LINEAR ALGEBRA AND LINEAR OPERATORS IN ENGINEERING
with Applications in *Mathematica*

This is Volume 3 of
PROCESS SYSTEMS ENGINEERING
A Series edited by George Stephanopoulos and John Perkins

LINEAR ALGEBRA AND LINEAR OPERATORS IN ENGINEERING
with Applications in *Mathematica*

H. Ted Davis
Department of Chemical Engineering and Materials Science
University of Minnesota
Minneapolis, Minnesota

Kendall T. Thomson
School of Chemical Engineering
Purdue University
West Lafayette, Indiana

San Diego   San Francisco   New York   Boston   London   Sydney   Tokyo

This book is printed on acid-free paper.

Copyright © 2000 by ACADEMIC PRESS

All Rights Reserved.
No part of this publication may be reproduced or transmitted in any form or by any means, electronic or mechanical, including photocopy, recording, or any information storage and retrieval system, without permission in writing from the publisher.

Requests for permission to make copies of any part of the work should be mailed to: Permissions Department, Harcourt, Inc., 6277 Sea Harbor Drive, Orlando, Florida 32887-6777

Academic Press
A Harcourt Science and Technology Company
525 B Street, Suite 1900, San Diego, California 92101-4495, USA
http://www.academicpress.com

Academic Press
Harcourt Place, 32 Jamestown Road, London NW1 7BY, UK
http://www.hbuk.co.uk/ap/

Library of Congress Catalog Card Number: 00-100019

International Standard Book Number: 0-12-206349-X

PRINTED IN THE UNITED STATES OF AMERICA
00 01 02 03 04 05 QW 9 8 7 6 5 4 3 2 1

CONTENTS

PREFACE xi

1 Determinants

    1.1. Synopsis   1
    1.2. Matrices   2
    1.3. Definition of a Determinant   3
    1.4. Elementary Properties of Determinants   6
    1.5. Cofactor Expansions   9
    1.6. Cramer's Rule for Linear Equations   14
    1.7. Minors and Rank of Matrices   16
Problems   18
Further Reading   22

2 Vectors and Matrices

    2.1. Synopsis   25
    2.2. Addition and Multiplication   26
    2.3. The Inverse Matrix   28
    2.4. Transpose and Adjoint   33

2.5. Partitioning Matrices   35
2.6. Linear Vector Spaces   38
Problems   43
Further Reading   46

3 Solution of Linear and Nonlinear Systems

3.1. Synopsis   47
3.2. Simple Gauss Elimination   48
3.3. Gauss Elimination with Pivoting   55
3.4. Computing the Inverse of a Matrix   58
3.5. **LU**-Decomposition   61
3.6. Band Matrices   66
3.7. Iterative Methods for Solving $\mathbf{Ax} = \mathbf{b}$ 78
3.8. Nonlinear Equations   85
Problems   108
Further Reading   121

4 General Theory of Solvability of Linear Algebraic Equations

4.1. Synopsis   123
4.2. Sylvester's Theorem and the Determinants of Matrix Products   124
4.3. Gauss-Jordan Transformation of a Matrix   129
4.4. General Solvability Theorem for $\mathbf{Ax} = \mathbf{b}$ 133
4.5. Linear Dependence of a Vector Set and the Rank of Its Matrix   150
4.6. The Fredholm Alternative Theorem   155
Problems   159
Further Reading   161

5 The Eigenproblem

5.1. Synopsis   163
5.2. Linear Operators in a Normed Linear Vector Space   165
5.3. Basis Sets in a Normed Linear Vector Space   170
5.4. Eigenvalue Analysis   179
5.5. Some Special Properties of Eigenvalues   184
5.6. Calculation of Eigenvalues   189
Problems   196
Further Reading   203

6 Perfect Matrices

6.1. Synopsis   205
6.2. Implications of the Spectral Resolution Theorem   206
6.3. Diagonalization by a Similarity Transformation   213
6.4. Matrices with Distinct Eigenvalues   219
6.5. Unitary and Orthogonal Matrices   220
6.6. Semidiagonalization Theorem   225
6.7. Self-Adjoint Matrices   227
6.8. Normal Matrices   245
6.9. Miscellanea   249
6.10. The Initial Value Problem   254
6.11. Perturbation Theory   259
Problems   261
Further Reading   278

7 Imperfect or Defective Matrices

7.1. Synopsis   279
7.2. Rank of the Characteristic Matrix   280
7.3. Jordan Block Diagonal Matrices   282
7.4. The Jordan Canonical Form   288
7.5. Determination of Generalized Eigenvectors   294
7.6. Dyadic Form of an Imperfect Matrix   303
7.7. Schmidt's Normal Form of an Arbitrary Square Matrix   304
7.8. The Initial Value Problem   308
Problems   310
Further Reading   314

8 Infinite-Dimensional Linear Vector Spaces

8.1. Synopsis   315
8.2. Infinite-Dimensional Spaces   316
8.3. Riemann and Lebesgue Integration   319
8.4. Inner Product Spaces   322
8.5. Hilbert Spaces   324
8.6. Basis Vectors   326
8.7. Linear Operators   330
8.8. Solutions to Problems Involving k-term Dyadics   336
8.9. Perfect Operators   343
Problems   351
Further Reading   353

9 Linear Integral Operators in a Hilbert Space

9.1. Synopsis   355
9.2. Solvability Theorems   356
9.3. Completely Continuous and Hilbert-Schmidt Operators   366
9.4. Volterra Equations   375
9.5. Spectral Theory of Integral Operators   387
Problems   406
Further Reading   411

10 Linear Differential Operators in a Hilbert Space

10.1. Synopsis   413
10.2. The Differential Operator   416
10.3. The Adjoint of a Differential Operator   420
10.4. Solution to the General Inhomogeneous Problem   426
10.5. Green's Function: Inverse of a Differential Operator   439
10.6. Spectral Theory of Differential Operators   452
10.7. Spectral Theory of Regular Sturm-Liouville Operators   459
10.8. Spectral Theory of Singular Sturm-Liouville Operators   477
10.9. Partial Differential Equations   493
Problems   502
Further Reading   509

APPENDIX

A.1. Section 3.2: Gauss Elimination and the Solution to the Linear System $\mathbf{Ax} = \mathbf{b}$ 511
A.2. Example 3.6.1: Mass Separation with a Staged Absorber   514
A.3. Section 3.7: Iterative Methods for Solving the Linear System $\mathbf{Ax} = \mathbf{b}$ 515
A.4. Exercise 3.7.2: Iterative Solution to $\mathbf{Ax} = \mathbf{b}$—Conjugate Gradient Method   518
A.5. Example 3.8.1: Convergence of the Picard and Newton–Raphson Methods   519
A.6. Example 3.8.2: Steady-State Solutions for a Continuously Stirred Tank Reactor   521
A.7. Example 3.8.3: The Density Profile in a Liquid–Vapor Interface (Iterative Solution of an Integral Equation)   523
A.8. Example 3.8.4: Phase Diagram of a Polymer Solution   526
A.9. Section 4.3: Gauss–Jordan Elimination and the Solution to the Linear System $\mathbf{Ax} = \mathbf{b}$ 529
A.10. Section 5.4: Characteristic Polynomials and the Traces of a Square Matrix   531
A.11. Section 5.6: Iterative Method for Calculating the Eigenvalues of Tridiagonal Matrices   533
A.12. Example 5.6.1: Power Method for Iterative Calculation of Eigenvalues   534

A.13. Example 6.2.1: Implementation of the Spectral Resolution Theorem—Matrix Functions   535
A.14. Example 9.4.2: Numerical Solution of a Volterra Equation (Saturation in Porous Media)   537
A.15. Example 10.5.3: Numerical Green's Function Solution to a Second-Order Inhomogeneous Equation   540
A.16. Example 10.8.2: Series Solution to the Spherical Diffusion Equation (Carbon in a Cannonball)   542

INDEX   543

PREFACE

This textbook is aimed at first-year graduate students in engineering or the physical sciences. It is based on a course that one of us (H.T.D.) has given over the past several years to chemical engineering and materials science students.

The emphasis of the text is on the use of algebraic and operator techniques to solve engineering and scientific problems. Where the proof of a theorem can be given without too much tedious detail, it is included. Otherwise, the theorem is quoted along with an indication of a source for the proof. Numerical techniques for solving both nonlinear and linear systems of equations are emphasized. Eigenvector and eigenvalue theory, that is, the eigenproblem and its relationship to the operator theory of matrices, is developed in considerable detail.

Homework problems, drawn from chemical, mechanical, and electrical engineering as well as from physics and chemistry, are collected at the end of each chapter—the book contains over 250 homework problems. Exercises are sprinkled throughout the text. Some 15 examples are solved using *Mathematica,* with the *Mathematica* codes presented in an appendix. Partially solved examples are given in the text as illustrations to be completed by the student.

The book is largely self-contained. The first two chapters cover elementary principles. Chapter 3 is devoted to techniques for solving linear and nonlinear algebraic systems of equations. The theory of the solvability of linear systems is presented in Chapter 4. Matrices as linear operators in linear vector spaces are studied in Chapters 5 through 7. The last three chapters of the text use analogies between finite and infinite dimensional vector spaces to introduce the functional theory of linear differential and integral equations. These three chapters could serve as an introduction to a more advanced course on functional analysis.

H. Ted Davis

Kendall T. Thomson

1 DETERMINANTS

1.1. SYNOPSIS

For any square array of numbers, i.e., a square matrix, we can define a determinant—a scalar number, real or complex. In this chapter we will give the fundamental definition of a determinant and use it to prove several elementary properties. These properties include: determinant addition, scalar multiplication, row and column addition or subtraction, and row and column interchange. As we will see, the elementary properties often enable easy evaluation of a determinant, which otherwise could require an exceedingly large number of multiplication and addition operations.

Every determinant has cofactors, which are also determinants but of lower order (if the determinant corresponds to an $n \times n$ array, its cofactors correspond to $(n-1) \times (n-1)$ arrays). We will show how determinants can be evaluated as linear expansions of cofactors. We will then use these cofactor expansions to prove that a system of linear equations has a unique solution if the determinant of the coefficients in the linear equations is not 0. This result is known as Cramer's rule, which gives the analytic solution to the linear equations in terms of ratios of determinants. The properties of determinants established in this chapter will play (in the chapters to follow) a big role in the theory of linear and nonlinear systems and in the theory of matrices as linear operators in vector spaces.

1.2. MATRICES

A matrix $\mathbf{A}$ is an array of numbers, complex or real. We say $\mathbf{A}$ is an $m \times n$-dimensional matrix if it has m rows and n columns, i.e.,

$$\mathbf{A} = \begin{bmatrix} a_{11} & a_{12} & a_{13} & \cdots & a_{1n} \\ a_{21} & a_{22} & a_{23} & \cdots & a_{2n} \\ a_{31} & a_{32} & a_{33} & \cdots & a_{3n} \\ \vdots & \vdots & \vdots & \ddots & \vdots \\ a_{m1} & a_{m2} & a_{m3} & \cdots & a_{mn} \end{bmatrix}. \tag{1.2.1}$$

The numbers a_{ij} ($i = 1, \ldots, m$, $j = 1, \ldots, n$) are called the elements of $\mathbf{A}$ with the element a_{ij} belonging to the ith row and jth column of $\mathbf{A}$. An abbreviated notation for $\mathbf{A}$ is

$$\mathbf{A} = [a_{ij}]. \tag{1.2.2}$$

By interchanging the rows and columns of $\mathbf{A}$, the transpose matrix $\mathbf{A}^{\mathrm{T}}$ is generated. Namely,

$$\mathbf{A}^{\mathrm{T}} = \begin{bmatrix} a_{11} & a_{21} & \cdots & a_{m1} \\ a_{12} & a_{23} & \cdots & a_{m2} \\ \vdots & \vdots & \ddots & \vdots \\ a_{1n} & a_{2n} & \cdots & a_{mn} \end{bmatrix}. \tag{1.2.3}$$

The rows of $\mathbf{A}^{\mathrm{T}}$ are the columns of $\mathbf{A}$ and the ijth element of $\mathbf{A}^{\mathrm{T}}$ is a_{ji}, i.e., $(\mathbf{A}^{\mathrm{T}})_{ij} = a_{ji}$. If $\mathbf{A}$ is an $m \times n$ matrix, then $\mathbf{A}^{\mathrm{T}}$ is an $n \times m$ matrix.

When $m = n$, we say $\mathbf{A}$ is a square matrix. Square matrices figure importantly in applications of linear algebra, but non-square matrices are also encountered in common physical problems, e.g., in least squares data analysis. The $m \times 1$ matrix

$$\mathbf{x} = \begin{bmatrix} x_1 \\ x_2 \\ \vdots \\ x_m \end{bmatrix} \tag{1.2.4}$$

and the $1 \times m$ matrix

$$\mathbf{y}^{\mathrm{T}} = [y_1, \ldots, y_n] \tag{1.2.5}$$

are also important cases. They are called vectors. We say that $\mathbf{x}$ is an m-dimensional column vector containing m elements, and $\mathbf{y}^{\mathrm{T}}$ is an n-dimensional row vector containing n elements. Note that $\mathbf{y}^{\mathrm{T}}$ is the transpose of the $n \times 1$ matrix $\mathbf{y}$—the n-dimensional column vector $\mathbf{y}$.

DEFINITION OF A DETERMINANT

If **A** and **B** have the same dimensions, then they can be added. The rule of matrix addition is that corresponding elements are added, i.e.,

$$\mathbf{A} + \mathbf{B} = [a_{ij}] + [b_{ij}] = [a_{ij} + b_{ij}]$$

$$= \begin{bmatrix} a_{11} + b_{11} & a_{12} + b_{12} & \cdots & a_{1n} + b_{1n} \\ a_{21} + b_{21} & a_{22} + b_{22} & \cdots & a_{2n} + b_{2n} \\ \vdots & \vdots & \ddots & \vdots \\ a_{m1} + b_{m1} & a_{m2} + b_{m2} & \cdots & a_{mn} + b_{mn} \end{bmatrix}. \quad (1.2.6)$$

Consistent with the definition of $m \times n$ matrix addition, the multiplication of the matrix **A** by a complex number α (scalar multiplication) is defined by

$$\alpha \mathbf{A} = [\alpha a_{ij}]; \quad (1.2.7)$$

i.e., $\alpha \mathbf{A}$ is formed by replacing every element a_{ij} of **A** by αa_{ij}.

1.3. DEFINITION OF A DETERMINANT

A determinant is defined specifically for a square matrix. The various notations for the determinant of **A** are

$$D = D_A = \text{Det}\, \mathbf{A} = |\mathbf{A}| = |a_{ij}| = \begin{vmatrix} a_{11} & \cdots & a_{1n} \\ \vdots & \ddots & \vdots \\ a_{n1} & \cdots & a_{nn} \end{vmatrix}. \quad (1.3.1)$$

We define the determinant of **A** as follows:

$$D \equiv \sum_{l_1, \ldots, l_n}{}' (-1)^P a_{1l_1} a_{2l_2} \cdots a_{nl_n}, \quad (1.3.2)$$

where the summation is taken over all possible products of a_{ij} in which each product contains n elements and one and only one element from each row and each column. The indices $l_1, \ldots, l_n$ are permutations of the integers $1, \ldots, n$. We will use the symbol $\sum'$ to denote summation over all permutations. For a given set $\{l_1, \ldots, l_n\}$, the quantity P denotes the number of transpositions required to transform the sequence $l_1, l_2, \ldots, l_n$ into the ordered sequence $1, 2, \ldots, n$. A transposition is defined as an interchange of two numbers l_i and l_j. Note that there are $n!$ terms in the sum defining D since there are exactly $n!$ ways to reorder the set of numbers $\{1, 2, \ldots, n\}$ into distinct sets $\{l_1, l_2, \ldots, l_n\}$.

As an example of a determinant, consider

$$\begin{vmatrix} a_{11} & a_{12} & a_{13} \\ a_{21} & a_{22} & a_{23} \\ a_{31} & a_{32} & a_{33} \end{vmatrix} = a_{11}a_{22}a_{33} - a_{12}a_{21}a_{33} - a_{11}a_{23}a_{32} \\ - a_{13}a_{22}a_{31} + a_{13}a_{21}a_{32} + a_{12}a_{23}a_{31}. \quad (1.3.3)$$

The sign of the second term is negative because the indices $\{2, 1, 3\}$ are transposed to $\{1, 2, 3\}$ with the one transposition

$$2, 1, 3 \to 1, 2, 3,$$

and so $P = 1$ and $(-1)^P = -1$. However, the transposition also could have been accomplished with the three transpositions

$$2, 1, 3 \to 2, 3, 1 \to 1, 3, 2 \to 1, 2, 3,$$

in which case $P = 3$ and $(-1)^P = -1$. We see that the number of transpositions P needed to reorder a given sequence $l_1, \ldots, l_n$ is not unique. However, the evenness or oddness of P is unique and thus $(-1)^P$ is unique for a given sequence.

EXERCISE 1.3.1. Verify the signs in Eq. (1.3.3). Also, verify that the number of transpositions required for $a_{11}a_{25}a_{33}a_{42}a_{54}$ is even.

A definition equivalent to that in Eq. (1.3.2) is

$$D = \sum_{l_1, \ldots, l_n}{}' (-1)^P a_{l_1 1} a_{l_2 2} \cdots a_{l_n n}. \tag{1.3.4}$$

If the product $a_{l_1 1} a_{l_2 2} \cdots a_{l_n n}$ is reordered so that the first indices of the $a_{l_i i}$ are ordered in the sequence $1, \ldots, n$, the second indices will be in a sequence requiring P transpositions to reorder as $1, \ldots, n$. Thus, the $n!$ n-tuples in Eqs. (1.3.2) and (1.3.4) are the same and have the same signs.

The determinant in Eq. (1.3.3) can be expanded according to the defining equation (1.3.4) as

$$\begin{vmatrix} a_{11} & a_{12} & a_{13} \\ a_{21} & a_{22} & a_{23} \\ a_{31} & a_{32} & a_{33} \end{vmatrix} = a_{11}a_{22}a_{32} - a_{21}a_{12}a_{33} - a_{11}a_{32}a_{23} \tag{1.3.5}$$

$$- a_{31}a_{22}a_{13} + a_{21}a_{32}a_{13} + a_{31}a_{12}a_{23}.$$

It is obvious by inspection that the right-hand sides of Eqs. (1.3.3) and (1.3.5) are identical since the various terms differ only by the order in which the multiplication of each 3-tuple is carried out.

In the case of second- and third-order determinants, there is an easy way to generate the distinct n-tuples. For the second-order case,

$$\begin{vmatrix} a_{11} & a_{12} \\ a_{21} & a_{22} \end{vmatrix},$$

the product of the main diagonal, $a_{11}a_{22}$, is one of the 2-tuples and the product of the reverse main diagonal, $a_{12}a_{21}$, is the other. The sign of $a_{12}a_{21}$ is negative since $\{2, 1\}$ requires one transposition to reorder to $\{1, 2\}$. Thus,

$$\begin{vmatrix} a_{11} & a_{12} \\ a_{21} & a_{22} \end{vmatrix} = a_{11}a_{22} - a_{12}a_{21} \tag{1.3.6}$$

DEFINITION OF A DETERMINANT

since there are no other 2-tuples containing exactly one element from each row and column.

In the case of the third-order determinant, the six 3-tuples can be generated by multiplying the elements shown below by solid and dashed curves

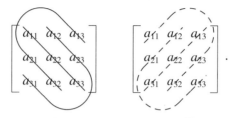

The products associated with solid curves require an even number of transpositions P and those associated with the dashed curves require an odd P. Thus, the determinant is given by

$$\begin{vmatrix} a_{11} & a_{12} & a_{13} \\ a_{21} & a_{22} & a_{23} \\ a_{31} & a_{32} & a_{33} \end{vmatrix} = a_{11}a_{22}a_{33} + a_{12}a_{23}a_{31} + a_{13}a_{21}a_{32} \\ - a_{13}a_{22}a_{31} - a_{11}a_{23}a_{32} - a_{12}a_{21}a_{33},$$
(1.3.7)

in agreement with Eq. (1.3.3), the defining expression. For example, the following determinant is 0:

$$\begin{vmatrix} 1 & 2 & 3 \\ 4 & 5 & 6 \\ 3 & 2 & 1 \end{vmatrix} = 1(5)1 + 2(6)3 + 3(2)4 - 3(5)3 - 2(6)1 - 1(2)4 \\ = 0.$$
(1.3.8)

The evaluation of a determinant by calculation of the $n!$ n-tuples requires $(n-1)(n!)$ multiplications. For a fourth-order determinant, this requires 72 multiplications, not many in the age of computers. However, if $n = 100$, the number of required multiplications would be

$$(n-1)\,n! \sim (n-1)\left(\frac{n}{e}\right)^n = 99\left(\frac{100}{2.718}\right)^{100} \\ \sim 3.7 \times 10^{158},$$
(1.3.9)

where Stirling's approximation, $n! \sim (n/e)^n$, has been used. If the time for one multiplication is 10^{-9} sec, then the required time to do the multiplications would be

$$3.7 \times 10^{149} \text{ sec}, \quad \text{or} \quad 1.2 \times 10^{142} \text{ years!} \tag{1.3.10}$$

Obviously, large determinants cannot be evaluated by direct calculation of the defining n-tuples. Fortunately, the method of Gauss elimination, which we will describe in Chapter 3, reduces the number of multiplications to n^3. For $n = 100$, this is 10^6 multiplications, as compared to 3.7×10^{158} by direct n-tuple evaluation. The Gauss elimination method depends on the application of some of the elementary properties of determinants given in the next section.

1.4. ELEMENTARY PROPERTIES OF DETERMINANTS

If the determinant of $\mathbf{A}$ is given by Eq. (1.3.2), then—because the elements of the transpose $\mathbf{A}^T$ are a_{ji}—it follows that

$$D_{\mathbf{A}^T} = \sum_{l_1,\ldots,l_n}{}' (-1)^P a_{l_1 1} a_{l_2 2} \cdots a_{l_n n}. \qquad (1.4.1)$$

However, according to Eq. (1.3.4), the right-hand side of Eq. (1.4.1) is also equal to the determinant D_A of $\mathbf{A}$. This establishes the property that

1. A determinant is invariant to the interchange of rows and columns; i.e., the determinant of $\mathbf{A}$ is equal to the determinant of $\mathbf{A}^T$.

For example,

$$\begin{vmatrix} 1 & 8 \\ 3 & 7 \end{vmatrix} = 7 - 24 = -17$$

$$\begin{vmatrix} 1 & 3 \\ 8 & 7 \end{vmatrix} = 7 - 24 = -17.$$

Another elementary property of a determinant is that

2. If two rows (columns) of a determinant are interchanged, then the determinant changes sign.

For example,

$$D = \begin{vmatrix} a_{11} & a_{12} \\ a_{21} & a_{22} \end{vmatrix} = a_{11}a_{22} - a_{21}a_{12},$$

$$D' = \begin{vmatrix} a_{21} & a_{22} \\ a_{11} & a_{12} \end{vmatrix} = a_{21}a_{12} - a_{11}a_{22} = -D.$$

From the definition of D in Eq. (1.3.2),

$$D = \sum{}' (-1)^P a_{1 l_1} a_{2 l_2} \cdots a_{i l_i} \cdots a_{j l_j} \cdots a_{n l_n},$$

it follows that the determinant D' formed by the interchange of rows i and j in D is

$$D' = \sum{}' (-1)^{P'} a_{1 l_1} a_{2 l_2} \cdots a_{j l_i} \cdots a_{i l_j} \cdots a_{n l_n}. \qquad (1.4.2)$$

Each term in D' corresponds to one in D if one transposition is carried out. Thus, P and P' differ by 1, and so $(-1)^{P'} = (-1)^{P+1} = -(-1)^P$. From this it follows that $D' = -D$. A similar proof that the interchange of two columns changes the sign of the determinant can be given using the definition of D in Eq. (1.3.4). Alternatively, from the fact that $D_A = D_{A^T}$, it follows that if the interchange of two rows changes the sign of the determinant, then the interchange of two columns does the same thing because the columns of $\mathbf{A}^T$ are the rows of $\mathbf{A}$.

The preceding property implies:

3. If any two rows (columns) of a matrix are the same, its determinant is 0.

If two rows (columns) are interchanged, $D = -D'$. However, if the rows (columns) interchanged are identical, then $D = D'$. The two equalities, $D = -D'$ and $D = D'$, are possible only if $D = D' = 0$.

Next, we note that

4. Multiplication of the determinant D by a constant k is the same as multiplying any row (column) by k.

This property follows from the commutative law of scalar multiplication, i.e., $kab = (ka)b = a(kb)$, or

$$kD = \sum{}'(-1)^P a_{1l_1} a_{2l_2} \cdots k a_{il_i} \cdots a_{nl_n}$$
$$= \sum{}'(-1)^P a_{l_1 1} a_{l_2 2} \cdots k a_{l_j j} \cdots a_{l_n n}$$
$$= \begin{vmatrix} a_{11} & a_{12} & \cdots & & a_{1n} \\ a_{21} & \cdots & \cdots & & a_{2n} \\ \vdots & & & & \vdots \\ ka_{i1} & \cdots & \cdots & & ka_{in} \\ a_{n1} & \cdots & \cdots & & a_{nn} \end{vmatrix} \quad (1.4.3)$$
$$= \begin{vmatrix} a_{11} & a_{12} & \cdots & ka_{1j} & \cdots & a_{1n} \\ a_{21} & a_{22} & \cdots & ka_{2j} & \cdots & a_{2n} \\ \vdots & & & \vdots & & \vdots \\ a_{n1} & \cdots & \cdots & ka_{nj} & \cdots & a_{nn} \end{vmatrix}.$$

Multiplication of the determinant

$$D = \begin{vmatrix} 1 & 2 & 3 \\ 2 & 4 & 6 \\ 1 & 2 & 8 \end{vmatrix}$$

by $\frac{1}{2}$ gives

$$\frac{1}{2}D = \begin{vmatrix} 1 & 1 & 3 \\ 2 & 2 & 6 \\ 1 & 1 & 8 \end{vmatrix}$$

from which we can conclude that $D/2 = 0$ and $D = 0$, since $D/2$ has two identical columns. Stated differently, the multiplication rule says that if a row (column) of D has a common factor k, then $D = kD'$, where D' is formed from D by replacing the row (column) with the common factor by the row (column) divided by the common factor. Thus, in the previous example,

$$D = 4 \begin{vmatrix} 1 & 1 & 3 \\ 1 & 1 & 3 \\ 1 & 1 & 8 \end{vmatrix}.$$

The fact that a determinant is 0 if two rows (columns) are the same yields the property:

5. The addition of a row (column) multiplied by a constant to any other row (column) does not change the value of D.

To prove this, note that

$$\begin{aligned} D' &= {\sum}'(-1)^P a_{1l_1} \cdots (a_{il_i} + k a_{jl_i}) \cdots a_{jl_j} \cdots a_{nl_n} \\ &= D + k {\sum}'(-1)^P a_{1l_1} \cdots a_{jl_i} \cdots a_{jl_j} \cdots a_{nl_n}. \end{aligned} \quad (1.4.4)$$

The second determinant on the right-hand-side of Eq. (1.4.4) is 0 since the elements of the ith and jth rows are the same. Thus, $D' = D$. The equality $D_A = D_{A^T}$ establishes the property for column addition. As an example,

$$\begin{vmatrix} 1 & 2 \\ 1 & 3 \end{vmatrix} = 3 - 2 = 1 = \begin{vmatrix} 1+2k & 2 \\ 1+3k & 3 \end{vmatrix} = 3 + 6k - 2 - 6k = 1.$$

Elementary properties can be used to simplify a determinant. For example,

$$\begin{vmatrix} 1 & 2 & 3 \\ 2 & 4 & 6 \\ 1 & 2 & 8 \end{vmatrix} = \begin{vmatrix} 1 & 2 & 3-2 \\ 2 & 4 & 6-4 \\ 1 & 2 & 8-2 \end{vmatrix} = \begin{vmatrix} 1 & 2 & 1 \\ 2 & 4 & 2 \\ 1 & 2 & 6 \end{vmatrix}$$

$$= 2(2) \begin{vmatrix} 1 & 1 & 1 \\ 1 & 1 & 1 \\ 1 & 1 & 6 \end{vmatrix} = 4 \begin{vmatrix} 1 & 1-1 & 1-1 \\ 1 & 1-1 & 1-1 \\ 1 & 1-1 & 6-1 \end{vmatrix} \quad (1.4.5)$$

$$= 4 \begin{vmatrix} 1 & 0 & 0 \\ 1 & 0 & 0 \\ 1 & 0 & 5 \end{vmatrix} = 4 \begin{vmatrix} 1 & 0 & 0 \\ 0 & 0 & 0 \\ 0 & 0 & 5 \end{vmatrix}.$$

The sequence of application of the elementary properties in Eq. (1.4.5) is, of course, not unique.

Another useful property of determinants is:

6. If two determinants differ only by one row (column), their sum differs only in that the differing rows (columns) are summed.

That is,

$$\begin{vmatrix} a_{11} & \cdots & a_{1i} & \cdots & a_{1n} \\ & & \vdots & & \\ a_{n1} & \cdots & a_{ni} & \cdots & a_{nn} \end{vmatrix} + \begin{vmatrix} a_{11} & \cdots & b_{1i} & \cdots & a_{1n} \\ & & \vdots & & \\ a_{n1} & \cdots & b_{ni} & \cdots & a_{nn} \end{vmatrix} = \begin{vmatrix} a_{11} & \cdots & a_{1i}+b_{1i} & \cdots & a_{1n} \\ & & \vdots & & \\ a_{n1} & \cdots & a_{ni}+b_{ni} & \cdots & a_{nn} \end{vmatrix}.$$

(1.4.6)

This property follows from the definition of determinants and the distributive law $(ca + cb = c(a+b))$ of scalar multiplication.

As the last elementary property of determinants to be given in this section, consider differentiation of D by the variable t:

$$\frac{dD}{dt} = \sum{}'(-1)^P \frac{da_{1l_1}}{dt} a_{2l_2} \cdots a_{nl_n} + \sum{}'(-1)^P a_{1l_1} \frac{da_{2l_2}}{dt} \cdots a_{nl_n}$$
$$+ \cdots + \sum{}'(-1)^P a_{1l_1} a_{2l_2} \cdots \frac{da_{nl_n}}{dt} \qquad (1.4.7)$$
$$= \sum_{i=1}^{n} D_i'$$

or

$$\frac{dD}{dt} = \sum{}'(-1)^P \frac{da_{l_1 1}}{dt} a_{l_2 2} \cdots a_{l_n n} + \sum{}'(-1)^P a_{l_1 1} a_{l_2 2} \cdots \frac{da_{l_n n}}{dt}$$
$$= \sum_{i=1}^{n} D_i''. \qquad (1.4.8)$$

The determinant D_i' is evaluated by replacing in D the elements of the ith row by the derivatives of the elements of the ith row. Similarly, for D_i'', replace in D the elements of the ith column by the derivatives of the elements of the ith column. For example,

$$\frac{d}{dt}\begin{vmatrix} a_{11} & a_{12} \\ a_{21} & a_{22} \end{vmatrix} = \begin{vmatrix} \frac{da_{11}}{dt} & \frac{da_{12}}{dt} \\ a_{21} & a_{22} \end{vmatrix} + \begin{vmatrix} a_{11} & a_{12} \\ \frac{da_{21}}{dt} & \frac{da_{22}}{dt} \end{vmatrix}$$
$$= \begin{vmatrix} \frac{da_{11}}{dt} & a_{12} \\ \frac{da_{21}}{dt} & a_{22} \end{vmatrix} + \begin{vmatrix} a_{11} & \frac{da_{12}}{dt} \\ a_{21} & \frac{da_{22}}{dt} \end{vmatrix}.$$

(1.4.9)

1.5. COFACTOR EXPANSIONS

We define the cofactor A_{ij} as the quantity $(-1)^{i+j}$ multiplied by the determinant of the matrix generated when the ith row and jth column of $\mathbf{A}$ are removed. For example, some of the cofactors of the matrix

$$\mathbf{A} = \begin{vmatrix} a_{11} & a_{12} & a_{13} \\ a_{21} & a_{22} & a_{23} \\ a_{31} & a_{32} & a_{33} \end{vmatrix} \qquad (1.5.1)$$

include

$$A_{11} = \begin{vmatrix} a_{22} & a_{23} \\ a_{32} & a_{33} \end{vmatrix}, \quad A_{21} = -\begin{vmatrix} a_{12} & a_{13} \\ a_{32} & a_{33} \end{vmatrix}, \quad A_{31} = \begin{vmatrix} a_{12} & a_{13} \\ a_{22} & a_{23} \end{vmatrix}. \quad (1.5.2)$$

In general,

$$A_{ij} = (-1)^{i+j} \begin{vmatrix} a_{11} & \cdots & a_{1,j-1} & a_{1,j+1} & \cdots & a_{1n} \\ \vdots & & \vdots & \vdots & & \vdots \\ a_{i-1,1} & \cdots & a_{i-1,j-1} & a_{i-1,j+1} & \cdots & a_{i-1,n} \\ a_{i+1,1} & \cdots & a_{i+1,j-1} & a_{i+1,j+1} & \cdots & a_{i+1,n} \\ \vdots & & \vdots & \vdots & & \vdots \\ a_{n1} & \cdots & a_{n,j-1} & a_{n,j+1} & \cdots & a_{nn} \end{vmatrix}. \quad (1.5.3)$$

Note that an $n \times n$ matrix has n^2 cofactors.

Cofactors are important because they enable us to evaluate an nth-order determinant as a linear combination of n $(n-1)$th-order determinants. The evaluation makes use of the following theorem:

COFACTOR EXPANSION THEOREM. *The determinant D of $\mathbf{A}$ can be computed from*

$$D = \sum_{j=1}^{n} a_{ij} A_{ij}, \quad \text{where } i \text{ is an arbitrary row}, \quad (1.5.4)$$

or

$$D = \sum_{i=1}^{n} a_{ij} A_{ij}, \quad \text{where } j \text{ is an arbitrary column}. \quad (1.5.5)$$

Equation (1.5.4) is called a cofactor expansion by the ith row and Eq. (1.5.5) is called a cofactor expansion by the jth column.

Before presenting the proof of the cofactor expansion, we will give an example. Let

$$\mathbf{A} = \begin{bmatrix} 4 & -1 & 0 \\ -1 & 4 & -1 \\ 0 & -1 & 4 \end{bmatrix}. \quad (1.5.6)$$

By the expression given in Eq. (1.3.7), it follows that

$$D = 64 + 0 + 0 + (-0) - 4 - 4 = 56.$$

The cofactor expansion by row 1 yields

$$D = 4 \begin{vmatrix} 4 & -1 \\ -1 & 4 \end{vmatrix} - (-1) \begin{vmatrix} -1 & -1 \\ 0 & 4 \end{vmatrix} + 0 \begin{vmatrix} -1 & 4 \\ 0 & -1 \end{vmatrix}$$

$$= 60 - 4 + 0 = 56,$$

and the cofactor expansion by column 2 yields

$$D_A = -1(-1)\begin{vmatrix} -1 & -1 \\ 0 & 4 \end{vmatrix} + 4\begin{vmatrix} 4 & 0 \\ 0 & 4 \end{vmatrix} - 1(-1)\begin{vmatrix} 4 & 0 \\ -1 & -1 \end{vmatrix} = 56.$$

To prove the cofactor expansion theorem, we start with the definition of the determinant given in Eq. (1.3.4). Choosing an arbitrary column j, we can rewrite this equation as

$$D = \sum_{i=1}^{n} \underset{l_1,\ldots,i,\ldots,l_n}{\sum\nolimits'} (-1)^P a_{l_1 1} a_{l_2 2} \cdots a_{ij} \cdots a_{l_n n}, \qquad (1.5.7)$$

where the primed sum now refers to the sum over all permutations in which $l_j = i$. For a given value of i in the first sum, we would like now to isolate the ijth cofactor of $\mathbf{A}$. To accomplish this, we must examine the factor $(-1)^P$ closely. First, we note that the permutations defined by P can be redefined in terms of permutations in which all elements *except* element i are in proper order plus the permutations required to put i in its place in the sequence $1, 2, \ldots, n$. For this new definition, the proper sequence, in general, would be

$$1, 2, 3, \ldots, i-1, i+1, i+2, \ldots, j-1, j+1, j+2, \ldots, n-1, n. \qquad (1.5.8)$$

We now define P'_{ij} as the number of permutations required to bring a sequence back to the proper sequence defined in Eq. (1.5.8). We now note that $|j-i|$ permutations are required to transform this new proper sequence back to the original proper sequence $1, 2, \ldots, n$. Thus, we can write $(-1)^P = (-1)^{P'_{ij}}(-1)^{i+j}$ and Eq. (1.5.7) becomes

$$D = \sum_{i=1}^{n}(-1)^{i+j}\left(\underset{l_1,\ldots,i,\ldots,l_n}{\sum\nolimits'} (-1)^{P'_{ij}} a_{l_1 1} a_{l_2 2} \cdots a_{l_{j-1} j-1} a_{l_{j+1} j+1} \cdots a_{l_n n}\right) a_{ij}, \qquad (1.5.9)$$

which we recognize using the definition of a cofactor as

$$D = \sum_{i=1}^{n} a_{ij} A_{ij}. \qquad (1.5.10)$$

A similar proof exists for Eq. (1.5.4).

With the aid of the cofactor expansion theorem, we see that the determinant of an upper triangular matrix, i.e.,

$$\mathbf{U} = \begin{bmatrix} u_{11} & u_{12} & u_{13} & \cdots & u_{1n} \\ 0 & u_{22} & u_{23} & \cdots & u_{2n} \\ 0 & 0 & u_{33} & \cdots & u_{3n} \\ \vdots & \vdots & \vdots & \ddots & \vdots \\ 0 & 0 & 0 & \cdots & u_{nn} \end{bmatrix}, \qquad (1.5.11)$$

where $u_{ij} = 0$, when $i > j$, is the product of the main diagonal elements of $\mathbf{U}$, i.e.,

$$|\mathbf{U}| = \prod_{i=1}^{n} u_{ii}. \qquad (1.5.12)$$

To derive Eq. (1.5.12), we use the cofactor expansion theorem with the first column of **U** to obtain

$$|\mathbf{U}| = u_{11} \begin{vmatrix} u_{22} & u_{23} & \cdots & u_{2n} \\ 0 & u_{33} & \cdots & u_{3n} \\ \vdots & \vdots & \ddots & \vdots \\ 0 & 0 & \cdots & u_{nn} \end{vmatrix} + \sum_{i=2}^{n} 0 \times U_{i1}$$

$$= u_{11} \begin{vmatrix} u_{22} & u_{23} & \cdots & u_{2n} \\ 0 & u_{33} & \cdots & u_{3n} \\ \vdots & \vdots & \ddots & \vdots \\ 0 & 0 & \cdots & u_{nn} \end{vmatrix},$$

where U_{i1} is the $i1$ cofactor of **U**. Repeat the process on the $(n-1)$th-order upper triangular determinant, then the $(n-2)$th one, etc., until Eq. (1.5.13) results. Similarly, the row cofactor expansion theorem can be used to prove that the determinant of the lower triangular matrix,

$$\mathbf{L} = \begin{bmatrix} l_{11} & 0 & \cdots & 0 \\ l_{21} & l_{22} & \cdots & 0 \\ \vdots & \vdots & \ddots & \vdots \\ l_{n1} & l_{n2} & \cdots & l_{nn} \end{bmatrix}, \qquad (1.5.13)$$

is

$$|\mathbf{L}| = \prod_{i=1}^{n} l_{ii}; \qquad (1.5.14)$$

i.e., it is again the product of the main diagonal elements. In **L**, $l_{ij} = 0$ when $j > i$.

The property of the row cofactor expansion is that the sum

$$\sum_{j=1}^{n} a_{ij} A_{ij}$$

replaces the ith row of D_A with the elements a_{ij} of the ith row; i.e., the sum puts in the ith row of D_A the elements $a_{i1}, a_{i2}, \ldots, a_{in}$. Thus, the quantity

$$\sum_{j=1}^{n} \alpha_j A_{ij}$$

puts the elements $\alpha_1, \alpha_2, \ldots, \alpha_n$ in the ith row of D_A, i.e.,

$$\sum_{j=1}^{n} \alpha_j A_{ij} = \begin{vmatrix} a_{11} & a_{12} & \cdots & a_{1n} \\ a_{21} & \cdots & \cdots & a_{2n} \\ a_{i-1,1} & \cdots & \cdots & a_{i-1,n} \\ \alpha_1 & \alpha_2 & \cdots & \alpha_n \\ a_{i+1,1} & \cdots & \cdots & a_{i+1,n} \\ \vdots & & & \vdots \\ a_{n1} & \cdots & \cdots & a_{nn} \end{vmatrix}. \qquad (1.5.15)$$

COFACTOR EXPANSIONS

Similarly, for the column expansion,

$$\sum_{i=1}^{n} \alpha_i A_{ij} = \begin{vmatrix} a_{11} & a_{1,j-1} & \alpha_1 & a_{1,j+1} & a_{1n} \\ \vdots & \vdots & \vdots & \vdots & \vdots \\ a_{n1} & a_{n,j-1} & \alpha_n & a_{n,j+1} & a_{nn} \end{vmatrix}. \tag{1.5.16}$$

■ **EXAMPLE 1.5.1.**

$$\mathbf{A} = \begin{bmatrix} 1 & 2 & 3 \\ 3 & 2 & 1 \\ 1 & 1 & 2 \end{bmatrix}$$

$$A_{11} = \begin{vmatrix} 2 & 1 \\ 1 & 2 \end{vmatrix}, \quad A_{21} = -\begin{vmatrix} 2 & 3 \\ 1 & 2 \end{vmatrix}, \quad A_{31} = \begin{vmatrix} 2 & 3 \\ 2 & 1 \end{vmatrix}$$

$$D_A = 1 \times A_{11} + 3 \times A_{21} + 1 \times A_{31}$$
$$= 1(3) + 3(-1) + 1(-4) = -4$$

$$\mathbf{A}' = \begin{bmatrix} \alpha_1 & 2 & 3 \\ \alpha_2 & 2 & 1 \\ \alpha_3 & 1 & 2 \end{bmatrix}$$

$$D_{A'} = \alpha_1 \times A_{11} + \alpha_2 \times A_{21} + \alpha_3 \times A_{31}$$
$$= 3\alpha_1 + \alpha_2 - 4\alpha_3.$$

The cofactor expansion of D_A by the first-column cofactors involves the same cofactors, A_{11}, A_{21}, and A_{31}, as the cofactor expansion of $D_{A'}$ by the first column. The difference between the two expansions is simply that multipliers of A_{11}, A_{21}, ■ ■ ■ and A_{31} differ since the elements of the first column differ.

Consider next the expansions

$$\sum_{i=1}^{n} a_{ik} A_{ij}, \quad k \neq j, \tag{1.5.17}$$

and

$$\sum_{j=1}^{n} a_{kj} A_{ij}, \quad k \neq i. \tag{1.5.18}$$

The determinant represented by Eq. (1.5.17) is the same as D_A, except that the jth column is replaced by the elements of the kth column of $\mathbf{A}$, i.e.,

$$\sum_{i=1}^{n} a_{ik} A_{ij} = \begin{vmatrix} & & \text{column } j & & \text{column } k & & \\ a_{11} & \cdots & a_{1k} & \cdots & a_{1k} & \cdots & a_{1n} \\ \vdots & \vdots & \vdots & \vdots & \vdots & \vdots & \vdots \\ a_{n1} & \cdots & a_{nk} & \cdots & a_{nk} & \cdots & a_{nn} \end{vmatrix}. \tag{1.5.19}$$

The determinant in Eq. (1.5.19) is 0 because columns j and k are identical. Similarly, the determinant represented by Eq. (1.5.18) is the same as D_A, except that the ith row is replaced by the elements of the kth row of **A**, i.e.,

$$\sum_{j=1}^{n} a_{kj} A_{ij} = \begin{vmatrix} a_{11} & \cdots & a_{1n} \\ \vdots & & \vdots \\ a_{k1} & \cdots & a_{kn} \\ \vdots & & \vdots \\ a_{k1} & \cdots & a_{kn} \\ \vdots & & \vdots \\ a_{n1} & \cdots & a_{nn} \end{vmatrix} \begin{matrix} \\ \\ \text{row } i \\ \\ \text{row } k \\ \\ \end{matrix} \qquad (1.5.20)$$

The determinant in Eq. (1.5.20) is 0 because rows i and k are identical.

Equations (1.5.19) and (1.5.20) embody the alien cofactor expansion theorem:

ALIEN COFACTOR EXPANSION THEOREM. *The alien cofactor expansions are 0*, i.e.,

$$\sum_{i=1}^{n} a_{ik} A_{ij} = 0, \qquad k \neq j,$$
$$\sum_{j=1}^{n} a_{kj} A_{ij} = 0, \qquad k \neq i. \qquad (1.5.21)$$

The cofactor expansion theorem and the alien cofactor expansion theorem can be summarized as

$$\sum_{i=1}^{n} a_{ik} A_{ij} = D_A \delta_{kj}, \qquad \sum_{j=1}^{n} a_{kj} A_{ij} = D_A \delta_{ki}, \qquad (1.5.22)$$

where δ_{kj} is the Kronecker delta function with the property

$$\begin{aligned} \delta_{kj} &= 1, & k &= j, \\ &= 0, & k &\neq j. \end{aligned} \qquad (1.5.23)$$

1.6. CRAMER'S RULE FOR LINEAR EQUATIONS

Frequently, in a practical situation, one wishes to know what values of the variables $x_1, x_2, \ldots, x_n$ satisfy the n linear equations

$$\begin{aligned} a_{11}x_1 + a_{12}x_2 + \cdots + a_{1n}x_n &= b_1 \\ a_{21}x_1 + a_{22}x_2 + \cdots + a_{2n}x_n &= b_2 \\ \vdots \quad \vdots \quad \vdots \quad \vdots \\ a_{n1}x_1 + a_{n2}x_2 + \cdots + a_{nn}x_n &= b_n \end{aligned} \qquad (1.6.1)$$

CRAMER'S RULE FOR LINEAR EQUATIONS

These equations can be summarized as

$$\sum_{j=1}^{n} a_{ij} x_j = b_i, \qquad i = 1, 2, \ldots, n. \tag{1.6.2}$$

Equation (1.6.2) is suggestive of a solution to the set of equations. Let us multiply Eq. (1.6.2) by the cofactor A_{ik} and sum over i. By interchanging the order of summation over i and j on the left-hand side of the resulting equation, we obtain

$$\sum_j \sum_i a_{ij} A_{ik} x_j = \sum_i b_i A_{ik}. \tag{1.6.3}$$

By the alien cofactor expansion, it follows that

$$\sum_i a_{ij} A_{ik} = 0 \tag{1.6.4}$$

unless $j = k$, whereas, when $j = k$, the cofactor expansion yields

$$\sum_i a_{ik} A_{ik} = D = \begin{vmatrix} a_{11} & a_{12} & \cdots & a_{1n} \\ a_{21} & a_{22} & \cdots & a_{2n} \\ \vdots & \vdots & & \vdots \\ a_{n1} & a_{n2} & \cdots & a_{nn} \end{vmatrix}. \tag{1.6.5}$$

Also, it follows from Eq. (1.5.19) that

$$D_k \equiv \sum_i b_i A_{ik} = \begin{vmatrix} a_{11} & a_{12} & \cdots & a_{1,k-1} & b_1 & a_{1,k+1} & \cdots & a_{1n} \\ \vdots & \vdots & & \vdots & \vdots & \vdots & & \vdots \\ a_{n1} & a_{n2} & \cdots & a_{n,k-1} & b_n & a_{n,k+1} & \cdots & a_{nn} \end{vmatrix}, \tag{1.6.6}$$

where D_k is the same as the determinant D except that the kth column of D has been replaced by the elements $b_1, b_2, \ldots, b_n$.

According to Eqs. (1.6.4)–(1.6.6), Eq. (1.6.3) becomes

$$D x_k = D_k. \tag{1.6.7}$$

Cramer's rule follows from the preceding result:

CRAMER'S RULE. *If the determinant D is not 0, then the solution to the linear system, Eq. (1.6.1), is*

$$x_k = \frac{D_k}{D}, \qquad k = 1, \ldots, n, \tag{1.6.8}$$

and the solution is unique.

To prove uniqueness, suppose x_i and y_i for $i = 1, \ldots, n$ are two solutions to Eq. (1.6.2). Then the difference between $\sum_j a_{ij} x_j = b_i$ and $\sum_j a_{ij} y_j = b_i$ yields

$$\sum_j a_{ij} (x_j - y_j) = 0, \qquad i = 1, \ldots, n. \tag{1.6.9}$$

Multiplication of Eq. (1.6.9) by A_{ik} and summation over i yields

$$D(x_k - y_k) = 0, \qquad (1.6.10)$$

or $x_k = y_k$, $k = 1, \ldots, n$, since $D \neq 0$. Incidentally, even if $D = 0$, the linear equations sometimes have a solution, but not a unique one. The full theory of the solution of linear systems will be presented in Chapter 4.

■ **EXAMPLE 1.6.1.** Use Cramer's rule to solve

$$2x_1 + x_2 = 7$$
$$x_1 + 2x_2 = 3.$$

Solution.

$$D = \begin{vmatrix} 2 & 1 \\ 1 & 2 \end{vmatrix} = 3$$

$$D_1 = \begin{vmatrix} 7 & 1 \\ 3 & 2 \end{vmatrix} = 11$$

$$D_2 = \begin{vmatrix} 2 & 7 \\ 1 & 3 \end{vmatrix} = -1$$

$$x_1 = \frac{D_1}{D} = \frac{11}{3}, \qquad x_2 = \frac{D_2}{D} = -\frac{1}{3}.$$

■ ■ ■

Even if the determinant found no other role, its utility in mathematics is assured by Cramer's rule. When $D \neq 0$, a unique solution exists for the linear equations in Eq. (1.6.1). We shall see later that, in the theory and applications of linear algebra, the determinant is important in a myriad of circumstances.

1.7. MINORS AND RANK OF MATRICES

Consider the $m \times n$ matrix

$$\mathbf{A} = \begin{bmatrix} a_{11} & a_{12} & \cdots & a_{1n} \\ a_{21} & a_{22} & \cdots & a_{2n} \\ \vdots & \vdots & & \vdots \\ a_{m1} & a_{m2} & \cdots & a_{mn} \end{bmatrix}. \qquad (1.7.1)$$

If $m - r$ rows and $n - r$ columns are struck from $\mathbf{A}$, the remaining elements form an $r \times r$ matrix whose determinant M_A^r is said to be an rth-order minor of $\mathbf{A}$. For example, striking the third row and the second and fourth columns of

$$\mathbf{A} = \begin{bmatrix} a_{11} & a_{12} & a_{13} & a_{14} & a_{15} \\ a_{21} & a_{22} & a_{23} & a_{24} & a_{25} \\ a_{31} & a_{32} & a_{33} & a_{34} & a_{35} \\ a_{41} & a_{42} & a_{43} & a_{44} & a_{45} \end{bmatrix} \qquad (1.7.2)$$

generates the minor

$$M_A^3 = \begin{vmatrix} a_{11} & a_{13} & a_{15} \\ a_{21} & a_{23} & a_{25} \\ a_{41} & a_{43} & a_{45} \end{vmatrix}. \tag{1.7.3}$$

We can now make the following important definition:

DEFINITION. *The rank r (or r_A) of a matrix $\mathbf{A}$ is the order of the largest nonzero minor of $\mathbf{A}$.*

For example, for

$$\mathbf{A} = \begin{bmatrix} 2 & -1 \\ -1 & 2 \end{bmatrix}, \quad |\mathbf{A}| = 3 \tag{1.7.4}$$

and so $r_A = 2$. On the other hand, all of the minors of

$$\mathbf{A} = \begin{bmatrix} 1 & 1 & 1 & 1 \\ 1 & 1 & 1 & 1 \\ 1 & 1 & 1 & 1 \end{bmatrix}, \tag{1.7.5}$$

except M_A^1, are 0. Thus, $r_A = 1$ for this 3×4 matrix. For an $m \times n$ matrix $\mathbf{A}$, it follows from the definition of rank that $r_A \leq \min(m, n)$.

Let us end this chapter by mentioning the principal minors and traces of a matrix. They are important in the analysis of the time dependence of systems of equations. The jth-order trace of a matrix $\mathbf{A}$, $\mathrm{tr}_j \mathbf{A}$, is defined as the sum of the jth-order minors generated by striking $n - j$ rows and columns intersecting on the main diagonal of $\mathbf{A}$. These minors are called the principal minors of $\mathbf{A}$. Thus, for a 3×3 matrix,

$$\mathrm{tr}_3 \mathbf{A} = \begin{vmatrix} a_{11} & a_{12} & a_{13} \\ a_{21} & a_{22} & a_{23} \\ a_{31} & a_{32} & a_{33} \end{vmatrix} = \mathrm{Det}\, \mathbf{A}, \tag{1.7.6}$$

$$\mathrm{tr}_2 \mathbf{A} = \begin{vmatrix} a_{11} & a_{12} \\ a_{21} & a_{22} \end{vmatrix} + \begin{vmatrix} a_{11} & a_{13} \\ a_{31} & a_{33} \end{vmatrix} + \begin{vmatrix} a_{22} & a_{23} \\ a_{32} & a_{33} \end{vmatrix}, \tag{1.7.7}$$

$$\mathrm{tr}_1 \mathbf{A} = a_{11} + a_{22} + a_{33}. \tag{1.7.8}$$

For an $n \times n$ matrix $\mathbf{A}$, the nth-order trace is just the determinant of $\mathbf{A}$ and $\mathrm{tr}_1 \mathbf{A}$ is the sum of the diagonal elements of $\mathbf{A}$. These are the most common traces encountered in practical situations. However, all the traces figure importantly in the theory of eigenvalues of $\mathbf{A}$. In some texts, the term trace of $\mathbf{A}$ is reserved for $\mathrm{tr}_1 \mathbf{A} = \sum_{j=1}^n a_{jj}$, and the objects $\mathrm{tr}_j \mathbf{A}$ are called the invariants of $\mathbf{A}$.

EXERCISE 1.7.1. Show that all of the traces of the matrix

$$A = \begin{bmatrix} -4 & 1 & 1 & 0 \\ 1 & -4 & 1 & 1 \\ 1 & 1 & -4 & 1 \\ 0 & 1 & 1 & -4 \end{bmatrix} \tag{1.7.9}$$

are positive. We will show in Chapter 6 that this implies that **A** has only negative eigenvalues. It also implies that, independently of initial conditions, the solution **x** to the equation

$$\frac{d\mathbf{x}}{dt} = \mathbf{A}\mathbf{x} \tag{1.7.10}$$

■■■ always vanishes with increasing time t.

PROBLEMS

1. Evaluate the determinant of the matrix given by Eq. (1.7.9) using the formulas

$$D = \sum_{l_1, \ldots, l_n} (-1)^P a_{1l_1} a_{2l_2} \cdots a_{nl_n} = \sum_{i=1}^{n} a_{ij} A_{ij}$$

2. Solve the following determinants by elementary operations

   (a) $\begin{vmatrix} 6 & 3 & 2 & 9 \\ 5 & 3 & 2 & 9 \\ 4 & 4 & 8 & 9 \\ 4 & 3 & 2 & 9 \end{vmatrix}.$

   (b) $\begin{vmatrix} a+b & c & c \\ a & b+c & a \\ b & b & c+a \end{vmatrix} = 4abc.$

   (c) Solve the following set of equations:

   $$3x_1 + x_2 + 2x_3 = 1$$
   $$-x_1 + 4x_2 + 5x_3 = 1$$
   $$-7x_1 + 2x_2 + x_3 = -1.$$

3. Evaluate the determinants

   $\begin{vmatrix} 0 & 0 & 0 & a_1 \\ 0 & 0 & b_1 & a_2 \\ 0 & c_1 & b_2 & a_3 \\ d_1 & c_2 & b_3 & a_4 \end{vmatrix}$

and

$$\begin{vmatrix} b_2 & b_3 & b_4 & b_5 \\ 0 & c_3 & 0 & 0 \\ 0 & d_3 & d_4 & d_5 \\ 0 & e_3 & 0 & e_5 \end{vmatrix}.$$

4. Using Cramer's rule, find x, y, and z for the following system of equations:

$$3x - 4y + 2z = 1$$
$$2x + 3y - 3z = -1$$
$$5x - 5y + 4z = 7.$$

5. Using Cramer's rule, find x, y, and z for the following system of equations:

$$6/x - 2/y + 1/z = 4$$
$$2/x + 5/y - 2/z = 3/4$$
$$5/x - 1/y + 3/z = 63/4.$$

6. Show that

$$\begin{vmatrix} 0 & a & b & c \\ a & 0 & c & b \\ b & c & 0 & a \\ c & b & a & 0 \end{vmatrix} = \begin{vmatrix} 0 & 1 & 1 & 1 \\ 1 & 0 & c^2 & b^2 \\ 1 & c^2 & 0 & a^2 \\ 1 & b^2 & a^2 & 0 \end{vmatrix}.$$

7. Use the determinant properties to evaluate

$$\begin{vmatrix} b^2 + c^2 & a^2 & a^2 \\ b^2 & c^2 + a^2 & b^2 \\ c^2 & c^2 & a^2 + b^2 \end{vmatrix}.$$

8. Using Cramer's rule, find x, y, and z, where

$$3x + 4y - 2z = 3$$
$$2x + 2y - 3z = 1$$
$$-x + y - 2z = -2.$$

9. Using Cramer's rule, find x, y, and z for the system of equations

$$4x + 7y - z = 7$$
$$3x + 2y + 2z = 9$$
$$x + 5y - 3z = 3.$$

10. Using Cramer's rule, find x, y, and z for the system of equations
$$x + 3y = 0$$
$$2x + 6y + 4z = 0$$
$$-x + 2z = 0.$$

11. Using Cramer's rule, find x, y, and z for the system of equations
$$x + y + z = 1$$
$$x + 1.0001y + 2z = 2$$
$$x + 2y + 2z = 1.$$

12. Let
$$D_1 = \begin{vmatrix} a_{11} & a_{12} \\ a_{21} & a_{22} \end{vmatrix} \quad \text{and} \quad D_2 = \begin{vmatrix} b_{11} & b_{12} \\ b_{21} & b_{22} \end{vmatrix}.$$

Show that
$$D = D_1 D_2 = \begin{vmatrix} a_{11} & a_{12} & 0 & 0 \\ a_{21} & a_{22} & 0 & 0 \\ -1 & 0 & b_{11} & b_{12} \\ 0 & -1 & b_{21} & b_{22} \end{vmatrix}.$$

13. What is the rank of
$$\mathbf{A} = \begin{bmatrix} 2 & 1 & 3 & 4 \\ -1 & 1 & -2 & -1 \\ 0 & 3 & -1 & 2 \end{bmatrix}?$$

14. Evaluate the determinant
$$D = \begin{vmatrix} 3 & 5 & 2 & 4 \\ 1 & 1 & -1 & 6 \\ 2 & 3 & 5 & 1 \\ 2 & 1 & 4 & 8 \end{vmatrix}$$

by first generating zero entries where you can and then using a cofactor expansion.

15. Show that
$$\begin{vmatrix} 1+c_1 & 1 & 1 & 1 \\ 1 & 1+c_2 & 1 & 1 \\ 1 & 1 & 1+c_3 & 1 \\ 1 & 1 & 1 & 1+c_4 \end{vmatrix} = c_1 c_2 c_3 c_4 \left(1 + \frac{1}{c_1} + \frac{1}{c_2} + \frac{1}{c_3} + \frac{1}{c_4}\right).$$

16. What is the rank of

$$\mathbf{A} = \begin{bmatrix} 1 & 2 & 3 & 4 \\ 1 & 4 & 6 & 8 \\ 5 & 7 & 9 & 1 \end{bmatrix} ?$$

17. Give all of the minors of

$$\mathbf{A} = \begin{bmatrix} a_{11} & a_{12} & a_{13} \\ a_{21} & a_{22} & a_{23} \end{bmatrix}.$$

18. Give all of the traces of the matrix whose determinant is shown in Problem 15.

19. Solve the equation

$$\begin{vmatrix} 1-x & 7+2x & 0+3x \\ 4+2x & 10-4x & 6-6x \\ 2 & 4 & 5 \end{vmatrix} = 0.$$

20. Without expanding the determinant, show that

$$\begin{vmatrix} x^2 & x & 1 \\ y^2 & y & 1 \\ z^2 & y & 1 \end{vmatrix} = (x-y)(y-z)(x-z).$$

21. Consider the set of matrices

$$\mathbf{A}_n = \begin{bmatrix} 1 & b & & & & & & \\ b & a & b & & & & 0 & \\ & b & a^2 & b & & & & \\ & & b & a^3 & b & & & \\ & & & b & a^4 & \ddots & & \\ & & & & \ddots & \ddots & b & \\ 0 & & & & & b & a^{n-1} \end{bmatrix}.$$

(a) Defining the determinants D_n as

$$D_n = |\mathbf{A}_n|,$$

find a recursion relation for D_n (i.e., $D_n = f(D_{n-1}, D_{n-2}, \ldots; a, b)$).

(b) Letting $a = 0.5$ and $b = 1$, write a computer program to evaluate D_{99}.

22. Consider the n-dimensional matrix

$$\begin{bmatrix} x & a & & & & & \\ a & x & a & & & \mathbf{0} & \\ & a & x & a & & & \\ & & a & x & a & & \\ & & & a & x & \ddots & \\ & \mathbf{0} & & & \ddots & \ddots & \end{bmatrix}.$$

For the case where $a = 1$ and $x = 2\cos\theta$, prove that the determinant is given by $D = \sin(n+1)\theta / \sin\theta$ as long as θ is restricted to $0 < \theta < \pi$.

23. Find the determinant of the $n \times n$ matrix whose diagonal elements are 0 and whose off-diagonal elements are a, i.e.,

$$\begin{vmatrix} 0 & a & a & \cdots & a & a \\ a & 0 & a & \cdots & a & a \\ a & a & 0 & \cdots & a & a \\ \vdots & \vdots & \vdots & & \vdots & \vdots \\ a & a & a & \cdots & 0 & a \\ a & a & a & \cdots & a & 0 \end{vmatrix}.$$

24. Find the following determinant:

$$\begin{vmatrix} 1+a_1 & a_2 & a_3 & \cdots & a_n \\ a_1 & 1+a_2 & a_3 & \cdots & a_n \\ a_1 & a_2 & 1+a_3 & \cdots & a_n \\ \vdots & \vdots & \vdots & & \vdots \\ a_1 & a_2 & a_3 & \cdots & 1+a_n \end{vmatrix}.$$

25. Prove the following relation for the *Vandermonde determinant*:

$$\begin{vmatrix} 1 & 1 & \cdots & 1 \\ x_1 & x_2 & \cdots & x_n \\ x_1^2 & x_2^2 & \cdots & x_n^2 \\ \vdots & \vdots & & \vdots \\ x_1^{n-1} & x_2^{n-1} & \cdots & x_n^{n-1} \end{vmatrix} = \prod_{i=1}^{n-1} \prod_{j=i+1}^{n} (x_j - x_i).$$

FURTHER READING

Aitken, A. C. (1948). "Determinants and Matrices." Oliver and Boyd, Edinburgh.
Aitken, A. C. (1964). "Determinants and Matrices." Interscience, New York.
Amundson, A. R. (1964). "Mathematical Methods in Chemical Engineering." Prentice-Hall, New Jersey.
Bronson, R. (1995). "Linear Algebra: an Introduction." Academic Press, San Diego.

FURTHER READING

Muir, T. (1960). "A Treatise on the Theory of Determinants." Dover, New York.
Muir, T. (1930). "Contributions to the History of Determinants, 1900-1920." Blackie & Son, London/Glasgow.
Nomizu, K. (1966). "Fundamentals of Linear Algebra." McGraw-Hill, New York.
Stigant, S. A. (1959). "The Elements of Determinants, Matrices and Tensors for Engineers." Macdonald, London.
Turnbull, H. W. (1928). "The Theory of Determinants, Matrices and Invariants." Blackie, London/Glasgow.
Vein, R. "Determinants and Their Applications in Mathematical Physics," Springer, New York.

2
VECTORS AND MATRICES

2.1. SYNOPSIS

In this chapter we will define the properties of matrix addition and multiplication for the general $m \times n$ matrix containing m rows and n columns. We will show that a vector is simply a special class of matrices: a column vector is an $m \times 1$ matrix and a row vector is a $1 \times n$ matrix. Thus, vector addition, scalar or inner products, and vector dyadics are defined by matrix addition and multiplication.

The inverse $\mathbf{A}^{-1}$ of the square matrix $\mathbf{A}$ is the matrix such that $\mathbf{A}\mathbf{A}^{-1} = \mathbf{A}^{-1}\mathbf{A} = \mathbf{I}$, where $\mathbf{I}$ is the unit matrix. We will show that when the inverse exists it can be evaluated in terms of the cofactors of $\mathbf{A}$ through the adjugate matrix

$$\text{adj}\,\mathbf{A} = \begin{bmatrix} A_{11} & A_{21} & \cdots & A_{n1} \\ A_{12} & A_{22} & \cdots & A_{n2} \\ \vdots & \vdots & & \vdots \\ A_{1n} & A_{2n} & \cdots & A_{nn} \end{bmatrix}.$$

Specifically, by using the cofactor expansion theorems of Chapter 1, we will prove that the inverse can be evaluated as

$$\mathbf{A}^{-1} = \frac{1}{D}\,\text{adj}\,\mathbf{A}.$$

We will also derive relations for evaluating the inverse, transpose, and adjoint of the product of matrices. The inverse of a product of matrices **AB** can be computed from the product of the inverses $\mathbf{B}^{-1}$ and $\mathbf{A}^{-1}$. Similar expressions hold for the transpose and adjoint of a product. The concept of matrix partitioning and its utility in computing the inverse of a matrix will be discussed.

Finally, we will introduce *linear vector spaces* and the important concept of linear independence of vectors sets. We will also expand upon the concept of vector norms, which are required in defining *normed linear vector spaces*. Matrix norms based on the length or norm of a vector are then defined and several very general properties of norms are derived. The utility of matrix norms will be demonstrated in analyzing the solvability of linear equations.

2.2. ADDITION AND MULTIPLICATION

The rules of matrix addition were given in Eq. (1.2.6). To be conformable for addition (i.e., for addition to be defined), the matrices **A** and **B** must be of the same dimension $m \times n$. The elements of $\mathbf{A} + \mathbf{B}$ are then $a_{ij} + b_{ij}$; i.e., corresponding elements are added to make the matrix sum. Using this rule for addition, the product of a matrix **A** with a scalar (complex or real number) α was defined as

$$\alpha \mathbf{A} = \begin{bmatrix} \alpha a_1 & \cdots & \alpha a_{1n} \\ \vdots & & \vdots \\ \alpha a_{m1} & \cdots & \alpha a_{mn} \end{bmatrix}. \tag{2.2.1}$$

Using the properties of addition and scalar multiplication, and the definition of the derivative of **A**,

$$\frac{d\mathbf{A}}{dt} = \lim_{\Delta t \to 0} \frac{\mathbf{A}(t + \Delta t) - \mathbf{A}(t)}{\Delta t}, \tag{2.2.2}$$

we find that

$$\frac{d\mathbf{A}}{dt} = \lim_{\Delta t \to 0} \begin{bmatrix} \dfrac{a_{11}(t+\Delta t) - a_{11}(t)}{\Delta t} & \cdots & \dfrac{a_{1n}(t+\Delta t) - a_{1n}(t)}{\Delta t} \\ \vdots & & \vdots \\ \dfrac{a_{m1}(t+\Delta t) - a_{m1}(t)}{\Delta t} & \cdots & \dfrac{a_{mn}(t+\Delta t) - a_{mn}(t)}{\Delta t} \end{bmatrix}$$

$$= \begin{bmatrix} \dfrac{da_{11}}{dt} & \cdots & \dfrac{da_{1n}}{dt} \\ \vdots & & \vdots \\ \dfrac{da_{m1}}{dt} & \cdots & \dfrac{da_{mn}}{dt} \end{bmatrix}. \tag{2.2.3}$$

We can therefore conclude that the derivative of a matrix $d\mathbf{A}/dt$ is a matrix whose elements are the derivatives of the elements of **A**, i.e.,

$$\frac{d\mathbf{A}}{dt} = \left[\frac{da_{ij}}{dt} \right]. \tag{2.2.4}$$

ADDITION AND MULTIPLICATION

Note that $|d\mathbf{A}/dt| \neq d|\mathbf{A}|/dt$. The determinant of $d\mathbf{A}/dt$ is a nonlinear function of the derivatives of a_{ij}, whereas the derivative of the determinant $|\mathbf{A}|$ is linear.

If $\mathbf{A}$ and $\mathbf{B}$ are conformable for matrix multiplication, i.e., if $\mathbf{A}$ is an $m \times n$ matrix and $\mathbf{B}$ is an $n \times p$ matrix, then the product

$$\mathbf{C} = \mathbf{AB} \tag{2.2.5}$$

is defined. $\mathbf{C}$ is then an $m \times p$ matrix with elements

$$c_{ij} = \sum_{k=1}^{n} a_{ik} b_{kj}. \tag{2.2.6}$$

Thus, the ijth element of $\mathbf{C}$ is the product of the ith row of $\mathbf{A}$ and the jth column of $\mathbf{B}$, and so $\mathbf{A}$ and $\mathbf{B}$ are conformable for the product $\mathbf{AB}$ if the number of columns of $\mathbf{A}$ equals the number of rows of $\mathbf{B}$. For example, if

$$\mathbf{A} = \begin{bmatrix} 1 & 1 \\ 2 & 1 \\ 3 & 1 \end{bmatrix} \quad \text{and} \quad \mathbf{B} = \begin{bmatrix} 1 & 2 \\ 2 & 1 \end{bmatrix},$$

then

$$\mathbf{AB} = \begin{bmatrix} 3 & 3 \\ 4 & 5 \\ 5 & 7 \end{bmatrix}, \tag{2.2.7}$$

whereas $\mathbf{BA}$ is not defined.

EXERCISE 2.2.1. Solve the linear system of equations

$$\frac{d\mathbf{A}}{dt} = \mathbf{BA}, \tag{2.2.8}$$

where

$$\mathbf{A}(t=0) = \begin{bmatrix} 1 & 1 \\ 2 & 1 \end{bmatrix} \tag{2.2.9}$$

and

$$\mathbf{B} = \begin{bmatrix} -2 & 1 \\ 1 & -2 \end{bmatrix}. \tag{2.2.10}$$

If $\mathbf{x}$ and $\mathbf{y}$ are n-dimensional vectors, then the product $\mathbf{x}^T\mathbf{y}$ is defined since the transpose $\mathbf{x}^T$ of $\mathbf{x}$ is a $1 \times n$ matrix and $\mathbf{y}$ is an $n \times 1$ matrix. The product is a 1×1 matrix (a scalar) given by

$$\mathbf{x}^T\mathbf{y} = \sum_{i=1}^{n} x_i y_i. \tag{2.2.11}$$

$\mathbf{x}^T\mathbf{y}$ is sometimes called the scalar or inner product of $\mathbf{x}$ and $\mathbf{y}$. The scalar product is only defined if $\mathbf{x}$ and $\mathbf{y}$ have the same dimension. If the vector $\mathbf{x}$ is real, then the scalar product $\mathbf{x}^T\mathbf{x} = \sum_{i=1}^{n} x_i^2$ is positive as long as $\mathbf{x}$ is not 0. We define the "length" $\|\mathbf{x}\|$ of $\mathbf{x}$ as

$$\|\mathbf{x}\| = \sqrt{\mathbf{x}^T\mathbf{x}} \tag{2.2.12}$$

for real vectors. For a complex vector, $\sum_i x_i^2$ is not necessarily positive—or even real. In this case, we define the inner product by

$$\mathbf{x}^\dagger \mathbf{y} = \sum_{i=1}^{n} x_i^* y_i \tag{2.2.13}$$

and the length of $\mathbf{x}$ by

$$\|\mathbf{x}\| = \sqrt{\mathbf{x}^\dagger \mathbf{x}} = \sqrt{\sum_{i=1}^{n} |x_i|^2}, \tag{2.2.14}$$

where the quantity $|x_i|^2$ ($\equiv x_i^* x_i$) is the square of the modulus of x_i. Here, again, $\mathbf{x}^\dagger$ denotes the *adjoint* of $\mathbf{x}$, namely, the complex conjugate of the transpose of $\mathbf{x}$,

$$\mathbf{x}^\dagger = [x_1^*, x_2^*, \ldots, x_n^*], \tag{2.2.15}$$

where x_i^* is the complex conjugate of x_i. The length $\|\mathbf{x}\|$ has the desired property that it is 0 *if and only if* every component of $\mathbf{x}$ is 0 (i.e., if $\mathbf{x} = \mathbf{0}$).

By definition, the matrix product $\mathbf{AB}$ exists only if $\mathbf{A}$ is an $m \times p$ matrix and $\mathbf{B}$ is a $p \times n$ matrix. The product $\mathbf{AB}$ is then an $m \times n$ matrix. Thus, if $\mathbf{x}$ is an m-dimensional vector ($m \times 1$ matrix) and $\mathbf{y}$ is an n-dimensional vector ($n \times 1$ matrix), the vector $\mathbf{y}^T$ is a $1 \times n$ matrix. The product $\mathbf{xy}^T$ therefore exists as an $m \times n$ matrix given by

$$\mathbf{xy}^T = \begin{bmatrix} x_1 y_1 & x_1 y_2 & \cdots & x_1 y_n \\ \vdots & \vdots & & \vdots \\ x_m y_1 & x_m y_2 & \cdots & x_m y_n \end{bmatrix}. \tag{2.2.16}$$

In the language of vectors and tensors, in Euclidean vector space the product $\mathbf{xy}^T$ is known as a dyadic. We will use this term in later chapters.

2.3. THE INVERSE MATRIX

We note that the product $\mathbf{Ax}$ is defined for an $m \times n$ matrix $\mathbf{A}$ and an n-dimensional vector $\mathbf{x}$ (i.e., an $n \times 1$ matrix). With the matrix product so defined, the set of linear equations

$$\begin{aligned} a_{11}x_1 + a_{12}x_2 + \cdots + a_{1n}x_n &= b_1 \\ \vdots \quad \vdots \quad \quad \vdots \quad \quad \vdots \\ a_{m1}x_1 + a_{m2}x_2 + \cdots + a_{mn}x_n &= b_m \end{aligned} \tag{2.3.1}$$

can be summarized by the single equation

$$\mathbf{Ax} = \mathbf{b}. \tag{2.3.2}$$

We can see that the multiplication of $\mathbf{A}$ by $\mathbf{x}$ transforms the n-dimensional vector $\mathbf{x}$ into the m-dimensional vector $\mathbf{b}$.

THE INVERSE MATRIX

When $n = m$ in Eq. (2.3.1) and $D_A \neq 0$, we know from Cramer's rule that Eq. (2.3.1) can be solved for **x** uniquely for any **b**. In this case, it seems natural to hunt for an $n \times n$ matrix $\mathbf{A}^{-1}$ that is the inverse of the $n \times n$ matrix **A**, i.e., a matrix such that

$$\mathbf{A}^{-1}\mathbf{A} = \mathbf{A}\mathbf{A}^{-1} = \mathbf{I}, \qquad (2.3.3)$$

where **I** is the identity matrix. The identity matrix is defined by the property that

$$\mathbf{I}\mathbf{y} = \mathbf{y}; \qquad (2.3.4)$$

i.e., the product of the $n \times n$ identity matrix **I** and an n-dimensional vector **y** produces the vector **y** again. With this property, Eq. (2.3.2) can be solved by multiplying it by $\mathbf{A}^{-1}$ to obtain

$$\mathbf{A}^{-1}\mathbf{A}\mathbf{x} = \mathbf{I}\mathbf{x} = \mathbf{x} = \mathbf{A}^{-1}\mathbf{b}. \qquad (2.3.5)$$

If $\mathbf{A}^{-1}$ can be found, then the solution **x** to $\mathbf{A}\mathbf{x} = \mathbf{b}$ is simply $\mathbf{A}^{-1}\mathbf{b}$.

The identity matrix **I** has the simple form

$$\mathbf{I} = \begin{bmatrix} 1 & 0 & \cdots & 0 \\ 0 & 1 & \cdots & 0 \\ \vdots & \vdots & \ddots & \vdots \\ 0 & 0 & \cdots & 1 \end{bmatrix} = [\delta_{ij}]; \qquad (2.3.6)$$

i.e., **I** is unity on the main diagonal and is 0 elsewhere.

We can use the cofactor expansion theorems proved in Chapter 1 to construct the inverse of a square, nonsingular (i.e., $|\mathbf{A}| \neq 0$) matrix **A**. We define the *adjugate* of a matrix **A** by

$$\operatorname{adj}\mathbf{A} = \begin{bmatrix} A_{11} & A_{21} & \cdots & A_{n1} \\ A_{12} & A_{22} & \cdots & A_{n2} \\ \vdots & \vdots & & \vdots \\ A_{1n} & A_{2n} & \cdots & A_{nn} \end{bmatrix}, \qquad (2.3.7)$$

where A_{ij} is the cofactor of a_{ij}. Note that the ijth element of $\operatorname{adj}\mathbf{A}$ is the cofactor of the jith element of **A**. The elements of the matrix product

$$\mathbf{A}\operatorname{adj}\mathbf{A} \qquad (2.3.8)$$

are then

$$\sum_{k=1}^{n} a_{ik} A_{jk}. \qquad (2.3.9)$$

But, according to the cofactor theorems (Eq. (1.5.18)),

$$\sum_{k=1}^{n} a_{ik} A_{jk} = D\delta_{ij}, \qquad (2.3.10)$$

where D is the determinant of $\mathbf{A}$. This means that $\mathbf{A} \operatorname{adj} \mathbf{A}$ is a diagonal matrix having the determinant D on the main diagonal (and, of course, 0's elsewhere), i.e.,

$$\mathbf{A} \operatorname{adj} \mathbf{A} = \begin{bmatrix} D & 0 & 0 & \cdots & 0 \\ 0 & D & 0 & \cdots & 0 \\ 0 & 0 & D & \cdots & 0 \\ \vdots & \vdots & \vdots & \ddots & \vdots \\ 0 & 0 & 0 & \cdots & D \end{bmatrix} = D\mathbf{I}. \qquad (2.3.11)$$

Likewise, the ijth element of the product

$$(\operatorname{adj} \mathbf{A})\, \mathbf{A} \qquad (2.3.12)$$

is

$$\sum_{k=1}^{n} A_{ki} a_{kj}, \qquad (2.3.13)$$

which again, by the cofactor expansion theorems, obeys

$$\sum_{k=1}^{n} A_{ki} a_{kj} = D \delta_{ij} \qquad (2.3.14)$$

and implies

$$(\operatorname{adj} \mathbf{A})\, \mathbf{A} = D\mathbf{I}. \qquad (2.3.15)$$

In summary, the properties of the $\operatorname{adj} \mathbf{A}$ are that

$$\mathbf{A} \operatorname{adj} \mathbf{A} = (\operatorname{adj} \mathbf{A})\, \mathbf{A} = D\mathbf{I}. \qquad (2.3.16)$$

If $D \neq 0$, then Eqs. (2.3.11) and (2.3.15) can be rearranged to give

$$\mathbf{A}\left(\frac{1}{D} \operatorname{adj} \mathbf{A}\right) = \left(\frac{1}{D} \operatorname{adj} \mathbf{A}\right) \mathbf{A} = \mathbf{I}, \qquad (2.3.17)$$

which, when compared with Eq. (2.3.3), shows that the inverse of $\mathbf{A}$ can be computed as

$$\mathbf{A}^{-1} = \frac{1}{D} \operatorname{adj} \mathbf{A}. \qquad (2.3.18)$$

Since the elements of $(\operatorname{adj} \mathbf{A})\, \mathbf{b}$ are

$$\sum_{j=1}^{n} A_{ji} b_j = D_i, \qquad i = 1, \ldots, n, \qquad (2.3.19)$$

where D_i is defined in Eq. (1.6.6), it follows that the elements of $\mathbf{x} = \mathbf{A}^{-1}\mathbf{b}$ are

$$x_i = \frac{D_i}{D}, \qquad i = 1, \ldots, n. \qquad (2.3.20)$$

THE INVERSE MATRIX

Not surprisingly, the solution $\mathbf{x} = \mathbf{A}^{-1}\mathbf{b}$ to $\mathbf{A}\mathbf{x} = \mathbf{b}$ is exactly the one given by Cramer's rule. What is important, however, is that Eq. (2.3.8) gives the inverse of $\mathbf{A}$ once and for all. We can therefore generate a solution for a particular $\mathbf{b}$ by taking the matrix product $\mathbf{A}^{-1}\mathbf{b}$.

It should be noted that true inverses are only defined for square matrices. Otherwise, $\mathbf{A}^{-1}\mathbf{A}$ and $\mathbf{A}\mathbf{A}^{-1}$ cannot be equal to the same square matrix. We can, however, define left and right inverses or pseudo-inverses. If there exists a matrix $\mathbf{G}$ such that

$$\mathbf{G}\mathbf{A} = \mathbf{I}, \qquad (2.3.21)$$

then we say $\mathbf{G}$ is a left inverse of $\mathbf{A}$. Accordingly, if there exists a matrix $\mathbf{H}$ such that

$$\mathbf{A}\mathbf{H} = \mathbf{I}, \qquad (2.3.22)$$

then we say $\mathbf{H}$ is a right inverse of $\mathbf{A}$. Note that left and right inverses need not be unique.

EXAMPLE 2.3.1. Find the inverse of

$$\mathbf{A} = \begin{bmatrix} 2 & 1 \\ 2 & 2 \end{bmatrix}. \qquad (2.3.23)$$

Since $|\mathbf{A}| = 2$, the adjugate formula gives

$$\mathbf{A}^{-1} = \frac{1}{2}\begin{bmatrix} 2 & -1 \\ -2 & 2 \end{bmatrix}, \qquad (2.3.24)$$

such that

$$\mathbf{A}^{-1}\mathbf{A} = \frac{1}{2}\begin{bmatrix} 2 & 0 \\ 0 & 2 \end{bmatrix} = \begin{bmatrix} 1 & 0 \\ 0 & 1 \end{bmatrix} \equiv \mathbf{I} = \mathbf{A}\mathbf{A}^{-1}.$$

EXAMPLE 2.3.2. Find the left or right inverse of

$$\mathbf{A} = \begin{bmatrix} 1 & -1 \\ -1 & 1 \end{bmatrix}. \qquad (2.3.25)$$

Since $|\mathbf{A}| = 0$, the adjugate formula will not give an inverse. To find a left inverse, we must find

$$\mathbf{G} = \begin{bmatrix} g_{11} & g_{12} \\ g_{21} & g_{22} \end{bmatrix}, \qquad (2.3.26)$$

such that

$$\mathbf{G}\mathbf{A} = \begin{bmatrix} g_{11} - g_{12} & -g_{11} + g_{12} \\ g_{21} - g_{22} & -g_{21} + g_{22} \end{bmatrix} = \begin{bmatrix} 1 & 0 \\ 0 & 1 \end{bmatrix} \qquad (2.3.27)$$

or

$$g_{11} - g_{12} = 1$$
$$-g_{11} + g_{12} = 0$$
$$g_{21} - g_{22} = 0$$
$$-g_{21} + g_{22} = 1. \tag{2.3.28}$$

These equations imply that $g_{11} = g_{12}$, $g_{21} = g_{22}$, and $0 = 1$, which is impossible. ■ ■ ■ Thus, **G** does not exist. Similarly, we can show that **H** does not exist.

Examples 2.3.1 and 2.3.2 illustrate two general results for a square matrix. Namely, if a square matrix is nonsingular ($|\mathbf{A}| \neq 0$), its inverse always exists. We have shown this by the above adjugate constructions. Also, if a square matrix is singular ($|\mathbf{A}| = 0$), it has neither a right nor a left inverse. The proof of this property requires a bit more than we have learned so far. We need to know that the determinant of a product of square matrices obeys the formula

$$|\mathbf{AB}| = |\mathbf{A}| \, |\mathbf{B}|. \tag{2.3.29}$$

From this property, which is proved in Chapter 4, and the property $|\mathbf{I}| = 1$, it follows that $|\mathbf{GA}| = 0$ is always true if $|\mathbf{A}| = 0$. Therefore, the equations $\mathbf{GA} = \mathbf{I}$ and $\mathbf{AH} = \mathbf{I}$, upon taking the determinants, imply that $0 = 1$, which is impossible.

EXAMPLE 2.3.3. Find the right inverse of

$$\mathbf{A} = \begin{bmatrix} 1 & -1 & 1 \\ 1 & 1 & 2 \end{bmatrix}. \tag{2.3.30}$$

The pertinent equation is $\mathbf{AH} = \mathbf{I}$,

$$\mathbf{AH} = \begin{bmatrix} 1 & -1 & 1 \\ 1 & 1 & 2 \end{bmatrix} \begin{bmatrix} h_{11} & h_{12} \\ h_{21} & h_{22} \\ h_{31} & h_{32} \end{bmatrix} = \begin{bmatrix} 1 & 0 \\ 0 & 1 \end{bmatrix}. \tag{2.3.31}$$

The corresponding set of equations is

$$h_{11} - h_{21} + h_{31} = 1, \quad h_{12} - h_{22} + h_{32} = 0$$
$$h_{11} + h_{21} + 2h_{31} = 0, \quad h_{12} + h_{22} + 2h_{32} = 1. \tag{2.3.32}$$

For arbitrary h_{31} and h_{32}, these equations yield

$$\mathbf{H} = \frac{1}{2} \begin{bmatrix} 1 - 3h_{31} & 1 - 3h_{32} \\ -1 - h_{31} & 1 - h_{32} \\ 2h_{31} & 2h_{32} \end{bmatrix}. \tag{2.3.33}$$

■ ■ ■ Thus, the right inverse of **A** exists, but is not unique.

EXERCISE 2.3.1. Show that the matrix defined by Eq. (2.3.30) does not have ■ ■ ■ a left inverse.

■ **EXERCISE 2.3.2.** Show that the matrix

$$A = \begin{bmatrix} 1 & -1 & 1 \\ 1 & -1 & 1 \end{bmatrix} \qquad (2.3.34)$$

■ ■ ■ has neither a left nor a right inverse.

An important property of the inverses of square matrices is that

$$(AB)^{-1} = B^{-1}A^{-1}, \qquad (ABC)^{-1} = C^{-1}B^{-1}A^{-1}, \qquad \text{etc.;} \qquad (2.3.35)$$

that is, the inverse of a string of products of matrices is the product of the inverse of the matrices written in the opposite sequence. To prove Eq. (2.3.35), let D denote the inverse of AB, i.e., $D = (AB)^{-1}$. Then

$$DAB = ABD = I. \qquad (2.3.36)$$

However, $B^{-1}A^{-1}(ABD) = B^{-1}(A^{-1}A)BD = B^{-1}IBD = (B^{-1}B)D = ID = D$. Since $ABD = I$ and $B^{-1}A^{-1}I = B^{-1}A^{-1}$, it follows that $D = B^{-1}A^{-1}$, or

$$(AB)^{-1} = B^{-1}A^{-1}. \qquad (2.3.37)$$

To prove the property for longer strings, set

$$B' = BC. \qquad (2.3.38)$$

Then

$$(ABC)^{-1} = (AB')^{-1} = (B')^{-1}A^{-1} = C^{-1}B^{-1}A^{-1}, \qquad (2.3.39)$$

where the result in Eq. (2.3.37) is used.

2.4. TRANSPOSE AND ADJOINT

We saw in Chapter 1 that the transpose A^T of the $m \times n$ matrix A is the $n \times m$ matrix formed by interchanging the rows and columns of A, i.e.,

$$A^T = \begin{bmatrix} a_{11} & a_{21} & \cdots & a_{m1} \\ a_{12} & a_{22} & \cdots & a_{m2} \\ \vdots & \vdots & \ddots & \vdots \\ a_{1n} & a_{2n} & \cdots & a_{mn} \end{bmatrix}. \qquad (2.4.1)$$

Accordingly, the adjoint $A^\dagger$ of A is the complex conjugate of the transpose of A, i.e.,

$$A^\dagger = \begin{bmatrix} a_{11}^* & \cdots & a_{m1}^* \\ \vdots & & \vdots \\ a_{1n}^* & \cdots & a_{mn}^* \end{bmatrix}, \qquad (2.4.2)$$

where a_{ij}^* is the complex conjugate of the element a_{ij}. In the shorter notation,

$$\mathbf{A} = [a_{ij}], \qquad \mathbf{A}^T = [a_{ji}], \qquad \mathbf{A}^\dagger = [a_{ji}^*]. \qquad (2.4.3)$$

Since a scalar α is a 1×1 matrix, the adjoint of a scalar is just its complex conjugate, i.e., $\alpha^\dagger = \alpha^*$. Of course, if $\mathbf{A}$ is real the adjoint and the transpose are the same since $(\mathbf{A}^T)^* = \mathbf{A}^T$.

In many cases, one needs the transpose or adjoint of matrix products such as $\mathbf{AB}$, $\mathbf{ABC}$, etc., for which the following property is useful:

$$(\mathbf{AB})^T = \mathbf{B}^T \mathbf{A}^T, \qquad (\mathbf{ABC})^T = \mathbf{C}^T \mathbf{B}^T \mathbf{A}^T, \qquad \text{etc.,} \qquad (2.4.4a)$$

$$(\mathbf{AB})^\dagger = \mathbf{B}^\dagger \mathbf{A}^\dagger, \qquad (\mathbf{ABC})^\dagger = \mathbf{C}^\dagger \mathbf{B}^\dagger \mathbf{A}^\dagger, \qquad \text{etc.;} \qquad (2.4.4b)$$

that is, the transpose (adjoint) of a product string of matrices is the product of the transposes (adjoints) of the matrices taken in reverse order. The proof of Eq. (2.4.4a) is straightforward. The ijth element of $\mathbf{B}^T \mathbf{A}^T$ is $[\mathbf{B}^T \mathbf{A}^T]_{ij} = \sum_k b_{ik}^T a_{kj}^T = \sum_k a_{jk} b_{ki}$, proving that $(\mathbf{AB})^T = \mathbf{B}^T \mathbf{A}^T$. If we define $\mathbf{D} = \mathbf{BC}$, then we have just proved that $(\mathbf{AD})^T = \mathbf{D}^T \mathbf{A}^T$ and $\mathbf{D}^T = \mathbf{C}^T \mathbf{B}$, so that $(\mathbf{ABC})^T = \mathbf{C}^T \mathbf{B}^T \mathbf{A}^T$. A similar proof holds for a product stream of any length. Moreover, since the adjoint is simply the complex conjugate of the transpose, Eq. (2.4.4b) follows immediately from Eq. (2.4.4a). The matrices in Eq. (2.4.4) must be conformable to multiplication but, in general, do not have to be square.

EXAMPLE 2.4.1. If $\mathbf{A}$ is an $m \times n$ matrix, $\mathbf{x}$ an m-dimensional vector ($m \times 1$ matrix), and $\mathbf{y}$ an n-dimensional vector ($n \times 1$ matrix), then the product $\mathbf{x}^\dagger \mathbf{A} \mathbf{y}$ is a 1×1 matrix or a scalar. The adjoint of $\mathbf{x}^\dagger \mathbf{A} \mathbf{y}$, namely, $(\mathbf{x}^\dagger \mathbf{A} \mathbf{y})^\dagger = \mathbf{y}^\dagger \mathbf{A}^\dagger \mathbf{x}$, is simply its complex conjugate, $(\mathbf{x}^\dagger \mathbf{A} \mathbf{y})^*$.

In many practical situations (such as reactor stability analysis or process control), one would like to know whether the equation

$$\frac{d\mathbf{x}}{dt} = -\mathbf{A}\mathbf{x} \qquad (2.4.5)$$

predicts that an arbitrary perturbation $\mathbf{x}$ will decay to 0 or diverge in time. In such a problem, the matrix $\mathbf{A}$ is square, say $n \times n$. The answer to the question is: if the scalar

$$\mathbf{z}^\dagger (\mathbf{A} + \mathbf{A}^\dagger) \mathbf{z}$$

is positive for every nonzero n-dimensional vector $\mathbf{z}$, then $\mathbf{x} \to 0$ as $t = \infty$ for any initial condition. To prove this, note that

$$\frac{d\|\mathbf{x}\|^2}{dt} = \frac{d}{dt}(\mathbf{x}^\dagger \mathbf{x}) = \left(\frac{d\mathbf{x}^\dagger}{dt}\right)\mathbf{x} + \mathbf{x}^\dagger \frac{d\mathbf{x}}{dt}. \qquad (2.4.6)$$

If $\mathbf{x}$ obeys Eq. (2.4.5), then

$$\frac{d\|\mathbf{x}\|^2}{dt} = -\mathbf{x}^\dagger (\mathbf{A} + \mathbf{A}^\dagger) \mathbf{x}. \qquad (2.4.7)$$

If $\mathbf{x}^\dagger(\mathbf{A} + \mathbf{A}^\dagger)\mathbf{x}$ is positive for every vector $\mathbf{x}$, then Eq. (2.4.7) implies that

$$\frac{d\|\mathbf{x}\|^2}{dt} \leq 0 \quad \text{for all time,} \tag{2.4.8}$$

or the magnitude of all the components of $\mathbf{x}$ must approach 0. If, for any $\mathbf{x}$, the product $\mathbf{x}^\dagger(\mathbf{A} + \mathbf{A}^\dagger)\mathbf{x}$ is negative, then there will be some initial conditions for which the length of $\mathbf{x}$, and hence the magnitude of its components, grow in time. In such a case, the reactor or control system would be unstable.

Let us close this section by noting that the inverse of a transpose (adjoint) is the transpose (adjoint) of the inverse, i.e.,

$$(\mathbf{A}^T)^{-1} = (\mathbf{A}^{-1})^T \quad \text{and} \quad (\mathbf{A}^\dagger)^{-1} = (\mathbf{A}^{-1})^\dagger. \tag{2.4.9}$$

The properties $\mathbf{I}^T = \mathbf{I}$ and $\mathbf{I} = (\mathbf{A}\mathbf{A}^{-1})^T = (\mathbf{A}^{-1})^T \mathbf{A}^T$ show that $(\mathbf{A}^{-1})^T = (\mathbf{A}^T)^{-1}$. Similarly, it follows from $\mathbf{I}^\dagger = \mathbf{I}$ that $(\mathbf{A}^{-1})^\dagger = (\mathbf{A}^\dagger)^{-1}$.

2.5. PARTITIONING MATRICES

Any matrix can be partitioned into an array of matrices and/or vectors. As long as the matrices $\mathbf{A}$ and $\mathbf{B}$ are conformable for addition or multiplication, they can be partitioned into conformable matrices of matrices. Consider, for example, the partition of the 3×6 matrix $\mathbf{A}$ denoted by the lines:

$$\mathbf{A} = \begin{bmatrix} a_{11} & a_{12} & a_{13} & a_{14} & a_{15} & a_{16} \\ a_{21} & a_{22} & a_{23} & a_{24} & a_{25} & a_{26} \\ a_{31} & a_{32} & a_{33} & a_{34} & a_{35} & a_{36} \end{bmatrix} = \begin{bmatrix} \mathbf{A}_{11} & \mathbf{A}_{12} & \mathbf{A}_{13} \\ \mathbf{A}_{21} & \mathbf{A}_{22} & \mathbf{A}_{23} \end{bmatrix}, \tag{2.5.1}$$

where

$$\mathbf{A}_{11} = \begin{bmatrix} a_{11} & a_{12} & a_{13} \\ a_{21} & a_{22} & a_{23} \end{bmatrix}, \quad \mathbf{A}_{12} = \begin{bmatrix} a_{14} \\ a_{24} \end{bmatrix}, \quad \mathbf{A}_{13} = \begin{bmatrix} a_{15} & a_{16} \\ a_{25} & a_{26} \end{bmatrix}, \tag{2.5.2}$$

$$\mathbf{A}_{21} = [a_{31}, a_{32}, a_{33}], \quad \mathbf{A}_{22} = a_{34}, \quad \mathbf{A}_{23} = [a_{35}, a_{36}].$$

When $\mathbf{A}$ and $\mathbf{B}$ are conformable for addition, and are conformably partitioned, then their sum can be given by

$$\mathbf{A} + \mathbf{B} = \begin{bmatrix} \mathbf{A}_{11} & \mathbf{A}_{12} & \cdots & \mathbf{A}_{1q} \\ \vdots & \vdots & & \vdots \\ \mathbf{A}_{p1} & \mathbf{A}_{p2} & \cdots & \mathbf{A}_{pq} \end{bmatrix} + \begin{bmatrix} \mathbf{B}_{11} & \mathbf{B}_{12} & \cdots & \mathbf{B}_{1q} \\ \vdots & \vdots & & \vdots \\ \mathbf{B}_{p1} & \mathbf{B}_{p2} & \cdots & \mathbf{B}_{pq} \end{bmatrix}$$

$$= \begin{bmatrix} \mathbf{A}_{11} + \mathbf{B}_{11} & \cdots & \mathbf{A}_{1q} + \mathbf{B}_{1q} \\ \vdots & & \vdots \\ \mathbf{A}_{p1} + \mathbf{B}_{p2} & \cdots & \mathbf{A}_{rq} + \mathbf{B}_{pq} \end{bmatrix}. \tag{2.5.3}$$

By "conformably partitioned" we mean the elements $\mathbf{A}_{ij}$ and $\mathbf{B}_{ij}$ have to have the same dimension for addition.

When **A** and **B** are conformable for multiplication (i.e., the number of columns of **A** equals the number of rows of **B**), the product of the partitioned matrices **A** and **B** is

$$\mathbf{AB} = \begin{bmatrix} \mathbf{A}_{11} & \cdots & \mathbf{A}_{1p} \\ \vdots & & \vdots \\ \mathbf{A}_{q1} & \cdots & \mathbf{A}_{qp} \end{bmatrix} \begin{bmatrix} \mathbf{B}_{11} & \cdots & \mathbf{B}_{1r} \\ \vdots & & \vdots \\ \mathbf{B}_{p1} & \cdots & \mathbf{B}_{qr} \end{bmatrix}$$

$$= \begin{bmatrix} \sum_{k=1}^{p} \mathbf{A}_{1k}\mathbf{B}_{k1} & \cdots & \sum_{k=1}^{p} \mathbf{A}_{1k}\mathbf{B}_{kr} \\ \vdots & & \vdots \\ \sum_{k=1}^{p} \mathbf{A}_{qk}\mathbf{B}_{k1} & \cdots & \sum_{k=1}^{p} \mathbf{A}_{qk}\mathbf{B}_{kr} \end{bmatrix}. \quad (2.5.4)$$

Note that the product $\mathbf{A}_{ik}\mathbf{B}_{kj}$ has to be conformable for each k and for all ij pairs in question.

A partitioning that will often be useful in later chapters is the partitioning of a matrix **x** into its column vectors, i.e.,

$$\mathbf{X} = \begin{bmatrix} x_{11} & x_{12} & \cdots & x_{1n} \\ \vdots & \vdots & & \vdots \\ x_{m1} & x_{m2} & \cdots & x_{mn} \end{bmatrix} = [\mathbf{x}_1, \mathbf{x}_2, \ldots, \mathbf{x}_n], \quad (2.5.5)$$

where $\mathbf{x}_j$ is an m-dimensional vector with components x_{ij}. To see how this is useful, consider the equation

$$\mathbf{AX} = \mathbf{I} \quad (2.5.6)$$

for the $n \times n$ square matrices **A**, **X**, and **I**. The unit matrix **I** can be partitioned as

$$\mathbf{I} = [\mathbf{e}_1, \mathbf{e}_2, \ldots, \mathbf{e}_n], \quad (2.5.7)$$

where $\mathbf{e}_i$ is a vector with 1 in the ith row and 0's elsewhere, i.e.,

$$\mathbf{e}_i = \begin{bmatrix} 0 \\ \vdots \\ 0 \\ 1 \\ 0 \\ \vdots \\ 0 \end{bmatrix}. \quad (2.5.8)$$

The vectors $\mathbf{e}_i$ are orthonormal, which means

$$\mathbf{e}_i^\dagger \mathbf{e}_j = \delta_{ij}. \quad (2.5.9)$$

PARTITIONING MATRICES

If in Eq. (2.5.6) $\mathbf{X}$ is partitioned as shown in Eq. (2.5.5) and $\mathbf{I}$ as in Eq. (2.5.7), then multiplication of $\mathbf{A}$ and $\mathbf{X}$ using the rule shown in Eq. (2.5.4) yields

$$[\mathbf{A}\mathbf{x}_1, \mathbf{A}\mathbf{x}_2, \ldots, \mathbf{A}\mathbf{x}_n] = [\mathbf{e}_1, \mathbf{e}_2, \ldots, \mathbf{e}_n,\,], \qquad (2.5.10)$$

or, equating matrix components,

$$\mathbf{A}\mathbf{x}_i = \mathbf{e}_i, \qquad i = 1, \ldots, n. \qquad (2.5.11)$$

When $\mathbf{A}$ is not singular ($|\mathbf{A}| \neq 0$), Cramer's rule assures the solution to the equations of (2.5.11). In this case, the matrix $\mathbf{X}$ is the inverse $\mathbf{A}^{-1}$ of $\mathbf{A}$. Thus, by using matrix partitioning, we find yet another way to obtain the inverse of $\mathbf{A}$, namely, to solve the linear equations $\mathbf{A}\mathbf{x}_i = \mathbf{e}_i$ for each unit vector $\mathbf{e}_i$. The efficient solution of such equations is the topic of the next chapter.

■ **EXAMPLE 2.5.1.** Find the inverse of

$$\mathbf{A} = \begin{bmatrix} 2 & 3 \\ 8 & 7 \end{bmatrix}. \qquad (2.5.12)$$

$\mathbf{A}\mathbf{x}_1 = \mathbf{e}_1$ yields the equations

$$\begin{aligned} 2x_{11} + 3x_{21} &= 1 \\ 8x_{11} + 7x_{21} &= 0 \end{aligned} \qquad (2.5.13)$$

or

$$\mathbf{x}_1 = \begin{bmatrix} -\frac{7}{10} \\ \frac{4}{5} \end{bmatrix}. \qquad (2.5.14)$$

$\mathbf{A}\mathbf{x}_2 = \mathbf{e}_2$ yields

$$\begin{aligned} 2x_{12} + 3x_{22} &= 0 \\ 8x_{12} + 7x_{22} &= 1 \end{aligned} \qquad (2.5.15)$$

or

$$\mathbf{x}_2 = \begin{bmatrix} \frac{3}{10} \\ -\frac{1}{5} \end{bmatrix} \qquad (2.5.16)$$

and so

$$\mathbf{A}^{-1} = [\mathbf{x}_1, \mathbf{x}_2] = \begin{bmatrix} -\frac{7}{10} & \frac{3}{10} \\ \frac{4}{5} & -\frac{1}{5} \end{bmatrix}. \qquad (2.5.17)$$

■ ■ ■ It is easy to verify that $\mathbf{A}\mathbf{A}^{-1} = \mathbf{I}$.

■ **EXERCISE 2.5.1.** (a) Give the solution to

$$\mathbf{A}\mathbf{x} = \mathbf{b} \qquad (2.5.18)$$

for arbitrary $\mathbf{b}$ for the matrix $\mathbf{A}$ in the preceding example.
■ ■ ■ (b) Calculate the transpose of the inverse of $\mathbf{A}$.

2.6. LINEAR VECTOR SPACES

In the analysis of linear systems, it is frequently useful to define a vector space S and linear operators in that space. Generalization of the concept of *vector length* to the concept of *operator norm* can lead to some very strong statements about the existence of solutions and the convergence of some solution techniques.

We say S is a linear vector space if it is a collection of elements $\mathbf{x}, \mathbf{y}, \ldots$ for which addition is defined and which obey the following properties:

1. If $\mathbf{x}$ and $\mathbf{y}$ belong to S, then $\mathbf{x} + \mathbf{y}$ belongs to S; i.e., if $\mathbf{x}, \mathbf{y} \in S$, then $\mathbf{x} + \mathbf{y} \in S$.
2. $\mathbf{x} + \mathbf{y} = \mathbf{y} + \mathbf{x}$.
3. There exists the zero element $\mathbf{0}$ such that

$$\mathbf{x} + \mathbf{0} = \mathbf{x}.$$

4. For every element $\mathbf{x}$, there exists an element $-\mathbf{x}$ such that

$$\mathbf{x} + (-\mathbf{x}) = \mathbf{0}.$$

5. If α and β are complex numbers, then

$$\alpha(\beta \mathbf{x}) = (\alpha\beta)\mathbf{x}$$
$$(\alpha + \beta)\mathbf{x} = \alpha\mathbf{x} + \beta\mathbf{x}$$
$$\alpha(\mathbf{x} + \mathbf{y}) = \alpha\mathbf{x} + \alpha\mathbf{y}.$$

The collection E_n of all n-dimensional vectors obeying the previously given rules of addition and multiplication by a scalar obviously has the properties 1–5. Thus, E_n is a linear vector space. In much of this book, E_n will be the linear vector space of interest. However, the properties of a linear vector space admit much more general objects than n-dimensional vectors. A pertinent example is a collection of all functions sharing common properties, such as continuity, differentiability, integrability, and the like. Another example is the collection of all $n \times m$ matrices.

An important vector concept is *linear independence*. If $\mathbf{x}_1$ and $\mathbf{x}_2$ are linearly dependent, then there exists a complex (or real) number c such that $\mathbf{x}_1 + c\mathbf{x}_2 = 0$. For example, if

$$\mathbf{x}_1 = \begin{bmatrix} 1 \\ 2 \end{bmatrix} \quad \text{and} \quad \mathbf{x}_2 = \begin{bmatrix} \frac{1}{2} \\ 1 \end{bmatrix}, \tag{2.6.1}$$

then $\mathbf{x}_1 = 2\mathbf{x}_2$ or $c = -2$. If $\mathbf{x}_1$ and $\mathbf{x}_2$ are linearly independent, then there is no number c such that $\mathbf{x}_1 + c\mathbf{x}_2 = 0$. For example, the vectors

$$\mathbf{x}_1 = \begin{bmatrix} 1 \\ 1 \end{bmatrix} \quad \text{and} \quad \mathbf{x}_2 = \begin{bmatrix} 2 \\ 0 \end{bmatrix} \tag{2.6.2}$$

are linearly independent because there is no multiplier of $\mathbf{x}_2$ that will place 1 in the second component of $\mathbf{x}_2$. The general definition of linear independence is that the set of p n-dimensional vectors $\{\mathbf{x}_1, \mathbf{x}_2, \ldots, \mathbf{x}_p\}$ are linearly independent if there exists no nonzero set of numbers $\{c_1, c_2, \ldots, c_p\}$ (not all of which are 0) such that

$$\sum_{i=1}^{p} c_i \mathbf{x}_i = \mathbf{0}. \tag{2.6.3}$$

EXAMPLE 2.6.1. Prove that the vectors

$$\mathbf{x}_1 = \begin{bmatrix} 3 \\ 2 \\ 1 \end{bmatrix}, \quad \mathbf{x}_2 = \begin{bmatrix} 1 \\ 2 \\ 3 \end{bmatrix}, \quad \text{and} \quad \mathbf{x}_3 = \begin{bmatrix} 5 \\ 6 \\ 7 \end{bmatrix} \quad (2.6.4)$$

are linearly dependent.

We seek a solution to

$$\sum_{i=1}^{3} c_i \mathbf{x}_i = \mathbf{0} \quad (2.6.5)$$

or

$$\begin{aligned} 3c_1 + c_2 + 5c_3 &= 0 \\ 2c_1 + 2c_2 + 6c_3 &= 0 \\ c_1 + 3c_2 + 7c_3 &= 0. \end{aligned} \quad (2.6.6)$$

Assume $c_3 \neq 0$ and solve the first two equations for c_1 and c_2. Multiply the first equation by 2 and subtract the second equation

$$\begin{aligned} 2(3c_1 + c_2 &= -5c_3) \\ \underline{2c_1 + 2c_2 &= -6c_3} \\ 4c_1 &= -4c_3 \quad \text{or} \quad c_1 = -c_3 \text{ and } c_2 = -2c_3. \end{aligned} \quad (2.6.7)$$

The values $c_1 = -c_3$ and $c_2 = -2c_3$ satisfy the third equation in Eq. (2.6.6). Therefore, $c_3 = 1$, $c_1 = -1$, and $c_2 = -2$ are a solution to Eq. (2.6.5), and so the vectors $\mathbf{x}_1, \mathbf{x}_2, \mathbf{x}_3$ are linearly dependent.

We create what is known as a *normed linear vector space* by assigning to every vector $\mathbf{x}$ in S the norm (or length) $\|\mathbf{x}\|$. By definition, the norm is required to satisfy the properties:

1. $\|\mathbf{x}\| \geq 0$ and $\|\mathbf{x}\| = 0$ if and only if $\mathbf{x} = \mathbf{0}$.
2. $\|\alpha \mathbf{x}\| = |\alpha| \|\mathbf{x}\|$ for any complex number α and any $\mathbf{x}$. (2.6.8)
3. $\|\mathbf{x} + \mathbf{y}\| \leq \|\mathbf{x}\| + \|\mathbf{y}\|$.

Here, $|\alpha|$ denotes the magnitude of the complex number α.

Property 3 is known as the "triangle inequality" after the well-known property of Euclidean vectors as illustrated in Figure 2.6.1. In ordinary Euclidean space,

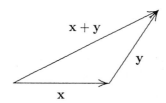

FIGURE 2.6.1

this inequality represents the property that the shortest distance between points a and b is a straight line (which lies along $\mathbf{x} + \mathbf{y}$ and is of length $\|\mathbf{x}+\mathbf{y}\| = [\sum_{i=1}^{3}(x_i + y_i)^2]^{1/2}$, where x_i and y_i denote the Cartesian coordinates of $\mathbf{x}$ and $\mathbf{y}$).

Clearly, if the norm $\|\mathbf{x}\|$ is defined as the square root of the inner product $\mathbf{x}^\dagger \mathbf{x}$, then properties 1–3 follow. However, this definition of the norm is only one among many, and it is not always the most convenient one. For example, if $\mathbf{x} \in E_n$, the quantity

$$\|\mathbf{x}\|_p = \left(\sum_{i=1}^{n} |x_i|^p\right)^{1/p} \tag{2.6.9}$$

for any positive real number p is also an acceptable norm. The special choice of $p = 2$ corresponds to the familiar length $\sqrt{\mathbf{x}^\dagger \mathbf{x}}$ introduced earlier. For the case $p = \infty$, it follows that

$$\begin{aligned}
\|\mathbf{x}\|_\infty &= \lim_{p\to\infty}\left[|x_m|^p \sum_{i=1}^{n}(|x_i|/|x_m|)^p\right]^{1/p} \\
&= |x_m| \lim_{p\to\infty}(\nu)^{1/p} = |x_m|\nu^0 \\
&= |x_m|,
\end{aligned} \tag{2.6.10}$$

where $|x_m|$ is the magnitude of x_i of maximum value, i.e.,

$$|x_m| = \max |x_i|, \qquad 1 \geq i \geq n, \tag{2.6.11}$$

and ν is the number of components x_i having the same maximum magnitude. For example, if

$$\mathbf{x} = \begin{bmatrix} \frac{1}{2} \\ e^{-i\theta} \\ e^{i\theta} \end{bmatrix} \tag{2.6.12}$$

where θ is real and $i = \sqrt{-1}$, then

$$|x_1| = 1/2, \qquad |x_2| = 1, \quad \text{and} \quad |x_3| = 1,$$

and so $|x_m| = 1$ and $\nu = 2$.

There are still other norms that we can define. For example, consider a self-adjoint matrix $\mathbf{A}$ (i.e., $\mathbf{A}^\dagger = \mathbf{A}$) with the additional property that

$$\mathbf{x}^\dagger \mathbf{A} \mathbf{x} > 0 \qquad \text{for all } \mathbf{x} \neq \mathbf{0} \text{ in } E_n.$$

Such a matrix is called positive definite. The quantity $\mathbf{x}^\dagger \mathbf{A} \mathbf{x}$ is real since $(\mathbf{x}^\dagger \mathbf{A} \mathbf{x})^* = \mathbf{x}^\dagger \mathbf{A}^\dagger \mathbf{x} = \mathbf{x}^\dagger \mathbf{A} \mathbf{x}$, and thus the quantity

$$\|\mathbf{x}\| \equiv \sqrt{\mathbf{x}^\dagger \mathbf{A} \mathbf{x}} \tag{2.6.13}$$

obeys the requisite properties of a norm. Similarly, if **A** is positive definite, the quantity

$$\|\mathbf{x}\|_{p,\mathbf{A}} = \left\{ \sum_{i=1}^{n} \left[\left(\sum_{j=1}^{n} a_{ij}^* x_j^* \right) \left(\sum_{j=1}^{n} a_{ij} x_j \right) \right]^{p/2} \right\}^{1/p} \quad (2.6.14)$$

obeys the conditions for a norm for any positive real number p.

Next, we want to define linear operators in linear vector spaces. An operator **A** is a linear operator in S if it obeys the properties:

1. If $\mathbf{x} \in S$, then $\mathbf{Ax} \in S$. $\quad$ (2.6.15)

2. If $\mathbf{x}$ and $\mathbf{y} \in S$, then $\mathbf{A}(\alpha\mathbf{x} + \beta\mathbf{y}) = \alpha\mathbf{Ax} + \beta\mathbf{Ay} \in S$, $\quad$ (2.6.16)

where α and β are complex numbers.

If $S = E_n$, an n-dimensional vector space, then all $n \times n$ matrices are linear operators in E_n when the product $\mathbf{Ax}$ obeys the rules of matrix multiplication. Square matrices as operators in vector spaces will preoccupy us in much of this book. However, if S is a function space, then the most common linear operators are differential or integral operators (more on this in Chapters 8–10).

We define the *norm* of a linear operator as

$$\|\mathbf{A}\| = \max_{\mathbf{x} \neq \mathbf{0}} \|\mathbf{Ax}\| / \|\mathbf{x}\|. \quad (2.6.17)$$

Thus, $\|\mathbf{A}\|$ is the largest value that the ratio on the right-hand side of Eq. (2.6.17) achieves for any nonzero vector in S. From properties 1–3 of the norm of a vector, the following relations hold for the norm of a linear operator:

1. $\|\mathbf{A}\| \geq 0$ and $\|\mathbf{A}\| = 0$ if and only if $\mathbf{A} = \mathbf{0}$. $\quad$ (2.6.18)

2. $\|\alpha\mathbf{A}\| = |\alpha|\,\|\mathbf{A}\|$ for any complex number α. $\quad$ (2.6.19)

3. $\|\mathbf{A} + \mathbf{B}\| \leq \|\mathbf{A}\| + \|\mathbf{B}\|$. $\quad$ (2.6.20)

4. $\|\mathbf{AB}\| \leq \|\mathbf{A}\|\,\|\mathbf{B}\|$. $\quad$ (2.6.21)

Conditions 1 and 2 are easy to establish from the properties of $\|\mathbf{x}\|$. To prove condition 3, note that

$$\|(\mathbf{A} + \mathbf{B})\mathbf{x}\| = \|\mathbf{Ax} + \mathbf{Bx}\| \leq \|\mathbf{Ax}\| + \|\mathbf{Bx}\|, \quad (2.6.22)$$

and so

$$\begin{aligned}\|(\mathbf{A} + \mathbf{B})\| &= \max_{\mathbf{x} \neq \mathbf{0}} \frac{\|(\mathbf{A} + \mathbf{B})\mathbf{x}\|}{\|\mathbf{x}\|} \\ &\leq \max_{\mathbf{x} \neq \mathbf{0}} \frac{\|\mathbf{Ax}\|}{\|\mathbf{x}\|} + \max_{\mathbf{x} \neq \mathbf{0}} \frac{\|\mathbf{Bx}\|}{\|\mathbf{x}\|} \\ &\leq \|\mathbf{A}\| + \|\mathbf{B}\|. \end{aligned} \quad (2.6.23)$$

For condition 4, note that

$$\|\mathbf{Ay}\| \leq \|\mathbf{A}\|\,\|\mathbf{y}\| \quad (2.6.24)$$

follows from the definition of $\|\mathbf{A}\|$. But if we set $\mathbf{y} = \mathbf{Bx}$ and use again the property

$$\|\mathbf{y}\| = \|\mathbf{Bx}\| \leq \|\mathbf{B}\| \, \|\mathbf{x}\|, \tag{2.6.25}$$

we find

$$\|\mathbf{ABx}\| \leq \|\mathbf{A}\| \, \|\mathbf{B}\| \, \|\mathbf{x}\| \tag{2.6.26}$$

or

$$\|\mathbf{AB}\| = \max_{\mathbf{x} \neq 0} \frac{\|\mathbf{ABx}\|}{\|\mathbf{x}\|} \leq \|\mathbf{A}\| \, \|\mathbf{B}\|, \tag{2.6.27}$$

which proves condition 4.

EXERCISE 2.6.1. Show that, for the space E_n and the norm $\|\mathbf{x}\|_p$,

$$\|\mathbf{A}\|_\infty = \max_{1 \leq i \leq n} \sum_{j=1}^{n} |a_{ij}| \tag{2.6.28}$$

and

$$\|\mathbf{A}\|_1 = \max_{1 \leq j \leq n} \sum_{i=1}^{n} |a_{ij}|, \tag{2.6.29}$$

i.e., that the $p = \infty$ norm of the matrix $\mathbf{A}$ is the maximum of the sum of the magnitudes of the row elements and the $p = 1$ norm is the maximum of the sum of the magnitudes of the column elements.

Let us close this chapter with an example of how operator norms can be useful. Suppose we want to solve the linear equation

$$\mathbf{x} - \mathbf{Ax} = \mathbf{b}. \tag{2.6.30}$$

The above equation can be rearranged to

$$\mathbf{x} = \mathbf{Ax} + \mathbf{b}. \tag{2.6.31}$$

Letting $\mathbf{x}^{(0)}$ denote an initial guess of the solution, the next estimate to the solution can be obtained from

$$\mathbf{x}^{(1)} = \mathbf{Ax}^{(0)} + \mathbf{b}. \tag{2.6.32}$$

Continuing this process, known as Picard iteration, gives for the kth estimate to the solution

$$\mathbf{x}^{(k)} = \mathbf{Ax}^{(k-1)} + \mathbf{b}. \tag{2.6.33}$$

The question is: will this solution converge? Suppose $\mathbf{x}$ is the true solution to Eq. (2.6.30). Then subtraction of Eq. (2.6.33) from the successive iteration equations yields

$$\mathbf{x} - \mathbf{x}^{(1)} = \mathbf{A}(\mathbf{x} - \mathbf{x}^{(0)}) \tag{2.6.34}$$

$$\mathbf{x} - \mathbf{x}^{(k)} = \mathbf{A}(\mathbf{x} - \mathbf{x}^{(k-1)}) = \mathbf{A}^k(\mathbf{x} - \mathbf{x}^{(0)}). \tag{2.6.35}$$

Taking the norm of Eq. (2.6.35) and using the properties of the norm, we obtain

$$\|\mathbf{x} - \mathbf{x}^{(k)}\| \leq \|\mathbf{A}\|^k \|\mathbf{x} - \mathbf{x}^{(0)}\|. \tag{2.6.36}$$

Thus, if $\|\mathbf{A}\| < 1$, Eq. (2.6.36) guarantees that $\|\mathbf{x} - \mathbf{x}^{(k)}\| \to 0$ as $k \to \infty$ and so in Picard iteration $\mathbf{x}^{(k)}$ converges to the solution of Eq. (2.6.30). If $\|\mathbf{A}\| > 1$, Picard iteration may or may not converge to a solution.

In the next chapter, we will introduce iterative solution schemes whose convergence can be analyzed in terms of vector and matrix norms.

PROBLEMS

1. Show that the unit vectors $\mathbf{e}_i$ in an n-dimensional space are similar to Cartesian unit vectors in three-dimensional Euclidean space.
2. Find the inverse of

$$\mathbf{A} = \begin{bmatrix} -1 & 1 & 0 \\ 1 & -2 & 1 \\ 0 & 1 & -2 \end{bmatrix}.$$

   Calculate the determinant of the inverse of $\mathbf{A}$.
3. Find the maximum value of the function

$$f = \mathbf{x}^T \mathbf{A} \mathbf{x} - 2\mathbf{b}^T \mathbf{x},$$

   where $\mathbf{A}$ is the matrix defined in Problem 2 and

$$\mathbf{b} = \begin{bmatrix} 3 \\ 2 \\ 1 \end{bmatrix}.$$

   Recall that $\partial f / \partial x_i = 0$, $i = 1, 2$, and 3, at the maximum.
4. Calculate the scalar or inner product $\mathbf{y}^\dagger \mathbf{x}$, the lengths $\|\mathbf{x}\|$ and $\|\mathbf{y}\|$, and the dyadic $\mathbf{x}\mathbf{y}^\dagger$, where

$$\mathbf{x} = \begin{bmatrix} 1 \\ 2 \\ 1 \\ 2 \end{bmatrix} \quad \text{and} \quad \mathbf{y} = \begin{bmatrix} i \\ 2 \\ 1 \\ 2i \end{bmatrix},$$

   and $i = \sqrt{-1}$.
5. Compute the adjugate and inverse of the matrix defined in Problem 2.
6. Find the right inverse of

$$\mathbf{A} = \begin{bmatrix} 1 & 2 & 2 \\ 3 & 1 & 1 \end{bmatrix}.$$

7. Prove that a two-dimensional space in which the quantity $\|\mathbf{x}\|$ is defined by

$$\|\mathbf{x}\| = \mathbf{x}^\dagger \mathbf{A} \mathbf{x}$$

is a normed linear vector space if

$$\mathbf{A} = \begin{bmatrix} 2 & -1 \\ -1 & 2 \end{bmatrix}.$$

Hint: Show that the properties in Eq. (2.6.8) are obeyed.

8. Prove that the vectors

$$\mathbf{x}_1 = \begin{bmatrix} 1 \\ 2 \\ 3 \end{bmatrix}, \quad \mathbf{x}_2 = \begin{bmatrix} 5 \\ 1 \\ 2 \end{bmatrix}, \quad \text{and} \quad \mathbf{x}_2 = \begin{bmatrix} 3 \\ 7 \\ 1 \end{bmatrix}$$

are linearly independent.

9. Consider the equation

$$\mathbf{x} = \mathbf{A}\mathbf{x} + \mathbf{b}, \tag{1}$$

where

$$\mathbf{A} = \begin{bmatrix} \frac{1}{2} & \frac{1}{4} & \frac{1}{5} \\ \frac{1}{6} & \frac{1}{3} & \frac{1}{7} \\ \frac{1}{5} & \frac{1}{4} & \frac{1}{4} \end{bmatrix} \quad \text{and} \quad \mathbf{b} = \begin{bmatrix} 5 \\ 6 \\ 3 \end{bmatrix}.$$

Prove that Eq. (1) can be solved by Picard iteration, i.e., if

$$\mathbf{x}^{(k+1)} = \mathbf{A}\mathbf{x}^{(k)} + \mathbf{b}, \quad k = 1, 2, \ldots,$$

then $\mathbf{x}^{(k)} \to \mathbf{x}$, the solution of Eq. (1), as $k \to \infty$.

10. Calculate the adjugate and inverse of the matrix

$$\mathbf{A} = \begin{bmatrix} 1 & 5 & 3 \\ 2 & 1 & 7 \\ 3 & 2 & 1 \end{bmatrix}.$$

11. Show that if

$$\mathbf{A}(\theta) = \begin{bmatrix} \cos\theta & -\sin\theta \\ \sin\theta & -\cos\theta \end{bmatrix}$$

then

$$\mathbf{A}(\theta_1)\mathbf{A}(\theta_2) = \mathbf{A}(\theta_2)\mathbf{A}(\theta_1) = \mathbf{A}(\theta_1 + \theta_2). \tag{1}$$

Consider the two-dimensional Euclidean vector represented as

$$\mathbf{x} = \begin{bmatrix} x \\ y \end{bmatrix}.$$

Examine $\mathbf{A}(\theta)\mathbf{x}$ and give a geometric interpretation of the result in Eq. (1).

12. Suppose that $\mathbf{A} = [a_{ij}]$, where

$$a_{ij} = (-1)^{n-j} \frac{(n-j)!}{(n-j-i+1)!(i-1)!}, \qquad i, j = 1, \ldots, n.$$

Compute $\mathbf{A}^3$.

13. Find all 2×2 solutions of the equation

$$\mathbf{A}^n = \mathbf{A},$$

where n is a positive integer. Consider first the case $n = 2$ and then solve for the general case.

14. Prove that, in general,

$$e^{(\mathbf{A}+\mathbf{B})t} \neq e^{\mathbf{A}t} e^{\mathbf{B}t}.$$

Under what conditions would $e^{(\mathbf{A}+\mathbf{B})t} = e^{\mathbf{A}t} e^{\mathbf{B}t}$ be true?

15. Suppose

$$\frac{d\mathbf{x}}{dt} = \mathbf{A}(t)\mathbf{x}, \qquad \mathbf{x}(t=0) = \mathbf{x}_0, \tag{1}$$

where the elements a_{ij} of $\mathbf{A}$ depend on t.

(a) Use the result in Problem 14 to prove that

$$\mathbf{x} = \exp\left(\int_0^t \mathbf{A}(\tau)\, d\tau\right) \mathbf{x}_0$$

is not a solution to Eq. (1).

(b) If $\mathbf{A}$ is independent of time, prove that

$$\mathbf{x} = e^{\mathbf{A}t} \mathbf{x}_0$$

is the solution to Eq. (1).

16. Prove that the *Frobenius norm*,

$$\|\mathbf{A}\|_F = \left(\sum_{i,j}^n |a_{ij}|^2\right)^{1/2},$$

obeys the requisite conditions of a norm, namely, Eqs. (2.6.18)–(2.6.21).

17. (a) Define a normed linear vector space for the linear vector space which consists of all the $n \times m$ matrices.
(b) What are the linear operators in the space?

FURTHER READING

Bellman, R. (1970). "Introduction to Matrix Analysis." McGraw–Hill, New York.
Bronson, R. (1995). "Linear Algebra: an Introduction." Academic Press, San Diego.
Hoffman, K. and Kunze, R. (1971). "Linear Algebra." 2nd Ed., Prentice Hall International, Englewood Cliffs, NJ.
Householder, A. S. (1965). "The Theory of Matrices in Numerical Analysis." Blaisdell, New York.
Nomizu, K. (1966). "Fundamentals of Linear Algebra." McGraw–Hill, New York.
Noble B. and Daniel, J. W. (1977). "Applied Linear Algebra." Prentice Hall International, Englewood Cliffs, NJ.
Smiley, M. F. (1951). "Algebra of Matrices." Allyn & Bacon, Needham Heights, MA.
Wade, T. L. (1951). "The Algebra of Vectors and Matrices." Addison-Wesley, Reading, MA.

3
SOLUTION OF LINEAR AND NONLINEAR SYSTEMS

3.1. SYNOPSIS

In this chapter we will deal exclusively with problems involving square matrices. We know from Cramer's rule that if $|\mathbf{A}| \neq 0$ the linear system $\mathbf{Ax} = \mathbf{b}$ has a unique solution. However, we learned in Chapter 1 that a straightforward application of Cramer's rule for a large system can be prohibitively expensive. In the next section we shall discover that a method due to Gauss (Gauss elimination) greatly reduces the cost of solving linear equations.

We will explain how Gauss elimination can be used to find the inverse of matrices and the **LU**-decomposition when it exists. The **LU**-decomposition of a matrix can lead to reduced costs of solving sequential linear equations involving the same matrix. It also can be the most economical way to store the matrix $\mathbf{A}$ for eventual computer solutions to linear equations. In each row of a band matrix, there are nonzero elements only in p columns to the left and q columns to the right of the main diagonal element. This structure leads to very efficient **LU**-decomposition if $p + q$ is small compared to the dimension of the vector space.

Despite the power of Gauss elimination, it is often more efficient to use iterative methods for solving linear equations. We will consider three such methods—the Jacobi, the Gauss–Seidel, and the successive overrelaxation (SOR) methods—and explore the conditions for the convergence of each. We will close the chapter by introducing the Picard and Newton–Raphson methods for solving nonlinear algebraic systems. The Newton–Raphson method is the more powerful, albeit more complicated, of the two methods. Getting a good first guess is essential to the success of both methods and thus continuation methods are discussed. We will see that, when the Newton–Raphson method converges, it does so quadratically with the number of iterations. This quadratic convergence is not only very efficient, but it also provides a fingerprint whereby programming errors can be detected.

3.2. SIMPLE GAUSS ELIMINATION

Consider the equation $\mathbf{Ux} = \mathbf{b}$, where $\mathbf{U}$ is an upper triangular matrix; i.e., $\mathbf{U}$ has nonzero elements only on and above the main diagonal. The equations corresponding to $\mathbf{Ux} = \mathbf{b}$ are

$$\begin{aligned}
u_{11}x_1 + u_{12}x_2 + \cdots + u_{1n}x_n &= b_1 \\
u_{22}x_2 + \cdots + u_{2n}x_n &= b_2 \\
&\vdots \\
u_{n-1,n-1}x_{n-1} + u_{n-1,n}x_n &= b_{n-1} \\
u_{nn}x_n &= b_n.
\end{aligned} \quad (3.2.1)$$

Symbolically, these equations can be represented in Fig. 3.2.1, where the triangle represents the matrix $\mathbf{U}$ and the vertical lines represent the vectors $\mathbf{x}$ and $\mathbf{b}$.

Expansion by cofactors shows that the determinant of $\mathbf{U}$ is given by $|\mathbf{U}| = \prod_{i=1}^{n} u_{ii}$. Thus, if none of the diagonal elements of $\mathbf{U}$ is 0, then $|\mathbf{U}| \neq 0$ and Eq. (3.2.1) has a unique solution. The solution can be determined by *back substitution*, namely,

$$x_n = \frac{b_n}{u_{nn}}$$

$$x_{n-1} = \frac{b_{n-1} - u_{n-1,n}x_n}{u_{n-1,n-1}}$$

$$\vdots$$

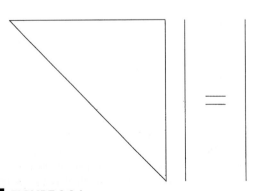

FIGURE 3.2.1

or

$$x_n = \frac{b_n}{u_{nn}}$$

$$x_i = \frac{b_i - \sum_{k=i+1}^{n} u_{ik} x_k}{u_{ii}}, \quad i = n-1, n-2, \ldots, 1. \tag{3.2.2}$$

From x_n we compute x_{n-1}, from which we compute x_{n-2}, etc. until $\{x_n, \ldots, x_1\}$ have been determined.

To estimate the time it takes to evaluate the x_i, we ignore the faster addition and subtraction operations and count only the slower multiplication and division operations. We note that at the ith step there are $(n-i)$ multiplications and one division. Thus, the total number of multiplications and divisions can be determined from the series

$$\sum_{i=n}^{1}(n-i+1) = \sum_{i=1}^{n}(n-i+1) = \frac{1}{2}n(n+1). \tag{3.2.3}$$

The way to evaluate a sum such as that in Eq. (3.2.3) is to note that in analogy with the integral

$$\sum_{i=1}^{n}(n-i+1)\,\Delta i \leftrightarrow \int_{1}^{n}(n-i+1)\,di$$

$$= n(n-1) - \frac{1}{2}(n^2-1) + n - 1, \tag{3.2.4}$$

we expect a sum of first-order terms in n and summed to n terms to be a quadratic function of n. Thus, we try

$$\sum_{i=1}^{n}(n-i+1) = an^2 + bn + c. \tag{3.2.5}$$

By taking the three special cases, $n = 1, 2$, and 3, we generate three equations for a, b, and c. The solution to these equations is $a = b = \frac{1}{2}$ and $c = 0$, thus yielding Eq. (3.2.3).

By blind computation with Cramer's rule, we saw in Eq. (1.3.9) that the number of multiplications and divisions of an nth-order system is $(n-1)(n/e)^n$ or 3.7×10^{158} if $n = 100$. By back substitution, Eq. (3.2.1) requires $50(101) = 5.05 \times 10^3$ such operations if $n = 100$. Quite a savings!

Similarly, if $\mathbf{L}$ is a lower triangular matrix, the linear system $\mathbf{Lx} = \mathbf{b}$ is

$$\begin{aligned}
l_{11} x_1 & = b_1 \\
l_{21} x_1 + l_{22} x_2 & = b_2 \\
\vdots \qquad \vdots \qquad\qquad \vdots & \\
l_{n1} x_1 + l_{n2} x_2 + \cdots + l_{nn} x_n & = b_n,
\end{aligned} \tag{3.2.6}$$

represented symbolically in Fig. 3.2.2. The determinant of $\mathbf{L}$ is $|\mathbf{L}| = \prod_{i=1}^{n} l_{ii}$. Thus, if no diagonal elements are 0, $|\mathbf{L}| \neq 0$ and Eq. (3.2.6) has a unique solution,

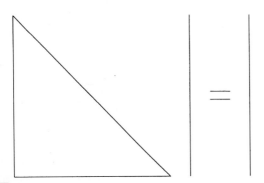

FIGURE 3.2.2

which can be computed by *forward substitution*, namely,

$$x_1 = \frac{b_1}{l_{11}}$$

$$x_2 = \frac{b_2 - l_{21}x_1}{l_{22}}$$

$$\vdots$$

or

$$x_1 = \frac{b_1}{l_{11}}$$

$$x_i = \frac{b_i - \sum_{k=1}^{i-1} l_{ik}x_k}{l_{ii}}, \quad i = 2, \ldots, n. \quad (3.2.7)$$

We note that at the ith step there are $i - 1$ multiplications and one division. Thus, solution of Eq. (3.2.5) by forward substitution requires

$$\sum_{i=1}^{n} i = \frac{1}{2}n(n+1)$$

multiplications and divisions, which is the same as solution of Eq. (3.2.1) by back substitution.

Clearly, if the matrix **A** is an upper or lower triangular matrix, the cost of solving $\mathbf{Ax} = \mathbf{b}$ is much smaller than what is expected by Cramer's rule. The method of Gauss takes advantage of such savings by transforming the problem to an upper triangular problem and solving it by back substitution. One could, equally easily, transform the problem to a lower triangular matrix and solve it by forward substitution.

Consider the set of equations

$$2x_1 + x_2 + 3x_3 = 1$$
$$3x_1 + 2x_2 + x_3 = 2 \quad (3.2.8)$$
$$4x_1 + x_2 + 2x_3 = 3.$$

We can multiply the first equation by $-\frac{3}{2}$ and add it to the second equation and then multiply the first equation by $-\frac{4}{2}$ and add it to the third equation. The result is

$$2x_1 + x_2 + 3x_3 = 1$$
$$\frac{1}{2}x_2 - \frac{7}{2}x_3 = \frac{1}{2} \qquad (3.2.9)$$
$$-x_2 - 4x_3 = 1.$$

Now we multiply the second equation by 2 and add it to the third one to obtain

$$2x_1 + x_2 + 3x_3 = 1$$
$$\frac{1}{2}x_2 - \frac{7}{2}x_3 = \frac{1}{2} \qquad (3.2.10)$$
$$-11x_3 = 2.$$

Note that the equations have been rearranged into upper triangular form. By back substitution,

$$x_3 = -\frac{2}{11}$$
$$x_2 = \frac{\left(\frac{1}{2} - \frac{7}{11}\right)}{\left(\frac{1}{2}\right)} = -\frac{3}{11} \qquad (3.2.11)$$
$$x_1 = \frac{\left(1 + \frac{3}{11} + \frac{6}{11}\right)}{2} = \frac{10}{11}.$$

The preceding steps constitute an example of solution of an algebraic system by Gauss elimination. In the general case, one wants to solve the system $\mathbf{Ax} = \mathbf{b}$, or

$$a_{11}x_1 + a_{12}x_2 + a_{13}x_3 + \cdots + a_{1n}x_n = b_1$$
$$a_{21}x_1 + a_{22}x_2 + a_{23}x_3 + \cdots + a_{2n}x_n = b_2$$
$$\vdots \qquad \vdots \qquad (3.2.12)$$
$$a_{n1}x_1 + a_{n2}x_2 + a_{n3}x_3 + \cdots + a_{nn}x_n = b_n.$$

As the first step in the elimination process, we multiply the first equation by $-a_{i1}/a_{11}$ and add it to the ith equation for $i = 2, \ldots, n$. The result is

$$a_{11}^{(1)}x_1 + a_{12}^{(1)}x_2 + a_{13}^{(1)}x_3 + \cdots + a_{1n}^{(1)}x_n = b_1^{(1)}$$
$$a_{22}^{(2)}x_2 + a_{23}^{(2)}x_3 + \cdots + a_{2n}^{(2)}x_n = b_2^{(2)}$$
$$\vdots \qquad \vdots \qquad \vdots \qquad (3.2.13)$$
$$a_{n2}^{(2)}x_2 + a_{n3}^{(2)}x_3 + \cdots + a_{nn}^{(2)}x_n = b_n^{(2)}.$$

The coefficients in the first equation are unchanged and are denoted as $a_{ij}^{(1)}$ to distinguish them from the coefficients in the equations $i = 2, \ldots, n$, which have changed. In fact,

$$a_{ij}^{(2)} = a_{ij}^{(1)} - \frac{a_{i1}^{(1)}}{a_{11}^{(1)}} a_{1j}^{(1)}, \quad i, j = 2, \ldots, n,$$

$$b_i^{(2)} = b_i^{(1)} - \frac{a_{i1}^{(1)}}{a_{11}^{(1)}} b_1^{(1)}, \quad i = 2, \ldots, n. \tag{3.2.14}$$

In the next step, the second equation is multiplied by $-a_{i2}^{(2)}/a_{22}^{(2)}$ and the result is added to the ith equation for $i = 3, \ldots, n$. This eliminates the terms in x_2 in equations $i = 3, \ldots, n$ and generates the set of equations

$$\begin{aligned}
a_{11}^{(1)} x_1 + a_{12}^{(1)} x_2 + a_{13}^{(1)} x_3 + \cdots + a_{1n}^{(1)} x_n &= b_1^{(1)} \\
a_{22}^{(2)} x_2 + a_{23}^{(2)} x_3 + \cdots + a_{2n}^{(2)} x_n &= b_2^{(2)} \\
a_{33}^{(3)} x_3 + \cdots + a_{3n}^{(3)} x_n &= b_3^{(3)} \\
&\vdots \\
a_{n3}^{(3)} x_3 + \cdots + a_{nn}^{(3)} x_n &= b_n^{(3)}.
\end{aligned} \tag{3.2.15}$$

Continuation of this process eventually transforms Eq. (3.2.12) into the upper triangular problem

$$\begin{aligned}
a_{11}^{(1)} x_1 + a_{12}^{(1)} x_2 + a_{13}^{(1)} x_3 + \cdots + a_{1n}^{(1)} x_n &= b_1^{(1)} \\
a_{22}^{(2)} x_2 + a_{23}^{(2)} x_3 + \cdots + a_{2n}^{(2)} x_n &= b_2^{(2)} \\
a_{33}^{(3)} x_3 + \cdots + a_{3n}^{(3)} x_n &= b_3^{(3)} \\
&\vdots \\
a_{nn}^{(n)} x_n &= b_n^{(n)},
\end{aligned} \tag{3.2.16}$$

which can be solved by back substitution. This procedure is Gauss elimination in its simplest form.

Since the operations used in transforming Eq. (3.2.12) to Eq. (3.2.16) involve only the additions to or subtractions from various equations by multiples of other equations, the determinant of the matrix in Eq. (3.2.16) is the same as the determinant of $\mathbf{A}$, i.e.,

$$|\mathbf{A}| = \prod_{k=1}^{n} a_{kk}^{(k)}. \tag{3.2.17}$$

Thus, a by-product of Gauss elimination is the evaluation of the determinant of $\mathbf{A}$.

Let us now calculate the number of multiplications and divisions needed for Gauss elimination. At the kth elimination step, there are $i = k+1, \ldots, n$ ratios $a_{ik}^{(k)}/a_{kk}^{(k)}$ to compute and $j = k+1, \ldots, n+1$ products $(a_{ik}^{(k)}/a_{kk}^{(k)}) a_{kj}^{(k)}$ to compute for each i. For the purpose of counting, we call $b_i^{(k)}$ the $(n+1)$th element. Thus, it takes

$$\sum_{k=1}^{n-1}(n-k) + \sum_{k=1}^{n-1}(n-k)(n+1-k) = \frac{1}{2}n(n-1) + \frac{1}{3}n(n^2-1) \quad (3.2.18)$$

multiplications and divisions to carry out the upper triangularization of the system of equations. To this we must add $\frac{1}{2}n(n+1)$ multiplications and divisions to carry out the back substitution to get a final solution. Thus, the total number of operations for Gauss elimination is $\frac{1}{3}n(n^2+3n-1)$. With $n = 100$ and a computer taking 10^{-9} sec for an arithmetic operation, a Gauss elimination solution costs about $\frac{1}{3}(100)^3 \times 10^{-9}$ sec $= 3.3 \times 10^{-4}$ sec compared to 3.7×10^{142} years for Cramer's rule.

In carrying out the steps of Gauss elimination, the variables $x_1, x_2, \ldots$ play only a passive role. What is happening is that the augmented matrix

$$\text{aug } \mathbf{A} = [\mathbf{A}, \mathbf{b}] \quad (3.2.19)$$

is being rearranged by addition and subtraction of multiples of rows. In the case of the problem in Eq. (3.2.8), the augmented matrix is

$$\text{aug } \mathbf{A} = \begin{bmatrix} 2 & 1 & 3 & 1 \\ 3 & 2 & 1 & 2 \\ 4 & 1 & 2 & 3 \end{bmatrix}. \quad (3.2.20)$$

The first step in Gauss elimination is the subtraction of $\frac{3}{2}$ times the first row from the second row and twice the first row from the third row to obtain

$$\begin{bmatrix} 2 & 1 & 3 & 1 \\ 0 & \frac{1}{2} & -\frac{7}{2} & \frac{1}{2} \\ 0 & -1 & -4 & 1 \end{bmatrix}. \quad (3.2.21)$$

In the next step, twice the second row is added to the third row to obtain

$$\begin{bmatrix} 2 & 1 & 3 & 1 \\ 0 & \frac{1}{2} & -\frac{7}{2} & \frac{1}{2} \\ 0 & 0 & -11 & 2 \end{bmatrix} = [\mathbf{A}_{tr}, \mathbf{b}_{tr}]. \quad (3.2.22)$$

Once Gauss elimination has transformed the augmented matrix to the form $[\mathbf{A}_{tr}, \mathbf{b}_{tr}]$, where $\mathbf{A}_{tr}$ is an upper triangular matrix, the solution to $\mathbf{Ax} = \mathbf{b}$ can be computed from

$$\mathbf{A}_{tr}\mathbf{x} = \mathbf{b}_{tr} \quad (3.2.23)$$

by backward substitution.

Gauss elimination as just carried out above was based on the assumption that the coefficients $a_{kk}^{(k)}$ were always nonzero. This is not always the case, even when $|\mathbf{A}| \neq 0$ and a solution is assured. Consider the equations

$$x_1 + 2x_2 + 3x_3 = 1$$
$$2x_1 + 4x_2 + 2x_3 = 1 \tag{3.2.24}$$
$$x_1 + x_2 + x_3 = 3.$$

After the first step in Gauss elimination, the set becomes

$$x_1 + 2x_2 + 3x_3 = 1$$
$$-4x_3 = -1 \tag{3.2.25}$$
$$-x_2 - 2x_3 = 2.$$

In this case, $a_{22}^{(2)} = 0$ and so the next step would fail since $a_{23}^{(2)}/a_{22}^{(2)} = \infty$. On the other hand, if the second and third equations are interchanged, we obtain

$$x_1 + 2x_2 + 3x_3 = 1$$
$$-x_2 - 2x_3 = 2 \tag{3.2.26}$$
$$-4x_3 = -1,$$

which can be solved by backward substitution.

As long as $|\mathbf{A}| \neq 0$, at any step k in Gauss elimination there will exist at least one equation $i \geq k$ such that the coefficient $a_{ik}^{(k)} \neq 0$. Otherwise, the determinant of $\mathbf{A}$ would be 0, in contradiction to the nonsingular case we are considering in this chapter. A strategy of implementation of Gauss elimination that lends itself to an easy computer program can now be described.

We want to solve the $n \times n$ problem $\mathbf{Ax} = \mathbf{b}$. First, we define the components of b_i as $b_i = a_{i,n+1}$ and the equations as rows R_i, i.e.,

$$R_1: \quad a_{11}x_1 + a_{12}x_2 + \cdots + a_{1n}x_n = a_{1,n+1}$$
$$R_1: \quad a_{21}x_1 + a_{22}x_2 + \cdots + a_{2n}x_n = a_{2,n+1}$$
$$\vdots \qquad \vdots \tag{3.2.27}$$
$$R_n: \quad a_{n1}x_1 + a_{n2}x_2 + \cdots + a_{nn}x_n = a_{n,n+1}.$$

The elements of the augmented matrix, $\text{aug}\,\mathbf{A}$, are a_{ij}, $i = 1, \ldots, n$, $j = 1, \ldots, n+1$. In a computer code to execute Gauss elimination, we set the dimension of the augmented matrix to be $n \times (n+1)$, input the elements a_{ij}, and set $DET = 1$. Then we carry out the following sequence of operations:

Elimination

Step 1. For $i = 1, \ldots, n-1$, do steps 2–5.

Step 2. Find the first-row p, $i \leq p \leq n$, with nonzero entry in the ith column ($a_{pi} \neq 0$). If none is found, then NO SOLUTION, $|\mathbf{A}| = 0$, and STOP. Otherwise,

Step 3. If $p \neq i$, then perform a row swap

$$R_p \leftrightarrow R_i \tag{3.2.28}$$

and set

$$DET = (-1)DET. \tag{3.2.29}$$

Step 4. For $j = i+1, \ldots, n$, do step 5.
Step 5. Set

$$R_j \to R_j - \frac{a_{ji}}{a_{ii}} R_i. \tag{3.2.30}$$

Step 6. If $a_{nn} = 0$, then NO SOLUTION, $|\mathbf{A}| = 0$, and STOP.

Backward Substitution

Step 7. Set

$$x_n = \frac{a_{n,n+1}}{a_{nn}}. \tag{3.2.31}$$

Step 8. For $i = n-1, \ldots, 1$, set

$$x_i = \left(a_{i,n+1} - \sum_{j=i+1}^{n} a_{ij} x_j\right)/a_{ii}. \tag{3.2.32}$$

Step 9.

$$|\mathbf{A}| = DET \prod_{i=1}^{n} a_{ii}. \tag{3.2.33}$$

Mathematica routines are provided in the Appendix for the Gauss elimination algorithm described above and in Section 3.3.

In the computer program just outlined, we assume that the elements of the augmented matrix are written over at every elimination step. Thus, at the end of the elimination process, the augmented matrix is composed of the elements $a_{ij}^{(i)}$, $i = 1, \ldots, n$, $j = i, \ldots, n+1$. Note that an interchange of rows changes the sign of the determinant, and so the role of Eq. (3.2.29) is to keep track of the net sign change. Subtracting a multiple of a row from other rows does not change the sign of a determinant, and so there is no effect on the determinant in Eq. (3.2.30). The determinant of $\mathbf{A}$ is then the product $\prod_i a_{ii}^{(i)}$ times the net sign change from row swaps; hence Eq. (3.2.33) follows.

3.3. GAUSS ELIMINATION WITH PIVOTING

If all our arithmetic is done with infinite precision, Gauss elimination with row swaps as described in the preceding section will always yield the correct solution to $\mathbf{Ax} = \mathbf{b}$ whenever $|\mathbf{A}| \neq 0$. However, with finite-figure floating-point arithmetic, round-off errors can cause problems. For example, consider the equations

$$x_1 + x_2 + x_3 = 1$$
$$x_1 + 1.00001x_2 + 2x_3 = 2 \qquad (3.3.1)$$
$$x_1 + 2x_2 + 2x_3 = 1.$$

Gauss elimination reduces these equations to

$$x_1 + x_2 + x_3 = 1$$
$$0.00001x_2 + x_3 = 1 \qquad (3.3.2)$$
$$99{,}999 x_3 = 100{,}000.$$

Let us now perform back substitution with a computer (admittedly a little fellow, but it makes the point) that does four-figure floating-point arithmetic. We find

$$x_3 = 1.0000, \qquad x_2 = 0, \qquad x_1 = 0. \qquad (3.3.3)$$

The actual solution is, of course,

$$x_3 = \frac{100{,}000}{99{,}999} = -x_2, \qquad x_1 = 1. \qquad (3.3.4)$$

One way to try to avoid round-off errors is to introduce so-called "pivoting" strategies in executing Gauss elimination. Consider the general elimination problem at the kth step. The equations are of the form

$$\begin{aligned}
a_{11}^{(1)} x_1 + a_{12}^{(1)} x_2 + \cdots &\phantom{{}+{}} + a_{1n}^{(1)} x_n = b_1^{(1)} \\
a_{22}^{(2)} x_2 + \cdots &\phantom{{}+{}} + a_{2n}^{(2)} x_n = b_2^{(2)} \\
&\vdots \\
a_{k-1,\,k-1}^{(k-1)} x_{k-1} + \cdots &\phantom{{}+{}} + a_{k-1,\,n}^{(k-1)} x_n = b_{k-1}^{(k-1)} \\
a_{kk}^{(k)} x_k + \cdots &\phantom{{}+{}} + a_{kn}^{(k)} x_n = b_k^{(k)} \\
&\vdots \\
a_{nk}^{(k)} x_k + \cdots &\phantom{{}+{}} + a_{nn}^{(k)} x_n = b_n^{(k)}.
\end{aligned} \qquad (3.3.5)$$

These equations can be represented schematically as in Fig. 3.3.1.

When the row swap as described in the previous section is not numerically stable, *partial pivoting* is frequently used to increase stability. In this pivoting strategy, we determine the largest magnitude of the elements $a_{jk}^{(k)}$, $j = k, k+1, \ldots, n$. Suppose this is the element $a_{pk}^{(k)}$, i.e.,

$$|a_{pk}^{(k)}| = \max_{k \leq j \leq n} |a_{jk}^{(k)}|, \qquad (3.3.6)$$

i.e., $a_{pk}^{(k)}$ is the largest magnitude element of the kth column lying below the $(k-1)$th row. In the partial pivoting scheme, rows k and p are swapped and the next Gauss elimination step is carried out. The interchange process of partial pivoting is shown in Fig. 3.3.2.

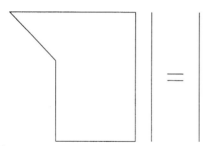

FIGURE 3.3.1

EXERCISE 3.3.1. Show that partial pivoting eliminates the round-off problem encountered in solving Eq. (3.3.1).

Note that the only difference between partial pivoting and the Gauss elimination process outlined in Eqs. (3.2.27)–(3.2.33) is that in the former one hunts the first nonzero $a_{ik}^{(k)}$, $i \geq k$, whereas in the latter one seeks the largest of $|a_{ik}^{(k)}|$, $i \geq k$. In either case, once the row of the object element has been determined, say row p, one interchanges rows k and p and continues the Gauss elimination process. Thus, in partial pivoting the only change in the process given by Eqs. (3.2.27)–(3.2.33) is to replace the search for the first-row p, $k \leq p \leq n$, with nonzero a_{pk} by the search for the row p with the maximum of $|a_{ik}^{(k)}|$, $i \geq k$.

Although not as often used, *complete pivoting* gives even further numerical stability. In this scheme, one hunts the row p and column q containing the largest value of $|a_{ij}^{(k)}| i = k, \ldots, n$, $j = k, \ldots, n$, i.e.,

$$|a_{pq}^{(k)}| = \max_{k \leq i,\, j \leq n} |a_{ij}^{(k)}|. \tag{3.3.7}$$

The rows k and p and columns k and q are then interchanged so that element $a_{pk}^{(k)}$ becomes the pivot element for the next step of Gauss elimination. Schematically, the interchange is shown in Fig. 3.3.3. Complete pivoting costs substantially more to compute and is more complicated to program, and so it is not usually employed in canned computer programs. The column interchange amounts to renaming the elements of the vector **x**, and so this feature needs to be added to a computer program executing complete pivoting.

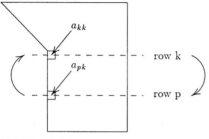

FIGURE 3.3.2

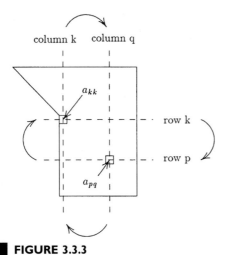

FIGURE 3.3.3

In closing, we note that the simplest Gauss elimination procedure, without row interchange or pivoting, frequently works and is guaranteed to be successful in two cases: (1) when **A** is strictly diagonally dominant, i.e.,

$$|a_{ii}| > \sum_{\substack{j=1 \\ j \neq i}}^{n} |a_{ij}|, \qquad i = 1, \ldots, n, \tag{3.3.8}$$

and (2) when **A** is positive definite, i.e., $\mathbf{x}^\dagger \mathbf{A} \mathbf{x} > 0$ for all $\mathbf{x} \neq \mathbf{0}$.

3.4. COMPUTING THE INVERSE OF A MATRIX

In Chapter 1 we saw that the inverse of a nonsingular matrix ($|\mathbf{A}| \neq 0$) is given by

$$\mathbf{A}^{-1} = \frac{1}{|\mathbf{A}|} \operatorname{adj} \mathbf{A}, \tag{3.4.1}$$

where the ij element of adj **A** is the ji cofactor of **A**. In retrospect, this would be an expensive way to compute $\mathbf{A}^{-1}$. A cheaper method can be devised using Gauss elimination. Consider the p problems

$$\mathbf{A}\mathbf{x}_i = \mathbf{b}_i, \qquad i = 1, \ldots, p. \tag{3.4.2}$$

These can be summarized in partitioned form as

$$[\mathbf{A}\mathbf{x}_1, \mathbf{A}\mathbf{x}_2, \ldots, \mathbf{A}\mathbf{x}_p] = [\mathbf{b}_1, \mathbf{b}_2, \ldots, \mathbf{b}_p]$$

or

$$\mathbf{A}[\mathbf{x}_1, \ldots, \mathbf{x}_p] = [\mathbf{b}_1, \ldots, \mathbf{b}_p]. \tag{3.4.3}$$

The augmented matrix corresponding to Eq. (3.4.3) is

$$[\mathbf{A}, \mathbf{b}_1, \mathbf{b}_2, \ldots, \mathbf{b}_p]. \tag{3.4.4}$$

If Gauss elimination is carried out on this augmented matrix, the number of division and multiplication operations is

$$\sum_{k=1}^{n-1}(n-k)+\sum_{k=1}^{n-1}(n-k)(n+p-k)=\frac{1}{3}n(n^2-1)+\frac{1}{2}pn(n-1). \tag{3.4.5}$$

Programming Gauss elimination for Eq. (3.4.4) is the same as that which was outlined in Eqs. (3.2.26)–(3.2.32), or its variation with pivoting, except that the augmented matrix is of dimension $n \times (n + p)$ and it has elements a_{ij}, $i, j = 1, \ldots, n$, and $a_{i,n+j} = b_{ij}$, $i = 1, \ldots, n$, $j = 1, \ldots, p$.

Gauss elimination transforms Eq. (3.4.4) into

$$[\mathbf{A}_{tr}, \mathbf{b}_{1,tr}, \ldots, \mathbf{b}_{p,tr}], \tag{3.4.6}$$

where $\mathbf{A}_{tr}$ is an upper triangular matrix. The solutions to the equations in Eq. (3.4.2) can now be computed by backward substitution from

$$\mathbf{A}_{tr}\mathbf{x}_i = \mathbf{b}_{i,tr}, \quad i = 1, \ldots, p. \tag{3.4.7}$$

The number of division and multiplication operations required to solve these equations is

$$\frac{p}{2}n(n-1), \tag{3.4.8}$$

and so the total number of operations needed to solve Eq. (3.4.2) for $\mathbf{x}_i$ is the sum of Eqs. (3.4.5) and (3.4.8), i.e.,

$$\frac{1}{3}n(n^2-1)+pn(n-1). \tag{3.4.9}$$

Let us now consider the problem of finding the inverse of $\mathbf{A}$. Define $\mathbf{x}_i$ to be the solution of

$$\mathbf{A}\mathbf{x}_i = \mathbf{e}_i, \quad i = 1, \ldots, n, \tag{3.4.10}$$

where $\mathbf{e}_i$ is the unit vector defined in Chapter 2:

$$\mathbf{e}_i = \begin{bmatrix} 0 \\ \vdots \\ 0 \\ 1 \\ 0 \\ \vdots \\ 0 \end{bmatrix}, \tag{3.4.11}$$

where only the ith element of $\mathbf{e}_i$ is nonzero. Defining the matrix

$$\mathbf{X} = [\mathbf{x}_1, \ldots, \mathbf{x}_r], \qquad (3.4.12)$$

we see that

$$\mathbf{AX} = [\mathbf{Ax}_1, \ldots, \mathbf{Ax}_n] = [\mathbf{e}_1, \ldots, \mathbf{e}_n] = \mathbf{I}, \qquad (3.4.13)$$

where $\mathbf{I} = [\delta_{ij}]$ is the unit matrix.

We now assume that $\mathbf{A}$ is nonsingular; i.e., $\mathbf{A}^{-1}$ exists. Multiplying Eq. (3.4.13) by $\mathbf{A}^{-1}$, we find

$$\mathbf{X} = [\mathbf{x}_1, \ldots, \mathbf{x}_n] = \mathbf{A}^{-1}; \qquad (3.4.14)$$

i.e., the inverse of $\mathbf{A}$ is a matrix whose column vectors are the solution of the set of equations in Eq. (3.4.10). We have just shown that the number of divisions and multiplications for solving n such equations is

$$\frac{1}{3}n(n^2 - 1) + n^2(n - 1). \qquad (3.4.15)$$

Thus, for large n, the inverse can be computed with $\frac{4}{3}n^3$ operations.

EXAMPLE 3.4.1. Find the inverse of

$$\mathbf{A} = \begin{bmatrix} -2 & 1 & 0 \\ 1 & -2 & 1 \\ 0 & 1 & -2 \end{bmatrix}. \qquad (3.4.16)$$

The appropriate augmented matrix is

$$\begin{bmatrix} -2 & 1 & 0 & 1 & 0 & 0 \\ 1 & -2 & 1 & 0 & 1 & 0 \\ 0 & 1 & -2 & 0 & 0 & 1 \end{bmatrix}. \qquad (3.4.17)$$

After the first elimination step, the matrix is

$$\begin{bmatrix} -2 & 1 & 0 & 1 & 0 & 0 \\ 0 & -\frac{3}{2} & 1 & \frac{1}{2} & 1 & 0 \\ 0 & 1 & -2 & 0 & 0 & 1 \end{bmatrix}.$$

After the second step,

$$[\mathbf{A}_{\text{tr}}, \mathbf{e}_{1,\text{tr}}, \mathbf{e}_{2,\text{tr}}, \mathbf{e}_{3,\text{tr}}] = \begin{bmatrix} -2 & 1 & 0 & 1 & 0 & 0 \\ 0 & -\frac{3}{2} & 1 & \frac{1}{2} & 1 & 0 \\ 0 & 0 & -\frac{4}{3} & \frac{1}{3} & \frac{2}{3} & 1 \end{bmatrix}. \qquad (3.4.18)$$

LU-DECOMPOSITION

Solving $\mathbf{A}_{tr}\mathbf{x}_i = \mathbf{e}_{i,tr}$ by backward substitution, we obtain

$$\mathbf{A}^{-1} = [\mathbf{x}_1, \mathbf{x}_2, \mathbf{x}_3] = \begin{bmatrix} -\frac{3}{4} & -\frac{1}{2} & -\frac{1}{4} \\ -\frac{1}{2} & -1 & -\frac{1}{2} \\ -\frac{1}{4} & -\frac{1}{2} & -\frac{3}{4} \end{bmatrix}. \tag{3.4.19}$$

■ ■ ■

3.5. LU-DECOMPOSITION

Under certain conditions, a matrix $\mathbf{A}$ can be factored into a product $\mathbf{LU}$ in which $\mathbf{L}$ is a lower triangular matrix and $\mathbf{U}$ is an upper triangular matrix. The following theorem identifies a class of matrices that can be decomposed into a product $\mathbf{LU}$.

LU-Decomposition Theorem. *If $\mathbf{A}$ is an $n \times n$ matrix such that*

$$|\mathbf{A}_p| \neq 0, \qquad p = 1, \ldots, n - 1, \tag{3.5.1}$$

where $\mathbf{A}_p$ is the matrix formed by the elements at the intersections of the first p rows and columns of $\mathbf{A}$, then there exists a unique lower triangular matrix $\mathbf{L} = [l_{ij}]$, $l_{ii} = 1$, and a unique upper triangular matrix $\mathbf{U} = [u_{ij}]$ such that

$$\mathbf{A} = \mathbf{LU}. \tag{3.5.2}$$

The determinants $|\mathbf{A}_p|$, $p = 1, \ldots, n$, are known as principal minors of $\mathbf{A}$. Note that since $|\mathbf{A}| = |\mathbf{A}_n| = 0$ is allowed by the theorem, an **LU**-decomposition can exist for a singular matrix (i.e., for a matrix for which $\mathbf{A}^{-1}$ does not exist), but not too singular since $|\mathbf{A}_{n-1}| \neq 0$ implies the rank of $\mathbf{A}$ is greater than or equal to $n - 1$.

An example of a matrix obeying the hypothesis of the theorem is

$$\mathbf{A} = \begin{bmatrix} -2 & 1 & 0 \\ 1 & -2 & 1 \\ 0 & 0 & 0 \end{bmatrix}. \tag{3.5.3}$$

In this case,

$$\mathbf{L} = \begin{bmatrix} 1 & 0 & 0 \\ -\frac{1}{2} & 1 & 0 \\ 0 & 0 & 1 \end{bmatrix} \tag{3.5.4}$$

and

$$\mathbf{U} = \begin{bmatrix} -2 & 1 & 0 \\ 0 & -\frac{3}{2} & 1 \\ 0 & 0 & 1 \end{bmatrix}. \tag{3.5.5}$$

The reader can readily verify that $\mathbf{LU} = \mathbf{A}$ and that $|\mathbf{A}| = 0$. Of course, $\mathbf{A}$ need not be singular, as the example

$$\mathbf{A} = \begin{bmatrix} -3 & 1 & 1 \\ 1 & -3 & 1 \\ 1 & 1 & -3 \end{bmatrix} \tag{3.5.6}$$

illustrates. In this case,

$$\mathbf{L} = \begin{bmatrix} 1 & 0 & 0 \\ -\dfrac{1}{3} & 1 & 0 \\ -\dfrac{1}{3} & -\dfrac{1}{2} & 1 \end{bmatrix} \tag{3.5.7}$$

and

$$\mathbf{U} = \begin{bmatrix} -3 & 1 & 1 \\ 0 & -\dfrac{8}{3} & \dfrac{4}{3} \\ 0 & 0 & -2 \end{bmatrix}. \tag{3.5.8}$$

It is worthwhile to note that the hypothesis of the theorem only constitutes a sufficient condition for **LU**-decomposition. For example, if

$$\mathbf{A} = \begin{bmatrix} 0 & 1 \\ 0 & 1 \end{bmatrix}, \tag{3.5.9}$$

then $|\mathbf{A}_1| = 0$ and $|\mathbf{A}_2| = |\mathbf{A}| = 0$. Nevertheless, there exist $\mathbf{L}$ and $\mathbf{U}$ such that

$$\mathbf{LU} = \mathbf{A}. \tag{3.5.10}$$

In particular, if

$$\mathbf{L} = \begin{bmatrix} 1 & 0 \\ 1 & 1 \end{bmatrix} \quad \text{and} \quad \mathbf{U} = \begin{bmatrix} 0 & 1 \\ 0 & 0 \end{bmatrix}, \tag{3.5.11}$$

then

$$\mathbf{LU} = \begin{bmatrix} 0 & 1 \\ 0 & 1 \end{bmatrix} = \mathbf{A}. \tag{3.5.12}$$

If $\mathbf{A} = \mathbf{LU}$, it follows from the rules of matrix multiplication that

$$a_{ij} = \sum_{k=1}^{\min(i,\,j)} l_{ik} u_{kj}. \tag{3.5.13}$$

LU-DECOMPOSITION

The upper limit on the summation over k comes from the fact that $\mathbf{L}$ has only zero elements above the main diagonal and $\mathbf{U}$ has only zero elements below the main diagonal.

When a matrix can be upper triangularized by Gauss elimination without interchange of rows (which is true, for example, for positive-definite matrices), LU-decomposition is generated as a by-product. Recall that at the kth step of Gauss elimination we eliminate the coefficients $a_{ik}^{(k)}$, $i > k$, by replacing the elements $a_{ij}^{(k)}$, $i, j = k+1, \ldots, n$, by

$$a_{ij}^{(k+1)} = a_{ij}^{(k)} - m_{ik}^{(k)} a_{kj}^{(k)}, \qquad i = k+1, \ldots, n, \ j = k, \ldots, n, \qquad (3.5.14)$$

where

$$m_{ik}^{(k)} \equiv \frac{a_{ik}^{(k)}}{a_{kk}^{(k)}}, \qquad i = k, \ldots, n. \qquad (3.5.15)$$

Let us define a lower triangular matrix $\mathbf{L} = [l_{ij}]$ by

$$l_{ij} = \begin{cases} m_{ij}^{(j)}, & i \geq j, \\ 0, & i < j. \end{cases} \qquad (3.5.16)$$

Recalling that $a_{ij}^{(1)} = a_{ij}$, the original elements of $\mathbf{A}$, and expressing Eq. (3.4.14) for the sequence of k's ranging from 1 to r, we obtain

$$\begin{aligned}
a_{ij}^{(2)} &= a_{ij}^{(1)} - l_{i1} a_{1j}^{(1)}, \qquad i = 2, \ldots, n, \ j = 1, \ldots, n, \\
a_{ij}^{(3)} &= a_{ij}^{(2)} - l_{i2} a_{2j}^{(2)} \\
a_{ij}^{(4)} &= a_{ij}^{(3)} - l_{i3} a_{3j}^{(3)} \\
&\vdots \\
a_{ij}^{(r+1)} &= a_{ij}^{(r)} - l_{ir} a_{rj}^{(r)}
\end{aligned} \qquad (3.5.17)$$

or

$$a_{ij} = a_{ij}^{(r+1)} + \sum_{k=1}^{r} l_{ik} a_{kj}^{(k)}. \qquad (3.5.18)$$

Gauss elimination generates an upper triangular matrix with elements $a_{ij}^{(i)}$ on and above the main diagonal and 0 below the main diagonal. With $r = i - 1$, Eq. (3.5.18) becomes

$$\begin{aligned}
a_{ij} &= a_{ij}^{(i)} + \sum_{k=1}^{i-1} l_{ik} a_{kj}^{(k)} \\
&= \sum_{k=1}^{i} l_{ik} a_{kj}^{(k)},
\end{aligned} \qquad (3.5.19)$$

since $l_{ii} = 1$. Since $a_{kj}^{(k)} = 0$ if $k > j$, Eq. (3.5.19) becomes

$$a_{ij} = \sum_{k=1}^{\min(i,j)} l_{ik} a_{kj}^{(k)}. \qquad (3.5.20)$$

Comparison of Eqs. (3.5.13) and (3.5.20) yields the conclusion that the **LU**-decomposition of **A** is

$$\mathbf{L} = \begin{bmatrix} 1 & 0 & \cdots & 0 & 0 \\ m_{21} & 1 & \cdots & 0 & 0 \\ \vdots & \vdots & & \vdots & \vdots \\ m_{n1} & m_{n2} & \cdots & m_{n,n-1} & 1 \end{bmatrix} \qquad (3.5.21)$$

and

$$\mathbf{U} = \begin{bmatrix} a_{11}^{(1)} & a_{12}^{(1)} & \cdots & a_{1n}^{(1)} \\ 0 & a_{22}^{(2)} & \cdots & a_{2n}^{(2)} \\ \vdots & \vdots & \ddots & \vdots \\ 0 & 0 & \cdots & a_{nn}^{(n)} \end{bmatrix}, \qquad (3.5.22)$$

where $m_{ij} = a_{ij}^{(j)}/a_{jj}^{(j)}$ and the elements $a_{ij}^{(i)}$ of **U** are the elements resulting from the upper triangularization of **A** by Gauss elimination *without* row interchange.

As an example of **LU**-decomposition by Gauss elimination, consider

$$\mathbf{A} = \begin{bmatrix} -2 & 1 & 0 \\ 1 & -2 & 1 \\ 0 & 1 & -2 \end{bmatrix}. \qquad (3.5.23)$$

Multiplying the first row by $m_{21} = -\frac{1}{2}$ and subtracting from the second row, and multiplying the first row by $m_{31} = 0$ and subtracting from the third row, we obtain

$$\begin{bmatrix} -2 & 1 & 0 \\ 0 & -\frac{3}{2} & 1 \\ 0 & 1 & -2 \end{bmatrix}. \qquad (3.5.24)$$

Multiplying the second row by $m_{32} = -\frac{2}{3}$ and subtracting from the third row then yields

$$\mathbf{U} = \begin{bmatrix} -2 & 1 & 0 \\ 0 & -\frac{3}{2} & 1 \\ 0 & 0 & -\frac{4}{3} \end{bmatrix}. \qquad (3.5.25)$$

With $m_{ii} = 1$, $\mathbf{L} = [m_{ij}]$ becomes

$$\mathbf{L} = \begin{bmatrix} 1 & 0 & 0 \\ -\frac{1}{2} & 1 & 0 \\ 0 & -\frac{2}{3} & 1 \end{bmatrix}. \qquad (3.5.26)$$

The reader can verify that $\mathbf{A} = \mathbf{LU}$. This example serves to illustrate how the **LU**-decomposition of $\mathbf{A}$ can be obtained by simply storing the multipliers m_{ij} when the simplest Gauss elimination process (no row interchanges) works.

As an example of how **LU**-decomposition might be useful, consider the equations

$$\mathbf{A}\mathbf{x}_1 = \mathbf{b}_1 \qquad (3.5.27)$$

$$\mathbf{A}\mathbf{x}_2 = \mathbf{b}_2(\mathbf{x}_1) \qquad (3.5.28)$$

in which $\mathbf{b}_2$ depends on the solution of Eq. (3.5.27). Performing Gauss elimination twice to solve these equations requires twice $n(n^2 - 1)/3 + n(n+1)/2$ divisions and multiplications, whereas carrying out Gauss elimination to solve Eq. (3.5.27) costs $n(n^2 - 1)/3 + n(n+1)/2$ and yields $\mathbf{L}$ and $\mathbf{U}$. Equation (3.5.28) can then be written as

$$\mathbf{LU}\mathbf{x}_2 = \mathbf{b}_2, \qquad (3.5.29)$$

and setting $\mathbf{y} = \mathbf{U}\mathbf{x}_2$, the equation

$$\mathbf{L}\mathbf{y} = \mathbf{b}_2$$

can be solved by forward substitution for $\mathbf{y}$. Subsequently, the equation

$$\mathbf{U}\mathbf{x}_2 = \mathbf{y} \qquad (3.5.30)$$

can be solved by backward substitution for $\mathbf{x}_2$ at a cost of $n(n+1)$. Thus, without using **LU**-decomposition, the solution for $\mathbf{x}_1$ and $\mathbf{x}_2$ costs

$$\frac{2}{3}n(n^2 - 1) + n(n+1), \qquad (3.5.31)$$

whereas by using **LU**-decomposition it only costs

$$\frac{1}{3}n(n^2 - 1) + \frac{1}{2}n(n+1), \qquad (3.5.32)$$

a savings on the order of $n^3/3$ for large n.

As pointed out at the end of Section 3.3, the simplest Gauss elimination process is guaranteed to work for strictly diagonally dominant matrices and for positive-definite matrices. Thus, **LU**-decomposition is assured for these two classes.

3.6. BAND MATRICES

If the matrix **A** is such that $a_{ij} = 0$ for $j \geq i + p$ and $i \geq j + q$, we say **A** is a *band matrix* of bandwidth $w = p + q + 1$. Examples are

$$\begin{bmatrix} 1 & 0 & 0 & 0 \\ 0 & 1 & 0 & 0 \\ 0 & 0 & 1 & 0 \\ 0 & 0 & 0 & 1 \end{bmatrix}, \quad a_{ij} = 0, \; j \geq i+1, \; i \geq j+1, \; w = 1, \quad (3.6.1)$$

and

$$\begin{bmatrix} 2 & 2 & 0 & 0 \\ 5 & 2 & 4 & 0 \\ 0 & 7 & 2 & 9 \\ 0 & 0 & 6 & 2 \end{bmatrix}, \quad a_{ij} = 0, \; j \geq i+2, \; i \geq j+2, \; w = 3. \quad (3.6.2)$$

Symbolically, we represent band matrices as in Fig. 3.6.1. In the part of the matrix labeled A, there are nonzero elements. In the parts labeled 0, all elements are 0. When **A** is banded, n large, and p and q small, the number of operations in the Gauss elimination solution to $\mathbf{Ax} = \mathbf{b}$ is on the order of npq, which is much smaller than the n^3 required for a full matrix.

One of the most important properties of a band matrix is that its **LU**-decomposition, when it exists, decomposes **A** into a product of more narrowly banded matrices. In particular, if the **LU**-decomposition can be achieved with Gauss elimination (without row interchange), it follows that the structure of **L** and **U** is as shown in Eq. (3.6.3) (see Fig. 3.6.2). Only in the parts of **L** and **U** labeled **L**$'$ and **U**$'$ are there nonzero elements. As an example of the simplification resulting

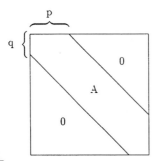

FIGURE 3.6.1

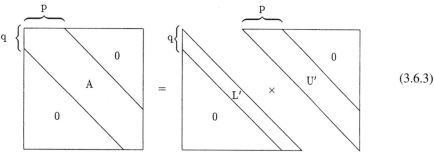

FIGURE 3.6.2 Equation (3.6.3).

$$\tag{3.6.3}$$

from Eq. (3.6.3), consider the tridiagonal matrix and its **LU**-decomposition:

$$\begin{bmatrix} a_1 & c_1 & & & & \\ b_2 & a_2 & c_2 & & \mathbf{0} & \\ & b_3 & a_3 & c_3 & & \\ & & \ddots & \ddots & \ddots & \\ \mathbf{0} & & & b_{n-1} & a_{n-1} & c_{n-1} \\ & & & & b_n & c_n \end{bmatrix} \tag{3.6.4}$$

$$= \begin{bmatrix} 1 & & & & & \\ \beta_2 & 1 & & \mathbf{0} & & \\ & \beta_3 & 1 & & & \\ & \mathbf{0} & \ddots & \ddots & & \\ & & & & \beta_n & 1 \end{bmatrix} \begin{bmatrix} \alpha_1 & \delta_1 & & & \\ & \alpha_2 & \delta_2 & \mathbf{0} & \\ & & \ddots & \ddots & \\ & \mathbf{0} & & \alpha_{n-1} & \delta_{n-1} \\ & & & & \alpha_n \end{bmatrix}.$$

Carrying out the matrix multiplications on the right-hand side of Eq. (3.6.4) and comparing with the corresponding elements in **A**, we find

$$\delta_i = c_i \tag{3.6.5}$$

and

$$\alpha_1 = a_1, \quad \beta_k = \frac{b_k}{\alpha_{k-1}}, \quad \alpha_k = a_k - \beta_k c_{k-1}, \quad k = 2, \ldots, n. \tag{3.6.6}$$

The number of operations needed to find the coefficients α_i, β_i, and δ_i are only $3(n-1)$ additions, multiplications, and divisions.

As a special case of a tridiagonal matrix, consider

$$\mathbf{A} = \begin{bmatrix} 1 & -1 & 0 & 0 \\ -1 & 2 & -1 & 0 \\ 0 & -1 & 2 & -1 \\ 0 & 0 & -1 & 2 \end{bmatrix}. \tag{3.6.7}$$

With the rules given in Eq. (3.6.6), we find

$$\mathbf{L} = \begin{bmatrix} 1 & 0 & 0 & 0 \\ -1 & 1 & 0 & 0 \\ 0 & -1 & 1 & 0 \\ 0 & 0 & -1 & 1 \end{bmatrix} \qquad (3.6.8)$$

and

$$\mathbf{U} = \begin{bmatrix} 1 & -1 & 0 & 0 \\ 0 & 1 & -1 & 0 \\ 0 & 0 & 1 & -1 \\ 0 & 0 & 0 & 1 \end{bmatrix}. \qquad (3.6.9)$$

Note that $\mathbf{U} = \mathbf{L}^{\mathrm{T}}$ for this case.

The inverse of $\mathbf{L}$ is easy to find. Consider the augmented matrix, $[\mathbf{L}, \mathbf{I}]$, corresponding to $\mathbf{L}\mathbf{x}_i = \mathbf{e}_i$, $i = 1, \ldots, 4$,

$$\begin{bmatrix} 1 & 0 & 0 & 0 & 1 & 0 & 0 & 0 \\ -1 & 1 & 0 & 0 & 0 & 1 & 0 & 0 \\ 0 & -1 & 1 & 0 & 0 & 0 & 1 & 0 \\ 0 & 0 & -1 & 1 & 0 & 0 & 0 & 1 \end{bmatrix}. \qquad (3.6.10)$$

In this case, Gauss elimination gives

$$\begin{bmatrix} 1 & 0 & 0 & 0 & 1 & 0 & 0 & 0 \\ 0 & 1 & 0 & 0 & 1 & 1 & 0 & 0 \\ 0 & 0 & 1 & 0 & 1 & 1 & 1 & 0 \\ 0 & 0 & 0 & 1 & 1 & 1 & 1 & 1 \end{bmatrix}. \qquad (3.6.11)$$

The solutions to $\mathbf{L}\mathbf{x}_i = \mathbf{e}_i$, $i = 1, \ldots, 4$, are

$$\mathbf{x}_1 = \begin{bmatrix} 1 \\ 1 \\ 1 \\ 1 \end{bmatrix}, \quad \mathbf{x}_2 = \begin{bmatrix} 0 \\ 1 \\ 1 \\ 1 \end{bmatrix}, \quad \mathbf{x}_3 = \begin{bmatrix} 0 \\ 0 \\ 1 \\ 1 \end{bmatrix}, \quad \mathbf{x}_4 = \begin{bmatrix} 0 \\ 0 \\ 0 \\ 1 \end{bmatrix}, \qquad (3.6.12)$$

and so the inverse of $\mathbf{L}$, which is just the matrix $[\mathbf{x}_1, \mathbf{x}_2, \mathbf{x}_3, \mathbf{x}_4]$, is the lower triangular matrix

$$\mathbf{L}^{-1} = \begin{bmatrix} 1 & 0 & 0 & 0 \\ 1 & 1 & 0 & 0 \\ 1 & 1 & 1 & 0 \\ 1 & 1 & 1 & 1 \end{bmatrix}. \qquad (3.6.13)$$

Since $\mathbf{U} = \mathbf{L}^{\mathrm{T}}$, it follows that $\mathbf{U}^{-1} = (\mathbf{L}^{-1})^{\mathrm{T}}$, or

$$\mathbf{U}^{-1} = \begin{bmatrix} 1 & 1 & 1 & 1 \\ 0 & 1 & 1 & 1 \\ 0 & 0 & 1 & 1 \\ 0 & 0 & 0 & 1 \end{bmatrix}. \tag{3.6.14}$$

By the same procedure, it is easy to show that the **LU**-decomposition of the $n \times n$ tridiagonal matrix

$$\mathbf{A} = \begin{bmatrix} 1 & -1 & & & & \mathbf{0} \\ -1 & 2 & -1 & & & \\ & \ddots & \ddots & \ddots & & \\ & & & -1 & 2 & -1 \\ \mathbf{0} & & & & -1 & 2 \end{bmatrix} \tag{3.6.15}$$

is

$$\mathbf{L} = \begin{bmatrix} 1 & & & \mathbf{0} \\ -1 & 1 & & \\ & \ddots & \ddots & \\ \mathbf{0} & & -1 & 1 \end{bmatrix}, \quad \mathbf{U} = \mathbf{L}^{\mathrm{T}}, \tag{3.6.16}$$

and

$$\mathbf{L}^{-1} = \begin{bmatrix} 1 & & & \mathbf{0} \\ 1 & 1 & & \\ \vdots & & \ddots & \\ 1 & \cdots & 1 & 1 \end{bmatrix}, \quad \mathbf{U}^{-1} = (\mathbf{L}^{-1})^{\mathrm{T}}. \tag{3.6.17}$$

Since the inverse of $\mathbf{A}$ is $\mathbf{U}^{-1}\mathbf{L}^{-1}$, we find

$$\mathbf{A}^{-1} = \begin{bmatrix} n & n-1 & n-2 & n-3 & \cdots & 2 & 1 \\ n-1 & n-1 & n-2 & n-3 & \cdots & 2 & 1 \\ n-2 & n-2 & n-2 & n-3 & \cdots & 2 & 1 \\ \vdots & \vdots & \vdots & \vdots & \ddots & \vdots & \vdots \\ 2 & 2 & 2 & 2 & \cdots & 2 & 1 \\ 1 & 1 & 1 & 1 & \cdots & 1 & 1 \end{bmatrix}.$$

For the 4×4 case, this becomes

$$\mathbf{A}^{-1} = \begin{bmatrix} 5 & 4 & 3 & 2 & 1 \\ 4 & 4 & 3 & 2 & 1 \\ 3 & 3 & 3 & 2 & 1 \\ 2 & 2 & 2 & 2 & 1 \\ 1 & 1 & 1 & 1 & 1 \end{bmatrix}.$$

An important point for band matrices can be made by comparing $\mathbf{A}^{-1}$ to $\mathbf{L}$ and $\mathbf{U}$. $\mathbf{A}^{-1}$ is a full matrix, requiring storage of n^2 elements. $\mathbf{L}$ and $\mathbf{U}$ require storage of only $4n - 2$ elements. To solve $\mathbf{LUx} = \mathbf{b}$, we could solve $\mathbf{Ly} = \mathbf{b}$ and $\mathbf{Ux} = \mathbf{y}$ in $n(n - 1)$ multiplications and divisions, and so evaluation of $\mathbf{A}^{-1}\mathbf{b}$ requires only n^2 divisions and multiplications. Thus, when storage is a problem, it is better to find the LU-decomposition of a banded matrix, store $\mathbf{L}$ and $\mathbf{U}$, and solve $\mathbf{LUx} = \mathbf{b}$ as needed.

As another example, consider the $n \times n$ tridiagonal matrix

$$\mathbf{A} = \begin{bmatrix} 2 & -1 & & & 0 \\ -1 & 2 & -1 & & \\ & \ddots & \ddots & \ddots & \\ & & -1 & 2 & -1 \\ 0 & & & -1 & 2 \end{bmatrix}. \quad (3.6.18)$$

The main diagonal elements are 2, whereas the elements on the diagonals just above and just below the main diagonal are -1. For this case, we find

$$\mathbf{L} = \begin{bmatrix} 1 & & & & & \\ -\dfrac{1}{2} & 1 & & & 0 & \\ & -\dfrac{2}{3} & 1 & & & \\ & & \ddots & & & \\ & & & -\dfrac{j-1}{j} & 1 & \\ 0 & & & & -\dfrac{n-1}{n} & 1 \end{bmatrix} \quad (3.6.19)$$

and

$$\mathbf{U} = \begin{bmatrix} 2 & -1 & & & & 0 \\ & \dfrac{3}{2} & -1 & & & \\ & & \ddots & \ddots & & \\ & 0 & & \dfrac{j+1}{j} & -1 & \\ & & & & & \dfrac{n+1}{n} \end{bmatrix}. \quad (3.6.20)$$

The inverses of $\mathbf{L}$ and $\mathbf{U}$ are not as easy to compute as in the previous example. However, the solution of $\mathbf{LUx} = \mathbf{b}$ only requires $n(n - 1)$ divisions and multiplications.

Matrix theory plays an important and often critical role in the analysis of differential equations. Although we will explore the general theory and formal solutions to specific differential equations in Chapter 10, we are nonetheless confronted with the cold hard fact that most practical applications in engineering and applied science offer no analytical solutions. We must, therefore, rely on numerical procedures in solving such problems. These procedures, as with the finite-difference method, usually involve discretizing coordinates into some finite grid in which differential quantities can be approximated.

As a physical example, consider the change in concentration c of a single-component solution under diffusion. The equation obeyed by c in time t and direction y is

$$\frac{\partial c}{\partial t} = D \frac{\partial^2 c}{\partial y^2}, \tag{3.6.21}$$

where D is the diffusion coefficient. At $t = 0$, we are given the initial condition

$$c(y, t) = f(y), \tag{3.6.22}$$

and at the boundaries of the system, $y = 0$ and $y = L$, the concentration is fixed at the values

$$\begin{aligned} c(y = 0, t) &= \alpha \\ c(y = L, t) &= \beta. \end{aligned} \tag{3.6.23}$$

Using the finite-difference approximation, we estimate c only at a set of nodal points $y_i = i \, \Delta y$, $i = 1, \ldots, n$, where $(n + 1) \Delta y = L$. In other words, we divide the interval $(0, L)$ into $n + 1$ identical intervals as shown in Fig. 3.6.3. If we let c_i denote $c(y_i, t)$ (the value of c at y_i), then the finite-difference approximation of $\partial^2 c / \partial y^2$ at y_i is given by

$$\frac{\partial^2 c(y_i, t)}{\partial y^2} = \frac{c_{i-1} - 2c_i + c_{i+1}}{(\Delta y)^2}. \tag{3.6.24}$$

With this approximation, Eq. (3.6.21), at the nodal points, reduces to the set of equations

$$\begin{aligned} \frac{dc_1}{d\tau} &= \alpha - 2c_1 + c_2 \\ \frac{dc_i}{d\tau} &= c_{i-1} - 2c_i + c_{i+1}, \quad i = 2, \ldots, n - 1, \\ \frac{dc_n}{d\tau} &= c_{n-1} - 2c_n + \beta, \end{aligned} \tag{3.6.25}$$

where $\tau \equiv t \, \Delta y^2 / D$.

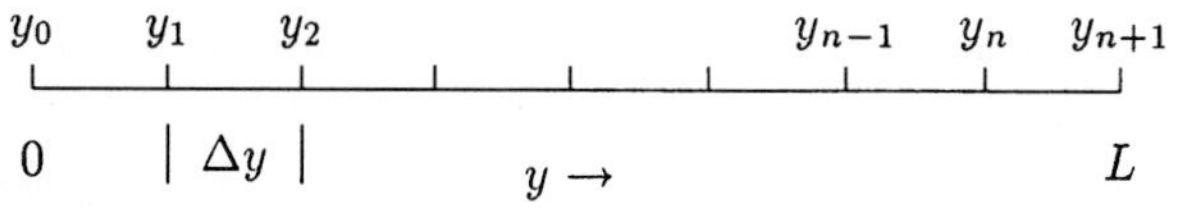

FIGURE 3.6.3

With the definitions

$$\mathbf{c} = \begin{bmatrix} c_1 \\ c_2 \\ \vdots \\ c_{n-1} \\ c_n \end{bmatrix} \quad \text{and} \quad \mathbf{b} = \begin{bmatrix} \alpha \\ 0 \\ \vdots \\ 0 \\ \beta \end{bmatrix}, \qquad (3.6.26)$$

the equations in Eq. (3.6.25) can be summarized as

$$\frac{d\mathbf{c}}{d\tau} = -\mathbf{A}\mathbf{c} + \mathbf{b}, \qquad (3.6.27)$$

where $\mathbf{A}$ is the tridiagonal matrix defined by Eq. (3.6.18). The initial condition corresponding to Eq. (3.6.22) is then given by

$$\mathbf{c}(\tau = 0) = \begin{bmatrix} f(y_1) \\ f(y_2) \\ \vdots \\ f(y_n) \end{bmatrix}. \qquad (3.6.28)$$

If we replace the boundary condition $c(y=0,t) = \alpha$ by the boundary condition $\partial c/\partial y = 0$ at $y = 0$ (barrier to diffusion), it gives for the finite-element approximation

$$c_1 = c(y=0, t) \qquad (3.6.29)$$

and changes the first equation in Eq. (3.6.25) to

$$\frac{dc_1}{d\tau} = -c_1 + c_2. \qquad (3.6.30)$$

Once again, we obtain Eq. (3.6.27), but the matrix $\mathbf{A}$ in this case is the one given by Eq. (3.6.15).

We note that, at steady state, $d\mathbf{c}/dt = 0$, and the concentration at the nodal points is given by $\mathbf{A}\mathbf{c} = \mathbf{b}$. The formal solution to Eq. (3.6.27) can, therefore, be written as

$$\mathbf{c}(\tau) = \mathbf{c}(0)\exp(-\tau\mathbf{A}) + \left[\int_0^\tau \exp(-(\tau-\xi)\mathbf{A})\,d\xi\right]\mathbf{b}. \qquad (3.6.31)$$

To compute $\mathbf{c}(\tau)$ from this formula, we need to know how to calculate the exponential of a matrix. Fortunately, the eigenvalue analysis discussed in Chapters 5 and 6 will make this easy.

As a further example, consider a one-dimensional problem in which the concentration $c(x,t)$ of some species is fixed at $x = 0$ and $x = a$, i.e.,

$$c(0,t) = \alpha, \qquad c(a,t) = \beta. \qquad (3.6.32)$$

Further, the concentration initially has the distribution

$$c(x, t = 0) = h(x), \tag{3.6.33}$$

and in the interval $0 < x < a$ is described by the equation

$$\frac{\partial c}{\partial t} = D\frac{\partial^2 c}{\partial x^2} + f(c), \tag{3.6.34}$$

where D is the diffusion coefficient and $f(c)$ represents the production or consumption rate of the species by chemical reaction.

In general, $f(c)$ will not be linear so that analytical solutions to Eq. (3.6.34) are not available. So, again, we solve this problem by introducing the finite-difference approximation. Again, the first and second spatial derivatives of c at the nodal points x_j are approximated by the difference formulas:

$$\frac{\partial c(x_j, t)}{\partial x} \approx \frac{c(x_{j+1}, t) - c(x_{j-1}, t)}{2\Delta x} \tag{3.6.35}$$

and

$$\frac{\partial^2 c(x_j, t)}{\partial x^2} \approx \frac{c(x_{j+1}, t) - 2c(x_j, t) + c(x_{j-1}, t)}{(\Delta x)^2}. \tag{3.6.36}$$

These are the so-called centered difference formulas, which approximate $\partial c(x_j)/\partial x$ and $\partial^2 c(x_j)/\partial x^2$ to order $(\Delta x)^2$.

With the notation

$$c_j = c(x_j, t), \tag{3.6.37}$$

the finite-difference approximation to Eq. (3.6.34) at the nodal points $j = 1, \ldots, n$ is given by

$$\frac{dc_j}{dt} = D\frac{c_{j+1} - 2c_j + c_{j-1}}{(\Delta x)^2} + f(c_j). \tag{3.6.38}$$

With the boundary conditions $c_0 = \alpha$ and $c_{n+1} = \beta$, Eq. (3.6.38) can be expressed as the matrix equation

$$\frac{d\mathbf{c}}{d\tau} = \mathbf{Ac} + \frac{(\Delta x)^2}{D}\mathbf{f(c)} + \mathbf{b}, \tag{3.6.39}$$

where $\tau \equiv tD/(\Delta x)^2$,

$$\mathbf{c} = \begin{bmatrix} c_1 \\ \vdots \\ c_n \end{bmatrix}, \quad \mathbf{b} = \begin{bmatrix} \alpha \\ 0 \\ \vdots \\ 0 \\ \beta \end{bmatrix}, \quad \mathbf{f} = \begin{bmatrix} f(c_1) \\ \vdots \\ f(c_n) \end{bmatrix}, \tag{3.6.40}$$

and

$$A = \begin{bmatrix} 2 & -1 & 0 & \cdots & & 0 \\ -1 & 2 & -1 & \cdots & & 0 \\ 0 & -1 & 2 & \cdots & & 0 \\ \vdots & \vdots & & \ddots & & \vdots \\ 0 & \cdots & & & 2 & -1 \\ 0 & \cdots & & 0 & -1 & 2 \end{bmatrix}. \qquad (3.6.41)$$

With the initial condition

$$\mathbf{c}(0) = \begin{bmatrix} h(x_1) \\ \vdots \\ h(x_n) \end{bmatrix}, \qquad (3.6.42)$$

Eq. (3.6.39) can be solved to determine the time evolution of the concentration at the nodal points. If the nodal points are sufficiently close together, $\mathbf{c}(t)$ provides a good approximation to $c(x, t)$.

For a two-dimensional problem with square boundaries, the above problem becomes

$$\frac{\partial c}{\partial t} = D\left(\frac{\partial^2 c}{\partial x^2} + \frac{\partial^2 c}{\partial y^2}\right) + f(c). \qquad (3.6.43)$$

The initial condition is given by

$$c(x, y, t = 0) = h(x, y), \qquad (3.6.44)$$

and the boundary conditions are indicated in Fig. 3.6.4. The finite-difference approximation for the square is generated by dividing the square into $(n+1)^2$ identical cells of area $(\Delta x)^2$. Numbering the nodal points, indicated by solid circles in the figure, from 1 to n^2 as indicated above, we can express the finite-difference approximations of the second derivatives at the ith node as

$$\frac{\partial^2 c_i}{\partial x^2} = \frac{c_{i+1} - 2c_i + c_{i-1}}{(\Delta x)^2} \qquad (3.6.45)$$

$$\frac{\partial^2 c_i}{\partial y^2} = \frac{c_{i+n} - 2c_i + c_{i-n}}{(\Delta x)^2}. \qquad (3.6.46)$$

With the boundary conditions given above, the matrix form of the finite-difference approximation to Eq. (3.6.43) is

$$\frac{d\mathbf{c}}{d\tau} = -\mathbf{A}\mathbf{c} + \frac{(\Delta x)^2}{D}\mathbf{f}(\mathbf{c}) + \mathbf{b}. \qquad (3.6.47)$$

$\mathbf{c}$ and $\mathbf{f}(\mathbf{c})$ are as defined in Eq. (3.6.40), but $\mathbf{A}$ in this case has five diagonals with nonzero elements and $\mathbf{b}$ is more complicated than before. For the special

BAND MATRICES

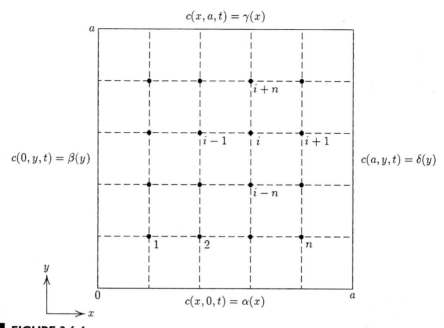

FIGURE 3.6.4

case $n = 3$, the formulas for $\mathbf{A}$ and $\mathbf{b}$ are

$$\mathbf{A} = \begin{bmatrix} 4 & -1 & 0 & -1 & 0 & 0 & 0 & 0 & 0 \\ -1 & 4 & -1 & 0 & -1 & 0 & 0 & 0 & 0 \\ 0 & -1 & 4 & 0 & 0 & -1 & 0 & 0 & 0 \\ -1 & 0 & 0 & 4 & -1 & 0 & -1 & 0 & 0 \\ 0 & -1 & 0 & -1 & 4 & -1 & 0 & -1 & 0 \\ 0 & 0 & -1 & 0 & -1 & 4 & 0 & 0 & -1 \\ 0 & 0 & 0 & -1 & 0 & 0 & 4 & -1 & 0 \\ 0 & 0 & 0 & 0 & -1 & 0 & -1 & 4 & -1 \\ 0 & 0 & 0 & 0 & 0 & -1 & 0 & -1 & 4 \end{bmatrix} \quad (3.6.48)$$

and

$$\mathbf{b} = \begin{bmatrix} \alpha(x_1) + \beta(y_1) \\ \alpha(x_2) \\ \alpha(x_3) + \delta(y_3) \\ \beta(y_4) \\ 0 \\ \delta(y_6) \\ \gamma(x_7) + \beta(y_7) \\ \gamma(x_8) \\ \gamma(x_9) + \delta(y_9) \end{bmatrix}. \quad (3.6.49)$$

Although the two-dimensional problem is more complicated to reduce to matrix form, the resulting matrix equation is about as simple as that of the one-dimensional problem.

EXAMPLE 3.6.1 (Mass Separation with a Staged Absorber). It is common in engineering practice to strip some solute from a fluid by contacting it through countercurrent flow with another fluid that absorbs the solute. Schematically, the device for this can be represented as a sequence of stages, at each stage of which the solute reaches equilibrium between the fluid streams. This equilibrium is frequently characterized by a constant K. Thus, at stage i, the concentration x_i in one fluid is related to the concentration y_i in the other fluid by the equation

$$y_i = K x_i. \tag{3.6.50}$$

A schematic diagram of the absorber is shown in Fig. 3.6.5.

L is the flow rate of inert carrier fluid in stream 1 and G is the corresponding quantity in stream 2. Lx_j is the rate at which solute leaves stage j with stream 1 and Gy_j is the rate at which it leaves stage j with stream 2. Thus, if hx_j and gy_j represent the amount of solute in stage j in streams 1 and 2, h and g being the "hold-up" capacities of each stream, then a mass balance on stage j yields the equation of change

$$h\frac{dx_j}{dt} + g\frac{dy_j}{dt} = Lx_{j-1} + Gy_{j+1} - Lx_j - Gy_j, \qquad j=1,\ldots,n. \tag{3.6.51}$$

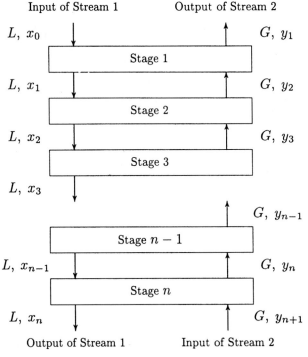

FIGURE 3.6.5

BAND MATRICES

Combined with Eq. (3.6.50), this expression becomes

$$\frac{dx_j}{d\tau} = Lx_{j-1} - (L + KG)x_j + KGx_{j+1}, \tag{3.6.52}$$

where $\tau \equiv t/(h+gK)$. In matrix form, this set of equations can be summarized as

$$\frac{d\mathbf{x}}{d\tau} = \mathbf{Ax} + \mathbf{b}, \tag{3.6.53}$$

where

$$\mathbf{x} = \begin{bmatrix} x_1 \\ \vdots \\ x_n \end{bmatrix}, \quad \mathbf{b} = \begin{bmatrix} Lx_0 \\ 0 \\ \vdots \\ 0 \\ Gy_{n+1} \end{bmatrix}, \tag{3.6.54}$$

and

$$\mathbf{A} = \begin{bmatrix} -(L+GK) & GK & 0 & 0 & \cdots & 0 \\ L & -(L+GK) & GK & 0 & \cdots & \vdots \\ 0 & L & -(L+GK) & GK & \cdots & \\ \vdots & & & & \ddots & \\ 0 & & \cdots & 0 & L & -(L+GK) \end{bmatrix}. \tag{3.6.55}$$

A is referred to as a tridiagonal matrix because it has nonzero elements only along the main diagonal and along two other diagonals parallel to the main diagonal. Such matrix problems are especially easy to handle.

In a typical problem L, G, K, x_0, and y_{n+1} are the givens. Questions asked might be: (1) How many stages n are needed for a given separation x_n? (2) What are the steady-state values of x_i and y_i? (3) Is the steady state stable? (4) What is the start-up behavior of the absorber? (5) Can cyclic behavior occur?

The countercurrent separation process described above leads to an equation like Eq. (3.6.27) with a tridiagonal matrix **A** in which $a_i = -(L + GK)$, $b_i = L$, and $c_i = GK$. At steady state, $\mathbf{Ax} = -\mathbf{b}$, where

$$\mathbf{b} = \begin{bmatrix} Lx_0 \\ 0 \\ \vdots \\ 0 \\ Gy_{n+1} \end{bmatrix}.$$

Consider a process where a liquid stream of $3500\,\text{lb}_\text{m}/\text{hr}$ water containing 1% nicotine is contacted with pure kerosene ($4000\,\text{lb}_\text{m}/\text{hr}$) in a countercurrent

TABLE 3.6.1 Existing Nicotine Concentration versus Number of Theoretical Trays

n	x_n
7	0.00125
8	0.00111
9	0.00099
10	0.00090
11	0.00083

liquid–liquid extractor at room temperature. The dilute partition coefficient for nicotine is estimated as $K = 0.876$ (weight fraction nicotine in kerosene)/(weight fraction nicotine in H_2O). If the target nicotine concentration in water is $x_n < 0.001$, how many theoretical trays are required?

To solve this problem, the matrix equation $\mathbf{Ax} = -\mathbf{b}$ must be solved at different values of n (the number of stages). The matrix $\mathbf{A}$ and the vectors $\mathbf{x}$ and $\mathbf{b}$ are defined in Eqs. (3.6.54) and (3.6.55). Given $L = 3500$, $G = 4000$, $K = 0.876$, $x_0 = 0.01$, and $y_{n+1} = 0$, the Gauss elimination procedure can be applied to find $\mathbf{x}(n)$ as a function of n. The correct number of theoretical stages is then the minimum value of n such that $x_n < 0.001$. Table 3.6.1 shows the values of x_n for different values of n. For this system, the required number of theoretical trays is 9.

■ ■ ■   A Mathematica program to solve this problem is given in the Appendix.

3.7. ITERATIVE METHODS FOR SOLVING Ax = b

Cramer's rule and Gauss elimination are direct methods for solving linear equations. Gauss elimination is the preferred method for large problems when sufficient computer memory is available for matrix storage. However, iterative methods are sometimes used because they are easier to program or require less computer memory for storage.

Jacobi's method is perhaps the simplest iterative method. Consider the equations

$$\sum_{j=1}^{n} a_{ij} x_j = b_i, \qquad i = 1, \ldots, n, \qquad (3.7.1)$$

for the case $a_{ii} \neq 0$, $i = 1, \ldots, n$. Note that Eq. (3.7.1) can be rewritten as

$$x_i = \left(-\sum_{\substack{j=1 \\ j \neq i}}^{n} a_{ij} x_j + b_i \right) \Big/ a_{ii}, \qquad i = 1, \ldots, n. \qquad (3.7.2)$$

ITERATIVE METHODS FOR SOLVING Ax = b

If we guess the solution $\{x_1^{(0)}, \ldots, x_n^{(0)}\}$ and insert it on the right-hand side of Eq. (3.7.2), we obtain the new estimate $\{x_1^{(1)}, \ldots, x_n^{(1)}\}$ of the solution from

$$x_i^{(1)} = \left(-\sum_{\substack{j=1 \\ j \neq i}}^{n} a_{ij} x_j^{(0)} + b_i\right) \Big/ a_{ii}, \qquad i = 1, \ldots, n. \tag{3.7.3}$$

Continuing the process, we find that at the $(k+1)$th iteration the estimate is given by

$$x_i^{(k+1)} = \left(-\sum_{\substack{j=1 \\ j \neq i}}^{n} a_{ij} x_j^{(k)} + b_i\right) \Big/ a_{ii}, \qquad i = 1, \ldots, n. \tag{3.7.4}$$

If $\mathbf{x}^{(k)}$ converges to a limit $\mathbf{x}$ as $k \to \infty$, then $\mathbf{x}$ is the solution to the linear equation set in Eq. (3.7.4); i.e., $\mathbf{x}$ is the solution to $\mathbf{Ax} = \mathbf{b}$. This is called the Jacobi iteration method. If $|\mathbf{A}| = 0$, the method, of course, fails. However, even if $|\mathbf{A}| \neq 0$, the method does not necessarily converge to a solution. We will discuss the convergence problem further, later in this section.

In computing $x_1^{(k+1)}, x_2^{(k+1)}, \ldots$ from Eq. (3.7.4), we note that when calculating $x_i^{(k+1)}$ from the ith equation the estimates $x_1^{(k+1)}, \ldots, x_{i-1}^{(k+1)}$ are already known. Thus, we could hope to accelerate convergence by updating the quantities already known on the right-hand side of Eq. (3.7.4). In other words, we can estimate $x_i^{(k+1)}$ from

$$x_i^{(k+1)} = \left(-\sum_{j=1}^{i-1} a_{ij} x_j^{(k+1)} - \sum_{j=i+1}^{n} a_{ij} x_j^{(k)} + b_i\right) \Big/ a_{ii}, \qquad i = 1, \ldots, n. \tag{3.7.5}$$

This iteration scheme is called the Gauss–Seidel method and it usually converges faster than the Jacobi method.

Still another method, which is a variation of the Gauss–Seidel method, is the successive overrelaxation (SOR) method. If we define the "residual" $r_i^{(k)}$ as

$$\begin{aligned} r_i^{(k)} &\equiv x_i^{(k+1)} - x_i^{(k)} \\ &= \left(-\sum_{j=1}^{i-1} a_{ij} x_j^{(k+1)} - \sum_{j=i}^{n} a_{ij} x_j^{(k)} + b_i\right) \Big/ a_{ii}, \end{aligned} \tag{3.7.6}$$

where $x_i^{(k+1)}$ and $x_i^{(k)}$ are Gauss–Seidel estimates, then $r_i^{(k)}$ is a measure of how fast the iterations are converging. For example, if successive estimates oscillate a lot, then $r_i^{(k)}$ will be rather large. The SOR method "relaxes" the $(k+1)$th oscillation somewhat by averaging the estimate $x_i^{(k)}$ and the residual. Namely, the $(k+1)$th SOR estimate is defined by

$$x_i^{(k+1)} = x_i^{(k)} + \omega r_i^{(k)} \tag{3.7.7}$$

or

$$x_i^{(k+1)} = (1-\omega)x_i^{(k)} + \omega\left(-\sum_{j=1}^{i-1} a_{ij}x_j^{(k+1)} - \sum_{j=i+1}^{n} a_{ij}x_j^{(k)} + b_i\right)\bigg/a_{ii}, \quad i=1,\ldots,n. \tag{3.7.8}$$

Thus, the SOR estimate of $x_i^{(k+1)}$ is a weighted average of the kth estimate $x_i^{(k)}$ (weighted by $1-\omega$) and the estimate of $x_i^{(k+1)}$ computed by the Gauss–Seidel method (the coefficient of ω in Eq. (3.7.8)). Note that ω is an arbitrary weighting parameter—although, as we shall claim shortly, it cannot lie outside the range $0 < \omega < 2$ if the method is to converge. For every matrix $\mathbf{A}$, there is a "best value" $\tilde{\omega}$ of the SOR parameter for which the method converges the most rapidly. With $\omega = 1$, the SOR method is the same as the Gauss–Seidel method. The SOR method will converge faster than the Gauss–Seidel method if the optimal $\tilde{\omega}$ is used.

To illustrate the three iterative methods just described, consider the problem $\mathbf{Ax} = \mathbf{b}$, where

$$\mathbf{A} = \begin{bmatrix} 4 & -1 & -1 & 0 & 0 \\ -1 & 4 & 0 & -1 & 0 \\ -1 & 0 & 4 & -1 & -1 \\ 0 & -1 & -1 & 4 & 0 \\ 0 & 0 & -1 & 0 & 4 \end{bmatrix} \tag{3.7.9a}$$

and

$$\mathbf{b} = \begin{bmatrix} 1 \\ 2 \\ 0 \\ 1 \\ 2 \end{bmatrix}. \tag{3.7.9b}$$

Table 3.7.1 shows $x_i^{(k)}$ versus iteration k for the three methods using as the zeroth-order estimate $\mathbf{x}^{(0)} = \mathbf{0}$. The solution found by Gauss elimination is given at the bottom of the table and we will use this as the correct solution $\mathbf{x}$ and define the error of the iterative result as

$$\text{error} = \|\mathbf{x}^{(k)} - \mathbf{x}\|_{p=2} = \sqrt{\sum_{i=1}^{n}(x_i^{(k)} - x_i)^2}. \tag{3.7.10}$$

We see that, in order to find a solution with error less than 10^{-4}, it takes 15 iterations for the Jacobi method, 8 for the Gauss–Seidel method, and 5 for the SOR method. The value of ω (1.1035) used is optimal for the matrix $\mathbf{A}$. For this matrix, SOR is the best method, Gauss–Seidel the next best, and Jacobi the poorest. Mathematica programs have been included in the Appendix for all three of these methods.

ITERATIVE METHODS FOR SOLVING $\mathbf{Ax}=\mathbf{b}$

TABLE 3.7.1 Comparison of Rates of Convergence of Three Iterative Solutions of $\mathbf{Ax}=\mathbf{b}$ for A and b Given in Eqs. (3.7.9a) and (3.7.9b)

k	$x_1^{(k)}$	$x_2^{(k)}$	$x_3^{(k)}$	$x_4^{(k)}$	$x_5^{(k)}$
			Jacobi method		
1	0.25	0.5	0	0.25	0.5
2	0.375	0.625	0.25	0.375	0.5
3	0.46875	0.6875	0.3125	0.46875?	0.5625
4	0.5	0.734375	0.375	0.5	0.578125
5	0.527344	0.75	0.394531	0.527394	0.59375
6	0.536133	0.763672	0.412109	0.536133	0.598633
7	0.543945	0.768066	0.417725	0.543945	0.603027
8	0.546448	0.771973	0.422729	0.546448	0.604431
9	0.548676	0.773224	0.424332	0.548676	0.605682
10	0.549389	0.774338	0.425758	0.549389	0.606083
11	0.550024	0.774694	0.426215	0.550024	0.60644
12	0.550227	0.775012	0.426622	0.550227	0.606554
			Gauss–Seidel method		
1	0.25	0.5625	0.0625	0.40625	0.515625
2	0.40625	0.703125	0.332031	0.508789	0.583008
3	0.508789	0.754395	0.400146	0.538635	0.600037
4	0.538635	0.769318	0.419327	0.547161	0.604832
5	0.547161	0.773581	0.424788	0.549592	0.606197
6	0.549592	0.774796	0.426345	0.550285	0.606586
7	0.550285	0.775143	0.426789	0.550483	0.606697
			SOR method		
1	0.275875	0.627857	0.076107	0.470081	0.572746
2	0.441528	0.738257	0.401619	0.541685	0.603268
3	0.54464?	0.77503?	0.424549	0.550745	0.606434
4	0.550439	0.775323	0.427148	0.550605	0.606824
5	0.550636	0.775309	0.427003	0.550575	0.606743
			Gauss elimination method		
	0.550562	0.775281	0.426966	0.550562	0.606742

The defining equations for Jacobi iteration, Eq. (3.7.2), can be written in vector form as

$$\mathbf{x} = \mathbf{B}_J \mathbf{x} + \mathbf{c}_J, \tag{3.7.11}$$

where

$$\mathbf{B}_J = -\mathbf{D}^{-1}(\mathbf{L}+\mathbf{U}) \tag{3.7.12}$$

and

$$\mathbf{c}_J = \mathbf{D}^{-1}\mathbf{b}. \tag{3.7.13}$$

The quantities **D**, **L**, and **U** are given by

$$\mathbf{D} = \begin{bmatrix} a_{11} & & \mathbf{0} \\ & \ddots & \\ \mathbf{0} & & a_{nn} \end{bmatrix}, \tag{3.7.14}$$

$$\mathbf{L} = \begin{bmatrix} 0 & & & & \\ a_{21} & 0 & & \mathbf{0} & \\ a_{31} & a_{32} & 0 & & \\ \vdots & \vdots & \ddots & \ddots & \\ a_{n1} & a_{n2} & \cdots & a_{n,n-1} & 0 \end{bmatrix}, \tag{3.7.15}$$

and

$$\mathbf{U} = \begin{bmatrix} 0 & a_{12} & a_{13} & \cdots & a_{1n} \\ & 0 & a_{23} & \cdots & a_{2n} \\ & & 0 & \cdots & a_{3n} \\ & \mathbf{0} & & \ddots & \vdots \\ & & & & 0 \end{bmatrix}. \tag{3.7.16}$$

D is a diagonal matrix with the diagonal elements of **A** along the main diagonal. **L** is a lower diagonal matrix having 0's on the main diagonal and the elements of **A** below the main diagonal. **U** is an upper diagonal matrix having 0's on the main diagonal and the elements of **A** above the main diagonal. Thus, it follows that

$$\mathbf{A} = \mathbf{D} + \mathbf{L} + \mathbf{U}. \tag{3.7.17}$$

Clearly, Eq. (3.7.11) is a rearrangement of $\mathbf{Ax} = \mathbf{b}$, which can be accomplished only if none of the diagonal elements is 0. Note that **L** and **U** defined here have nothing to do with the **L** and **U** in our discussion of **LU**-decomposition. The iterative solution of Eq. (3.7.11) is given by

$$\mathbf{x}^{(k+1)} = \mathbf{B}_J \mathbf{x}^{(k)} + \mathbf{c}_J, \qquad k = 0, 1, \ldots. \tag{3.7.18}$$

The vector form of the defining equations of the Gauss–Seidel iteration is

$$(\mathbf{I} + \mathbf{D}^{-1}\mathbf{L})\mathbf{x} = -\mathbf{D}^{-1}\mathbf{U}\mathbf{x} + \mathbf{D}^{-1}\mathbf{b}$$

or

$$\mathbf{x} = \mathbf{B}_{GS}\mathbf{x} + \mathbf{c}_{GS}, \tag{3.7.19}$$

where

$$\mathbf{B}_{GS} = -(\mathbf{D} + \mathbf{L})^{-1}\mathbf{U} \tag{3.7.20}$$

and

$$\mathbf{c}_{GS} = (\mathbf{D} + \mathbf{L})^{-1}\mathbf{b}. \qquad (3.7.21)$$

Equation (3.7.19) is again just a formal rearrangement of $\mathbf{Ax} = \mathbf{b}$, which is allowed if $\mathbf{A}$ has no zero elements on the main diagonal. $\mathbf{D} + \mathbf{L}$ is a lower diagonal matrix, which is nonsingular if $|\mathbf{D}| \neq 0$. The corresponding iterative solution of Eq. (3.7.19) becomes

$$\mathbf{x}^{(k+1)} = \mathbf{B}_{GS}\mathbf{x}^{(k)} + \mathbf{c}_{GS}, \qquad k = 0, 1, \ldots. \qquad (3.7.22)$$

Finally, the SOR iteration begins with the vector equation

$$\mathbf{x} = \mathbf{B}_{SOR}\mathbf{x} + \mathbf{c}_{SOR}, \qquad (3.7.23)$$

where

$$\mathbf{B}_{SOR} = (\mathbf{D} + \omega\mathbf{L})^{-1}\left[(1 - \omega)\mathbf{D} - \omega\mathbf{U}\right] \qquad (3.7.24)$$

and

$$\mathbf{c}_{SOR} = (\mathbf{D} + \omega\mathbf{L})^{-1}\mathbf{b}. \qquad (3.7.25)$$

Equation (3.7.23) is again a formal rearrangement of $\mathbf{Ax} = \mathbf{b}$, which is allowed if $|\mathbf{D}| \neq 0$ or $a_{ii} \neq 0$, $i = 1, \ldots, n$, and the iterative solution to Eq. (3.7.23) is

$$\mathbf{x}^{(k+1)} = \mathbf{B}_{SOR}\mathbf{x}^{(k)} + \mathbf{c}_{SOR}, \qquad k = 0, 1, \ldots. \qquad (3.7.26)$$

The components of Eqs. (3.7.18), (3.7.22), and (3.7.26) can be shown to correspond to Eqs. (3.7.4), (3.7.5), and (3.7.7), which are the Jacobi, Gauss–Seidel, and SOR equations, respectively.

We now need some mechanism to predict the convergence properties of iterative methods. It turns out that a sufficient condition for convergence of the iteration process to the solution $\mathbf{x}$ of $\mathbf{Ax} = \mathbf{b}$ is provided by the following theorem:

THEOREM. *If*

$$\|\mathbf{B}\| < 1,$$

then the process

$$\mathbf{x}^{(k+1)} = \mathbf{B}\mathbf{x}^{(k)} + \mathbf{c}, \qquad k = 0, 1, \ldots, \qquad (3.7.27)$$

generates the solution $\mathbf{x}$ *of*

$$\mathbf{x} = \mathbf{B}\mathbf{x} + \mathbf{c}. \qquad (3.7.28)$$

To prove the theorem, suppose $\mathbf{x}$ is the solution of Eq. (3.7.28). Subtracting Eq. (3.7.28) from Eq. (3.7.27), we obtain

$$\mathbf{x}^{(k+1)} - \mathbf{x} = \mathbf{B}(\mathbf{x}^{(k)} - \mathbf{x}). \tag{3.7.29}$$

Note that

$$\begin{aligned} \mathbf{x}^{(1)} - \mathbf{x} &= \mathbf{B}(\mathbf{x}^{(0)} - \mathbf{x}) \\ \mathbf{x}^{(2)} - \mathbf{x} &= \mathbf{B}(\mathbf{x}^{(1)} - \mathbf{x}) = \mathbf{B}^2(\mathbf{x}^{(0)} - \mathbf{x}) \\ &\vdots \\ \mathbf{x}^{(k)} - \mathbf{x} &= \mathbf{B}^k(\mathbf{x}^{(0)} - \mathbf{x}). \end{aligned} \tag{3.7.30}$$

From the properties of matrix norms, it follows that

$$\|\mathbf{x}^{(k)} - \mathbf{x}\| \leq \|\mathbf{B}^k\| \, \|\mathbf{x}^{(0)} - \mathbf{x}\| \leq \|\mathbf{B}\|^k \|\mathbf{x}^{(0)} - \mathbf{x}\|. \tag{3.7.31}$$

Thus, if the norm $\|\mathbf{B}\|$ of $\mathbf{B}$ is less than 1, Eq. (3.7.31) implies that

$$\lim_{k \to \infty} \|\mathbf{x}^{(k)} - \mathbf{x}\| = 0; \tag{3.7.32}$$

i.e., the sequence $\mathbf{x}^{(k)}$ generated by Eq. (3.7.27) converges to the solution of Eq. (3.7.28).

We know that if $\mathbf{B} = \mathbf{B}_J$, $\mathbf{B}_{GS}$, or $\mathbf{B}_{SOR}$, Eq. (3.7.27) is equivalent to $\mathbf{A}\mathbf{x} = \mathbf{b}$, and so the Jacobi, Gauss–Seidel, and SOR processes will generate the solution to $\mathbf{A}\mathbf{x} = \mathbf{b}$ if the norm of the corresponding matrix $\mathbf{B}$ is less than 1. Interestingly, the value of $\mathbf{c}$, and therefore of $\mathbf{b}$, does not affect convergence of the iteration process.

■ ■ ■ **EXERCISE 3.7.1.** Use the $p = \infty$ norm to prove that Jacobi's method converges when $\mathbf{A}$ is a diagonally dominant matrix.

Using the eigenvalue techniques that will be developed in Chapters 5 and 6, we can prove that the SOR method converges *if and only if* $\mathbf{B}_{SOR}$ exists ($|\mathbf{D}| \neq 0$) and the parameter ω lies in the range

$$0 < \omega < 2. \tag{3.7.33}$$

■ **EXERCISE 3.7.2.** If $|\mathbf{A}| \neq 0$ and $\mathbf{A}$ is a real $n \times n$ matrix, then the problem $\mathbf{A}\mathbf{x} = \mathbf{b}$ can be solved by the *conjugate gradient method*. According to this method, we start with the initial guess $\mathbf{x}_0$, compute the vector $\mathbf{r}_0 = \mathbf{b} - \mathbf{A}\mathbf{x}_0$, and set $\mathbf{p}_0 = \mathbf{r}_0$. We then iterate

$$\alpha_i = \frac{\|\mathbf{r}_i\|^2}{\mathbf{p}_i^T \mathbf{A} \mathbf{p}_i} \tag{3.7.34a}$$

$$\mathbf{x}_{i+1} = \mathbf{x}_i + \alpha_i \mathbf{p}_i \tag{3.7.34b}$$

$$\mathbf{r}_{i+1} = \mathbf{r}_i - \alpha_i \mathbf{A} \mathbf{p}_i \tag{3.7.34c}$$

$$\beta_i = \frac{\|\mathbf{r}_{i+1}\|^2}{\|\mathbf{r}_i\|^2} \tag{3.7.34d}$$

$$\mathbf{p}_{i+1} = \mathbf{r}_{i+1} + \beta_i \mathbf{p}_i \tag{3.7.34e}$$

for $i = 1, 2, \ldots, n-1$. A Mathematica program implementing the above conjugate gradient method is included in the Appendix. Show that the vector $\mathbf{x}_n$ converges to the solution to $\mathbf{Ax} = \mathbf{b}$ for a variety of matrices $\mathbf{A}$.

3.8. NONLINEAR EQUATIONS

Most "real-world" problems are nonlinear. Given this fact, it might appear a bit surprising that linear equations receive so much attention in numerical analysis and applied mathematics. The reason is simple: the most common method for solving nonlinear systems is to do so iteratively through linearized equations. The Newton–Raphson method is frequently the method of choice for such iterative solutions. Another iterative approach is the Picard method, which is easier than, but not as powerful as, the Newton–Raphson method. Both methods require an initial guess of the solution and, as we will see, their success depends on the "goodness" of the initial guess.

Although linear equations occasionally admit multiple solutions, they very often have unique solutions. Nonlinear equations, on the other hand, normally admit multiple solutions. For example, the equation

$$x^4 - 5x^2 + 4 = 0 \tag{3.8.1}$$

has four solutions, namely, $x = 1, -1, 2$, and -2. Similarly, the equation

$$\cos^2 x - 1 = 0 \tag{3.8.2}$$

has an infinite number of solutions, namely, $x = n\pi$, $n = 0, \pm 1, \pm 2, \ldots$. There are, of course, nonlinear equations that have unique solutions. For example,

$$e^{-x} - x = 0 \tag{3.8.3}$$

has the unique solution $x = 0.567\ldots$. Furthermore, the equation $e^{-ax} - x = 0$ has a unique solution for any positive real value of a and no solution at all for any negative real value of a. When there are multiple solutions, the particular solution to which an iterative scheme converges will depend on the initial guess.

In this section we will concern ourselves with a system of n equations and n unknowns, i.e.,

$$f_1(x_1, \ldots, x_n) = 0$$
$$\vdots \tag{3.8.4}$$
$$f_n(x_1, \ldots, x_n) = 0,$$

which we can express in vector form as

$$\mathbf{f}(\mathbf{x}) = \mathbf{0}, \tag{3.8.5}$$

where $\mathbf{f}$ is a column vector whose components f_i are nonlinear functions of the components x_i of the column vector $\mathbf{x}$.

For Picard iteration, we rewrite $\mathbf{f}$ as $\mathbf{f}(\mathbf{x}) = \mathbf{x} - \mathbf{g}(\mathbf{x})$, and Eq. (3.8.5) is rearranged to get

$$\mathbf{x} = \mathbf{g}(\mathbf{x}). \tag{3.8.6}$$

If we choose the initial estimate of the solution to be $\mathbf{x}^{(1)}$, subsequent estimates are then computed from

$$\mathbf{x}^{(k+1)} = \mathbf{g}(\mathbf{x}^{(k)}), \qquad k = 1, 2, \ldots. \tag{3.8.7}$$

If this iteration converges, i.e., if

$$\lim_{k \to \infty} \|\mathbf{x}^{(k+1)} - \mathbf{x}^{(k)}\| = 0, \tag{3.8.8}$$

then $\lim_{k \to \infty} \mathbf{x}^{(k)} = \mathbf{x}$, where $\mathbf{x}$ is a solution of Eq. (3.8.5). In practical applications, the criterion for convergence is usually that one accepts $\mathbf{x}^{(k+1)}$ as a solution when $\|\mathbf{x}^{(k+1)} - \mathbf{x}^{(k)}\| < \epsilon$, where the tolerance ϵ is a small number chosen by physical considerations.

The Newton–Raphson method also begins with an initial guess $\mathbf{x}^{(1)}$. However, this method is derived by expanding the equations in Eq. (3.8.4) in a multivariable Taylor series about $\mathbf{x}^{(1)}$, obtaining

$$f_i(x_1, \ldots, x_n) = f_i(x_1^{(1)}, \ldots, x_n^{(1)}) + \sum_{j=1}^n (x_j - x_j^{(1)}) \frac{\partial f_i}{\partial x_j}(x_1, \ldots, x_n)\bigg|_{\mathbf{x} = \mathbf{x}^{(1)}}$$
$$+ \mathcal{O}(\mathbf{x} - \mathbf{x}^{(1)})^2, \tag{3.8.9}$$

or, in vector notation,

$$\mathbf{f}(\mathbf{x}) = \mathbf{f}(\mathbf{x}^{(1)}) + \mathbf{J}(\mathbf{x}^{(1)})(\mathbf{x} - \mathbf{x}^{(1)}) + \mathcal{O}(\mathbf{x} - \mathbf{x}^{(1)})^2. \tag{3.8.10}$$

The matrix $\mathbf{J}(\mathbf{x})$ is called the Jacobian matrix (or just the Jacobian) and has the elements

$$j_{ij} = \frac{\partial f_i}{\partial x_j}(x_1, \ldots, x_n). \tag{3.8.11}$$

In the Newton–Raphson method, we assume the higher order terms in the Taylor expansion are small and we obtain the next guess $\mathbf{x}^{(2)}$ for the solution to $\mathbf{f}(\mathbf{x}) = \mathbf{0}$ from the expression

$$\mathbf{0} = \mathbf{f}(\mathbf{x}^{(1)}) + \mathbf{J}(\mathbf{x}^{(2)} - \mathbf{x}^{(1)}), \tag{3.8.12}$$

which can be solved using Gauss elimination or any of the iterative methods discussed in the previous section. Iteration of this process yields

$$\mathbf{0} = \mathbf{f}(\mathbf{x}^{(k)}) + \mathbf{J}(\mathbf{x}^{(k)})(\mathbf{x}^{(k+1)} - \mathbf{x}^{(k)}), \tag{3.8.13}$$

where $k = 1, 2, \ldots$. When the process converges, the result $\mathbf{x} = \lim_{k \to \infty} \mathbf{x}^{(k)}$ is a solution to $\mathbf{f}(\mathbf{x}) = \mathbf{0}$.

In practice, the quantity $\mathbf{x}^{(k+1)}$ is accepted as a solution when

$$\|\mathbf{x}^{(k+1)} - \mathbf{x}^{(k)}\| < \epsilon_x \quad \text{and} \quad \|\mathbf{f}(\mathbf{x}^{(k+1)})\| < \epsilon_f, \tag{3.8.14}$$

where the tolerances ϵ_x and ϵ_f are small quantities chosen by physical considerations. For example, if $\|\mathbf{x}^{(k+1)}\| = 10^8$, then $\epsilon_x = 1$ could be sufficiently small for practical purposes. On the other hand, if $\|\mathbf{x}^{(k+1)}\| = 10^{-8}$, then ϵ_x should be considerably smaller than 10^{-8}. One way to choose ϵ_x would be to require $\epsilon_x < 10^{-6}\|\mathbf{x}^{(1)}\|$, and likewise $\epsilon_f < 10^{-6}\|\mathbf{f}(\mathbf{x}^{(1)})\|$. Of course, 10^{-6} could be replaced by a smaller or larger number, depending on the tolerance one is willing to accept.

The Newton–Raphson method will always converge to a solution when the guess $\mathbf{x}^{(1)}$ is sufficiently close to a root of the equation $\mathbf{f}(\mathbf{x}) = \mathbf{0}$. This is not so for the Picard method. For instance, for the equation

$$x = -\sqrt{3x - e^x + 2}, \tag{3.8.15a}$$

there is no guess for which the Picard method will converge to the solution, which is $x = -0.390272$. Interestingly, if Eq. (3.8.15a) is squared and rearranged to get

$$x = \frac{x^2 + e^x - 2}{3}, \tag{3.8.15b}$$

then the guess $x^{(0)} = 1$ will lead to the convergence of a Picard iteration to the root $x = -0.390272$. In general, for $x = g(x)$, when $|dg/dx| < 1$ for $x^{(0)}$ in the vicinity of the root x, the Picard method will converge. Otherwise, it is likely to fail.

■ **EXERCISE 3.8.1.** Show that the value of dg/dx for x near -0.390272 guarantees convergence of the Picard method for Eq. (3.8.15b) but not for Eq. (3.8.15a). Explore how robust the Picard method is for Eq. (3.8.15b) by trying different ini-
■ ■ ■ tial guesses.

An important feature of the Newton–Raphson method is that, when it starts to converge, the convergence is quadratic. In other words, if convergence begins and

$$\|\mathbf{x}^{(k+1)} - \mathbf{x}^{(k)}\| = 10^{-1}C, \tag{3.8.16}$$

where C is some scale factor, then after ν more iterations

$$\|\mathbf{x}^{(k+1+\nu)} - \mathbf{x}^{(k+\nu)}\| = 10^{-2^\nu}C. \tag{3.8.17}$$

Thus, at three iterations beyond the kth iteration, the difference $\|\mathbf{x}^{(k+1+\nu)} - \mathbf{x}^{(k+\nu)}\|$ is a factor of 10^{-8} smaller than at the kth iteration. This remarkable result actually can be used as a test of the correctness of the evaluation of the Jacobian in a problem. Any error in $\mathbf{J}$ will reduce the rate of convergence below the theoretical quadratic rate.

Picard iteration usually converges at a much slower rate (usually arithmetically instead of quadratically) than the Newton–Raphson method, as illustrated in the following example.

EXAMPLE 3.8.1 (Illustration of Quadratic Convergence of the Newton–Raphson Method). Consider the system of equations

$$0 = x_1 \tan(x_1) - \sqrt{x_2^2 - x_1^2}$$
$$0 = -\frac{(x_1 + \pi/4)}{\tan(x_1 + \pi/4)} - \sqrt{x_2^2 - \left(x_1 + \frac{\pi}{4}\right)^2}. \quad (3.8.18)$$

Starting with the Picard method, the residual expressions in Eq. (3.8.18) are rewritten in the proper form as

$$x_1 = g_1(x_1, x_2) \equiv x_1 - x_1 \tan(x_1) + \sqrt{x_2^2 - x_1^2}$$
$$x_2 = g_2(x_1, x_2) \equiv x_2 + \frac{(x_1 + \pi/4)}{\tan(x_1 + \pi/4)} + \sqrt{x_2^2 - \left(x_1 + \frac{\pi}{4}\right)^2}. \quad (3.8.19)$$

It is now just a matter of choosing an initial guess and iterating as in Eq. (3.8.7).

It turns out that the only choice of initial values that yields a convergent solution using the Picard method is the exact solution (within the requested tolerance). Here is another example of the inadequacy of the Picard method in certain applications.

Using the Newton–Raphson method, we define the residual expressions $\mathbf{f}(\mathbf{x}) = \mathbf{0}$ from Eq. (3.8.18)

$$f_1(x_1, x_2) = x_1 \tan(x_1) - \sqrt{x_2^2 - x_1^2}$$
$$f_2(x_1, x_2) = -\frac{(x_1 + \pi/4)}{\tan(x_1 + \pi/4)} - \sqrt{x_2^2 - \left(x_1 + \frac{\pi}{4}\right)^2}. \quad (3.8.20)$$

The Jacobian elements are then

$$\frac{\partial f_1}{\partial x_1} = \frac{x_1}{\sqrt{x_2^2 - x_1^2}} + x_1 \sec^2(x_1) + \tan(x_1)$$

$$\frac{\partial f_1}{\partial x_2} = -\frac{x_2}{\sqrt{x_2^2 - x_1^2}}$$

$$\frac{\partial f_2}{\partial x_1} = \frac{(x_1 + \pi/4)}{\sqrt{x_2^2 - (x_1 + \pi/4)^2}} - \cot\left(x_1 + \frac{\pi}{4}\right) + \left(x_1 + \frac{\pi}{4}\right)^2 \left(x_1 + \frac{\pi}{4}\right)$$

$$\frac{\partial f_2}{\partial x_2} = -\frac{x_2}{\sqrt{x_2^2 - (x_1 + \pi/4)^2}}.$$

With a tolerance of 10^{-14} (i.e., $\|\mathbf{x}^{(k+1)} - \mathbf{x}^{(k)}\| < 10^{-14}$ and $\|\mathbf{f}\| < 10^{-14}$), the Newton–Raphson method yields the solutions $x_1 = 0.99279$ and $x_2 = 1.81713$. Table 3.8.1 illustrates the convergence properties of the Newton–Raphson procedure for the final five iterations.

A Mathematica program for solving this problem is given in the Appendix.

TABLE 3.8.1 Residuals and Jacobian Elements

	Residuals		Jacobian		
$\|\mathbf{x}^{(k+1)} - \mathbf{x}^{(k)}\|$	$\|\mathbf{f}\|$	j_{11}	j_{12}	j_{21}	j_{22}
0.6185×10^{-1}	0.2130×10^{0}	6.38067	-1.20243	6.64350	-4.55258
0.6540×10^{-2}	0.1696×10^{-1}	5.58271	-1.19462	6.77445	-4.80056
0.5087×10^{-4}	0.1518×10^{-3}	5.51184	-1.19396	6.81964	-4.85622
0.3951×10^{-8}	0.1013×10^{-7}	5.51125	-1.19395	6.81972	-4.85644
0.1513×10^{-15}	0.1391×10^{-14}	5.51125	-1.19395	6.81972	-4.85644

EXAMPLE 3.8.2. A common model of a chemical reactor is the continuously stirred tank reactor (CSTR). A feed stream with volumetric flow rate q pours into a well-stirred tank holding a liquid volume V. The concentration of the reactant (denoted here as A) and the temperature in the feed stream are c_{A_0} (moles/volume) and T_0, respectively. We define the concentration and temperature in the tank as c_A and T, respectively, and note that the product stream will—in the ideal limit considered here—have the same values. The exiting volumetric flow rate is q. The situation is illustrated by the diagram in Fig. 3.8.1.

We assume that the reaction of A to its products is first order and that the rate of consumption of A in the tank is

$$-k_0 \exp\left(\frac{-E}{RT}\right) c_A V, \quad (3.8.21)$$

where R is the gas constant (Avogadro's number times k_B) and $k_0 \exp(-E/RT)$ is the Arrhenius formula for the rate constant—k_0 and E being constants characteristic of the chemical species. Mass and energy balances on the contents of the tank yield, under appropriate conditions, the rates of change of c_A and T with time t:

$$V\frac{dc_A}{dt} = q(c_{A_0} - c_A) - k_0 \exp\left(\frac{-E}{RT}\right) c_A V \equiv f_1(T, c_A) \quad (3.8.22)$$

and

$$V\rho\hat{C}_p \frac{dT}{dt} = q\rho\hat{C}_p(T_0 - T) - k_0 \exp\left(\frac{-E}{RT}\right) c_A V (\Delta H) - AU(T - T_b)$$

$$\equiv f_2(T, c_A). \quad (3.8.23)$$

The symbol ρ represents the density and $\hat{C}_p$ the specific heat per unit mass of the contents of the tank. ΔH denotes the molar heat of reaction and the term $-AU(T - T_b)$ accounts for heat lost from the tank to its surroundings (which is at temperature T_b). U is the overall heat transfer coefficient and A is the area through which heat is lost.

The first question to ask regarding the stirred tank reactor is: What are the steady states of the reactor under conditions of constant q, c_{A_0}, T_0, and T_b? The answer is the roots of the equations

$$0 = f_1(T^s, c_A^s)$$
$$0 = f_2(T^s, c_A^s). \quad (3.8.24)$$

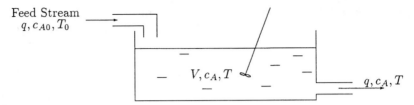

FIGURE 3.8.1

The Newton–Raphson technique reduces finding these roots to a matrix problem.

The next question is whether a given steady state (T^s, c_A^s) is stable to small fluctuations. For small deviations of T and c_A from T^s and c_A^s, $f_1(T, c_A)$ and $f_2(T, c_A)$ can be expanded in a Taylor series and truncated after linear terms. For such a situation, Eqs. (3.8.22) and (3.8.23) become

$$\frac{dx_1}{dt} = a_{11} x_1 + a_{12} x_2$$

$$\frac{dx_2}{dt} = a_{21} x_1 + a_{22} x_2$$

or

$$\frac{d\mathbf{x}}{dt} = \mathbf{J}(c_A^s, T^s)\mathbf{x}, \tag{3.8.25}$$

where

$$\mathbf{x} = \begin{bmatrix} c_A - c_A^s \\ T - T^s \end{bmatrix} \tag{3.8.26}$$

and

$$\mathbf{J} \equiv \begin{bmatrix} \dfrac{\partial f_1}{\partial c_A} & \dfrac{\partial f_1}{\partial T} \\ \dfrac{\partial f_2}{\partial c_A} & \dfrac{\partial f_2}{\partial T} \end{bmatrix}. \tag{3.8.27}$$

The steady state is stable if, for an arbitrary value of $\mathbf{x}$ at $t = 0$, Eq. (3.8.24) predicts that $\mathbf{x} \to \mathbf{0}$ as $t \to \infty$. As we shall learn later, the asymptotic behavior of $\mathbf{x}$ can be deduced from the properties of the matrix $\mathbf{J}$. Indeed, if all the eigenvalues of $\mathbf{J}$ have negative real parts, then $\mathbf{x}$ will go to 0 as t goes to ∞. Otherwise, $\mathbf{x}$ will not approach 0.

It is worth noting that the matrix generated in solving Eq. (3.8.27) by the Newton–Raphson technique is the matrix $\mathbf{A}$ needed for stability analysis.

■ ■ ■ **EXAMPLE 3.8.3** (The Density Profile in a Liquid–Vapor Interface). According to an approximate molecular theory of a planar interface, the density $n(x)$ between the liquid and vapor phase obeys the nonlinear integral equation

$$\int_{-\infty}^{\infty} K(|x' - x|)[n(x') - n(x)] \, dx' = \mu_0(n(x)) - \mu, \tag{3.8.28}$$

where $K(|x|)$ is a function of order unity for x on the order of a molecular interaction distance σ and which is vanishingly small when $|x| \gg \sigma$. μ is the chemical potential of the system and $\mu_0(n)$ is the chemical potential of a homogeneous fluid at density n.

For the van der Waals model, often used in qualitatively modeling fluid behavior,

$$\mu_0(n) = \mu^\dagger(T) - k_B T \ln\left(\frac{1}{nb} - 1\right) + \frac{nbk_B T}{1 - nb} - 2na, \qquad (3.8.29)$$

where a and b ($b \approx \sigma^3$) are parameters of the model. The van der Waals equation of state for the pressure of a homogeneous fluid is

$$P_0(n) = \frac{nk_B T}{1 - nb} - n^2 a, \qquad (3.8.30)$$

and the value of the chemical potential at the liquid and vapor coexistence densities, n_l and n_g, are determined by the thermodynamic equilibrium conditions

$$\begin{aligned} P_0(n_l) &= P_0(n_g) \\ \mu_0(n_l) &= \mu_0(n_g) = \mu. \end{aligned} \qquad (3.8.31)$$

Once μ is fixed, the density versus positions x in a liquid–vapor interface can be determined by solving Eq. (3.8.28). Of course, K must be specified for actual computations. The formula

$$K(|x|) = \frac{2a}{b^{1/3}} \exp\left(\frac{-|x|}{b^{1/3}}\right) \qquad (3.8.32)$$

yields qualitatively correct results for a planar interface.

To solve Eq. (3.8.28), we assume that, for $x > L$, $n(x) = n_g$ and, for $x < -L$, $n(x) = n_l$ so that the equation becomes

$$\int_{-L}^{L} K(|x' - x|)n(x')\,dx' - \left[\int_{-\infty}^{\infty} K(|x'|)\,dx'\right]n(x) - \mu_0(n(x))$$
$$= -\mu - n_g \int_{L}^{\infty} K(|x' - x|)\,dx' - n_l \int_{-\infty}^{-L} K(|x' - x|)\,dx' \equiv g(x). \qquad (3.8.33)$$

Dividing the interval $(-L, L)$ into N intervals of width Δx and introducing the approximation (the trapezoidal rule)

$$\int_{-L}^{L} K(|x' - x_i|)n(x')\,dx' = \sum_{j=1}^{N-1} \Delta x\, K(|x_j - x_i|)n(x_j)$$
$$+ \frac{1}{2}\Delta x\left[K(|x_0 - x_i|)n(x_0) + K(|x_N - x_i|)n(x_N)\right], \qquad (3.8.34)$$

we obtain for Eq. (3.8.33)

$$\sum_{j=1}^{N-1} K_{ij}n_j - \alpha n_i + \beta_i - \mu_0(n_i) = g_i, \qquad (3.8.35)$$

where $K_{ij} \equiv \Delta x\, K(|x_j - x_i|)$, $\alpha \equiv \int_{-\infty}^{\infty} K(|x'|)\,dx'$, $\beta_i \equiv \tfrac{1}{2}(K_{0i}n_1 + K_{Ni}n_g)$, $n_i \equiv n(x_i)$, and $g_i \equiv g(x_i)$. The matrix formula for Eq. (3.8.35) is

$$\mathbf{Kn} - \alpha\mathbf{n} + \boldsymbol{\beta} - \boldsymbol{\mu}_0(\mathbf{n}) = \mathbf{g}. \tag{3.8.36}$$

Since $\boldsymbol{\mu}_0(\mathbf{n})$ is a nonlinear function of $\mathbf{n}$, an iterative method must be used. The Appendix has a Mathematica program to solve Eq. (3.8.36).

Frequently, one is interested in the solution $\mathbf{x}$ to the problem

$$\mathbf{f}(\alpha, \mathbf{x}) = \mathbf{0} \tag{3.8.37}$$

for various values of a parameter α. Suppose one has found the solution $\mathbf{x}(\alpha)$ for the parameter value α. If α' is another value that is not far removed from α, we can take $\mathbf{x}_1 = \mathbf{x}(\alpha)$ as the first guess in a Newton–Raphson scheme and obtain the solution $\mathbf{x}(\alpha')$ by iteration of the equation

$$\mathbf{f}(\alpha', \mathbf{x}^{(k)}) + \mathbf{J}(\alpha', \mathbf{x}^{(k)})(\mathbf{x}^{(k+1)} - \mathbf{x}^{(k)}) = 0. \tag{3.8.38}$$

This is called *zeroth-order continuation* of the Newton–Raphson technique.

First-order continuation is an improvement over zeroth-order continuation. In this case, suppose we have found the solution $\mathbf{x}(\alpha)$ from the Newton–Raphson method. Then consider a new value $\alpha' = \alpha + \delta\alpha$, where $\delta\alpha$ is a small displacement from α. We then estimate an initial value of $\mathbf{x}^{(1)}$ of $\mathbf{x}(\alpha')$ from the equation

$$\begin{aligned}\mathbf{f}(\alpha, \mathbf{x}) + \mathbf{J}(\alpha, \mathbf{x}(\alpha))(\mathbf{x}^{(1)}(\alpha') - \mathbf{x}(\alpha)) \\ + (\alpha' - \alpha)\frac{\partial \mathbf{f}}{\partial \alpha'}(\alpha', \mathbf{x})\bigg|_{\alpha'=\alpha,\, \mathbf{x}=\mathbf{x}(\alpha)} = \mathbf{0}.\end{aligned} \tag{3.8.39}$$

Since $\mathbf{x}(\alpha)$ is a solution for the parameter value α, $\mathbf{f}(\alpha, \mathbf{x}(\alpha)) = \mathbf{0}$, and since $\mathbf{J}(\alpha, \mathbf{x}(\alpha))$ has been evaluated already in finding $\mathbf{x}(\alpha)$, the only work to do in Eq. (3.8.39) to obtain $\mathbf{x}^{(1)}$ is to take the derivatives $\partial \mathbf{f}/\partial \alpha$ and solve the resulting linear equation for $\mathbf{x}^{(1)}(\alpha')$. From here the solution $\mathbf{x}(\alpha')$ is obtained by the Newton–Raphson procedure in Eq. (3.8.38).

If, instead of a single parameter α, there is a collection of parameters denoted by $\boldsymbol{\alpha}$, Eq. (3.8.37) is replaced by

$$\mathbf{J}(\boldsymbol{\alpha}, \mathbf{x}(\boldsymbol{\alpha}))(\mathbf{x}^{(1)}(\boldsymbol{\alpha}') - \mathbf{x}(\boldsymbol{\alpha})) + \sum_{i=1}^{p}(\alpha'_i - \alpha_i)\frac{\partial \mathbf{f}}{\partial \alpha'_i}\bigg|_{\boldsymbol{\alpha}'=\boldsymbol{\alpha},\, \mathbf{x}=\mathbf{x}(\boldsymbol{\alpha})} = \mathbf{0}, \tag{3.8.40}$$

where the components of $\boldsymbol{\alpha}'$ are $\alpha_i + \delta\alpha_i$. Again the solution $\mathbf{x}(\boldsymbol{\alpha}')$ is found by iteration of Eq. (3.8.38).

Continuation strategy is often used when there is no good first guess $\mathbf{x}^{(1)}$ of the solution at the parameter value(s) $\boldsymbol{\alpha}$ of interest. If, however, at $\boldsymbol{\alpha}$, a good guess exists, then the solution $\mathbf{x}(\boldsymbol{\alpha})$ is found and zeroth- or first-order continuation is used repeatedly by advancing the parameter(s) by some small value $\delta\boldsymbol{\alpha}$ until the desired value $\boldsymbol{\alpha}$ is reached. The following example illustrates the value of using continuation with the Newton–Raphson technique.

NONLINEAR EQUATIONS

EXAMPLE 3.8.4 (Phase Diagram of a Polymer Solution). In the Flory–Huggins theory of polymeric solutions, the chemical potentials of solvent molecules and polymer molecules obey the equations

$$\mu_1 = kT \ln x_1 + kT\left(1 - \frac{1}{\nu}\right)x_2 + \epsilon x_2^2 \qquad (3.8.41)$$

$$\mu_2 = kT \ln x_2 + kT(1 - \nu)x_1 + \nu \epsilon x_1^2, \qquad (3.8.42)$$

where x_1 is the volume fraction of solvent, x_2 is the volume fraction of polymer, and ν is the molecular weight of the polymer measured in units of the solvent molecular weight. ϵ is a polymer–solvent interaction parameter, k is Boltzmann's constant, and T is the absolute temperature. Note that the volume fractions sum to unity, i.e.,

$$x_1 + x_2 = 1. \qquad (3.8.43)$$

Solution properties are determined by the values of the dimensionless parameters ν and $\chi = kT/\epsilon$. At certain overall concentrations and temperatures, the solution splits into two coexisting phases at different concentrations. From thermodynamics, the conditions of phase coexistence are that

$$\begin{aligned}\mu_1(x_1^\alpha, x_2^\alpha) &= \mu_1(x_1^\beta, x_2^\beta) \\ \mu_2(x_1^\alpha, x_2^\alpha) &= \mu_2(x_1^\beta, x_2^\beta),\end{aligned} \qquad (3.8.44)$$

where x_i^α and x_i^β are the volume fractions of component i in phases α and β. As the parameter χ increases (temperature increases), the compositions of the phases become more and more similar until, at the critical point,

$$x_{2c} = \frac{1}{1+\sqrt{\nu}} \quad \text{and} \quad \frac{1}{\chi_c} = \frac{1}{2}\left(1 + \frac{1}{\sqrt{\nu}}\right)^2, \qquad (3.8.45)$$

and the phases are indistinguishable. Above x_{2c}, there are no coexisting phases.

For the polymer scientist, the problem is to solve Eq. (3.8.44) to determine the two-phase coexistence curve as a function of χ. If the variables $x = x_1^\alpha$, $y = x_1^\beta$, and

$$f_i = \frac{1}{kT}[\mu_i(x, 1-x) - \mu_i(y, 1-y)] \qquad (3.8.46)$$

are defined, the problem becomes

$$\mathbf{f} = \mathbf{0}, \qquad (3.8.47)$$

where the components of $\mathbf{f}$ are f_1 and f_2. The range of values of molecular weights ν of interest to the polymer scientist are typically $\nu > 100$. Solving Eq. (3.8.44) requires a good first guess, which is difficult to obtain for $\nu > 100$. On the other hand, for $\nu = 1$, Eq. (3.8.44) admits the analytical solution

$$\chi = \frac{1 - 2x}{\ln[(1-x)/x]} \qquad (3.8.48)$$
$$y = 1 - x.$$

This suggests using the Newton–Raphson technique with continuation from $v = 1$ to find the coexistence curve for values of v of interest to the polymer scientist.

The following routine solves for the coexistence curve at some given v (in this example, the goal will be to find the phase diagram for $v = 500$). The routine can be broken down into two parts, each of which uses first-order continuation. The first part uses the solution at $v = 1$ (at a particular choice of y) to determine the first set of points (x, y, χ) for the coexistence curve at $v = 500$. This is accomplished by a succession of first-order continuation steps, followed by the Newton–Raphson technique at incrementing values of v. The incrementation of v must be chosen small enough to assure convergence of the Newton–Raphson steps. Once a solution is found at $v = 500$, the second part of the program uses the Newton–Raphson technique to solve (3.8.44) at several values of y, again using first-order continuation to construct the initial guesses x and χ for each value of y. Alternatively, we could have chosen values of χ and then computed values of x and y. However, since the y values sometimes approach unity, where the Jacobian is singular, the problem is better behaved when y is fixed instead.

The derivatives needed for the calculation are

$$\frac{\partial f_1}{\partial x} = \frac{1}{x} - \left(1 - \frac{1}{v}\right) - \frac{2(1-x)}{\chi}$$

$$\frac{\partial f_2}{\partial x} = -\frac{1}{1-x} + (1-v) + \frac{2vx}{\chi}$$

$$\frac{\partial f_1}{\partial \chi} = -\frac{(1-x)^2 - (1-y)^2}{\chi^2}$$

$$\frac{\partial f_2}{\partial \chi} = -\frac{v(x^2 - y^2)}{\chi^2}$$

$$\frac{\partial f_1}{\partial y} = -\frac{1}{y} + \left(1 - \frac{1}{v}\right) + \frac{2(1-y)}{\chi} \qquad (3.8.49)$$

$$\frac{\partial f_2}{\partial y} = \frac{1}{1-y} - (1-v) - \frac{2vy}{\chi}$$

$$\frac{\partial f_1}{\partial v} = -\frac{x-y}{v^2}$$

$$\frac{\partial f_2}{\partial v} = -(x-y) + \frac{x^2 - y^2}{\chi}.$$

The solution is shown in Fig. 3.8.2, where we present a plot of χ versus solvent volume fraction in the coexisting phases for $v = 1$, 50, and 500. At a given value of χ below the critical point, there are two values of x_1, corresponding to the left and right branches of the curve in Fig. 3.8.2. These values of x_1 are the volume fractions of solvent in the coexisting phases α and β. It is interesting to note that for $v = 500$ not far below the critical point are two distinct coexisting phases having very little polymer.

A Mathematica program to solve this problem is given in the Appendix. Here we outline the program logic in case the reader wishes to program the problem in some other language.

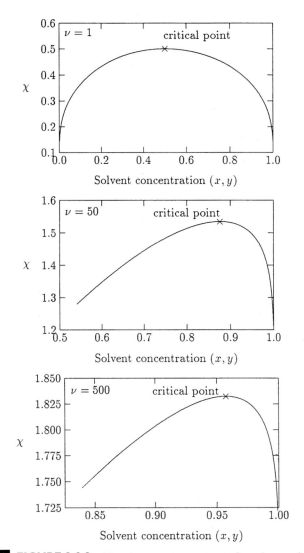

FIGURE 3.8.2 Two-phase coexistence curves for polymer solutions.

We begin the program by defining the user input parameters. These are the variables of the program that both define the problem and establish the accuracy desired. The variable *nucalc* is the target value of ν used for the calculation of the coexistence curve and *deltanu* is the step size used in the first continuation sequence from $\nu = 1$ to $\nu = nucalc$. The value of *ymax* defines the maximum desired value of y on the coexistence curve and *ydelta* is the step size used for the incrementation of y (in the second continuation sequence):

$$nucalc = 500$$
$$deltanu = 0.005$$
$$ymax = 0.999$$
$$ydelta = 0.001.$$

Next, the parameters specific to the Newton–Raphson routine are defined. *TOLERANCE* is used in checking for the convergence of the solution and *MAXITER* sets the maximum number of iterations allowed in the calculation:

$$CONVERGE = 10^{-7}$$

$$MAXITER = 100.$$

The solution arrays are then initialized to their values for $\nu = 1$. Here we are defining $y[1]$ as *ymax* and when the coexistence curve is eventually calculated, $y[i]$ will take on values of y as it is deincremented by *ydelta*. The values of $x[i]$ and $\chi[i]$ are the corresponding solution values for $y[i]$. Initially, the array need only be defined for $i = 1$:

$$y[1] = ymax$$

$$x[1] = 1 - y[1]$$

$$\chi[1] = (1 - 2x[1])/\ln\bigl[(1 - x[1])/x[1]\bigr].$$

Before beginning the main program body, the initial value of ν must be set to 1 and the number of ν steps (*nustep*) must be calculated for control of the loop:

$$\nu = 1$$

$$nustep = (nucalc - 1.0)/deltanu.$$

The Jacobian elements for the first continuation step must also be calculated (at $\nu = 1$):

$$jacobian[1, 1] = \frac{\partial f_1}{\partial x}$$

$$jacobian[1, 2] = \frac{\partial f_1}{\partial \chi}$$

$$jacobian[2, 1] = \frac{\partial f_2}{\partial x}$$

$$jacobian[2, 2] = \frac{\partial f_2}{\partial \chi}.$$

The loop for the first continuation sequence begins. Here we let k range from 1 to *nustep*:

$$\text{For } k = 1, \ nustep.$$

We now compute the residual vector for the next continuation step:

$$residual[1] = -\frac{\partial f_1}{\partial \nu} deltanu$$

$$residual[2] = -\frac{\partial f_2}{\partial \nu} deltanu,$$

and then compute the solution to the linear equations in Eq. (3.8.40) using a Gaussian elimination routine. *xdel* is the solution array containing $(\chi^{(k+1)} - \chi^{(k)})$ and $(x^{(k+1)} - x^{(k)})$ as elements:

$$xdel = LinearSolve(jacobian, residual).$$

The values of $\chi[1]$ and $x[1]$ are then updated and the value of ν is incremented in preparation for the Newton–Raphson loop:

$$\chi[1] = \chi[1] + xdel[1]$$
$$x[1] = x[1] + xdel[2].$$

Before beginning the Newton–Raphson loop, some internal variables need to be initialized. The variable *normdel* will be used to store the value of the modulus of *xdel* and *normres* will be the modulus of the residual vector *residual*. Both will be compared to *TOLERANCE* in order to determine if convergence has been obtained.

$$normdel = 0$$
$$normres = 0.$$

Also, the value of ν must be updated:

$$\nu = \nu + deltanu.$$

The Newton–Raphson routine first computes the updated components of the Jacobian matrix and the residual vector. Then it solves for the vector *xdel* using the Gaussian elimination subroutine. The respective moduli *normdel* and *normres* are then computed and the solutions $\chi[1]$ and $x[1]$ are updated. The *if-statement* checks for convergence using the criterion that both *normdel* and *normres* must be less than *TOLERANCE*. The loop is repeated until convergence is obtained or the maximum number of iterations is reached:

$$For\ i = 2, MAXITER$$

$$jacobian[1, 1] = \frac{\partial f_1}{\partial x}$$

$$jacobian[1, 2] = \frac{\partial f_1}{\partial \chi}$$

$$jacobian[2, 1] = \frac{\partial f_2}{\partial x}$$

$$jacobian[2, 2] = \frac{\partial f_2}{\partial \chi}$$

$$residual[1] = -f_1(x[1], y[1], \chi[1], \nu)$$

$$residual[2] = -f_1(x[1], y[1], \chi[1], \nu)$$

$$xdel = LinearSolve(jacobian, residual)$$

$$normdel = \sqrt{(xdel[1]^2 + xdel[2]^2)}$$

$$normdel = \sqrt{(residual[1]^2 + residual[2]^2)}$$

$$\chi[1] = \chi[1] + xdel[1]$$

$$x[1] = x[1] + xdel[2]$$

If $(normdel < TOLERANCE)$ and

$(normres < TOLERANCE)$, then $\langle exit\ loop \rangle$

end loop

End loop over k:

end loop

At this point, a solution has been found at the target value of $\nu = nucalc$. The second continuation sequence now begins for the calculation of the coexistence curve. The first step is to find the minimum value of y for our deincrementation ($ymin$). This value corresponds to the critical point values given in Eq. (3.8.45).

$$ymin = 1 - 1/(1 + \sqrt{\nu}).$$

Now we need to find the number of deincrementation steps in y to take ($nmax$). The function Floor is used to find the maximum integer less than its argument and subtracting the small number 10^{-10} is required to ensure that the critical point is not included in our count:

$$nmax = \text{Floor}((ymax - ymin)/ydelta - 10^{-10}) + 1.$$

Next, the calculation of the coexistence curve proceeds by utilizing successive first-order continuation steps with respect to y. This is done by looping over the $nmax$ values of $y[i]$.

$$\text{For } i = 1, nmax.$$

Calculate the current value of y:

$$y[i] = ymax - (i-1)ydelta.$$

Conduct first-order continuation and update of x and χ:

$$residual[1] = -\frac{\partial f_1}{\partial y_{i-1}} ydelta$$

$$residual[2] = -\frac{\partial f_2}{\partial y_{i-1}} ydelta$$

$$xdel = LinearSolve(jacobian, residual)$$

$$\chi[i] = \chi[i-1] + xdel[1]$$

$$x[i] = x[i-1] + xdel[2].$$

Begin the Newton–Raphson loop:

$$normdel = 0$$
$$normres = 0$$
For $i = j, MAXITER$
$$jacobian[1, 1] = \frac{\partial f_1}{\partial x}$$
$$jacobian[1, 2] = \frac{\partial f_1}{\partial \chi}$$
$$jacobian[2, 1] = \frac{\partial f_2}{\partial x}$$
$$jacobian[2, 2] = \frac{\partial f_2}{\partial \chi}$$
$$residual[1] = -f_1(x[i], y[i], \chi[i], \nu)$$
$$residual[2] = -f_1(x[i], y[i], \chi[i], \nu)$$
$$xdel = LinearSolve(jacobian, residual)$$
$$normdel = \sqrt{(xdel[1]^2 + xdel[2]^2)}$$
$$normdel = \sqrt{(residual[1]^2 + residual[2]^2)}$$
$$\chi[i] = \chi[i] + xdel[1]$$
$$x[i] = x[i] + xdel[2]$$
If $(normdel < TOLERANCE)$ and
$(normres < TOLERANCE)$, then$\langle exit\ loop \rangle$
end loop
end loop

Finally, calculate the values of x, y, and χ at the critical point:

$$x[nmax + 1] = ymin$$
$$y[nmax + 1] = ymin$$
$$\chi[nmax + 1] = \left(1 + 1/\sqrt{(\nu + deltanu)}\right)^2/2.$$

ILLUSTRATION 3.8.1 (Liquid–Vapor Phase Diagram). In dimensionless units, the van der Waals equation of state is

$$x = \frac{8y}{3x - 1} - \frac{3}{x^2}, \qquad (3.8.50)$$

where

$$z = \frac{P}{P_c}, \qquad y = \frac{T}{T_c}, \qquad x = \frac{V}{V_c}.$$

P, T, and V are pressure, temperature, and molar volume, respectively; and P_c, T_c, and V_c are the respective values of these quantities at the liquid–vapor critical point (the point at which liquid and vapor become indistinguishable). The conditions for the critical point are

$$\frac{\partial z}{\partial x} = \frac{\partial^2 z}{\partial x^2} = 0. \quad (3.8.51)$$

(i) Verify that these equations imply that $x = y = z = 1$ at the critical point.

The dimensional van der Waals equation is given by

$$P = \frac{RT}{V-b} - \frac{a}{V^2}, \quad (3.8.52)$$

where R is the universal gas constant and b and a are molecular constants characteristic of a given fluid.

(ii) Find P_c, T_c, and V_c in terms of R, b, and a.

The chemical potential μ of a van der Waals fluid is given by

$$\mu = \frac{RTV}{V-b} - RT\ln\left(\frac{V}{b} - 1\right) - \frac{2a}{V}. \quad (3.8.53)$$

At liquid–vapor equilibrium, the pressure and chemical potential of each phase are equal, i.e.,

$$\begin{aligned} P(V_l, T) - P(V_v, T) &= 0 \\ \mu(V_l, T) - \mu(V_v, T) &= 0, \end{aligned} \quad (3.8.54)$$

where V_l and V_v are the molar volumes of each phase.

(iii) From these equilibrium conditions, find V_l and V_v as a function of temperature for temperatures less than T_c.

The problem is easier to solve in dimensionless variables. First, define $w \equiv \mu/\mu_c$ and

$$\mathbf{x} \equiv \begin{bmatrix} x_l \\ x_v \end{bmatrix}, \quad (3.8.55)$$

where $x_l = V_l/V_c$ and $x_v = V_v/V_c$, and

$$\begin{aligned} f_1 &= z(x_l, y) - z(x_v, y) \\ f_2 &= w(x_l, y) - w(x_v, y). \end{aligned} \quad (3.8.56)$$

Now let

$$\mathbf{f} \equiv \begin{bmatrix} f_1 \\ f_2 \end{bmatrix}, \quad (3.8.57)$$

and thus the equilibrium equation to solve is

$$\mathbf{f}(\mathbf{x}) = \mathbf{0}. \quad (3.8.58)$$

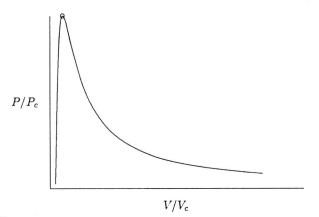

FIGURE 3.8.3 Two-phase coexistence curves for van der Waals fluids.

Use the Newton–Raphson method to solve this equation for values of $0.50 \leq y < 1.0$. As a first guess, take $y = 0.99$ and try $x_l = 0.85$ and $x_v = 1.3$. If that guess does not work, try others. Once you have a converged solution for $y = 0.99$, you can either use zeroth- or first-order continuation to solve the problem for the above range of y values.

In dimensionless form, the results look like Fig. 3.8.3. The point on the curve denotes the critical point. The left branch of the curve represents the molar volume of coexisting liquid and the right branch represents the molar volume of the vapor. The points defining the curve are generated by solving the equilibrium equation for various temperatures. Plot the coexistence curve in dimensionless form for the van der Waals equation of state.

(iv) Outside the liquid–vapor coexistence region, only one phase is stable. Plot P versus V for various temperatures. These curves are called pressure isotherms. Three such isotherms are illustrated in Fig. 3.8.4. Give a quantitative plot of P/P_c versus V/V_c for several isotherms.

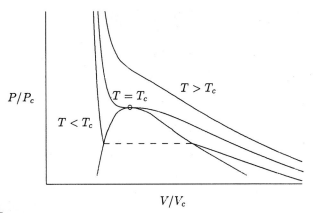

FIGURE 3.8.4 P–V isotherms for van der Waals fluids.

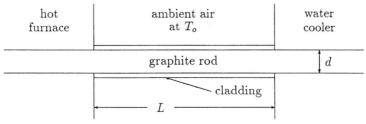

FIGURE 3.8.5

(v) In the van der Waals model, b denotes molecular volume and a/b denotes the strength of pair intermolecular attraction. Explain qualitatively, from your dimensionless phase diagram, how an increase or decrease in molecular size or intermolecular interaction affects the phase diagram of a van der Waals fluid.

(vi) The fact that all phase diagrams of all fluids obeying the van der Waals equation of state reduce to the same dimensionless phase diagram is known as obeying the *law of corresponding states*. State this law physically and explain how one would use it.

■ ■ ■

ILLUSTRATION 3.8.2 (Cooling a Graphite Electrode). Consider the graphite electrode illustrated in Fig. 3.8.5. The electrode, of diameter d, passes from a hot furnace into a water cooler to help control the electrode temperature. The temperature t where the electrode leaves the furnace (at position $x = 0$) is $T(0) = 1600°$C. The rod resides for a distance L in ambient air at temperature $T_0 = 20°$C. In this space, the electrode is covered with an insulating cladding, which gives an overall effective heat transfer coefficient of U (between the air and the electrode). It is desired that the water cooler keep the temperature of the rod at 160°C at the point where it enters the cooler (at $x = L$). What is the rate that heat must be absorbed by the water cooler to hold $T(L) = 160°$C.

Assuming that the temperature of the electrode varies only along its axis, it can be shown (see p. 41 of V. G. Jenson and G. V. Jeffreys, *Mathematical Methods in Chemical Engineering*, Academic Press, 1977) that it obeys the heat equation

$$(k_0 - \alpha T)\frac{d^2 T}{dx^2} - \alpha \left(\frac{dT}{dx}\right)^2 - \beta(T - T_0) = 0, \tag{3.8.59}$$

where $k_0 - \alpha T$ is the temperature-dependent thermal conductivity of graphite, and $\beta = 4U/d$. The parameters of the problem are

$$U = 1.7 \text{ W}/°\text{C m}^2$$
$$k_0 = 152.6 \text{ W}/°\text{C m}^2$$
$$\alpha = 0.056 \text{ W}/(°\text{C})^2 \text{ m}$$
$$d = 10 \text{ cm}$$
$$L = 25 \text{ cm}$$
$$T_0 = 20°\text{C}$$

$$T(0) = 1600°C$$
$$T(L) = 160°C.$$

The rate of heat transfer to the water cooler is

$$Q = -(k_0 - \alpha T)\left(\frac{\pi d}{4}\right)^2 \frac{dT}{dx} \quad \text{at } x = L. \tag{3.8.60}$$

The heat equation can be solved by two different methods. The first is accomplished by transforming the differential equation to an integral equation. Define u by

$$u \equiv \frac{dT}{dx}$$

so that, by using the relation $d(dT/dx) = (du/dT)dT$, we obtain

$$\frac{d^2 T}{dx^2} = u \frac{du}{dx}.$$

Then the heat equation becomes

$$\frac{1}{2\alpha}(k_0 - \alpha T)\frac{d(u^2)}{dT} - u^2 - \frac{\beta}{\alpha}(T - T_0) = 0, \tag{3.8.61}$$

which can be rearranged first to

$$\frac{d(u^2)}{dT} - \frac{2\alpha}{k_0 - \alpha T} u^2 = 2\beta \frac{T - T_0}{k_0 - \alpha T} \tag{3.8.62}$$

and then to

$$\frac{1}{(k_0 - \alpha T)^2} \frac{d}{dT}\left((k_0 - \alpha T)^2 u^2\right) = 2\beta \frac{T - T_0}{k_0 - \alpha T}. \tag{3.8.63}$$

This expression integrates to

$$(k_0 - \alpha T)^2 u^2 = \beta(T - T_0)^2(k_0 - \alpha T_0) - \frac{2\alpha\beta}{3}(T - T_0)^3 + C, \tag{3.8.64}$$

where C is a constant of integration. Recall that $u = dT/dx$. With this substitution, the above expression rearranges to $dx = f(x)\,dT$, which can be integrated to obtain

$$x = \int_{T(0)}^{T(x)} \frac{(k_0 - \alpha T)}{\left(C + \beta(k_0 - \alpha T)^2 - \frac{2}{3}\alpha\beta(T - T_0)^3\right)^{1/2}} dT = 0. \tag{3.8.65a}$$

The constant of integration, C, is determined from

$$f(C) = L - \int_{T(0)}^{T(L)} \frac{(k_0 - \alpha T)}{\left(C + \beta(k_0 - \alpha T)^2 - \frac{2}{3}\alpha\beta(T - T_0)^3\right)^{1/2}} dT = 0. \tag{3.8.65b}$$

(i) Calculate C by first plotting $f(C)$ versus C and finding the interval of C in which $f(c)$ switches from negative to positive. Then use the Picard method or the Newton–Raphson method to precisely determine C such that $f(C) = 0$. In evaluating $f(C)$, use the trapezoidal rule for computing an integral.

(ii) Once C is determined, calculate and plot x versus T, again using the trapezoidal rule. Determine the required cooling rate for the water cooler.

Another way to solve the heat equation is to discretize it using the finite-difference approximation. Let $x_i = i\,\Delta x$, where $\Delta x = L/(N+1)$ and $i = 0, 1, \ldots, N+1$. Next, let $T_i = T(x_i)$, $i = 0, \ldots, N+1$. The midpoint approximations to the first and second derivatives of T, evaluated at x_i, are

$$\left.\frac{dT}{dx}\right|_{x_i} = \frac{T_{i+1} - T_{i-1}}{2\Delta x} \qquad (3.8.66)$$

and

$$\left.\frac{d^2T}{dx^2}\right|_{x_i} = \frac{T_{i+1} - 2T_i + T_{i-1}}{(\Delta x)^2}. \qquad (3.8.67)$$

With these approximations, the heat equation becomes

$$(k_0 - \alpha T_i)\left(\frac{T_{i+1} - 2T_i + T_{i-1}}{(\Delta x)^2}\right) - \frac{\alpha}{4(\Delta x)^2}(T_{i+1} - T_{i-1})^2 - \beta(T_i - T_0) = 0 \qquad (3.8.68)$$

for $i = 1, \ldots, N$. The boundary points are given by

$$(k_0 - \alpha T_1)\left(\frac{T_2 - 2T_1 + T(0)}{(\Delta x)^2}\right) - \frac{\alpha}{4(\Delta x)^2}(T_2 - T(0))^2 - \beta(T_1 - T_0) = 0 \qquad (3.8.69)$$

and

$$(k_0 - \alpha T_N)\left(\frac{T(L) - 2T_N + T_{N-1}}{(\Delta x)^2}\right) - \frac{\alpha}{4(\Delta x)^2}(T(L) - T_{N-1})^2 - \beta(T_N - T_0) = 0. \qquad (3.8.70)$$

(iii) Solve this nonlinear algebraic system for T_i using the Newton–Raphson method and the above parameters. Compare the computation time for each method.

ILLUSTRATION 3.8.3 (Multireaction CSTR at Constant Temperature). Consider the CSTR analyzed in Example 3.8.2. We saw how the steady-state concentration could be determined by using the Newton–Raphson scheme. In the following problem we will solve for the steady-state output concentrations of a multicomponent mixture of reactants and products in a constant-temperature CSTR. The chemical species (A, B, C, and D) are assumed to be in dilute aqueous solution with input concentrations c_{A_0}, c_{B_0}, c_{C_0}, and c_{D_0} (mol/L). The total volumetric flow rate is Q (L/s) and the reactor volume is V_r (L). The following reaction set applies:

$$A + B \rightarrow C \qquad (1)$$
$$B + C \rightarrow D \qquad (2)$$
$$D \rightarrow B + C, \qquad (3)$$

with corresponding rate equations

$$r_1 = k_1 c_A c_B$$
$$r_2 = k_2 c_B c_C \qquad (3.8.71)$$
$$r_3 = k_3 c_D,$$

where r_i is the rate in moles/second per volume of reaction i.

For an ideal CSTR, the output concentrations are assumed to be equal to the mixture concentrations within the reactor (perfect mixing). Mole balances for each of the species can then be approximated as

Component A: $\quad 0 = Q(c_{A_0} - c_A) + (-k_1 c_A c_B) V_r$

Component B: $\quad 0 = Q(c_{B_0} - c_B) + (-k_1 c_A c_B - k_2 c_B c_C + k_3 c_D) V_r$

Component C: $\quad 0 = Q(c_{C_0} - c_C) + (k_1 c_A c_B - k_2 c_B c_C + k_3 c_D) V_r$

Component D: $\quad 0 = Q(c_{D_0} - c_D) + (k_2 c_B c_C - k_3 c_D) V_r.$

$$(3.8.72)$$

The problem can be recast into dimensionless form by introducing the following quantities:

$$Q^* \equiv \frac{Q}{V_r k_1 c_{A_0}} \qquad (3.8.73)$$

$$k_i^* \equiv \frac{k_i}{k_1} \qquad (3.8.74)$$

$$c_\nu^* \equiv \frac{c_\nu}{c_{A_0}}. \qquad (3.8.75)$$

Only A and B are supplied to the reactor. Thus, the mole balance equations can be rewritten in dimensionless form as

$$\begin{aligned} f_1 &= 0 = Q^*(1 - c_A^*) + (-c_A^* c_B^*) \\ f_2 &= 0 = Q^*(c_{B_0}^* - c_B^*) + (-c_A^* c_B^* - k_2^* c_B^* c_C^* + k_3^* c_D^*) \\ f_3 &= 0 = Q^*(-c_C^*) + (c_A^* c_B^* - k_2^* c_B^* c_C^* + k_3^* c_D^*) \\ f_4 &= 0 = Q^*(-c_D^*) + (k_2^* c_B^* c_C^* - k_3^* c_D^*). \end{aligned} \qquad (3.8.76)$$

The next step is to define the problem in vector notation. By relabeling the dimensionless reactor concentrations as $\{x_1, x_2, x_3, x_4\}$, i.e.,

$$\mathbf{x} \equiv \begin{bmatrix} c_A^* \\ c_B^* \\ c_C^* \\ c_D^* \end{bmatrix}, \qquad (3.8.77)$$

we can recast the problem as

$$\mathbf{f}(\mathbf{x}) = \mathbf{0}, \qquad (3.8.78)$$

where $\mathbf{f}$ is defined in Eq. (3.8.76).

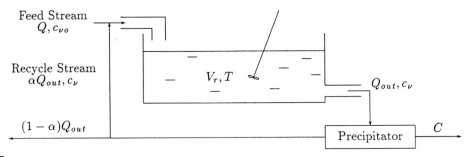

FIGURE 3.8.6

(i) Use the Newton–Raphson method to solve for the output concentrations of the reactor when $Q^* = 1$, $k_2^* = 5$, $k_3^* = 10$, and $c_{B_0}^* = 1$.

(ii) The desired product is C, which is removed through precipitation from the reactor output stream. Note that D may be recycled to the reactor, possibly increasing the yield of C. Figure 3.8.6 illustrates the flow diagram. The precipitator is assumed to remove component C from the output stream without changing the total volumetric flow rate. We define the dimensionless output flow rate by $Q_{\text{out}}^* \equiv Q_{\text{out}}/(V_r k_1 c_{A_0})$ and the recycle fraction α as the fraction of the reactor output flow rate that is recycled. Reformulate the problem in dimensionless parameters and assume the same input values of Q^* and c_{B_0}. Solve the new system of equations for various values of $0 < \alpha \leq 1$ using zeroth- or first-order continuation. Plot the dimensionless production rate of C ($Q_{\text{out}}^* c_C^*$) as a function of α.

ILLUSTRATION 3.8.4 (Temperature Profile in a Plug Flow Reactor). It is desired to calculate the adiabatic temperature profile of a plug flow reactor for the gaseous phase reaction

$$A + B \to C. \tag{3.8.79}$$

The reactor, illustrated in Fig. 3.8.7, is loaded with catalyst beads and has an effective cross-sectional area of S_e. We define the total molar flow rate through the reactor as N and the individual component flow rates as N_i such that

$$N = N_A + N_B + N_C.$$

The initial flow rates are designated as $\{N_{A_0}, N_{B_0}, N_{C_0} = 0\}$. We define the steady-state conversion rate, which is a function of position along the catalyst bed, as $\xi(x)$. Mole balances on each species yield

$$\begin{aligned} N_A(x) &= N_{A_0} - \xi(x) \\ N_B(x) &= N_{B_0} - \xi(x) \\ N_C(x) &= \xi(x) \end{aligned} \tag{3.8.80}$$

and

$$N(x) = N_0 - \xi(x), \tag{3.8.81}$$

where the total initial molar flow rate is $N_0 = N_{A_0} + N_{B_0}$.

NONLINEAR EQUATIONS

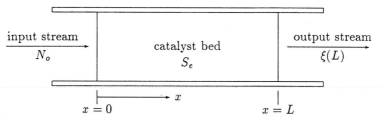

FIGURE 3.8.7

The overall rate of reaction at a point x in the catalyst bed is

$$r(x) = \frac{1}{S_e}\frac{d\xi}{dx} = k\frac{N_A(x)}{\tilde{V}(x)N(x)}\frac{N_B(x)}{\tilde{V}(x)N(x)}, \qquad (3.8.82)$$

where $\tilde{V}(x)$ is the molar volume of the gas mixture given by

$$\tilde{V}(x) = \frac{RT(x)}{P}, \qquad (3.8.83)$$

P is the pressure (assumed constant), R is the universal gas constant, and $T(x)$ is the reactor temperature. The differential equation describing ξ is, therefore,

$$\frac{d\xi}{dx} = S_e k \left(\frac{P}{RT(x)}\right)^2 \frac{(N_{A_0} - \xi)(N_{B_0} - \xi)}{(N_0 - \xi)^2}, \qquad (3.8.84)$$

with the initial condition that $\xi(0) = 0$.

If the reactor were operating isothermally, we would simply integrate Eq. (3.8.84) to generate an expression for $\xi(x)$. However, the temperature dependence must be determined through an enthalpy balance. Defining the molar heat capacity of component i as $\tilde{C}_i$, the initial mixture temperature as T_i, and the molar heat of reaction as $\tilde{H}_r$ determined from the reference temperature T_r, the enthalpy balance can be expressed as

$$\begin{aligned} 0 = & (T_i - T_r)(\tilde{C}_A N_{A_0} + \tilde{C}_B N_{B_0}) \\ & - (T - T_r)(\tilde{C}_A(N_{A_0} - \xi) + \tilde{C}_B(N_{B_0} - \xi)) \\ & - (T - T_r)\tilde{C}_C \xi - \tilde{H}_r \xi. \end{aligned} \qquad (3.8.85)$$

We can recast these equations in dimensionless form by introducing

$$x^* = \frac{x}{L} \qquad (3.8.86)$$

$$\xi^* = \frac{\xi}{N_{A_0}} \qquad (3.8.87)$$

$$\alpha = \frac{N_{B_0}}{N_{A_0}} \qquad (3.8.88)$$

$$\kappa = \frac{kS_e L}{N_{A_0}}\left(\frac{P}{RT_i}\right)^2 \qquad (3.8.89)$$

$$\theta = \frac{T}{T_i} \tag{3.8.90}$$

$$\tilde{C}_v^* = \frac{\tilde{C}_v}{\tilde{C}_A} \tag{3.8.91}$$

$$\tilde{H}_r^* = \frac{\tilde{H}_r}{C_A(T_i - T_r)}, \tag{3.8.92}$$

where L is the total length of the catalyst bed. Equation (3.8.84) is then

$$\frac{d\xi^*}{dx^*} = \kappa \left(\frac{1}{\theta}\right)^2 \frac{(1-\xi^*)(\alpha - \xi^*)}{(1+\alpha - \xi^*)^2}, \tag{3.8.93}$$

(i) For the isothermal case, θ is a constant and equal to 1. Equation (3.8.93) can be integrated directly to yield the following implicit expression for ξ^*:

$$\kappa x^* = \xi^* + \frac{\alpha^2}{1-\alpha}\ln|1-\xi^*| - \left(\frac{\alpha^3}{1-\alpha} + (1+\alpha)^2 + \alpha\right)\ln\left|\frac{\alpha - \xi^*}{\alpha}\right|, \tag{3.8.94}$$

where $\alpha \neq 1$. Plot ξ^* versus κx^* on the interval $0 < x^* \leq 1$ for several values of α.

(ii) For the nonisothermal case, Eq. (3.8.85) can be written in dimensionless form as

$$0 = 1 + \alpha \tilde{C}_B^* - \frac{\theta - \theta_r}{1-\theta_r}\left(1 - \xi^* + \tilde{C}_B^*(\alpha - \xi^*)\right)$$
$$- \frac{\theta - \theta_r}{1-\theta_r}\tilde{C}_C^*\xi^* - \tilde{H}_r^*\xi^*, \tag{3.8.95}$$

where $\theta_r = T_r/T_i \neq 1$. Solve the coupled equations (3.8.93) and (3.8.95) by either the trapezoidal rule or the finite-difference approximation for the dimensionless parameters $\kappa = 0.1$, $\alpha = 0.5$, $\tilde{C}_B^* = 1.2$, $\tilde{C}_C^* = 0.85$, $\theta_r = 0.25$, and $\tilde{H}_r^* = 2.8$. Plot the dimensionless conversion ξ^* and reactor temperature profile θ for $0 < x^* \leq 1$. Compare the output conversion to the isothermal case.

(iii) Derive a solution for Eq. (3.8.93) for the isothermal case when $\alpha = 1$. Plot ξ^* versus κx^* for this case.

■ ■ ■

PROBLEMS

1. Use Gauss elimination to solve

$$\mathbf{Ax} = \mathbf{b},$$

where

$$\mathbf{A} = \begin{bmatrix} 1 & 2 & 3 \\ 2 & 3 & 4 \\ 3 & 1 & 2 \end{bmatrix} \quad \text{and} \quad \mathbf{b} = \begin{bmatrix} -4 \\ -5 \\ 0 \end{bmatrix}.$$

PROBLEMS

2. Find the **LU**-decomposition of the matrix **A** in Problem 1.
3. Consider the matrix

$$\mathbf{A} = \begin{bmatrix} 3 & -2 & 0 & 0 & 0 \\ -1 & 3 & -2 & 0 & 0 \\ 0 & -1 & 3 & -2 & 0 \\ 0 & 0 & -1 & 3 & -2 \\ 0 & 0 & 0 & -1 & 3 \end{bmatrix}.$$

Without the aid of a computer:

(a) Find the inverse $\mathbf{A}^{-1}$ of $\mathbf{A}$.
(b) Evaluate the determinants of $\mathbf{A}$ and $\mathbf{A}^{-1}$.
(c) Solve $\mathbf{A}\mathbf{x} = \mathbf{b}$, where $b_i = 1, i = 1, \ldots, 5$.

4. Write a computer program to carry out Gauss elimination with partial pivoting.
5. Consider heat transfer in a square slab of length 10 in. on a side. At steady state, the temperature obeys the equation

$$\frac{\partial^2 T}{\partial x^2} + \frac{\partial^2 T}{\partial y^2} = 0.$$

Assume that the temperatures of the boundaries of the plates are controlled so that

$$T(x, 0) = 370 \text{ K}(x/10 \text{ in.})$$
$$T(0, y) = 100 \text{ K}(y/10 \text{ in.})$$
$$T(x, 10 \text{ in.}) = 100 \text{ K}(1 - x/10 \text{ in.})^2$$
$$T(10 \text{ in.}, y) = 370 \text{ K}(1 - y/10 \text{ in.})^3.$$

(a) Verify the finite-difference forms given for **A** and **b** in Eqs. (3.6.40) and (3.6.41). Generalize the results to an arbitrary subdivision of the square into n cells.
(b) Using your Gauss elimination program, compute the temperature distribution by the method of finite differences. Estimate the discretization n for which the computed temperature obeys the accuracy criterion

$$\left| \int_0^{10} \int_0^{10} T^{(n)}(x, y) \, dx \, dy - \int_0^{10} \int_0^{10} T^{(n-1)}(x, y) \, dx \, dy \right|$$
$$< 10^{-8} \int_0^{10} \int_0^{10} T^{(n-1)}(x, y) \, dx \, dy,$$

where $T^{(n)}$ and $T^{(n-1)}$ are temperatures computed from successive discretization.

6. Solve the problem

$$\mathbf{Ax} = \mathbf{b},$$

where $\mathbf{A}$ is a 100×100 tridiagonal matrix with $+5$ on the main diagonal, -3 on the diagonal just above the main diagonal, and -2 on the diagonal just below the main diagonal, and $\mathbf{b}$ is a vector with all elements equal to 1. Calculate $\mathbf{x}$ using the following methods:

(a) Simple Gauss elimination.
(b) The Jacobi method.
(c) The Gauss–Seidel method.
(d) The SOR method.

Compare convergence properties and total evaluation times for each method.

7. Consider the system $\mathbf{Ax} = \mathbf{b}$, where

$$10^{-5}x_1 + 10^{-5}x_2 + x_3 = 2$$
$$10^{-5}x_1 - 10^{-5}x_2 + x_3 = -2 \times 10^{-5}$$
$$x_1 + x_2 + 2x_3 = 1.$$

(a) Does the system have a solution? Why?

Using three-digit floating-point arithmetic, solve the system by:

(b) Simple Gauss elimination without pivoting.
(c) Gauss elimination with complete pivoting.
(d) What is the exact solution to six-digit floating-point arithmetic?

8. (Least Squares Method). It is frequently known that some quantity p depends linearly on a set of variables, say $s_1, \ldots, s_n$, i.e.,

$$p = \sum_{j=1}^{n} x_j s_j, \qquad (1)$$

where the x_j are constant coefficients. In the language of data fitting (statistics), Eq. (1) is referred to as a linear regression formula.

Suppose that in an experiment run m times (where m does not necessarily equal n), the values of $p, s_1, \ldots, s_n$ are measured. We would like to use these data to estimate the values of the coefficients $\{x_1, \ldots, x_n\}$. For this, we define

$$L = \sum_{i=1}^{m}\left(p_i - \sum_{j=1}^{n} s_{ij}x_j\right)^2. \qquad (2)$$

If the measurements could be done with infinite accuracy and if the regression formula (1) were indeed obeyed by p and the s_j, then for each run i the equation

$$p_i - \sum_{j=1}^{n} s_{ij}x_j = 0 \qquad (3)$$

would hold and the quantity L would be identically 0. Any error in measurement would give rise to a violation of Eq. (3) and yield a positive value of L. L provides a collective measure of the error or scatter of data from all the runs.

To determine the "best" set $\{x_i, \ldots, x_n\}$ of coefficients from m experimental runs, we choose the coefficients so that L is a minimum. This procedure is known as the method of least squares for fitting a linear regression formula.

(a) To minimize L, the coefficients must be chosen such that

$$\frac{\partial L}{\partial x_k} = 0, \qquad k = 1, \ldots, n, \tag{4}$$

By defining the $m \times n$ matrix $\mathbf{C}$ such that c_{ij} is the value of s_j measured in the ith experiment, prove that the condition in Eq. (4) leads to the matrix expression

$$\mathbf{C}^{\mathrm{T}}\mathbf{C}\mathbf{x} = \mathbf{C}^{\mathrm{T}}\mathbf{p},$$

where

$$\mathbf{x} = \begin{bmatrix} x_1 \\ \vdots \\ x_n \end{bmatrix}, \qquad \mathbf{p} = \begin{bmatrix} p_1 \\ \vdots \\ p_m \end{bmatrix}.$$

(b) Use your Gauss elimination program to solve the equation when $n = 5$, $m = 7$, and the experimental data are as listed in Table 3.8. Indicate the degree of scattering of the data.

9. Consider the following three equations of state for single-component fluids:

Redlich–Kwong:
$$P = \frac{RT}{V-b} - \frac{a}{\sqrt{T}V(V+b)}.$$

Soave:
$$P = \frac{RT}{V-b} - \frac{a\alpha}{V(V+b)}.$$

Peng–Robinson:
$$P = \frac{RT}{V-b} - \frac{a\alpha}{V^2 + 2Vb - b^2}.$$

Using the least squares method and the Newton–Raphson method, generate the "best fit" parameters for methane for each of the above equations of state using the experimental data given in Table 3.P.9. Use your calculated parameters to plot P versus V for methane at $T = 350$ K for each equation of state. Include the above experimental values in your plot.

10. Consider the structural framework shown in Fig. 3.P.10, consisting of n rods attached at a common pin joint at A. The rods are free swiveling at their attachment points and at the pin joint. We assume an external force $\mathbf{f}$ is applied at the pin joint, and denote $\{t_1, \ldots, t_n\}$ as the tension of rods $1, \ldots, n$ and $\{\theta_1, \ldots, \theta_n\}$ as their orientations with respect to the x axis.

TABLE 3.P.9

Data	T (K)	P (bar)	V (L/g mol)
1	200	1.0	16.52
2	200	10.0	1.553
3	200	20.0	0.7149
4	300	1.0	24.90
5	300	10.0	2.452
6	300	20.0	1.206
7	400	1.0	33.24
8	400	10.0	3.310
9	400	20.0	1.648

Letting f_1 and f_2 denote the x and y components of **f**, a force balance at A yields

$$\mathbf{St} = \mathbf{f}, \qquad (1)$$

where **S** is the $2 \times n$ structure matrix

$$\mathbf{S} = \begin{bmatrix} \cos\theta_1 & \cos\theta_2 & \cos\theta_3 & \cdots & \cos\theta_n \\ \sin\theta_1 & \sin\theta_2 & \sin\theta_3 & \cdots & \sin\theta_n \end{bmatrix}, \qquad (2)$$

and **t** is an n-dimensional vector with components $t_1, \ldots, t_n$. With no applied force ($\mathbf{f} = \mathbf{0}$), when none of the rods is under any tension, each of the rods is of some equilibrium length $\{l_1^0, l_2^0, \ldots, l_n^0\}$ and equilibrium orientation $\{\theta_1^0, \theta_2^0, \ldots, \theta_n^0\}$.

(a) Defining the two-dimensional displacement vector **d** as the x- and y-coordinate values of the displacement of the pin joint (A) from its equilibrium position, and the n-dimensional strain vector **e** as the

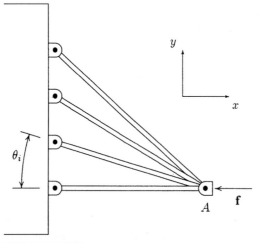

FIGURE 3.P.10

elongation of the rods under tension ($e_i = l_i - l_i^0$), prove that

$$\mathbf{e} = \mathbf{S}^T\mathbf{d}$$

in the "small strain" approximation (i.e., $\theta_i \approx \theta_i^0$).

(b) If the rods are assumed to be Hookean substances, then the strain $e_i \equiv l_i - l_i^0$ obeys the linear law $e_i = k_i t_i$, where k_i is the flexibility constant (k_i^{-1} is the elastic modulus) of the ith rod. In matrix notation, $\mathbf{e} = \mathbf{Kt}$, where $\mathbf{K}$ is the matrix of materials constants

$$\mathbf{K} = \begin{bmatrix} k_1 & 0 & 0 & \cdots & 0 \\ 0 & k_2 & 0 & \cdots & 0 \\ 0 & 0 & k_3 & & 0 \\ \vdots & \vdots & & \ddots & \vdots \\ 0 & 0 & 0 & \cdots & k_n \end{bmatrix}.$$

Show that the strain $\mathbf{d}$ can be related to the external force $\mathbf{f}$ by $\mathbf{Ad} = \mathbf{f}$, where $\mathbf{A} = \mathbf{SK}^{-1}\mathbf{S}^T$.

(c) Consider the case of five rods all made of steel of elastic modulus 207×10^9 N/m² (note the units). The above elastic modulus is defined by the relation $\sigma = E\epsilon$, where σ has units of stress (N/m²), ϵ is the strain ($\epsilon_i = (l_i - l_i^0)/l_i^0$) in dimensionless form, and E is the elastic modulus. Assume that all of the rods have 0.1 cm² cross-sectional area, and the attachment pins are separated by 2 cm (i.e., 2 cm apart) along the wall. When $f = 0$, the point A is at (8 cm, 7 cm) where the origin is located on the wall at the point of attachment of the lowest rod and the y axis is along the wall. If each of the components of the displacement vector $\mathbf{d}$ undergoes a 1% increase, find the force required and the tensions in each of the rods.

(d) If the force applied at point A is given, set up the algorithm to find the displacement vector. *Hint*: Note that $\mathbf{SK}^{-1}\mathbf{S}^T$ is set up in terms of the θ_i's, which are unknown quantities.

(e) Assume that a force of 50 kN is applied at A at 45° to the horizontal in the downward direction. Use the algorithm set up in part (d) to obtain the displacement vector $\mathbf{d}$. Also find the new lengths of the rods and the tensions in each of them. Use the Newton–Raphson method to solve the system.

11. Solve the following equation:

$$x = 1 + \exp(-a\sqrt{x})$$

by

(a) Picard iteration (error criterion: $(|x^{(k+1)} - x^{(k)}| \leq 10^{-5})$).
(b) The Newton–Raphson method (error criteria: $|x^{(k+1)} - x^{(k)}| \leq 10^{-5}$ and $|f(x^{(k)})| \leq 10^{-6}$). (Let $a = \frac{1}{2}$.)

12. Use the Newton–Raphson method with first-order continuation to find x as a function of a from the equation

$$x = 1 + \exp(-a\sqrt{x}).$$

13. Consider the system of equations

$$0 = -x + h^2\left(\exp(y\sqrt{x}) + 3x^2\right)$$
$$0 = -y + 0.5 + h^2 \tan(\exp(x) + y^2).$$

Use the Newton–Raphson method to solve these equations for $h = 0.1$ and 0.2.

14. Consider the concentric cylinders shown in Fig. 3.P.14. Two fluids, labeled A and B, are separated by a meniscus, which is pinned on the tops of the inner ($z = a, r = R_1$) and outer ($z = b, r = R_2$) cylinders.

At equilibrium, the meniscus position $z = z(r)$ is determined by the Young–Laplace equation

$$-\Delta P + \Delta \rho\, gz = 2H\gamma, \tag{1}$$

where γ is the surface tension, ΔP is the capillary pressure (pressure of phase A at $z = 0$ minus pressure that would exist at $z = 0$ if phase A were replaced by phase B), $\Delta \rho$ is the difference between the density of fluids A and B, g is the acceleration of gravity, and H is the mean curvature,

$$2H = \frac{1}{\left[1 + (dz/dr)^2\right]^{3/2}} \frac{d^2z}{dr^2} + \frac{1}{r\left[1 + (dz/dr)^2\right]^{1/2}} \frac{dz}{dr}. \tag{2}$$

One often wants to find the shape of the meniscus, i.e., the function $z(r)$, by solving Eq. (1) subject to the boundary conditions

$$z(r = R_1) = a \quad \text{and} \quad z(r = R_2) = b.$$

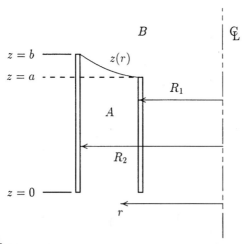

FIGURE 3.P.14

(a) By introducing the finite-difference approximation, show that Eq. (1) takes the form

$$z_{j+1} - 2z_j + z_{j-1} + \frac{[(\Delta r)^2 + (z_{j+1} - z_{j-1})^2/4]}{j \Delta r}(z_{j+1} - z_{j-1})/2$$
$$+ \frac{1}{\gamma \Delta r}[(\Delta r)^2 + (z_{j+1} - z_{j-1})^2/4]^{3/2}[\Delta P - \Delta \rho g z_j] = 0 \quad (3)$$

for $j = 1, \ldots, n$.

(b) Use the Newton–Raphson method to solve for the meniscus shape when fluid A is liquid water (at 293 K), fluid B is air at atmospheric pressure, $R_1 = 1$ cm, $R_2 = 1.5$ cm, $a = 1$ cm, and $b = 1.2$ cm. At 293 K, $\rho_{water} = 0.998$ g/cm^3, $\rho_{air} = 0.0012$ g/cm^3, and $\gamma = 72.75$ dyn/cm. Plot the resulting curve $z(r)$.

15. In dimensionless units, the continuous stirred tank reactor equations can be expressed as

$$\frac{dx_1}{dt} = 1 - x_1 - \alpha \exp\left(-\frac{\beta}{x_2}\right) x_1$$

$$\frac{dx_2}{dt} = 1 - x_2 + \gamma \exp\left(-\frac{\beta}{x_2}\right) x_1 - \delta x_2.$$

Assume that $\alpha = 100$, $\beta = 5$, $\gamma = 149.39$, and $\delta = 0.553$.

(a) Find the steady-state solutions.
(b) Which steady-state solutions are stable to small perturbations? Why?
(c) With the initial value $\mathbf{x}_0 = (1 \pm 0.1)\mathbf{x}^s$, compute and plot $\mathbf{x}(t)$ versus time for each of the steady states $\mathbf{x}^s$. Use the Runge–Kutta method to solve the transient problem.

The Runge–Kutta method to use for solving

$$\frac{d\mathbf{x}}{dt} = \mathbf{f}(\mathbf{x}), \qquad \mathbf{x}_0 = \mathbf{x}(t = 0),$$

is as follows:

$$\mathbf{x}_{n+1} = \mathbf{x}_n + \frac{1}{6}[\mathbf{k}_1 + 2\mathbf{k}_2 + 2\mathbf{k}_3 + \mathbf{k}_4],$$

with

$$\mathbf{k}_1 = h\mathbf{f}(\mathbf{x}_n)$$
$$\mathbf{k}_2 = h\mathbf{f}\left(\mathbf{x}_n + \frac{1}{2}\mathbf{k}_1\right)$$
$$\mathbf{k}_3 = h\mathbf{f}\left(\mathbf{x}_n + \frac{1}{2}\mathbf{k}_2\right)$$
$$\mathbf{k}_4 = h\mathbf{f}(\mathbf{x}_n + \mathbf{k}_3),$$

where $\mathbf{x}_n = \mathbf{x}(t = nh)$.

16. Consider the set of equations

$$f_1 = 1 + h^2(\exp(y\sqrt{x}) + 3x^2) - kz - x = 0$$
$$f_2 = 0.5 + h^2 \tan(\exp(x) + y^2) + k \sin z - y = 0$$
$$f_3 = 1 + h^2 xyz + kz^2 + k^2 \cos y - z = 0.$$

When $h = k = 0$, the solution to these equations is $x = 1$, $y = 0.5$, and $z = 1$. Use the Newton–Raphson method and continuation to explore solutions to the set as a function of the parameters h and k. In particular, find the solutions as a function of h when $k = 0$ and of k when $h = 0$. Then look along the lines of constant values of $h + k$.

17. (a) Use Newton–Raphson method to determine the nonzero root of $x = 1 - \exp(-\alpha x)$ to four correct decimal places (i.e., $|\epsilon_{n+1}| = |x_{n+1} - x_n| < 10^{-5}$) for $\alpha = 2$. Choose $x_0 = 3$ to show that the solution converges quadratically.

   (b) Given a new α, set up the first-order continuation scheme for the above problem. Give a stepwise complete algorithm to obtain the new solution. Give all necessary equations.

   (c) Starting with $\alpha = 2$ and the solution obtained from part (a), perform the first-order continuation and one Newton–Raphson iteration for $\alpha = 3.0$.

18. The time-dependent reaction–diffusion equation in Eq. (3.6.34) can be solved iteratively by noting that the time derivative in the finite-difference approximation can be written as

$$\frac{\partial c}{\partial t} = \frac{c(x, t + \Delta t) - c(x, t)}{\Delta t}.$$

By defining $c_i^{(k)}$ as the value of c at the ith nodal point and the kth time step (i.e., $c_i^{(k)} = c(i \Delta x, k \Delta t)$), the matrix form of the time-dependent problem becomes

$$\mathbf{c}^{(k+1)} - \mathbf{c}^{(k)} = \frac{\Delta t\, D}{(\Delta x)^2} \left[\mathbf{A}\mathbf{c}^{(k)} + f(\mathbf{c}^{(k)}) \right] + \Delta t \mathbf{b}, \tag{1}$$

where $\mathbf{A}$ is defined in Eq. (3.6.41) (this is often referred to as the *Euler method*).

Consider the reaction–diffusion problem for the second-order reaction $f(c) = -kc^2$ over the spatial interval $(0 \leq x \leq 1)$. Using the dimensionless values $D = 0.001$ and $k = 0.25$, plot the dimensionless concentration profile at $t = 10$ for the following initial and boundary conditions:

(a)

$$c(0, t) = 0$$
$$c(1, t) = 1$$
$$c(x, 0) = x^2.$$

(b)
$$c(0, t) = 0$$
$$\frac{\partial}{\partial x} c(1, t) = 0$$
$$c(x, 0) = x^2.$$

(c) An important improvement to the Euler method is to replace the right-hand side of Eq. (1) with its future $(k + 1)$ values, giving

$$\mathbf{c}^{(k+1)} - \mathbf{c}^{(k)} = \frac{\Delta t\, D}{(\Delta x)^2} \left[\mathbf{A} \mathbf{c}^{(k+1)} + f\left(\mathbf{c}^{(k+1)}\right) \right] + \Delta t\, \mathbf{b} \qquad (2)$$

(referred to as the *implicit* or *backward Euler method*). Write a program to solve Eq. (2) and repeat parts (a) and (b) above using the implicit Euler method.

19. Find the inverse of the $n \times n$ Hilbert matrix $\mathbf{A}$ whose components are
$$a_{ij} = \frac{1}{i + j},$$
for the case $n = 3$ and 10.

20. For $n = 10$, solve by Gauss–Seidel iteration the equation $\mathbf{A}\mathbf{x} = \mathbf{b}$ for $b_1 = 1$, $b_6 = 3$, and $b_i = 0$, $i \neq 1$ or 6, where $\mathbf{A}$ is the Hilbert matrix defined in Problem 19.

21. Find the **LU**-decomposition of the 10×10 matrix whose elements are
$$a_{ii} = 2i, \qquad a_{i,i+1} = -i, \qquad a_{i,i-1} = -i$$
and $a_{ij} = 0$ otherwise.

22. The pressure and chemical potential of a van der Waals fluid are given by
$$P = \frac{nkT}{1 - nb} - n^2 a$$
$$\mu = -kT \ln\left(\frac{1}{nb} - 1\right) + \frac{nbkT}{1 - nb} - 2na,$$
where n and T are the density and the temperature, respectively, k is Boltzmann's constant, and a and b are parameters characteristic of the fluid. A vapor of density n_v is in equilibrium with a liquid at density n_l if
$$P(n_v) = P(n_l)$$
$$\mu(n_v) = \mu(n_l). \qquad (1)$$

Find the densities n_v and n_l of coexisting vapor and liquid densities as a function of temperature T. The problem should be solved in terms of the dimensionless variables
$$n^* = nb \quad \text{and} \quad T^* = \frac{kTb}{a}.$$

In terms of these dimensionless variables, the critical point is $n_c^* = \frac{1}{3}$ and $T_c^* = \frac{8}{27}$. Above T_c^*, liquid and vapor do not coexist (i.e., $n_v^* = n_l^*$ is the only solution to Eq. (1)), but for T^* below T_c^* the equations in Eq. (1) admit the solutions $n_v^* \neq n_l^*$ appropriate for coexistence. In terms of the dimensionless variables, the equations in Eq. (1) become

$$\frac{n_v^* T^*}{1 - n_v^*} - n_v^{*2} = \frac{n_l^* T^*}{1 - n_l^*} - n_l^{*2} - T^* \ln\left(\frac{1}{n_v^*} - 1\right) + \frac{n_v^* T^*}{1 - n_v^*} - 2n_v^*$$

$$= -T^* \ln\left(\frac{1}{n_l^*} - 1\right) + \frac{n_l^* T^*}{1 - n_l^*} - 2n_l^*.$$

Use the Newton–Raphson method to solve these equations for n_v^* and n_l^* as a function of temperature. A good first guess is needed.

23. Consider the van der Waals equation of state in Example 3.8.3:

$$P = \frac{nkT}{1 - nb} - n^2 a$$

$$\mu = -kT \ln\left(\frac{1}{nb} - 1\right) + \frac{nbkT}{1 - nb} - 2na.$$

(a) Using the Newton–Raphson method, calculate the boiling-point temperature (T) and the equilibrium liquid and vapor densities (n_l and n_g) for ethanol ($a = 12.016$ atm-L^2/g-mol^2, $b = 0.08405$ L/g-mol) at atmospheric pressure ($P = 1$ atm).

(b) Calculate and plot the planar interfacial density for ethanol using values of $L/\sqrt[3]{b} = 1, 5,$ and 10.

24. (Reversible Anionic Polymerization). A reaction scheme sometimes used to model polymerization is

$$I + M \xrightarrow{k_1} P_1 \quad \text{initiation step}$$

$$\left.\begin{array}{c} P_1 + M \underset{k_2}{\overset{k_1}{\rightleftharpoons}} P_2 \\ \\ P_n + M \underset{k_2}{\overset{k_1}{\rightleftharpoons}} P_{n+1} \end{array}\right\} \text{polymerization steps}$$

where I denotes the concentration of the initiator, M the monomer, and P_n the polymer containing n monomers. The rate equations corresponding to this scheme are

$$\frac{dI}{dt} = -k_1 I M \tag{1}$$

$$\frac{dM}{dt} = -k_1 I M - k_1 \sum_{n=1}^{\infty} P_n + k_2 \sum_{n=2}^{\infty} P_n \tag{2}$$

$$\frac{dP_1}{dt} = k_1 I M - k_1 M P_1 + k_2 P_2 \tag{3}$$

$$\frac{dP_n}{dt} = k_1 M P_{n-1} - (k_1 M + k_2) P_n + k_2 P_{n+1}, \quad n > 2. \quad (4)$$

In matrix form, the rate equations for $P_1, P_2, \ldots$ become

$$\frac{d\mathbf{P}}{dt} = \mathbf{AP} + \mathbf{b}, \quad (5)$$

where

$$\mathbf{P} = \begin{bmatrix} P_1 \\ P_2 \\ \vdots \end{bmatrix}, \quad \mathbf{b} = \begin{bmatrix} -k_1 I M \\ 0 \\ \vdots \end{bmatrix} \quad (6)$$

and $\mathbf{A}$ is the tridiagonal matrix

$$\mathbf{A} = \begin{bmatrix} -k_1 M & k_2 & 0 & 0 & \cdots & 0 & \cdots \\ k_1 M & -(k_1 M + k_2) & k_2 & 0 & \cdots & 0 & \cdots \\ 0 & k_1 M & -(k_1 M + k_2) & k_2 & 0 & \cdots \\ \vdots & \vdots & \vdots & \vdots & \vdots \end{bmatrix}. \quad (7)$$

This problem is special in that the vectors and matrices have an infinite number of entries and that $\mathbf{A}$ and $\mathbf{b}$ depend on time (through I and M, which must satisfy Eqs. (1) and (2)).

(a) Solve for the steady-state solution $\mathbf{P}^*$ in terms of k_1, k_2, M_0, and ξ^*, where M_0 is the initial monomer concentration and ξ^* is the steady-state conversion defined by $\xi^* = (M_0 - M^*)/M_0$ (M^* is the steady-state monomer concentration).

(b) Describe how you would go about solving for the steady-state conversion ξ^* for a given set of initial monomer and initiator concentrations.

(c) In terms of the rate constants, what is the maximum possible value for the conversion ξ^* for a given initial monomer concentration M_0? What initial conditions would be needed to achieve this conversion?

25. (Heat Exchanger Performance). As a design engineer, you have been asked to evaluate the performance of a heat exchanger and offer your advice. Two liquid streams (labeled "cold" and "hot") are contacted in a single-pass heat exchanger as shown in Fig. 3.P.25.

Currently, the cold stream enters the exchanger at 95°F and exits at 110°F, while the hot stream enters at 150°F and exits at 115°F. However, recent changes in the reactor temperature specification requires the cold stream to exit at 120°F (10° hotter than the current specs). Furthermore, the flow rates of both streams (Q_c and Q_h) are fixed by other process constraints and cannot be changed.

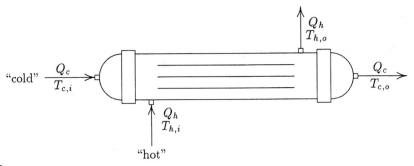

FIGURE 3.P.25

Rather than design and install a new heat exchanger (a costly and time-consuming job), you observe that the two streams in the current configuration are contacted "co-currently." Thus, you suspect that simply swapping the input and output points of one of the streams (i.e., changing to a "countercurrent" configuration) may be all that is needed to meet the new specifications.

Your goal is to estimate what the output temperatures of both streams would be ($t_{c,o}$ and $t_{h,o}$) if you just rearranged the exchanger to be "countercurrent." (Assume that the heat transfer coefficient is unaffected.)

(a) Use the following design equations to derive two independent equations for the two unknowns ($t_{c,o}$ and $t_{h,o}$). *Hint*: Define new variables $\alpha_c = Q_c C_{v,c}/(UA)$ and $\alpha_h = Q_h C_{v,h}/(UA)$ and eliminate them from the system of equations.

"Cocurrent" configuration:

$$Q_c C_{v,c}(T_{c,o} - T_{c,i}) = Q_h C_{v,h}(T_{h,i} - T_{h,o})$$

$$Q_c C_{v,c}(T_{c,o} - T_{c,i}) = UA \frac{[(T_{h,i} - T_{c,i}) - (T_{h,o} - T_{c,o})]}{\ln[(T_{h,i} - T_{c,i})/(T_{h,o} - T_{c,o})]}.$$

"Countercurrent" configuration:

$$Q_c C_{v,c}(t_{c,o} - T_{c,i}) = Q_h C_{v,h}(T_{h,i} - t_{h,o})$$

$$Q_c C_{v,c}(t_{c,o} - T_{c,i}) = UA \frac{[(t_{h,o} - T_{c,i}) - (T_{h,i} - t_{c,o})]}{\ln[(t_{h,o} - T_{c,i})/(T_{h,i} - t_{c,o})]}.$$

Variables:

Q_c = flow rate of "cold" stream
Q_h = flow rate of "hot" stream
$C_{v,c}$ = heat capacity of "cold" stream
$C_{v,h}$ = heat capacity of "hot" stream
U = overall heat transfer coefficient
A = effective contact area of exchanger
$T_{c,i}$ = inlet temperature of "cold" stream

$T_{h,i}$ = inlet temperature of "hot" stream
$T_{c,o}$ = outlet temperature of "cold" stream (original configuration)
$T_{h,o}$ = outlet temperature of "hot" stream (original configuration)
$t_{c,o}$ = outlet temperature of "cold" stream (new configuration)
$t_{h,o}$ = outlet temperature of "hot" stream (new configuration)

(b) Calculate the Jacobian matrix elements for the Newton–Raphson procedure in terms of the current guesses ($t_{c,o}$ and $t_{h,o}$).
(c) Solve this problem using the Newton–Raphson method.
(d) The process engineers inform you that the "cold" stream inlet temperature occasionally drops from 95 to 85°F. Using first-order continuation, calculate new initial guesses for a Newton–Raphson calculation at this new inlet temperature ($T'_{c,i} = 85°F$) based on the converged solutions ($t^*_{c,o}$ and $t^*_{h,o}$) from the old inlet temperature ($T_{c,i} = 95°F$). Use the Newton–Raphson method to solve for the new outlet temperature.

FURTHER READING

Adby, P. R. (1980). "Applied Circuit Theory: Matrix and Computer Methods." Halsted, New York.
Chapra, S. C., and Canale, R. P. (1988). "Numerical Methods for Engineers." McGraw-Hill, New York.
Dahlquist, G., and Bjorck, A. (1974). "Numerical Methods." Prentice Hall International, Englewood Cliffs, NJ.
Faddeeva, V. N. (1959). "Computational Methods of Linear Algebra." Dover, New York.
Gere, J. M., and Weaver, W. (1965). "Analysis of Framed Structures." Van Nostrand, Princeton, NJ.
Golub, G. H., and Van Loan, C. F. (1989). "Matrix Computations," 2nd Ed. Johns Hopkins Press, Baltimore.
Hackbush, W. (1994). "Iterative Solution of Large Sparse Systems of Equations." Springer-Verlag, New York.
Hoffman, J. (1992). "Numerical Methods for Engineers and Scientists." McGraw-Hill, New York.
Householder, A. S. (1965). "The Theory of Matrices in Numerical Analysis." Blaisdell, Boston.
Lewis, W. E., and Pryce, D. G. (1966). "Application of Matrix Theory to Electrical Engineering." Spon, London.
Magid, A. R. (1985). "Applied Matrix Models: a Second Course in Linear Algebra with Computer Applications," Wiley, New York.
Press, W. H., Teukolsky, S. A., Vetterling. W. T., and Flannery, B. P. (1988). "Numerical Recipes in Fortran," 2nd Ed. Cambridge University Press, Cambridge, UK.
Saad, Y. (1996). "Iterative Methods for Sparse Linear Systems." PWS, Boston.
Strang, G. (1988). "Linear Algebra and Its Applications." Academic Press, New York.
Varga, R. S. (1962). "Matrix Iterative Analysis." Prentice Hall International, Englewood Cliffs, NJ.
Watkins, S. W. (1991). "Fundamentals of Matrix Computations," Wiley, New York.
Wolfram, S. (1996). "The Mathematica Book." 3rd Ed. Cambridge University Press, Cambridge, UK.

4
GENERAL THEORY OF SOLVABILITY OF LINEAR ALGEBRAIC EQUATIONS

4.1. SYNOPSIS

If in the linear system $\mathbf{Ax} = \mathbf{b}$ the matrix $\mathbf{A}$ is square and nonsingular, then Cramer's rule establishes the existence of a unique solution for any vector $\mathbf{b}$. The only problem remaining in this case is to find the most efficient way to solve for $\mathbf{x}$. This was the topic of the previous chapter. On the other hand, if $\mathbf{A}$ is singular or non-square, the linear system may or may not have a solution. The goals of this chapter are to find the conditions under which the linear system has a solution and to determine what kind of solution (or solutions) there is.

In pursuit of these goals, we will first prove *Sylvester's theorem*, namely, that the rank r_C of $\mathbf{C} = \mathbf{AB}$ obeys the inequality $r_C \leq \min(r_A, r_B)$, where r_A and r_B are the ranks of $\mathbf{A}$ and $\mathbf{B}$, respectively. This theorem can be used to establish that $r_C = r_A$ (or r_B) if $\mathbf{A}$ (or $\mathbf{B}$) is a square nonsingular matrix. A by-product of our proof of Sylvester's theorem is that $|\mathbf{C}| = |\mathbf{A}||\mathbf{B}|$ if $\mathbf{A}$ and $\mathbf{B}$ are square matrices of the same dimension. Thus, the determinant of the product of square matrices equals the product of their determinants.

We will then prove the *solvability theorem*, which states that $\mathbf{Ax} = \mathbf{b}$ *if and only if* the rank of the augmented matrix $[\mathbf{A}, \mathbf{b}]$ equals the rank of $\mathbf{A}$. When a solution does exist for an $m \times n$ matrix, the general solution will be of the form $\mathbf{x} = \mathbf{x}_p + \sum_{i=1}^{n-r} c_i \mathbf{x}_h^{(i)}$, where $\mathbf{x}_p$ is a particular solution, i.e., $\mathbf{Ax} = \mathbf{b}$, and the vectors $\mathbf{x}_h^{(1)}, \ldots, \mathbf{x}_h^{(n-r)}$ are $n-r$ linearly independent solutions of the homogeneous equation $\mathbf{Ax}_h = \mathbf{0}$ (the c_i are arbitrary numbers).

As a by-product of the solvability theorem, we will also prove that if $\mathbf{A}$ is an $m \times n$ matrix, of rank r_A, then $\mathbf{A}$ contains r linearly independent row vectors. The remaining $n - r$ column vectors and $m - r$ row vectors are linear combinations of the r linearly independent ones. It follows from these properties that the number of linearly independent column vectors in the set $\{\mathbf{a}_1, \ldots, \mathbf{a}_n\}$ is equal to the rank of the matrix $\mathbf{A} = [\mathbf{a}_1, \ldots, \mathbf{a}_n]$. Gauss or Gauss–Jordan transformation of $\mathbf{A}$ provides an efficient way to determine the rank of $\mathbf{A}$ or the number of linearly independent vectors in the set $\{\mathbf{a}_1, \ldots, \mathbf{a}_n\}$.

In the last section, we will restate the solvability theorem as the *Fredholm alternative theorem*. That is, $\mathbf{A}\mathbf{x} = \mathbf{b}$ *if and only if* $\mathbf{b}$ is orthogonal to $\mathbf{z}$ (i.e., $\mathbf{b}^\dagger \mathbf{z} = 0$) for any solution of the homogeneous equation $\mathbf{A}^\dagger \mathbf{z} = \mathbf{0}$, where $\mathbf{A}^\dagger$ is the adjoint of $\mathbf{A}$. The advantage of the Fredholm alternative form of the solvability theorem is that it is generalizable to abstract linear vector spaces (e.g., function spaces in which integral or differential operations replace matrices) where the concept of the rank of a matrix does not arise.

4.2. SYLVESTER'S THEOREM AND THE DETERMINANTS OF MATRIX PRODUCTS

Recall from Chapter 1 that an rth-order determinant $M_A^{(r)}$ of the matrix formed by striking $m - r$ rows and $n - r$ columns of an $m \times n$ matrix $\mathbf{A}$ is called an rth-order minor of $\mathbf{A}$. For example, consider the matrix

$$\mathbf{A} = \begin{bmatrix} a_{11} & a_{12} & a_{13} & a_{14} \\ a_{21} & a_{22} & a_{23} & a_{24} \\ a_{31} & a_{32} & a_{33} & a_{34} \end{bmatrix}. \tag{4.2.1}$$

If we strike the second row and the second and third columns

$$\begin{bmatrix} a_{11} & & & a_{14} \\ \hline a_{31} & & & a_{34} \end{bmatrix}, \tag{4.2.2}$$

we obtain the second-order minor

$$M_A^{(2)} = \begin{vmatrix} a_{11} & a_{14} \\ a_{31} & a_{34} \end{vmatrix}. \tag{4.2.3}$$

Striking just the fourth column, we obtain the third-order minor

$$M_A^{(3)} = \begin{vmatrix} a_{11} & a_{12} & a_{13} \\ a_{21} & a_{22} & a_{23} \\ a_{31} & a_{32} & a_{33} \end{vmatrix}. \tag{4.2.4}$$

We previously defined the rank r_A of a matrix $\mathbf{A}$ as the highest order minor of $\mathbf{A}$ that has a nonzero determinant. For instance, the rank of

$$\begin{bmatrix} 1 & 2 & 1 \\ 1 & 1 & 1 \end{bmatrix} \tag{4.2.5}$$

SYLVESTER'S THEOREM AND THE DETERMINANTS OF MATRIX PRODUCTS

is 2 since striking the third column gives the minor

$$M^{(2)} = \begin{vmatrix} 1 & 2 \\ 1 & 1 \end{vmatrix} = -1. \tag{4.2.6}$$

The rank of

$$\begin{bmatrix} 1 & 1 & 1 \\ 1 & 1 & 1 \end{bmatrix}, \tag{4.2.7}$$

on the other hand, is 1 since striking one column gives

$$M^{(2)} = \begin{vmatrix} 1 & 1 \\ 1 & 1 \end{vmatrix} = 0, \tag{4.2.8}$$

and striking two columns and one row gives

$$M^{(1)} = |1| = 1. \tag{4.2.9}$$

One of the objectives of this section will be to relate the rank of the product of two matrices to the ranks of the matrices; i.e., if $C = AB$, what is the relationship among r_C and r_A and r_B? If A is a $p \times n$ matrix with elements a_{ij} and B is an $n \times q$ matrix with elements b_{ij}, then C is a $p \times q$ matrix with elements

$$c_{ij} = \sum_{k=1}^{n} a_{ik} b_{kj}, \qquad i = 1, \ldots, p, \ j = 1, \ldots, q. \tag{4.2.10}$$

Since only a square matrix has a determinant, we know from the outset that $r_A \leq \min(p, n)$, $r_B \leq \min(n, q)$, and $r_C \leq \min(p, q)$. There are, of course, cases where $r_C = r_A = r_B$; e.g., if

$$A = \begin{bmatrix} 2 & 0 \\ 0 & 1 \end{bmatrix} \quad \text{and} \quad B = \begin{bmatrix} 1 & 0 \\ 0 & 2 \end{bmatrix}, \quad \text{then} \quad C = \begin{bmatrix} 2 & 0 \\ 0 & 2 \end{bmatrix}, \tag{4.2.11}$$

and so $r_C = r_A = r_B = 2$. However, there are also cases where $r_C \leq r_A$ and r_B; e.g., if

$$A = \begin{bmatrix} 1 & 1 \\ 0 & 0 \end{bmatrix} \quad \text{and} \quad B = \begin{bmatrix} 1 & 0 \\ -1 & 0 \end{bmatrix}, \quad \text{then} \quad C = \begin{bmatrix} 0 & 0 \\ 0 & 0 \end{bmatrix}, \tag{4.2.12}$$

and so $r_C = 0$, $r_A = r_B = 1$. Similarly, if

$$A = \begin{bmatrix} 1 & 1 \\ 0 & 0 \end{bmatrix} \quad \text{and} \quad B = \begin{bmatrix} 1 & 1 & 1 \\ -1 & 0 & -1 \end{bmatrix}, \quad \text{then} \quad C = \begin{bmatrix} 0 & 1 & 0 \\ 0 & 0 & 0 \end{bmatrix}, \tag{4.2.13}$$

and so $r_C = 1$, $r_A = 1$, and $r_B = 2$. The question is whether there are other possibilities. The answer is provided by

SYLVESTER'S THEOREM. *If the rank of* $\mathbf{A}$ *is* r_A *and the rank of* $\mathbf{B}$ *is* r_B, *then the rank* r_C *of* $\mathbf{C} = \mathbf{AB}$ *obeys the inequality*

$$r_C \leq \min(r_A, r_B). \tag{4.2.14}$$

The proof of the theorem is straightforward though somewhat tedious. If $\mathbf{A}$ and $\mathbf{B}$ are $p \times n$ and $n \times q$ matrices, then

$$\mathbf{C} = \begin{bmatrix} c_{11} & c_{12} & \cdots & c_{1q} \\ \vdots & & & \vdots \\ c_{p1} & c_{p2} & \cdots & c_{pq} \end{bmatrix}, \tag{4.2.15}$$

where the elements c_{ij} are given by Eq. (4.2.10). Let the indices $h_1, \ldots, h_r$ and $k_1, \ldots, k_r$ denote the rows and columns of $M_C^{(r)}$, an rth-order minor of $\mathbf{C}$ given as

$$M_C^{(r)} = \begin{vmatrix} c_{h_1 k_1} & c_{h_1 k_2} & \cdots & c_{h_1 k_r} \\ \vdots & \vdots & & \vdots \\ c_{h_r k_1} & c_{h_r k_2} & \cdots & c_{h_r k_r} \end{vmatrix}. \tag{4.2.16}$$

Since $c_{h_i k_1} = \sum_{j_1=1}^n a_{h_i j_1} b_{j_1 k_1}$, $M_C^{(r)}$ can be rewritten as

$$\begin{aligned}
M_C^{(r)} &= \begin{vmatrix} \sum_{j_1} a_{h_1 j_1} b_{j_1 k_1} & c_{h_1 k_2} & \cdots & c_{h_1 k_r} \\ \vdots & \vdots & & \vdots \\ \sum_{j_1} a_{h_r j_1} b_{j_1 k_1} & c_{h_r k_2} & \cdots & c_{h_r k_r} \end{vmatrix} \\
&= \sum_{j_1} \begin{vmatrix} a_{h_1 j_1} b_{j_1 k_1} & c_{h_1 k_2} & \cdots & c_{h_1 k_r} \\ \vdots & \vdots & & \vdots \\ a_{h_r j_1} b_{j_1 k_1} & c_{h_r k_2} & \cdots & c_{h_r k_r} \end{vmatrix} \\
&= \sum_{j_1} b_{j_1 k_1} \begin{vmatrix} a_{h_1 j_1} & c_{h_1 k_2} & \cdots & c_{h_1 k_r} \\ \vdots & \vdots & & \vdots \\ a_{h_r j_1} & c_{h_r k_2} & \cdots & c_{h_r k_r} \end{vmatrix}.
\end{aligned} \tag{4.2.17}$$

In accomplishing the succession of equations in Eq. (4.2.17), we have utilized the elementary properties 6 and 4 of determinants discussed in Chapter 1. We can continue the process by setting $c_{h_1 k_2} = \sum_{j_2=1}^n a_{h_1 j_2} b_{j_2 k_2}$ and carrying out the same elementary operations to obtain

$$M_C^{(r)} = \sum_{j_1} \sum_{j_2 \neq j_1} b_{j_1 k_1} b_{j_2 k_2} \begin{vmatrix} a_{h_1 j_1} & a_{h_1 j_2} & c_{h_1 k_3} & \cdots & c_{h_1 k_r} \\ \vdots & \vdots & \vdots & & \vdots \\ a_{h_r j_1} & a_{h_r j_2} & c_{h_r k_3} & \cdots & c_{h_r k_r} \end{vmatrix}. \tag{4.2.18}$$

Continuation of the process finally yields

$$M_C^{(r)} = \sum_{j_1} \sum_{j_2 \neq 1} \cdots \sum_{j_r \neq j_1 \cdots j_{r-1}} b_{j_1 k_1} b_{j_2 k_2} \cdots b_{j_r k_r}$$

$$\times \begin{vmatrix} a_{h_1 j_1} & a_{h_1 j_2} & \cdots & a_{h_1 j_r} \\ \vdots & \vdots & & \vdots \\ a_{h_r j_1} & a_{h_r j_2} & \cdots & a_{h_r j_r} \end{vmatrix} \quad (4.2.19)$$

$$= \sum_{j_1} \cdots \sum_{j_r \neq j_1 \cdots j_{r-1}} b_{j_1 k_1} \cdots b_{j_r k_r} M_A^{(r)},$$

where we have identified that the remaining determinant is an rth-order minor $M_A^{(r)}$ of $\mathbf{A}$. Note that if $r > r_A$, then the minor $M_A^{(r)}$ is 0 by the definition of r_A. Thus, Eq. (4.2.19) proves that

$$r_C \leq r_A. \quad (4.2.20)$$

Beginning again with Eq. (4.2.16), we can express the elements of the first *row* of $M_C^{(r)}$ as

$$c_{h_1 k_j} = \sum_{j_1=1}^{n} a_{h_1 j_1} b_{j_1 k_j} \quad (4.2.21)$$

and then use elementary properties to obtain

$$M_C^{(r)} = \sum_{j_1} a_{h_1 j_1} \begin{vmatrix} b_{j_1 k_1} & b_{j_1 k_2} & \cdots & b_{j_1 k_r} \\ c_{h_2 k_1} & c_{h_2 k_2} & \cdots & c_{h_2 k_r} \\ \vdots & \vdots & & \vdots \\ c_{h_r k_1} & c_{h_r j_2} & \cdots & c_{h_r k_r} \end{vmatrix}. \quad (4.2.22)$$

Continuation of this process with the other rows eventually yields

$$M_C^{(r)} = \sum_{j_1} \cdots \sum_{j_r \neq j_1 \cdots j_{r-1}} a_{h_1 j_1} \cdots a_{h_r j_r} \begin{vmatrix} b_{j_1 k_1} & \cdots & b_{j_1 k_r} \\ \vdots & & \vdots \\ b_{j_r k_1} & \cdots & b_{j_r k_r} \end{vmatrix} \quad (4.2.23)$$

$$= \sum_{j_1} \cdots \sum_{j_r \neq j_1 \cdots j_{r-1}} a_{h_1 j_1} \cdots a_{h_r j_r} M_B^{(r)}.$$

Again, the rth-order minor $M_B^{(r)}$ of $\mathbf{B}$ will be 0 if $r > r_B$, and so we conclude

$$r_C \leq r_B. \quad (4.2.24)$$

The combination of Eqs. (4.2.20) and (4.2.24) implies Sylvester's theorem, Eq. (4.2.14).

In the special case that **A** and **B** are square $n \times n$ matrices

$$\mathbf{C} = \begin{bmatrix} \sum_{j_1} a_{1j_1} b_{j_1 1} & \sum_{j_2} a_{1j_2} b_{j_2 2} & \cdots & \sum_{j_n} a_{1j_n} b_{j_n n} \\ \vdots & \vdots & & \vdots \\ \sum_{j_1} a_{nj_1} b_{j_1 1} & \sum_{j_2} a_{nj_2} b_{j_2 2} & \cdots & \sum_{j_n} a_{nj_n} b_{j_n n} \end{bmatrix}, \qquad (4.2.25)$$

and elementary operations similar to those yielding Eq. (4.2.23) give

$$|\mathbf{C}| = \sum_{j_1} \cdots \sum_{j_r \neq j_1 \cdots j_{r-1}} a_{1j_1} \cdots a_{nj_n} \begin{vmatrix} b_{j_1 1} & b_{j_1 2} & \cdots & b_{j_1 n} \\ \vdots & \vdots & \ddots & \vdots \\ b_{j_n 1} & b_{j_n 2} & \cdots & b_{j_n n} \end{vmatrix}. \qquad (4.2.26)$$

The determinant above involving the b_{ij} elements will be nonzero only if the integers $j_1, j_2, \ldots, j_n$ are a permutation of $1, 2, \ldots, n$. Otherwise, two rows of the determinant would be the same. Moreover, the determinant in Eq. (4.2.26) is equal to $(-1)^P |\mathbf{B}|$, where P is the number of transpositions necessary to reorder the integers $j_1, j_2, \ldots, j_n$ to the sequence $1, 2, \ldots, n$ (thereby rearranging the columns to the proper order to get $|\mathbf{B}|$). Thus, Eq. (4.2.26) can be rewritten as

$$|\mathbf{C}| = |\mathbf{B}| \sum_{j_1} \cdots \sum_{j_n} (-1)^P a_{1j_1} \cdots a_{nj_n}. \qquad (4.2.27)$$

Noting that the factor to the right of $|\mathbf{B}|$ is, by definition, the determinant $|\mathbf{A}|$, we have proved the following:

THEOREM. *If **A** and **B** are square matrices, the determinant of their product **C** is the product of their determinants, i.e.,*

$$|\mathbf{C}| = |\mathbf{A}| |\mathbf{B}|. \qquad (4.2.28)$$

This result is not only useful in evaluating products of determinants, but it is also frequently employed in proving theorems concerning matrices and linear equations. For example, if **A** is nonsingular, i.e., $|\mathbf{A}| \neq 0$ and $\mathbf{A}^{-1}$ exists, then $\mathbf{A}\mathbf{A}^{-1} = \mathbf{I}$ and Eq. (4.2.28) implies $|\mathbf{A}\mathbf{A}^{-1}| = |\mathbf{A}||\mathbf{A}^{-1}| = |\mathbf{I}| = 1$, or $|\mathbf{A}^{-1}| = 1/|\mathbf{A}|$. This greatly simplifies finding the determinant of the inverse of **A**.

Before leaving this section, let us establish a corollary to Sylvester's theorem that is very useful in the theory of solvability of $\mathbf{A}\mathbf{x} = \mathbf{b}$.

COROLLARY. *The multiplication of a matrix **B** by a square nonsingular matrix **A** does not change the rank of **B**; i.e., if $\mathbf{C} = \mathbf{A}\mathbf{B}$, and if **A** is square and nonsingular, then*

$$r_C = r_B. \qquad (4.2.29)$$

The proof is simple. If **A** is a nonsingular $n \times n$ matrix, then its rank r_A is n since $|\mathbf{A}| \neq 0$, and so is the rank of $\mathbf{A}^{-1}$ since $|\mathbf{A}^{-1}| \neq 0$. According to Sylvester's theorem,

$$r_C \leq \min(r_A, r_B) = \min(n, r_B). \qquad (4.2.30)$$

But
$$\mathbf{B} = \mathbf{A}^{-1}\mathbf{C} \tag{4.2.31}$$

for which Sylvester's theorem requires

$$r_B \leq \min(r_{A^{-1}}, r_C) = \min(n, r_C). \tag{4.2.32}$$

Equations (4.2.30) and (4.2.32) combine to imply that $r_C \leq r_B$ and $r_B \leq r_C$, which can only be true if $r_C = r_B$, thus proving the corollary. A similar proof establishes that the rank of $\mathbf{BA}$ is the same as that of $\mathbf{B}$ if $\mathbf{A}$ is a square nonsingular matrix.

EXAMPLE 4.2.1. Consider the matrices

$$\mathbf{A} = \begin{bmatrix} -2 & 1 \\ 1 & -2 \end{bmatrix}, \quad \mathbf{B} = \begin{bmatrix} 1 & -2 & 1 \\ -1 & 2 & -1 \end{bmatrix}, \tag{4.2.33}$$

and

$$\mathbf{AB} = \begin{bmatrix} -3 & 6 & -3 \\ 3 & -6 & 3 \end{bmatrix}. \tag{4.2.34}$$

We note that since $|\mathbf{A}| = 3$, $r_A = 2$, and since the order of the only nonzero minor of $\mathbf{B}$ is 1, $r_B = 1$. The same is true of $\mathbf{AB}$ and so $r_{AB} = 1$. This is admittedly a rather trivial example, but in the next sections we will flex the true muscle of the corollary.

4.3. GAUSS–JORDAN TRANSFORMATION OF A MATRIX

Consider the nonsingular matrix

$$\mathbf{A} = \begin{bmatrix} a_{11} & a_{12} & a_{13} \\ a_{21} & a_{22} & a_{23} \\ a_{31} & a_{32} & a_{33} \end{bmatrix}. \tag{4.3.1}$$

Assuming $a_{11} \neq 0$, the first step in Gauss elimination yields

$$\begin{bmatrix} a_{11} & a_{12} & a_{13} \\ 0 & b_{22} & b_{23} \\ 0 & b_{32} & b_{33} \end{bmatrix}, \tag{4.3.2}$$

where $b_{ij} = a_{ij} - (a_{i1}/a_{11})a_{1j}$, $i = 2, 3$, $j = 1, 2, 3$. If $b_{22} = 0$, we interchange rows 2 and 3 (one of b_{22} and b_{32} have to be nonzero since $r_A = 3$) to obtain

$$\begin{bmatrix} a_{11} & a_{12} & a_{13} \\ 0 & b'_{22} & b'_{23} \\ 0 & b'_{32} & b'_{33} \end{bmatrix}. \tag{4.3.3}$$

Next, we multiply row 2 by a_{12}/b'_{22} and subtract it from the first row to obtain

$$\begin{bmatrix} a_{11} & 0 & c_{13} \\ 0 & b'_{22} & b'_{23} \\ 0 & b'_{32} & b'_{33} \end{bmatrix}. \qquad (4.3.4)$$

We then carry out Gauss elimination on the third row to get

$$\begin{bmatrix} a_{11} & 0 & c_{13} \\ 0 & b'_{22} & b'_{23} \\ 0 & 0 & c_{33} \end{bmatrix}. \qquad (4.3.5)$$

Multiplying the third row by c_{13}/c_{33} and subtracting it from the first row, and multiplying the third row by b'_{23}/c_{33} and subtracting it from the second row, we finally arrive at

$$\begin{bmatrix} a_{11} & 0 & 0 \\ 0 & b'_{22} & 0 \\ 0 & 0 & c_{33} \end{bmatrix}. \qquad (4.3.6)$$

The process we just described leading to the result in Eq. (4.3.6) is known as Gauss–Jordan elimination.

If we wish to solve the system of equations

$$\begin{aligned} a_{11}x_1 + a_{12}x_2 + a_{13}x_3 &= b_1 \\ a_{21}x_1 + a_{22}x_2 + a_{23}x_3 &= b_2 \\ a_{31}x_1 + a_{32}x_2 + a_{33}x_3 &= b_3 \end{aligned} \qquad (4.3.7)$$

represented by

$$\mathbf{Ax} = \mathbf{b}, \qquad (4.3.8)$$

where $r_A = 3$, then the Gauss–Jordan transformation of the augmented matrix $[\mathbf{A}, \mathbf{b}]$ yields

$$\begin{bmatrix} a_{11} & 0 & 0 & \alpha_1 \\ 0 & b'_{22} & 0 & \alpha_2 \\ 0 & 0 & c_{33} & \alpha_3 \end{bmatrix}, \qquad (4.3.9)$$

where α_i are the values of b_i obtained through the transformation process. The linear equations corresponding to Eq. (4.3.9) are then

$$\begin{aligned} a_{11}x_1 &= \alpha_1 \\ b'_{22}x_2 &= \alpha_2 \\ c_{33}x_3 &= \alpha_3, \end{aligned} \qquad (4.3.10)$$

which leads to the simple solution $x_1 = \alpha_1/a_{11}$, $x_2 = \alpha_2/b'_{22}$, and $x_3 = \alpha_3/c_{33}$.

For the general nonsingular problem $\mathbf{Ax} = \mathbf{b}$, the Gauss–Jordan transformation converts the equations

$$\begin{aligned}
a_{11}x_1 + a_{12}x_2 + \cdots + a_{1n}x_n &= b_1 \\
a_{21}x_1 + a_{22}x_2 + \cdots + a_{2n}x_n &= b_2 \\
&\vdots \\
a_{n1}x_1 + a_{n2}x_2 + \cdots + a_{nn}x_n &= b_n
\end{aligned} \qquad (4.3.11)$$

into the system

$$\begin{aligned}
a_{11}x_1 &= \alpha_1 \\
a'_{22}x_2 &= \alpha_2 \\
&\ddots \vdots \\
& a'_{nn}x_n = \alpha_n
\end{aligned} \qquad (4.3.12)$$

for which the solution is simply $x_i = \alpha_i/a'_{ii}$. In Eq. (4.3.12) the quantities a'_{ii} and α_i are values of the original a_{ij} and b_i obtained through the transformation process. We see that for nonsingular problems the advantage of Gauss–Jordan elimination over Gauss elimination is that the backward substitution step is eliminated (or rendered trivial since $x_i = \alpha_i/a'_{ii}$). A Mathematica routine is provided in the Appendix for the Gauss–Jordan algorithm.

If we use a complete pivoting strategy for numerical stability, then column interchange, as well as row interchange, could occur. Suppose a column exchange occurs at the second step. Then the matrix in Eq. (4.3.2) would be

$$\begin{bmatrix} a_{11} & a'_{12} & a'_{13} \\ 0 & b'_{22} & b'_{23} \\ 0 & b'_{32} & b'_{33} \end{bmatrix}, \qquad (4.3.13)$$

with $a'_{12} = a_{13}$, $a'_{13} = a_{12}$, $b'_{22} = b_{23}$, etc. The corresponding system of equations can be expressed as

$$\begin{aligned}
a_{11}y_1 + a'_{12}y_2 + a'_{13}y_3 &= \alpha_1 \\
b'_{22}y_2 + b'_{23}y_3 &= \alpha_2 \\
b'_{32}y_2 + b'_{33}y_3 &= \alpha_3,
\end{aligned} \qquad (4.3.14)$$

where $y_1 = x_1$, $y_2 = x_3$, and $y_3 = x_2$. Thus, Gauss–Jordan elimination with row and column interchange transforms Eq. (4.3.7) into the system

$$\mathbf{A}_{\text{tr}}\mathbf{y} = \boldsymbol{\alpha} \qquad (4.3.15)$$

or

$$
\begin{aligned}
a_{11}y_1 &= \alpha_1 \\
a'_{22}y_2 &= \alpha_2 \\
a'_{33}y_3 &= \alpha_3.
\end{aligned}
\qquad (4.3.16)
$$

The components y_1, y_2, y_3 of $\mathbf{y}$ are simply a reordering of the components x_1, x_2, x_3 of $\mathbf{x}$.

The correspondence between the components of $\mathbf{x}$ and $\mathbf{y}$ is easily determined by keeping track of the column interchanges. To do this symbolically, let us define a matrix $\mathbf{I}_{ij}$, which, when multiplied by $\mathbf{x}$, will interchange components i and j. For a four-dimensional problem, an example is

$$
\mathbf{I}_{23} = \begin{bmatrix} 1 & 0 & 0 & 0 \\ 0 & 0 & 1 & 0 \\ 0 & 1 & 0 & 0 \\ 0 & 0 & 0 & 1 \end{bmatrix}
\qquad (4.3.17)
$$

from which it is easy to show that

$$
\mathbf{I}_{23}\mathbf{x} = \begin{bmatrix} x_1 \\ x_3 \\ x_2 \\ x_4 \end{bmatrix}.
\qquad (4.3.18)
$$

For this four-dimensional case, it is easy to see that the matrix $\mathbf{I}_{ij}$, which interchanges components x_i and x_j through the product $\mathbf{I}_{ij}\mathbf{x}$, is obtained by interchanging columns i and j of the unit matrix $\mathbf{I}$. This turns out to be generally true. That is, for an n-dimensional case, the matrix $\mathbf{I}_{ij}$, obtained by interchanging columns i and j of the $n \times n$ unit matrix $\mathbf{I}$, interchanges rows i and j of $\mathbf{x}$ through the multiplication $\mathbf{I}_{ij}\mathbf{x}$.

To see how the $\mathbf{I}_{ij}$ might be useful, imagine a Gauss–Jordan elimination process in which columns 1 and 3 are interchanged followed by columns 2 and 5 and then columns 3 and 6. In the first interchange $\mathbf{x}$ is transformed into $\mathbf{I}_{13}\mathbf{x}$, in the second into $\mathbf{I}_{25}\mathbf{I}_{13}\mathbf{x}$, and in the third into $\mathbf{I}_{36}\mathbf{I}_{25}\mathbf{I}_{13}\mathbf{x}$, so that, in the equation $\mathbf{A}_{\mathrm{tr}}\mathbf{y} = \boldsymbol{\alpha}$,

$$
\mathbf{y} = \mathbf{Q}\mathbf{x}. \qquad (4.3.19)
$$

We can see that the determinant of $\mathbf{I}_{ij}$ is equal to -1, since $|\mathbf{I}_{ij}|$ differs from $|\mathbf{I}|$ only by the interchange of two columns. This implies that the matrix $\mathbf{Q}$ is also nonsingular. For the $\mathbf{Q}$ defined in Eq. (4.3.19), $|\mathbf{Q}| = |\mathbf{I}_{36}||\mathbf{I}_{25}||\mathbf{I}_{13}| = -1$, and since $\mathbf{Q}$ is nonsingular, the solution $\mathbf{x}$ to $\mathbf{A}\mathbf{x} = \mathbf{b}$ is uniquely determined as $\mathbf{x} = \mathbf{Q}^{-1}\mathbf{y}$ from the solution to $\mathbf{A}_{\mathrm{tr}}\mathbf{y} = \boldsymbol{\alpha}$.

4.4. GENERAL SOLVABILITY THEOREM FOR Ax = b

Although up to this point we have concentrated mostly on solving $\mathbf{Ax} = \mathbf{b}$ for nonsingular matrices $\mathbf{A}$, the more general problem is of substantial practical and theoretical interest. Consider the system of equations

$$\begin{aligned} a_{11}x_1 + a_{12}x_2 + \cdots + a_{1n}x_n &= b_1 \\ a_{21}x_1 + a_{22}x_2 + \cdots + a_{2n}x_n &= b_2 \\ &\vdots \\ a_{m1}x_1 + a_{m2}x_2 + \cdots + a_{mn}x_n &= b_m \end{aligned} \qquad (4.4.1)$$

or

$$\mathbf{Ax} = \mathbf{b}, \qquad (4.4.2)$$

where $\mathbf{A}$ is an $m \times n$ matrix with elements a_{ij}, $\mathbf{x}$ is an n-dimensional vector, and $\mathbf{b}$ is an m-dimensional vector. We will consider the general situation in which m and n need not be the same and the rank r of $\mathbf{A}$ will be less than or equal to $\min(m, n)$. Square nonsingular and singular matrices constitute special cases of the general theory.

A non-square problem may or may not have a solution and when it does have a solution it is may not be unique. For instance, the system

$$\begin{aligned} x_1 + x_2 + x_3 &= 1 \\ 2x_1 + x_2 + 2x_3 &= 2 \end{aligned} \qquad (4.4.3)$$

has the solution

$$x_1 = -1 - x_3, \qquad x_2 = 0 \qquad (4.4.4)$$

for arbitrary x_3, and thus a solution exists but is not unique. On the other hand, the system

$$\begin{aligned} x_1 + 2x_2 + x_3 &= 1 \\ 2x_1 + 4x_2 + 2x_3 &= 1 \end{aligned} \qquad (4.4.5)$$

has no solution since multiplication of the first equation by 2 and subtraction from the second equation yields the contradiction

$$0 = -1. \qquad (4.4.6)$$

Thus, the equations in Eq. (4.4.3) are compatible, whereas those in Eq. (4.4.5) are not. The theory of solvability of $\mathbf{Ax} = \mathbf{b}$ must address this compatability issue for the general case.

As an aid in examining the general case, we will first discuss Gauss–Jordan elimination for the 3×4 matrix

$$\mathbf{A} = \begin{bmatrix} a_{11} & a_{12} & a_{13} & a_{14} \\ a_{21} & a_{22} & a_{23} & a_{24} \\ a_{31} & a_{32} & a_{33} & a_{34} \end{bmatrix}. \qquad (4.4.7)$$

At this point, we know nothing about $\mathbf{A}$ except the values of a_{ij}. For the elimination process, we note that the matrix $\mathbf{I}_{ij}$ defined in the previous section has the useful properties that pre-multiplication of $\mathbf{A}$ by $\mathbf{I}_{ij}$ (i.e., $\mathbf{I}_{ij}\mathbf{A}$) interchanges rows i and j and post-multiplication of $\mathbf{A}$ by $\mathbf{I}_{ij}$ (i.e., $\mathbf{A}\mathbf{I}_{ij}$) interchanges columns i and j. For example, if

$$\mathbf{I}_{12} = \begin{bmatrix} 0 & 1 & 0 \\ 1 & 0 & 0 \\ 0 & 0 & 1 \end{bmatrix}, \qquad (4.4.8)$$

then it is easy to see that

$$\begin{aligned}\mathbf{I}_{12}\mathbf{A} &= \begin{bmatrix} 0 & 1 & 0 \\ 1 & 0 & 0 \\ 0 & 0 & 1 \end{bmatrix} \begin{bmatrix} a_{11} & a_{12} & a_{13} & a_{14} \\ a_{21} & a_{22} & a_{23} & a_{24} \\ a_{31} & a_{32} & a_{33} & a_{34} \end{bmatrix} \\ &= \begin{bmatrix} a_{21} & a_{22} & a_{23} & a_{24} \\ a_{11} & a_{12} & a_{13} & a_{14} \\ a_{31} & a_{32} & a_{33} & a_{34} \end{bmatrix}.\end{aligned} \qquad (4.4.9)$$

And if

$$\mathbf{I}_{12} = \begin{bmatrix} 0 & 1 & 0 & 0 \\ 1 & 0 & 0 & 0 \\ 0 & 0 & 1 & 0 \\ 0 & 0 & 0 & 1 \end{bmatrix}, \qquad (4.4.10)$$

then

$$\begin{aligned}\mathbf{A}\mathbf{I}_{12} &= \begin{bmatrix} a_{11} & a_{12} & a_{13} & a_{14} \\ a_{21} & a_{22} & a_{23} & a_{24} \\ a_{31} & a_{32} & a_{33} & a_{34} \end{bmatrix} \begin{bmatrix} 0 & 1 & 0 & 0 \\ 1 & 0 & 0 & 0 \\ 0 & 0 & 1 & 0 \\ 0 & 0 & 0 & 1 \end{bmatrix} \\ &= \begin{bmatrix} a_{12} & a_{11} & a_{13} & a_{14} \\ a_{22} & a_{21} & a_{23} & a_{24} \\ a_{32} & a_{31} & a_{33} & a_{34} \end{bmatrix}.\end{aligned} \qquad (4.4.11)$$

Note that, for an $m \times n$ matrix $\mathbf{A}$, the matrix $\mathbf{I}_{ij}$ has to be an $m \times m$ matrix for pre-multiplication and an $n \times n$ matrix for post-multiplication.

GENERAL SOLVABILITY THEOREM FOR Ax = b

For Gauss–Jordan elimination, one other matrix is needed. This is the square matrix $\mathbf{J}_{ij}(k)$, $i \neq j$, which has elements $j_{ll} = 1$ for $l = 1, \ldots, n$, $j_{ij} = k$, and $j_{lm} = 0$ if $l \neq m$ and $lm \neq ij$. To generate $\mathbf{J}_{ij}(k)$, we can start with the unit matrix $\mathbf{I}$ and add k as the $\{ij\}$th element. For example, the 3×3 version of $\mathbf{J}_{23}(k)$ is

$$\mathbf{J}_{23}(k) = \begin{bmatrix} 1 & 0 & 0 \\ 0 & 1 & k \\ 0 & 0 & 1 \end{bmatrix}. \qquad (4.4.12)$$

Notice that the product of $\mathbf{J}_{23}(k)$ with the matrix defined by Eq. (4.4.7) is

$$\mathbf{J}_{23}(k)\mathbf{A} = \begin{bmatrix} a_{11} & a_{12} & a_{13} & a_{14} \\ a_{21} + ka_{31} & a_{22} + ka_{32} & a_{23} + ka_{33} & a_{24} + ka_{34} \\ a_{31} & a_{32} & a_{33} & a_{34} \end{bmatrix}. \qquad (4.4.13)$$

Thus, pre-multiplying $\mathbf{A}$ by $\mathbf{J}_{ij}(k)$ produces a matrix in which the product of k and the elements of the jth row are added to the elements of the ith row. According to the elementary properties of determinants, the determinants of $\mathbf{A}$ and $\mathbf{J}_{ij}(k)\mathbf{A}$ are the same.

Since $\mathbf{J}_{ij}(k)$ can be generated by multiplying the ith column of $\mathbf{I}$ by k and adding this to the jth column, it follows that the determinant of $\mathbf{J}_{ij}(k)$ is 1. We also note that the inverse of $\mathbf{J}_{ij}(k)$ is $\mathbf{J}_{ij}(-k)$ because

$$\mathbf{J}_{ij}(-k)\mathbf{J}_{ij}(k) = \mathbf{I}, \qquad (4.4.14)$$

by inspection. Since $\mathbf{I}_{ij}\mathbf{I}_{ij} = \mathbf{I}$, the matrix $\mathbf{I}_{ij}$ is its own inverse. The matrices $\mathbf{I}_{ij}$ and $\mathbf{J}_{ij}(k)$ are the tools we need to establish the solvability of any linear system represented by $\mathbf{A}\mathbf{x} = \mathbf{b}$.

In Gauss–Jordan elimination, the matrix $\mathbf{A}$ is transformed into $\mathbf{A}_{\text{tr}}$ by multiplication of the square nonsingular matrices $\mathbf{I}_{ij}$ and $\mathbf{J}_{ij}$. It is important to remember that this will not change the rank of $\mathbf{A}$. Consider the matrix defined by Eq. (4.4.7). If the rank r of $\mathbf{A}$ is not 0, a nonzero element can be put at the $\{11\}$ position by the interchange of rows and/or columns. As was shown above, this can be accomplished by pre- and/or post-multiplying $\mathbf{A}$ by the appropriate $\mathbf{I}_{ij}$ matrices. For simplicity of discussion, suppose that $r > 0$ (otherwise, $\mathbf{A} = [0]$) and that $a_{11} \neq 0$. Then the first step of Gauss elimination is accomplished by the matrix operation

$$\mathbf{J}_{21}\left(-\frac{a_{21}}{a_{11}}\right)\mathbf{A} = \begin{bmatrix} a_{11} & a_{12} & a_{13} & a_{14} \\ 0 & b_{22} & b_{23} & b_{24} \\ a_{31} & a_{32} & a_{33} & a_{34} \end{bmatrix}, \qquad (4.4.15)$$

where

$$\mathbf{J}_{21} = \begin{bmatrix} 1 & 0 & 0 \\ -\dfrac{a_{21}}{a_{11}} & 1 & 0 \\ 0 & 0 & 1 \end{bmatrix} \qquad (4.4.16)$$

and $b_{22} = a_{22} - a_{12}a_{21}/a_{11}$, $b_{23} = a_{23} - a_{13}a_{21}/a_{11}$, and $b_{24} = a_{24} - a_{14}a_{21}/a_{11}$. In the next step, Eq. (4.4.15) is multiplied by $\mathbf{J}_{31}(-a_{31}/a_{11})$ to obtain

$$\mathbf{J}_{31}\left(-\frac{a_{31}}{a_{11}}\right)\mathbf{J}_{21}\left(-\frac{a_{21}}{a_{11}}\right)\mathbf{A} = \begin{bmatrix} a_{11} & a_{12} & a_{13} & a_{14} \\ 0 & b_{22} & b_{23} & b_{24} \\ 0 & b_{32} & b_{33} & b_{34} \end{bmatrix}. \quad (4.4.17)$$

If the rank $r_A = 1$, then all the elements b_{ij} in Eq. (4.4.17) are 0 and the transformed matrix is simply

$$\mathbf{A}_{tr} = \begin{bmatrix} a_{11} & a_{12} & a_{13} & a_{14} \\ 0 & 0 & 0 & 0 \\ 0 & 0 & 0 & 0 \end{bmatrix}. \quad (4.4.18)$$

If $r_A > 1$, then at least one of the elements b_{ij} has to be nonzero. By pre- and/or post-multiplication of Eq. (4.4.17) by the appropriate $\mathbf{I}_{ij}$ matrices, row and/or column interchanges can be carried out to place a nonzero element at the {22} position. For illustration, assume that $b_{22} = b_{32} = b_{23} = 0$ and $b_{33} \neq 0$. Then

$$\mathbf{I}_{23}\mathbf{J}_{31}\left(-\frac{a_{31}}{a_{11}}\right)\mathbf{J}_{21}\left(-\frac{a_{21}}{a_{11}}\right)\mathbf{A}\mathbf{I}_{23} = \begin{bmatrix} a_{11} & a_{13} & a_{12} & a_{14} \\ 0 & b_{33} & 0 & b_{34} \\ 0 & 0 & 0 & b_{24} \end{bmatrix}. \quad (4.4.19)$$

Pre-multiplication of the above matrix by $\mathbf{J}_{12}(-a_{13}/b_{33})$ yields

$$\begin{bmatrix} a_{11} & 0 & a_{12} & c_{14} \\ 0 & b_{33} & 0 & b_{34} \\ 0 & 0 & 0 & b_{24} \end{bmatrix}. \quad (4.4.20)$$

If the rank r_A of $\mathbf{A}$ is 2, then $b_{24} = 0$; i.e., the transformed matrix is

$$\mathbf{A}_{tr} = \begin{bmatrix} a_{11} & 0 & a_{12} & c_{14} \\ 0 & b_{33} & 0 & b_{34} \\ 0 & 0 & 0 & 0 \end{bmatrix}. \quad (4.4.21)$$

If, however, $r_A = 3$, then $b_{24} \neq 0$, and post-multiplication of Eq. (4.4.20) by $\mathbf{I}_{34}$ yields

$$\begin{bmatrix} a_{11} & 0 & c_{14} & a_{12} \\ 0 & b_{33} & b_{34} & 0 \\ 0 & 0 & b_{24} & 0 \end{bmatrix}. \quad (4.4.22)$$

Pre-multiplication by $\mathbf{J}_{13}(-c_{14}/b_{24})$ followed by $\mathbf{J}_{23}(-b_{34}/b_{24})$ gives the transformed matrix

$$\mathbf{A}_{tr} = \begin{bmatrix} a_{11} & 0 & 0 & a_{12} \\ 0 & b_{33} & 0 & 0 \\ 0 & 0 & b_{34} & 0 \end{bmatrix}. \quad (4.4.23)$$

Equations (4.4.18), (4.4.21), and (4.4.23) are the Gauss–Jordan transformations of $\mathbf{A}$ for the cases $r_A = 1, 2,$ and 3, respectively (with the special situation conditions $b_{22} = b_{32} = b_{23} = 0$ introduced for illustration purposes). The relationship between $\mathbf{A}_{tr}$, say in the case of Eq. (4.4.23), is

$$\mathbf{A}_{tr} = \mathbf{PAQ}, \tag{4.4.24}$$

where

$$\mathbf{P} = \mathbf{J}_{23}\left(-\frac{b_{34}}{b_{24}}\right)\mathbf{J}_{13}\left(-\frac{c_{14}}{b_{24}}\right)\mathbf{J}_{12}\left(-\frac{a_{13}}{b_{33}}\right)\mathbf{I}_{23}\mathbf{J}_{31}\left(-\frac{a_{31}}{a_{11}}\right)\mathbf{J}_{21}\left(-\frac{a_{21}}{a_{11}}\right)$$

$$= \begin{bmatrix} 1 + \dfrac{a_{13}a_{31}}{a_{11}b_{33}} + \dfrac{a_{21}c_{14}}{a_{11}b_{34}} & -\dfrac{c_{14}}{b_{34}} & -\dfrac{a_{13}}{b_{33}} \\ \dfrac{a_{21}b_{24}}{a_{11}b_{34}} - \dfrac{a_{31}}{a_{11}} & -\dfrac{b_{24}}{b_{34}} & 1 \\ -\dfrac{a_{21}}{a_{11}} & 1 & 0 \end{bmatrix} \tag{4.4.25}$$

and

$$\mathbf{Q} = \mathbf{I}_{23}\mathbf{I}_{34} = \begin{bmatrix} 1 & 0 & 0 & 0 \\ 0 & 0 & 0 & 1 \\ 0 & 1 & 0 & 0 \\ 0 & 0 & 1 & 0 \end{bmatrix}. \tag{4.4.26}$$

$\mathbf{P}$ is a product of nonsingular 3×3 matrices, and so it is a nonsingular 3×3 matrix itself. Likewise, $\mathbf{Q}$ is a nonsingular 4×4 matrix. For the particular cases resulting in Eqs. (4.4.25) and (4.4.26), we find $|\mathbf{P}| = -1, |\mathbf{Q}| = -1$.

From this 3×4 example, the matrix transformation in the Gauss–Jordan elimination method becomes clear for the general matrix

$$\mathbf{A} = \begin{bmatrix} a_{11} & a_{12} & \cdots & a_{1n} \\ a_{21} & a_{22} & \cdots & a_{2n} \\ \vdots & \vdots & & \vdots \\ a_{m1} & a_{m2} & \cdots & a_{mn} \end{bmatrix}. \tag{4.4.27}$$

If the rank of $\mathbf{A}$ is r, then, through a sequence of pre- and post-multiplications by the $\mathbf{I}_{ij}$ and $\mathbf{J}_{ij}$ matrices, the matrix can be transformed into

$$\mathbf{A}_{tr} = \mathbf{PAQ} = \begin{bmatrix} \gamma_{11} & 0 & \cdots & 0 & \gamma_{1,r+1} & \cdots & \gamma_{1n} \\ 0 & \gamma_{22} & \cdots & 0 & \gamma_{2,r+1} & \cdots & \gamma_{2n} \\ \vdots & \vdots & & \vdots & \vdots & & \vdots \\ 0 & 0 & \cdots & \gamma_{rr} & \gamma_{r,r+1} & \cdots & \gamma_{rn} \\ 0 & 0 & & 0 & 0 & & 0 \\ \vdots & \vdots & & \vdots & \vdots & & \vdots \\ 0 & 0 & & 0 & 0 & & 0 \end{bmatrix} \tag{4.4.28}$$

in which **P** is a product of the $\mathbf{I}_{ij}$ and $\mathbf{J}_{ij}$ matrices and **Q** is a product of the $\mathbf{I}_{ij}$ matrices, and where $\gamma_{ij} = 0$ for $i \neq j$ and $i, j = 1, \ldots, r$ and $\gamma_{ij} = 0$ for $i > r$. The values of γ_{ij} for $i \leq r$, $j > r$ may or may not be 0. Thus, the matrix $\mathbf{A}_{tr}$ can be partitioned as

$$\mathbf{A}_{tr} = \begin{bmatrix} \mathbf{A}_1 & \mathbf{A}_2 \\ \mathbf{0}_3 & \mathbf{0}_4 \end{bmatrix}, \quad (4.4.29)$$

where $\mathbf{A}_1$ is an $r \times r$ matrix having nonzero elements γ_{ii} only on the main diagonal; $\mathbf{A}_2$ is an $r \times (n-r)$ matrix with the elements γ_{ij}, $i = 1, \ldots, r$ and $j = r+1, \ldots, n$; $\mathbf{0}_3$ is an $(m-r) \times r$ matrix with only zero elements; and $\mathbf{0}_4$ is an $(m-r) \times (n-r)$ matrix with only zero elements.

Since the determinants $|\mathbf{P}|$ and $|\mathbf{Q}|$ must be equal to ± 1, the rank of $\mathbf{A}_{tr}$ is the same as the rank of **A**. Thus, the Gauss–Jordan elimination must lead to exactly r nonzero diagonal elements γ_{ii}. If there were fewer than r, then the rank of **A** would be greater than r. Either case would contradict the hypothesis that the rank of **A** is r.

The objective of this section is to determine the solvability of the system of equations

$$\begin{aligned} a_{11}x_1 + a_{12}x_2 + \cdots + a_{1n}x_n &= b_1 \\ a_{21}x_1 + a_{22}x_2 + \cdots + a_{2n}x_n &= b_2 \\ \vdots \quad \vdots \quad \quad \vdots \\ a_{m1}x_1 + a_{m2}x_2 + \cdots + a_{mn}x_n &= b_m \end{aligned} \quad (4.4.30)$$

represented by

$$\mathbf{Ax} = \mathbf{b}. \quad (4.4.31)$$

If **P** and **Q** are the matrices in Eq. (4.4.28), it follows that Eq. (4.4.31) can be transformed into

$$\mathbf{PAQQ}^{-1}\mathbf{x} = \mathbf{Pb} \quad (4.4.32)$$

or

$$\mathbf{A}_{tr}\mathbf{y} = \boldsymbol{\alpha}, \quad (4.4.33)$$

where

$$\mathbf{y} = \mathbf{Q}^{-1}\mathbf{x} \quad \text{and} \quad \boldsymbol{\alpha} = \mathbf{Pb}. \quad (4.4.34)$$

In tableau form Eq. (4.4.33) reads

$$\gamma_{11}y_1 \qquad\qquad +\gamma_{1,r+1}y_{r+1}+\cdots+\gamma_{1n}y_n=\alpha_1$$
$$\gamma_{22}y_2 \qquad +\gamma_{2,r+1}y_{r+1}+\cdots+\gamma_{2n}y_n=\alpha_2$$
$$\vdots \qquad\qquad \vdots \qquad \vdots$$
$$\gamma_{rr}y_r+\gamma_{r,r+1}y_{r+1}+\cdots+\gamma_{rn}y_n=\alpha_r \qquad (4.4.35)$$
$$0=\alpha_{r+1}$$
$$\vdots$$
$$0=\alpha_m.$$

If the vector **b** is such that not all α_i, where $i > r$, are 0, then the equations in Eq. (4.4.35) are inconsistent and so no solution exists. If $\alpha_i = 0$ for all $i > r$, then the solution to Eq. (4.4.35) is

$$y_1 = -\beta_{1,r+1}y_{r+1} - \cdots - \beta_{1n}y_n + \alpha_1/\gamma_{11}$$
$$y_2 = -\beta_{2,r+1}y_{r+1} - \cdots - \beta_{2n}y_n + \alpha_2/\gamma_{22}$$
$$\vdots \qquad\qquad (4.4.36)$$
$$y_r = -\beta_{r,r+1}y_{r+1} - \cdots - \beta_{rn}y_n + \alpha_r/\gamma_{rr}$$

for arbitrary values of $y_{r+1}, \ldots, y_n$ and where

$$\beta_{ij} \equiv \frac{\gamma_{ij}}{\gamma_{ii}}. \qquad (4.4.37)$$

Of course, if Eq. (4.4.35) or Eq. (4.4.33) has a solution **y**, then the vector $\mathbf{x} = \mathbf{Qy}$ is a solution to $\mathbf{Ax} = \mathbf{b}$ (i.e., to Eq. (4.4.31)).

Let us explore further the conditions of solvability of Eq. (4.4.33). The augmented matrix corresponding to this equation is

$$[\mathbf{A}_{\text{tr}}, \boldsymbol{\alpha}] = \begin{bmatrix} \gamma_{11} & 0 & 0 & \cdots & 0 & \gamma_{1,r+1} & \cdots & \gamma_{1n} & \alpha_1 \\ 0 & \gamma_{22} & 0 & \cdots & 0 & \gamma_{2,r+1} & \cdots & \gamma_{2n} & \alpha_2 \\ \vdots & & & & \vdots & \vdots & & \vdots & \vdots \\ 0 & & & & \gamma_{rr} & \gamma_{r,r+1} & \cdots & \gamma_{rn} & \alpha_r \\ 0 & & & & 0 & 0 & \cdots & 0 & \alpha_{r+1} \\ \vdots & & & & \vdots & \vdots & & \vdots & \vdots \\ 0 & & & & 0 & 0 & \cdots & 0 & \alpha_m \end{bmatrix}. \qquad (4.4.38)$$

As indicated above, Eq. (4.4.33) has a solution *if and only if* $\alpha_i = 0$, $i > r$. This is equivalent to the condition that the rank of the augmented matrix $[\mathbf{A}_{\text{tr}}, \boldsymbol{\alpha}]$ is the same as the rank of the matrix $\mathbf{A}_{tr}$—which is the same as the rank r of $\mathbf{A}$. If $\mathbf{A}_{tr}\mathbf{y} = \boldsymbol{\alpha}$ has a solution, so does $\mathbf{A}\mathbf{x} = \mathbf{b}$.

When the rank of $[\mathbf{A}_{tr}, \boldsymbol{\alpha}]$ is r, what is the rank of the augmented matrix $[\mathbf{A}, \mathbf{b}]$? From the corollary of the previous section, it follows that the rank of $\mathbf{C}$, where

$$\mathbf{C} = \mathbf{P}[\mathbf{A}, \mathbf{b}] = [\mathbf{PA}, \mathbf{Pb}] = [\mathbf{PA}, \boldsymbol{\alpha}], \tag{4.4.39}$$

is the same as the rank of $[\mathbf{A}, \mathbf{b}]$ since $\mathbf{P}$ is a square nonsingular matrix. Note also that the augmented matrix

$$[\mathbf{A}_{tr}, \boldsymbol{\alpha}] = [\mathbf{PAQ}, \boldsymbol{\alpha}] \tag{4.4.40}$$

has the same rank as $\mathbf{C}$ since $\mathbf{PAQ}$ and $\mathbf{PA}$ only differ in the interchange of columns—an elementary operation that does not change the rank of a matrix. Thus, the rank of $[\mathbf{A}, \mathbf{b}]$ is identical to the rank of $[\mathbf{A}_{tr}, \boldsymbol{\alpha}]$.

Collectively, what we have shown above implies the solvability theorem:

THEOREM. *The equation $\mathbf{A}\mathbf{x} = \mathbf{b}$ has a solution if and only if the rank of the augmented matrix $[\mathbf{A}, \mathbf{b}]$ is the same as the rank of $\mathbf{A}$.*

Since the solution is not unique when $\mathbf{A}$ is a singular square matrix or a non-square matrix, we need to explore further the nature of the solutions in these cases. First, we assume that the solvability condition is obeyed ($\alpha_i = 0$, $i > r$) and consider again Eq. (4.4.36). A particular solution to this set of equations is

$$\begin{aligned} y_i &= \frac{\alpha_i}{\gamma_{ii}}, \quad i = 1, \ldots, r, \\ y_i &= 0, \quad i = r+1, \ldots, n, \end{aligned} \tag{4.4.41}$$

or

$$\mathbf{y}_p = \begin{bmatrix} \dfrac{\alpha_1}{\gamma_{11}} \\ \vdots \\ \dfrac{\alpha_r}{\gamma_{rr}} \\ 0 \\ \vdots \\ 0 \end{bmatrix}. \tag{4.4.42}$$

However, if $\mathbf{y}_p$ is a solution to Eq. (4.4.36), then so is $\mathbf{y}_p + \mathbf{y}_h$, where $\mathbf{y}_h$ is any solution to the homogeneous equations

$$\begin{aligned} y_1 &= -\beta_{1,r+1} y_{r+1} - \cdots - \beta_{1n} y_n \\ &\vdots \\ y_r &= -\beta_{r,r+1} y_{r+1} - \cdots - \beta_{rn} y_n. \end{aligned} \tag{4.4.43}$$

One simple solution to this set of equations can be obtained by letting $y_{r+1} = 1$ and $y_i = 0$, $i > r + 1$. The solution is

$$\mathbf{y}_h^{(1)} = \begin{bmatrix} -\beta_{1,r+1} \\ \vdots \\ -\beta_{r,r+1} \\ 1 \\ 0 \\ \vdots \\ 0 \end{bmatrix}. \qquad (4.4.44)$$

Similarly, choosing $y_{r+1} = 0$, $y_{r+2} = 1$, and $y_i = 0$, $i > r + 2$, gives the solution

$$\mathbf{y}_h^{(2)} = \begin{bmatrix} -\beta_{1,r+2} \\ \vdots \\ -\beta_{r,r+2} \\ 0 \\ 1 \\ 0 \\ \vdots \\ 0 \end{bmatrix}. \qquad (4.4.45)$$

Using this method, the set of solutions $\mathbf{y}_h^{(1)}, \ldots, \mathbf{y}_h^{(n-r)}$ can be generated in which

$$\mathbf{y}_h^{(j)} = \begin{bmatrix} -\beta_{1,r+j} \\ \vdots \\ -\beta_{r,r+j} \\ 0 \\ \vdots \\ 1 \\ \vdots \\ 0 \end{bmatrix} \quad (r+j)\text{th row} \qquad (4.4.46)$$

for $j = 1, \ldots, n - r$. There are two important aspects of these homogeneous solutions. First, they are linearly independent since the $(r + j)$th element of $\mathbf{y}_h^{(j)}$ is 1, whereas the $(r + j)$th element of all other $\mathbf{y}_h^{(k)}$, $k \neq j$, is 0, and so any linear combination of $\mathbf{y}_h^{(k)}$, $k \neq j$, will still have a zero $(r + j)$th element. The second aspect is that *any* solution to Eq. (4.4.43) can be expressed as a linear combination of the vectors $\mathbf{y}_h^{(j)}$. To see this, suppose that $y_{r+1}, \ldots, y_n$ are given arbitrary values in Eq. (4.4.43). The solution can be written as

$$\mathbf{y}_h = \begin{bmatrix} -\beta_{1,r+1}y_{r+1} & -\beta_{1,r+2}y_{r+2} & \cdots & -\beta_{1n}y_n \\ \vdots & & & \\ -\beta_{r,r+1}y_{r+1} & -\beta_{r,r+2}y_{r+2} & \cdots & -\beta_{rn}y_n \\ y_{r+1} & + \quad 0 & \cdots & + \quad 0 \\ 0 & + \quad y_{r+2} & \cdots & + \quad 0 \\ \vdots & \vdots & & \vdots \\ 0 & + \quad 0 & \cdots & + \quad y_n \end{bmatrix}, \tag{4.4.47}$$

which, in turn, can be expressed as

$$\mathbf{y}_h = \sum_{j=1}^{n-r} y_{r+j} \mathbf{y}_h^{(j)}. \tag{4.4.48}$$

The essence of this result is that the equation $\mathbf{A}_{tr}\mathbf{y} = \boldsymbol{\alpha}$ has exactly $n-r$ linearly independent homogeneous solutions. Any other homogeneous solution will be a linear combination of these linearly independent solutions.

In general, the solution to $\mathbf{A}_{tr}\mathbf{y} = \boldsymbol{\alpha}$ can be expressed as

$$\mathbf{y} = \mathbf{y}_p + \sum_{j=1}^{n-r} c_j \mathbf{y}_h^{(j)}, \tag{4.4.49}$$

where $\mathbf{y}_p$ is a particular solution obeying $\mathbf{A}_{tr}\mathbf{y}_p = \boldsymbol{\alpha}$, $\mathbf{y}_h^{(j)}$ are the linearly independent solutions obeying $\mathbf{A}_{tr}\mathbf{y}_h^{(j)} = \mathbf{0}$, and c_j are arbitrary complex numbers.

Because $\mathbf{y}_h^{(j)}$ is a solution to the homogeneous equation $\mathbf{A}_{tr}\mathbf{y}_h = \mathbf{0}$, the vectors $\mathbf{x}_h^{(j)} = \mathbf{Q}\mathbf{y}_h^{(j)}$ are solutions to the homogeneous equation $\mathbf{A}\mathbf{x}_h^{(j)} = \mathbf{0}$. The set $\{\mathbf{x}_h^{(j)}\}$, $j = 1, \ldots, n-r$, is also linearly independent. To prove this, assume that the set is linearly dependent. Then there exist numbers $\alpha_1, \ldots, \alpha_{n-r}$, not all of which are 0, such that

$$\sum_{j=1}^{n-r} \alpha_j \mathbf{x}_h^{(j)} = \mathbf{0}. \tag{4.4.50}$$

But multiplying Eq. (4.4.50) by $\mathbf{Q}^{-1}$ and recalling that $\mathbf{y}_h^{(j)} = \mathbf{Q}^{-1}\mathbf{x}_h^{(j)}$ yields

$$\sum_{j=1}^{n-r} \alpha_j \mathbf{y}_h^{(j)} = \mathbf{0}. \tag{4.4.51}$$

Since the vectors $\mathbf{y}_h^{(j)}$ are linearly independent, the only set $\{\alpha_j\}$ obeying (4.4.51) is $\alpha_1 = \cdots = \alpha_{n-r} = 0$, which is a contradiction to the hypothesis that the $\mathbf{x}_h^{(j)}$'s are linearly dependent. Thus, the vectors $\mathbf{x}_h^{(j)}$, $j = 1, \ldots, n-r$, must be linearly independent.

We summarize the findings of this section with the complete form of the *solvability theorem:*

SOLVABILITY THEOREM. *The equation*

$$\mathbf{A}\mathbf{x} = \mathbf{b} \qquad (4.4.52)$$

has a solution if and only if the rank of the augmented matrix $[\mathbf{A}, \mathbf{b}]$ *is equal to the rank r of the matrix* $\mathbf{A}$. *The general solution has the form*

$$\mathbf{x} = \mathbf{x}_p + \sum_{j=1}^{n-r} c_j \mathbf{x}_h^{(j)}, \qquad (4.4.53)$$

where $\mathbf{x}_p$ *is a particular solution satisfying the inhomogeneous equation* $\mathbf{A}\mathbf{x}_p = \mathbf{b}$, *the set* $\mathbf{x}_h^{(j)}$ *consists of* $n - r$ *linearly independent vectors satisfying the homogeneous equation* $\mathbf{A}\mathbf{x}_h = \mathbf{0}$, *and the coefficients* c_j *are arbitrary complex numbers.*

For those who find proofs tedious, this theorem is the "take-home lesson" of this section. Its beauty is its completeness and generality. $\mathbf{A}$ is an $m \times n$ matrix, m need not equal n, whose rank r can be equal to or less than the smallest of the number of rows m or the number of columns n. If the rank of $[\mathbf{A}, \mathbf{b}]$ equals that of $\mathbf{A}$, we know exactly how many solutions to look for; if not, we know not to look for any; or if not, we know to find where the problem lies if we thought we had posed a solvable physical problem.

EXAMPLE 4.4.1. Consider the electric circuit shown in Fig. 4.4.1. The V's denote voltages at the conductor junctions indicated by solid circles and the conductance c of each of the conductors (straight lines) is given. The current i across a conductor is given by Ohm's law, $i = c\Delta V$, where ΔV is the voltage drop between conductor junctions. A current i enters the circuit where the voltage is V_I and leaves where it is V_O. The values of V_I and V_O are set by external conditions.

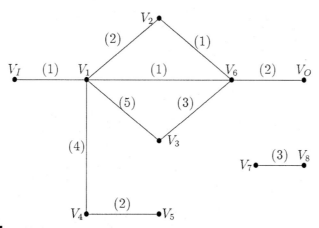

FIGURE 4.4.1

CHAPTER 4 GENERAL THEORY OF SOLVABILITY OF LINEAR ALGEBRAIC EQUATIONS

From the conservation of current at each junction and from the values of the conductances, we can determine the voltages $V_1, \ldots, V_8$.

The conservation conditions at each junction are

$$2(V_2 - V_1) + (V_6 - V_1) + 5(V_3 - V_1) + 4(V_4 - V_1) + (V_1 - V_1) = 0$$
$$2(V_1 - V_2) + (V_6 - V_2) = 0$$
$$3(V_6 - V_3) + 5(V_1 - V_3) = 0$$
$$4(V_1 - V_4) + 2(V_5 - V_4) = 0 \quad (4.4.54)$$
$$2(V_4 - V_5) = 0$$
$$(V_1 - V_6) + (V_2 - V_6) + 3(V_3 - V_6) + 2(V_O - V_6) = 0$$
$$3(V_8 - V_7) = 0$$

or

$$\begin{aligned}
-13V_1 + 2V_2 + 5V_3 + 4V_4 \quad\quad + V_6 \quad\quad\quad &= -V_1 \\
2V_1 - 3V_2 \quad\quad\quad\quad\quad + V_6 \quad\quad\quad &= 0 \\
5V_1 \quad\quad - 8V_3 \quad\quad\quad + 3V_6 \quad\quad\quad &= 0 \\
4V_1 \quad\quad\quad\quad - 6V_4 + 2V_5 \quad\quad\quad &= 0 \quad (4.4.55) \\
2V_4 - 2V_5 \quad\quad\quad &= 0 \\
V_1 + V_2 + 3V_3 \quad\quad\quad - 7V_6 \quad\quad\quad &= -2V_O \\
- 3V_7 + 3V_8 &= 0,
\end{aligned}$$

which, in matrix form, is written as

$$\mathbf{AV} = \mathbf{b}, \quad (4.4.56)$$

where

$$\mathbf{A} = \begin{bmatrix} -13 & 2 & 5 & 4 & 0 & 1 & 0 & 0 \\ 2 & -3 & 0 & 0 & 0 & 1 & 0 & 0 \\ 5 & 0 & -8 & 0 & 0 & 3 & 0 & 0 \\ 4 & 0 & 0 & -6 & 2 & 0 & 0 & 0 \\ 0 & 0 & 0 & 2 & -2 & 0 & 0 & 0 \\ 1 & 1 & 3 & 0 & 0 & -7 & 0 & 0 \\ 0 & 0 & 0 & 0 & 0 & 0 & -3 & 3 \end{bmatrix} \quad (4.4.57)$$

and

$$\mathbf{b} = \begin{bmatrix} -V_\text{I} \\ 0 \\ 0 \\ 0 \\ 0 \\ 0 \\ -2V_\text{O} \\ 0 \end{bmatrix}. \tag{4.4.58}$$

We will assume that $V_\text{I} = 1$ and $V_\text{O} = 0$ and use Gauss–Jordan elimination to find the voltages $V_1, \ldots, V_8$. Note that since $\mathbf{A}$ is a 7×8 matrix, its rank is less than or equal to 7, and so if the system has a solution it is not unique. The reason is that the conductor between junctions 7 and 8 is "floating." Our analysis will tell us what we know physically, namely, that voltages V_7 and V_8 are equal to each other but are otherwise unknown from the context of the problem.

Solution. We perform Gauss–Jordan elimination on $\mathbf{A}$, transforming the problem into the form of Eq. (4.4.32). The resulting transformation matrices $\mathbf{P}$ and $\mathbf{Q}$ are

$$\mathbf{P} = \begin{bmatrix} \frac{1729}{303} & \frac{507}{101} & \frac{1495}{303} & \frac{1729}{303} & \frac{1729}{303} & \frac{1105}{303} & 0 \\ \frac{105}{101} & \frac{2450}{1313} & \frac{1260}{1313} & \frac{105}{101} & \frac{105}{101} & \frac{1085}{1313} & 0 \\ \frac{4715}{2121} & \frac{1476}{707} & \frac{5945}{2121} & \frac{4715}{2121} & \frac{4715}{2121} & \frac{3854}{2121} & 0 \\ \frac{37506}{20705} & \frac{32994}{20705} & \frac{6486}{4141} & \frac{23547}{8282} & \frac{23547}{8282} & \frac{4794}{4141} & 0 \\ \frac{57988}{128169} & \frac{5668}{14241} & \frac{50140}{128169} & \frac{91015}{128169} & \frac{157069}{128169} & \frac{37060}{128169} & 0 \\ \frac{85}{109} & \frac{93}{109} & \frac{94}{109} & \frac{85}{109} & \frac{85}{109} & 1 & 0 \\ 0 & 0 & 0 & 0 & 0 & 0 & 1 \end{bmatrix} \tag{4.4.59}$$

and

$$\mathbf{Q} = \mathbf{I}_8,$$

the $n = 8$ identity matrix. The matrix $\mathbf{A}$ then becomes

$$\mathbf{A}_{tr} = \mathbf{PAQ} = \begin{bmatrix} -13 & 0 & 0 & 0 & 0 & 0 & 0 & 0 \\ 0 & -\frac{35}{13} & 0 & 0 & 0 & 0 & 0 & 0 \\ 0 & 0 & -\frac{41}{7} & 0 & 0 & 0 & 0 & 0 \\ 0 & 0 & 0 & -\frac{846}{205} & 0 & 0 & 0 & 0 \\ 0 & 0 & 0 & 0 & -\frac{436}{423} & 0 & 0 & 0 \\ 0 & 0 & 0 & 0 & 0 & -\frac{303}{109} & 0 & 0 \\ 0 & 0 & 0 & 0 & 0 & 0 & -3 & 3 \end{bmatrix}, \quad (4.4.60)$$

and the vector $\mathbf{b}$ is transformed into

$$\boldsymbol{\alpha} = \mathbf{Pb} = \begin{bmatrix} -\left(\frac{2210}{303} V_O + \frac{1729}{303} V_I\right) \\ -\left(\frac{2170}{1313} V_O + \frac{105}{101} V_I\right) \\ -\left(\frac{7708}{2121} V_O + \frac{4715}{2121} V_I\right) \\ -\left(\frac{9588}{4141} V_O + \frac{37506}{20705} V_I\right) \\ -\left(\frac{74120}{128169} V_O + \frac{57988}{128169} V_I\right) \\ -\left(2 V_O + \frac{85}{109} V_I\right) \\ 0 \end{bmatrix}. \quad (4.4.61)$$

Since for this example $\mathbf{y} = \mathbf{x}$, the voltages can be obtained by inspection. Substituting $V_I = 1$ and $V_O = 0$, we find

$$\begin{aligned} V_1 &= 0.439 \\ V_2 &= 0.386 \\ V_3 &= 0.380 \\ V_4 &= 0.439 \\ V_5 &= 0.439 \\ V_6 &= 0.281. \end{aligned} \quad (4.4.62)$$

The last equation in $\mathbf{A}_{tr}\mathbf{y} = \boldsymbol{\alpha}$ reduces to $V_7 = V_8$. Thus, the values of V_7 and V_8 cannot be determined uniquely from the equation system. By Gauss–Jordan elimination, we found that the rank of $\mathbf{A}$ is equal to the number of rows (7) and by augmenting $\mathbf{A}$ with $\mathbf{b}$ the rank does not change. Therefore, although a solution does exist, it is not a unique one since $\mathbf{A}$ is a non-square matrix (and $m < n$).

■■■ EXAMPLE 4.4.2 (Chemical Reaction Equilibria). We, of course, know that molecules are made up of atoms. For example, water is composed of two hydrogen atoms, H, and one oxygen atom, O. We say that the chemical formula for water is H_2O. Likewise, the formula for methane is CH_4, indicating that methane is composed of one carbon atom, C, and four hydrogen atoms. We also know that chemical reactions can interconvert molecules; e.g., in the reaction

$$2H_2 + O_2 = 2H_2O, \tag{4.4.63}$$

two hydrogen molecules combine with one molecule of oxygen to form two molecules of water. And in the reaction

$$2CH_4 + O_2 = 2CH_3OH, \tag{4.4.64}$$

two molecules of methane combine with one molecule of oxygen to form two molecules of methanol.

Suppose that there are m atoms, labeled $\alpha_1, \ldots, \alpha_m$, some or all of which are contained in molecule M_j. The chemical formula for M_j is then

$$M_j = (\alpha_1)_{a_{ij}} (\alpha_2)_{a_{ij}}, \ldots, (\alpha_m)_{a_{mj}}. \tag{4.4.65}$$

M_j is thus totally specified by the column vector

$$\mathbf{a}_j = \begin{bmatrix} a_{1j} \\ a_{2j} \\ \vdots \\ a_{mj} \end{bmatrix}. \tag{4.4.66}$$

For example, if H, O, and C are the atoms 1, 2, and 3, and methane is designated as molecule 1, then the corresponding vector

$$\mathbf{a}_1 = \begin{bmatrix} 4 \\ 0 \\ 1 \end{bmatrix} \tag{4.4.67}$$

tells us that the formula for methane is $(H)_4(O)_0(C)_1$ or CH_4.

If we are interested in reactions involving the n molecules $M_1, \ldots, M_n$, the vectors specifying the atomic compositions of the molecules form the atomic matrix

$$\mathbf{A} = [\mathbf{a}_1, \ldots, \mathbf{a}_n] = \begin{bmatrix} a_{11} & a_{12} & \cdots & a_{1n} \\ a_{21} & a_{22} & \cdots & a_{2n} \\ \vdots & \vdots & & \vdots \\ a_{m1} & a_{m2} & \cdots & a_{mn} \end{bmatrix}. \tag{4.4.68}$$

We know from the solvability theory developed above that if the rank of A is r, then only r of the vectors of $\mathbf{A}$ are linearly independent. The remaining $n - r$ vectors are linear combinations of these r vectors; i.e., if $\{\mathbf{a}_1, \ldots, \mathbf{a}_r\}$ denotes the set of linearly independent vectors, then there exist numbers β_{kj} such that

$$\mathbf{a}_k = \sum_{j=1}^{r} \beta_{kj} \mathbf{a}_j, \quad k = r+1, \ldots, n. \tag{4.4.69}$$

These equations represent chemical reactions among the molecular species. Since each vector represents a different molecule, Eq. (4.4.69) implies that the number of independent molecular components is r and a minimum of $n-r$ reactions exists for all of the different molecular species since each equation in Eq. (4.4.69) contains a species not present in the other equations.

As an application, consider the atoms H, O, and C and the molecules H_2, O_2, H_2O, CH_4 and CH_3OH. The atomic matrix is given by

	H_2	O_2	H_2O	CH_4	CH_3OH
H	2	0	2	4	4
O	0	2	1	0	1
C	0	0	0	1	1

The rank of this matrix is 3, and so there are three independent molecular components and there have to be at least $5 - 3 = 2$ reactions to account for all the molecular species. H_2, O_2, and CH_4 can be chosen to be the independent components (because $\mathbf{a}_1$, $\mathbf{a}_2$, and $\mathbf{a}_4$ are linearly independent) and the equilibrium of the two reactions in Eqs. (4.4.63) and (4.4.64) suffice to account thermodynamically for reactions among the independent species. H_2, O_2, and H_2O cannot be chosen since $\mathbf{a}_3 = \mathbf{a}_1 + \mathbf{a}_2/2$, reflecting the physical fact that carbon is missing from these three molecules.

■ ■ ■

ILLUSTRATION 4.4.1 (Virial Coefficients of a Gas Mixture). Statistical mechanics provides a rigorous set of mixing rules when describing gas mixtures with the virial equation of state. The compressibility of a mixture at low density can be written as

$$z = 1 + B_{mix}\rho + C_{mix}\rho^2 + \cdots, \quad (4.4.70)$$

where ρ is the molar density of the fluid and the virial coefficients, B_{mix}, C_{mix}, etc., are functions only of temperature and composition. At low enough density, we can truncate the series after the second term. From statistical mechanics, we can define pair coefficients that are only functions of temperature. The second virial coefficient for an N-component mixture is then given by

$$B_{mix} = \sum_{i=1}^{N}\sum_{j=1}^{N} y_i y_j B_{ij}(T), \quad (4.4.71)$$

where y_i refers to the mole fraction of component i.

We desire to find the virial coefficients for a three-component gas mixture from the experimental values of B_{mix} given in Table 4.4.1. For a three-component system, the relevant coefficients are B_{11}, B_{22}, B_{33}, B_{12}, B_{13}, and B_{23}. The mixing rule is given by

$$\begin{aligned}B_{mix} =& y_1^2 B_{11} + y_2^2 B_{22} + y_3^2 B_{33} \\ &+ 2y_1 y_2 B_{12} + 2y_1 y_3 B_{13} + 2y_2 y_3 B_{23}.\end{aligned} \quad (4.4.72)$$

GENERAL SOLVABILITY THEOREM FOR Ax = b

TABLE 4.4.1 Second Virial Coefficient at 200 K for Ternary Gas Mixture

B_{mix}	y_1	y_2	y_3
10.3	0	0.25	0.75
17.8	0	0.50	0.50
26.2	0	0.75	0.25
13.7	0.50	0.25	0.25
7.64	0.75	0	0.25

We can recast this problem in matrix form as follows. We define the vectors

$$\mathbf{b} \equiv \begin{bmatrix} B_{mix,1} \\ B_{mix,2} \\ B_{mix,3} \\ B_{mix,4} \\ B_{mix,5} \end{bmatrix}, \quad \mathbf{x} \equiv \begin{bmatrix} B_{11} \\ B_{22} \\ B_{33} \\ B_{12} \\ B_{13} \\ B_{23} \end{bmatrix}, \quad (4.4.73)$$

and the 5×6 matrix $\mathbf{A}$ by

$$\mathbf{A} \equiv \begin{bmatrix} y_{1,1}^2 & y_{2,1}^2 & y_{3,1}^2 & y_{1,1}y_{2,1} & y_{1,1}y_{3,1} & y_{2,1}y_{3,1} \\ y_{1,2}^2 & y_{2,2}^2 & y_{3,2}^2 & y_{1,2}y_{2,2} & y_{1,2}y_{3,2} & y_{2,2}y_{3,2} \\ y_{1,3}^2 & y_{2,3}^2 & y_{3,3}^2 & y_{1,3}y_{2,3} & y_{1,3}y_{3,3} & y_{2,3}y_{3,3} \\ y_{1,4}^2 & y_{2,4}^2 & y_{3,4}^2 & y_{1,4}y_{2,4} & y_{1,4}y_{3,4} & y_{2,4}y_{3,4} \\ y_{1,5}^2 & y_{2,5}^2 & y_{3,5}^2 & y_{1,5}y_{2,5} & y_{1,5}y_{3,5} & y_{2,5}y_{3,5} \end{bmatrix}, \quad (4.4.74)$$

where $y_{i,j}$ refers to the ith component and the jth measurement in Table 4.4.1. Similarly, the subscripts in the components of $\mathbf{b}$ refer to the measurements. Solving for the virial coefficients has then been reduced to solving the linear system

$$\mathbf{A}\mathbf{x} = \mathbf{b}. \quad (4.4.75)$$

(i) Using the solvability theorems of this chapter, determine if a solution exists for Eq. (4.4.75) using the data in Table 4.4.1. If a solution exists, is it unique? Find the most general solution to the equation if one exists. If a solution does not exist, explain why.

We can find the "best fit" solution to Eq. (4.4.75) by applying the least squares analysis from Chapter 3. We define the quantity

$$L = \sum_{i=1}^{5} \left(b_i - \sum_{j=1}^{6} A_{ij} x_j \right)^2, \quad (4.4.76)$$

which represents a measure of the relative error in the parameters x_i. The vector $\mathbf{x}$ that minimizes L can be found from the requirement that each of the derivatives of L with respect to x_i be equal to 0. The solution (see Chapter 3, Problem 8) is

$$\mathbf{A}^T \mathbf{A} \mathbf{x} = \mathbf{A}^T \mathbf{b}. \quad (4.4.77)$$

Here **x** contains the best fit parameters B_{ij} for the given data.

(ii) Show that Eq. (4.4.77) has a unique solution for the data given in Table 4.4.1. Find the best fit virial coefficients. How does this best fit solution compare to the general solution (if it exists) found in (i)?

4.5. LINEAR DEPENDENCE OF A VECTOR SET AND THE RANK OF ITS MATRIX

Consider the set $\{\mathbf{a}_1, \ldots, \mathbf{a}_n\}$ of m-dimensional vectors. Whether this set is linearly dependent depends on whether a set of numbers $\{x_1, \ldots, x_n\}$, not all of which are 0, can be found such that

$$\sum_{j=1}^n x_j \mathbf{a}_j = \mathbf{0}. \qquad (4.5.1)$$

We can pose the problem in a slightly different manner by defining a matrix $\mathbf{A}$ whose column vectors are $\mathbf{a}_1, \ldots, \mathbf{a}_n$. In partition form, this is

$$[\mathbf{a}_1, \mathbf{a}_2, \ldots, \mathbf{a}_n]. \qquad (4.5.2)$$

Next, we define the vector **x** by

$$\mathbf{x} = \begin{bmatrix} x_1 \\ x_2 \\ \vdots \\ x_n \end{bmatrix}. \qquad (4.5.3)$$

Now we ask the question: Does the equation

$$\mathbf{A}\mathbf{x} = \mathbf{0} \qquad (4.5.4)$$

have a nontrivial ($\mathbf{x} \neq \mathbf{0}$) solution? This equation can be expressed as

$$[\mathbf{a}_1, \mathbf{a}_2, \ldots, \mathbf{a}_n] \begin{bmatrix} x_1 \\ x_2 \\ \vdots \\ x_n \end{bmatrix} = \mathbf{0} \qquad (4.5.5)$$

or, carrying out the vector product,

$$\sum_{j=1}^n \mathbf{a}_j x_j = \mathbf{0}, \qquad (4.5.6)$$

which is the same as Eq. (4.5.1).

$$\begin{aligned} a_{11} x_1 + \cdots + a_{1n} x_n &= 0 \\ &\vdots \\ a_{m1} x_1 + \cdots + a_{mn} x_n &= 0. \end{aligned} \qquad (4.5.7)$$

The question of whether the vectors $\mathbf{a}_1, \ldots, \mathbf{a}_n$ are linearly dependent is thus seen to be the question of whether the homogeneous equation $\mathbf{Ax} = \mathbf{0}$ has a solution, and the answer to this latter question depends on the rank of $\mathbf{A}$.

From what we have learned from the solvability theory of $\mathbf{Ax} = \mathbf{b}$, we can immediately draw several conclusions: (1) If $\mathbf{A}$ is a square nonsingular matrix, then $\mathbf{x} = \mathbf{0}$ is the only solution to Eq. (4.5.4). Thus, a set of n n-dimensional vectors $\mathbf{a}_j$ are linearly independent if and only if their matrix $\mathbf{A}$ has a rank n, i.e., $|\mathbf{A}| \neq 0$. (2) This conclusion is a subcase of a more general one: a set of n m-dimensional vectors $\mathbf{a}_j$ are linearly independent *if and only if* their matrix $\mathbf{A}$ has a rank n. This follows from the fact that $\mathbf{Ax} = \mathbf{0}$ admits $n - r$ solutions. (3) A set of n m-dimensional vectors $\mathbf{a}_j$ are linearly dependent *if and only if* the rank r of their matrix $\mathbf{A}$ is less than n. This also follows from the fact that $\mathbf{Ax} = \mathbf{0}$ admits $n - r$ solutions. (4) From this it follows that if $n > m$, then the set of n m-dimensional vectors $\mathbf{a}_j$ will always be linearly dependent since $r \leq (m, n)$.

We are familiar with these properties from Euclidean vector space. We know that no three coplanar vectors can be used as a basis set for an arbitrary three-dimensional vector. What is useful here is the fact that analysis of the rank of the matrix whose Cartesian vector components form the column vectors of the matrix will establish whether a given set of three vectors is coplanar. Coplanar means that one of the vectors is a linear combination of the other two (linear dependence). From the solvability theory developed in this chapter, we can use the rank of $\mathbf{A} = [\mathbf{a}_1, \ldots, \mathbf{a}_n]$ not only to determine whether the set $\{\mathbf{a}_1, \ldots, \mathbf{a}_n\}$ is linearly dependent but also to find how many of the $\mathbf{a}_j$ are linearly dependent on a subset of the vectors in the set.

The rank of $\mathbf{A}$ is equal to the number of linearly independent column vectors and to the number of linearly independent row vectors that $\mathbf{A}$ contains. To prove this, consider first an $m \times n$ matrix $\mathbf{A}$ of rank r in which the upper left corner of $\mathbf{A}$ contains an rth-order nonzero minor. Proof of this special case will be shown to suffice in establishing the general case. We can rearrange the homogeneous equation $\mathbf{Ax} = \mathbf{0}$ into the form

$$a_{11}x_1 + \cdots + a_{1r}x_r = -a_{1,r+1}x_{r+1} - \cdots - a_{1n}x_n$$
$$\vdots \tag{4.5.8}$$
$$a_{m1}x_1 + \cdots + a_{mr}x_r = -a_{m,r+1}x_{r+1} - \cdots - a_{mn}x_n.$$

Since the minor

$$\begin{vmatrix} a_{11} & a_{12} & \cdots & a_{1r} \\ a_{21} & a_{22} & \cdots & a_{2r} \\ \vdots & \vdots & & \vdots \\ a_{r1} & a_{r2} & \cdots & a_{rr} \end{vmatrix} \tag{4.5.9}$$

is nonzero and the rank of $\mathbf{A}$ is r, the equations in Eq. (4.5.8) have a solution $\{x_1, \ldots, x_r\}$ for arbitrary $x_{r+1}, \ldots, x_n$. Note that $\mathbf{Ax} = \mathbf{0}$ always has a nontrivial solution for $r < \min(m, n)$ because the rank of $[\mathbf{A}, \mathbf{0}]$ is the same as the rank of $\mathbf{A}$.

One solution to Eq. (4.5.8) is obtained by setting $x_{r+1} = 1$ and $x_j = 0$ and solving for $x_1^{(1)}, \ldots, x_r^{(1)}$. With this solution, Eq. (4.5.8) can be rearranged to get

$$a_{1,r+1} = -\left(a_{11}x_1^{(1)} + a_{12}x_2^{(1)} + \cdots + a_{1r}x_r^{(1)}\right)$$
$$\vdots \qquad \vdots \qquad (4.5.10)$$
$$a_{m,r+1} = -\left(a_{m1}x_1^{(1)} + a_{m2}x_2^{(1)} + \cdots + a_{mr}x_r^{(1)}\right)$$

or, in vector notation,

$$\mathbf{a}_{r+1} = -\sum_{j=1}^{r} x_j^{(1)} \mathbf{a}_j. \qquad (4.5.11)$$

This proves that the $(r+1)$th column vector of $\mathbf{A}$ is a linear combination of the set $\{\mathbf{a}_1, \ldots, \mathbf{a}_r\}$. In general, if $x_i = 1$ for $i > r$ and $x_j = 0$ for $j > r$ and $j \neq i$, then the solution $\{x_1^{(i)}, \ldots, x_r^{(i)}\}$ of Eq. (4.5.8) can be found, and so

$$a_{1i} = -\left(a_{11}x_1^{(i)} + \cdots + a_{1r}x_r^{(i)}\right)$$
$$\vdots \qquad \vdots \qquad (4.5.12)$$
$$a_{mi} = -\left(a_{m1}x_1^{(i)} + \cdots + a_{mr}x_r^{(i)}\right)$$

or

$$\mathbf{a}_i = -\sum_{j=1}^{r} x_j^{(i)} \mathbf{a}_j, \qquad i > r. \qquad (4.5.13)$$

Thus, we have proven that all the column vectors $\mathbf{a}_{r+1}, \ldots, \mathbf{a}_n$ are linear combinations of the first r column vectors $\mathbf{a}_1, \ldots, \mathbf{a}_r$. The vectors $\mathbf{a}_1, \ldots, \mathbf{a}_r$ are linearly independent because, otherwise, there would exist a set of numbers $\{c_1, \ldots, c_r\}$, not all 0, such that

$$\sum_{i=1}^{r} c_i \mathbf{a}_i = 0$$

or

$$a_{11}c_i + \cdots + a_{1r}c_r = 0$$
$$\vdots \qquad \vdots \qquad (4.5.14)$$
$$a_{m1}c_i + \cdots + a_{mr}c_r = 0.$$

If this set of equations has a nontrivial solution, then the rank of the matrix $[\mathbf{a}_1, \ldots, \mathbf{a}_r]$ has to be less than r, which contradicts our hypothesis.

In summary, for a matrix of rank r and of the form considered here, the last $n - r$ column vectors are linearly dependent on the first r column vectors, which themselves are linearly independent. Since the rank of the transpose $\mathbf{A}^T$ of $\mathbf{A}$ is also r, it follows that the last $m - r$ column vectors of $\mathbf{A}^T$ are linearly dependent on the first r column vectors, which themselves are linearly dependent. But the column vectors of $\mathbf{A}^T$ are simply the row vectors of $\mathbf{A}$, and so we conclude that the last $m-r$ row vectors of $\mathbf{A}$ are linearly dependent on the first r row vectors, which are themselves linearly independent.

Next, consider the general case, i.e., an $m \times n$ matrix $\mathbf{A}$ of rank r, but in which the upper left corner does not contain a nonzero minor. By the interchange of columns and rows, however, a matrix $\mathbf{A}'$ can be obtained that does have a nonzero rth-order minor in the upper left corner. We have already shown that the rank of $\mathbf{A}'$ is also r, and so if

$$\mathbf{A}' = [\mathbf{a}'_1, \mathbf{a}'_2, \ldots, \mathbf{a}'_n], \tag{4.5.15}$$

then the r m-dimensional column vectors $\mathbf{a}'_1, \ldots, \mathbf{a}'_r$ are linearly independent and the $n-r$ column vectors $\mathbf{a}'_{r+1}, \ldots, \mathbf{a}'_n$ are linear combinations of the first r column vectors. Also, from what we presented above, the r n-dimensional row vectors $[a'_{i1}, \ldots, a'_{in}]$, $i = 1, \ldots, r$, are linearly independent and the remaining $m-r$ row vectors are linear combinations of the first r row vectors.

To prove what was just stated, note first that the relationship between $\mathbf{A}$ and $\mathbf{A}'$ is

$$\mathbf{A}' = \mathbf{Q}^{(1)} \mathbf{A} \mathbf{Q}^{(2)}, \tag{4.5.16}$$

where the square matrices $\mathbf{Q}^{(1)}$ and $\mathbf{Q}^{(2)}$ are products of the $\mathbf{I}_{ij}$ matrices that accomplish the appropriate row interchanges and column interchanges, respectively. Since the determinants of $\mathbf{Q}^{(1)}$ and $\mathbf{Q}^{(2)}$ are ± 1, it follows that the ranks of $\mathbf{A}'$ and $\mathbf{A}$ are the same. From the property $\mathbf{I}_{ij}\mathbf{I}_{ij} = \mathbf{I}$, it follows that

$$\mathbf{Q}^{(i)}\mathbf{Q}^{(i)} = \mathbf{I} \tag{4.5.17}$$

or that $\mathbf{Q}^{(i)}$ equals its own inverse. Consequently,

$$\begin{aligned}
\mathbf{A} &= \mathbf{Q}^{(1)} \mathbf{A}' \mathbf{Q}^{(2)} \\
&= \mathbf{Q}^{(1)} [\mathbf{a}'_1, \ldots, \mathbf{a}'_n] \mathbf{Q}^{(2)} \\
&= [\mathbf{Q}^{(1)}\mathbf{a}'_1, \ldots, \mathbf{Q}^{(1)}\mathbf{a}'_n] \begin{bmatrix} q^{(2)}_{11} & \cdots & q^{(2)}_{1n} \\ \vdots & & \vdots \\ q^{(2)}_{n1} & \cdots & q^{(2)}_{nn} \end{bmatrix} \\
&= \left[\sum_{k=1}^{n} q^{(2)}_{k1} \mathbf{Q}^{(1)} \mathbf{a}'_k, \ldots, \sum_{k=1}^{n} q^{(2)}_{kn} \mathbf{Q}^{(1)} \mathbf{a}'_k \right]
\end{aligned} \tag{4.5.18}$$

and, therefore,

$$\mathbf{a}_i = \sum_{k=1}^{n} q^{(2)}_{ki} \mathbf{Q}^{(1)} \mathbf{a}'_k. \tag{4.5.19}$$

We proved already that

$$\mathbf{a}'_k = \sum_{j=1}^{r} \alpha_{kj} \mathbf{a}'_j, \quad k > r, \tag{4.5.20}$$

which, when inserted into Eq. (4.5.19), yields

$$\mathbf{a}_i = \sum_{j=1}^{r} \beta_{ij} \mathbf{Q}^{(1)} \mathbf{a}'_j, \qquad (4.5.21)$$

where $\beta_{ij} = \sum_{k=1}^{n} q_{ki}^{(2)} \alpha_{kj}$. The vector $\mathbf{a}'_j$ is related to one of the set $\{\mathbf{a}_1, \ldots, \mathbf{a}_n\}$, say $\mathbf{a}_{l_j}$, by the row interchange operation $\mathbf{Q}^{(1)}$, i.e.,

$$\mathbf{a}'_j = \mathbf{Q}^{(1)} \mathbf{a}_{l_j}, \qquad (4.5.22)$$

and it follows that

$$\mathbf{Q}^{(1)} \mathbf{a}'_j = \mathbf{Q}^{(1)} \mathbf{Q}^{(1)} \mathbf{a}_{l_j} = \mathbf{a}_{l_j}. \qquad (4.5.23)$$

Thus, Eq. (4.5.21) reads

$$\mathbf{a}_i = \sum_{\{l_j\}} \beta_{ij} \mathbf{a}_{l_j}, \qquad (4.5.24)$$

where $\{l_j\}$ indicates the indices of the r column vectors of $\mathbf{A}$ that were moved to the columns $1, \ldots, r$ in $\mathbf{A}$ to put a nonzero rth-order minor in its upper left corner. This proves that any column vector in $\mathbf{A}$ is a linear combination of the r column vectors $\mathbf{a}_{l_1}, \mathbf{a}_{l_2}, \ldots, \mathbf{a}_{l_r}$.

These r vectors are linearly independent. To prove this, assume the contrary; i.e., assume that there exists a set of numbers $\{c_1, \ldots, c_r\}$, not all 0, such that

$$\sum_{\{l_j\}} c_j \mathbf{a}_{l_j} = \mathbf{0}. \qquad (4.5.25)$$

By multiplying Eq. (4.5.24) by $\mathbf{Q}^{(1)}$, it follows that Eq. (4.5.24) implies

$$\sum_{j=1}^{r} c_j \mathbf{a}'_j = \mathbf{0},$$

or that the set $\{\mathbf{a}'_1, \ldots, \mathbf{a}'_r\}$ is linearly dependent. However, this is a contradiction, and so the vectors $\mathbf{a}_{l_1}, \ldots, \mathbf{a}_{l_r}$ must be linearly independent.

Similarly, by considering the transpose of $\mathbf{A}'$, we can prove that r of the row vectors of $\mathbf{A}$ are linearly independent and that the other $m - r$ row vectors are linear combinations of these r row vectors.

The "take-home" lesson of this section is as follows:

THEOREM. *If the rank of the m × n matrix* **A** *is r, then* (a) *there are r m-dimensional column vectors (and r n-dimensional row vectors) that are linearly independent and* (b) *the remaining n − r column vectors (and m − r row vectors) are linear combinations of the r linearly independent vectors.*

■ EXAMPLE 4.5.1. How many of the vectors

$$\mathbf{a}_1 = \begin{bmatrix} 1 \\ 2 \\ 3 \end{bmatrix}, \quad \mathbf{a}_2 = \begin{bmatrix} 1 \\ 1 \\ 1 \end{bmatrix}, \quad \mathbf{a}_3 = \begin{bmatrix} 2 \\ 3 \\ 4 \end{bmatrix}, \quad \mathbf{a}_4 = \begin{bmatrix} 3 \\ 5 \\ 7 \end{bmatrix} \qquad (4.5.26)$$

are linearly independent? Since the rank r of **A**,

$$\mathbf{A} = [\mathbf{a}_1, \mathbf{a}_2, \mathbf{a}_3, \mathbf{a}_4] = \begin{bmatrix} 1 & 1 & 2 & 3 \\ 2 & 1 & 3 & 5 \\ 3 & 1 & 4 & 7 \end{bmatrix}, \qquad (4.5.27)$$

is less than or equal to 3, we know at most three vectors are linearly independent. By Gauss elimination, we transform **A** to

$$\mathbf{A}_{\mathrm{tr}} = \begin{bmatrix} 1 & 1 & 2 & 3 \\ 0 & -1 & -1 & -1 \\ 0 & 0 & 0 & 0 \end{bmatrix}. \qquad (4.5.28)$$

Thus, the rank of **A** is 2. Therefore, only two of the vectors are linearly independent. Indeed, the pair $\mathbf{a}_1$ and $\mathbf{a}_2$ are linearly independent and

■ ■ ■ $$\mathbf{a}_3 = \mathbf{a}_1 + \mathbf{a}_2 \quad \text{and} \quad \mathbf{a}_4 = 2\mathbf{a}_3 - \mathbf{a}_2. \qquad (4.5.29)$$

4.6. THE FREDHOLM ALTERNATIVE THEOREM

For algebraic equations, the condition of solvability, stated as the equality of the ranks of the augmented matrix [**A**, **b**] and the matrix **A**, is especially attractive. Straightforward Gauss–Jordan elimination establishes the ranks of both [**A**, **b**] and **A** and results in a final set of equations, which, when solvable, require very little further work to obtain the solution or solutions to **Ax** = **b**. There is, however, another way to state the solvability theorem, known as the Fredholm alternative theorem. While not suggestive of a method of solution, it is powerful because its form carries over to much more general vector spaces (e.g., function spaces whose operators are differential or integral operators instead of matrices) where the concept of the rank of a determinant is not defined.

Before stating the theorem, some additional properties of the matrix **A** need to be established. Recall that the adjoint $\mathbf{A}^\dagger$ is the complex conjugate of the transpose

$\mathbf{A}^T$ of $\mathbf{A}$; i.e., if

$$\mathbf{A} = \begin{bmatrix} a_{11} & a_{12} & \cdots & a_{1n} \\ a_{21} & a_{22} & \cdots & a_{2n} \\ \vdots & \vdots & & \vdots \\ a_{m1} & a_{m2} & \cdots & a_{mn} \end{bmatrix}, \qquad (4.6.1)$$

then

$$\mathbf{A}^\dagger = \begin{bmatrix} a_{11}^* & a_{21}^* & \cdots & a_{m1}^* \\ a_{12}^* & a_{22}^* & \cdots & a_{m2}^* \\ \vdots & \vdots & & \vdots \\ a_{1n}^* & a_{2n}^* & \cdots & a_{mn}^* \end{bmatrix}. \qquad (4.6.2)$$

Since the interchange of rows and columns does not change the value of a determinant, it follows that the rank of the adjoint $\mathbf{A}^\dagger$ is the same as the rank of $\mathbf{A}$. This can be seen explicitly by recalling the form of $\mathbf{A}_{tr}$ in Eq. (4.4.39). The adjoint of $\mathbf{A}_{tr}$ is

$$\mathbf{A}_{tr}^\dagger = \begin{bmatrix} \gamma_{11}^* & 0 & \cdots & 0 & 0 & \cdots & 0 \\ 0 & \gamma_{22}^* & \cdots & 0 & 0 & \cdots & 0 \\ 0 & 0 & \cdots & 0 & 0 & \cdots & 0 \\ \vdots & \vdots & & \vdots & \vdots & & \vdots \\ 0 & 0 & \cdots & \gamma_{rr}^* & 0 & \cdots & 0 \\ \gamma_{1,r+1}^* & \gamma_{2,r+1}^* & \cdots & \gamma_{r,r+1}^* & 0 & \cdots & 0 \\ \vdots & \vdots & & \vdots & \vdots & & \vdots \\ \gamma_{1n}^* & \gamma_{2n}^* & \cdots & \gamma_{rn}^* & 0 & \cdots & 0 \end{bmatrix}. \qquad (4.6.3)$$

Since only r columns of $\mathbf{A}_{tr}^\dagger$ have nonzero elements and since $\prod_{i=1}^r \gamma_{ii}^* \neq 0$, it follows that the rank of $\mathbf{A}_{tr}^\dagger$ is r, the rank of $\mathbf{A}$. Recall, however, that $\mathbf{A}_{tr} = \mathbf{PAQ}$, where $|\mathbf{P}|$ and $|\mathbf{Q}|$ are ± 1. Thus, $\mathbf{A}_{tr}^\dagger = \mathbf{Q}^\dagger \mathbf{A}^\dagger \mathbf{P}^\dagger$ from which it follows that the rank of $\mathbf{A}_{tr}^\dagger$ is the same as the rank of $\mathbf{A}^\dagger$, which proves our claim that the rank of $\mathbf{A}^\dagger$ is the same as the rank r of $\mathbf{A}$.

The matrix $\mathbf{A}^\dagger$ has m column vectors (which are the complex conjugates of the transpose of the row vectors of $\mathbf{A}$), and so, according to the solvability theorem in Eqs. (4.4.52) and (4.4.53), the homogeneous equation

$$\mathbf{A}^\dagger \mathbf{z} = \mathbf{0} \qquad (4.6.4)$$

has $m - r$ linearly independent solutions; i.e., there exist $m - r$ n-dimensional vectors $\mathbf{z}_1, \mathbf{z}_2, \ldots, \mathbf{z}_{m-r}$ satisfying Eq. (4.6.4). Recall that the homogeneous equation

$$\mathbf{A}\mathbf{x} = \mathbf{0} \qquad (4.6.5)$$

has $n - r$ linearly independent m-dimensional vector solutions $\mathbf{x}_1, \ldots, \mathbf{x}_{n-r}$. Thus, only if $\mathbf{A}$ is a square matrix do $\mathbf{A}$ and $\mathbf{A}^\dagger$ have the same number of solutions to their homogeneous equations.

THE FREDHOLM ALTERNATIVE THEOREM

FREDHOLM ALTERNATIVE THEOREM. *The equation*

$$\mathbf{A}\mathbf{x} = \mathbf{b} \tag{4.6.6}$$

has a solution if and only if $\mathbf{b}$ *is orthogonal to the solutions of Eq.* (4.6.4)

$$\mathbf{b}^\dagger \mathbf{z}_j = 0, \tag{4.6.7}$$

where $\mathbf{z}_j$ *is any of the* $m - r$ *linearly independent solutions of the homogeneous adjoint equation* (4.6.4).

The solvability condition required by the Fredholm alternative theorem places $m - r$ conditions on $\mathbf{b}$, namely, $\mathbf{b}^\dagger \mathbf{z}_j = \sum_{i=1}^{m} b_i^* z_{ij} = 0, j = 1, \ldots, m - r$. The conditions required to ensure that the rank of $[\mathbf{A}, \mathbf{b}]$ is the same as the rank of $\mathbf{A}$ are that

$$\alpha_{r+1} = \alpha_{r+2} = \cdots = \alpha_m = 0, \tag{4.6.8}$$

where $\alpha_i = \sum_{k=1}^{m} p_{ik} b_k$ and p_{ik} are elements of the matrix $\mathbf{P}$ in the transformation $\mathbf{A}_{\mathrm{tr}} = \mathbf{PAQ}$. Thus, the solvability conditions in Eq. (4.6.8) also place $m - r$ conditions on $\mathbf{b}$ and must, of course, be equivalent to the conditions of the Fredholm alternative theorem.

The proof of the necessity ("only if") on the conditions of the Fredholm alternative theorem is quite simple. Suppose the solution $\mathbf{x}$ to Eq. (4.6.6) exists and take the inner product of $\mathbf{z}_i$ and Eq. (4.6.6) to obtain

$$\mathbf{z}_i^\dagger \mathbf{A}\mathbf{x} = \mathbf{z}_i^\dagger \mathbf{b}, \tag{4.6.9}$$

where $\mathbf{z}_i$ is any solution to Eq. (4.6.4). Taking the adjoint of each side of Eq. (4.6.9)—using the rule (Eq. (2.4.4b)) for forming adjoints of products of matrices—we obtain

$$\mathbf{x}^\dagger \mathbf{A}^\dagger \mathbf{z}_i = \mathbf{b}^\dagger \mathbf{z}_i. \tag{4.6.10}$$

But $\mathbf{A}^\dagger \mathbf{z}_i = \mathbf{0}$, or $\mathbf{b}^\dagger \mathbf{z}_i = 0$, proving that Eq. (4.6.7) is a necessary condition for $\mathbf{A}\mathbf{x} = \mathbf{b}$ to have a solution. To prove the sufficiency condition, we must assume that the conditions in Eq. (4.5.7) are true and prove that this implies the existence of $\mathbf{x}$. This part of the proof is somewhat tedious and will not be given here.

■ **EXAMPLE 4.6.1.** Under what conditions does

$$\begin{bmatrix} 2 & 1 \\ 1 & 2 \\ 3 & 4 \end{bmatrix} \begin{bmatrix} x_1 \\ x_2 \end{bmatrix} = \begin{bmatrix} b_1 \\ b_2 \\ b_3 \end{bmatrix} \quad \text{or} \quad \mathbf{A}\mathbf{x} = \mathbf{b} \tag{4.6.11}$$

have a solution? The homogeneous equation $\mathbf{A}^\dagger \mathbf{z} = \mathbf{0}$ is given by

$$\begin{bmatrix} 2 & 1 & 3 \\ 1 & 2 & 4 \end{bmatrix} \begin{bmatrix} z_1 \\ z_2 \\ z_3 \end{bmatrix} = \mathbf{0} \tag{4.6.12}$$

or

$$2z_1 + z_2 + 3z_3 = 0$$
$$z_1 + 2z_2 + 4z_3 = 0 \qquad (4.6.13)$$

and has the solution

$$\mathbf{z} = \begin{bmatrix} -\dfrac{2}{3} \\ -\dfrac{5}{3} \\ 1 \end{bmatrix}, \qquad (4.6.14)$$

where we have set z_3 equal to 1 (but a solution exists for arbitrary z_3).

The solvability condition, $\mathbf{b}^\dagger \mathbf{z} = \mathbf{0}$, is then

$$-\frac{2}{3}b_1^* - \frac{5}{3}b_2^* + b_3^* = 0. \qquad (4.6.15)$$

If, for example, $b_1 = 1, b_2 = 1$, and $b_3 = \frac{7}{3}$, then Eq. (4.6.11) has a solution. In this case, $\mathbf{Ax} = \mathbf{b}$ is

$$2x_1 + x_2 = 1$$
$$x_1 + 2x_2 = 1$$
$$3x_1 + 4x_2 = \frac{7}{3}. \qquad (4.6.16)$$

Gauss–Jordan elimination yields

$$2x_1 = \frac{2}{3}$$
$$\frac{3}{2}x_2 = \frac{1}{2} \qquad (4.6.17)$$
$$0 = 0,$$

or $x_1 = \frac{1}{3}$ and $x_2 = \frac{1}{3}$. Since the rank of $\mathbf{A}$ is 2 and $n = 2$, $n - r = 0$, there are no homogeneous solutions to $\mathbf{Ax} = \mathbf{b}$, and so

$$\mathbf{x} = \begin{bmatrix} \dfrac{1}{3} \\ \dfrac{1}{3} \end{bmatrix} \qquad (4.6.18)$$

■ ■ ■ is a unique solution to $\mathbf{Ax} = \mathbf{b}$, even though $\mathbf{A}$ is not a square matrix.

PROBLEMS

EXERCISE 4.6.1. Show the solvability conditions that the Fredholm alternative theorem requires for **b** when

$$A = \begin{bmatrix} 1 & 1 \\ 1 & 1 \\ 1 & 1 \end{bmatrix}. \tag{4.6.19}$$

Pick a **b** satisfying the conditions and find the most general solution to $Ax = b$ for this case. ∎ ∎ ∎

PROBLEMS

1. **A** and **B** are defined as

$$A = \begin{bmatrix} 1 & -2 \\ -2 & 3 \end{bmatrix} \quad \text{and} \quad B = \begin{bmatrix} -2 & 1 \\ 1 & 1 \end{bmatrix}.$$

   (a) Compute the determinants $|A|$, $|B|$, and $|AB|$ and verify that $|AB| = |A|\,|B|$.
   (b) Compute A^{-1} and B^{-1} and verify that $|A^{-1}| = 1/|A|$ and $|B^{-1}| = 1/|B|$.

2. If **B** and **C** are of ranks r_B and r_C, show that the rank of **A**, where

$$A = \begin{bmatrix} B & 0 \\ 0 & C \end{bmatrix},$$

   is $r_A + r_C$.

3. Find the *general* solutions to $Ax = b$ for the following:

   (a)
$$A = \begin{bmatrix} 1 & -1 & 2 & 1 \\ 6 & -10 & 10 & 4 \\ -2 & 4 & -3 & -1 \\ 2 & -2 & 4 & 2 \end{bmatrix}, \quad b = \begin{bmatrix} 1 \\ -3 \\ 2 \\ 3 \end{bmatrix}.$$

   (b)
$$A = \begin{bmatrix} 3 & -1 & 1 \\ 0 & 2 & 2 \\ 6 & -4 & 0 \\ 3 & 1 & 3 \\ 3 & -3 & -1 \end{bmatrix}, \quad b = \begin{bmatrix} 6 \\ 4 \\ 8 \\ 10 \\ 2 \end{bmatrix}.$$

   (c)
$$A = \begin{bmatrix} 2 & 0 & 1 & 0 & 2 \\ 0 & 1 & 1 & 1 & 0 \\ 2 & 1 & 2 & 1 & 1 \end{bmatrix}, \quad b = \begin{bmatrix} 0 \\ 0 \\ 0 \end{bmatrix}.$$

4. Consider the system of equations

$$2x + y + \alpha z = \beta$$
$$2x - \alpha y + 2z = \beta$$
$$x - 2y + 2\alpha z = 1.$$

(a) For what values of α and β does the system have a unique solution?
(b) Use Cramer's rule to obtain the unique solution in terms of α and β.
(c) Are there any other nonunique solutions for other values of α and β? If so, give a single example (i.e., choose specific values for α and β and give the general form of the solution.)

5. Consider the set of equations

$$x - 3y = -2$$
$$2x + y = 3$$
$$3x - 2y = \alpha.$$

(a) Are there values of α for which this set has no solution? If so, what are they?
(b) Are there values of α for which this set has a solution? If so, give an example.

6. For what values of k will the system

$$2x + ky + z = 0$$
$$(k - 1)x - y - 2z = 0$$
$$4x + y + 4z = 0$$

have nontrivial solutions?

7. Prove that the equations

$$x + (\cos \gamma)y + (\cos \beta)z = 0$$
$$(\cos \gamma)x + y + (\cos \alpha)z = 0$$
$$(\cos \beta)x + (\cos \alpha)y + z = 0$$

have a nontrivial solution if $\alpha + \beta + \gamma = 0$.

8. Consider the equations

$$ax + by + cz = b_1$$
$$a^2x + b^2y + c^2z = b_2$$
$$a^3x + b^3y + c^3z = b_3.$$

Give the solution to these equations when a, b, and c are different. Give the conditions for a solution when $a = b \neq c$ and give the most general solution in this case.

9. Consider the augmented matrix

$$[\mathbf{A}, \mathbf{b}] = \begin{bmatrix} 1 & 3 & -8 & 2 \\ 1 & -9 & -10 & -3 \\ -1 & 3 & 9 & 0 \end{bmatrix}.$$

(a) Use simple Gauss elimination to find the rank of $[\mathbf{A}, \mathbf{b}]$ and $\mathbf{A}$.
(b) How many of the column vectors of $[\mathbf{A}, \mathbf{b}]$ are linearly independent? Why? How many row vectors are linearly independent? Why?
(c) If the problem

$$\mathbf{A}\mathbf{x} = \mathbf{b} \tag{1}$$

has a solution, find the most general one. If there is no solution, why not?
(d) Is there a solution to

$$\mathbf{A}^\dagger \mathbf{z} = \mathbf{0}?$$

What are the implications of the answer to this question to the solvability of Eq. (1)?

10. Repeat parts (a)–(d) in Problem 9 using the following augmented matrix:

$$[\mathbf{A}, \mathbf{b}] = \begin{bmatrix} 3 & -1 & 2 & -1 \\ 6 & 0 & 2 & -2 \\ 0 & 2 & 2 & 0 \\ -9 & 3 & 0 & 3 \\ 3 & 2 & 3 & -1 \end{bmatrix}.$$

11. Prove that

$$(\mathbf{I} + \mathbf{B})\mathbf{x} = \mathbf{b}$$

has a solution for arbitrary real $\mathbf{b}$ if $\mathbf{B}$ is a real skew symmetric matrix ($\mathbf{B}^T = -\mathbf{B}$).

12. In the pyrolysis of a low-molecular-weight hydrocarbon, the following species are present: C_2H_6, H, C_2H_5, CH_3, CH_4, H_2, C_2H_4, C_3H_8, and C_4H_{10}. Determine the number of independent reactions among these species.

13. A reaction mixture is found to consist of O_2, H_2, CO, CO_2, H_2CO, CH_3OH, C_2H_5OH, $(CH_3)_2CO$, CH_3CHO, CH_4, and H_2O. How many independent components are there in the mixture? Which ones can they be? What is the minimum number of reactions possible to produce this mixture?

FURTHER READING

Amundson, A. R. (1964). "Mathematical Methods in Chemical Engineering." Prentice Hall, Englewood Cliffs, NJ.
Bellman, R. (1970). "Introduction to Matrix Analysis." McGraw-Hill, New York.
Noble B., and Daniel, J. W. (1977). "Applied Linear Algebra." Prentice Hall, Englewood Cliffs, NJ.

5
THE EIGENPROBLEM

5.1. SYNOPSIS

In this chapter we will present the concept of an $n \times n$ matrix as a *linear operator* in an n-dimensional linear vector space E_n. We can, as shown in Chapter 2, define an inner product from which we can create an *inner product space*. The inner product provides us with convenient vector and matrix norms. We will show that the inner product $\langle \mathbf{x}, \mathbf{y} \rangle$ obeys the *Schwarz inequality*, $|\langle \mathbf{x}, \mathbf{y} \rangle| \leq \|\mathbf{x}\| \|\mathbf{y}\|$, where $\|\mathbf{x}\|$ denotes the norm or length of the vector $\mathbf{x}$. The Schwarz inequality, in turn, implies the *triangle inequality* $\|\mathbf{x} + \mathbf{y}\| \leq \|\mathbf{x}\| + \|\mathbf{y}\|$ for the norms (lengths) of the vectors, and $\|\mathbf{A} + \mathbf{B}\| \leq \|\mathbf{A}\| + \|\mathbf{B}\|$ for the norms of the matrices. The inequality also implies that $\|\mathbf{AB}\| \leq \|\mathbf{A}\| \|\mathbf{B}\|$.

Any set of n linearly independent vectors $\{\mathbf{x}_1, \ldots, \mathbf{x}_n\}$ forms a basis set in E_n such that any vector $\mathbf{x}$ belonging to E_n can be expanded in the form

$$\mathbf{x} = \sum_{i=1}^{n} \alpha_i \mathbf{x}_i,$$

where α_i are complex scalar numbers. We can subsequently define the reciprocal basis set $\{\mathbf{z}_1, \ldots, \mathbf{z}_n\}$ by the properties $\mathbf{x}_i^\dagger \mathbf{z}_j = \delta_{ij}$. We will show that the vectors $\{\mathbf{z}_i\}$ can be formed from the column vectors of $(\mathbf{X}^{-1})^\dagger$, where $\mathbf{X} = [\mathbf{x}_1, \ldots, \mathbf{x}_n]$. We say that $\{\mathbf{x}_i\}$ and $\{\mathbf{z}_i\}$, satisfying the orthonormality conditions $\mathbf{x}_i^\dagger \mathbf{z}_j = \delta_{ij}$, form biorthogonal sets.

The concept of basis and reciprocal basis sets is immensely important in the analysis of *perfect* matrices. We say that the $n \times n$ matrix $\mathbf{A}$ is a *perfect* matrix if $\mathbf{A}$ has n eigenvectors, where an eigenvector $\mathbf{x}_i$ of $\mathbf{A}$ obeys the eigenequation

$$\mathbf{A}\mathbf{x}_i = \lambda_i \mathbf{x}_i,$$

and λ_i is a scalar (real or complex number) called the eigenvalue. Specifically, we say $\mathbf{x}_i$ is an eigenvector of $\mathbf{A}$ with eigenvalue λ_i. We will show that the $n \times n$ identity matrix $\mathbf{I}_n$ can be expressed by the equation

$$\mathbf{I}_n = \sum_{i=1}^{n} \mathbf{x}_i \mathbf{z}_i^\dagger$$

for any linearly independent basis set $\{\mathbf{x}_i\}$ and its corresponding reciprocal basis set $\{\mathbf{z}_i\}$. Since $\mathbf{A} = \mathbf{A}\mathbf{I}$, it follows that a *perfect* matrix $\mathbf{A}$ obeys the *spectral resolution theorem*

$$\mathbf{A} = \sum_{i=1}^{n} \lambda_i \mathbf{x}_i \mathbf{z}_i^\dagger.$$

From this we will show that the reciprocal vectors $\{\mathbf{z}_i\}$ are the eigenvectors of the adjoint matrix $\mathbf{A}^\dagger$ and that the eigenvalues of the adjoint matrix are λ_i^* (the complex conjugate of λ_i).

The spectral resolution theorem enables us to express the function $f(\mathbf{A})$ as $f(\mathbf{A}) = \sum_{i=1}^{n} f(\lambda_i) \mathbf{x}_i \mathbf{z}_i^\dagger$ for any function $f(t)$ that can be expanded as a Taylor or Laurent series in t near the eigenvalues λ_i. As we will show, the spectral resolution theorem for $\exp(\alpha \mathbf{A})$ provides a simple solution to the differential equation $d\mathbf{x}/dt = \mathbf{A}\mathbf{x}$, $\mathbf{x}(t=0) = \mathbf{x}_0$.

We will see that the eigenvalues of $\mathbf{A}$ obey the characteristic equation $|\mathbf{A} - \lambda \mathbf{I}| = 0$, which is an nth-degree polynomial $P_n(\lambda) = \sum_{j=0}^{n} a_j (-\lambda)^{n-j}$, where $a_j = \text{tr}_j \mathbf{A}$. Note that $\text{tr}_j \mathbf{A}$ is the jth trace of $\mathbf{A}$, which is the sum of the jth-order principal minors of $\mathbf{A}$. The traces of $\mathbf{A}$ are invariant to a similarity transformation; i.e., $\text{tr}_j \mathbf{A} = \text{tr}_j (\mathbf{S}^{-1} \mathbf{A} \mathbf{S})$, where $\mathbf{S}$ is any nonsingular matrix. This invariance implies that

$$\text{tr}_j \mathbf{A} = \sum_{i_j > i_{j-1}}^{n} \sum^{n} \cdots \sum_{i_2 > i_1 = 1}^{n} \sum^{n} \lambda_{i_j} \lambda_{i_{j-1}} \cdots \lambda_{i_1} = \text{tr}_j \mathbf{\Lambda};$$

i.e., the jth trace of $\mathbf{A}$ is the sum of all distinct j-tuples of the eigenvalues $\lambda_1, \ldots, \lambda_n$. $\mathbf{\Lambda}$ is a diagonal matrix with the eigenvalues of $\mathbf{A}$ on the main diagonal, i.e., $\mathbf{\Lambda} = [\lambda_i \delta_{ij}]$. We will also prove that the traces obey the property $\text{tr}_j(\mathbf{AB}) = \text{tr}_j(\mathbf{BA})$. This implies that the eigenvalues of $\mathbf{AB}$ are the same as the eigenvalues of $\mathbf{BA}$.

We can determine the degree of degeneracy of an eigenvalue without actually solving for the eigenvectors. The number n_i of linearly independent eigenvectors of $\mathbf{A}$ that correspond to a given eigenvalue λ_i (called the degeneracy of λ_i) is equal to $n - r_{\lambda_i}$, where r_{λ_i} is the rank of the characteristic determinant $|\mathbf{A} - \lambda_i \mathbf{I}|$. If $\mathbf{A}$ has n_i eigenvectors $\{\mathbf{x}_i\}$ corresponding to the eigenvalue λ_i, then the adjoint $\mathbf{A}^\dagger$ has n_i eigenvectors $\{\mathbf{z}_i\}$ corresponding to the eigenvalue λ_i^*. If $\lambda_i \neq \lambda_j$, then the eigenvectors of $\mathbf{A}$ corresponding to λ_i are orthogonal to the eigenvectors of $\mathbf{A}^\dagger$ corresponding to λ_j^*.

LINEAR OPERATORS IN A NORMED LINEAR VECTOR SPACE

Solving the characteristic polynomial equation $P_n(\lambda) = 0$ can be numerically tricky, and so other methods are sometimes sought for finding the eigenvalues. For instance, tridiagonal matrices are often handled via a recurrence method. The power method is an iterative method that is especially easy to implement for computing the eigenvalue of maximum magnitude, but the method can also be used to compute any eigenvalue. We will end the chapter by presenting Gerschgorin's theorem, which gives a domain in complex space containing all the eigenvalues of a given matrix. As we will see, this is sometimes useful for estimating eigenvalues.

5.2. LINEAR OPERATORS IN A NORMED LINEAR VECTOR SPACE

We asserted in Section 2.6 that S is a *linear vector space* if it has the properties:
1. If $\mathbf{x}, \mathbf{y} \in S$, then

$$\mathbf{x} + \mathbf{y} \in S. \tag{5.2.1}$$

2.
$$\mathbf{x} + \mathbf{y} = \mathbf{y} + \mathbf{x}. \tag{5.2.2}$$

3. There exists a zero vector $\mathbf{0}$ such that

$$\mathbf{x} + \mathbf{0} = \mathbf{x}. \tag{5.2.3}$$

4. For every $\mathbf{x} \in S$, there exists $-\mathbf{x}$ such that

$$\mathbf{x} + (-\mathbf{x}) = \mathbf{0}. \tag{5.2.4}$$

5. If α and β are complex numbers, then

$$\begin{aligned} \alpha(\beta \mathbf{x}) &= (\alpha\beta)\mathbf{x} \\ (\alpha + \beta)\mathbf{x} &= \alpha\mathbf{x} + \beta\mathbf{x} \\ \alpha(\mathbf{x} + \mathbf{y}) &= \alpha\mathbf{x} + \alpha\mathbf{y}. \end{aligned} \tag{5.2.5}$$

To produce the *inner product space* S, we add to the above properties of S an *inner product*, $\langle \mathbf{x}, \mathbf{y} \rangle$, which is a scalar object defined by the following properties: if $\mathbf{x}, \mathbf{y} \in S$, then

$$\begin{aligned} \langle \mathbf{x}, \alpha\mathbf{y} \rangle &= \alpha \langle \mathbf{x}, \mathbf{y} \rangle, \\ \langle \mathbf{y}, \mathbf{x} \rangle &= \langle \mathbf{x}, \mathbf{y} \rangle^*, \\ \langle \mathbf{x}, \mathbf{x} \rangle &> 0 \quad \text{if } \mathbf{x} \neq \mathbf{0}, \end{aligned}$$

and

$$\langle \mathbf{x}, \mathbf{x} \rangle = 0 \quad \text{if and only if } \mathbf{x} = \mathbf{0}, \tag{5.2.6}$$

where $\langle \mathbf{x}, \mathbf{y} \rangle^*$ denotes the complex conjugate of $\langle \mathbf{x}, \mathbf{y} \rangle$.

We define the *norm* or *length* $\|\mathbf{x}\|$ of a vector $\mathbf{x}$ in an inner product space by

$$\|\mathbf{x}\| \equiv \sqrt{\langle \mathbf{x}, \mathbf{x}\rangle}. \tag{5.2.7}$$

The properties of the inner product given in Eq. (5.2.6) yield a vector norm (or length) obeying the defining conditions of a *normed linear vector space* (Eq. (2.6.8)), which are the physical conditions commonly associated with the concept of length in the three-dimensional Euclidean vector space we occupy. Namely,

1. $\|\mathbf{x}\| \geq 0$ and $\|\mathbf{x}\| = 0$ only if $\mathbf{x} = \mathbf{0}$.
2. $\|\alpha \mathbf{x}\| = |\alpha|\, \|\mathbf{x}\|$ for any complex number. (5.2.8)
3. $\|\mathbf{x} + \mathbf{y}\| \leq \|\mathbf{x}\| + \|\mathbf{y}\|$.

Here $|\alpha|$ denotes the absolute value of the complex number α, i.e., $|\alpha| = \sqrt{\alpha_R^2 + \alpha_I^2}$, where α_R and α_I are the real and imaginary parts of α ($\alpha = \alpha_R + i\alpha_I$).

The proof of item (3), called the *triangle inequality*, proceeds by first proving the *Schwarz inequality*. To prove this, we note that $J \geq 0$, where

$$J \equiv \langle \mathbf{x} + \alpha \mathbf{y}, \mathbf{x} + \alpha \mathbf{y}\rangle = \|\mathbf{x}\|^2 + \alpha^* \langle \mathbf{x}, \mathbf{y}\rangle^* + \alpha \langle \mathbf{x}, \mathbf{y}\rangle + |\alpha|^2 \|\mathbf{y}\|^2, \tag{5.2.9}$$

and $\alpha (= \alpha_R + i\alpha_I)$ is an arbitrary complex number. Since $J \geq 0$, we seek the value of α that minimizes the value of J for a given $\mathbf{x}$ and $\mathbf{y}$. At the minimum, α_R and α_I obey the equations

$$\frac{\partial J}{\partial \alpha_R} = \langle \mathbf{x}, \mathbf{y}\rangle^* + \langle \mathbf{x}, \mathbf{y}\rangle + 2\alpha_R \|\mathbf{y}\|^2 = 0 \tag{5.2.10}$$

$$\frac{\partial J}{\partial \alpha_I} = -\langle \mathbf{x}, \mathbf{y}\rangle^* i + \langle \mathbf{x}, \mathbf{y}\rangle i + 2\alpha_I \|\mathbf{y}\|^2 = 0 \tag{5.2.11}$$

whose solutions yield

$$\alpha_R = \frac{-\mathrm{Re}\langle \mathbf{x}, \mathbf{y}\rangle}{\|\mathbf{y}\|^2}, \qquad \alpha_I = \frac{\mathrm{Im}\langle \mathbf{x}, \mathbf{y}\rangle}{\|\mathbf{y}\|^2} \tag{5.2.12}$$

or

$$\alpha = -\frac{\langle \mathbf{x}, \mathbf{y}\rangle^*}{\|\mathbf{y}\|^2}. \tag{5.2.13}$$

The notation $\mathrm{Re}\langle \mathbf{x}, \mathbf{y}\rangle$ and $\mathrm{Im}\langle \mathbf{x}, \mathbf{y}\rangle$ denotes the real and imaginary parts of $\langle \mathbf{x}, \mathbf{y}\rangle$. Insertion of Eq. (5.2.13) into Eq. (5.2.9) yields

$$J_{\min} = \|\mathbf{x}\|^2 - \frac{2|\langle \mathbf{x}, \mathbf{y}\rangle|^2}{\|\mathbf{y}\|^2} + \frac{|\langle \mathbf{x}, \mathbf{y}\rangle|^2}{\|\mathbf{y}\|^4}\|\mathbf{y}\|^2 \geq 0 \tag{5.2.14}$$

or, upon rearranging,

$$|\langle \mathbf{x}, \mathbf{y}\rangle| \leq \|\mathbf{x}\|\, \|\mathbf{y}\|, \tag{5.2.15}$$

which is the Schwarz inequality.

LINEAR OPERATORS IN A NORMED LINEAR VECTOR SPACE

In the ordinary Euclidean vector space, $\langle \mathbf{a}, \mathbf{b}\rangle \equiv \mathbf{a} \cdot \mathbf{b} = \|\mathbf{a}\| \|\mathbf{b}\| \cos\theta$, where θ is the angle between the vectors $\mathbf{a}$ and $\mathbf{b}$. If $\mathbf{a}$ and $\mathbf{b}$ are colinear, then $|\cos\theta| = 1$ and $|\mathbf{a} \cdot \mathbf{b}| = \|\mathbf{a}\| \|\mathbf{b}\|$. Otherwise, $|\cos\theta| < 1$ since the projection of $\mathbf{a}$ onto the direction of $\mathbf{b}$ is shorter than the length of $\mathbf{a}$, as shown in Fig. 5.2.1. In an abstract linear vector space, the Schwarz inequality is geometrically equivalent to the projection of the vector $\mathbf{x}$ onto the direction $\hat{\mathbf{y}}$ of $\mathbf{y}$, where $\hat{\mathbf{y}}$ is the unit vector

$$\hat{\mathbf{y}} = \frac{\mathbf{y}}{\|\mathbf{y}\|}. \tag{5.2.16}$$

In terms of this unit vector, Eq. (5.2.15) reads

$$|\langle \mathbf{x}, \hat{\mathbf{y}}\rangle| \le \|\mathbf{x}\|, \tag{5.2.17}$$

which is analogous to $|\mathbf{a} \cdot \hat{\mathbf{b}}| = \|\mathbf{a}\| |\cos\theta| \le \|\mathbf{a}\|$. We see that only if $\mathbf{x}$ and $\mathbf{y}$ are colinear will the equality hold in Eq. (5.2.15).

To prove the triangle inequality, note that

$$\|\mathbf{x} + \mathbf{y}\|^2 = \|\mathbf{x}\|^2 + 2|\langle \mathbf{x}, \mathbf{y}\rangle| + \|\mathbf{y}\|^2, \tag{5.2.18}$$

which, when combined with Eq. (5.2.15), yields

$$\|\mathbf{x} + \mathbf{y}\|^2 \le \|\mathbf{x}\|^2 + 2\|\mathbf{x}\| \|\mathbf{y}\| + \|\mathbf{y}\|^2 = (\|\mathbf{x}\| + \|\mathbf{y}\|)^2$$

or

$$\|\mathbf{x} + \mathbf{y}\| \le \|\mathbf{x}\| + \|\mathbf{y}\|. \tag{5.2.19}$$

Linear operators in a vector space were also defined in Section 2.6. We say that $\mathbf{A}$ is a linear operator in S if it has the properties

1. If $\mathbf{x} \in S$, then $\mathbf{A}\mathbf{x} \in S$. (5.2.20)
2. If $\mathbf{x}, \mathbf{y} \in S$, then $\mathbf{A}(\alpha\mathbf{x} + \beta\mathbf{y}) = \alpha\mathbf{A}\mathbf{x} + \beta\mathbf{A}\mathbf{y}$. (5.2.21)

We saw in Section 2.6 that if the norm of a linear operator is defined by

$$\|\mathbf{A}\| = \max_{\mathbf{x}\ne 0} \frac{\|\mathbf{A}\mathbf{x}\|}{\|\mathbf{x}\|}, \tag{5.2.22}$$

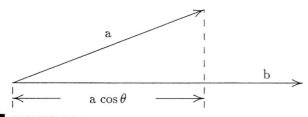

FIGURE 5.2.1

then the vector norm properties listed in Eq. (5.2.8) imply the following properties for the norms of linear operators:

1. $\|\mathbf{A}\| \geq 0$ and $\|\mathbf{A}\| = 0$ *if and only if* $\mathbf{A} = \mathbf{0}$. (5.2.23)
2. $\|\alpha \mathbf{A}\| = |\alpha| \|\mathbf{A}\|$, α for any complex number. (5.2.24)
3. $\|\mathbf{A} + \mathbf{B}\| \leq \|\mathbf{A}\| + \|\mathbf{B}\|$. (5.2.25)
4. $\|\mathbf{AB}\| \leq \|\mathbf{A}\| \|\mathbf{B}\|$. (5.2.26)

Any $n \times n$ matrix $\mathbf{A}$ obeying the previously defined properties of multiplication and addition is a linear operator in the n-dimensional linear vector space E_n. If we define the inner product by

$$\langle \mathbf{x}, \mathbf{y} \rangle = \mathbf{x}^\dagger \mathbf{y} = \sum_{i=1}^{n} x_i^* y_i, \qquad (5.2.27)$$

then $\mathbf{A}$ becomes a linear operator in a *normed linear vector space* with the norm $\sqrt{\mathbf{x}^\dagger \mathbf{x}}$. As an operator, $\mathbf{A}$ transforms a vector $\mathbf{x}$ in E_n into another vector in E_n. Geometrically, this transformation can involve rotation, stretching (or shrinking), or both. The possibilities are illustrated in Fig. 5.2.2 for vectors in E_3. Note that the vector $\mathbf{y}$ is the same length as $\mathbf{x}$, i.e., $\|\mathbf{x}\| = \|\mathbf{y}\|$, and so in this case the action of $\mathbf{A}$ on $\mathbf{x}$ is purely a rotation since $\mathbf{x}$ and $\mathbf{y}$ are not colinear. The vector $\mathbf{z}$ is neither colinear with $\mathbf{x}$ nor of the same length, and so in this case the action of $\mathbf{A}$ is to rotate and stretch (if $\|\mathbf{z}\| > \|\mathbf{x}\|$) or shrink (if $\|\mathbf{z}\| < \|\mathbf{x}\|$) the vector $\mathbf{x}$. In the third case, $\mathbf{Ax} = \lambda \mathbf{x}$, where λ is a scalar (real or complex). Thus, $\lambda \mathbf{x}$ is colinear with $\mathbf{x}$ and so $\mathbf{A}$ merely stretches ($|\lambda| > 1$) or shrinks ($|\lambda| < 1$) the vector $\mathbf{x}$. In this last case, i.e., *when*

$$\mathbf{Ax} = \lambda \mathbf{x}, \qquad (5.2.28)$$

we say that $\mathbf{x}$ *is an eigenvector of* $\mathbf{A}$ *and* λ *is an eigenvalue of* $\mathbf{A}$. Determining what vectors $\mathbf{x}$ and numbers λ satisfy Eq. (5.2.28) is referred to as the *eigenproblem*.

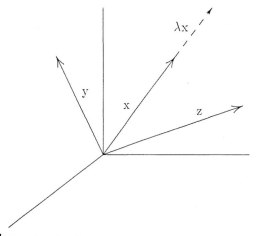

FIGURE 5.2.2

We will soon learn that a matrix can be usefully classified in terms of how many linearly independent eigenvectors it has. If an $n \times n$ matrix has n eigenvectors, it is called a *perfect* matrix. Otherwise, it is called an *imperfect* or *defective* matrix. The value of this classification will become clear later.

The infinite-dimensional vector space E_∞, whose vectors $\mathbf{x}$ have an infinite number of components, occurs in physical problems such as those encountered in Chapters 9 and 10. With the inner product defined by

$$\mathbf{x}^\dagger \mathbf{y} = \sum_{i=1}^\infty x_i^* y_i, \tag{5.2.29}$$

the requirement that vectors $\mathbf{x}$ and $\mathbf{y}$ belong to a *normed linear vector space* is that

$$\|\mathbf{x}\|^2 = \sum_{i=1}^\infty |x_i|^2 < \infty \quad \text{and} \quad \|\mathbf{y}\|^2 = \sum_{i=1}^\infty |y_i|^2 < \infty. \tag{5.2.30}$$

Thus, the vector

$$\mathbf{x} = \begin{bmatrix} 1 \\ \dfrac{1}{2} \\ \dfrac{1}{3} \\ \vdots \end{bmatrix} \tag{5.2.31}$$

with the norm

$$\|\mathbf{x}\| = \left[\sum_{i=1}^\infty \frac{1}{i^2}\right]^{1/2} \simeq 1.26 \tag{5.2.32}$$

belongs to a *normed linear vector space*, whereas the vector

$$\mathbf{y} = \begin{bmatrix} 1 \\ \dfrac{1}{\sqrt{2}} \\ \dfrac{1}{\sqrt{3}} \\ \vdots \end{bmatrix} \tag{5.2.33}$$

with the norm

$$\|\mathbf{y}\| = \left[\sum_{i=1}^\infty \frac{1}{i}\right]^{1/2} = \infty \tag{5.2.34}$$

does not belong to a *normed linear vector space*.

We can define inner products other than $\mathbf{x}^\dagger\mathbf{y}$, which are sometimes useful. For example, if $\mathbf{A}$ is a positive-definite matrix (i.e., if $\mathbf{A} = \mathbf{A}^\dagger$ and $\mathbf{x}^\dagger\mathbf{A}\mathbf{x} > 0$ for all $\mathbf{x} \neq \mathbf{0}$ in E_n), then the inner products

$$\langle \mathbf{y}, \mathbf{x} \rangle = \mathbf{y}^\dagger \mathbf{A} \mathbf{x} \tag{5.2.35}$$

and

$$\langle \mathbf{y}, \mathbf{x} \rangle = \mathbf{y}^\dagger \mathbf{A}^\dagger \mathbf{A} \mathbf{x} \tag{5.2.36}$$

satisfy the defining criteria in Eq. (5.2.6). The inner product in either Eq. (5.2.35) or (5.2.36) with the norm defined by $\|\mathbf{x}\| = \sqrt{\langle \mathbf{x}^\dagger, \mathbf{x} \rangle}$ will yield a *normed linear vector space* obeying the conditions in Eq. (5.2.8).

We can, of course, define a normed linear vector space in terms of the conditions in Eq. (5.2.8) without defining an inner product. For example, the norm $\|\mathbf{x}\| = \left[\sum_{i=1}^n |x_i|^p\right]^{1/p}$, for any positive real p, defines an acceptable normed linear vector space in E_n. However, $\|\mathbf{x}\|_p$ is not necessarily related to an inner product.

In the remainder of this and the next two chapters, we will concentrate primarily on $n \times n$ matrices as linear operators in the finite-dimensional linear vector space E_n in which we will usually define the inner product as $\mathbf{y}^\dagger\mathbf{x}$. We will, however, demonstrate the utility of alternative inner products such as Eq. (5.2.35) or (5.2.36) for symmetrizing certain problems and for expediting certain proofs. In the last three chapters, we will explore linear operators in infinite-dimensional spaces. These include integral or differential operators in function spaces.

■ **EXERCISE 5.2.1.** Consider the matrix

$$\mathbf{A} = \begin{bmatrix} 2 & -1 \\ -1 & 2 \end{bmatrix}.$$

Show that the inner product

$$\langle \mathbf{y}, \mathbf{x} \rangle = \mathbf{y}^\dagger \mathbf{A} \mathbf{x}$$

■ ■ ■ obeys all the conditions in Eq. (5.2.6).

5.3. BASIS SETS IN A NORMED LINEAR VECTOR SPACE

If $\mathbf{x}$ is a vector in E_n, it can always be expressed in the form

$$\mathbf{x} = \sum_{i=1}^n x_i \mathbf{e}_i, \tag{5.3.1}$$

where

$$\mathbf{e}_i = \begin{bmatrix} 0 \\ \vdots \\ 0 \\ 1 \\ 0 \\ \vdots \\ 0 \end{bmatrix} i\text{th row}. \qquad (5.3.2)$$

The set $\{\mathbf{e}_1, \ldots, \mathbf{e}_n\}$ is thus a basis set in E_n in the sense that any vector $\mathbf{x}$ can be written as a linear combination of $\{\mathbf{e}_i\}$. Since the set obeys the condition

$$\mathbf{e}_i^\dagger \mathbf{e}_j = \delta_{ij}, \qquad (5.3.3)$$

where $\delta_{ij} = 1$ for $i = j$ and $\delta_{ij} = 0$ for $i \neq j$, we say that the set $\{\mathbf{e}_i\}$ forms an *orthonormal basis set*. That is to say, the vectors $\mathbf{e}_i$ are mutually orthogonal ($\mathbf{e}_i^\dagger \mathbf{e}_j = 0$, $i \neq j$) and of unit length ($\|\mathbf{e}_i\|^2 = \mathbf{e}_i^\dagger \mathbf{e}_i = 1$).

Although the set $\{\mathbf{e}_i\}$ is perhaps the simplest basis set, it is by no means the only one. Suppose $\{\mathbf{x}_1, \mathbf{x}_2, \ldots, \mathbf{x}_n\}$ is any linearly independent set in E_n. Then this set provides a basis set in E_n. The requirement for a basis set is that for any vector $\mathbf{x}$ there exists a unique set of numbers $\{\alpha_1, \ldots, \alpha_n\}$ such that

$$\mathbf{x} = \sum_{i=1}^n \alpha_i \mathbf{x}_i. \qquad (5.3.4)$$

This equation can be rewritten as

$$\mathbf{x} = \mathbf{X}\boldsymbol{\alpha}, \qquad (5.3.5)$$

where $\mathbf{X}$ is a square matrix whose column vectors are $\mathbf{x}_i$, i.e.,

$$\mathbf{X} = [\mathbf{x}_1, \ldots, \mathbf{x}_n] \qquad (5.3.6)$$

and $\boldsymbol{\alpha}$ is the vector

$$\boldsymbol{\alpha} = \begin{bmatrix} \alpha_1 \\ \vdots \\ \alpha_n \end{bmatrix}. \qquad (5.3.7)$$

Since the vectors $\{\mathbf{x}_1, \ldots, \mathbf{x}_n\}$ are linearly independent, $\mathbf{X}$ is nonsingular ($|\mathbf{X}| \neq 0$) and Eq. (5.3.5) has the unique solution

$$\boldsymbol{\alpha} = \mathbf{X}^{-1}\mathbf{x}, \qquad (5.3.8)$$

proving that the set $\{\mathbf{x}_1, \ldots, \mathbf{x}_n\}$ forms a basis set in E_n.

We say $\{\mathbf{x}_1, \ldots, \mathbf{x}_n\}$ is an *orthonormal basis set* if

$$\mathbf{x}_i^\dagger \mathbf{x}_j = \delta_{ij}. \qquad (5.3.9)$$

As we have seen, one example of such a set is $\{\mathbf{e}_1, \mathbf{e}_2, \ldots, \mathbf{e}_n\}$. However, there are numerous other orthonormal sets. In fact, we can use the *Gram–Schmidt procedure* to construct an orthonormal set from any linearly independent set $\{\mathbf{z}_1, \ldots, \mathbf{z}_n\}$. The Gram–Schmidt procedure constructs an orthogonal set $\{\mathbf{y}_1, \ldots, \mathbf{y}_n\}$ from the set $\{\mathbf{z}_i\}$ as follows:

$$\mathbf{y}_1 = \mathbf{z}_1$$

$$\mathbf{y}_2 = \mathbf{z}_2 - \frac{(\mathbf{y}_1^\dagger \mathbf{z}_2)}{\|\mathbf{y}_1\|^2} \mathbf{y}_1$$

$$\mathbf{y}_3 = \mathbf{z}_3 - \frac{(\mathbf{y}_1^\dagger \mathbf{z}_3)}{\|\mathbf{y}_1\|^2} \mathbf{y}_1 - \frac{(\mathbf{y}_2^\dagger \mathbf{z}_3)}{\|\mathbf{y}_2\|^2} \mathbf{y}_2$$

$$\vdots \qquad (5.3.10)$$

$$\mathbf{y}_i = \mathbf{z}_i - \sum_{j=1}^{i-1} \frac{(\mathbf{y}_j^\dagger \mathbf{z}_i)}{\|\mathbf{y}_j\|^2} \mathbf{y}_j$$

$$\vdots$$

$$\mathbf{y}_n = \mathbf{z}_n - \sum_{j=1}^{n-1} \frac{(\mathbf{y}_j^\dagger \mathbf{z}_n)}{\|\mathbf{y}_j\|^2} \mathbf{y}_j.$$

Note that

$$\mathbf{y}_1^\dagger \mathbf{y}_2 = 0$$
$$\mathbf{y}_1^\dagger \mathbf{y}_3 = \mathbf{y}_2^\dagger \mathbf{y}_3 = 0$$
$$\vdots \qquad (5.3.11)$$
$$\mathbf{y}_j^\dagger \mathbf{y}_i = 0, \qquad i \neq j, \; j = 1, \ldots, i-1, \; i = 1, \ldots, n,$$

i.e., each vector $\mathbf{y}_i$ is orthogonal to every other vector $\mathbf{y}_j$, $j \neq i$. There are exactly n orthogonal vectors $\mathbf{y}_i$ corresponding to the n linearly independent vectors $\mathbf{z}_i$. If there were fewer $\mathbf{y}_i$, say $n-1$, then $\mathbf{y}_n = \mathbf{0}$, which implies that a linear combination of $\mathbf{z}_i$ is 0. This, of course, is a contradiction of our hypothesis that the $\mathbf{z}_i$ are linearly independent. If we now define

$$\mathbf{x}_i = \frac{\mathbf{y}_i}{\|\mathbf{y}_i\|}, \qquad i = 1, \ldots, n, \qquad (5.3.12)$$

then the vectors $\mathbf{x}_i$ obey

$$\mathbf{x}_i^\dagger \mathbf{x}_j = \delta_{ij}, \qquad (5.3.13)$$

and so they form an orthonormal set. The Gram–Schmidt procedure is the same for a more general inner product—simply replace $\mathbf{y}_i^\dagger \mathbf{z}_j$ in Eq. (5.3.10) by $\langle \mathbf{y}, \mathbf{z} \rangle$.

BASIS SETS IN A NORMED LINEAR VECTOR SPACE

EXAMPLE 5.3.1. Consider the set

$$\mathbf{z}_1 = \begin{bmatrix} 2 \\ 1 \\ 1 \end{bmatrix}, \quad \mathbf{z}_2 = \begin{bmatrix} 1 \\ 2 \\ 1 \end{bmatrix}, \quad \mathbf{z}_3 = \begin{bmatrix} 1 \\ 1 \\ 2 \end{bmatrix}. \tag{5.3.14}$$

Since the determinant of

$$[\mathbf{z}_1, \mathbf{z}_2, \mathbf{z}_3] = \begin{bmatrix} 2 & 1 & 1 \\ 1 & 2 & 1 \\ 1 & 1 & 2 \end{bmatrix} \tag{5.3.15}$$

is nonzero (D equals 4), the $\mathbf{z}_i$ are linearly independent. By the Gram–Schmidt procedure, we obtain

$$\mathbf{y}_1 = \begin{bmatrix} 2 \\ 1 \\ 1 \end{bmatrix}$$

$$\mathbf{y}_2 = \begin{bmatrix} 1 \\ 2 \\ 1 \end{bmatrix} - \frac{5}{6} \begin{bmatrix} 2 \\ 1 \\ 1 \end{bmatrix} = \begin{bmatrix} -\frac{2}{3} \\ \frac{7}{6} \\ \frac{1}{6} \end{bmatrix} \tag{5.3.16}$$

$$\mathbf{y}_3 = \begin{bmatrix} 1 \\ 1 \\ 2 \end{bmatrix} - \frac{5}{6} \begin{bmatrix} 2 \\ 1 \\ 1 \end{bmatrix} - \frac{5}{11} \begin{bmatrix} -\frac{2}{3} \\ \frac{7}{6} \\ \frac{1}{6} \end{bmatrix} = \begin{bmatrix} -\frac{4}{11} \\ -\frac{4}{11} \\ \frac{12}{11} \end{bmatrix}.$$

Next, defining $\mathbf{x}_i = \mathbf{y}_i / \|\mathbf{y}_i\|$, we find the orthonormal set

$$\mathbf{x}_1 = \begin{bmatrix} \sqrt{\frac{2}{3}} \\ \sqrt{\frac{1}{6}} \\ \sqrt{\frac{1}{6}} \end{bmatrix}, \quad \mathbf{x}_2 = \begin{bmatrix} -2\sqrt{\frac{2}{33}} \\ 7\sqrt{\frac{1}{66}} \\ \sqrt{\frac{1}{66}} \end{bmatrix}, \quad \mathbf{x}_3 = \begin{bmatrix} -\sqrt{\frac{1}{11}} \\ -\sqrt{\frac{1}{11}} \\ 3\sqrt{\frac{1}{11}} \end{bmatrix} \tag{5.3.17}$$

■ ■ ■ for which $\mathbf{x}_i^\dagger \mathbf{x}_j = \delta_{ij}$.

When $\{\mathbf{x}_1, \ldots, \mathbf{x}_n\}$ is an orthonormal set, the coefficients α_i in the expansion

$$\mathbf{x} = \sum_{i=1}^{n} \alpha_i \mathbf{x}_i \tag{5.3.18}$$

are especially simple to calculate. If we multiply Eq. (5.3.18) by $\mathbf{x}_j^\dagger$, we find

$$\mathbf{x}_j^\dagger \mathbf{x} = \sum_{i=1}^{n} \alpha_i \mathbf{x}_j^\dagger \mathbf{x}_i = \sum_{i=1}^{n} \alpha_i \delta_{ij} \tag{5.3.19}$$

or

$$\alpha_j = \mathbf{x}_j^\dagger \mathbf{x}. \tag{5.3.20}$$

Thus, for an orthonormal basis set, the coefficients α_i in Eq. (5.3.18) are just the inner products of $\mathbf{x}_i$ with $\mathbf{x}$. Since the product $\mathbf{x}_i \alpha_i$ is the same as $\alpha_i \mathbf{x}_i$, insertion of Eq. (5.3.20) into Eq. (5.3.18) allows the rearrangement

$$\mathbf{x} = \sum_{i=1}^{n} \mathbf{x}_i \mathbf{x}_i^\dagger \mathbf{x} = \left(\sum_{i=1}^{n} \mathbf{x}_i \mathbf{x}_i^\dagger \right) \mathbf{x}. \tag{5.3.21}$$

This relationship is valid for any vector $\mathbf{x}$ in E_n. Therefore, the matrix $\sum_{i=1}^{n} \mathbf{x}_i \mathbf{x}_i^\dagger$ has the property that it maps any vector $\mathbf{x}$ in E_n onto itself. This is, by definition, the unit matrix $\mathbf{I}$, i.e.,

$$\mathbf{I} = \sum_{i=1}^{n} \mathbf{x}_i \mathbf{x}_i^\dagger. \tag{5.3.22}$$

Equation (5.3.22) is called a *resolution of the identity matrix*. Application of the right-hand side of Eq. (5.3.22) to $\mathbf{x}$ yields $\mathbf{x}$ again, but expands it as a linear combination of the basis set $\{\mathbf{x}_i\}$. Different orthonormal basis sets are analogous to different Cartesian coordinate frames in a three-dimensional Euclidean vector space.

EXERCISE 5.3.1. Show by direct summation that

$$\sum_{i=1}^{3} \mathbf{e}_i \mathbf{e}_i^\dagger = \sum_{i=1}^{3} \mathbf{x}_i \mathbf{x}_i^\dagger = \begin{bmatrix} 1 & 0 & 0 \\ 0 & 1 & 0 \\ 0 & 0 & 1 \end{bmatrix}, \tag{5.3.23}$$

where the $\mathbf{x}_i$ are given by Eq. (5.3.17).

When $\{\mathbf{x}_1, \ldots, \mathbf{x}_n\}$ is a non-orthonormal basis set, finding the coefficients in the expansion $\mathbf{x} = \sum_i \alpha_i \mathbf{x}_i$ is not quite so simple. However, all we have to do is find the reciprocal basis $\{\mathbf{z}_i, \ldots, \mathbf{z}_n\}$, which is related to $\{\mathbf{x}_1, \ldots, \mathbf{x}_n\}$ by the conditions

$$\mathbf{z}_i^\dagger \mathbf{x}_j = \delta_{ij}. \tag{5.3.24}$$

Those familiar with crystallography or other areas of physics might recall that the three non-coplanar Euclidean vectors $\mathbf{a}$, $\mathbf{b}$, and $\mathbf{c}$ have the reciprocal vectors

$$\hat{\mathbf{a}} = \frac{\mathbf{b} \times \mathbf{c}}{(\mathbf{a} \cdot \mathbf{b} \times \mathbf{c})}, \quad \hat{\mathbf{b}} = \frac{\mathbf{a} \times \mathbf{c}}{(\mathbf{b} \cdot \mathbf{a} \times \mathbf{c})}, \quad \hat{\mathbf{c}} = \frac{\mathbf{a} \times \mathbf{b}}{(\mathbf{c} \cdot \mathbf{a} \times \mathbf{b})}, \tag{5.3.25}$$

with the properties

$$\mathbf{a} \cdot \hat{\mathbf{a}} = 1, \qquad \mathbf{b} \cdot \hat{\mathbf{b}} = 1, \qquad \mathbf{c} \cdot \hat{\mathbf{c}} = 1$$
$$\mathbf{a} \cdot \hat{\mathbf{b}} = \mathbf{a} \cdot \hat{\mathbf{c}} = \mathbf{b} \cdot \hat{\mathbf{a}} = \mathbf{b} \cdot \hat{\mathbf{c}} = \mathbf{c} \cdot \hat{\mathbf{a}} = \mathbf{c} \cdot \hat{\mathbf{b}} = 0, \tag{5.3.26}$$

where $\mathbf{a} \cdot \mathbf{b}$ and $\mathbf{c} \times \mathbf{b}$ denote the dot and cross product of Euclidean vectors. By definition of the cross product, the direction of $\mathbf{a} \times \mathbf{b}$ is perpendicular to the plane defined by $\mathbf{a}$ and $\mathbf{b}$.

In any case, let us first assume the existence of the reciprocal basis set, explore its implications, and then prove its existence. Since $\{\mathbf{x}_1, \ldots, \mathbf{x}_n\}$ is a linearly independent set, it is a basis set, and so a unique set of $\{\alpha_i\}$ exists for a given $\mathbf{x}$ such that

$$\mathbf{x} = \sum_{i=1}^{n} \alpha_i \mathbf{x}_i. \tag{5.3.27}$$

Multiplying Eq. (5.3.27) by $\mathbf{z}_j^\dagger$ and using Eq. (5.3.24), we find

$$\alpha_i = \mathbf{z}_i^\dagger \mathbf{x}. \tag{5.3.28}$$

Thus, if the reciprocal set $\{\mathbf{z}_i\}$ is known, finding $\{\alpha_i\}$ for the expansion of $\mathbf{x}$ in the basis set $\{\mathbf{x}_i\}$ is as easy as it is for an orthonormal basis set.

Also, insertion of Eq. (5.3.28) into Eq. (5.3.27) yields

$$\mathbf{x} = \sum_{i=1}^{n} \mathbf{x}_i \mathbf{z}_i^\dagger \mathbf{x} = \left(\sum_{i=1}^{n} \mathbf{x}_i \mathbf{z}_i^\dagger \right) \mathbf{x}. \tag{5.3.29}$$

Again, this equation is valid for any vector $\mathbf{x}$ in E_n, and so we again conclude

$$\mathbf{I} = \sum_{i=1}^{n} \mathbf{x}_i \mathbf{z}_i^\dagger. \tag{5.3.30}$$

Thus, the *resolution of the identity* is a sum of the dyadics $\mathbf{x}_i \mathbf{z}_i^\dagger$ for an arbitrary basis set and an orthonormal basis set, for which $\mathbf{z}_i = \mathbf{x}_i$, is just a special case.

Let us now return to the problem of proving the existence of and finding the reciprocal set $\{\mathbf{z}_i\}$ for the set $\{\mathbf{x}_i\}$. Let

$$\mathbf{X} = [\mathbf{x}_1, \ldots, \mathbf{x}_n]. \tag{5.3.31}$$

Since the vectors $\mathbf{x}_i$ are linearly independent, $\mathbf{X}^{-1}$ exists and

$$\mathbf{X}^{-1}\mathbf{X} = \mathbf{I} = [\delta_{ij}]. \tag{5.3.32}$$

Next, we define

$$\mathbf{Z}^\dagger = \mathbf{X}^{-1} \qquad \text{or} \qquad \mathbf{Z} = (\mathbf{X}^{-1})^\dagger, \tag{5.3.33}$$

which we can write in partitioned form with the column vectors $\{\mathbf{z}_i\}$ as

$$\mathbf{Z} = [\mathbf{z}_1, \ldots, \mathbf{z}_n], \tag{5.3.34}$$

and so

$$\mathbf{Z}^\dagger = \begin{bmatrix} \mathbf{z}_1^\dagger \\ \mathbf{z}_2^\dagger \\ \vdots \\ \mathbf{z}_n^\dagger \end{bmatrix} = \mathbf{X}^{-1}. \tag{5.3.35}$$

However,

$$\mathbf{Z}^\dagger \mathbf{X} = \begin{bmatrix} \mathbf{z}_1^\dagger \\ \mathbf{z}_2^\dagger \\ \vdots \\ \mathbf{z}_n^\dagger \end{bmatrix} [\mathbf{x}_1, \ldots, \mathbf{x}_n] = [\mathbf{z}_i^\dagger \mathbf{x}_j] = \mathbf{X}^{-1} \mathbf{X} = [\delta_{ij}], \tag{5.3.36}$$

proving that

$$\mathbf{z}_i^\dagger \mathbf{x}_j = \delta_{ij},$$

which is the property we sought for the reciprocal set $\{\mathbf{z}_i\}$ of the set $\{\mathbf{x}_i\}$.

We have just proved that the set $\{\mathbf{z}_i\}$ exists and consists simply of the column vectors of $(\mathbf{X}^{-1})^\dagger$.

EXAMPLE 5.3.2. Consider the set

$$\mathbf{x}_1 = \begin{bmatrix} 2 \\ 1 \\ 1 \end{bmatrix}, \quad \mathbf{x}_2 = \begin{bmatrix} 1 \\ 2 \\ 1 \end{bmatrix}, \quad \mathbf{x}_3 = \begin{bmatrix} 1 \\ 1 \\ 2 \end{bmatrix}. \tag{5.3.37}$$

The inverse of $\mathbf{X} = [\mathbf{x}_1, \mathbf{x}_2, \mathbf{x}_3]$ is

$$\mathbf{X}^{-1} = \begin{bmatrix} \frac{3}{4} & -\frac{1}{4} & -\frac{1}{4} \\ -\frac{1}{4} & \frac{3}{4} & -\frac{1}{4} \\ -\frac{1}{4} & -\frac{1}{4} & \frac{3}{4} \end{bmatrix}, \tag{5.3.38}$$

and

$$\mathbf{Z} = (\mathbf{X}^{-1})^\dagger = \begin{bmatrix} \frac{3}{4} & -\frac{1}{4} & -\frac{1}{4} \\ -\frac{1}{4} & \frac{3}{4} & -\frac{1}{4} \\ -\frac{1}{4} & -\frac{1}{4} & \frac{3}{4} \end{bmatrix}. \tag{5.3.39}$$

BASIS SETS IN A NORMED LINEAR VECTOR SPACE

Therefore, the reciprocal vectors of $\mathbf{x}_1$, $\mathbf{x}_2$, and $\mathbf{x}_3$ are

$$\mathbf{z}_1 = \begin{bmatrix} \frac{3}{4} \\ -\frac{1}{4} \\ -\frac{1}{4} \end{bmatrix}, \quad \mathbf{z}_2 = \begin{bmatrix} -\frac{1}{4} \\ \frac{3}{4} \\ -\frac{1}{4} \end{bmatrix}, \quad \mathbf{z}_3 = \begin{bmatrix} -\frac{1}{4} \\ -\frac{1}{4} \\ \frac{3}{4} \end{bmatrix}. \tag{5.3.40}$$

■ ■ ■ **EXERCISE 5.3.2.** Show that $\mathbf{x}_i^\dagger \mathbf{z}_j = \mathbf{z}_j^\dagger \mathbf{x}_i = \delta_{ij}$ and that

$$\sum_{i=1}^{3} \mathbf{x}_i \mathbf{z}_i^\dagger = \begin{bmatrix} 1 & 0 & 0 \\ 0 & 1 & 0 \\ 0 & 0 & 1 \end{bmatrix} \tag{5.3.41}$$

■ ■ ■ for the preceding example.

We can now demonstrate why perfect matrices are attractive. Suppose that the $n \times n$ matrix $\mathbf{A}$ is perfect and that $\mathbf{x}$ obeys the equation

$$\frac{d\mathbf{x}}{dt} = \mathbf{A}\mathbf{x}, \quad \mathbf{x}_0 = \mathbf{x}(t=0). \tag{5.3.42}$$

Since $\mathbf{A}$ is perfect, it has n linearly independent eigenvectors $\mathbf{x}_i$ ($\mathbf{A}\mathbf{x}_i = \lambda_i \mathbf{x}_i$, $i = 1, \ldots, n$). The identity $\mathbf{I}$ can, therefore, be resolved as in Eq. (5.3.30) as a sum of the dyadic $\mathbf{x}_i \mathbf{z}_i^\dagger$, where the set $\{\mathbf{z}_i\}$ is the reciprocal of the set $\{\mathbf{x}_i\}$. It follows that

$$\mathbf{A} = \mathbf{A}\mathbf{I} = \sum_{i=1}^{n} \mathbf{A}\mathbf{x}_i \mathbf{z}_i^\dagger = \sum_{i=1}^{n} \lambda_i \mathbf{x}_i \mathbf{z}_i^\dagger. \tag{5.3.43}$$

Thus, we have proven

THEOREM. *When $\mathbf{A}$ is perfect, the spectral resolution $\mathbf{A} = \sum_{i=1}^{n} \lambda_i \mathbf{x}_i \mathbf{z}_i^\dagger$ exists, where $\mathbf{x}_i$ and $\mathbf{z}_i$, $i = 1, \ldots, n$, are the eigenvectors of $\mathbf{A}$ and their reciprocal vectors.*

The eigenvalues are called the spectra of $\mathbf{A}$ in a tradition going back to quantum mechanics.

Taking the adjoint of $\mathbf{A}$, we find

$$\mathbf{A}^\dagger = \sum_{i=1}^{n} \lambda_i^* \mathbf{z}_i \mathbf{x}_i^\dagger \tag{5.3.44}$$

from which it follows that

$$\mathbf{A}^\dagger \mathbf{z}_j = \lambda_j^* \mathbf{z}_j, \quad j = 1, \ldots, n. \tag{5.3.45}$$

This implies the following:

1. If $\mathbf{A}$ is perfect, so is its adjoint $\mathbf{A}^\dagger$.
2. The eigenvalues of $\mathbf{A}^\dagger$ are complex conjugates of the eigenvalues of $\mathbf{A}$.

3. The eigenvectors $\mathbf{z}_i$ of $\mathbf{A}^\dagger$ are the reciprocal vectors of the eigenvectors $\mathbf{x}_i$ of $\mathbf{A}$.

From repeated multiplication of Eq. (5.3.43) by $\mathbf{A}$, it follows that

$$\mathbf{A}^k = \sum_{i=1}^n \lambda_i^k \mathbf{x}_i \mathbf{z}_i^\dagger \qquad \text{for any positive integer } k, \qquad (5.3.46)$$

or that any positive power of a perfect matrix has a spectral resolution into a linear combination of the dyadics $\mathbf{x}_i \mathbf{z}_i^\dagger$. We can use this result to expand the matrix function $\exp(t\mathbf{A})$ in a Taylor series as

$$\begin{aligned}
\exp(t\mathbf{A}) &= \sum_{k=0}^\infty \frac{(t\mathbf{A})^k}{k!} = \sum_{k=0}^\infty \sum_{i=1}^n \frac{t^k \lambda_i^k}{k!} \mathbf{x}_i \mathbf{z}_i^\dagger \\
&= \sum_{i=1}^n \left(\sum_{k=0}^\infty \frac{t^k \lambda_i^k}{k!} \right) \mathbf{x}_i \mathbf{z}_i^\dagger \qquad (5.3.47) \\
&= \sum_{i=1}^n \exp(t\lambda_i) \mathbf{x}_i \mathbf{z}_i^\dagger,
\end{aligned}$$

thus obtaining a *spectral resolution* of $\exp(t\mathbf{A})$. In fact, for any function $f(t)$ that can be expanded in a Taylor or Laurent series for t near λ_i, $i = 1, \ldots, n$, it follows from Eq. (5.3.45) that

$$f(\mathbf{A}) = \sum_{i=1}^n f(\lambda_i) \mathbf{x}_i \mathbf{z}_i^\dagger. \qquad (5.3.48)$$

Clearly, when $\mathbf{A}$ is *perfect*, the spectral resolution theorem greatly simplifies evaluating functions of $\mathbf{A}$. As an example, consider the formal solution of Eq. (5.3.42)

$$\mathbf{x} = \exp(t\mathbf{A})\mathbf{x}_0, \qquad (5.3.49)$$

which, with the spectral resolution in Eq. (5.3.46), becomes

$$\mathbf{x} = \sum_{i=1}^n \exp(t\lambda_i) \mathbf{x}_i (\mathbf{z}_i^\dagger \mathbf{x}_0). \qquad (5.3.50)$$

EXAMPLE 5.3.3. Suppose that the vectors in Eq. (5.3.37) are eigenvectors of $\mathbf{A}$ with eigenvalues $\lambda_1 = -1$, $\lambda_2 = -2$, and $\lambda_3 = -3$. Assume that

$$\mathbf{x}_0 = \begin{bmatrix} 1 \\ 1 \\ 1 \end{bmatrix} \qquad (5.3.51)$$

in Eq. (5.3.42). The reciprocal vectors in Eq. (5.3.40) yield

$$\mathbf{z}_1^\dagger \mathbf{x}_0 = \frac{1}{4}, \qquad \mathbf{z}_2^\dagger \mathbf{x}_0 = \frac{1}{4}, \qquad \mathbf{z}_3^\dagger \mathbf{x}_0 = \frac{1}{4}, \qquad (5.3.52)$$

and so

$$\mathbf{x} = \frac{1}{4}e^{-t}\begin{bmatrix} 2 \\ 1 \\ 1 \end{bmatrix} + \frac{1}{4}e^{-2t}\begin{bmatrix} 1 \\ 2 \\ 1 \end{bmatrix} + \frac{1}{4}e^{-3t}\begin{bmatrix} 1 \\ 1 \\ 2 \end{bmatrix}$$

$$= \frac{1}{4}\begin{bmatrix} 2e^{-t} & + & e^{-2t} & + & e^{-3t} \\ e^{-t} & + & 2e^{-2t} & + & e^{-3t} \\ e^{-t} & + & e^{-2t} & + & 2e^{-3t} \end{bmatrix}. \qquad (5.3.53)$$

■ ■ ■

In the next section we will address the problem of finding the eigenvalues and eigenvectors of a matrix. In Chapter 8 we will see that, in more general normed linear vector spaces, *perfect operators* exist whose eigenvectors form a basis set for the space. We will see that these operators yield a powerful spectral resolution theorem and, eventually, we will discover that, even for *imperfect* or *defective* matrices, eigenanalysis will be of great benefit in solving matrix problems.

5.4. EIGENVALUE ANALYSIS

An $n \times n$ square matrix $\mathbf{A}$ has an eigenvalue if there exists a number λ such that the eigenequation

$$(\mathbf{A} - \lambda \mathbf{I})\mathbf{x} = \mathbf{0} \qquad (5.4.1)$$

has a solution. From solvability theory, we know that this homogeneous equation has a nontrivial solution *if and only if* the determinant of $(\mathbf{A} - \lambda \mathbf{I})$ is 0, i.e.,

$$|\mathbf{A} - \lambda \mathbf{I}| = 0. \qquad (5.4.2)$$

Equation (5.4.2) is known as the characteristic equation of $\mathbf{A}$. Such a determinant will yield an nth-degree polynomial in λ, i.e.,

$$|\mathbf{A} - \lambda \mathbf{I}| = P_n(\lambda) = (-\lambda)^n + a_1(-\lambda)^{n-1} + a_2(-\lambda)^{n-2} + \cdots + a_n$$

$$= \sum_{j=0}^{n} a_j(-\lambda)^{n-j}, \qquad (5.4.3)$$

where $a_0 = 1$ and a_j, $j > 0$, are functions of the elements of $\mathbf{A}$. Thus, the eigenvalues of $\mathbf{A}$ are roots of the polynomial equation (known as the characteristic polynomial equation)

$$P_n(\lambda) = \sum_{j=0}^{n} a_j(-\lambda)^{n-\lambda} = 0. \qquad (5.4.4)$$

From the theory of equations, we know that an nth-degree polynomial will have at least one distinct root and can have as many as n distinct roots.

For example, the polynomial

$$(1 - \lambda)^n = 0 \qquad (5.4.5)$$

has one root, $\lambda_1 = 1$, whereas the polynomial

$$\prod_{i=1}^{n}(i - \lambda) = 0 \tag{5.4.6}$$

has n distinct roots, $\lambda_i = i$, $i = 1, \ldots, n$. We say that the root $\lambda_1 = 1$ of Eq. (5.4.5) is a root of multiplicity n. In another example,

$$(1 - \lambda)^3(2 - \lambda)^3 \prod_{i=3}^{n-6}(i - \lambda) = 0, \tag{5.4.7}$$

we say that the root $\lambda_1 = 1$ is of multiplicity $p_{\lambda_1} = 3$, the root $\lambda_2 = 2$ is of multiplicity $p_{\lambda_2} = 3$, and the roots $\lambda_3 = 3$, $\lambda_4 = 4, \ldots, \lambda_{n-6} = n - 6$ are distinct roots or roots of multiplicity $p_{\lambda_i} = 1$. The multiplicities always sum to n, i.e.,

$$\sum_{i=1}^{n} p_{\lambda_i} = n. \tag{5.4.8}$$

The coefficients a_i can be related to the matrix $\mathbf{A}$ through the polynomial property

$$a_j = \frac{(-1)^{n-j}}{(n-j)!}\left[\frac{d^{n-j}P_n(\lambda)}{d\lambda^{n-j}}\right]_{\lambda=0}, \tag{5.4.9}$$

which follows from the definition $P_n(\lambda) = \sum_{j=0}^{n} a_j(-\lambda)^{n-j}$. We note from the elementary properties of determinants that

$$\frac{dP_n}{d\lambda} = \frac{d}{d\lambda}\begin{vmatrix} a_{11} - \lambda & a_{12} & \cdots & a_{1n} \\ a_{21} & a_{22} - \lambda & \cdots & a_{2n} \\ \vdots & \vdots & & \vdots \\ a_{n1} & a_{n2} & \cdots & a_{nn} - \lambda \end{vmatrix}$$

$$= \begin{vmatrix} -1 & a_{12} & \cdots & a_{1n} \\ 0 & a_{22} - \lambda & \cdots & a_{2n} \\ 0 & a_{32} & \cdots & a_{3n} \\ \vdots & \vdots & & \vdots \\ 0 & a_{n2} & \cdots & a_{nn} - \lambda \end{vmatrix}$$

$$+ \begin{vmatrix} a_{11} - \lambda & 0 & a_{13} & \cdots & a_{1n} \\ a_{21} & -1 & a_{23} & \cdots & a_{2n} \\ a_{31} & 0 & a_{33} - \lambda & \cdots & a_{3n} \\ \vdots & \vdots & \vdots & & \vdots \\ a_{n1} & 0 & a_{n3} & \cdots & a_{nn} - \lambda \end{vmatrix} + \cdots \tag{5.4.10}$$

EIGENVALUE ANALYSIS

$$+\begin{vmatrix} a_{11}-\lambda & a_{12} & \cdots & 0 \\ a_{21} & a_{22}-\lambda & \cdots & 0 \\ a_{31} & a_{32} & \cdots & 0 \\ \vdots & \vdots & & \vdots \\ a_{n1} & a_{n2} & \cdots & -1 \end{vmatrix}.$$

Thus, differentiation of $|\mathbf{A} - \lambda\mathbf{I}|$ by λ generates a sum of n determinants, where the ith column vector of the ith determinant is replaced by $-\mathbf{e}_i$, whereas the other columns are the same as those of $|\mathbf{A} - \lambda\mathbf{I}|$. Cofactor expansion yields

$$\frac{dP_n}{d\lambda} = (-1)\begin{vmatrix} a_{22}-\lambda & a_{23} & \cdots & a_{2n} \\ a_{32} & a_{33}-\lambda & \cdots & a_{3n} \\ \vdots & \vdots & & \vdots \\ a_{n2} & a_{n3} & \cdots & a_{nn}-\lambda \end{vmatrix}$$
$$+ (-1)\begin{vmatrix} a_{11}-\lambda & a_{13} & \cdots & a_{1n} \\ a_{31} & a_{33}-\lambda & \cdots & a_{3n} \\ \vdots & \vdots & & \vdots \\ a_{n1} & a_{n3} & \cdots & a_{nn}-\lambda \end{vmatrix} + \cdots \quad (5.4.11)$$
$$+ (-1)\begin{vmatrix} a_{11}-\lambda & a_{12} & \cdots & a_{1,n-1} \\ a_{21} & a_{22}-\lambda & \cdots & a_{2,n-1} \\ \vdots & \vdots & & \vdots \\ a_{n1} & a_{n2} & \cdots & a_{n-1,n-1}-\lambda \end{vmatrix}$$
$$= (-1)\text{tr}_{n-1}(\mathbf{A} - \lambda\mathbf{I}),$$

where $\text{tr}_i \mathbf{A}$ denotes the ith trace of $\mathbf{A}$ as defined by Eqs. (1.7.6)–(1.7.8) for a 3×3 matrix. Recall that, for an $n \times n$ matrix, the trace $\text{tr}_i \mathbf{A}$ is the sum of the ith-order minors generated by striking $n - i$ rows and columns intersecting on the main diagonal of $\mathbf{A}$.

The second derivative of $P_n(\lambda)$ is obtained by differentiating Eq. (5.4.10). The result is

$$\frac{d^2 P_n}{d\lambda^2} = \begin{vmatrix} -1 & 0 & a_{13} & \cdots & a_{1n} \\ 0 & -1 & a_{23} & & a_{2n} \\ 0 & 0 & a_{33}-\lambda & & a_{3n} \\ \vdots & \vdots & \vdots & & \vdots \\ 0 & 0 & a_{n3} & & a_{nn}-\lambda \end{vmatrix}$$

$$+ \begin{vmatrix} -1 & 0 & a_{13} & \cdots & a_{1n} \\ 0 & -1 & a_{23} & & a_{2n} \\ 0 & 0 & a_{33}-\lambda & & a_{3n} \\ \vdots & \vdots & \vdots & & \vdots \\ 0 & 0 & a_{n3} & & a_{nn}-\lambda \end{vmatrix}$$

$$+ \begin{vmatrix} -1 & a_{12} & 0 & a_{14} & \cdots & a_{1n} \\ 0 & a_{22}-\lambda & 0 & a_{24} & \cdots & a_{2n} \\ 0 & a_{32} & -1 & a_{34} & \cdots & a_{3n} \\ \vdots & \vdots & \vdots & \vdots & & \vdots \\ 0 & a_{n2} & 0 & a_{n4} & \cdots & a_{nn}-\lambda \end{vmatrix}$$

$$+ \begin{vmatrix} -1 & a_{12} & 0 & a_{14} & \cdots & a_{1n} \\ 0 & a_{22}-\lambda & 0 & a_{24} & \cdots & a_{2n} \\ 0 & a_{32} & -1 & a_{34} & \cdots & a_{3n} \\ \vdots & \vdots & \vdots & \vdots & & \vdots \\ 0 & a_{n2} & 0 & a_{n4} & \cdots & a_{nn}-\lambda \end{vmatrix} + \cdots . \quad (5.4.12)$$

Successive cofactor expansions by the columns containing the vectors $-\mathbf{e}_i$ yield

$$\frac{d^2 P_n}{d\lambda^n} = (-1)^2 2\, \mathrm{tr}_{n-2}(\mathbf{A} - \lambda \mathbf{I}). \quad (5.4.13)$$

Continuing this process results in the following expression for the general case

$$\frac{d^i P_n}{d\lambda^i} = (-1)^i\, i!\, \mathrm{tr}_{n-i}(\mathbf{A} - \lambda \mathbf{I}). \quad (5.4.14)$$

The factor of $i!$ comes from the fact that, by taking the ith derivative of $|\mathbf{A} - \lambda \mathbf{I}|$, the same i columns are differentiated $i!$ ways by changing the order of differentiation.

Setting $i = n - j$ and using Eq. (5.4.13) in Eq. (5.4.9), we find

$$a_j = \mathrm{tr}_j\, \mathbf{A}. \quad (5.4.15)$$

The coefficient a_j, $j > 0$, in the characteristic polynomial $P_n(\lambda)$ is simply the jth trace of $\mathbf{A}!$. Since

$$\mathrm{tr}_1\, \mathbf{A} = \sum_{i=1}^{n} a_{ii} \quad \text{and} \quad \mathrm{tr}_n\, \mathbf{A} = |\mathbf{A}|, \quad (5.4.16)$$

we see that the coefficient a_1 is equal to the sum of the diagonal elements of $\mathbf{A}$, often simply called the trace of $\mathbf{A}$, and the coefficient a_n is equal to the determinant of $\mathbf{A}$. Mathematica routines are provided in the Appendix for calculating the jth-order trace and characteristic polynomial of a square matrix.

EIGENVALUE ANALYSIS

■ **EXAMPLE 5.4.1.** Find the eigenvalues, eigenvectors, and reciprocal eigenvectors of

$$\mathbf{A} = \begin{bmatrix} 2 & -1 & -1 \\ -1 & 2 & -1 \\ -2 & 2 & -1 \end{bmatrix}. \quad (5.4.17)$$

We find that

$$a_3 = \mathrm{tr}_3\, \mathbf{A} = |\mathbf{A}| = -3 \quad (5.4.18)$$

$$a_2 = \mathrm{tr}_2\, \mathbf{A} = \begin{vmatrix} 2 & -1 \\ 2 & -1 \end{vmatrix} + \begin{vmatrix} 2 & -1 \\ -2 & -1 \end{vmatrix} + \begin{vmatrix} 2 & -1 \\ -1 & 2 \end{vmatrix}$$

$$= 0 - 4 + 3 = -1 \quad (5.4.19)$$

$$a_1 = \mathrm{tr}_1\, \mathbf{A} = 2 + 2 - 1 = 3. \quad (5.4.20)$$

Thus,

$$P_3(\lambda) = -\lambda^3 + 3\lambda^2 + \lambda - 3 = 0. \quad (5.4.21)$$

The eigenvalues are the roots of Eq. (5.3.20), which are

$$\lambda_1 = -1, \quad \lambda_2 = 1, \quad \lambda_3 = 3. \quad (5.4.22)$$

The eigenvector corresponding to λ_i obeys the homogeneous equation

$$(\mathbf{A} - \lambda_i \mathbf{I})\mathbf{x}_i = \mathbf{0} \quad (5.4.23)$$

and the solutions to these equations are

$$\mathbf{x}_1 = \begin{bmatrix} 1 \\ 1 \\ 2 \end{bmatrix}, \quad \mathbf{x}_2 = \begin{bmatrix} 1 \\ 1 \\ 0 \end{bmatrix}, \quad \mathbf{x}_3 = \begin{bmatrix} -3 \\ 1 \\ 2 \end{bmatrix}. \quad (5.4.24)$$

The reciprocal vectors are the column vectors of $(\mathbf{X}^{-1})^\dagger$, where $\mathbf{X} = [\mathbf{x}_1, \mathbf{x}_2, \mathbf{x}_3]$. They are

$$\mathbf{z}_1 = \begin{bmatrix} -\frac{1}{4} \\ -\frac{1}{4} \\ \frac{1}{2} \end{bmatrix}, \quad \mathbf{z}_2 = \begin{bmatrix} 0 \\ 1 \\ -\frac{1}{2} \end{bmatrix}, \quad \mathbf{z}_3 = \begin{bmatrix} -\frac{1}{4} \\ \frac{1}{4} \\ 0 \end{bmatrix}. \quad (5.4.25)$$

■ ■ ■

■ **EXERCISE 5.4.1.** Verify that the vectors $\mathbf{z}_i$ in Eq. (5.4.25) are eigenvectors of the transpose $\mathbf{A}^\dagger$ of $\mathbf{A}$, where $\mathbf{A}$ is given by Eq. (5.4.17). Also, verify the spectral resolution theorem for $\mathbf{A}$; i.e., show by direct calculation that

$$\sum_{i=1}^{n} \lambda_i \mathbf{x}_i \mathbf{z}_i^\dagger = \mathbf{A}. \quad (5.4.26)$$

■ ■ ■

EXERCISE 5.4.2. Use the spectral resolution theorem to prove the Hamilton–Cayley theorem for perfect matrices. The Hamilton–Cayley theorem claims that $\mathbf{A}$ obeys its characteristic equation, i.e.,

$$P_n(\mathbf{A}) = \sum_{j=0}^{n} a_j(-\mathbf{A})^{n-j} = \mathbf{0}. \tag{5.4.27}$$

The theorem is valid for any square matrix, but the proof is harder for the general case.

If the multiplicity p_{λ_i} of the eigenvalue λ_i is greater than 1, then the eigenvalue might have multiple linearly independent eigenvectors. The number of eigenvectors, however, is not necessarily equal to p_{λ_i}. For example, if $\mathbf{A} = \mathbf{I}$, then any basis set $\{\mathbf{x}_1, \ldots, \mathbf{x}_n\}$ is a linearly independent set of eigenvectors of $\mathbf{I}$, i.e., $\mathbf{I}\mathbf{x}_i = \mathbf{x}_i$, and all n eigenvectors have the same eigenvalue $\lambda_i = 1$. Consider the eigenequation

$$(\mathbf{A} - \lambda_i \mathbf{I})\mathbf{x}_i = \mathbf{0} \tag{5.4.28}$$

for an $n \times n$ matrix $\mathbf{A}$. The number of linearly independent solutions of Eq. (5.4.28) is $n - r_{\lambda_i}$, where r_{λ_i} is the rank of the characteristic determinant $|\mathbf{A} - \lambda_i \mathbf{I}|$.

5.5. SOME SPECIAL PROPERTIES OF EIGENVALUES

From the general theory of polynomial equations, we know that a polynomial, $f(x) = \sum_{j=0}^{n} c_j x^{n-j}$, $c_0 = 1$, can be written in the factored form

$$f(x) = \prod_{j=1}^{n}(x - x_j), \tag{5.5.1}$$

where the x_i are the roots of $f(x)$; i.e., they obey the equation

$$f(x_i) = 0. \tag{5.5.2}$$

Thus, the characteristic polynomial

$$P_n(\lambda) = (-\lambda)^n + a_1(-\lambda)^{n-1} + \cdots + a_{n-1}(-\lambda) + a_n \tag{5.5.3}$$

can be expressed in the factored form

$$P_n(\lambda) = \prod_{j=1}^{n}(\lambda_j - \lambda), \tag{5.5.4}$$

where λ_j are the roots of $P_n(\lambda) = 0$. The λ_j are, of course, just the eigenvalues of $\mathbf{A}$, and the coefficients a_j were shown in the previous section to be related to $\mathbf{A}$ by the equation

$$a_j = \operatorname{tr}_j \mathbf{A}. \tag{5.5.5}$$

Expanding the factors in Eq. (5.5.4), we obtain

$$P_n(\lambda) = (-\lambda)^n + \left(\sum_{j=1}^{n} \lambda_j\right)(-\lambda)^{n-1} + \left(\sum_{i>j=1}^{n}\sum \lambda_i \lambda_j\right)(-\lambda)^{n-2} + \cdots + \prod_{j=1}^{n}\lambda_j. \tag{5.5.6}$$

Comparison of Eqs. (5.5.3) and (5.5.6) leads to the relationships

$$\begin{aligned} a_1 &= \operatorname{tr}_1 \mathbf{A} = \sum_{j=1}^{n} \lambda_j \\ a_2 &= \operatorname{tr}_2 \mathbf{A} = \sum_{i>j=1}^{n}\sum^{n} \lambda_i \lambda_j \\ &\vdots \\ a_k &= \operatorname{tr}_k \mathbf{A} = \sum_{i_k>i_{k-1}}^{n}\sum^{n} \cdots \sum_{i_2>i_1=1}^{n}\sum^{n} \lambda_{i_k}\lambda_{i_{k-1}}\cdots \lambda_{i_1} \\ &\vdots \\ a_n &= \operatorname{tr}_n \mathbf{A} = \prod_{j=1}^{n} \lambda_j. \end{aligned} \tag{5.5.7}$$

The first and last of these relationships are encountered rather frequently in matrix analysis. They can be usefully restated as

$$\sum_{j=1}^{n} a_{jj} = \sum_{j=1}^{n} \lambda_j \tag{5.5.8}$$

and

$$|\mathbf{A}| \prod_{j=1}^{n} \lambda_j; \tag{5.5.9}$$

i.e., the sum of the eigenvalues of $\mathbf{A}$ equals the sum of the diagonal elements of $\mathbf{A}$, and the product of the eigenvalues of $\mathbf{A}$ equals the determinant of $\mathbf{A}$.

The kth trace of $\mathbf{A}$ is equal to the sum of all distinct k-tuples of products of the eigenvalues of $\mathbf{A}$, i.e., every product of k eigenvalues that contains a different combination of eigenvalues. For example, if $n = 3$, then

$$\operatorname{tr}_2 \mathbf{A} = \lambda_2\lambda_1 + \lambda_3\lambda_1 + \lambda_3\lambda_2. \tag{5.5.10}$$

The result in Eq. (5.5.7) implies a very special relationship between $\mathbf{A}$ and the diagonal eigenvalue matrix

$$\mathbf{\Lambda} = \begin{bmatrix} \lambda_1 & 0 & 0 & \cdots & 0 \\ 0 & \lambda_2 & 0 & \cdots & 0 \\ \vdots & \vdots & \vdots & & \vdots \\ 0 & 0 & 0 & \cdots & \lambda_n \end{bmatrix} = [\lambda_i \delta_{ij}]. \tag{5.5.11}$$

From the definition of the kth trace, it follows that

$$\operatorname{tr}_k \mathbf{\Lambda} = \sum_{i_k > i_{k-1}}^{n}\sum^{n} \cdots \sum_{i_2 > i_1 = 1}^{n}\sum^{n} \lambda_{i_k}\lambda_{i_{k-1}}\cdots\lambda_{i_1}, \tag{5.5.12}$$

which, when combined with Eq. (5.5.7), yields the conclusion that $\mathbf{A}$ and $\mathbf{\Lambda}$ have identical traces. Namely,

$$\operatorname{tr}_k \mathbf{A} = \operatorname{tr}_k \mathbf{\Lambda}. \tag{5.5.13}$$

This result can be summarized in the following theorem:

THEOREM. *The traces of $\mathbf{A}$ are equal to the corresponding traces of the diagonal matrix $\mathbf{\Lambda}$ whose diagonal elements are the eigenvalues of the matrix $\mathbf{A}$.*

To give a related theorem, let us first define a *similarity transformation*.

DEFINITION. *We say $\mathbf{A}$ and $\mathbf{B}$ are related by a similarity transformation if*

$$\mathbf{B} = \mathbf{S}^{-1}\mathbf{A}\mathbf{S}, \tag{5.5.14}$$

where $\mathbf{S}$ is a nonsingular matrix.

From this definition, we can now prove the following theorem:

THEOREM. *The traces of $\mathbf{A}$ are unchanged by a similarity transformation, i.e.,*

$$\operatorname{tr}_j \mathbf{A} = \operatorname{tr}_j (\mathbf{S}^{-1}\mathbf{A}\mathbf{S}). \tag{5.5.15}$$

The theorem follows from the fact that $\mathbf{A}$ and $\mathbf{S}^{-1}\mathbf{A}\mathbf{S}$ have the same eigenvalues and, therefore, from Eq. (5.5.13), they have the same traces. Note that the eigenvalues of $\mathbf{S}^{-1}\mathbf{A}\mathbf{S}$ obey the characteristic equation

$$|\mathbf{S}^{-1}\mathbf{A}\mathbf{S} - \lambda\mathbf{I}| = 0. \tag{5.5.16}$$

But since $\mathbf{I} = \mathbf{S}^{-1}\mathbf{I}\mathbf{S}$, Eq. (5.5.16) can be written as

$$|\mathbf{S}^{-1}(\mathbf{A} - \lambda\mathbf{I})\mathbf{S}| = |\mathbf{S}^{-1}|\,|\mathbf{A} - \lambda\mathbf{I}|\,|\mathbf{S}|$$
$$= |\mathbf{S}^{-1}\mathbf{S}|\,|\mathbf{A} - \lambda\mathbf{I}| \tag{5.5.17}$$
$$= |\mathbf{A} - \lambda\mathbf{I}| = 0,$$

and thus it follows that

THEOREM. *The eigenvalues of $\mathbf{A}$ are unchanged by a similarity transformation.*

In Section 5.2 we showed that the eigenvalues of the adjoint, $\mathbf{A}^\dagger$, of $\mathbf{A}$ are the complex conjugates, λ_i^*, of $\mathbf{A}$ if $\mathbf{A}$ is *perfect*. In fact, whether or not $\mathbf{A}$ is perfect, the eigenvalues of $\mathbf{A}^\dagger$ are λ_i^*. To prove this, consider the characteristic polynomial of $\mathbf{A}^\dagger$:

$$Q_n(\nu) = |\mathbf{A}^\dagger - \nu\mathbf{I}|. \tag{5.5.18}$$

The eigenvalues ν_i of $\mathbf{A}^\dagger$ obey the equation $Q_n(\nu) = 0$, and since the complex conjugate of 0 is 0, it follows that

$$0 = Q_n^*(\nu) = |\mathbf{A}^\dagger - \nu\mathbf{I}|^* = |\mathbf{A}^T - \nu^*\mathbf{I}| \tag{5.5.19}$$

or

$$0 = \sum_{j=0}^{n} b_j(-\nu^*)^{n-j}, \tag{5.5.20}$$

where $b_j = \operatorname{tr}_j \mathbf{A}^T$. But since the minors of $\mathbf{A}^T$ are the same as the minors of $\mathbf{A}$, it follows that

$$b_j = \operatorname{tr}_j \mathbf{A} = a_j, \tag{5.5.21}$$

and so the roots ν_i^* of Eq. (5.5.20) are the roots of $P_n(\nu^*) = \sum_{j=0}^{n} a_j(-\nu^*)^{n-j} = 0$. Thus, $\nu_i^* = \lambda_i$ or $\nu_i = \lambda_i^*$, proving that

THEOREM. *The eigenvalues of the adjoint $\mathbf{A}^\dagger$ of $\mathbf{A}$ are the complex conjugates of the eigenvalues of $\mathbf{A}$ for any square matrix $\mathbf{A}$.*

There are two other general relationships between the eigenproperties of $\mathbf{A}$ and its adjoint $\mathbf{A}^\dagger$. The first is expressed in the following theorem:

THEOREM. *The number of eigenvectors that $\mathbf{A}$ has corresponding to λ_i is equal to the number of eigenvectors that $\mathbf{A}^\dagger$ has corresponding to λ_i^*.*

This conclusion follows from the fact that the rank of $|\mathbf{A} - \lambda_i\mathbf{I}|$ equals the rank of $|\mathbf{A}^\dagger - \lambda_i^*\mathbf{I}|$. If the rank is r_{λ_i}, and $\mathbf{A}$ is an $n \times n$, then there are $n - r_{\lambda_i}$ linearly independent solutions to $\mathbf{A}\mathbf{x}_i = \lambda_i\mathbf{x}_i$ and $n - r_{\lambda_i}$ linearly independent solutions to $\mathbf{A}^\dagger\mathbf{z}_i = \lambda_i^*\mathbf{z}_i$.

The other property is that if $\lambda_i \neq \lambda_j$, then $\langle \mathbf{z}_j, \mathbf{x}_i \rangle = \langle \mathbf{z}_i, \mathbf{x}_j \rangle = 0$. To show this, consider the eigenequations $\mathbf{A}\mathbf{x}_i = \lambda_i\mathbf{x}_i$ and $\mathbf{A}^\dagger\mathbf{z}_j = \lambda_j^*\mathbf{z}_j$. It follows that $\langle \mathbf{z}_j, \mathbf{A}\mathbf{x}_i \rangle = \lambda_i\langle \mathbf{z}_j, \mathbf{x}_i \rangle$ and $\langle \mathbf{A}^\dagger\mathbf{z}_j, \mathbf{x}_i \rangle = \lambda_j\langle \mathbf{z}_j, \mathbf{x} \rangle$. However, $\langle \mathbf{A}^\dagger\mathbf{z}_j, \mathbf{x}_i \rangle = \langle \mathbf{z}_j, \mathbf{A}\mathbf{x}_i \rangle$, which follows from the definition of the adjoint, and so we find that $(\lambda_i - \lambda_j)\langle \mathbf{z}_j, \mathbf{x}_i \rangle = 0$, or $\langle \mathbf{z}_j, \mathbf{x}_i \rangle = 0$ if $\lambda_i \neq \lambda_j$. Thus, we have shown that

THEOREM. *The eigenvectors $\mathbf{x}_i$ of $\mathbf{A}$ corresponding to λ_i are orthogonal to the eigenvectors $\mathbf{z}_j$ of $\mathbf{A}^\dagger$ corresponding to λ_j^* if $\lambda_i \neq \lambda_j$.*

∎ EXERCISE 5.5.1. Define the nonsingular matrix

$$\mathbf{S} = \begin{bmatrix} -2 & 11 & 0 \\ 1 & -2 & 1 \\ 0 & 1 & -2 \end{bmatrix}. \tag{5.5.22}$$

(a) Find the inverse $\mathbf{S}^{-1}$.
(b) Calculate the matrix $\mathbf{B}$, where

$$\mathbf{B} = \mathbf{S}^{-1}\mathbf{A}\mathbf{S} \tag{5.5.23}$$

and $\mathbf{A}$ is the matrix defined in Eq. (5.3.17).

(c) Compute the eigenvalues of **B**. These should be the same as the eigenvalues of **A**.

■ ■ ■ (d) Find the eigenvectors of **B**. Are they eigenvectors of **A**?

EXAMPLE 5.5.1. The Cartesian components of the stress tensor are

$$\mathbf{T} = \begin{bmatrix} 4 & -2 & -1 \\ -2 & 5 & 2 \\ -1 & 2 & 4 \end{bmatrix} \tag{5.5.24}$$

in a convenient set of units.

(a) Find the values of the maximum and minimum traction force for the mechanical loading giving rise to the state of stress represented by **T**.

(b) Find the directions of the maximum and minimum traction forces. These forces are eigenvalues of the equation

$$\mathbf{T}\hat{\mathbf{n}} = \lambda\hat{\mathbf{n}}, \tag{5.5.25}$$

and the corresponding eigenvectors are their corresponding directions.

The eigenvalues of Eq. (5.5.25) are

$$\lambda_1 = 3, \qquad \lambda_2 = 5 - 2\sqrt{2}, \qquad \lambda_3 = 5 + 2\sqrt{2} \tag{5.5.26}$$

or

$$\lambda_1 = 3, \qquad \lambda_2 = 2.17157, \qquad \lambda_3 = 7.82943. \tag{5.5.27}$$

Thus, the maximum and minimum traction forces are $\lambda_3 = 7.82943$ and $\lambda_2 = 2.17157$, respectively. The normalized eigenvectors of Eq. (5.5.25) are

$$\hat{\mathbf{n}}_1 = \frac{1}{\sqrt{2}} \begin{bmatrix} 1 \\ 0 \\ 1 \end{bmatrix}, \qquad \hat{\mathbf{n}}_2 = \frac{1}{2} \begin{bmatrix} -1 \\ -\sqrt{2} \\ 1 \end{bmatrix}, \qquad \hat{\mathbf{n}}_3 = \frac{1}{2} \begin{bmatrix} -1 \\ \sqrt{2} \\ 1 \end{bmatrix}. \tag{5.5.28}$$

Thus, if $\hat{i}, \hat{j}$, and $\hat{k}$ are the mutually orthogonal unit vectors of the Cartesian coordinate system in which **T** is expressed, the direction of the maximum in three-dimensional Euclidean space is

$$\hat{n}_3 = -\frac{1}{2}\hat{i} + \frac{\sqrt{2}}{2}\hat{j} + \frac{1}{2}\hat{k} \tag{5.5.29}$$

and for the minimum traction force it is

$$\hat{n}_2 = -\frac{1}{2}\hat{i} - \frac{\sqrt{2}}{2}\hat{j} + \frac{1}{2}\hat{k}. \tag{5.5.30}$$

What are the physical angles between the $\hat{n}_2$ and $\hat{n}_3$ directions and the directions
■ ■ ■ of $\hat{i}, \hat{j}$, and $\hat{k}$?

5.6. CALCULATION OF EIGENVALUES

The most obvious way to compute the eigenvalues of $\mathbf{A}$ is to find the roots of the characteristic polynomial $P_n(\lambda)$. This is, indeed, the preferred method for small matrices, say $n \leq 5$. However, for large n, the polynomial can be ill-conditioned; i.e., its roots can be very sensitive to round-off errors. For example, suppose that the exact eigenvalues of $\mathbf{A}$ are $\lambda_i = i$, $i = 1, \ldots, 20$. Since $a_i = \text{tr}_i \mathbf{A}$, the characteristic polynomial for $\mathbf{A}$ is

$$P_n(\lambda) = \lambda^{20} - 210\lambda^{19} + 20615\lambda^{18} + \cdots + 20! \tag{5.6.1}$$

If $a_1 = -210$ is replaced by $a_1 = -210 - 2^{-23} = -210.000000119\ldots$ and all of the other coefficients are unchanged, the roots $\lambda_{16} = 16$ and $\lambda_{17} = 17$ become (to nine decimals)

$$\begin{aligned}\lambda_{16} &= 16.730737466 - 2.812624894i \\ \lambda_{17} &= 16.730737466 - 2.812624894i.\end{aligned} \tag{5.6.2}$$

As we can see, a change in a_1 on the order of $10^{-10}\%$ changes the eigenvalues substantially—even resulting in complex values instead of real ones.

EXERCISE 5.6.1. You can verify by plotting the polynomial

$$P_3(\lambda) = -\lambda^3 + 8\lambda^2 - 13\lambda + 5 \tag{5.6.3}$$

that it has three real roots. However, if you use Mathematica to solve $P_3(\lambda) = 0$ at various levels of precision, you will see that one of the roots has a small spurious imaginary part that decreases with increasing precision in the calculation.

For tridiagonal matrices, there is a simple iterative method for computing eigenvalues. Let

$$P_n(\lambda) = |\mathbf{A} - \lambda\mathbf{I}| = \begin{vmatrix} a_1 - \lambda & b_1 & 0 & 0 & \cdots & 0 \\ c_1 & a_2 - \lambda & b_2 & 0 & \cdots & 0 \\ 0 & c_2 & a_3 - \lambda & b_3 & \cdots & 0 \\ \vdots & \vdots & \vdots & \vdots & & \vdots \\ 0 & 0 & 0 & c_{n-2} & a_{n-1} - \lambda & b_{n-1} \\ 0 & 0 & 0 & 0 & c_{n-1} & a_n - \lambda \end{vmatrix}. \tag{5.6.4}$$

If the last row and last column of this determinant are struck, the remaining determinant is $P_{n-1}(\lambda)$, the characteristic polynomial of the $(n-1) \times (n-1)$ tridiagonal matrix. Similarly, if the last two rows and two columns are struck, the result is $P_{n-2}(\lambda)$, the characteristic polynomial of the $(n-2) \times (n-2)$ tridiagonal matrix. Thus, by using the cofactor expansion of $|\mathbf{A} - \lambda\mathbf{I}|$ by its last column followed by the cofactor expansion on the coefficient b_{n-1} in the first cofactor expansion, we obtain

$$P_n(\lambda) = (a_n - \lambda)P_{n-1}(\lambda) - b_{n-1}c_{n-1}P_{n-2}(\lambda). \tag{5.6.5}$$

This result is valid for any n, and so the expansion process generates the recurrence formulas

$$P_r(\lambda) = (a_r - \lambda)P_{r-1}(\lambda) - b_{r-1}c_{r-1}P_{r-2}(\lambda), \qquad r = n, \ldots, 3$$

$$\vdots$$

$$P_2(\lambda) = (a_2 - \lambda)P_1(\lambda) - b_1c_1P_0(\lambda)$$

$$P_1(\lambda) = (a_1 - \lambda)P_0(\lambda),$$

(5.6.6)

where $P_0(\lambda) = 1$.

The eigenvalues can now be computed by guessing a value λ' and computing $P_1(\lambda'), P_2(\lambda'), \ldots, P_n(\lambda')$. If $P_n(\lambda') = 0$, λ' is an eigenvalue. If not, we make another guess and continue the process. As we can see from Eq. (5.6.6), the calculation of $P_n(\lambda')$ for each guess takes $3(n-1)$ operations. A Mathematica program implemeting this method has been included in the Appendix.

In the next chapter we will prove that any *self-adjoint* matrix can be transformed into a tridiagonal matrix and thus the recurrence method for computing eigenvalues can be used.

EXERCISE 5.6.2. Use Eq. (5.6.6) to find the eigenvalues of

$$\mathbf{A} = \begin{bmatrix} 1 & 2 & 0 \\ 1 & 3 & 2 \\ 0 & 1 & 4 \end{bmatrix}. \tag{5.6.7}$$

Once an eigenvalue λ of a tridiagonal matrix is found, the corresponding eigenvector satisfies $(\mathbf{A} - \lambda\mathbf{I})\mathbf{x} = \mathbf{0}$ and can be easily computed (if $b_i \neq 0$) from

$$x_2 = -\frac{1}{b_1}(a_1 - \lambda)x_1$$

$$x_3 = -\frac{1}{b_2}\big[(a_2 - \lambda)x_2 + c_1x_1\big]$$

$$\vdots$$

$$x_n = -\frac{1}{b_{n-1}}\big[(a_{n-1} - \lambda)x_{n-1} + c_{n-2}x_{n-2}\big].$$

(5.6.8)

The initial value x_1 can be given any nonzero value or can be chosen to normalize $\mathbf{x}$, i.e., chosen such that $\|\mathbf{x}\| = 1$.

An easy method to program for computing the maximum-magnitude eigenvalue of a matrix $\mathbf{A}$ is the *power method*. The method works if the eigenvalues can be ordered as

$$|\lambda_1| \leq |\lambda_2| \leq |\lambda_3| \leq \cdots \leq |\lambda_{n-1}| < |\lambda_n| \tag{5.6.9}$$

such that λ_n is the eigenvalue of maximum magnitude. If λ_n is of multiplicity 1, i.e., $|\lambda_n| \neq |\lambda_i|$, $i = 1, \ldots, n-1$, then the power method works. We will prove

CALCULATION OF EIGENVALUES

the method only for *perfect* matrices; however, it works for any matrix as long as the conditions in Eq. (5.6.9) are valid.

For a perfect matrix, the spectral resolution theorem holds, namely,

$$\mathbf{A} = \sum_{i=1}^{n} \lambda_i \mathbf{x}_i \mathbf{z}_i^\dagger, \tag{5.6.10}$$

where $\mathbf{A}\mathbf{x}_i = \lambda_i \mathbf{x}_i$ and $\mathbf{z}_i^\dagger \mathbf{x}_j = \delta_{ij}$. We begin with an arbitrary vector $\mathbf{u}_0$. Then we define

$$\begin{aligned}\mathbf{u}_k \equiv \mathbf{A}^k \mathbf{u}_0 &= \sum_{i=1}^{n} \lambda_i^k (\mathbf{z}_i^\dagger \mathbf{x}_0) \mathbf{x}_i \\ &= \lambda_n^k \left[(\mathbf{z}_n^\dagger \mathbf{u}_0) \mathbf{x}_n + \sum_{i=1}^{n-1} \left(\frac{\lambda_i}{\lambda_n}\right)^k (\mathbf{z}_i^\dagger \mathbf{u}_0) \mathbf{x}_i \right].\end{aligned} \tag{5.6.11}$$

As the value of k increases, the terms in Eq. (5.6.11) multiplied by $(\lambda_i/\lambda_n)^k$ contribute less and less since $(\lambda_i/\lambda_n)^k \to 0$ as $k \to \infty$. Thus, at large enough k,

$$\mathbf{u}_k \simeq \lambda_n^k (\mathbf{z}_n^\dagger \mathbf{u}_0) \mathbf{x}_n. \tag{5.6.12}$$

We do have to assume that the arbitrarily chosen $\mathbf{u}_0$ is not orthogonal to $\mathbf{z}_n$. Usually, round-off errors prevent this coincidence. If we define

$$\mathbf{w}_k = \frac{\mathbf{u}_k}{\|\mathbf{u}_k\|}, \tag{5.6.13}$$

then

$$\mathbf{w}_k = \left(\frac{\lambda_n}{|\lambda_n|}\right)^k \frac{\mathbf{z}_n^\dagger \mathbf{u}_0}{|\mathbf{z}_n^\dagger \mathbf{u}_0|} \frac{\mathbf{x}_n}{\|\mathbf{x}_n\|}, \tag{5.6.14}$$

and so

$$\frac{\mathbf{w}_k^\dagger \mathbf{A} \mathbf{w}_k}{\mathbf{w}_k^\dagger \mathbf{w}_k} \to \lambda_n \qquad \text{as } k \to \infty \tag{5.6.15}$$

and

$$\mathbf{w}^{(k)} \to \alpha \mathbf{x}_n \qquad \text{as } k \to \infty. \tag{5.6.16}$$

Since any multiple of an eigenvector of $\mathbf{A}$ is also an eigenvector of $\mathbf{A}$, Eqs. (5.6.13) and (5.6.14) yield the eigenvalue of maximum magnitude and its eigenvector. A simple computer program utilizing the power method is given below.

Choose an initial eigenvector guess $\mathbf{u}_0$:

$$\mathbf{u} = \mathbf{u}_0.$$

Initialize the variable *TOLERANCE* used in the convergence criterion:

$$TOLERANCE = \langle \text{some value} \rangle.$$

Initialize the variable ϵ to some value greater than *TOLERANCE*:

$\epsilon = \langle\text{some value greater than tolerance}\rangle$.

Choose an initial guess for the eigenvalue and set it equal to λ_{old}:

$\lambda_{\text{old}} = \langle\text{some value}\rangle$.

Begin iteration loop and continue until solution converges:

While ($\epsilon > TOLERANCE$)

$\mathbf{u}_{\text{new}} = \mathbf{A}\mathbf{u}$

$\lambda = \mathbf{u}^\dagger \mathbf{u}_{\text{new}} / \mathbf{u}^\dagger \mathbf{u}$

$\mathbf{u} = \mathbf{u}_{\text{new}}$

$\epsilon = |\lambda - \lambda_{\text{old}}|$

$\lambda_{\text{old}} = \lambda$

Continue

With sufficiently stringent tolerance, the power method will give a good approximation to λ_n and $\mathbf{x}_n$, which, if normalized, is $\mathbf{x}_n = \mathbf{u}_{\text{new}} / \|\mathbf{u}_{\text{new}}\|$.

The power method can be used to find the eigenvalue of minimum magnitude as well as all of the distinct eigenvalues. However, the matrix inverses must be computed in this case. Again, we begin with an arbitrary guess $\mathbf{x}_0$. For a perfect matrix with only nonzero eigenvalues,

$$\mathbf{A}^{-1}\mathbf{x}_i = \lambda_i^{-1}\mathbf{x}_i, \qquad i = 1, \ldots, n, \tag{5.6.17}$$

and so the spectral resolution of $\mathbf{A}^{-1}$ is

$$\mathbf{A}^{-1} = \sum_{i=1}^{n} \lambda_i^{-1} \mathbf{x}_i \mathbf{z}_i^\dagger, \tag{5.6.18}$$

where $\mathbf{x}_i$ and $\mathbf{z}_i$ are the eigenvalues of $\mathbf{A}$ and $\mathbf{A}^\dagger$. From this it follows that

$$\mathbf{A}^{-k}\mathbf{x}_0 \rightarrow (\lambda_1)^{-k} \mathbf{x}_1 (\mathbf{z}_1 \mathbf{x}_0), \tag{5.6.19}$$

where $|\lambda_1| < |\lambda_2| \leq \cdots \leq |\lambda_n|$.

Thus, with the iterative sequence

$$\mathbf{A}\mathbf{u}_1 = \mathbf{x}_0 \text{ or } \mathbf{u}_1 = \mathbf{A}^{-1}\mathbf{x}_0, \qquad \mathbf{w}_1 = \frac{\mathbf{u}_1}{\|\mathbf{u}_1\|} \tag{5.6.20}$$

$$\mathbf{u}_2 = \mathbf{A}^{-1}\mathbf{w}_1, \qquad \mathbf{w}_2 = \frac{\mathbf{u}_2}{\|\mathbf{u}_2\|}, \tag{5.6.21}$$

and so on until

$$\mathbf{u}_k = \mathbf{A}^{-1}\mathbf{w}_{k-1}, \qquad \mathbf{w}_k = \frac{\mathbf{u}_k}{\|\mathbf{u}_k\|}, \tag{5.6.22}$$

CALCULATION OF EIGENVALUES

it follows that

$$\frac{\mathbf{w}_k^\dagger \mathbf{A} \mathbf{w}_k}{\mathbf{w}_k^\dagger \mathbf{w}_k} \to \lambda_1 \quad \text{as } k \to \infty \tag{5.6.23}$$

and

$$\mathbf{w}_k \to \mathbf{x}_1.$$

Similarly, if λ_i is distinct, i.e., if no other eigenvalue is equal to λ_i, and if β is closer to λ_i than any other eigenvalue, then the inverse power method for $\mathbf{A} - \beta \mathbf{I}$ generates

$$(\mathbf{A} - \beta \mathbf{I})^{-k} \mathbf{x}_0 \to (\lambda_i - \beta)^{-k} \mathbf{x}_i (\mathbf{z}_i^\dagger \mathbf{x}_0). \tag{5.6.24}$$

Thus, if $\mathbf{u}_1 = (\mathbf{A} - \beta \mathbf{I})^{-1} \mathbf{x}_0$ and $\mathbf{w}_1 = \mathbf{u}_1 / \|\mathbf{u}_1\|$, the iterative sequence

$$\mathbf{u}_k = (\mathbf{A} - \beta \mathbf{I})^{-1} \mathbf{w}_{k-1}, \quad \mathbf{w}_k = \frac{\mathbf{u}_k}{\|\mathbf{u}_k\|}, \quad k = 2, 3, \ldots, \tag{5.6.25}$$

yields

$$\frac{\mathbf{w}_k^\dagger \mathbf{A} \mathbf{w}_k}{\mathbf{w}_k^\dagger \mathbf{w}_k} \to (\lambda_i)^{-1} \quad \text{as } k \to \infty \tag{5.6.26}$$

and

$$\mathbf{w}_k \to \mathbf{x}_i. \tag{5.6.27}$$

If λ_i has a multiplicity greater than 1, the power method will still generate the eigenvalue and one of the eigenvectors, but not all of them.

EXAMPLE 5.6.1. Use the power method to find the eigenvectors and eigenvalues of the matrix

$$\mathbf{A} = \begin{bmatrix} 4+i & \frac{1}{2}(3+2i) & -(1+i) & -\frac{1}{2} \\ -(1+i) & \frac{1}{2}(3-2i) & 1+i & \frac{1}{2} \\ 1 & \frac{1}{2}(3+i) & 2 & -\frac{1}{2}(1+i) \\ 1+i & \frac{1}{2}(3+2i) & -(1+i) & \frac{5}{2} \end{bmatrix}.$$

We choose a tolerance of 10^{-15}, an initial eigenvalue guess of 2, and an initial eigenvector guess of

$$\mathbf{x}_0 = \begin{bmatrix} 1 \\ 1 \\ 1 \\ 1 \end{bmatrix}.$$

Application of the power method outlined above results in convergence after 406 steps (i.e., $\|\lambda^{(k+1)} - \lambda^{(k)}\| < 10^{-15}$). The resulting solution is $\lambda_1 = 3 + i$ and

$$\mathbf{x}_1 = -(0.0335351 + 0.49874i) \begin{bmatrix} 1 \\ -1 \\ -1 \\ 1 \end{bmatrix},$$

which can be normalized to

$$\mathbf{x}_1 = \frac{1}{2} \begin{bmatrix} 1 \\ -1 \\ -1 \\ 1 \end{bmatrix}.$$

Next, we apply the inverse power method on the matrix $\mathbf{A}^{-1}$ given by

$$\mathbf{A}^{-1} = \begin{bmatrix} \frac{1}{30}(7-9i) & -\frac{1}{15}(2+6i) & \frac{1}{10}(1+3i) & \frac{1}{30}(1+3i) \\ \frac{1}{10}(1+3i) & \frac{1}{15}(7+6i) & -\frac{1}{10}(1+3i) & -\frac{1}{30}(1+3i) \\ -\frac{1}{15}(1+3i) & -\frac{1}{60}(7+21i) & \frac{1}{5}(2+i) & \frac{1}{20}(1+3i) \\ -\frac{1}{10}(1+3i) & -\frac{1}{15}(2+6i) & \frac{1}{10}(1+3i) & \frac{1}{30}(11+3i) \end{bmatrix}.$$

Using the same tolerance and initial guesses, the solution converges after 33 iterations, yielding $\lambda_2 = 1 - i$ and

$$\mathbf{x}_2 = \frac{1}{2} \begin{bmatrix} 1 \\ -1 \\ 1 \\ 1 \end{bmatrix}.$$

For the intermediate eigenvalues, we use the inverse power method on the matrix $(\mathbf{A} - b\mathbf{I})^{-1}$. Choosing $b = 2$ gives

$$(\mathbf{A} - b\mathbf{I})^{-1} = \begin{bmatrix} \frac{1}{2}(3-i) & \frac{1}{2}(2-i) & -\frac{1}{2}(1-i) & -\frac{1}{2} \\ -\frac{1}{2}(1-i) & \frac{1}{2}i & \frac{1}{2}(1-i) & \frac{1}{2} \\ 1 & \frac{1}{4}(5-i) & 0 & -\frac{1}{4}(1-i) \\ \frac{1}{2}(1-i) & \frac{1}{2}(2-i) & -\frac{1}{2}(1-i) & \frac{1}{2} \end{bmatrix},$$

and by using the original values of the tolerance, eigenvalue, and eigenvector, the solution converges after 68 steps, giving $\lambda_3 = 3$ and

$$\mathbf{x}_3 = 0.37794 \begin{bmatrix} 1 \\ 1 \\ 2 \\ 1 \end{bmatrix}.$$

By changing the initial eigenvector guess to

$$\mathbf{x}_0 = \begin{bmatrix} -1 \\ 1 \\ -1 \\ 1 \end{bmatrix}$$

and recomputing, we find that the solution again converges after 68 steps and the resulting eigenvalue is $\lambda_4 = 3$, as before. However, we find that the eigenvector is different:

$$\mathbf{x}_4 = 0.57735 \begin{bmatrix} -1 \\ 1 \\ 0 \\ 1 \end{bmatrix},$$

and, by inspection, we see that it is not colinear with $\mathbf{x}_3$ and is, in fact, a separate solution. This is an example of a degenerate eigenvalue. More specifically, we say that $\lambda_3 = 3$ with multiplicity $p_{\lambda_3} = 2$. The two eigenvectors, $\mathbf{x}_3$ and $\mathbf{x}_4$, form a linearly independent basis in E_2 and, after some careful examination, we see that any linear combination of these vectors in also an eigenvector of $\mathbf{A}$ with eigenvalue $\lambda = 3$. A Mathematica program for this example has been included in the Appendix.

■ ■ ■

A way of roughly estimating the neighborhood of eigenvalues is provided by *Gerschgorin's theorem*:

GERSCHGORIN'S THEOREM. *Each of the eigenvalues of* $\mathbf{A}$ *lies in the union of circles*

$$|\lambda - a_{ii}| \leq r_i, \quad r_i = \sum_{\substack{j=1 \\ j \neq i}}^{n} |a_{ij}|. \tag{5.6.28}$$

The proof of the theorem is straightforward. If λ is an eigenvalue of $\mathbf{A}$, then there exists an eigenvector $\mathbf{x}$ such that $\mathbf{A}\mathbf{x} = \lambda \mathbf{x}$. We can rewrite this as

$$(\lambda - a_{ii})x_i = \sum_{\substack{j=1 \\ j \neq i}}^{n} a_{ij} x_j, \quad i = 1, \ldots, n, \tag{5.6.29}$$

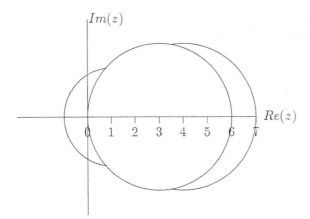

FIGURE 5.6.1

where we have isolated the diagonal terms from the nondiagonal terms. But if we choose i such that $|x_i| = \|\mathbf{x}\|_{p=\infty}$, then

$$|\lambda - a_{ii}| \leq \sum_{\substack{j=1 \\ j \neq i}} |a_{ij}| \frac{|x_j|}{|x_i|} \leq r_i, \qquad (5.6.30)$$

which proves the theorem.

EXAMPLE 5.6.2. Consider the matrix

$$\mathbf{A} = \begin{bmatrix} 1 & 1 & 1 \\ 1 & 3 & 2 \\ 2 & 1 & 4 \end{bmatrix}. \qquad (5.6.31)$$

The circles in Eq. (5.6.28) are thus

$$|z - 1| = 2, \qquad |z - 3| = 3, \qquad |z - 4| = 3, \qquad (5.6.32)$$

■ ■ ■ and the eigenvalues lie in their union.

The eigenvalues for this example turn out to be real and lie between $\frac{1}{2}$ and 6, an interval that is, indeed, contained in the union of the three circles shown in Fig. 5.6.1.

PROBLEMS

1. Consider the following inner product candidates:

    (a) Does

    $$\langle \mathbf{x}, \mathbf{y} \rangle = \mathbf{x}^\dagger \mathbf{A} \mathbf{y},$$

    where

    $$\mathbf{A} = \begin{bmatrix} 4 & 1 \\ 2 & -4 \end{bmatrix},$$

    form an inner product? Why?

(b) Does
$$\langle x, y \rangle = x^\dagger A^\dagger A y$$
form an inner product? Why?

(c) Does
$$\langle x, y \rangle = x^\dagger B y,$$
where
$$B = \begin{bmatrix} 4 & 1 \\ 1 & 4 \end{bmatrix},$$
form an inner product? Why?

(d) Find the eigenvalues of A, $A^\dagger A$, and B.

2. Use the Gram–Schmidt procedure to construct an orthonormal set from the vectors
$$x_1 = \begin{bmatrix} 1 \\ 0 \\ 1 \end{bmatrix}, \quad x_2 = \begin{bmatrix} 2 \\ 1 \\ 0 \end{bmatrix}, \quad x_3 = \begin{bmatrix} 1 \\ 2 \\ 3 \end{bmatrix}.$$

3. (a) Compute $\exp(tA)$ for
$$A = \begin{bmatrix} 2 & 1 & 2 \\ 1 & 2 & 1 \\ 1 & 1 & 2 \end{bmatrix}.$$

(b) Compute $\sin(At)$.

4. Consider the matrix
$$A = \begin{bmatrix} 1 & 2 & 1 \\ 1 & 3 & 2 \\ 2 & 1 & 4 \end{bmatrix}.$$

(a) Evaluate $\text{tr}_j A$, $j = 1, 2, 3$.
(b) Find the eigenvalues of A.
(c) How many eigenvectors does A have? Find them.
(d) Find the reciprocal vectors of the eigenvectors and give the eigenvalues and eigenvectors of the adjoint $A^\dagger$.

5. Consider the polynomial
$$(t+2)(t^2-1)^6 - 2 \times 10^{-8} t^{11} = 0.$$

(a) What would be the solution to this equation if solved on a machine using floating-point arithmetic and nine decimals?
(b) Use the Newton–Raphson method to solve the polynomial to 12-place accuracy to test the solution found in part (a).

6. Consider a 10×10 matrix $\mathbf{A}$ whose eigenvalues are $\lambda_i = i$. Suppose that the jth trace is evaluated with a systematic error of $10^{-5}(-1)^j$; i.e., the characteristic polynomial is evaluated as

$$\tilde{P}_n(\lambda) = \sum_{j=0}^{n} [a_j + (-1)^j 10^{-5}](-\lambda)^{n-j}$$

instead of

$$P_n(\lambda) = \sum_{j=0}^{n} a_j (-\lambda)^{n-j},$$

where $a_j = \text{tr}_j \mathbf{A}$.

(a) Calculate a_j from the eigenvalues of $\mathbf{A}$.
(b) Find the roots of $\tilde{P}_n(\lambda)$ and compare them to the roots of $P_n(\lambda)$.

7. Does the infinite set

$$\mathbf{e}_2, \mathbf{e}_3, \ldots, \mathbf{e}_j, \quad j = 2, 3, \ldots,$$

where the ith component of $\mathbf{e}_j$ is δ_{ij}, form a basis set in E_∞? Why?

8. Prove that $\mathbf{AB}$ and $\mathbf{BA}$ have the same eigenvalues, where $\mathbf{A}$ and $\mathbf{B}$ are both perfect nonsingular matrices. How are the eigenvectors of $\mathbf{AB}$ and $\mathbf{BA}$ related?

9. Prove that $\mathbf{A}^\dagger \mathbf{A}$ is positive semidefinite, i.e., $\mathbf{x}^\dagger \mathbf{A}^\dagger \mathbf{A} \mathbf{x} \geq 0$.

10. (a) Show that the matrix

$$\mathbf{A} = \begin{bmatrix} 0 & 1 & 0 & 0 & \cdots & 0 \\ 0 & 0 & 1 & 0 & \cdots & 0 \\ \vdots & \vdots & \vdots & \vdots & & \vdots \\ 0 & 0 & 0 & 0 & \cdots & 1 \\ -a_n & -a_{n-1} & -a_{n-2} & -a_{n-3} & \cdots & -a_1 \end{bmatrix}$$

has eigenvectors

$$\mathbf{x}_i = \begin{bmatrix} 1 \\ \lambda_i \\ \lambda_i^2 \\ \vdots \\ \lambda_i^{n-1} \end{bmatrix},$$

where λ_i is an eigenvalue of $\mathbf{A}$. Show also that the characteristic equation becomes

$$\lambda^n + a_1 \lambda^{n-1} + a_1 \lambda^{n-1} + \cdots + a_n = 0.$$

$\mathbf{A}$ is known as the companion matrix because of its relationship to an $(n+1)$th-order ordinary differential equation.

(b) Suppose $n = 10$, $a_1 = a_2 = \cdots = a_9 = 0$, and $a_{10} = 10^{-10}$. Find the eigenvalues and eigenvectors of $\mathbf{A}$.

11. Consider the matrix
$$A = \begin{bmatrix} 2 & -2 & 3 \\ 1 & 1 & 1 \\ 1 & 3 & -1 \end{bmatrix}.$$

(a) Compute the eigenvalues of A.
(b) Compute the eigenvectors x_i of A and z_i of $A^\dagger$. Show by computation that x_i and z_i are a biorthogonal set, i.e., $x_i^\dagger z_j \neq 0$ if $i = j$ and $= 0$ if $i \neq j$. Construct the reciprocal set for $\{x_i\}$.
(c) Show by direct calculation that A obeys the Hamilton–Cayley theorem, i.e., that A satisfies its characteristic polynomial, i.e., $P_3(A) = 0$.
(d) Compute A^4 using the Hamilton–Cayley theorem.
(e) Compute A^{-1}.
(f) Show that the eigenvectors of A are linearly independent.
(g) Calculate the eigenvalues of
$$2A^6 + 5A^{-1} + 3A^2 - 7A^{-1} - 8I.$$

12. Consider the space S of square integrable functions $f(t)$ defined on the domain $-1 \leq t \leq 1$. Define the inner product as
$$\langle f, g \rangle = \int_{-1}^{1} f^*(t) g(t)\, dt.$$
Suppose
$$z_1(t) = 1, \qquad z_2(t) = t, \qquad z_3(t) = t^2.$$
Use the Gram–Schmidt procedure to construct an orthonormal set of functions $x_1(t), x_2(t), x_3(t)$ from these functions.

13. Suppose
$$A = \begin{bmatrix} -2 & 1 \\ 1 & -2 \end{bmatrix}.$$

(a) Find the normalized eigenvectors x_i and the eigenvalues λ_i of A.
(b) Show by direct substitution that
$$I = x_1 x_1^\dagger + x_2 x_2^\dagger.$$

(c) Using part (b), show that
$$A = \lambda_1 x_1 x_1^\dagger + \lambda_2 x_2 x_2^\dagger$$
and
$$\exp(tA) = \exp(\lambda_1 t) x_1 x_1^\dagger + \exp(\lambda_2 t) x_2 x_2^\dagger.$$

(d) Solve
$$\frac{dx}{dt} = Ax,$$
where
$$x(t = 0) = \begin{bmatrix} 1 \\ 2 \end{bmatrix}.$$

14. (a) Is the inner product

$$\mathbf{x}^T \mathbf{A} \mathbf{x}$$

positive for all nonzero, real vectors **x**, where

$$\mathbf{A} = \begin{bmatrix} -2 & 1 & 0 \\ 0 & -2 & 1 \\ 0 & 0 & -2 \end{bmatrix}?$$

(b) Is **A** positive definite? Why?
(c) What are the eigenvalues of **A**? Find the eigenvectors of **A**.

15. Find the eigenvalues and eigenvectors of the matrices

(a)
$$\mathbf{A} = \begin{bmatrix} 2 & 1 \\ 1 & 2 \end{bmatrix}.$$

(b)
$$\mathbf{A} = \begin{bmatrix} 4 & 1 \\ 0 & 4 \end{bmatrix}.$$

(c)
$$\mathbf{A} = \begin{bmatrix} 2 & 1 & 0 \\ 0 & 2 & 1 \\ 0 & 0 & 2 \end{bmatrix}.$$

(d)
$$\mathbf{A} = \begin{bmatrix} 2 & 1 & 0 & 0 & 0 & 0 \\ 0 & 2 & 1 & 0 & 0 & 0 \\ 0 & 0 & 2 & 0 & 0 & 0 \\ 0 & 0 & 0 & 4 & 1 & 0 \\ 0 & 0 & 0 & 0 & 4 & 0 \\ 0 & 0 & 0 & 0 & 0 & 3 \end{bmatrix}.$$

16. According to the Hamilton–Caley theorem, a square matrix satisfies its characteristic equation, i.e.,

$$P_n(\mathbf{A}) = \mathbf{0}.$$

For the matrix

$$\mathbf{A} = \begin{bmatrix} -1 & 0 & 1 \\ 5 & -1 & -5 \\ 2 & -4 & -3 \end{bmatrix}:$$

(a) Find the traces $\text{tr}_j \mathbf{A}$, $j = 1, 2,$ and 3.
(b) Show by direct substitution that **A** satisfies its characteristic equation.
(c) Find the eigenvalues of **A**.
(d) Is **A** perfect? Why?
(e) Find the eigenvectors of **A**.

(f) Find the matrix **X** that diagonalizes **A** under a similarity transformation.
(g) Find the inverse of the matrix **X**.
(h) Find the set of vectors z_i that are the reciprocal of the set of eigenvectors of **A**.
(i) Evaluate A^4 using the Hamilton–Cayley theorem and the spectral resolution theorem.

17. According to the power method, the maximum-magnitude eigenvalue and its eigenvector can be found by starting with an arbitrary vector x_0 and hunting the vector

$$x_k = Ax_{k-1} = A^k x_0, \quad k = 1, 2, \ldots.$$

An estimate of the maximum-magnitude eigenvalue is

$$\lambda = \frac{x_k^\dagger A x_k}{x_k^\dagger x_k}.$$

Consider the matrix

$$A = \begin{bmatrix} -2 & 1 & 0 \\ 1 & -2 & 1 \\ 0 & 1 & -2 \end{bmatrix}.$$

(a) Find the eigenvalues and eigenvectors of **A** by a direct method.
(b) Use the power method to find the maximum-magnitude eigenvalue and its eigenvector of **A**. How large must k be to find the estimated eigenvalue $\tilde{\lambda}$ and its estimated normalized eigenvector $\tilde{x}$ to a tolerance

$$\frac{|\tilde{\lambda} - \lambda|}{|\lambda|} < 10^{-6}$$

and

$$\|\tilde{x} - x\| < 10^{-6},$$

where λ is the maximum-magnitude eigenvalue and **x** its normalized eigenvalue. It is probably best to do this part of the exercise on a computer, although it is not impossible to do by hand.
(c) Calculate the adjugate of **A** and from it the inverse of **A**. Use the power method to determine the minimum-magnitude eigenvalue of **A** to the same tolerance as given in part (b).

18. The traction vector $t_{\hat{n}}$ on a surface whose normal is in the direction $\hat{n}$ is related to the stress tensor **T** of a material by

$$t_{\hat{n}} = T \cdot \hat{n}, \tag{1}$$

where $\hat{n}$ is a unit vector. In Cartesian coordinates, Eq. (1) can be expressed as

$$\begin{bmatrix} t_1 \\ t_2 \\ t_3 \end{bmatrix} = \begin{bmatrix} T_{11} & T_{12} & T_{13} \\ T_{21} & T_{22} & T_{23} \\ T_{31} & T_{32} & T_{33} \end{bmatrix} \begin{bmatrix} n_1 \\ n_2 \\ n_3 \end{bmatrix},$$

or as $\mathbf{t} = \mathbf{Tn}$ in matrix notation. From physical principles it is known that the stress tensor is real and that $T_{ij} = T_{ji}$.

If the material is loaded so that $T_{11} = T_{22} = 3$, $T_{33} = 4$, $T_{12} = 2$, and $T_{13} = T_{23} = 0$, we wish to find the principal directions, i.e., the directions in which the traction vector is the largest and the smallest. To find these directions, we want to find the extrema of $Q - \lambda \mathbf{n}^T \mathbf{n}$, where $Q = \mathbf{n}^T \mathbf{t} = \mathbf{n}^T \mathbf{Tn} = \sum T_{ij} n_i n_j$ and λ is the Lagrange multiplier resulting from the constraint that $\hat{n}$ be a unit vector, i.e., that $\mathbf{n}^T \mathbf{n} = \mathbf{I}$. The extremum condition is

$$\frac{\partial}{\partial n_i}[Q - \lambda \mathbf{n}^T \mathbf{n}] = 0, \quad i = 1, 2, \text{ and } 3. \tag{2}$$

(a) How many linearly independent solutions does Eq. (2) have? Why?

(b) Find the direction vectors $\mathbf{n}$ and the values of the traction vectors in these directions (principal directions and values).

(c) It turns out that the material is nonlinear and a pair of points in the material originally separated by $\mathbf{x}$ is separated by $\mathbf{y}$,

$$\mathbf{y} = \mathbf{T}^{3/4} \mathbf{x}$$

after the load is applied. If $\mathbf{x}^T = [1, 1, 1]$, find the length of $\mathbf{y}$ after loading.

19. Consider the matrix equation

$$\mathbf{A}^2 - \mathbf{BA} + \mathbf{C} = \mathbf{0},$$

where

$$\mathbf{B} = \begin{bmatrix} 2 & 1 \\ 1 & 2 \end{bmatrix}, \quad \mathbf{C} = \begin{bmatrix} 1 & 2 \\ 2 & 1 \end{bmatrix}.$$

Find $\mathbf{A}$ with the aid of the spectral resolution theorem for $\mathbf{B}$ and $\mathbf{C}$.

20. Prove that $\text{tr}_k(\mathbf{BAB}^{-1}) = \text{tr}_k(\mathbf{A})$.

21. Prove that $\text{tr}_1((\mathbf{AB})^k) = \text{tr}_1((\mathbf{BA})^k)$.

22. Use the result from Problem 21 to prove that $\text{tr}_k(\mathbf{AB}) = \text{tr}_k(\mathbf{BA})$.

23. Prove that if $|a_{ii}| > \sum_{j \neq i} |a_{ij}|$ for all $i = 1, 2, \ldots, n$, then the $n \times n$ matrix $\mathbf{A}$ is nonsingular.

24. Prove Perrin's theorem, which states that, for an $n \times n$ real-valued matrix $\mathbf{A}$ with $a_{ij} > 0$ for all i and j, the eigenvalue of the largest magnitude is:

(a) Real and positive.

(b) Nondegenerate.

25. Use the recurrence formulas in Eq. (5.6.6) to compute the eigenvalues of the tridiagonal matrix

$$\mathbf{A} = \begin{bmatrix} -5 & 1 & 0 & 0 & 0 \\ 3 & -5 & 1 & 0 & 0 \\ 0 & 3 & -6 & 2 & 0 \\ 0 & 0 & 4 & -6 & 1 \\ 0 & 0 & 0 & 5 & -7 \end{bmatrix}.$$

Compute the eigenvectors of **A**.

26. Use the power method to find the maximum eigenvalue of

$$\mathbf{A} = \begin{bmatrix} 1 & 2 & 1 \\ 1 & 3 & 2 \\ 2 & 1 & 4 \end{bmatrix}.$$

27. Use the power method to find the maximum eigenvalue of an $n \times n$ matrix with elements

$$a_{ij} = \frac{1}{i+j-1}.$$

Find the eigenvalue as a function of n for n between 1 and 50. Plot the results.

28. Evaluate the contour integral:

$$\frac{1}{2\pi i} \int_C f(z)(z\mathbf{I} - \mathbf{A})^{-1} \, dz,$$

where **A** is an $n \times n$ nonsingular matrix and $i = \sqrt{-1}$.

FURTHER READING

Amundson, A. R. (1964). "Mathematical Methods in Chemical Engineering." Prentice Hall, Englewood Cliffs, NJ.

Bellman, R. (1970). "Introduction to Matrix Analysis." McGraw-Hill, New York.

Golub, G. H., and Van Loan, C. F. (1989). "Matrix Computations." 2nd Ed. Johns Hopkins University Press, Baltimore.

Noble, B. (1969). "Applied Linear Algebra." Prentice Hall, Englewood Cliffs, NJ.

Press, W. H., Teukolsky, S. A., Vetterling, W. T., and Flannery, B. P. (1988). "Numerical Recipes in Fortran." 2nd Ed. Cambridge University Press, Cambridge, UK.

Wilkinson, J. H. (1965). "The Algebraic Eigenvalue Problem." Clarendon Press, Oxford.

Noble, B., and Daniel, J. W. (1977). "Applied Linear Algebra." Prentice Hall, Englewood Cliffs, NJ.

Smith, B. T. (1974). "Matrix Eigensystem Routines, EISPACK Guide," Springer-Verlag, New York.

6
PERFECT MATRICES

6.1. SYNOPSIS

The most important property of a *perfect* matrix is that it obeys the spectral resolution theorem, $\mathbf{A} = \sum_{i=1}^{n} \lambda_i \mathbf{x}_i \mathbf{z}_i^\dagger$, where λ_i and $\mathbf{x}_i$, $i = 1, \ldots, n$, are the eigenvalues and eigenvectors of $\mathbf{A}$ and $\mathbf{z}_i$, $i = 1, \ldots, n$, are the eigenvectors of the adjoint $\mathbf{A}^\dagger$. We say the vectors $\{\mathbf{x}_i\}$ and $\{\mathbf{z}_i\}$ form a biorthogonal set, i.e., $\mathbf{x}_i^\dagger \mathbf{z}_j = \mathbf{z}_j^\dagger \mathbf{x}_i = \delta_{ij}$. The set $\{\mathbf{z}_i\}$ are the column vectors of the matrix $(\mathbf{X}^{-1})^\dagger$, where $\mathbf{X} = [\mathbf{x}_1, \ldots, \mathbf{x}_n]$. Thus, the reciprocal vectors $\{\mathbf{z}\}$ can easily be computed from the eigenvectors $\{\mathbf{x}_i\}$. We will show that the spectral decomposition of $\mathbf{A}$ leads to the result $f(\mathbf{A}) = \sum_{i=1}^{n} f(\lambda_i) \mathbf{x}_n \mathbf{z}_i^\dagger$, when the function $f(t)$ is defined for values of t equal to the eigenvalues of $\mathbf{A}$.

The spectral resolution theorem for a perfect matrix is equivalent to the following theorem: if $\mathbf{A}$ is *perfect*, it can be diagonalized by a similarity transformation; i.e., there exists a matrix $\mathbf{X}$ such that $\mathbf{X}^{-1} \mathbf{A} \mathbf{X} = \mathbf{\Lambda}$, where $\mathbf{\Lambda}$ is a diagonal matrix. The column vectors of $\mathbf{X}$ are the eigenvalues of $\mathbf{A}$ and the diagonal elements of $\mathbf{\Lambda}$ are the eigenvalues of $\mathbf{A}$. Furthermore, if the eigenvalues of $\mathbf{A}$ are distinct, i.e., if all the roots of the characteristic polynomial $P_n(\lambda) = |\mathbf{A} - \lambda \mathbf{I}|$ are of multiplicity 1, then $\mathbf{A}$ is a perfect matrix, but a perfect matrix need not have distinct eigenvalues.

We say a matrix $\mathbf{U}$ is unitary if it has the property $\mathbf{U}^\dagger = \mathbf{U}^{-1}$. This property is equivalent to the property $\|\mathbf{x}\| = \|\mathbf{y}\|$ for every $\mathbf{x} \in E_n$, where $\mathbf{y} = \mathbf{U}\mathbf{x}$. In other words, as a linear operator in E_n, a unitary matrix rotates a vector without changing its length. We will prove in Section 6.6 that *any* square matrix can be semidiagonalized by a unitary transformation; i.e., for any square matrix $\mathbf{A}$, there exists a unitary matrix $\mathbf{U}$ such that $\mathbf{U}^\dagger \mathbf{A} \mathbf{U}$ has only zero elements below the main diagonal.

A *normal* matrix is defined by the property $AA^\dagger = A^\dagger A$. The class of normal matrices includes unitary matrices ($A^\dagger = A^{-1}$), orthogonal matrices (real matrices with the property $A^T = A^{-1}$), and self-adjoint matrices ($A^\dagger = A$). With the aid of the semidiagonalization theorem, we will prove that a normal matrix has the spectral decomposition $A = \sum_i \lambda_i x_i x_i^\dagger$, where λ_i and x_i, $i = 1, \ldots, n$, are the eigenvalues and the eigenvectors of A. The eigenvectors $\{x_i\}$, therefore, form an orthonormal basis set, i.e., $x_i^\dagger x_j = \delta_{ij}$. The spectral decomposition theorem for normal matrices is equivalent to the following theorem: if A is normal, there exists a unitary matrix X such that $X^\dagger A X = \Lambda$; i.e., A can be diagonalized by a unitary transformation. The column vectors of X are the eigenvectors of A and the elements of the diagonal matrix Λ are its eigenvalues.

If A is a unitary or orthogonal matrix, its eigenvalues are of modulus 1, i.e., $|\lambda_i| = 1$. If A is self-adjoint, its eigenvalues are real. If all of the eigenvalues of a self-adjoint matrix are positive, then it is positive definite. In the most general case of a normal matrix, the eigenvalues can be arbitrary complex numbers.

We will find that the spectral resolution theorem for perfect matrices greatly simplifies the solution of the differential equations $dx/dt = Ax$ and $d^2x/dt^2 = Ax$. For self-adjoint and normal matrices, the theorem can be used to find a coordinate system in which coupled harmonic oscillators are decoupled. The theorem can also be used in the analysis of the extrema of a multivariable function. For positive-definite matrices, the theorem will be used to find the covariance and mean square deviation for a multivariate Gaussian distribution. As an example of a negative-definite matrix, the problem of one-dimensional diffusion or heat transfer will be analyzed. The spectral resolution theorem for normal matrices will also be used to prove that, in the $p = 2$ norm, the norm $\|A\|$ of a normal matrix A is $|\lambda_{\max}|$. Furthermore, the condition number $\kappa(A)$ equals $|\lambda_{\max}|/|\lambda_{\min}|$, where $\lambda_{\max}$ is the eigenvalue of maximum absolute value and $\lambda_{\min}$ is the eigenvalue of minimum absolute value.

We will prove that the product of a positive-definite (or negative-definite) and a self-adjoint matrix is a perfect matrix and has real eigenvalues. We will also show that if A is perfect, and has only distinct eigenvalues, and if A and B commute ($AB = BA$), then A and B have the same eigenvectors. Also, if A and B are self-adjoint and commute, then they have the same eigenvectors. If A and B commute and are self-adjoint, they can be diagonalized by the same unitary transformation. We can prove that if A is positive definite and B is self-adjoint, the bilinear function $x^\dagger(A + B)x$ can in a coordinate transformation be transformed into $w^\dagger(I + M)w$, where M is a diagonal matrix with real elements along the diagonal. We will also show that if A is perfect, and has only distinct eigenvalues, it can be represented by Sylvester's formula in terms of eigenvalues and the adjugates of $A - \lambda_i I$.

Finally, we will show that a pth-order initial value problem can be reduced to the solution of the problem $dx/dt = Ax$ for one or more equations. For a single pth-order differential equation with constant coefficients, we will derive the eigenvectors of the companion matrix, and the necessary and sufficient conditions for A to be a perfect matrix are given.

6.2. IMPLICATIONS OF THE SPECTRAL RESOLUTION THEOREM

In Chapter 5 we saw that a perfect $n \times n$ matrix A has n linearly independent eigenvectors x_i, $i = 1, \ldots, n$. These form a basis set and imply the *spectral*

IMPLICATIONS OF THE SPECTRAL RESOLUTION THEOREM

resolution theorem

$$\mathbf{A} = \sum_{i=1}^{n} \lambda_i \mathbf{x}_i \mathbf{z}_i^\dagger. \tag{6.2.1}$$

Equation (6.2.1) is known as the spectral resolution or spectral decomposition of $\mathbf{A}$. The vector set $\{\mathbf{z}_i\}$ is composed of the n eigenvectors of $\mathbf{A}^\dagger$ (with eigenvalues $\{\lambda_i^*\}$), which are also the reciprocal vectors to the set $\{\mathbf{x}_i\}$, i.e.,

$$\mathbf{x}_i^\dagger \mathbf{z}_j = \mathbf{z}_j^\dagger \mathbf{x}_i = \delta_{ij}. \tag{6.2.2}$$

If $\{\mathbf{x}_i\}$ and $\{\mathbf{z}_i\}$ are calculated independently as eigenvectors of $\mathbf{A}$ and $\mathbf{A}^\dagger$, respectively, one must binormalize the vectors $\mathbf{z}_i$ so that $\mathbf{x}_i^\dagger \mathbf{z}_i = 1$. An implication of Eq. (6.2.1) is that to generate the class of all perfect matrices in E_n, we would first construct every linearly independent set of n vectors $\{\mathbf{x}_i\}$, then construct their reciprocal vectors $\{\mathbf{z}_i\}$, and, finally, generate every linear combination $\sum_i \lambda_i \mathbf{x}_i \mathbf{z}_i^\dagger$ for every possible set $\{\lambda_i\}$ of complex numbers.

Since

$$\mathbf{A}^k \mathbf{x}_i = \lambda_i^k \mathbf{x}_i \quad \text{and} \quad (\mathbf{A}^\dagger)^k \mathbf{z}_i = (\lambda_i^*)^k \mathbf{z}_i, \tag{6.2.3}$$

where k is any positive integer, it follows that the eigenvectors $\mathbf{x}_1, \ldots, \mathbf{x}_n$ are also eigenvectors of $\mathbf{A}^k$ with eigenvalues $\lambda_1^k, \ldots, \lambda_n^k$, and thus the spectral decomposition of $\mathbf{A}^k$ is

$$\mathbf{A}^k = \sum_{i=1}^{n} \lambda_i^k \mathbf{x}_i \mathbf{z}_i^\dagger. \tag{6.2.4}$$

If $\mathbf{A}$ is nonsingular, then $\mathbf{A}^{-1}$ exists and $\lambda_i \neq 0$, $i = 1, \ldots, n$. If $\lambda_j = 0$, then $|\mathbf{A}| = \prod_{i=1}^{n} \lambda_i = 0$ and the matrix would then be singular. However, since $\mathbf{A}^{-1}$ exists, the equation $\mathbf{A}\mathbf{x}_i = \lambda_i \mathbf{x}_i$ can be rearranged to get

$$\mathbf{A}^{-1} \mathbf{x}_i = \lambda_i^{-1} \mathbf{x}_i, \quad i = 1, \ldots, n, \tag{6.2.5}$$

where k is any positive integer. We see that the eigenvectors $\mathbf{x}_i, \ldots, \mathbf{x}_n$ are also eigenvectors of $\mathbf{A}^{-k}$ with eigenvalues $\lambda_1^{-k}, \ldots, \lambda_n^{-k}$. The spectral decomposition of $\mathbf{A}^{-k}$ is, therefore,

$$\mathbf{A}^{-k} = \sum_{i=1}^{n} \lambda_i^{-k} \mathbf{x}_i \mathbf{z}_i^\dagger. \tag{6.2.6}$$

If $f(t)$ is a function that can be expressed in the form

$$f(t) = \sum_{k=0}^{\infty} c_k t^k + \sum_{k=1}^{\infty} d_k t^{-k} \tag{6.2.7}$$

for t equal to the eigenvalues $\{\lambda_i\}$, then Eqs. (6.2.4) and (6.2.6) lead to the spectral decomposition

$$f(\mathbf{A}) = \sum_{i=1}^{n} f(\lambda_i) \mathbf{x}_i \mathbf{z}_i^\dagger. \tag{6.2.8}$$

Thus, as stated in the previous chapter, if $f(t)$ is expressible in a Laurent series or a Taylor series ($d_k = 0$), the matrix function $f(\mathbf{A})$ has eigenvectors $\mathbf{x}_i$ and eigenvalues $f(\lambda_i)$ and obeys the spectral decomposition theorem given by Eq. (6.2.8). Note that if $f(t)$ is a Laurent series, then $\mathbf{A}$ must be nonsingular for Eq. (6.2.8) to hold. If $f(t)$ is a Taylor series, then $\mathbf{A}$ simply has to be perfect for Eq. (6.2.8) to be valid.

EXAMPLE 6.2.1. Consider the matrix

$$\mathbf{A} = \begin{bmatrix} 0 & 1 \\ 1 & 0 \end{bmatrix}. \qquad (6.2.9)$$

The eigenvalues of $\mathbf{A}$ are $\lambda_1 = -1$ and $\lambda_2 = 1$, and the corresponding eigenvectors are

$$\mathbf{x}_1 = \begin{bmatrix} -\dfrac{1}{\sqrt{2}} \\ \dfrac{1}{\sqrt{2}} \end{bmatrix} \quad \text{and} \quad \mathbf{x}_2 = \begin{bmatrix} \dfrac{1}{\sqrt{2}} \\ \dfrac{1}{\sqrt{2}} \end{bmatrix}. \qquad (6.2.10)$$

For this matrix, it turns out that the reciprocal basis vectors are identical to the eigenvectors ($\mathbf{z}_i = \mathbf{x}_i$, $i = 1, 2$) since the $\mathbf{x}_i$ are orthogonal. From Eq. (6.2.10) we can construct the dyads

$$\mathbf{x}_1 \mathbf{x}_1^\dagger = \begin{bmatrix} \dfrac{1}{2} & -\dfrac{1}{2} \\ -\dfrac{1}{2} & \dfrac{1}{2} \end{bmatrix} \quad \text{and} \quad \mathbf{x}_2 \mathbf{x}_2^\dagger = \begin{bmatrix} \dfrac{1}{2} & \dfrac{1}{2} \\ \dfrac{1}{2} & \dfrac{1}{2} \end{bmatrix}, \qquad (6.2.11)$$

and using the spectral resolution theorem of Eq. (6.2.8), the matrices $\exp(\mathbf{A}t)$, $\sin \mathbf{A}t$, and $\cos \mathbf{A}t$ can be evaluated as

$$\exp(\mathbf{A}t) = \exp(-t) \begin{bmatrix} \dfrac{1}{2} & -\dfrac{1}{2} \\ -\dfrac{1}{2} & \dfrac{1}{2} \end{bmatrix} + \exp(t) \begin{bmatrix} \dfrac{1}{2} & \dfrac{1}{2} \\ \dfrac{1}{2} & \dfrac{1}{2} \end{bmatrix} = \begin{bmatrix} \cosh t & \sinh t \\ \sinh t & \cosh t \end{bmatrix},$$

$$(6.2.12)$$

$$\sin \mathbf{A}t = \sin(-t) \begin{bmatrix} \dfrac{1}{2} & -\dfrac{1}{2} \\ -\dfrac{1}{2} & \dfrac{1}{2} \end{bmatrix} + \sin(t) \begin{bmatrix} \dfrac{1}{2} & \dfrac{1}{2} \\ \dfrac{1}{2} & \dfrac{1}{2} \end{bmatrix} = \begin{bmatrix} 0 & \sin t \\ \sin t & 0 \end{bmatrix},$$

$$(6.2.13)$$

$$\cos \mathbf{A}t = \cos(-t) \begin{bmatrix} \dfrac{1}{2} & -\dfrac{1}{2} \\ -\dfrac{1}{2} & \dfrac{1}{2} \end{bmatrix} + \cos(t) \begin{bmatrix} \dfrac{1}{2} & \dfrac{1}{2} \\ \dfrac{1}{2} & \dfrac{1}{2} \end{bmatrix} = \begin{bmatrix} \cos t & 0 \\ 0 & \cos t \end{bmatrix}.$$

$$(6.2.14)$$

For the case of simple perfect matrices, the spectral resolution theorem can be applied ■ ■ ■ using Mathematica for most functions (built-in or user defined)—see the Appendix.

IMPLICATIONS OF THE SPECTRAL RESOLUTION THEOREM

Exercise 6.2.1. Using the spectral resolution theorem, derive expressions for the matrices $\exp(\mathbf{A}t)$, $\sin \mathbf{A}t$, and $\cos \mathbf{A}t$ for the matrix

$$\mathbf{A} = \begin{bmatrix} 0 & -i \\ i & 0 \end{bmatrix},$$

where $i = \sqrt{-1}$.

The spectral resolution theorem for perfect matrices can be used to compute $f(\mathbf{A})$ even if $f(t)$ cannot be expanded in a series in t. For example, consider the square root of $\mathbf{A}$, i.e., $f(\mathbf{A})$ when $f(t) = t^{1/2}$. If we set

$$f(\mathbf{A}) = \sum_{i=1}^{n} \lambda_i^{1/2} \mathbf{x}_i \mathbf{z}_i^{\dagger}, \tag{6.2.15}$$

we see by direct calculation that

$$f(\mathbf{A}) f(\mathbf{A}) = \sum_i \sum_j \lambda_i^{1/2} \lambda_j^{1/2} \mathbf{x}_i \mathbf{z}_i^{\dagger} \mathbf{x}_j \mathbf{z}_j^{\dagger}$$

$$= \sum_{i,j} \lambda_i^{1/2} \lambda_j^{1/2} \delta_{ij} \mathbf{x}_i \mathbf{z}_j^{\dagger} = \sum_i \lambda_i \mathbf{x}_i \mathbf{z}_i^{\dagger} \tag{6.2.16}$$

$$= \mathbf{A}.$$

The property we want of the square root $\sqrt{\mathbf{A}}$ is that, when it is squared, the result equals $\mathbf{A}$. The matrix defined by Eq. (6.2.15) does just this. Analogously, the νth power of $\mathbf{A}$ is $\sum_i \lambda_i^{\nu} \mathbf{x}_i \mathbf{z}_i^{\dagger}$, where $-\infty < \nu < \infty$ if $|\mathbf{A}| \neq 0$ and $0 < \nu < \infty$ otherwise.

Another quantity of interest is the logarithm of $\mathbf{A}$, i.e., $\ln \mathbf{A}$. We require that the matrix $\ln \mathbf{A}$ have the property

$$\exp(\ln \mathbf{A}) = \mathbf{A}; \tag{6.2.17}$$

i.e., the exponential of the logarithm of $\mathbf{A}$ equals $\mathbf{A}$. Let us define

$$\ln \mathbf{A} = \sum_{i=1}^{n} \ln \lambda_i \, \mathbf{x}_i \mathbf{z}_i^{\dagger}. \tag{6.2.18}$$

The eigenvectors $\mathbf{x}_1, \ldots, \mathbf{x}_n$ of $\mathbf{A}$ are thus eigenvectors in Eq. (6.2.18) and form a basis set. According to the spectral resolution theorem, if $f(t) = \exp(t)$, then

$$f(\ln \mathbf{A}) = \sum_{i=1}^{n} \exp(\ln \lambda_i) \mathbf{x}_i \mathbf{z}_i^{\dagger} = \sum_{i=1}^{n} \lambda_i \mathbf{x}_i \mathbf{z}_i^{\dagger} = \mathbf{A}. \tag{6.2.19}$$

Thus, the exponential of Eq. (6.2.18) yields $\mathbf{A}$, the desired property of the logarithm of $\mathbf{A}$. Since the logarithm of 0 is $-\infty$, the spectral decomposition fails when $\lambda_i = 0$. Also, when an eigenvalue λ_i is negative or complex, one has to decide what meaning to give $\ln \lambda_i$.

The spectral resolution theorem is especially useful for investigating the stability of nonlinear systems. Suppose

$$\frac{d\mathbf{x}}{dt} = \mathbf{f}(\mathbf{x}), \tag{6.2.20}$$

where $\mathbf{x} \in E_n$ and the components $f_i(\mathbf{x})$ of $\mathbf{f}(\mathbf{x})$ are nonlinear functions of $\mathbf{x}$. At steady state,

$$\mathbf{f}(\mathbf{x}^s) = \mathbf{0}. \tag{6.2.21}$$

To find the steady states, we can use the Newton–Raphson scheme; i.e., we begin with a guess, $\mathbf{x}^{(0)}$, and iteratively solve

$$\mathbf{J}(\mathbf{x}^{(k)})(\mathbf{x}^{(k+1)} - \mathbf{x}^{(k)}) = -\mathbf{f}(\mathbf{x}^{(k)}), \tag{6.2.22}$$

$k = 0, 1, \ldots$, as described in Chapter 3. The elements of the Jacobian $\mathbf{J}(\mathbf{x})$ are just $j_{ij} = \partial f_i / \partial x_j$.

Suppose we have found a steady-state solution $\mathbf{x}^s$ and we would like to know if the solution is stable to small perturbations, i.e., small values of $\mathbf{x} - \mathbf{x}^s$. For $\mathbf{x}$ near $\mathbf{x}^s$, one can linearize the right-hand side of Eq. (6.2.20) to obtain

$$\frac{d\mathbf{x}}{dt} = \mathbf{f}(\mathbf{x}^s) + \mathbf{J}(\mathbf{x}^s)(\mathbf{x} - \mathbf{x}^s). \tag{6.2.23}$$

Defining $\mathbf{y} \equiv \mathbf{x} - \mathbf{x}^s$ and recalling that $\mathbf{f}(\mathbf{x}^s) = 0$, we find

$$\frac{d\mathbf{y}}{dt} = \mathbf{J}(\mathbf{x}^s)\mathbf{y}. \tag{6.2.24}$$

Conveniently, the Jacobian $\mathbf{J}(\mathbf{x}^s)$ would already have been evaluated in the Newton–Raphson solution for $\mathbf{x}^s$. The formal solution to Eq. (6.2.24) is

$$\mathbf{y}(t) = \exp(t\mathbf{J})\mathbf{y}_0, \tag{6.2.25}$$

where $\mathbf{y}_0 = \mathbf{y}(t = 0)$ is the initial perturbation. If the matrix $\mathbf{J}$ is perfect and if $\{\mathbf{x}_i\}$ and $\{\mathbf{z}_i\}$ are the eigenvectors of $\mathbf{J}$ and $\mathbf{J}^\dagger$, then the spectral resolution theorem yields

$$\mathbf{y}(t) = \sum_{i=1}^{n} \exp(\lambda_i t)(\mathbf{z}_i^\dagger \mathbf{y}_0)\mathbf{x}_i. \tag{6.2.26}$$

Thus, if the real part of the eigenvalues λ_i of $\mathbf{J}$ are negative, then $\mathbf{y} \to 0$ as $t \to \infty$ and so the steady state is stable to small perturbations. Otherwise, it is not. If stability of the steady state is all we want to know, we only need to know the eigenvalues of the Jacobian.

EXAMPLE 6.2.2. In a chemical reaction process, the concentrations obey the kinetic equations

$$\frac{dx}{dt} = 4x^2y^2 - 2xy^2 + 5y - 9 \tag{6.2.27}$$

$$\frac{dy}{dt} = -3x^3y - 12xy^2 - 3y + 9. \tag{6.2.28}$$

The Jacobian for the system is

$$\mathbf{J} = \begin{bmatrix} 8xy^2 - 2y^2 & 8x^2y - 4xy + 5 \\ -9x^2y - 12y^2 & -3x^3 - 24xy - 3 \end{bmatrix}. \tag{6.2.29}$$

A steady-state solution to the problem is

$$x = 0.07709, \qquad y = 1.89351 \tag{6.2.30}$$

for which the corresponding eigenvalues of $\mathbf{J}$ are

$$\lambda_1 = -5.73216 + 13.9189i, \qquad \lambda_2 = -5.73216 - 13.9189i. \tag{6.2.31}$$

Thus, the steady state is stable. Since λ_1 and λ_2 are different, we know that the matrix $\mathbf{J}$ is perfect.

EXERCISE 6.2.2. Give the spectral decomposition of $\exp(t\mathbf{J})$ for the above example and plot x and y versus t for $x_0 - x^s = y_0 - y^s = 1$.

It is interesting to think of the spectral resolution of the matrix $\mathbf{A}$ as a decomposition into a weighted sum of special dyads. If we define the dyad

$$\mathbf{E}_i \equiv \mathbf{x}_i \mathbf{z}_i^\dagger, \tag{6.2.32}$$

then spectral decomposition yields

$$\mathbf{A} = \sum_{i=1}^n \lambda_i \mathbf{E}_i, \tag{6.2.33}$$

a dyadic representation of $\mathbf{A}$. It is easy to show that, since $\mathbf{x}_i^\dagger \mathbf{z}_j = \delta_{ij}$, the dyads $\mathbf{E}_i$ have the properties

$$\mathbf{E}_i^2 = \mathbf{E}_i, \qquad \mathbf{E}_i \mathbf{E}_j = \mathbf{0} \text{ if } i \neq j, \qquad \mathbf{I} = \sum_{i=1}^n \mathbf{E}_i. \tag{6.2.34}$$

As an operator, $\mathbf{E}_i$ projects a vector $\mathbf{y}$ onto the direction of the eigenvector $\mathbf{x}_i$, i.e.,

$$\mathbf{E}_i \mathbf{y} = (\mathbf{z}_i^\dagger \mathbf{y}) \mathbf{x}_i. \tag{6.2.35}$$

The $\mathbf{E}_i$ are said to be projection operators and the set $\{\mathbf{E}_i\}$ for a perfect matrix is said to form a complete set of projection operators because

$$\sum_{i=1}^n \mathbf{E}_i \mathbf{y} = \mathbf{y}; \tag{6.2.36}$$

i.e., the sum $\sum_i \mathbf{E}_i$ projects any vector $\mathbf{y} \in E_n$ onto itself. A given $\mathbf{E}_i$ projects its eigenvector $\mathbf{x}_i$ onto itself, i.e., $\mathbf{E}_i \mathbf{x}_i = \mathbf{x}_i$. Furthermore, $\mathbf{E}_i \mathbf{x}_j = \mathbf{0}$ if $i \neq j$.

ILLUSTRATION 6.2.1 (Process Control of a Phase Separator). A process stream of benzene is partially condensed and flashed in a phase separator at $P = 14$ atm and $T = 200°C$. The flow rates to the separator for the liquid and vapor streams are $W_l = 300$ lb$_m$/min and $W_v = 160$ lbm/min, respectively. A common strategy for controlling the process environment is illustrated in Fig. 6.2.1. The liquid level in the separator is maintained at a constant set point value of $z = z^*$ by a liquid level controller (LLC) and a control valve on the exiting liquid stream. Similarly, the pressure in the vessel is maintained at $P = P^*$ by a pressure controller (PC) and a control valve on the the exiting vapor stream.

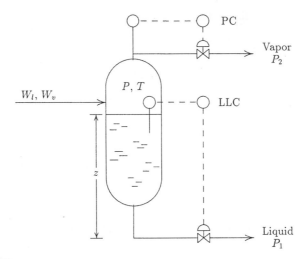

FIGURE 6.2.1

We can derive expressions for the time derivative of our control variables (z and P) by performing mass balances on both the liquid and the vapor streams. This liquid stream mass balance yields

$$\rho_l S \frac{dz}{dt} = W_l - \rho_l Q_l, \tag{6.2.37}$$

where S is the internal cross-sectional area of the separator vessel, ρ_l is the liquid benzene density, and Q_l is the volumetric flow rate of the exiting liquid stream. If we assume ideal gas behavior, we can write the vapor mass balance as

$$\frac{d}{dt}\left(\frac{MP}{RT}(L-z)S\right) = W_v - \frac{MP}{RT} Q_v, \tag{6.2.38}$$

where M is the molecular weight of benzene, R is the universal gas constant, and Q_v is the volumetric flow rate of the exiting vapor stream measured at P and T in the vessel. $L - z$ is the total vapor phase volume.

The volumetric flow rate of the liquid can be estimated using Bernoulli's equation as

$$Q_l = \pi D_l^2 \sqrt{\frac{\rho_l z + (P - P_1)}{8\alpha}}, \tag{6.2.39}$$

where P_1 is the downstream liquid pressure and D_l is the inner pipe diameter for the liquid stream. The parameter α represents the collective energy-dispersive terms due to friction within the pipe, bends in the pipe, entrance and exit effects, and, most important, the dispersive factor corresponding to the control valve. The value of α is adjusted by opening and closing the control valve and hence regulating the liquid flow rate out of the vessel.

A similar expression can be derived for the vapor flow rate by assuming adiabatic expansion through the vapor stream control valve. For an ideal gas, we get

$$Q_v = \pi D_2^2 \sqrt{\frac{\gamma RT\left[(P/P_2)^{(\gamma-1)/\gamma} - 1\right]}{8M(\gamma - 1)\beta}}, \quad (6.2.40)$$

where P_2 is the downstream vapor pressure, D_2 is the inner pipe diameter for the vapor stream, γ is the ratio of heat capacities, and β again is the collective energy-dispersive terms for the vapor stream.

Substituting these expressions into the mass balances gives the following differential equations for the control variables:

$$\frac{dz}{dt} = \frac{W_1}{S\rho_1} - \frac{\pi D_1^2}{S}\sqrt{\frac{\rho_1 z + (P - P_1)}{8\alpha}} \quad (6.2.41)$$

and

$$\frac{dP}{dt} = \frac{PW_v}{S\rho_1(L-z)} - \frac{P\pi D_1^2}{S(L-z)}\sqrt{\frac{\rho_1 z + (P - P_1)}{8\alpha}}$$
$$+ \frac{RTW_v}{SM(L-z)} - \frac{P\pi D_2^2}{S(L-z)}\sqrt{\frac{\gamma RT\left[(P/P_2)^{(\gamma-1)/\gamma} - 1\right]}{8M(\gamma - 1)\beta}}. \quad (6.2.42)$$

(i) For the process variables, $P^* = 14$ atm, $T = 200°$C, $D_1 = 3$ in., $D_2 = 4$ in., $P_1 = 5$ atm, $P_2 = 5$ atm, $M = 78.11$, $\rho_1 = 54$ lb$_m$/ft^3, $S = 113.1$ in.^{2}, $z^* = 5$ ft, $L = 7$ ft, and $\gamma = 1.21$, solve for the equilibrium values α^* and β^*. By constructing the appropriate Jacobian matrix, determine whether this equilibrium is stable or not.

(ii) To account for fluctuations in W_v and W_1, as well as both downstream pressures, the parameters α and β are to be controlled with the following proportional control schemes:

$$\frac{d\alpha}{dt} = -k_1(z - z^*) \quad (6.2.43)$$

$$\frac{d\beta}{dt} = -k_2(P - P^*). \quad (6.2.44)$$

Determine whether this system is stable for the control parameter values $k_1 = 1.0$ sec^{-1} ft^{-1} and $k_2 = 0.5$ sec^{-1} psi^{-1}. Will the control scheme produce oscillatory behavior?

6.3. DIAGONALIZATION BY A SIMILARITY TRANSFORMATION

We have previously defined a perfect $n \times n$ matrix as one that has n linearly independent eigenvectors. An equivalent definition of a perfect matrix is that $\mathbf{A}$ is perfect if it can be diagonalized by a similarity transformation. To see the equivalence, consider the matrix

$$\mathbf{X} = [\mathbf{x}_1, \mathbf{x}_2, \ldots, \mathbf{x}_n], \quad (6.3.1)$$

where the column vectors are the eigenvectors of **A**. Taking the product

$$\mathbf{AX} = [\mathbf{Ax}_1, \mathbf{Ax}_2, \ldots, \mathbf{Ax}_n] = [\lambda_1\mathbf{x}_1, \lambda_2\mathbf{x}_2, \ldots, \lambda_n\mathbf{x}_n] \qquad (6.3.2)$$

and noting that if **Λ** is the diagonal matrix

$$\mathbf{\Lambda} = \begin{bmatrix} \lambda_1 & 0 & \cdots & 0 \\ 0 & \lambda_2 & \cdots & 0 \\ \vdots & \vdots & & \vdots \\ 0 & 0 & \cdots & \lambda_n \end{bmatrix}, \qquad (6.3.3)$$

then

$$\mathbf{X\Lambda} = [\mathbf{x}_1, \ldots, \mathbf{x}_n] \begin{bmatrix} \lambda_1 & 0 & \cdots & 0 \\ 0 & \lambda_2 & \cdots & 0 \\ \vdots & \vdots & & \vdots \\ 0 & 0 & \cdots & \lambda_n \end{bmatrix} = [\lambda_1\mathbf{x}_1, \ldots, \lambda_n\mathbf{x}_n], \qquad (6.3.4)$$

and so we find

$$\mathbf{AX} = \mathbf{X\Lambda} \quad \text{or} \quad \mathbf{X}^{-1}\mathbf{AX} = \mathbf{\Lambda}. \qquad (6.3.5)$$

Thus,

THEOREM. *If **A** is perfect, it can be diagonalized by a similarity transformation. The column vectors of the transformation matrix **X** are the eigenvectors of **A** and the elements of the diagonal matrix are the eigenvalues of **A**.*

When the matrix function $f(\mathbf{A})$ can be spectrally decomposed, i.e., when $f(\lambda_i)$ exists for $i = 1, \ldots, n$, it can also be diagonalized by a similarity transformation. This follows from the fact that $f(\mathbf{A})\mathbf{x}_i = f(\lambda_i)\mathbf{x}_i, i = 1, \ldots, n$, and so the argument leading to Eq. (6.3.5) yields the result

$$\mathbf{X}^{-1} f(\mathbf{A})\mathbf{X} = f(\mathbf{\Lambda}) = \begin{bmatrix} f(\lambda_1) & 0 & \cdots & 0 \\ 0 & f(\lambda_2) & \cdots & 0 \\ \vdots & \vdots & & \vdots \\ 0 & 0 & \cdots & f(\lambda_n) \end{bmatrix}. \qquad (6.3.6)$$

Rearranging Eq. (6.3.6) in the form

$$f(\mathbf{A}) = \mathbf{X} f(\mathbf{\Lambda}) \mathbf{X}^{-1}, \qquad (6.3.7)$$

we see that if **A** is perfect, then the solution to

$$\frac{d\mathbf{y}}{dt} = \mathbf{A}\mathbf{y}, \qquad \mathbf{y}_0 = \mathbf{y}(t=0), \qquad (6.3.8)$$

is

$$\mathbf{y} = \mathbf{X}\exp(t\mathbf{\Lambda})\mathbf{X}^{-1}\mathbf{y}_0 = \mathbf{X}\begin{bmatrix} \exp(t\lambda_1) & 0 & \cdots & 0 \\ 0 & \exp(t\lambda_2) & \cdots & 0 \\ \vdots & \vdots & & \vdots \\ 0 & 0 & \cdots & \exp(t\lambda_n) \end{bmatrix}\mathbf{X}^{-1}\mathbf{y}_0, \tag{6.3.9}$$

again illustrating that the asymptotic behavior of $\mathbf{y}$ can be understood by knowing only the eigenvalues of $\mathbf{A}$.

ILLUSTRATION 6.3.1 (Heat Exchanger Profile). Countercurrent heat exchangers are used extensively to cool hot process fluids. Consider a process stream at temperature T_p entering a water-cooled heat exchanger illustrated in Fig. 6.3.1. It is desired to cool the stream down to a temperature T_1 shown exiting the exchanger on the right. Cooling water at T_w is supplied to the exchanger jacket and exits at temperature T_2. The process stream has a mass flow rate of W_p (kg/s) and water is supplied at a rate of W_w.

We would like to determine the temperature profiles of each stream within the heat exchanger. The overall length of the fluid "contact zone" is given by L and we represent the position along this contact zone by x. We can then write enthalpy balances at any arbitrary position x along the exchanger for each stream. The process stream gives

$$0 = W_p \hat{C}_p (T_p - T_{\text{ref}}) - W_p \hat{C}_p \big(t_p(x) - T_{\text{ref}}\big) \\ - \int_0^x U A_1 \big(t_p(x') - t_w(x')\big) dx', \tag{6.3.10}$$

where $t_p(x)$ and $t_w(x)$ represent the process and water stream temperatures, T_{ref} is the reference temperature for calculating enthalpies, $\hat{C}_p$ and $\hat{C}_w$ are the respective heat capacities, U is the local overall heat transfer coefficient, and A_1 is the area per unit length of heat transfer surface along the exchanger. The first two terms in Eq. (6.3.10) represent the net rate of enthalpy (we are neglecting pressure drops) entering a control volume encompassing the range $x = 0$ to $x = x$ within the exchanger. The last term

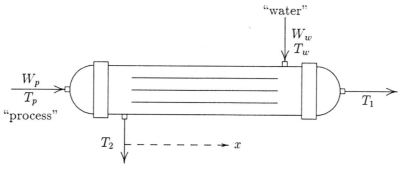

FIGURE 6.3.1

represents the total heat transferred to the water stream from the process stream. A similar equation can be derived for the water stream:

$$0 = W_w \hat{C}_w (t_w(x) - T_{\text{ref}}) - W_w \hat{C}_w (T_2 - T_{\text{ref}})$$
$$+ \int_0^x U A_1 (t_p(x') - t_w(x')) \, dx'. \quad (6.3.11)$$

Differentiating both balance equations with respect to x, we arrive at the following coupled set of differential equations:

$$\frac{dt_p}{dx} = -\frac{U A_1}{W_p \hat{C}_p} (t_p - t_w) \quad (6.3.12)$$

$$\frac{dt_w}{dx} = -\frac{U A_1}{W_w \hat{C}_w} (t_p - t_w). \quad (6.3.13)$$

We can express the above system of equations in matrix form using the following definitions:

$$\mathbf{t} \equiv \begin{bmatrix} t_p \\ t_w \end{bmatrix} \quad (6.3.14)$$

and

$$\mathbf{A} \equiv \begin{bmatrix} -\alpha_p & \alpha_p \\ -\alpha_w & \alpha_w \end{bmatrix}, \quad (6.3.15)$$

with

$$\alpha_p \equiv \frac{U A_1}{W_p \hat{C}_p} \quad \text{and} \quad \alpha_w \equiv \frac{U A_1}{W_w \hat{C}_w}. \quad (6.3.16)$$

Equations (6.3.12) and (6.3.13) then become

$$\frac{d\mathbf{t}}{dx} = \mathbf{A}\mathbf{t}, \quad (6.3.17)$$

with boundary conditions

$$\mathbf{t}(x = 0) = \begin{bmatrix} T_p \\ T_2 \end{bmatrix}. \quad (6.3.18)$$

(i) We now make use of the theorems in this section by defining the matrix $\mathbf{S}$ in the similarity transformation that diagonalizes $\mathbf{A}$. Namely,

$$\mathbf{S}^{-1} \mathbf{A} \mathbf{S} = \mathbf{\Lambda}, \quad (6.3.19)$$

where $\mathbf{\Lambda}$ is the diagonal matrix containing the eigenvalues of $\mathbf{A}$. We can then make the linear transformation

$$\mathbf{s} = \mathbf{S}^{-1} \mathbf{t}. \quad (6.3.20)$$

Multiplying Eq. (6.3.17) by $\mathbf{S}^{-1}$ and making use of the identity $\mathbf{I} = \mathbf{SS}^{-1}$, show that Eq. (6.3.17) can be rewritten as

$$\frac{d\mathbf{s}}{dx} = \mathbf{\Lambda s}. \tag{6.3.21}$$

(ii) For the general case where $\alpha_p \neq \alpha_w$, find the eigenvalues of $\mathbf{A}$ and matrices $\mathbf{S}$ and $\mathbf{S}^{-1}$.

(iii) Solve the decoupled system in Eq. (6.3.21) using Eq. (6.3.20) to transform the boundary conditions into the new representation. How are the temperature profiles different when $\alpha_p > \alpha_w$ as opposed to when $\alpha_p < \alpha_w$?

(iv) Determine the outlet temperatures T_1 and T_2 given the inlet conditions $T_p = 95$ K, $T_w = 10$ K, $\alpha_p = 1.2$ ft^{-1}, $\alpha_w = 2.5$ ft^{-1}, and $L = 3$ ft. Plot both temperature profiles over the range $0 \leq x \leq 3$.

ILLUSTRATION 6.3.2 (Multicomponent Reaction System). Consider a system of n chemical components C_i interconnected by first-order reversible reactions. The reactions are represented by

$$C_i \underset{k_{ij}}{\overset{k_{ji}}{\rightleftharpoons}} C_j, \quad i \neq j, \; i, j = 1, 2, \ldots, n, \tag{6.3.22}$$

where k_{ji} is the rate constant for conversion of component i to j and k_{ij} is the rate constant for conversion of component j to i. If ω_i represents the mole fraction of component i, the rate equations for the reaction system are

$$\frac{d\omega_i}{dt} = \sum_{j \neq i}^{n} k_{ij} \omega_j - \sum_{j \neq i}^{n} k_{ji} \omega_i \tag{6.3.23}$$

or

$$\frac{d\boldsymbol{\omega}}{dt} = \mathbf{K}\boldsymbol{\omega}, \tag{6.3.24}$$

where

$$\mathbf{K} = \begin{bmatrix} -\sum_{j \neq 1}^{n} k_{j1} & k_{12} & \cdots & k_{1n} \\ k_{21} & -\sum_{j \neq 2}^{n} k_{j2} & \cdots & k_{2n} \\ \vdots & \vdots & \ddots & \vdots \\ k_{n1} & k_{n2} & \cdots & -\sum_{j=1}^{n-1} k_{j2} \end{bmatrix}. \tag{6.3.25}$$

Note that $\sum_{i=1}^{n} \omega_i = 1$.

In general, $\mathbf{K}$ is not symmetric, but usually it is perfect. In this case, the concentrations of the chemical components as a function of time can be calculated from the formula

$$\boldsymbol{\omega}(t) = \exp(\mathbf{K}t)\boldsymbol{\omega}_0 = \sum_{i=1}^{n} \exp(\lambda_i t) \mathbf{x}_i \mathbf{z}_i^{\dagger} \boldsymbol{\omega}_0, \tag{6.3.26}$$

where λ_i and $\mathbf{x}_i$, $i = 1, \ldots, n$ are the eigenvalues and eigenvectors of $\mathbf{K}$ and $\mathbf{z}_i$ are the corresponding reciprocal vectors.

A system that has been studied carefully is the isomerization reactions of 1-butene, *trans*-2-butene, and *cis*-2-butene. The three reactions are

$$1\text{-butene} \underset{k_{12}}{\overset{k_{21}}{\rightleftharpoons}} cis\text{-2-butene}$$

$$1\text{-butene} \underset{k_{13}}{\overset{k_{31}}{\rightleftharpoons}} trans\text{-2-butene}$$

$$cis\text{-2-butene} \underset{k_{23}}{\overset{k_{32}}{\rightleftharpoons}} trans\text{-2-butene}$$

The chemical formulas for the isomers 1-butene and 2-butene are

$$\begin{array}{cccc} H & H & H & H \\ | & | & | & | \\ C= & C- & C- & C-H \\ | & & | & | \\ H & & H & H \end{array}$$

and

$$\begin{array}{cccc} H & H & H & H \\ | & | & | & | \\ H-C- & C= & C- & C-H \\ | & & & | \\ H & & & H \end{array}$$

where C and H denote carbon and hydrogen and "—" and "=" denote single and double chemical bonds between atoms. The *cis* and *trans* forms of 2-butene are distinguished by the spatial angles of the single bonds (not depicted here).

By comparing experiment and predictions, one can determine the rate constants k_{ij}. J. Wei and C. D. Prater report (*Adv. Catal.* **13** (1962), 256) the following values:

$$\mathbf{K} = \begin{bmatrix} -14.068 & 4.623 & 1.000 \\ 10.344 & -10.239 & 3.371 \\ 3.724 & 5.616 & -4.373 \end{bmatrix}. \tag{6.3.27}$$

Using these values, they give a comparison of calculated reaction paths with observed composition for butene isomerization shown in Fig. 6.3.2. Note that plotted in this form the units of t do not matter. $\mathbf{K}$ has been scaled so that $k_{13} = 1$. This amounts to choosing a particular set of units for the time t. Thus, one can calculate ω_i versus t, but, without the scale factor, the absolute values of t are unknown.

(i) Find the eigenvalues, eigenvectors, and reciprocal vectors for the matrix $\mathbf{K}$ given for the butene isomerization reaction.

(ii) Calculate as a function of time the concentrations of 1-butene and *cis*- and *trans*-2-butene for several initial concentrations.

(iii) Explain how to calculate the reaction paths shown in Fig. 6.3.2.

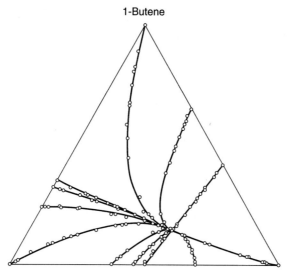

FIGURE 6.3.2 Reproduced from J. Wei and C. D. Prater (1962). "Advances in Catalysis and Related Subjects" Vol. 13, Academic Press.

(iv) Calculate the six reaction paths initially at the compositions $(\omega_1, \omega_2, \omega_3) =$ (0.25, 0.75, 0), (0.75, 0.25, 0), (0, 0.25, 0.75), (0, 0.75, 0.25), (0.25, 0, 0.75), and (0.75, 0, 0.25).

■ ■ ■ (v) Plot these new paths in a figure like that shown in Fig. 6.3.2.

6.4. MATRICES WITH DISTINCT EIGENVALUES

In Sections 5.3, 6.2, and 6.3, we have explored the implications of **A** being a perfect matrix. When can we know that **A** will be perfect? In subsequent sections we will show that normal, unitary, orthogonal, skew-symmetric, self-adjoint, and positive-definite matrices are always perfect. However, from the eigenvalue analysis presented in Section 5.4 we can already prove the following:

THEOREM. *A matrix* **A** *is perfect if all of its eigenvalues are distinct, i.e., if each root of* $P_n(\lambda) = |\mathbf{A} - \lambda \mathbf{I}| = 0$ *is of multiplicity* 1.

Recall from Eq. (5.4.11) that

$$\frac{dP_n}{d\lambda} = (-1)\mathrm{tr}_{n-1}(\mathbf{A} - \lambda \mathbf{I}). \qquad (6.4.1)$$

But the nth-degree characteristic polynomial can be expressed in the factored form

$$P_n(\lambda) = \prod_{i=1}^{n}(\lambda_i - \lambda), \qquad (6.4.2)$$

and so

$$\frac{dP_n(\lambda)}{d\lambda} = (-1) \sum_{j=1}^{n} \prod_{i \neq j}^{n} (\lambda_i - \lambda). \qquad (6.4.3)$$

If each eigenvalue is of multiplicity 1, it follows from Eqs. (6.4.3) and (6.4.1) that

$$\frac{dP_n(\lambda_k)}{d\lambda} = (-1) \prod_{i \neq k}^{n} (\lambda_i - \lambda_k) \neq 0, \qquad (6.4.4)$$

and so

$$\operatorname{tr}_{n-1}(\mathbf{A} - \lambda_k \mathbf{I}) \neq 0, \qquad k = 1, \ldots, n. \qquad (6.4.5)$$

According to Eq. (6.4.5), since $\operatorname{tr}_{n-1}(\mathbf{A} - \lambda_k \mathbf{I})$ is the sum of the principal minors of $\mathbf{A} - \lambda_k \mathbf{I}$, at least one principal minor of $\mathbf{A} - \lambda_k \mathbf{I}$ is not 0, and so the rank of $\mathbf{A} - \lambda_k \mathbf{I}$ is $n - 1$. From the solvability theorem, we know that the number of solutions of the homogeneous problem

$$(\mathbf{A} - \lambda_k \mathbf{I})\mathbf{x}_k = \mathbf{0} \qquad (6.4.6)$$

equals the difference between n and the rank of $\mathbf{A} - \lambda_k \mathbf{I}$, i.e., $n - (n - 1) = 1$. Thus, $\mathbf{A}$ is perfect if all of its eigenvalues are distinct.

It should be noted that having distinct eigenvalues is a sufficient but not a necessary condition for a matrix to be perfect. For instance, since $\mathbf{I}\mathbf{x} = \mathbf{x}$ for any vector $\mathbf{x}$, the eigenvalues of the unit matrix are all equal to 1 and, therefore, any basis set $\{\mathbf{x}_1, \ldots, \mathbf{x}_n\}$ in E_n is a set of eigenvectors for $\mathbf{I}$.

6.5. UNITARY AND ORTHOGONAL MATRICES

All perfect matrices can be diagonalized by a similarity transformation. However, there is a large class of matrices, namely, normal matrices, that can be diagonalized by a *unitary transformation*. Before discussing this class, let us describe unitary and orthogonal matrices. In a complex linear vector space E_n, the matrix $\mathbf{A}$ is a unitary matrix if it is a rotation operator in E_n, i.e., if the length of $\mathbf{y} = \mathbf{A}\mathbf{x}$ is the same as the length of $\mathbf{x}$. The equation representing this property is

$$\|\mathbf{x}\|^2 = \|\mathbf{A}\mathbf{x}\|^2 \qquad \text{for all } \mathbf{x} \in E_n. \qquad (6.5.1)$$

As an example of a rotation matrix, consider the following two Cartesian coordinate frames related by a rotation of angle θ in Fig. 6.5.1. If the coordinates of $\mathbf{r}$ in the frame $(\hat{i}, \hat{j})$ are x and y and in the frame $(\hat{i}', \hat{j}')$ they are x' and y', it follows that

$$\begin{aligned} x &= x' \cos \theta - y' \sin \theta \\ y &= x' \sin \theta + y' \cos \theta. \end{aligned} \qquad (6.5.2)$$

In matrix notation, this reads

$$\mathbf{x} = \mathbf{A}\mathbf{x}', \qquad (6.5.3)$$

UNITARY AND ORTHOGONAL MATRICES

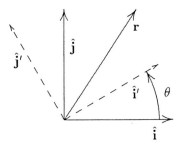

FIGURE 6.5.1

where

$$\mathbf{x} = \begin{bmatrix} x \\ y \end{bmatrix}, \quad \mathbf{x}' = \begin{bmatrix} x' \\ y' \end{bmatrix}, \quad \mathbf{A} = \begin{bmatrix} \cos\theta & -\sin\theta \\ \sin\theta & \cos\theta \end{bmatrix}. \quad (6.5.4)$$

It is easy to show that

$$x^2 + y^2 = x'^2 + y'^2 \quad \text{or} \quad \|\mathbf{x}\| = \|\mathbf{x}'\|, \quad (6.5.5)$$

which says that the length of **r** is unchanged—as required for the rotational transformation.

We call the transformation matrix of a rotation in a Euclidean vector space an orthogonal matrix instead of a unitary matrix. This is because the Euclidean vector space is real. In what follows we shall see that orthogonal matrices are a subclass of unitary matrices, the former being restricted to rotations in real linear vectors spaces and the latter including complex linear vector spaces.

To investigate the properties of unitary and orthogonal matrices, the following two theorems are useful. The first is:

THEOREM. *If*

$$\mathbf{x}^\dagger \mathbf{B} \mathbf{x} = 0 \quad \text{for all } \mathbf{x} \in E_n, \quad (6.5.6)$$

then $\mathbf{B} = \mathbf{0}$, *where* $\mathbf{0}$ *is the zero matrix, all of whose elements are 0.*

To prove this theorem, we write $\mathbf{x}^\dagger \mathbf{B} \mathbf{x} = 0$ as

$$\sum_{i,j}^n b_{ij} x_i x_j^* = 0. \quad (6.5.7)$$

Choosing $x_k \neq 0$ and $x_i = 0, i \neq k$, Eq. (6.5.7) implies

$$b_{kk} |x_k|^2 = 0 \quad \text{or} \quad b_{kk} = 0. \quad (6.5.8)$$

Since k is arbitrary, Eq. (6.5.8) proves that $b_{kk} = 0, k = 1, \ldots, n$. Next, we choose $x_k \neq 0$, $x_l \neq 0$, and $x_i = 0, i \neq k$ or l. Then Eq. (6.5.7) implies

$$b_{kl} x_k x_l^* + b_{lk} x_l x_k^* = 0 \quad (6.5.9)$$

or

$$(b_{kl} + b_{lk}) \operatorname{Re}(x_k x_l^*) + i(b_{kl} - b_{lk}) \operatorname{Im}(x_k x_l^*) = 0. \quad (6.5.10)$$

Since the real and imaginary parts of $x_k x_l^*$ can be varied independently, Eq. (6.5.10) implies

$$b_{kl} + b_{lk} = 0$$
$$b_{kl} - b_{lk} = 0 \qquad (6.5.11)$$

or $b_{kl} = b_{lk} = 0$ for arbitrary k and l. This completes the proof that Eq. (6.5.6) is true *if and only if* $\mathbf{B} = \mathbf{0}$.

The next theorem is:

THEOREM. *If*

$$\mathbf{B} = \mathbf{B}^T \text{ and } \mathbf{x}^T \mathbf{B} \mathbf{x} = 0 \qquad \text{for all real } \mathbf{x} \in E_n, \qquad (6.5.12)$$

then $\mathbf{B} = \mathbf{0}$.

The expression $\mathbf{x}^T \mathbf{B} \mathbf{x} = 0$ can be written as

$$\sum_{i,j} b_{ij} x_k x_j = 0. \qquad (6.5.13)$$

Again, we choose $x_k \neq 0$, $x_i = 0$, $i \neq k$, which leads to the conclusion $b_{kk} = 0$, $k = 1, \ldots, n$. And, if we choose $x_k \neq 0$, $x_l \neq 0$, $x_i = 0$, $i \neq k$ or l, we find

$$(b_{kl} + b_{lk}) x_k x_l = 0. \qquad (6.5.14)$$

Since $\mathbf{B} = \mathbf{B}^T$ by hypothesis ($b_{kl} = b_{lk}$), it follows that $b_{kl} = 0$ for arbitrary k and l, thus proving the theorem.

Note that if $\mathbf{B} \neq \mathbf{B}^T$, the condition $\mathbf{x}^T \mathbf{B} \mathbf{x} = 0$ for all real $\mathbf{x} \in E_n$ does not prove $\mathbf{B} = \mathbf{0}$. For example, if

$$\mathbf{B} = \begin{bmatrix} 0 & -1 \\ 1 & 0 \end{bmatrix}, \qquad (6.5.15)$$

then

$$\mathbf{x}^T \mathbf{B} \mathbf{x} = -x_1 x_2 + x_2 x_1 = 0 \qquad \text{for all } \mathbf{x} \in E_2. \qquad (6.5.16)$$

For this example, $\mathbf{B}$ has the property $\mathbf{B} = -\mathbf{B}^T$ for which

$$(\mathbf{x}^T \mathbf{B} \mathbf{x})^T = -\mathbf{x}^T \mathbf{B} \mathbf{x}. \qquad (6.5.17)$$

But $\mathbf{x}^T \mathbf{B} \mathbf{x}$ is a scalar (1×1 matrix) and so it must be equal to its transpose. Thus, Eq. (6.5.17) becomes

$$\mathbf{x}^T \mathbf{B} \mathbf{x} = -\mathbf{x}^T \mathbf{B} \mathbf{x} \qquad (6.5.18)$$

or $\mathbf{x}^T \mathbf{B} \mathbf{x} = 0$ for any $\mathbf{x} \in E_n$ if $\mathbf{B} = -\mathbf{B}^T$.

UNITARY AND ORTHOGONAL MATRICES

If $\mathbf{B} = \mathbf{B}^T$, we say that $\mathbf{B}$ is symmetric, whereas if $\mathbf{B} = -\mathbf{B}^T$, we say that $\mathbf{B}$ is asymmetric. Any arbitrary matrix can be decomposed into its symmetric and asymmetric parts:

$$\mathbf{B} = \mathbf{B}_{sy} + \mathbf{B}_{asy}$$
$$= \frac{1}{2}(\mathbf{B} + \mathbf{B}^T) + \frac{1}{2}(\mathbf{B} - \mathbf{B}^T). \tag{6.5.19}$$

Thus, in the quantity $\mathbf{x}^T \mathbf{B} \mathbf{x}$, only the symmetric part of $\mathbf{B}$ contributes, i.e., $\mathbf{x}^T \mathbf{B}_{asy} \mathbf{x} = 0$ for all $\mathbf{x} \in E_n$, and so

$$\mathbf{x}^T \mathbf{B} \mathbf{x} = \mathbf{x}^T \mathbf{B}_{sy} \mathbf{x} = \frac{1}{2} \mathbf{x}^T (\mathbf{B} + \mathbf{B}^T) \mathbf{x}. \tag{6.5.20}$$

This is, of course, not the case for the inner product $\mathbf{x}^\dagger \mathbf{B} \mathbf{x}$.

Let us return now to Eq. (6.5.1), the defining equation for a unitary matrix, and rewrite it as

$$\mathbf{x}^\dagger \mathbf{I} \mathbf{x} = \mathbf{x}^\dagger \mathbf{A}^\dagger \mathbf{A} \mathbf{x}$$

or

$$\mathbf{x}^\dagger (\mathbf{A}^\dagger \mathbf{A} - \mathbf{I}) \mathbf{x} = 0 \qquad \text{for all } \mathbf{x} \in E_n. \tag{6.5.21}$$

According to the theorem in Eq. (6.5.6), Eq. (6.5.21) establishes that a unitary matrix obeys the expression

$$\mathbf{A}^\dagger \mathbf{A} = \mathbf{I} \tag{6.5.22}$$

or, equivalently,

$$\mathbf{A}^{-1} = \mathbf{A}^\dagger; \tag{6.5.23}$$

the inverse of a unitary matrix is its adjoint. Also, since $|\mathbf{A}^\dagger \mathbf{A}| = |\mathbf{A}|^* |\mathbf{A}| = 1$, we see that the determinant of a unitary matrix is of unit magnitude; i.e., it is of modulus 1.

Equation (6.5.22) proves that the column vectors of a unitary matrix form an orthonormal set. To see this, let $\mathbf{A} = [\mathbf{a}_1, \ldots, \mathbf{a}_n]$ so that

$$\mathbf{A}^\dagger \mathbf{A} = \begin{bmatrix} \mathbf{a}_1^\dagger \\ \mathbf{a}_2^\dagger \\ \vdots \\ \mathbf{a}_n^\dagger \end{bmatrix} [\mathbf{a}_1, \mathbf{a}_2, \ldots, \mathbf{a}_n] = [\mathbf{a}_i^\dagger \mathbf{a}_j]. \tag{6.5.24}$$

However, $\mathbf{A}^\dagger \mathbf{A} = \mathbf{I} = [\delta_{ij}]$ and so it follows from Eq. (6.5.24) that

$$\mathbf{a}_i^\dagger \mathbf{a}_j = \delta_{ij}, \qquad i, j = 1, \ldots, n, m, \tag{6.5.25}$$

which is the defining property of an orthonormal basis set in E_n. In fact, Eqs. (6.5.23) and (6.5.25) are equivalent expressions for the defining property of a unitary matrix.

An orthogonal matrix is defined as a real matrix that rotates real vectors in E_n; i.e., if $\mathbf{A}$ is an orthogonal matrix, then

$$\|\mathbf{x}\|^2 = \|\mathbf{A}\mathbf{x}\|^2 \quad \text{or} \quad \mathbf{x}^T(\mathbf{A}^T\mathbf{A} - \mathbf{I})\mathbf{x} = 0 \quad \text{for all real } \mathbf{x} \in E_n. \quad (6.5.26)$$

Since the matrix $\mathbf{A}^T\mathbf{A} - \mathbf{I}$ is symmetric, the theorem in Eq. (6.5.12) is valid, and so an orthogonal matrix has the property

$$\mathbf{A}^T\mathbf{A} = \mathbf{I} \quad \text{or} \quad \mathbf{A}^{-1} = \mathbf{A}^T. \quad (6.5.27)$$

In this case, the determinant of $\mathbf{A}$ is equal to unity. Furthermore, the column vectors of $\mathbf{A}$ obey the conditions

$$\mathbf{a}_i^T \mathbf{a}_j = \delta_{ij}, \quad i, j = 1, \ldots, n. \quad (6.5.28)$$

Again, the column vectors of an orthogonal matrix form an orthonormal set of real vectors, which form a basis set in E_n. For the example given in Eq. (6.5.4),

$$\mathbf{a}_1 = \begin{bmatrix} \cos\theta \\ \sin\theta \end{bmatrix} \quad \text{and} \quad \mathbf{a}_2 = \begin{bmatrix} -\sin\theta \\ \cos\theta \end{bmatrix}. \quad (6.5.29)$$

It is easy to show that these are orthonormal vectors.

Examples of 2×2 unitary matrices are

$$\mathbf{a}_1 = \frac{1}{\sqrt{2}} \begin{bmatrix} i \\ -i \end{bmatrix}, \quad \mathbf{a}_2 = \frac{1}{\sqrt{2}} \begin{bmatrix} i \\ i \end{bmatrix}, \quad (6.5.30)$$

$$\mathbf{a}_1 = \frac{1}{\sqrt{2}} \begin{bmatrix} \exp(i\theta) \\ \exp(-i\theta) \end{bmatrix}, \quad \mathbf{a}_2 = \frac{1}{\sqrt{2}} \begin{bmatrix} \exp(i\theta) \\ -\exp(-i\theta) \end{bmatrix}. \quad (6.5.31)$$

The eigenvalues of the matrices with the column vectors in Eqs. (6.5.29), (6.5.30), and (6.5.31) are, respectively,

$$\lambda_1 = \cos\theta - i\sin\theta, \quad \lambda_2 = \cos\theta + i\sin\theta, \quad (6.5.32)$$

$$\lambda_1 = \frac{\sqrt{2}}{2}(i-1), \quad \lambda_2 = \frac{\sqrt{2}}{2}(i+1), \quad (6.5.33)$$

and

$$\lambda_1 = \frac{\exp(-i\theta)}{2\sqrt{2}}\left[-1 + \exp(2i\theta) - (1 + 6\exp(2i\theta) + (\exp(4i\theta))^{1/2})\right],$$
$$\lambda_2 = \frac{\exp(-i\theta)}{2\sqrt{2}}\left[-1 + \exp(2i\theta) - (1 + 6\exp(2i\theta) + (\exp(4i\theta))^{1/2})\right]. \quad (6.5.34)$$

All of these eigenvalues share the common property that their magnitude $|\lambda_i|$ is equal to 1. In a later section we will show that all of the eigenvalues of any unitary or orthogonal matrix have an absolute value of 1.

6.6. SEMIDIAGONALIZATION THEOREM

Suppose $\mathbf{U}$ is a unitary matrix and $\mathbf{A}$ is a square matrix of the same dimension. We say $\mathbf{B}$ is related to $\mathbf{A}$ by a *unitary transformation* if

$$\mathbf{B} = \mathbf{U}^\dagger \mathbf{A} \mathbf{U}. \tag{6.6.1}$$

Since $\mathbf{U}^\dagger = \mathbf{U}^{-1}$ for a unitary matrix, a unitary transformation is simply a similarity transformation with a unitary matrix. $\mathbf{B}$ and $\mathbf{A}$ have the same eigenvalues since the eigenvalues of $\mathbf{A}$ are unchanged under a similarity transformation.

The unitary transformation plays an important role in the powerful *semidiagonalization theorem*:

SEMIDIAGONALIZATION THEOREM. *Given any square matrix* $\mathbf{A}$, *there exists a unitary matrix* $\mathbf{U}$ *such that*

$$\mathbf{U}^\dagger \mathbf{A} \mathbf{U} = \begin{bmatrix} \lambda_1 & b_{12} & \cdots & b_{1n} \\ 0 & \lambda_2 & \cdots & b_{2n} \\ \vdots & \vdots & \ddots & \vdots \\ 0 & 0 & \cdots & \lambda_n \end{bmatrix}, \tag{6.6.2}$$

where the quantities λ_i *are the eigenvalues of* $\mathbf{A}$.

Thus, the theorem guarantees the existence of a unitary matrix that transforms $\mathbf{A}$ in a unitary transformation to an upper triangular matrix with the eigenvalues of $\mathbf{A}$ on the main diagonal.

To prove the semidiagonalization theorem, consider first a 2×2 matrix. Let $\mathbf{x}_1$ be a normalized ($\|\mathbf{x}_1\| = 1$) eigenvector of $\mathbf{A}$ and choose $\mathbf{x}_2$ to be a normalized vector orthogonal to $\mathbf{x}_1$, i.e., $\mathbf{x}_1^\dagger \mathbf{x}_2 = 0$ and $\|\mathbf{x}_2\| = 1$. We can choose any vector that is linearly independent of $\mathbf{x}_1$ and use the Gram–Schmidt orthogonalization procedure to find a vector orthogonal to $\mathbf{x}_1$. Define the unitary matrix $\mathbf{U} = [\mathbf{x}_1, \mathbf{x}_2]$. Then

$$\mathbf{A}\mathbf{U} = \mathbf{A}[\mathbf{x}_1, \mathbf{x}_2] = [\lambda_1 \mathbf{x}_1, \mathbf{A}\mathbf{x}_2] \tag{6.6.3}$$

and

$$\mathbf{U}^\dagger \mathbf{A} \mathbf{U} = \begin{bmatrix} \mathbf{x}_1^\dagger \\ \mathbf{x}_2^\dagger \end{bmatrix} [\lambda_1 \mathbf{x}_1, \mathbf{A}\mathbf{x}_2] = \begin{bmatrix} \lambda_1 & \mathbf{x}_1^\dagger \mathbf{A} \mathbf{x}_2 \\ 0 & \mathbf{x}_2^\dagger \mathbf{A} \mathbf{x}_2 \end{bmatrix}. \tag{6.6.4}$$

Since the eigenvalues of an upper triangular matrix are equal to its diagonal elements, $\mathbf{x}_2^\dagger \mathbf{A} \mathbf{x}_2$ is an eigenvalue of $\mathbf{U}^\dagger \mathbf{A} \mathbf{U}$, and, therefore, it must be an eigenvalue of $\mathbf{A}$ (since a unitary transformation does not affect the eigenvalues of $\mathbf{A}$). This proves the correctness of the semidiagonalization theorem for a 2×2 matrix. The rest of the proof will be accomplished by induction.

Assume that the theorem is true for an $n \times n$ matrix and consider the $(n+1) \times (n+1)$ matrix $\mathbf{A}$. Let $\mathbf{x}_1$ be the normalized eigenvector of $\mathbf{A}$ corresponding to the eigenvalue λ_1. Choose the vectors $\mathbf{x}_2, \ldots, \mathbf{x}_{n+1}$ such that

$$\mathbf{x}_i^\dagger \mathbf{x}_j = \delta_{ij}, \qquad i, j = 1, \ldots, n+1, \tag{6.6.5}$$

and let
$$\mathbf{U} = [\mathbf{x}_1, \mathbf{x}_2, \ldots, \mathbf{x}_{n+1}]. \qquad (6.6.6)$$

The unitary transform of $\mathbf{A}$ is

$$\mathbf{U}^\dagger \mathbf{A}\mathbf{U} = \begin{bmatrix} \lambda_1 & b_{12} & \cdots & b_{1,n+1} \\ 0 & & & \\ \vdots & & \mathbf{B} & \\ 0 & & & \end{bmatrix}, \qquad (6.6.7)$$

where $b_{1j} = \mathbf{x}_1^\dagger \mathbf{A}\mathbf{x}_j$ for $j = 2, \ldots, n+1$ and $\mathbf{B}$ is an $n \times n$ matrix with elements $\mathbf{x}_i^\dagger \mathbf{A}\mathbf{x}_j$ for $i, j = 2, \ldots, n+1$. The eigenvalues of $\mathbf{B}$ are the eigenvalues $\lambda_2, \ldots, \lambda_{n+1}$ of $\mathbf{A}$. This follows because

$$\prod_{i=1}^{n+1}(\lambda_i - \lambda) = |\mathbf{U}^\dagger \mathbf{A}\mathbf{U} - \lambda \mathbf{I}^{(n+1)}| = (\lambda_1 - \lambda)|\mathbf{B} - \lambda \mathbf{I}^{(n)}|, \qquad (6.6.8)$$

where $\mathbf{I}^{(n+1)}$ and $\mathbf{I}^{(n)}$ are the $(n+1) \times (n+1)$ and $n \times n$ unit matrices, respectively.

By hypothesis, there exists an $n \times n$ unitary matrix such that

$$\mathbf{V}^\dagger \mathbf{B}\mathbf{V} = \begin{bmatrix} \lambda_2 & b'_{22} & \cdots & b'_{2n} \\ 0 & \lambda_3 & \cdots & b'_{3n} \\ \vdots & \vdots & & \vdots \\ 0 & 0 & \cdots & \lambda_{n+1} \end{bmatrix}. \qquad (6.6.9)$$

We can, therefore, define the $(n+1) \times (n+1)$ unitary matrix

$$\mathbf{T} \equiv \begin{bmatrix} 1 & 0 & \cdots & 0 \\ 0 & & & \\ \vdots & & \mathbf{V} & \\ 0 & & & \end{bmatrix} \qquad (6.6.10)$$

from which Eq. (6.6.7) becomes

$$\mathbf{T}^\dagger \mathbf{U}^\dagger \mathbf{A}\mathbf{U}\mathbf{T} = \begin{bmatrix} \lambda_1 & b_{12} & \cdots & b_{1,n+1} \\ 0 & & & \\ \vdots & & \mathbf{V}^\dagger \mathbf{B}\mathbf{V} & \\ 0 & & & \end{bmatrix}$$

$$= \begin{bmatrix} \lambda_1 & b_{12} & \cdots & b_{1,n+1} \\ 0 & \lambda_2 & \cdots & b_{2,n+1} \\ \vdots & \vdots & & \vdots \\ 0 & 0 & \cdots & \lambda_{n+1} \end{bmatrix}. \qquad (6.6.11)$$

T and **U** are unitary, and so if $\mathbf{Q} \equiv \mathbf{TU}$, then $\mathbf{Q}^{-1} = \mathbf{U}^{-1}\mathbf{T}^{-1} = \mathbf{U}^\dagger \mathbf{T}^\dagger = \mathbf{Q}^\dagger$. Thus, **Q** is a unitary matrix and Eq. (6.6.11) proves that if the semidiagonalization theorem is true for $n \times n$ matrices, it is true for $(n+1) \times (n+1)$ matrices. Since we proved the theorem for the 2×2 case, this completes the proof for the general case.

6.7. SELF-ADJOINT MATRICES

Self-adjoint matrices, defined by the property

$$\mathbf{A} = \mathbf{A}^\dagger \quad \text{or} \quad a_{ij}^\dagger = a_{ji}^*, \tag{6.7.1}$$

are a very important class of matrices that appear in numerous physical problems.

According to the semidiagonalization theorem, there exists a unitary matrix such that

$$\mathbf{X}^\dagger \mathbf{A} \mathbf{X} = \begin{bmatrix} \lambda_1 & b_{12} & \cdots & b_{1n} \\ 0 & \lambda_2 & \cdots & b_{2n} \\ \vdots & \vdots & & \vdots \\ 0 & 0 & \cdots & \lambda_n \end{bmatrix}. \tag{6.7.2}$$

But taking the adjoint of Eq. (6.7.2) and using the property $(\mathbf{X}^\dagger \mathbf{A} \mathbf{X})^\dagger = \mathbf{X}^\dagger \mathbf{A}^\dagger \mathbf{X} = \mathbf{X}^\dagger \mathbf{A} \mathbf{X}$, we find that, for a self-adjoint matrix,

$$\mathbf{X}^\dagger \mathbf{A} \mathbf{X} = \begin{bmatrix} \lambda_1^* & 0 & \cdots & 0 \\ b_{12}^* & \lambda_2^* & \cdots & 0 \\ \vdots & \vdots & & \vdots \\ b_{1n}^* & b_{2n}^* & \cdots & \lambda_n^* \end{bmatrix}. \tag{6.7.3}$$

Comparison of Eqs. (6.7.2) and (6.7.3) leads to the conclusions that $b_{ij} = 0$, $\lambda_i = \lambda_i^*$, and

$$\mathbf{X}^\dagger \mathbf{A} \mathbf{X} = \mathbf{\Lambda}, \tag{6.7.4}$$

where $\mathbf{\Lambda} = [\lambda_i \delta_{ij}]$. Thus, the eigenvalues of a self-adjoint matrix are real, and the matrix can be diagonalized by a unitary transformation. Since $\mathbf{X}^\dagger = \mathbf{X}^{-1}$, Eq. (6.7.4) can be rearranged to give

$$\mathbf{A}\mathbf{X} = \mathbf{X}\mathbf{\Lambda} \quad \text{or} \quad [\mathbf{A}\mathbf{x}_1, \ldots, \mathbf{A}\mathbf{x}_n] = [\lambda_1 \mathbf{x}_1, \ldots, \lambda_n \mathbf{x}_n], \tag{6.7.5}$$

which yields

$$\mathbf{A}\mathbf{x}_i = \lambda_i \mathbf{x}_i, \quad i = 1, \ldots, n. \tag{6.7.6}$$

Since **X** is unitary, the $\mathbf{x}_i$ form an orthonormal set ($\mathbf{x}_i^\dagger \mathbf{x}_j = \delta_{ij}$). Stated in terms of the eigenproblem, we have just proved the following:

THEOREM. *A self-adjoint matrix is perfect, its eigenvalues are real, and its eigenvectors can always be chosen to form an orthonormal set.*

In Chapter 3 we defined a positive-definite matrix by the following property:

DEFINITION. *We say* **A** *is positive definite if, for any nonzero vector* **x**,

$$\mathbf{x}^\dagger \mathbf{A} \mathbf{x} = \rho > 0, \qquad (6.7.7)$$

where ρ is a real scalar quantity.

The requirement that ρ be real valued can be stated as $\rho = \rho^*$, and so by taking the adjoint of Eq. (6.7.7), we find

$$\mathbf{x}^\dagger \mathbf{A} \mathbf{x} = \mathbf{x}^\dagger \mathbf{A}^\dagger \mathbf{x} \qquad (6.7.8)$$

from which we immediately see that $\mathbf{A} = \mathbf{A}^\dagger$. Thus, a positive-definite matrix must be self-adjoint.

We can now use the above theorem and write $\mathbf{x} = \sum_{i=1}^n \alpha_i \mathbf{x}_i$ for any arbitrary n-dimensional vector **x**. We can choose the vectors $\{\mathbf{x}_i\}$ to be the eigenvectors of **A** since we have proven they form a complete basis set (or at least can be made to form one). Substitution into Eq. (6.7.7) yields

$$\begin{aligned}
\rho &= \sum_{i=1}^n \sum_{j=1}^n \alpha_i^* \mathbf{x}_i^\dagger \mathbf{A} \mathbf{x}_j \alpha_j \\
&= \sum_{i=1}^n \sum_{j=1}^n \alpha_i^* \alpha_j \lambda_j \mathbf{x}_i^\dagger \mathbf{x}_j \qquad (6.7.9) \\
&= \sum_{i=1}^n |\alpha_i|^2 \lambda_i,
\end{aligned}$$

where we have used the orthonormality conditions of $\{\mathbf{x}_i\}$ to obtain the last line. Since **x** is arbitrary, the coefficients α_i are arbitrary as well, and so in order to ensure that $\rho > 0$ for any nonzero vector **x**, all eigenvalues λ_i must be positive and nonzero.

We have just proven the following theorem:

THEOREM. *A positive-definite matrix is a self-adjoint matrix whose eigenvalues are positive and greater than* 0 ($\lambda_i > 0$).

A similar proof can be constructed to show that a negative-definite matrix is self-adjoint with all eigenvalues $\lambda_i < 0$. Furthermore, we say a matrix is positive (negative) semi-definite if it is self-adjoint and all of its eigenvalues are greater (less) than or equal to 0 (i.e., some of its eigenvalues are 0). The point here is that each of the above classes of matrices are simply subclasses of self-adjoint matrices.

It is important to realize that, although the eigenvectors of a self-adjoint matrix always form a basis set, they might not be orthonormal. Let us explore this point further. Suppose $\lambda_i \neq \lambda_j$. Then

$$\mathbf{x}_i^\dagger \mathbf{A} \mathbf{x}_j = \lambda_j \mathbf{x}_i^\dagger \mathbf{x}_j \qquad (6.7.10)$$

and

$$\mathbf{x}_j^\dagger \mathbf{A} \mathbf{x}_i = \lambda_i \mathbf{x}_j^\dagger \mathbf{x}_i. \qquad (6.7.11)$$

SELF-ADJOINT MATRICES

Taking the adjoint of Eq. (6.7.11) and using the properties $\mathbf{A}^\dagger = \mathbf{A}$ and $\lambda_i^* = \lambda_i$, we obtain

$$\mathbf{x}_i^\dagger \mathbf{A} \mathbf{x}_j = \lambda_i \mathbf{x}_i^\dagger \mathbf{x}_j, \qquad (6.7.12)$$

which, when subtracted from Eq. (6.7.10), yields

$$(\lambda_j - \lambda_i)\mathbf{x}_i^\dagger \mathbf{x}_j = 0 \quad \text{or} \quad \mathbf{x}_i^\dagger \mathbf{x}_j = 0. \qquad (6.7.13)$$

Thus, if the eigenvalues λ_i and λ_j are not equal, then $\mathbf{x}_i$ and $\mathbf{x}_j$ are orthogonal. However, if $\lambda_i = \lambda_j$, then we cannot conclude from Eq. (6.7.13) that $\mathbf{x}_i$ and $\mathbf{x}_j$ are orthogonal.

As an example, consider the matrix

$$\mathbf{A} = \begin{bmatrix} \dfrac{7}{2} & -\dfrac{1}{2} & 0 \\ -\dfrac{1}{2} & \dfrac{7}{2} & 0 \\ 0 & 0 & 4 \end{bmatrix}. \qquad (6.7.14)$$

The eigenvalues are $\lambda_1 = 3$, $\lambda_2 = 4$, and $\lambda_3 = 4$, with the corresponding eigenvectors

$$\mathbf{x}_1' = \begin{bmatrix} 1 \\ 1 \\ 0 \end{bmatrix}, \quad \mathbf{x}_2' = \begin{bmatrix} -1 \\ 1 \\ 1 \end{bmatrix}, \quad \mathbf{x}_3' = \begin{bmatrix} -1 \\ 1 \\ -1 \end{bmatrix}. \qquad (6.7.15)$$

A number of things are noteworthy here. The vector $\mathbf{x}_1'$ is orthogonal to $\mathbf{x}_2'$ and $\mathbf{x}_3'$, but $\mathbf{x}_2'$ and $\mathbf{x}_3'$ are not orthogonal and none of the vectors is normalized. Thus, the matrix $[\mathbf{x}_1', \mathbf{x}_2', \mathbf{x}_3']$ is not a unitary matrix. It will, however, diagonalize $\mathbf{A}$ in a similarity transformation.

We can transform $\mathbf{x}_2'$ and $\mathbf{x}_3'$ into orthogonal vectors using the Gram–Schmidt orthogonalization procedure. The resulting orthogonal vectors are linear combinations of $\mathbf{x}_2'$ and $\mathbf{x}_3'$, and, since $\mathbf{A}(c_2 \mathbf{x}_2' + c_3 \mathbf{x}_3') = 4(c_2 \mathbf{x}_2' + c_3 \mathbf{x}_3')$, these linear combinations are also eigenvectors of $\mathbf{A}$. Also, the eigenvectors can be normalized by multiplying by scalar factors and so they remain eigenvectors of $\mathbf{A}$. Orthogonalizing and normalizing $\mathbf{x}_1'$, $\mathbf{x}_2'$, and $\mathbf{x}_3'$ yields the orthonormal eigenvectors

$$\mathbf{x}_1 = \frac{1}{\sqrt{2}} \begin{bmatrix} 1 \\ 1 \\ 0 \end{bmatrix}, \quad \mathbf{x}_2 = \begin{bmatrix} 0 \\ 0 \\ 1 \end{bmatrix}, \quad \mathbf{x}_3 = \frac{1}{\sqrt{2}} \begin{bmatrix} -1 \\ 1 \\ 0 \end{bmatrix}. \qquad (6.7.16)$$

With these eigenvectors, $\mathbf{X} = [\mathbf{x}_1, \mathbf{x}_2, \mathbf{x}_3]$ is a unitary matrix and will diagonalize $\mathbf{A}$ in a unitary transformation.

The above example illustrates the facts that the eigenvectors of a self-adjoint matrix may be provided in an unnormalized form and that if the eigenvalue λ_i is of multiplicity p_{λ_i} greater than 1, then the eigenvectors corresponding to λ_i may not be orthogonal. However, since $\sum_i c_i \mathbf{x}_i$ is an eigenvector of $\mathbf{A}$, and since the p_{λ_i} eigenvectors have the same eigenvalue, the Gram–Schmidt procedure can be used to obtain

p_{λ_i} orthogonal eigenvectors. So, whereas we frequently say that the eigenvectors of a self-adjoint matrix form an orthonormal basis set, what we really mean is that the eigenvectors of a self-adjoint matrix can be arranged into an orthonormal basis set.

Since a self-adjoint matrix $\mathbf{A}$ is perfect and its eigenvectors can be chosen to be an orthonormal basis, the spectral decomposition of $\mathbf{A}$ can be expressed as

$$\mathbf{A} = \sum_{i=1}^{n} \lambda_i \mathbf{x}_i \mathbf{x}_i^\dagger. \tag{6.7.17}$$

Let us illustrate the usefulness of the unitary transformation theorem by applications to physical examples in the remainder of this section.

A good approximation to the Hamiltonian H of an atomic crystal far from its melting point is the harmonic oscillator or Hookean spring approximation

$$H = \sum_{i=1}^{3N} \frac{1}{2} m \left(\frac{dx_i}{dt} \right)^2 + \frac{1}{2} \sum_{i,j=1}^{3N} a_{ij} x_i x_j + U_0, \tag{6.7.18}$$

where m is the mass of the atoms, x_i is a Cartesian coordinate measuring the displacement along a Cartesian axis of an atom from its equilibrium lattice site, dx_i/dt is the velocity along the axis, a_{ij} is a component of the force restoring atoms to their lattice sites, and U_0 is the equilibrium potential energy (a constant). With this Hamiltonian (i.e., mechanical energy), Newton's law of motion yields the equations

$$m \frac{d^2 x_i}{dt^2} = -\sum_{j=1}^{3N} a_{ij} x_j, \quad i = 1, \ldots, 3N. \tag{6.7.19}$$

Since all of the degrees of freedom $\{x_i\}$ are coupled, the solution to these equations is usually difficult. However, if we rewrite Eqs. (6.7.18) and (6.7.19) in vector form, it becomes clear how to simplify them:

$$H = \frac{1}{2} m \frac{d\mathbf{x}^\dagger}{dt} \frac{d\mathbf{x}}{dt} + \frac{1}{2} \mathbf{x}^\dagger \mathbf{A} \mathbf{x} + U_0$$
$$m \frac{d^2 \mathbf{x}}{dt^2} = -\mathbf{A} \mathbf{x}, \tag{6.7.20}$$

where $\mathbf{x}$ is a $3N$-dimensional vector whose components are x_i and $\mathbf{A} = [a_{ij}]$. Since the coefficients a_{ij} are related to a potential energy function $U(x_1, \ldots, x_{3N})$ by

$$a_{ij} = \frac{\partial^2 U}{\partial x_i \, \partial x_j}, \tag{6.7.21}$$

they are real and invariant to the interchange of the order of partial differentiation (i.e., $a_{ij} = a_{ji}$). Thus, $\mathbf{A}$ is a self-adjoint matrix that can be diagonalized by a unitary transformation. Actually, since $\mathbf{x}$ and $\mathbf{A}$ are real, the orthonormalized eigenvectors of $\mathbf{A}$ form a real orthonormal set and the eigenvector matrix $\mathbf{X}$ is an orthogonal matrix.

SELF-ADJOINT MATRICES

To exploit the fact that $\mathbf{A}$ is self-adjoint, we note that $\mathbf{X}\mathbf{X}^\dagger = \mathbf{I}$ and we rewrite Eq. (6.7.20) as

$$H = \frac{1}{2}\frac{d\mathbf{x}^\dagger}{dt}\mathbf{X}\mathbf{X}^\dagger\frac{d\mathbf{x}}{dt} + \frac{1}{2}\mathbf{x}^\dagger\mathbf{X}\mathbf{X}^\dagger\mathbf{A}\mathbf{X}\mathbf{X}^\dagger\mathbf{x} \quad (6.7.22)$$

$$m\frac{d^2}{dt^2}\mathbf{X}^\dagger\mathbf{x} = -\mathbf{X}^\dagger\mathbf{A}\mathbf{X}\mathbf{X}^\dagger\mathbf{x}.$$

Defining the vector

$$\boldsymbol{\zeta} \equiv \mathbf{X}^\dagger\mathbf{x} \quad (6.7.23)$$

and using the relation $\mathbf{X}^\dagger\mathbf{A}\mathbf{X} = \boldsymbol{\Lambda}$, we obtain

$$H = \frac{m}{2}\frac{d\boldsymbol{\zeta}^\dagger}{dt}\frac{d\boldsymbol{\zeta}}{dt} + \frac{1}{2}\boldsymbol{\zeta}\boldsymbol{\Lambda}\boldsymbol{\zeta} + U_0 \quad (6.7.24)$$

or

$$H = \frac{1}{2}\sum_{i=1}^{3N}\left[m\left(\frac{d\zeta_i}{dt}\right)^2 + \lambda_i\zeta_i^2\right] + U_0 \quad (6.7.25)$$

and

$$m\frac{d^2\boldsymbol{\zeta}}{dt^2} = -\boldsymbol{\Lambda}\boldsymbol{\zeta}, \quad (6.7.26)$$

which in expanded form is

$$m\frac{d^2\zeta_i}{dt^2} = -\lambda_i\zeta_i, \quad i = 1, \ldots, 3N. \quad (6.7.27)$$

The unitary transformation has greatly simplified the equation of motion of the atoms of the crystal. It amounts to introducing new Cartesian coordinates $\zeta_i, \ldots, \zeta_{3N}$—related to the old Cartesian coordinates $(x_1, \ldots, x_{3N})$ by the rotational transformation $\mathbf{x} = \mathbf{X}\boldsymbol{\zeta}$—which are decoupled. By decoupled we mean that, in the new coordinate system, the degrees of freedom ζ_i behave as independent harmonic oscillators. The solution to Eq. (6.7.25) is simply

$$\zeta_i = a\sin\omega_i t + b\cos\omega_i t, \quad (6.7.28)$$

where $\omega_i = \sqrt{\lambda_i/m}$ is the characteristic frequency of vibration of the degree of freedom ζ_i. The quantities a and b are constants that depend on the initial conditions. Physicists refer to the transformation of Eq. (6.7.19) to Eq. (6.7.25) as finding the "normal modes" of a coupled harmonic oscillator system.

EXERCISE 6.7.1. Suppose the atomic masses are different so that

$$H = \frac{1}{2}\sum_{i=1}^{3N}m_i\left(\frac{dx_i}{dt}\right)^2 + \frac{1}{2}\sum_{i,j}^{N}a_{ij}x_ix_j \quad (6.7.29)$$

and

$$m\frac{d^2x_i}{dt^2} = -\sum_{j=1}^{3N} a_{ij}x_j. \tag{6.7.30}$$

Find the transformation that converts these equations to the forms of Eqs. (6.7.23) ■ ■ ■ and (6.7.24).

As another example, consider the problem of finding the minimum or maximum of a multivariable real function $f(x_1, \ldots, x_n)$ of real variables. At an extremum (minimum or maximum),

$$\frac{\partial f}{\partial x_i} = 0, \quad i = 1, \ldots, n. \tag{6.7.31}$$

We can expand the perturbed function $f(x_1 + \delta x_1, \ldots, x_n + \delta x_n)$ (near the extremum) in a Taylor series around the points $\{x_1, \ldots, x_n\}$, giving

$$f(\mathbf{x} + \delta \mathbf{x}) = f(\mathbf{x}) + \sum_{i=1}^{N} \delta x_i \frac{\partial f}{\partial x_i} + \frac{1}{2} \sum_{i,j=1}^{N} \delta x_i \delta x_j \frac{\partial^2 f}{\partial x_i \partial x_j} + \cdots. \tag{6.7.32}$$

Since $\partial f/\partial x_i = 0$, $i = 1, \ldots, n$, and defining the matrix $\mathbf{A}$ by its components

$$a_{ij} = \frac{\partial^2 f}{\partial x_i \partial x_j}, \tag{6.7.33}$$

Eq. (6.7.30) can be expressed in the form

$$f(\mathbf{x} + \delta \mathbf{x}) = f(\mathbf{x}) + \frac{1}{2} \delta \mathbf{x}^\dagger \mathbf{A} \, \delta \mathbf{x} + \cdots. \tag{6.7.34}$$

The matrix $\mathbf{A}$ is often referred to as the *Hessian matrix*. Let us say that we are hunting a minimum in f. If $\mathbf{x}$ is at a minimum, then, in a very small displacement $\delta \mathbf{x}$ away from $\mathbf{x}$, $f(\mathbf{x} + \delta \mathbf{x})$ will increase over its value $f(\mathbf{x})$. From Eq. (6.7.34) we see that this will be the case if

$$\delta \mathbf{x}^\dagger \mathbf{A} \, \delta \mathbf{x} > 0 \quad \text{for arbitrary } \delta \mathbf{x}. \tag{6.7.35}$$

Since $a_{ij} = a_{ji}$ and the values of a_{ij} are real, the matrix $\mathbf{A}$ is self-adjoint. Therefore, $\mathbf{A}$ can be diagonalized by a unitary transformation, or, equivalently, it has the spectral decomposition

$$\mathbf{A} = \sum_{i=1}^{N} \lambda_i \mathbf{x}_i \mathbf{x}_i^\dagger, \tag{6.7.36}$$

where the eigenvalues λ_i are real and the eigenvectors $\{\mathbf{x}_i\}$ form an orthonormal set. Insertion of Eq. (6.7.36) into Eq. (6.7.35) yields

$$\delta \mathbf{x}^\dagger \mathbf{A} \, \delta \mathbf{x} = \sum_{i=1}^{n} \lambda_i |\alpha_i|^2 > 0, \tag{6.7.37}$$

where

$$\alpha_i \equiv \mathbf{x}_i^\dagger \, \delta\mathbf{x}. \tag{6.7.38}$$

Since $|\alpha_i|^2$ is arbitrary and positive, we conclude from Eq. (6.7.35) that $f(\mathbf{x})$ is a minimum if all of the eigenvalues of $\mathbf{A}$ are positive (and is a maximum if all of the eigenvalues are negative). If some of the eigenvalues are 0 and the rest are positive, then $\mathbf{x}$ might be a point of inflection rather than a minimum. High-order derivatives of f would be necessary to analyze this situation.

Another way to express Eq. (6.7.35) is to use the relation $\mathbf{A} = \mathbf{X}\boldsymbol{\Lambda}\mathbf{X}^\dagger$ to obtain

$$\delta\mathbf{x}^\dagger \mathbf{A}\, \delta\mathbf{x} = \delta\mathbf{x}^\dagger \, \mathbf{X}\boldsymbol{\Lambda}\mathbf{X}^\dagger \, \delta\mathbf{x} = \boldsymbol{\alpha}^\dagger \boldsymbol{\Lambda}\boldsymbol{\alpha}, \tag{6.7.39}$$

where $\boldsymbol{\alpha} = \mathbf{X}^\dagger \delta\mathbf{x}$. This is equivalent to Eq. (6.7.37).

In the thermodynamic theory of phase equilibria, f is the Helmholtz free energy and x_i is the density of component i. Previously, we said that $\mathbf{A}$ is positive definite if it is self-adjoint and if

$$\mathbf{x}^\dagger \mathbf{A}\mathbf{x} > 0 \qquad \text{for all } \mathbf{x} \in E_n. \tag{6.7.40}$$

With the aid of Eq. (6.7.36), it can now be stated that

$$\mathbf{x}^\dagger \mathbf{A}\mathbf{x} = \sum_{i=1}^n \lambda_i |\alpha_i|^2, \qquad \alpha_i = \mathbf{x}_i^\dagger \mathbf{x}, \tag{6.7.41}$$

and so it follows that $\mathbf{A}$ is positive definite *if and only if* all of its eigenvalues are positive.

Because a self-adjoint operator can be diagonalized by a unitary transformation (or, equivalently, because the eigenvectors of a self-adjoint matrix can be arranged as an orthonormal basis set), the $p = 2$ norm and the condition number $\kappa(\mathbf{A})$ are simple functions of eigenvalues of $\mathbf{A}$. The $p = 2$ norm is given by

$$\|\mathbf{A}\|^2 = \max_{\mathbf{x} \neq 0} \frac{(\mathbf{A}\mathbf{x})^\dagger \mathbf{A}\mathbf{x}}{\|\mathbf{x}\|^2}. \tag{6.7.42}$$

With the aid of $\mathbf{I} = \sum_i \mathbf{x}_i \mathbf{x}_i^\dagger$ and $\mathbf{A} = \sum_i \lambda_i \mathbf{x}_i \mathbf{x}_i^\dagger$, we obtain

$$\mathbf{x} = \mathbf{I}\mathbf{x} = \sum_i \mathbf{x}_i \mathbf{x}_i^\dagger \mathbf{x} \qquad \text{and} \qquad \mathbf{A}\mathbf{x} = \sum_i \lambda_i \mathbf{x}_i \alpha_i, \tag{6.7.43}$$

where $\alpha_i = \mathbf{x}_i^\dagger \mathbf{x}$. Equation (6.7.42) then becomes

$$\begin{aligned}\|\mathbf{A}\|^2 &= \max_{\mathbf{x}} \frac{\sum_i \lambda_i^2 |\alpha_i|^2}{\sum_i |\alpha_i|^2} \\ &\leq |\lambda_{\max}|^2 \frac{\sum_i |\alpha_i|^2}{\sum_i |\alpha_i|^2} \leq |\lambda_{\max}|^2,\end{aligned} \tag{6.7.44}$$

where

$$|\lambda_{\max}| = \max_{1 \leq i \leq n} |\lambda_i|. \tag{6.7.45}$$

Denoting $\mathbf{x}_m$ as the eigenvector having the eigenvalue of the greatest absolute value, we can choose $\mathbf{x} = \mathbf{x}_m$. The equality in Eq. (6.7.44) can then be achieved and so

$$\|\mathbf{A}\| = |\lambda_{\max}|. \tag{6.7.46}$$

Thus, for the $p = 2$ norm of a self-adjoint matrix, we simply have to find the eigenvalues of maximum magnitude. The power method described in Chapter 5 is a way of finding this eigenvalue.

If $\mathbf{A}$ is not singular, then the eigenvalues of $\mathbf{A}^{-1}$ are λ_i^{-1} and the eigenvectors are the same as those of $\mathbf{A}$. The $p = 2$ norm of $\mathbf{A}^{-1}$ is then

$$\|\mathbf{A}^{-1}\| = \frac{1}{|\lambda_{\min}|}, \tag{6.7.47}$$

where

$$|\lambda_{\min}| = \min_{1 \le i \le n} |\lambda_i|. \tag{6.7.48}$$

Here we denote $\lambda_{\min}$ as the eigenvalue of minimum absolute value. The inverse power method applied to $\mathbf{A}^{-1}$ yields this eigenvalue. The condition number of a self-adjoint matrix in the $p = 2$ norm is simply

$$\kappa(\mathbf{A}) = \|\mathbf{A}\| \, \|\mathbf{A}^{-1}\| = \left|\frac{\lambda_{\max}}{\lambda_{\min}}\right|. \tag{6.7.49}$$

It is important to note that if $\mathbf{A}$ is perfect but is not self-adjoint, then the $p = 2$ norm of $\mathbf{A}$ is not in general $|\lambda_{\max}|$. This can be proved by a simple example:

$$\mathbf{A} = \begin{bmatrix} 1 & 1 \\ 0 & 0 \end{bmatrix},$$

$$\lambda_1 = 0, \qquad \lambda_2 = 1, \tag{6.7.50}$$

$$\mathbf{x}_1 = \begin{bmatrix} 1 \\ -1 \end{bmatrix}, \qquad \mathbf{x}_2 = \begin{bmatrix} 1 \\ 0 \end{bmatrix}.$$

Thus, $\mathbf{A}$ is perfect but

$$\mathbf{A}\mathbf{y} = \begin{bmatrix} y_1 + y_2 \\ 0 \end{bmatrix} \tag{6.7.51}$$

and

$$\frac{\|\mathbf{A}\mathbf{y}\|^2}{\|\mathbf{y}\|^2} = \frac{(y_1 + y_2)^2}{y_1^2 + y_2^2} = \frac{(1+\alpha)^2}{1+\alpha^2}. \tag{6.7.52}$$

The maximum of $(1+\alpha)^2/(1+\alpha^2)$ occurs when $\alpha = 1$, and so

$$\|\mathbf{A}\| = \sqrt{2}, \tag{6.7.53}$$

which is larger than $\lambda_{\max} = \lambda_2 = 1$.

SELF-ADJOINT MATRICES

In statistics and in the thermodynamic theory of density fluctuations, one frequently encounters the multivariate Gaussian distribution function

$$P = a \exp\left(-\sum_{i,j=1}^{n} a_{ij} x_i x_j\right) = a \exp(-\mathbf{x}^T \mathbf{A} \mathbf{x}), \quad (6.7.54)$$

where a_{ij} and x_i are real and $a_{ij} = a_{ji}$. The physical meaning of P is that

$$P(x_1, \ldots, x_n) \, dx_1 \ldots dx_n \quad (6.7.55)$$

is the probability that the variables will be observed to lie between $x_1, \ldots, x_n$ and $x_1 + dx_1, \ldots, x_n + dx_n$. Therefore, if we integrate P over all the variables $\{x_1, \ldots, x_n\}$ the result must be 1, i.e.,

$$\int_{-\infty}^{\infty} \cdots \int_{-\infty}^{\infty} P(x_1, \ldots, x_n) \, dx_1 \cdots dx_n = 1. \quad (6.7.56)$$

In using a probability distribution, we generally want to evaluate average quantities such as

$$\overline{g(\mathbf{x})} \equiv \int_{-\infty}^{\infty} \cdots \int_{-\infty}^{\infty} g(\mathbf{x}) P(\mathbf{x}) \, dx_1 \cdots dx_n. \quad (6.7.57)$$

The variables $\mathbf{x} = \{x_1, \ldots, x_n\}$ are defined as the deviations from a set of mean values (i.e., $x_i = \zeta_i - \overline{\zeta}_i$) and so $\overline{x_i} = 0$. The quantities $\overline{x_k x_l}$ are elements of the covariance matrix $\mathbf{G}$, which we define by $[\overline{x_i x_l}]$ or $\overline{\mathbf{x}\mathbf{x}^T}$. Note that the trace of $\mathbf{G}$, $\mathrm{tr}_1 \mathbf{G} = \sum_{k=1}^{n} \overline{x_k^2}$, is the mean square dispersion, σ^2, of the variables.

Evaluation of the various averages could be very difficult in general since all of the variables are coupled. However, for the Gaussian distribution, $\mathbf{A}$ is a real, self-adjoint matrix, and so it is diagonalizable by an orthogonal transformation; i.e., there exists an orthogonal matrix $\mathbf{X}$ such that

$$\mathbf{X}^T \mathbf{A} \mathbf{X} = \boldsymbol{\Lambda} \quad \text{and} \quad \mathbf{X}^T \mathbf{X} = \mathbf{I}. \quad (6.7.58)$$

As in the analysis of the coupled harmonic oscillators, the new variables

$$\mathbf{y} \equiv \mathbf{X}^T \mathbf{x} \quad (6.7.59)$$

can be introduced to simplify the problem. The relations $\mathbf{X}^T = \mathbf{X}^{-1}$ and $\mathbf{x} = \mathbf{X}\mathbf{y}$ yield

$$\mathbf{x}^T \mathbf{A} \mathbf{x} = \mathbf{y}^T \mathbf{X}^T \mathbf{A} \mathbf{X} \mathbf{y} = \mathbf{y}^T \boldsymbol{\Lambda} \mathbf{y} = \sum_{i=1}^{n} \lambda_i y_i^2. \quad (6.7.60)$$

To carry out the integrations in the variables y_i, the Jacobian $|\mathbf{J}|$ of the coordinate transformation must be calculated. The volume elements of the $\mathbf{x}$ and $\mathbf{y}$ coordinates are related by

$$dx_1 \cdots dx_n = |\mathbf{J}| \, dy_1 \cdots dy_n, \quad (6.7.61)$$

where

$$|\mathbf{J}| = \begin{vmatrix} \dfrac{\partial x_1}{\partial y_1} & \dfrac{\partial x_1}{\partial y_2} & \cdots & \dfrac{\partial x_1}{\partial y_n} \\ \vdots & \vdots & & \vdots \\ \dfrac{\partial x_n}{\partial y_1} & \dfrac{\partial x_n}{\partial y_2} & \cdots & \dfrac{\partial x_n}{\partial y_n} \end{vmatrix}. \qquad (6.7.62)$$

The relationships $x_i = \sum_j x_{ij} y_j$ and $\partial x_i / \partial y_j = x_{ij}$ show that $|\mathbf{J}| = |\mathbf{X}| = 1$ and thus the Jacobian of coordinate transformation is unity. Any student of mechanics will, of course, recall that in Cartesian coordinates $dx\,dy\,dz = dx'\,dy'\,dz'$ whenever the coordinates x, y, z are related to the coordinates x', y', z' by a rotation (i.e., by an orthogonal transformation). We have simply proven that this property is valid in any linear vector space regardless of its dimension.

We can now calculate averages in the $\mathbf{y}$ coordinate system. The first task is to compute the normalization constant a from Eq. (6.7.56)

$$a \int_{-\infty}^{\infty} \cdots \int_{-\infty}^{\infty} \exp\left(-\sum_{i=1}^{n} \lambda_i y_i^2\right) dy_1 \cdots dy_n = 1. \qquad (6.7.63)$$

Note that if any $\lambda_i \leq 0$, then the integrals in Eq. (6.7.62) are singular. This means that $\mathbf{A}$ has to be positive definite to properly define a probability distribution in which $\mathbf{A}$ and $\mathbf{x}$ are real. From mathematical tables (or using Mathematica), we can find the integral

$$\int_{-\infty}^{\infty} \exp(-\lambda_i y_i^2)\,dy_i = \sqrt{\dfrac{\pi}{\lambda_i}}, \qquad (6.7.64)$$

which gives for a the result

$$a = \dfrac{\sqrt{\prod_{i=1}^{n} \lambda_i}}{\pi^{n/2}} = \dfrac{\sqrt{|\mathbf{A}|}}{\pi^{n/2}}. \qquad (6.7.65)$$

Although we used eigenvalue analysis to obtain a, the relationship $|\mathbf{A}| = \prod_i \lambda_i$ leads to a result requiring only the evaluation of the determinant of $\mathbf{A}$.

Next, consider the average value of $\mathbf{x}\mathbf{x}^T$. This quantity is related to $\overline{\mathbf{y}\mathbf{y}^T}$ by $\mathbf{X}\overline{\mathbf{y}\mathbf{y}^T}\mathbf{X}^T$. It suffices then to determine $\overline{\mathbf{y}\mathbf{y}^T}$ or its elements $\overline{y_k y_l}$. We simply write

$$\overline{y_k y_l} = \dfrac{\sqrt{|\mathbf{A}|}}{\pi^{n/2}} \int_{-\infty}^{\infty} \cdots \int_{-\infty}^{\infty} y_k y_l \exp\left(-\sum_{i=1}^{n} \lambda_i y_i^2\right) dy_1 \cdots dy_n. \qquad (6.7.66)$$

If $k \neq l$, then $\overline{y_k y_l} = 0$ since the average value of every y_i is 0. If $k = l$, then we note that the integral in Eq. (6.7.66) is made up of the product of $n - 1$ integrals of the form in Eq. (6.7.64), and one integral of the form

$$\int_{-\infty}^{\infty} y_k^2 \exp\left(-\lambda_k y_k^2\right) dy_k = \dfrac{1}{2} \dfrac{\sqrt{\pi}}{\lambda_k^{3/2}}. \qquad (6.7.67)$$

The value of this integral was obtained by differentiating Eq. (6.7.64) with respect to λ_k. Equation (6.7.66) can now be solved

$$\overline{y_k y_l} = \frac{\sqrt{|\mathbf{A}|}}{\pi^{n/2}} \delta_{kl} \frac{1}{\lambda_k} \frac{\pi^{n/2}}{\sqrt{\prod_{i=1}^n \lambda_i}}$$

$$= \frac{1}{2} \frac{\delta_{kl}}{\lambda_k}. \tag{6.7.68}$$

The result for $\overline{\mathbf{yy}^T}$ is then

$$\overline{\mathbf{yy}^T} = \frac{1}{2} \mathbf{\Lambda}^{-1}. \tag{6.7.69}$$

The relationships $\mathbf{A}^{-1} = \mathbf{X}\mathbf{\Lambda}^{-1}\mathbf{X}^T$ and $\overline{\mathbf{xx}^T} = \mathbf{X}\overline{\mathbf{yy}^T}\mathbf{X}^T$, and Eq. (6.7.69) yield

$$\mathbf{G} \equiv \overline{\mathbf{xx}^T} = \frac{1}{2} \mathbf{A}^{-1}. \tag{6.7.70}$$

Thus, the ij elements $\overline{x_i x_j}$ of the covariance matrix $\mathbf{G}$ are simply one half of the ij elements of the inverse of the matrix $\mathbf{A}$. Again, although eigenanalysis was needed to obtain Eq. (6.7.69), the result is that computation of the covariance requires only calculation of the inverse of $\mathbf{A}$—no eigenvalue analysis is required.

■ **EXAMPLE 6.7.1.** As an example of a negative-definite matrix, let us consider a transport process. The concentration in Fickean diffusion or the temperature in heat transfer obeys the partial differential equation

$$\frac{\partial u}{\partial t} = D \frac{\partial^2 u}{\partial x^2}, \tag{6.7.71}$$

where t is time, x is position, and D is either the diffusion coefficient or the thermal diffusivity. With the initial condition

$$u(x, t = 0) = f(x) \tag{6.7.72}$$

and the boundary conditions

$$u(0, t) = u(l, t) = 0, \tag{6.7.73}$$

the equation can be solved by functional analysis, as will be shown in a later chapter. As described in Section 3.6, the finite-difference approximation

$$\frac{\partial^2 u(x_i)}{\partial x^2} \approx \frac{u_{i+1} - 2u_i + u_{i-1}}{(\Delta x)^2}, \tag{6.7.74}$$

where u_i is the value of u at position $x_i = i \Delta x$, approximates the differential equation as

$$\frac{d\mathbf{u}}{dt} = \frac{D}{(\Delta x)^2} \mathbf{A}\mathbf{u},$$

where **A** is the tridiagonal matrix

$$\mathbf{A} = \begin{bmatrix} -2 & 1 & 0 & 0 & \cdots & 0 & 0 & 0 \\ 1 & -2 & 1 & 0 & \cdots & 0 & 0 & 0 \\ 0 & 1 & -2 & 1 & \cdots & 0 & 0 & 0 \\ \vdots & \vdots & \vdots & \vdots & & \vdots & \vdots & \vdots \\ 0 & 0 & 0 & 0 & \cdots & 1 & -2 & 1 \\ 0 & 0 & 0 & 0 & \cdots & 0 & 1 & -2 \end{bmatrix} \quad (6.7.75)$$

and

$$\mathbf{u} = \begin{bmatrix} u_1 \\ u_2 \\ \vdots \\ u_n \end{bmatrix}. \quad (6.7.76)$$

Here $u_i = u(i\,\Delta x, t)$, $u_0 = u(0, t) = 0$, $u_{n+1} = u(l, t) = 0$, and $n = l/\Delta x - 1$. The initial condition yields

$$\mathbf{u}(t = 0) = \mathbf{u}_0 = \begin{bmatrix} f(x_1) \\ \vdots \\ f(x_n) \end{bmatrix}. \quad (6.7.77)$$

A is a self-adjoint operator and so its eigenvectors $\mathbf{x}_i$, $i = 1, \ldots, n$, can be chosen to be an orthonormal basis set. It can be shown that the eigenvalues of $\mathbf{A}$ are

$$\lambda_i = -2\left[1 - \cos\left(\frac{\pi i}{n+1}\right)\right], \quad i = 1, \ldots, n. \quad (6.7.78)$$

Since $\mathbf{A}^\dagger = \mathbf{A}$ and $\lambda_i < 0$, $i = 1, \ldots, n$, it follows that **A** is a negative-definite matrix. Thus, with the spectral resolution theorem

$$\exp\left(\frac{D}{(\Delta x)^2}\mathbf{A}t\right) = \sum_{i=1}^{n} \exp\left(\frac{D}{(\Delta x)^2}\lambda_i t\right)\mathbf{x}_i \mathbf{x}_i^\dagger, \quad (6.7.79)$$

the solution to the finite-difference equation is

$$\mathbf{u} = \sum_{l=1}^{n} \exp\left(-\frac{2Dt}{(\Delta x)^2}\left[1 - \cos\left(\frac{\pi i}{n+1}\right)\right]\right)\mathbf{x}_i(\mathbf{x}_i^\dagger \mathbf{u}_0). \quad (6.7.80)$$

At large t, the $i = 1$ term of the series dominates. Since $n \gg 1$ for a good finite-difference approximation, $1/(n+1) \ll 1$, $1 - \cos[\pi/(n+1)] \approx [(\pi/(n+1)]^2$, $\Delta x(n+1) = l$, and so

$$\mathbf{u} \approx \exp\left(-\frac{D\pi^2 t}{l^2}\right)(\mathbf{x}_1^\dagger \mathbf{u}_0)\mathbf{x}_1 \quad \text{for large } t. \quad (6.7.81)$$

Thus, $l^2/\pi^2 D$ is the characteristic time determining the approach of concentration or temperature to its asymptotic value. ■ ■ ■

SELF-ADJOINT MATRICES

■   **EXERCISE 6.7.2.** For the example just given, assume

$$f(x) = \sin\left(\frac{\pi x}{l}\right), \qquad (6.7.82)$$

$l = 1\,\text{cm}$ and $d/l^2 = 10^{-3}\,\text{s}^{-1}$. Consider the discretizations $n = 10$ and $n = 20$. Using the eigenvectors given in Problem 29 of this chapter, determine $\mathbf{u}$ for $t = 1, 10, 100,$ and $1000\,\text{s}$. The components of $\mathbf{u}$ approximate the profile $u(x, t)$ at $x_i = i\,\Delta x$, $i = 1, \ldots, n$. Plot the approximate profiles for $n = 10$ and $n = 20$ at
■ ■ ■ $t = 1, 10, 100,$ and $1000\,\text{s}$.

Another useful property of self-adjoint matrices, which is used extensively in quantum mechanics, is:

THEOREM. *If the self-adjoint matrices* $\mathbf{A}$ *and* $\mathbf{B}$ *commute, i.e., if*

$$\mathbf{AB} = \mathbf{BA}, \qquad (6.7.83)$$

then $\mathbf{A}$ *and* $\mathbf{B}$ *have the same eigenvectors.*

Consider the eigenvalue λ of $\mathbf{A}$ and suppose there are p eigenvectors $\mathbf{x}_1, \ldots, \mathbf{x}_p$ corresponding to λ. The eigenvectors can always be chosen to be orthonormal. Since

$$\mathbf{A}\mathbf{x}_i = \lambda \mathbf{x}_i, \qquad i = 1, \ldots, p, \qquad (6.7.84)$$

Eq. (6.7.83) yields

$$\mathbf{BA}\mathbf{x}_i = \mathbf{AB}\mathbf{x}_i = \lambda \mathbf{B}\mathbf{x}_i \qquad \text{or} \qquad \mathbf{A}\mathbf{y}_i = \lambda \mathbf{y}_i, \qquad (6.7.85)$$

where

$$\mathbf{y}_i = \mathbf{B}\mathbf{x}_i. \qquad (6.7.86)$$

Since $\mathbf{y}_i$ is an eigenvector of $\mathbf{A}$ corresponding to λ, it must be a linear combination of $\mathbf{x}_1, \ldots, \mathbf{x}_p$, i.e.,

$$\mathbf{B}\mathbf{x}_i = \sum_{j=1}^{p} c_{ji} \mathbf{x}_j, \qquad i = 1, \ldots, p. \qquad (6.7.87)$$

The orthonormality of the $\mathbf{x}_i$ ($\mathbf{x}_i^\dagger \mathbf{x}_j = \delta_{ij}$) enables us to deduce form Eq. (6.7.87) that

$$c_{ji} = \mathbf{x}_j^\dagger \mathbf{B}\mathbf{x}_i, \qquad (6.7.88)$$

and the fact that $\mathbf{B}$ is self-adjoint yields the result

$$c_{ij} = c_{ji}^*. \qquad (6.7.89)$$

Thus, the $p \times p$ matrix

$$\mathbf{C} = [c_{ij}] = \begin{bmatrix} c_{11} & c_{12} & \cdots & c_{1p} \\ c_{21} & c_{22} & \cdots & c_{2p} \\ \vdots & \vdots & & \vdots \\ c_{p1} & c_{p2} & \cdots & c_{pp} \end{bmatrix} \qquad (6.7.90)$$

is self-adjoint.

The equations in Eq. (6.7.87) can be expressed as

$$\mathbf{B}[\mathbf{x}_1,\ldots,\mathbf{x}_p] = [\mathbf{x}_1,\ldots,\mathbf{x}_p]\mathbf{C}. \tag{6.7.91}$$

Since $\mathbf{C}$ is self-adjoint, there exists a $p \times p$ unitary matrix $\mathbf{U}$ such that

$$\mathbf{U}^\dagger \mathbf{C} \mathbf{U} = \mathbf{\Lambda} = \begin{bmatrix} \alpha_1 & 0 & \cdots & 0 \\ 0 & \alpha_1 & \cdots & 0 \\ \vdots & \vdots & & \vdots \\ 0 & 0 & \cdots & \alpha_1 \end{bmatrix} \tag{6.7.92}$$

and Eq. (6.7.91) can be transformed into

$$\mathbf{B}[\mathbf{x}_1,\ldots,\mathbf{x}_p]\mathbf{U} = [\mathbf{x}_1,\ldots,\mathbf{x}_p]\mathbf{U}\mathbf{U}^\dagger \mathbf{C}\mathbf{U}, \tag{6.7.93}$$

which can be rewritten as

$$\begin{aligned}[] [\mathbf{B}\mathbf{z}_1,\ldots,\mathbf{B}\mathbf{z}_p] &= [\mathbf{z}_1,\ldots,\mathbf{z}_p]\mathbf{\Lambda} \\ &= [\alpha_1 \mathbf{z}_1,\ldots,\alpha_p \mathbf{z}_p] \end{aligned} \tag{6.7.94}$$

or

$$\mathbf{B}\mathbf{z}_i = \alpha_i \mathbf{z}_i, \qquad i = 1,\ldots,p, \tag{6.7.95}$$

where

$$\mathbf{z}_i = \sum_{j=1}^{p} u_{ji} \mathbf{x}_i. \tag{6.7.96}$$

Thus, the p vectors $\mathbf{z}_1,\ldots,\mathbf{z}_p$ are eigenvectors of $\mathbf{B}$, and since $\mathbf{A}\mathbf{x}_i = \lambda \mathbf{x}_i$, $i = 1,\ldots,p$, they are also eigenvectors of $\mathbf{A}$. This completes the proof of the theorem. The eigenvalues of $\mathbf{B}$ for the p eigenvectors $\{\mathbf{z}_i\}$ are eigenvalues of the matrix $\mathbf{C}$, and the column vectors of $\mathbf{U}$ are the orthonormalized eigenvectors of $\mathbf{C}$.

Since $\mathbf{B}$ is self-adjoint, the eigenvectors in Eq. (6.7.95) are orthogonal if the eigenvalues α_i are distinct or they can be orthogonalized by the Gram–Schmidt procedure if the multiplicity of a given eigenvalue α_i is greater than 1. Furthermore, they remain eigenvectors of $\mathbf{B}$ and of $\mathbf{A}$ in such an orthogonalization procedure. Thus, the spectral decompositions of $\mathbf{B}$ and $\mathbf{A}$ are

$$\mathbf{B} = \sum_{i=1}^{n} \alpha_i \mathbf{z}_i \mathbf{z}_i^\dagger, \qquad \mathbf{A} = \sum_{i=1}^{n} \lambda_i \mathbf{z}_i \mathbf{z}_i^\dagger, \tag{6.7.97}$$

where the eigenvectors $\mathbf{z}_i$ are orthonormalized (i.e., $\mathbf{z}_i^\dagger \mathbf{z}_j = \delta_{ij}$).

The theorem can be restated as follows:

THEOREM. *If the self-adjoint matrices* $\mathbf{A}$ *and* $\mathbf{B}$ *commute, then they can be diagonalized by the same unitary transformation; i.e., there exists a unitary matrix* $\mathbf{Y}$

SELF-ADJOINT MATRICES

such that

$$\mathbf{Y}^\dagger \mathbf{A} \mathbf{Y} = \begin{bmatrix} \lambda_1 & 0 & \cdots & 0 \\ 0 & \lambda_2 & \cdots & 0 \\ \vdots & \vdots & & \vdots \\ 0 & 0 & \cdots & \lambda_n \end{bmatrix}, \quad \mathbf{Y}^\dagger \mathbf{B} \mathbf{Y} = \begin{bmatrix} \alpha_1 & 0 & \cdots & 0 \\ 0 & \alpha_2 & \cdots & 0 \\ \vdots & \vdots & & \vdots \\ 0 & 0 & \cdots & \alpha_n \end{bmatrix}. \quad (6.7.98)$$

The column vectors of $\mathbf{Y}$ are orthonormal eigenvectors of $\mathbf{A}$ and $\mathbf{B}$.

Finally, in connection with evaluating averages by Gaussian distributions, we prove the following theorem:

THEOREM. *If $\mathbf{A}$ is positive definite and $\mathbf{B}$ is self-adjoint, then there exists a nonsingular matrix $\mathbf{T}$ such that if*

$$\mathbf{x} = \mathbf{T}\mathbf{w}, \quad (6.7.99)$$

then the bilinear form f,

$$f = \mathbf{x}^\dagger (\mathbf{A} + \mathbf{B}) \mathbf{x}, \quad (6.7.100)$$

becomes

$$f = \mathbf{w}^\dagger (\mathbf{I} + \mathbf{M}) \mathbf{w} = \sum_{i=1}^{n} (1 + \mu_i) |w_i|^2, \quad (6.7.101)$$

where $\mathbf{M}$ is a diagonal matrix with real diagonal elements μ_i, $i = 1, \ldots, n$.

The first step in the proof is to note that since $\mathbf{A}$ is positive definite, there exists a unitary matrix $\mathbf{U}$ such that $\mathbf{U}^\dagger \mathbf{A} \mathbf{U} = \mathbf{\Lambda}$, where $\mathbf{\Lambda}$ is a diagonal matrix with positive diagonal elements λ_i. If we define $\mathbf{y}$ by

$$\mathbf{x} = \mathbf{U}\mathbf{y}, \quad (6.7.102)$$

then f becomes

$$f = \mathbf{y}^\dagger \mathbf{\Lambda} \mathbf{y} + \mathbf{y}^\dagger \mathbf{C} \mathbf{y}, \quad (6.7.103)$$

where $\mathbf{C}$ is the self-adjoint matrix

$$\mathbf{C} = \mathbf{U}^\dagger \mathbf{B} \mathbf{U}. \quad (6.7.104)$$

Next, we introduce the transformation

$$\mathbf{y} = \mathbf{\Lambda}^{-1/2} \mathbf{z}, \quad (6.7.105)$$

where $\mathbf{\Lambda}^{-1/2}$ is a diagonal matrix with elements $\lambda_i^{-1/2}$. In terms of $\mathbf{z}$, f becomes

$$f = \mathbf{z}^\dagger \mathbf{z} + \mathbf{z}^\dagger \mathbf{D} \mathbf{z}, \quad (6.7.106)$$

where

$$\mathbf{D} = \boldsymbol{\Lambda}^{-1/2}\mathbf{C}\boldsymbol{\Lambda}^{-1/2}. \qquad (6.7.107)$$

$\mathbf{D}$ is self-adjoint, and so there exists a unitary matrix $\mathbf{V}$ such that

$$\mathbf{V}^\dagger \mathbf{D} \mathbf{V} = \mathbf{M} \equiv \begin{bmatrix} \mu_1 & 0 & \cdots & 0 \\ 0 & \mu_2 & \cdots & 0 \\ \vdots & \vdots & & \vdots \\ 0 & 0 & \cdots & \mu_n \end{bmatrix}. \qquad (6.7.108)$$

Inserting the transformation

$$\mathbf{z} = \mathbf{V}^\dagger \mathbf{w} \qquad (6.7.109)$$

into Eq. (6.7.106) leads to Eq. (6.7.101), thus proving the theorem. Combining Eqs. (6.7.102), (6.7.105), and (6.7.109), we find that $\mathbf{x} = \mathbf{U}\boldsymbol{\Lambda}^{-1/2}\mathbf{V}^\dagger\mathbf{w}$, or that the transformation matrix is

$$\mathbf{T} = \mathbf{U}\boldsymbol{\Lambda}^{-1/2}\mathbf{V}^\dagger \qquad (6.7.110)$$

and its inverse is $\mathbf{T}^{-1} = \mathbf{V}\boldsymbol{\Lambda}^{1/2}\mathbf{U}^\dagger$. The column vectors of $\mathbf{U}$ are the orthonormalized eigenvectors of $\mathbf{A}$. The elements of $\boldsymbol{\Lambda}$ are the eigenvalues of $\mathbf{A}$ and the column vectors of $\mathbf{V}$ are the orthonormalized eigenvectors of the matrix $\boldsymbol{\Lambda}^{-1/2}\mathbf{U}^\dagger\mathbf{B}\mathbf{U}\boldsymbol{\Lambda}^{-1/2}$.

∎ **ILLUSTRATION 6.7.1** (Conditions of Chemical Stability). The second law of thermodynamics, sometimes referred to as the *entropy maximum principle*, requires that, for a system in thermodynamic equilibrium, $\delta^2 S = 0$ for any and all fluctuations δU (internal energy), δV (volume), and δn_i (composition). As the principle implies, the entropy S, which is a function of U, V, and n_i (number of moles of component i), is in a local maximum at an equilibrium point.

It is often convenient to work with free energies when discussing fluids. The second law can be restated for a system at constant temperature and volume as requiring the Helmholtz free energy, $F(V, T, \{n_i\})$, be in a local minimum at equilibrium. We can expand F in a Taylor series around the equilibrium point at constant T and V, giving

$$F = \frac{1}{2}\sum_{i,j} \frac{\partial^2 F}{\partial n_i \, \partial n_j} \delta n_i \, \delta n_j. \qquad (6.7.111)$$

Recognizing that the chemical potential is defined as

$$\mu_i = \left(\frac{\partial F}{\partial n_i}\right)_{T, V, n_{k\neq i}}, \qquad (6.7.112)$$

the second law of thermodynamics requires that the matrix $\mathbf{A}$, with components

$$A_{ij} = \left(\frac{\partial \mu_i}{\partial n_j}\right)_{T, V}, \qquad (6.7.113)$$

be *positive definite*.

A similar expression can be derived for the Gibbs free energy $G(T, P, \{n_i\})$ by noting that, at constant T and P,

$$G = \frac{1}{2} \sum_{i,j} \frac{\partial^2 G}{\partial n_i \partial n_j} \delta n_i \delta n_j. \qquad (6.7.114)$$

The second law requires that, for a system in equilibrium at constant temperature and pressure, the Gibbs free energy must be in a local minimum. However, in this case, the volume is no longer constrained. We can imagine a fluctuation in which all of the components change in equal proportions such that the composition is constant and the total density of the system is unchanged. Such a fluctuation, in a bulk fluid, consists of an arbitrary change of the boundary of the system and should thus not affect stability.

(i) Use the Gibbs–Duhem equation,

$$\sum_i d\mu_i \, n_i = 0, \qquad (6.7.115)$$

to show that, for the fluctuation $\delta n_i = a n_i$, where a is a constant, the second-order fluctuation in the Gibbs free energy is given by

$$\delta^2 G = 0. \qquad (6.7.116)$$

The above result requires that we modify the stability conditions for the Gibbs free energy such that this one fluctuation is considered. We hence require that the matrix $\mathbf{A}'$ defined by

$$A'_{ij} = \left(\frac{\partial \mu_i}{\partial n_j}\right)_{T,P} \qquad (6.7.117)$$

be *positive semidefinite* with exactly one eigenvalue equal to 0.

(ii) A useful free-energy approximation for binary mixtures is the Wilson free energy given by

$$G(T, P, n_1, n_1) = n_1 g_1(T, P) + n_2 g_2(T, P)$$
$$+ RT\left(n_1 \ln x_1 + n_2 \ln x_2\right) \qquad (6.7.118)$$
$$- RT\left(n_1 \ln (x_1 + \Lambda_{12} x_2) + n_2 \ln (x_2 + \Lambda_{21} x_1)\right),$$

where x_i is the molar fraction of component i, g_i is the molar Gibbs free energy of pure component i, and Λ_{ij} are model parameters. Determine under what conditions a single-phase, binary mixture obeying the Wilson equation is thermodynamically stable.

■ ■ ■

ILLUSTRATION 6.7.2. (Reaction Diffusion in a Thin Film). The multi-component diffusion system offers a good example of the power of eigenanalysis. Consider a permeable, thin-film membrane of thickness δ. We wish to construct the steady-state concentration profile of a three-component system. Reagent A is in contact with one boundary of the film at concentration c_{A_0}. Similarly, reagent B contacts the other boundary with concentration c_{B_0}. Reagent A is converted to reagent C according to the first-order reaction $A \to C$. The rate expression is given by

$$r_A = -r_C = -kc_A. \tag{6.7.119}$$

We can write the multicomponent reaction–diffusion equation for this system in matrix form as

$$\frac{\partial \mathbf{c}}{\partial t} = \mathbf{D}\frac{\partial^2 \mathbf{c}}{\partial x^2} + \mathbf{R}\mathbf{c}, \tag{6.7.120}$$

where

$$\mathbf{c} \equiv \begin{bmatrix} c_A \\ c_B \\ c_C \end{bmatrix}, \tag{6.7.121}$$

and $\mathbf{D}$ contains the multicomponent diffusion coefficients

$$\mathbf{D} \equiv \begin{bmatrix} D_{AA} & D_{AB} & D_{AC} \\ D_{BA} & D_{BB} & D_{BC} \\ D_{CA} & D_{CB} & D_{CC} \end{bmatrix}. \tag{6.7.122}$$

The matrix $\mathbf{R}$ contains the first-order reaction terms and is defined for this system as

$$\mathbf{R} \equiv \begin{bmatrix} -k & 0 & 0 \\ 0 & 0 & 0 \\ k & 0 & 0 \end{bmatrix}. \tag{6.7.123}$$

The boundary conditions are

$$\begin{aligned} c_A(0) &= c_{A_0}, & c_A(\delta) &= 0, \\ c_B(0) &= 0, & c_B(\delta) &= c_{B_0}, \\ c_C(0) &= 0, & c_C(\delta) &= 0. \end{aligned} \tag{6.7.124}$$

We can recast Eq. (6.7.120) in a simpler form by making the following linear transformation:

$$\mathbf{u} = \mathbf{X}^{-1}\mathbf{c}, \tag{6.7.125}$$

where $\mathbf{X}$ diagonalizes $\mathbf{D}$ in the similarity transformation

$$\mathbf{X}^{-1}\mathbf{D}\mathbf{X} = \mathbf{\Lambda}. \tag{6.7.126}$$

(i) Show that Eq. (6.7.120) can be transformed in terms of $\mathbf{u}$ into

$$\frac{\partial \mathbf{u}}{\partial t} = \mathbf{\Lambda}\frac{\partial^2 \mathbf{u}}{\partial x^2} + \mathbf{Q}\mathbf{u}. \tag{6.7.127}$$

Express the matrix $\mathbf{Q}$ in terms of $\mathbf{X}$, $\mathbf{D}$, and $\mathbf{R}$.

(ii) Generate a fully decoupled system of equations to describe the steady-state concentrations by first multiplying Eq. (6.7.127) by $\boldsymbol{\Lambda}^{-1}$. Solve the equations and plot the steady-state values of c_A, c_B, and c_C within the membrane film ($0 \leq x \leq \delta$) for $\delta = 0.5$ mm, $c_{A_0} = 0.05$ mol/L, $c_{B_0} = 0.05$ mol/L, $k = 0.55$ s^{-1}, and

$$\mathbf{D} \equiv \begin{bmatrix} 2.05 & 1.96 & 1.10 \\ 1.96 & 1.75 & 1.25 \\ 1.10 & 1.25 & 0.96 \end{bmatrix} \times 10^{-3} \text{ cm}^2/\text{s}. \tag{6.7.128}$$

6.8. NORMAL MATRICES

We say that $\mathbf{A}$ is a *normal* matrix if

$$\mathbf{A}\mathbf{A}^\dagger = \mathbf{A}^\dagger \mathbf{A}. \tag{6.8.1}$$

Self-adjoint matrices are, of course, normal, but the class also includes unitary, orthogonal, and skew-symmetric ($\mathbf{A}^\dagger = -i\mathbf{A}$) matrices, as well as other matrices. For example, the 2×2 matrix

$$\mathbf{A} = \begin{bmatrix} 7 + 4i & 1 \\ 1 & 7 + 4i \end{bmatrix} \tag{6.8.2}$$

is a normal matrix, but is neither a unitary nor a self-adjoint matrix. Figure 6.8.1 gives a qualitative picture of the classes of matrices we have been discussing. In the figure, P.D. means positive definite.

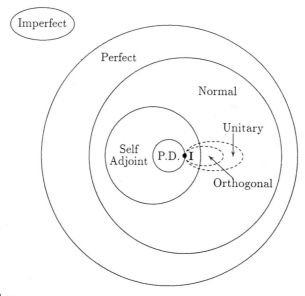

FIGURE 6.8.1

The normal matrices are, in fact, the class of all matrices whose eigenvectors form an orthonormal basis set. This fact follows from the theorem:

THEOREM. *If $\mathbf{A}$ is a normal matrix, there exists a unitary matrix $\mathbf{X}$ that diagonalizes $\mathbf{A}$ in a unitary transformation, i.e.,*

$$\mathbf{X}^\dagger \mathbf{A} \mathbf{X} = \mathbf{\Lambda} \quad \text{and} \quad \mathbf{X}^\dagger \mathbf{X} = \mathbf{I}. \qquad (6.8.3)$$

As stated in the previous section, this theorem is also equivalent to the spectral resolution theorem that there exists a basis set of orthonormal eigenvectors of $\mathbf{A}$ such that

$$\mathbf{A} = \sum_{i=1}^{n} \lambda_i \mathbf{x}_i \mathbf{x}_i^\dagger. \qquad (6.8.4)$$

The λ_i are not necessarily real. In fact, the eigenvalues of the matrix given by Eq. (6.8.2) are $\lambda_1 = 3 + 2i$ and $\lambda_2 = 4 + 2i$.

Normal matrices represent the largest class of matrices whose eigenvectors form an orthonormal basis set. In fact, the whole class can be constructed by selecting every orthonormal basis set $\{\mathbf{x}_i\}$ in E_n and taking linear combinations $\sum_i \lambda_i \mathbf{x}_i \mathbf{x}_i^\dagger$ with every possible set of complex numbers $\{\lambda_i\}$.

The proof of the above theorem makes use of the semidiagonalization theorem, which states that there exists a unitary matrix $\mathbf{Q}$ such that

$$\begin{aligned}
\mathbf{A} = \mathbf{Q} \mathbf{B}_n \mathbf{Q}^\dagger &= \mathbf{Q} \begin{bmatrix} \lambda_1 & b_{12} & \cdots & b_{1n} \\ 0 & \lambda_2 & \cdots & b_{2n} \\ \vdots & \vdots & & \vdots \\ 0 & 0 & \cdots & \lambda_n \end{bmatrix} \mathbf{Q}^\dagger \\
&= \mathbf{Q} \begin{bmatrix} \lambda_1 & b_{12} & \cdots & b_{1n} \\ 0 & & & \\ \vdots & & \mathbf{B}_{n-1} & \\ 0 & & & \end{bmatrix} \mathbf{Q}^\dagger,
\end{aligned} \qquad (6.8.5)$$

where $\mathbf{B}_{n-1}$ is an upper triangular matrix with eigenvalues $\lambda_2, \ldots, \lambda_n$ on the main diagonal. The adjoint of Eq. (6.8.5) yields the result

$$\mathbf{A}^\dagger = \mathbf{Q} \begin{bmatrix} \lambda_1^* & 0 & \cdots & 0 \\ b_{12}^* & & & \\ \vdots & & \mathbf{B}_{n-1}^\dagger & \\ b_{1n}^* & & & \end{bmatrix} \mathbf{Q}^\dagger. \qquad (6.8.6)$$

The product $\mathbf{A}^\dagger \mathbf{A}$, with the aid of the property $\mathbf{Q}\mathbf{B}_n^\dagger \mathbf{Q}^\dagger \mathbf{Q}\mathbf{B}_n \mathbf{Q}^\dagger = \mathbf{Q}\mathbf{B}_n^\dagger \mathbf{B}_n \mathbf{Q}^\dagger$, is

$$\mathbf{A}^\dagger \mathbf{A} = \mathbf{Q} \begin{bmatrix} |\lambda_1|^2 & \lambda_1^* b_{12} & \cdots & \lambda_1^* b_{1n} \\ \lambda_1 b_{12}^* & \vdots & & \vdots \\ \vdots & \vdots & & \vdots \\ \lambda_1 b_{1n}^* & & & \end{bmatrix} \mathbf{Q}^\dagger, \qquad (6.8.7)$$

whereas

$$\mathbf{A}\mathbf{A}^\dagger = \mathbf{Q} \begin{bmatrix} |\lambda_1|^2 + \sum_{j=2}^{n} |b_{1j}|^2 & \cdots & \cdots \\ \vdots & & \\ \vdots & & \end{bmatrix} \mathbf{Q}^\dagger. \qquad (6.8.8)$$

Using the property that $\mathbf{A}\mathbf{A}^\dagger - \mathbf{A}^\dagger \mathbf{A} = \mathbf{0}$ for a normal matrix, we find that

$$b_{1j} = 0, \qquad j = 2, \ldots, n. \qquad (6.8.9)$$

This reduces $\mathbf{A}$ to

$$\mathbf{A} = \mathbf{Q} \begin{bmatrix} \lambda_1 & 0 & \cdots & 0 \\ 0 & & & \\ \vdots & & \mathbf{B}_{n-1} & \\ 0 & & & \end{bmatrix} \mathbf{Q}^\dagger \qquad (6.8.10)$$

and gives the result

$$\mathbf{A}\mathbf{A}^\dagger = \mathbf{Q} \begin{bmatrix} |\lambda_1|^2 & 0 & \cdots & 0 \\ 0 & & & \\ \vdots & & \mathbf{B}_{n-1}\mathbf{B}_{n-1}^\dagger & \\ 0 & & & \end{bmatrix} \mathbf{Q}^\dagger \qquad (6.8.11)$$

and

$$\mathbf{B}_{n-1}\mathbf{B}_{n-1}^\dagger = \mathbf{B}_{n-1}^\dagger \mathbf{B}_{n-1} \qquad (6.8.12)$$

(since $\mathbf{A}\mathbf{A}^\dagger = \mathbf{A}^\dagger \mathbf{A}$).

$\mathbf{B}_{n-1}$ is an upper triangular matrix with elements $\lambda_2, \ldots, \lambda_n$ along the main diagonal and the elements of its first row are $\lambda_2, b_{23}, \ldots, b_{2n}$. Repeating the process we just used to obtain Eq. (6.8.11) leads to the conclusion that

$$b_{2j} = 0, \qquad j = 3, \ldots, n, \qquad (6.8.13)$$

and

$$A = Q \begin{bmatrix} \lambda_1 & 0 & \cdots & 0 \\ 0 & \lambda_2 & \cdots & 0 \\ \vdots & \vdots & & B_{n-2} \\ 0 & 0 & & \end{bmatrix} Q^\dagger, \qquad (6.8.14)$$

where B_{n-2} is an upper diagonal matrix with elements $\lambda_3, \ldots, \lambda_n$ along the main diagonal. Continuing the process yields the final result

$$A = Q\Lambda Q^\dagger, \qquad (6.8.15)$$

which completes the proof of the theorem. The orthonormal eigenvectors of A are, in fact, the column vectors of the unitary matrix Q that was guaranteed to exist by the semidiagonalization theorem.

If U is a unitary matrix, then $U^\dagger U = UU^\dagger = I$, and so, as mentioned earlier, a unitary matrix is a normal matrix. Thus, if $\mu_i, \ldots, \mu_n$ are the eigenvalues of U and $x_1, \ldots, x_n$ are the corresponding orthonormalized eigenvectors, the spectral decomposition is

$$U = \sum_{i=1}^n \mu_i x_i x_i^\dagger. \qquad (6.8.16)$$

From the fact that

$$\begin{aligned} UU^\dagger &= \sum_{i=1}^n \mu_i x_i x_i^\dagger \sum_{j=1}^n \mu_j^* x_j x_j^\dagger \\ &= \sum_{i=1}^n |\mu_i|^2 x_i x_i^\dagger = I = \sum_{i=1}^n x_i x_i^\dagger, \end{aligned} \qquad (6.8.17)$$

it follows that $|\mu_i|^2 = 1$; i.e., the eigenvalues of a unitary matrix are of unit magnitude. This leads to the conclusion that

THEOREM. *A unitary matrix U can always be expressed in the form*

$$U = \exp(iA), \qquad (6.8.18)$$

where A is a self-adjoint matrix.

To prove this, set

$$\mu_j = \exp(i\lambda_j), \qquad (6.8.19)$$

where λ_j is a real number. Note that, since $|\mu_j| = 1$, if we define $A = \sum_{j=1}^n \lambda_j x_j x_j^\dagger$, where the x_i are the orthonormalized eigenvectors of U, it follows that

$$\exp(iA) = \sum_{j=1}^n \exp(i\lambda_j) x_j x_j^\dagger = U. \qquad (6.8.20)$$

By definition, $\mathbf{A} = \mathbf{A}^\dagger$ if $\mathbf{U}$ is unitary, which proves the conclusion. The meaning of Eq. (6.8.18) is that unitary and self-adjoint matrices are related through an exponential mapping: for any self-adjoint matrix $\mathbf{A}$, there exists a unitary matrix such that $\mathbf{U} = \exp(i\mathbf{A})$ and for any unitary matrix $\mathbf{U}$ there exists a self-adjoint matrix such that $\mathbf{U} = \exp(i\mathbf{A})$.

As we did for self-adjoint matrices, we can use the spectral resolution theorem to prove that, for a normal matrix, in the $p = 2$ norm,

$$\|\mathbf{A}\| = |\lambda_{\max}| \tag{6.8.21}$$

and

$$\kappa(\mathbf{A}) = \frac{|\lambda_{\max}|}{|\lambda_{\min}|}, \tag{6.8.22}$$

where

$$|\lambda_{\max}| = \max_{1 \le i \le n} |\lambda_i| \quad \text{and} \quad |\lambda_{\min}| = \min_{1 \le i \le n} |\lambda_i|. \tag{6.8.23}$$

6.9. MISCELLANEA

The product $\mathbf{AB}$ of a self-adjoint matrix $\mathbf{A}$ and a positive-definite matrix $\mathbf{B}$ is not self-adjoint. Nevertheless, the following theorem can be proved:

THEOREM. *If $\mathbf{A}$ is self-adjoint and $\mathbf{B}$ is positive definite (or negative definite), then the product $\mathbf{AB}$ (and $\mathbf{BA}$) is a perfect matrix and its eigenvalues are real.*

The proof is straightforward. Consider the eigenproblem

$$\mathbf{AB}\mathbf{y}_i = \lambda_i \mathbf{y}_i. \tag{6.9.1}$$

Since $\mathbf{B}$ is positive definite, it is perfect and nonsingular and its square root $\mathbf{R} = \sum_i \mu_i^{1/2} \mathbf{x}_i \mathbf{x}_i^\dagger$ is also positive definite (μ_i is the eigenvalue of $\mathbf{B}$ and $\mathbf{x}_i$ is its orthonormal eigenvector). Putting $\mathbf{B} = \mathbf{RR}$ in Eq. (6.9.1), defining $\mathbf{z}_i \equiv \mathbf{R}\mathbf{y}_i$, and multiplying Eq. (6.9.1) by $\mathbf{R}$, we obtain

$$\mathbf{RAR}\mathbf{z}_i = \lambda_i \mathbf{z}_i. \tag{6.9.2}$$

Since the matrix $\mathbf{RAR}$ is self-adjoint, the eigenvalues λ_i are real and the eigenvectors $\{\mathbf{z}_i\}$ form an orthonormal basis set. Note that the eigenvectors $\mathbf{y}_i = \mathbf{R}^{-1}\mathbf{z}_i$, $i = 1, \ldots, n$, are linearly independent since the equation

$$\mathbf{0} = \sum_{i=1}^n \alpha_i \mathbf{y}_i = \sum_{i=1}^n \alpha_i \mathbf{R}^{-1} \mathbf{z}_i = \mathbf{R}^{-1} \sum_{i=1}^n \alpha_i \mathbf{z}_i \tag{6.9.3}$$

has only the solution $\alpha_i = 0$, $i = 1, \ldots, n$. This completes the proof that $\mathbf{AB}$ is a perfect matrix and its eigenvalues are real. Similarly, it is easy to prove that $\mathbf{BA}$ is perfect and has real eigenvalues.

If both **A** and **B** are positive definite, then an even more specific theorem can be proved:

THEOREM. *If **A** and **B** are positive definite, then the product **AB** (or **BA**) is a perfect matrix and its eigenvalues are real and positive.*

In the previous theorem, we established that **AB** is perfect and has real eigenvalues. According to Eq. (6.9.2), the eigenvalues of **AB** are the eigenvalues of **RAR**, where **R** is a positive definite matrix (more specifically, **R** is the square root of **B** and $\mathbf{R} = \mathbf{R}^\dagger$).

Consider the inner product $\mathbf{x}^\dagger \mathbf{RARx} = \mathbf{x}^\dagger \mathbf{R}^\dagger \mathbf{ARx}$. Defining $\mathbf{w} \equiv \mathbf{Rx}$, we note that if $\mathbf{x} \neq \mathbf{0}$, then $\mathbf{w} \neq \mathbf{0}$ since **R** is positive definite, and so the inner product

$$\mathbf{x}^\dagger \mathbf{RARx} = \mathbf{w}^\dagger \mathbf{Aw} > 0$$

for all vectors $\mathbf{x}$ in E_n (because **A** is, by choice, positive definite). Thus, from Eq. (6.9.2), it follows that $\mathbf{z}_i^\dagger \mathbf{RARz}_i = \lambda_i \|\mathbf{z}_i\|^2 > 0$, and so the eigenvalues of **AB** must be positive, proving the theorem. By simply interchanging the symbols **A** and **B** above, we can prove the same theorem for the product **BA**.

Note that if both **A** and **B** are self-adjoint but neither is positive definite (or negative definite), then their product might not be perfect. For example, if

$$\mathbf{A} = \begin{bmatrix} 1 & 1 \\ 1 & 0 \end{bmatrix} \quad \text{and} \quad \mathbf{B} = \begin{bmatrix} 0 & 2 \\ 2 & -1 \end{bmatrix}, \tag{6.9.4}$$

then

$$\mathbf{AB} = \begin{bmatrix} 2 & 1 \\ 0 & 2 \end{bmatrix}. \tag{6.9.5}$$

The eigenvalues of **AB** are $\lambda_1 = \lambda_2 = 2$ and the only eigenvector is

$$\mathbf{x}_1 = \begin{bmatrix} 1 \\ 0 \end{bmatrix}. \tag{6.9.6}$$

Thus, **AB** is not a perfect matrix.

■ **EXAMPLE 6.9.1.** In a multicomponent material, the concentration c_i of component i obeys the diffusion equation

$$\frac{\partial c_i}{\partial t} = \sum_{j=1}^n D_{ij} \nabla^2 c_j, \quad i = 1, \ldots, n, \tag{6.9.7}$$

where t is time, $\nabla^2 = \partial^2/\partial x^2 + \partial^2/\partial y^2 + \partial^2/\partial z^2$ is the Laplacian differential operator, D_{ij} is a diffusion coefficient through which a gradient in the concentration of component j causes diffusion of component i, and n is the number of components in the material. The quantity D_{ij} is related to the Onsager transport coefficients l_{ij} and the components of the Helmholtz free-energy curvature matrix $\mathbf{A} = [a_{ij}]$, where

$$a_{ij} = \frac{\partial^2 F}{\partial c_i \, \partial c_j}. \tag{6.9.8}$$

In particular,

$$D_{ij} = \sum_{k=1}^{n} l_{ik} a_{kj} \qquad (6.9.9)$$

or

$$\mathbf{D} = \mathbf{LA}, \qquad (6.9.10)$$

where

$$\mathbf{D} = [D_{ij}], \qquad \mathbf{L} = [l_{ij}], \qquad \mathbf{A} = [a_{ij}]. \qquad (6.9.11)$$

A relevant physical question is whether diffusion always gives rise to mixing in a closed multicomponent system; i.e., does diffusion always result in a homogeneous material? For present purposes, we assume that $\mathbf{D}$ is constant in space and time. Equation (6.9.7) can be summarized as

$$\frac{\partial \mathbf{c}}{\partial t} = \nabla^2 \mathbf{D} \mathbf{c}, \qquad (6.9.12)$$

where

$$\mathbf{c} = \begin{bmatrix} c_1 \\ \vdots \\ c_n \end{bmatrix}. \qquad (6.9.13)$$

The condition that the system is closed is

$$\sum_j D_{ij} \nabla c_j = 0 \qquad \text{or} \qquad \nabla \mathbf{D} \mathbf{c} = \mathbf{0} \qquad (6.9.14)$$

on the boundary ∂V of the system. The system is contained in the volume V.

From irreversible thermodynamics, it is known that the Onsager matrix $\mathbf{L}$ is positive definite. From thermodynamics, it is known that the free-energy curvature matrix is also positive definite if the material is thermodynamically stable. Thus, we can use the theorem proved above to show that the eigenvalues of $\mathbf{D}$ are all positive, and there exists a matrix $\mathbf{X}$, which diagonalizes $\mathbf{D}$ in a similarity transformation, i.e.,

$$\mathbf{X}^{-1} \mathbf{D} \mathbf{X} = \mathbf{\Lambda} = [\lambda_i \delta_{ij}], \qquad \lambda_i > 0, \qquad i = 1, \ldots, n. \qquad (6.9.15)$$

Defining

$$\boldsymbol{\psi} = \mathbf{X}^{-1} \mathbf{c} \qquad \text{or} \qquad \mathbf{c} = \mathbf{X} \boldsymbol{\psi} \qquad (6.9.16)$$

and multiplying Eqs. (6.9.12) and (6.9.14) by $\mathbf{X}^{-1}$, we find

$$\frac{\partial \boldsymbol{\psi}}{\partial t} = \nabla^2 \mathbf{\Lambda} \boldsymbol{\psi} \qquad (6.9.17a)$$

or

$$\frac{\partial \psi}{\partial t} = \lambda_i \nabla^2 \psi_i, \qquad i = 1, \ldots, n, \qquad (6.9.17b)$$

and

$$\nabla \mathbf{\Lambda} \boldsymbol{\psi} = \mathbf{0} \qquad \text{on } \partial V \qquad (6.9.18a)$$

or

$$\lambda_i \nabla \psi_i = 0, \qquad i = 1, \ldots, n, \text{ on } \partial V. \qquad (6.9.18b)$$

Multiplying Eq. (6.9.17b) by ψ_i and using the identity

$$\int_V \psi_i \nabla^2 \psi_i \, dV = -\int (\nabla \psi_i) \cdot \nabla \psi_i \, dV \\ + \int_V \nabla \cdot (\psi_i \nabla \psi_i) \, dV, \qquad (6.9.19)$$

with the property

$$\int_V \nabla \cdot (\psi_i \nabla \psi_i) \, dV = \int_{\partial V} \psi_i \nabla \psi_i \cdot d\mathbf{A}, \qquad (6.9.20)$$

where $d\mathbf{A}$ is an element of area on ∂V, we obtain

$$\frac{1}{2}\frac{d}{dt}\int \psi_i^2 \, dV = -\lambda_i \int (\nabla \psi_i)^2 \, dV. \qquad (6.9.21)$$

The right-hand side of Eq. (6.9.20) vanishes because of the closed-system boundary condition, Eq. (6.9.18b).

It follows from Eq. (6.9.21) that the positive quantity $\int \psi_i^2 \, dV$ decreases in time until it reaches 0, at which time $\nabla \psi_i = 0$ everywhere in V. Thus, we have proven that, in a thermodynamically stable material, diffusional transport in a closed system will always lead in time to a homogeneous system. If any eigenvalue λ_i changes sign, then the conclusion does not hold. Physically, this situation arises when the composition of the system is such that the material is thermodynamically unstable, in which case $|\mathbf{A}| < 0$ and $|\mathbf{D}| = |\mathbf{LA}| = |\mathbf{L}|\,|\mathbf{A}| < 0$. In this situation, the system "un-diffuses" and splits into two or more coexisting phases.

Another interesting property of a product of arbitrary square matrices is

THEOREM. *If* **A** *and* **B** *commute, i.e., if*

$$\mathbf{AB} = \mathbf{BA}, \qquad (6.9.22)$$

and if λ_i is an eigenvalue of **A** *of multiplicity* 1, *then the corresponding eigenvector*, $\mathbf{x}_i$, *is also an eigenvector of* **B**.

To prove this, multiply $\mathbf{A}\mathbf{x}_i = \lambda_i \mathbf{x}_i$ by **B** and use Eq. (6.9.22) to obtain

$$\mathbf{A}\mathbf{y}_i = \lambda_i \mathbf{y}_i, \qquad \text{where } \mathbf{y}_i = \mathbf{B}\mathbf{x}_i. \qquad (6.9.23)$$

Since $\mathbf{y}_i$ is an eigenvector of $\mathbf{A}$ of eigenvalue λ_i, then it differs from $\mathbf{x}_i$ only by a scalar factor, i.e., $\mathbf{y}_i = \alpha_i \mathbf{x}_i$, which yields the result

$$\mathbf{B}\mathbf{x}_i = \alpha_i \mathbf{x}_i, \tag{6.9.24}$$

proving the theorem. Note that this is a more general version of the theorem we proved in Section 6.7 for self-adjoint matrices.

We can also prove the following:

THEOREM. *If all of the eigenvalues are distinct or of multiplicity* 1, *i.e.*, $\lambda_i \neq \lambda_j$, $i \neq j$, *then it follows that*

$$\mathbf{A} = \sum_{i=1}^{n} \lambda_i \mathbf{x}_i \mathbf{z}_i^{\dagger} \quad \text{and} \quad \mathbf{B} = \sum_{i=1}^{n} \mu_i \mathbf{x}_i \mathbf{z}_i^{\dagger} \tag{6.9.25}$$

if $\mathbf{A}$ *and* $\mathbf{B}$ *commute.*

This also means that $\mathbf{A}$ and $\mathbf{B}$ can be diagonalized by the same similarity transformation; i.e., there exists a nonsingular matrix $\mathbf{X}$ such that

$$\mathbf{X}^{-1}\mathbf{A}\mathbf{X} = \begin{bmatrix} \lambda_1 & 0 & \cdots & 0 \\ 0 & \lambda_2 & \cdots & 0 \\ \vdots & \vdots & & \vdots \\ 0 & 0 & \cdots & \lambda_n \end{bmatrix} \quad \text{and} \quad \mathbf{X}^{-1}\mathbf{B}\mathbf{X} = \begin{bmatrix} \alpha_1 & 0 & \cdots & 0 \\ 0 & \alpha_2 & \cdots & 0 \\ \vdots & \vdots & & \vdots \\ 0 & 0 & \cdots & \alpha_n \end{bmatrix}. \tag{6.9.26}$$

The column vectors of $\mathbf{X}$ are the eigenvectors of $\mathbf{A}$ and $\mathbf{B}$.

Note that if $\mathbf{A}$ is not perfect, then $\mathbf{A}$ and $\mathbf{B}$ can commute and yet do not have the same eigenvectors. For example, the matrix

$$\mathbf{A} = \begin{bmatrix} 1 & 1 \\ 0 & 1 \end{bmatrix} \tag{6.9.27}$$

has only the eigenvector

$$\mathbf{x}_1 = \begin{bmatrix} 1 \\ 0 \end{bmatrix}. \tag{6.9.28}$$

However, the eigenvectors of $\mathbf{I}$ can be chosen to be

$$\mathbf{x}_1 = \begin{bmatrix} 1 \\ 0 \end{bmatrix}, \quad \mathbf{x}_2 = \begin{bmatrix} 0 \\ 1 \end{bmatrix}. \tag{6.9.29}$$

Since $\mathbf{IA} = \mathbf{AI}$, $\mathbf{I}$ and $\mathbf{A}$ commute, but $\mathbf{x}_2$ is not an eigenvector of $\mathbf{A}$.

An interesting property of a perfect matrix with distinct eigenvalues is that

$$\mathbf{x}_i \mathbf{z}_i^{\dagger} = \frac{\text{adj}(\lambda_i \mathbf{I} - \mathbf{A})}{\prod_{j \neq i}(\lambda_i - \lambda_j)}, \tag{6.9.30}$$

where adj$(\lambda \mathbf{I} - \mathbf{A})$ is the adjugate of $\lambda \mathbf{I} - \mathbf{A}$, $\mathbf{x}_i$ is the eigenvector corresponding to λ_i, and $\mathbf{z}_i$ is the reciprocal of $\mathbf{x}_i$ ($\mathbf{x}_i^\dagger \mathbf{z}_i = 1$). In Chapter 1 it was shown that

$$[\text{adj}(\lambda \mathbf{I} - \mathbf{A})](\lambda \mathbf{I} - \mathbf{A}) = |\lambda \mathbf{I} - \mathbf{A}|\,\mathbf{I}, \tag{6.9.31}$$

and so

$$[\text{adj}(\lambda \mathbf{I} - \mathbf{A})](\lambda \mathbf{I} - \mathbf{A}) = \prod_{j=1}^{n}(\lambda - \lambda_j)\mathbf{I}. \tag{6.9.32}$$

Multiplying Eq. (6.9.32) from the right by $\mathbf{x}_k \mathbf{z}_k^\dagger$ yields

$$[\text{adj}(\lambda_i \mathbf{I} - \mathbf{A})](\lambda_i - \lambda_k)\mathbf{x}_k \mathbf{z}_k^\dagger = \prod_{j=1}^{n}(\lambda_i - \lambda_j)\mathbf{x}_k \mathbf{z}_k^\dagger, \tag{6.9.33}$$

and dividing by $(\lambda_i - \lambda_k)$ gives

$$[\text{adj}(\lambda_i \mathbf{I} - \mathbf{A})]\mathbf{x}_k \mathbf{z}_k^\dagger = \prod_{j \neq k}^{n}(\lambda_i - \lambda_j)\mathbf{x}_k \mathbf{z}_k^\dagger. \tag{6.9.34}$$

Since $\sum_{k=1}^{n}\mathbf{x}_k \mathbf{z}_k^\dagger = \mathbf{I}$ and since $\sum_{k=1}^{n}\prod_{j\neq k}^{n}(\lambda_i - \lambda_j)\mathbf{x}\mathbf{z}_k^\dagger = \prod_{j\neq k}^{n}(\lambda_i - \lambda_j)\mathbf{x}_i \mathbf{z}_i^\dagger$, summing Eq. (6.9.33) over all k and dividing by $\prod_{j\neq i}^{n}(\lambda_i - \lambda_j)$ yields Eq. (6.9.29). Therefore, when all of the eigenvalues of a matrix are distinct, the spectral resolution theorem, $f(\mathbf{A}) = \sum_i f(\lambda_i)\mathbf{x}_i \mathbf{z}_i^\dagger$, can be expressed in the form

$$f(\mathbf{A}) = \sum_{I=1}^{n} f(\lambda_i) \frac{\text{adj}(\lambda_i \mathbf{I} - \mathbf{A})}{\prod_{j\neq i}^{n}(\lambda_i - \lambda_j)}, \tag{6.9.35}$$

which is known as *Sylvester's formula*.

6.10. THE INITIAL VALUE PROBLEM

The pth-order differential equation

$$\frac{d^p u}{dt^p} + a_p \frac{d^{p-1} u}{dt^{p-1}} + \cdots + a_2 \frac{du}{dt} + a_1 u = f$$

or

$$\frac{d^p u}{dt^p} + \sum_{i=1}^{p} a_{p+1-i} \frac{d^{p-i} u}{dt^{p-i}} = f \tag{6.10.1}$$

becomes an initial value problem (IVP) if the initial conditions

$$\frac{d^i u}{dt^i} = \gamma_i \qquad \text{at } t = 0 \text{ for } i = 0, 1, \ldots, p-1, \tag{6.10.2}$$

THE INITIAL VALUE PROBLEM

are given. If the values of a_i are constant and f is a known function of time, the IVP can be converted to a simple first-order equation as follows. Define $x_1, \ldots, x_p$ in the following way: $x_1 = u$ and

$$\frac{dx_1}{dt} = x_2 = \frac{du}{dt}$$
$$\frac{dx_2}{dt} = x_3 = \frac{d^2u}{dt^2} \qquad (6.10.3)$$
$$\vdots$$
$$\frac{dx_{p-1}}{dt} = x_p = \frac{d^{p-1}u}{dt^{p-1}}.$$

Then Eq. (6.10.1) becomes

$$\frac{dx_p}{dt} = -\sum_{i=1}^{p} a_i x_i + f. \qquad (6.10.4)$$

Combining Eqs. (6.10.3) and (6.10.4), we obtain

$$\frac{d\mathbf{x}}{dt} = \mathbf{A}\mathbf{x} + \mathbf{b}, \qquad (6.10.5)$$

where

$$\mathbf{x} = \begin{bmatrix} x_1 \\ \vdots \\ x_p \end{bmatrix}, \quad \mathbf{b} = \begin{bmatrix} 0 \\ \vdots \\ 0 \\ f \end{bmatrix}, \qquad (6.10.6)$$

and

$$\mathbf{A} = \begin{bmatrix} 0 & 1 & 0 & 0 & \cdots & 0 \\ 0 & 0 & 1 & 0 & \cdots & 0 \\ \vdots & \vdots & \vdots & \vdots & & \vdots \\ 0 & 0 & 0 & 0 & \cdots & 1 \\ -a_1 & -a_2 & -a_3 & -a_4 & \cdots & -a_p \end{bmatrix}, \qquad (6.10.7)$$

where $\mathbf{A}$ is known as the *companion matrix*.

The formal solution to Eq. (6.10.6) is

$$\mathbf{x} = \exp(\mathbf{A}t)\mathbf{x}_0 + \int_0^t \exp(\mathbf{A}(t-\tau))\mathbf{b}(\tau)\,d\tau, \qquad (6.10.8)$$

where

$$\mathbf{x}_0 = \begin{bmatrix} \gamma_0 \\ \gamma_1 \\ \vdots \\ \gamma_{p-1} \end{bmatrix}. \tag{6.10.9}$$

If $\mathbf{A}$ is perfect, the spectral decomposition of $\exp(\mathbf{A}t)$ allows us to express the solution in the form

$$\mathbf{x} = \sum_{i=1}^{p} \exp(\lambda_i t)(\mathbf{z}_i^\dagger \mathbf{x}_0)\mathbf{x}_i + \sum_{i=1}^{p} \int_0^t \exp(\lambda_i(t-\tau))[\mathbf{z}_i^\dagger \mathbf{b}(\tau)]\mathbf{x}_i \, d\tau. \tag{6.10.10}$$

Thus, when $\mathbf{A}$ is perfect, the solution to the IVP is a linear combination of exponential functions of time. This could have been found by trying exponential solutions. However, if $\mathbf{A}$ is not perfect, we shall learn in the next chapter that exponentials also arise in the solution of the IVP.

For the companion matrix given in Eq. (6.10.7), we can show that

$$\text{tr}_i \mathbf{A} = (-1)^i a_{p+1-i}. \tag{6.10.11}$$

The characteristic polynomial is given by

$$P_p(\lambda) = (-\lambda)^p + \sum_{i=1}^{p} (\text{tr}_i \mathbf{A})(-\lambda)^{p-i}, \tag{6.10.12}$$

and so the eigenvalues of $\mathbf{A}$ satisfy the equation

$$\lambda^p + a_p \lambda^{p-1} + a_{p-1} \lambda^{p-2} + \cdots + a_2 \lambda + a_1 = 0, \tag{6.10.13}$$

a result obtained by multiplying $P_p(\lambda)$ by $(-1)^p$ and noting that $(-1)^{2p} = (-1)^{2p-2i} = 1$. For a given eigenvalue λ, the eigenvector equation

$$\begin{bmatrix} -\lambda & 1 & 0 & \cdots & 0 & 0 \\ 0 & -\lambda & 1 & \cdots & 0 & 0 \\ \vdots & \vdots & \vdots & & \vdots & \vdots \\ 0 & 0 & 0 & \cdots & -\lambda & 1 \\ -a_1 & -a_2 & -a_3 & \cdots & -a_{p-1} & -(a_p + \lambda) \end{bmatrix} \begin{bmatrix} x_1 \\ x_2 \\ \vdots \\ \vdots \\ x_p \end{bmatrix} = \mathbf{0} \tag{6.10.14}$$

has the solution

$$\begin{aligned} -\lambda x_1 + x_2 &= 0 \\ -\lambda x_2 + x_3 &= 0 \\ &\vdots \\ -\lambda x_{p-1} + x_p &= 0 \\ -\sum_{i=1}^{p} a_i x_i - \lambda x_p &= 0 \end{aligned} \tag{6.10.15}$$

or

$$\mathbf{x} = x_1 \begin{bmatrix} 1 \\ \lambda \\ \lambda^2 \\ \vdots \\ \lambda^{p-1} \end{bmatrix}, \qquad (6.10.16)$$

where x_1 is arbitrary. From this result, we can conclude that there will be n linearly independent eigenvectors if all of the eigenvalues λ_i are distinct, i.e., if all of the roots of Eq. (6.10.13) are of multiplicity 1. In this case, $\mathbf{A}$ is perfect. If any root is of multiplicity greater than 1, then $\mathbf{A}$ is defective since it has fewer than p eigenvectors. Restated, we have just learned that

THEOREM. *The necessary and sufficient condition for the companion matrix to be perfect is that its eigenvalues be distinct.*

We will explore this further in Chapter 7.

■ ■ ■ **EXERCISE 6.10.1.** For $p = 3$, $a_1 = 4$, $a_2 = 4$, and $a_3 = 1$, find the eigenvalues and eigenvectors of the companion matrix and its adjoint.

A system of two linear differential equations (with initial conditions) can be expressed in vector form as

$$\begin{aligned} \frac{d\mathbf{x}}{dt} &= \mathbf{A}_1 \mathbf{x} + \mathbf{B}_1 \mathbf{y} + \mathbf{b}_1 \\ \frac{d\mathbf{y}}{dt} &= \mathbf{A}_2 \mathbf{x} + \mathbf{B}_2 \mathbf{y} + \mathbf{b}_2, \end{aligned} \qquad (6.10.17)$$

where the $p \times p$ matrices are

$$\mathbf{A}_i = \begin{bmatrix} 0 & 1 & 0 & 0 & \cdots & 0 \\ 0 & 0 & 1 & 0 & \cdots & 0 \\ \vdots & \vdots & \vdots & \vdots & & \vdots \\ 0 & 0 & 0 & 0 & \cdots & 1 \\ -a_{i1} & -a_{i2} & -a_{i3} & -a_{i4} & \cdots & -a_{ip} \end{bmatrix}, \quad i = 1, 2, \qquad (6.10.18)$$

$$\mathbf{B}_i = \begin{bmatrix} 0 & 1 & 0 & 0 & \cdots & 0 \\ 0 & 0 & 1 & 0 & \cdots & 0 \\ \vdots & \vdots & \vdots & \vdots & & \vdots \\ 0 & 0 & 0 & 0 & \cdots & 1 \\ -b_{i1} & -b_{i2} & -b_{i3} & -b_{i4} & \cdots & -b_{ip} \end{bmatrix}, \quad i = 1, 2, \qquad (6.10.19)$$

and the p-dimensional vectors are

$$\mathbf{b}_i = \begin{bmatrix} 0 \\ \vdots \\ 0 \\ f_i \end{bmatrix}, \quad i = 1, 2. \tag{6.10.20}$$

The problem can be further consolidated to the form

$$\frac{d}{dt}\begin{bmatrix} \mathbf{x} \\ \mathbf{y} \end{bmatrix} = \begin{bmatrix} \mathbf{A}_1 & \mathbf{B}_1 \\ \mathbf{A}_2 & \mathbf{B}_2 \end{bmatrix}\begin{bmatrix} \mathbf{x} \\ \mathbf{y} \end{bmatrix} + \begin{bmatrix} \mathbf{b}_1 \\ \mathbf{b}_2 \end{bmatrix} \tag{6.10.21}$$

or

$$\frac{d}{dt}\mathbf{w} = \mathbf{T}\mathbf{w} + \mathbf{b}, \tag{6.10.22}$$

where $\mathbf{w}$ and $\mathbf{b}$ are $2p$-dimensional vectors and $\mathbf{T}$ is a $2p \times 2p$ matrix.

For an initial value problem, the solution to Eq. (6.10.22) is

$$\mathbf{w} = \exp(\mathbf{T}t)\mathbf{w}_0 + \int_0^t \exp(\mathbf{T}(t-\tau))\mathbf{b}(\tau)\,d\tau, \tag{6.10.23}$$

where the vector $\mathbf{w}_0$ is composed of the values of u, v and their first $p-1$ derivatives at time $t = 0$.

Analogously, the three pth-order differential equations

$$\frac{d^p u}{dt^p} + \sum_{i=1}^p a_{1,p-i}\frac{d^{p-i}u}{dt^{p-i}} + \sum_{i=1}^p b_{1,p-i}\frac{d^{p-i}v}{dt^{p-i}} + \sum_{i=1}^p c_{1,p-i}\frac{d^{p-i}w}{dt^{p-i}} = f_1$$

$$\frac{d^p v}{dt^p} + \sum_{i=1}^p a_{2,p-i}\frac{d^{p-i}u}{dt^{p-i}} + \sum_{i=1}^p b_{2,p-i}\frac{d^{p-i}v}{dt^{p-i}} + \sum_{i=1}^p c_{2,p-i}\frac{d^{p-i}w}{dt^{p-i}} = f_2 \tag{6.10.24}$$

$$\frac{d^p w}{dt^p} + \sum_{i=1}^p a_{3,p-i}\frac{d^{p-i}u}{dt^{p-i}} + \sum_{i=1}^p b_{3,p-i}\frac{d^{p-i}v}{dt^{p-i}} + \sum_{i=1}^p c_{3,p-i}\frac{d^{p-i}w}{dt^{p-i}} = f_3$$

reduce to

$$\frac{d}{dt}\begin{bmatrix} \mathbf{x} \\ \mathbf{y} \\ \mathbf{z} \end{bmatrix} = \begin{bmatrix} \mathbf{A}_1 & \mathbf{B}_1 & \mathbf{C}_1 \\ \mathbf{A}_2 & \mathbf{B}_2 & \mathbf{C}_2 \\ \mathbf{A}_3 & \mathbf{B}_3 & \mathbf{C}_3 \end{bmatrix}\begin{bmatrix} \mathbf{x} \\ \mathbf{y} \\ \mathbf{z} \end{bmatrix} + \begin{bmatrix} \mathbf{b}_1 \\ \mathbf{b}_2 \\ \mathbf{b}_3 \end{bmatrix} \tag{6.10.25}$$

or

$$\frac{d}{dt}\mathbf{w} = \mathbf{T}\mathbf{w} + \mathbf{b},$$

where, this time, $\mathbf{w}$ and $\mathbf{b}$ are $3p$-dimensional vectors and $\mathbf{T}$ is a $3p \times 3p$ matrix. The generalization to an arbitrary number of pth-order differential equations ought to be obvious at this point.

6.11. PERTURBATION THEORY

In physical problems, one frequently encounters a matrix that can be split into a sum $\mathbf{A} + \epsilon\mathbf{B}$, where $\mathbf{A}$ is a matrix whose eigenproblem is solved and $\epsilon\mathbf{B}$ is small in the sense that

$$|\mathbf{x}^\dagger \mathbf{A}\mathbf{x}| \gg \epsilon |\mathbf{x}^\dagger \mathbf{B}\mathbf{x}| \tag{6.11.1}$$

for the vectors $\mathbf{x}$ of interest. When this is the case, the eigenvalue problem

$$(\mathbf{A} + \epsilon\mathbf{B})\mathbf{x}_i = \lambda_i \mathbf{x}_i \tag{6.11.2}$$

can be solved by perturbation theory. Perturbation theory has been especially useful in quantum mechanics, where $\mathbf{A}$ represents the Hamiltonian of some solvable system, $\mathbf{B}$ is some small external electric or magnetic field, and λ_i is the electronic or nuclear energy of the quantum state of interest.

We will restrict the analysis to the case of a distinct eigenvalue λ_i. From the characteristic equation

$$|\mathbf{A} + \epsilon\mathbf{B} - \lambda\mathbf{I}| = 0, \tag{6.11.3}$$

it follows that λ_i is a function of $\epsilon\mathbf{B}$. For small $\epsilon\mathbf{B}$, we expect to be able to expand a given eigenvalue λ in a series in ϵ (ϵ is a dimensionless index parameter that just keeps track of the power of $\mathbf{B}$ contributing to λ_i or $\mathbf{x}_i$). Thus, we expand

$$\lambda_i = \sum_{m=0}^{\infty} \epsilon^m \lambda_i^{(m)} \tag{6.11.4}$$

and

$$\mathbf{x}_i = \sum_{m=0}^{\infty} \epsilon^m \mathbf{x}_i^{(m)}, \tag{6.11.5}$$

where $\lambda^{(0)}$ and $\mathbf{x}^{(0)}$ are the eigenvalue and eigenvector of $\mathbf{A}$ in the absence of $\mathbf{B}$, i.e.,

$$\mathbf{A}\mathbf{x}_i^{(0)} = \lambda_i^{(0)} \mathbf{x}_i^{(0)}. \tag{6.11.6}$$

Inserting the expansions of Eqs. (6.11.4) and (6.11.5) into Eq. (6.11.2) gives

$$\mathbf{A} \sum_{m=0}^{\infty} \epsilon^m \mathbf{x}_i^{(m)} + \mathbf{B} \sum_{m=0}^{\infty} \epsilon^{m+1} \mathbf{x}_i^{(m)} = \sum_{m=0}^{\infty} \sum_{l=0}^{\infty} \epsilon^{m+l} \lambda_i^{(l)} x_i^{(m)} \tag{6.11.7}$$

or

$$\begin{aligned}\mathbf{A}\mathbf{x}_i^{(0)} + \epsilon(\mathbf{A}\mathbf{x}_i^{(1)} + \mathbf{B}\mathbf{x}_i^{(0)}) + \epsilon^2(\mathbf{A}\mathbf{x}_0^{(2)} + \mathbf{B}\mathbf{x}_i^{(1)}) + \cdots \\ = \lambda^{(0)}\mathbf{x}_i^{(0)} + \epsilon(\lambda^{(0)}\mathbf{x}_i^{(1)} + \lambda^{(1)}\mathbf{x}_i^{(0)}) \\ + \epsilon^2(\lambda^{(0)}\mathbf{x}_i^2 + \lambda^{(1)}\mathbf{x}_i^{(1)} + \lambda^{(2)}\mathbf{x}_i^{(0)}) + \cdots.\end{aligned} \tag{6.11.8}$$

Equating coefficients of like powers of ϵ on each side of Eq. (6.11.8), we obtain the perturbation sequence

$$\mathbf{A}\mathbf{x}_i^{(0)} = \lambda_i^{(0)}\mathbf{x}_i^{(0)}$$
$$(\mathbf{A} - \lambda^{(0)}\mathbf{I})\mathbf{x}_i^{(1)} = \lambda_i^{(1)}\mathbf{x}_i^{(0)} - \mathbf{B}\mathbf{x}_i^{(0)}$$
$$\vdots \qquad (6.11.9)$$
$$(\mathbf{A} - \lambda^{(0)}\mathbf{I})\mathbf{x}_i^{(l)} = \sum_{k=1}^{l}\lambda_i^{(k)}\mathbf{x}_i^{(l-k)} - \mathbf{B}\mathbf{x}_i^{(l-1)}, \qquad l = 1, 2, \cdots.$$

In quantum mechanics, and in the analysis of x-ray scattering, one is frequently only interested in first-order perturbation analysis, i.e. in the equations

$$\mathbf{A}\mathbf{x}_i^{(0)} = \lambda^{(0)}\mathbf{x}_i^{(0)} \qquad (6.11.10)$$

and

$$(\mathbf{A} - \lambda_i^{(0)}\mathbf{I})\mathbf{x}_i^{(1)} = \lambda_i^{(1)}\mathbf{x}_i^{(0)} - \mathbf{B}\mathbf{x}_i^{(0)}. \qquad (6.11.11)$$

By hypothesis, we assume that the eigenvalue $\lambda_i^{(0)}$ and eigenvector $\mathbf{x}_i^{(0)}$ of $\mathbf{A}$ are known. According to the Fredholm alternative theorem, Eq. (6.11.11) is solvable *if and only if*

$$\mathbf{z}_i^\dagger(\lambda_i^{(1)}\mathbf{x}_i^{(0)} - \mathbf{B}\mathbf{x}_i^{(0)}) = 0 \quad \text{or} \quad \lambda_i^{(1)} = \frac{\mathbf{z}_i^\dagger \mathbf{B}\mathbf{x}_i^{(0)}}{\mathbf{z}_i^\dagger \mathbf{x}_i^{(0)}}, \qquad (6.11.12)$$

where $\mathbf{z}_i$ is the solution to the equation

$$(\mathbf{A}_i^\dagger - \lambda_i^{(0)*}\mathbf{I})\mathbf{z}_i = \mathbf{0}. \qquad (6.11.13)$$

In quantum mechanics, $\mathbf{A}$ is always self-adjoint and so $\mathbf{A}^\dagger = \mathbf{A}$, $\mathbf{z}_i = \mathbf{x}_i^{(0)}$, and

$$\lambda_i^{(1)} = \frac{\mathbf{x}_i^{(0)\dagger}\mathbf{B}\mathbf{x}_i}{\mathbf{x}_i^{(0)\dagger}\mathbf{x}_i^{(0)}}. \qquad (6.11.14)$$

Thus, the first-order shift in the eigenvalue caused by a small perturbation can be computed directly from the perturbation operator $\mathbf{B}$ and the unperturbed eigenvector $\mathbf{x}_i^{(0)}$.

Expanding the unknown vector $\mathbf{x}_i^{(1)}$ in the basis set $\{\mathbf{x}_1^{(0)}, \ldots, \mathbf{x}_n^{(0)}\}$

$$\mathbf{x}_i^{(1)} = \sum_{j=1}^{n}\alpha_{ij}\mathbf{x}_j^{(0)}, \qquad (6.11.15)$$

we can write Eq. (6.11.11) as

$$\sum_j(\lambda_j^{(0)} - \lambda_i^{(0)})\alpha_{ij}\mathbf{x}_j = \lambda_i^{(1)}\mathbf{x}_i^{(0)} - \mathbf{B}\mathbf{x}_i^{(0)}. \qquad (6.11.16)$$

Multiplication of Eq. (6.11.16) by the reciprocal vector $\mathbf{z}_k$, where $\mathbf{z}_k^\dagger \mathbf{x}_j^{(0)} = \delta_{kj}$, leads to the result

$$(\lambda_k^{(0)} - \lambda_i^{(0)})\alpha_{ik} = \lambda_i^{(1)}\delta_{ik} - \mathbf{z}_k^\dagger \mathbf{B} \mathbf{x}_i^{(0)} \qquad (6.11.17)$$

or

$$\alpha_{ik} = \frac{-\mathbf{z}_k \mathbf{B} \mathbf{x}_i^{(0)}}{\lambda_k^{(0)} - \lambda_i^{(0)}} \qquad \text{for } k \neq i. \qquad (6.11.18)$$

For $k = i$, the left-hand side of Eq. (6.11.17) vanishes because $\lambda_i^{(0)} - \lambda_i^{(0)} = 0$ and the right-hand side vanishes owing to the solvability condition, $\lambda_i^{(1)} = \mathbf{z}_i^\dagger \mathbf{B} \mathbf{x}_i^{(0)}$ (where the eigenvectors of $\mathbf{A}^\dagger$ have been biorthonormalized with the eigenvectors $\{\mathbf{x}_i^{(0)}\}$). α_{ii} can be set to 0 without loss of generality and the first-order perturbation solution to the eigenproblem becomes

$$\lambda_i \approx \lambda_i^{(0)} + \mathbf{z}_i^\dagger \epsilon \mathbf{B} \mathbf{x}_i$$

$$\mathbf{x}_i \approx \mathbf{x}_i^{(0)} - \sum_{j \neq i} \frac{\mathbf{z}_j^\dagger \epsilon \mathbf{B} \mathbf{x}_i^{(0)}}{\lambda_j - \lambda_i} \mathbf{x}_j^{(0)}. \qquad (6.11.19)$$

■ **EXERCISE 6.11.1.** Develop the first-order perturbation theory for the case of degenerate eigenvalues, i.e., when more than one eigenvector has the same eigenvalue λ_i. The results of the first-order theory are

$$\lambda_{i_k} = \lambda_i + \epsilon \lambda_{i_k}^{(1)}$$

$$\mathbf{y}_{i_k} = \sum_{l=1}^r \beta_{i_l}^{(k)} \mathbf{x}_{i_l}^{(0)} + \epsilon \sum_{j \neq i_1, \ldots, i_r} \frac{[\mathbf{z}_{i_j}(\lambda_{i_k}^{(1)}\mathbf{I} - \mathbf{B}) \sum_{l=1}^r \beta_{i_l}^{(k)} \mathbf{x}_{i_l}^{(0)}] \mathbf{x}_j^{(0)}}{\lambda_j^{(0)} - \lambda_i^{(0)}} \qquad (6.11.20)$$

for $k = 1, \ldots, r$. Here the $\mathbf{x}_{i_1}^{(0)}, \ldots, \mathbf{x}_{i_r}^{(0)}$ are the r eigenvectors of $\mathbf{A}$ corresponding to the eigenvalue $\lambda_i^{(0)}$ and the $\mathbf{z}_{i_1}, \ldots, \mathbf{z}_{i_r}$ are the r eigenvectors of $\mathbf{A}^\dagger$ corresponding to the eigenvalue $\lambda_i^{(0)*}$. $\beta_{i_1}^{(k)}, \ldots, \beta_{i_r}^{(k)}$ and $\lambda_{i_k}^{(1)}$ are the eigenvectors and eigenvalues of

$$\mathbf{C}\beta = \lambda \beta, \qquad c_{kl} = \mathbf{z}_{i_k}^\dagger \mathbf{B} \mathbf{x}_{i_l}, \qquad k, l = 1, \ldots, r. \qquad (6.11.21)$$

Although the unperturbed eigenvectors $\mathbf{x}_{i_1}^{(0)}, \ldots, \mathbf{x}_{i_r}^{(0)}$ have the same eigenvalue, the perturbed eigenvectors $\mathbf{y}_{i_1}, \ldots, \mathbf{y}_{i_r}$ have different eigenvalues $\lambda_{i_1}, \ldots, \lambda_{i_r}$. Thus, the effect of the perturbation is to split or separate the degenerate eigenvalues. In quantum mechanics, it is well known that degenerate energy levels (different quantum states corresponding to the same energy level) can be split by imposing a small electric or magnetic field on the system. ■ ■ ■

PROBLEMS

1. Find the eigenvalues and eigenvectors of

$$\mathbf{U} = \frac{1}{2}\begin{bmatrix} 1+i & -1+i \\ 1+i & 1-i \end{bmatrix}$$

and verify that **U** is unitary. Verify that the eigenvalues are of modulus 1 (i.e., of unit magnitude).

2. (a) Form an orthonormal set x_1, x_2, x_3 from

$$\mathbf{y}_1 = \begin{bmatrix} 1 \\ 1 \\ 1 \\ -1 \end{bmatrix}, \quad \mathbf{y}_2 = \begin{bmatrix} 2 \\ -1 \\ -1 \\ 1 \end{bmatrix}, \quad \mathbf{y}_3 = \begin{bmatrix} -1 \\ 2 \\ 2 \\ 1 \end{bmatrix}.$$

(b) Find the orthonormal vector x_4 such that x_1, x_2, x_3, and x_4 form an orthonormal basis set.

(c) Find the reciprocal vectors of the set y_1, y_2, and y_3.

3. Find an orthonormal set of eigenvectors for

$$\mathbf{A} = \begin{bmatrix} 7 & -16 & -8 \\ -16 & 7 & 8 \\ -8 & 8 & -5 \end{bmatrix}.$$

Solve this problem manually, not with a computer. Show your work.

4. Suppose $\mathbf{A} = \mathbf{X}\mathbf{\Lambda}\mathbf{X}^{-1}$, where $\mathbf{\Lambda}$ is a diagonal matrix.

(a) Find the matrix **Y** that diagonalizes the matrix

$$\begin{bmatrix} \mathbf{0} & \mathbf{A} \\ \mathbf{A} & \mathbf{0} \end{bmatrix}$$

in a similar transformation.

(b) Show that the same matrix diagonalizes

$$\begin{bmatrix} f(\mathbf{A}) & g(\mathbf{A}) \\ g(\mathbf{A}) & f(\mathbf{A}) \end{bmatrix},$$

where $f(t)$ and $g(t)$ are expressible as Taylor series in t.

5. Show that the $n \times n$ Hilbert matrix

$$\mathbf{A} = [a_{ij}], \quad a_{ij} = \frac{1}{i+j-1}$$

is positive definite for $n = 10$. What about the general case?

6. Suppose

$$\mathbf{A}\frac{d^2\mathbf{x}}{dt^2} = -\mathbf{B}\mathbf{x},$$

where **A** and **B** are positive-definite matrices independent of t.

(a) Prove that the solutions to this equation are purely oscillatory.

(b) Solve the equation for

$$A = \begin{bmatrix} 2 & 1 \\ 1 & 2 \end{bmatrix}, \quad B = \begin{bmatrix} 4 & 3 \\ 3 & 4 \end{bmatrix}$$

and

$$x(0) = \begin{bmatrix} 1 \\ -1 \end{bmatrix}, \quad \frac{dx}{dt} = \begin{bmatrix} 1 \\ -1 \end{bmatrix}.$$

7. Let A be a real, symmetric $n \times n$ tridiagonal matrix such that $\det(A_k) \neq 0$, $k = 1, \ldots, n$, where the matrix A_k is formed by striking the last $n - k$ rows and columns of A.

   (a) Derive **LU**-decomposition formulas for A and use these to derive a recursion formula for computing $\det(A_k)$, $k = 1, \ldots, n$.
   (b) Determine the largest n for which A is positive definite if $a_{ii} = 2$, $a_{i,i+1} = 1.01$, $a_{i+1,i} = 1.01$, and $a_{ij} = 0$ otherwise.

8. If U is unitary, prove that, in the $p = 2$ norm,

$$\|UA\| = \|A\|.$$

9. Given

$$A = \begin{bmatrix} 0 & 1 & 0 \\ 1 & 0 & 0 \\ 0 & 0 & 1 \end{bmatrix},$$

find $\exp(tA)$.

10. Show that the $(n + m) \times (n + m)$ matrix

$$\begin{bmatrix} V & O \\ O & U \end{bmatrix}$$

is unitary, where V is an $n \times n$ unit matrix and U is an $m \times m$ unitary matrix.

11. The moments of inertia of a system of particles of masses m_i at positions (x_i, y_i, z_i) are defined by the matrix

$$J = \sum m_i \begin{bmatrix} y_i^2 + z_i^2 & -x_i y_i & -x_i z_i \\ -y_i x_i & x_i^2 + z_i^2 & -y_i z_i \\ -z_i x_i & -z_i y_i & x_i^2 + y_i^2 \end{bmatrix}.$$

Show that

$$J = -\sum m_i R_i^2,$$

where

$$R_i = \begin{bmatrix} 0 & -z_i & y_i \\ z_i & 0 & -x_i \\ -y_i & x_i & 0 \end{bmatrix}.$$

Deduce that **J** is a Cartesian tensor of order 2. If the body is continuous, the sums are replaced by integrals. Prove that the diagonal elements of the inertia tensor for a cube are all equal and that the off-diagonal elements are all 0, irrespective of the orientation of the coordinate axes with respect to the cube.

12. The moment of inertia of a rigid body is a second-rank tensor given by

$$\mathbf{J}_0 = \int_V \rho(\mathbf{r})(\mathbf{r} \cdot \mathbf{rI} - \mathbf{rr}) \, d^3r,$$

where $\rho(\mathbf{r})$ is the local density of the body and the subscript "0" on $\mathbf{J}_0$ indicates that the origin with respect to which the position vector $\mathbf{r}$ is measured is at point O. For a Cartesian coordinate system, $\mathbf{r} = x\hat{\mathbf{i}} + y\hat{\mathbf{j}} + z\hat{\mathbf{k}}$ and the unit matrix $\mathbf{I} = \hat{\mathbf{i}}\hat{\mathbf{i}} + \hat{\mathbf{j}}\hat{\mathbf{j}} + \hat{\mathbf{k}}\hat{\mathbf{k}}$.

Choosing the point O as the center of mass for any given body, the "principal directions" ($\hat{\mathbf{i}}'$, $\hat{\mathbf{j}}'$, and $\hat{\mathbf{k}}'$) are defined as the Cartesian unit vectors such that the moment of inertia is diagonal, i.e.,

$$\mathbf{J}_0 = J_1 \hat{\mathbf{i}}'\hat{\mathbf{i}}' + J_2 \hat{\mathbf{j}}'\hat{\mathbf{j}}' + J_3 \hat{\mathbf{k}}'\hat{\mathbf{k}}',$$

where the quantities J_1, J_2, and J_3 are called the principal moments of inertia.

Solve for the principal directions and principal moments of inertia for the following rigid bodies at constant density $\rho(\mathbf{r}) = \rho$:

(a) Sphere of radius R.
(b) Cylinder of radius R and length L.
(c) Rectangular parallelepiped of lengths L_1, L_2, and L_3.
(d) Spherical dumbbell consisting of two equal spheres of radius R held together by a rod of zero diameter and length L.
(e) Tetrahedral dumbbell consisting of four identical spheres of radius R held together at the center of mass by four equivalent cylinders of zero diameter and length L.

13. Prove that $|\exp(\mathbf{A})| > 1$ if $\mathbf{A}$ is positive definite.
14. Prove that for a perfect matrix $\mathbf{A}$ the determinant of $\exp(t\mathbf{A})$ is equal to $\exp(t \sum_{i=1}^{n} a_{ii})$.
15. $\mathbf{B}$ is positive definite and $\mathbf{C}$ is positive semidefinite ($\mathbf{x}^\dagger \mathbf{C} \mathbf{x} \leq 0$ for all $\mathbf{x} \in E_n$). Define

$$\mathbf{A} = \mathbf{B} + \mathbf{C}$$

and prove that (a) $\mathbf{A}$ is positive definite, (b) $|\mathbf{B}| \leq |\mathbf{A}|$, and (c) $\mathbf{B}^{-1} - \mathbf{A}^{-1}$ is positive semidefinite.

16. If $\mathbf{A}$ is positive definite, prove that

$$|\mathbf{A}| \leq a_{11} a_{22} \cdots a_{nn}.$$

Use this to prove that

$$|\mathbf{BB}^\dagger| \leq \prod_{i=1}^{n} \left(\sum_{j=1}^{n} |b_{ij}|^2 \right),$$

a relationship known as Hadamard's inequality.

17. If $\mathbf{A}$ is self-adjoint, prove that $\mathbf{I} + \epsilon \mathbf{A}$ is positive definite if $\epsilon > 0$ and sufficiently small. What condition must ϵ satisfy for the above to be true?

18. Prove that if $\mathbf{A}$ is anti-Hermitian, such that $\mathbf{A}^\dagger = -\mathbf{A}$, then

    (a) The eigenvalues of $\mathbf{A}$ (λ_i) are either purely imaginary or 0.
    (b) The eigenvectors of $\mathbf{A}$ satisfy $\mathbf{x}_i^\dagger \mathbf{x}_j = 0$ if $\lambda_i \neq \lambda_j$.

19. Prove that if $\mathbf{A}$ is skew-symmetric, such that $\mathbf{A}^\dagger = -i\mathbf{A}$, then

    (a) The eigenvalues of $\mathbf{A}$ (λ_i) are of the form $\alpha_i + i\alpha_i$, where α_i is a real scalar value.
    (b) The eigenvectors of $\mathbf{A}$ satisfy $\mathbf{x}_i^\dagger \mathbf{x}_j = 0$ if $\lambda_i \neq \lambda_j$.

20. Let $\mathbf{A}$ be a real, anti-Hermitian (anti-symmetric) matrix. If $\mathbf{z} = \mathbf{x} + i\mathbf{y}$ is an eigenvector of $\mathbf{A}$ with eigenvalue $i\mu$ (where $\mathbf{x}$, $\mathbf{y}$, and μ are all real and nonzero), then show that $\mathbf{x}^\dagger \mathbf{y} = 0$.

21. Prove that every matrix is uniquely expressible in the form $\mathbf{A} = \mathbf{H} + \mathbf{S}$, where $\mathbf{H}$ is self-adjoint and $\mathbf{S}$ is anti-Hermitian.

22. Consider the matrix $\mathbf{C} = \mathbf{A} + i\mathbf{B}$, where $\mathbf{A}$ and $\mathbf{B}$ are both n-dimensional square, normal matrices. *Note:* $i = \sqrt{-1}$.

    (a) Under what conditions is $\mathbf{C}$ a normal matrix?
    (b) If $\mathbf{C}$ is normal, what can we say about the eigenvalues of the commutator: $[\mathbf{A}, \mathbf{B}^\dagger]$?
    (c) Under what conditions is $\mathbf{C}$ a self-adjoint matrix?
    (d) If $\mathbf{C}$ is self-adjoint, do $\mathbf{A}$ and $\mathbf{B}$ necessarily commute? Explain why or why not.

23. Prove that *any* n-dimensional matrix $\mathbf{C}$ can be written in the form $\mathbf{C} = \mathbf{A} + i\mathbf{B}$, where $\mathbf{A}$ and $\mathbf{B}$ are both $n \times n$ self-adjoint matrices.

24. Consider the 2×2 matrix equation:

    $$\mathbf{A}^2 + \begin{bmatrix} 1 & 0 \\ 0 & 1 \end{bmatrix} = \mathbf{0}.$$

    (a) Find all *possible* eigenvalues of the matrix $\mathbf{A}$.
    (b) Do there exist any *real-valued* matrices $\mathbf{A}$ that satisfy the above equation? Explain why or why not and, if so, give an example.

25. Consider a point mass m moving in a three-dimensional force field whose potential energy is given by

    $$V = V_0 \exp\left(\frac{5x^2 + 5y^2 + 8z^2 - 8yz - 26ya - 8za}{a^2}\right),$$

    where a and V_0 are positive constants and x, y, and z are the Cartesian coordinates of the particle's position vector.

    (a) Show that V has a single minimum point. Find the position of that minimum (x^*, y^*, z^*).
    (b) Find the normal frequencies and modes of vibration about the minimum.

26. For the Gaussian distribution function given by Eq. (6.7.52), derive a formula for the average value of the product **xxxx**. Evaluate the average value of $x_1^2 x_2^2$ when

$$\mathbf{A} = \begin{bmatrix} 4 & -1 & 0 \\ -1 & 4 & -1 \\ 0 & -1 & 4 \end{bmatrix}.$$

27. Prove that the n-dimensional integral:

$$\frac{1}{\sqrt{\pi}^n} \int_{-\infty}^{\infty} \cdots \int_{-\infty}^{\infty} dx_1 \cdots dx_n \exp(-\mathbf{x}^\dagger \mathbf{A} \mathbf{x}) = \frac{1}{\sqrt{a^n - (-b)^n}},$$

where (elements not indicated are zero)

$$\mathbf{A} = \begin{bmatrix} a & & & & b \\ b & a & & & \\ & b & a & & \\ & & & \ddots & \ddots \\ & & & & b & a \end{bmatrix},$$

and a and b are real with $a > b$. *Hint:* Show that **A** is a normal matrix and can be diagonalized by a unitary transformation.

28. Evaluate the expression

$$\frac{\int_{-\infty}^{\infty} \cdots \int_{-\infty}^{\infty} dx_1 \cdots dx_n \exp(-\mathbf{x}^T \mathbf{A} \mathbf{x} + \mathbf{b}^T \mathbf{x})}{\int_{-\infty}^{\infty} \cdots \int_{-\infty}^{\infty} dx_1 \cdots dx_n \exp(-\mathbf{x}^T \mathbf{A} \mathbf{x})},$$

where **A** is a real, positive-definite matrix and **b** is a real vector.

29. Consider the tridiagonal matrix

$$\mathbf{A} = \begin{bmatrix} -(\alpha+\beta) & \alpha & 0 & 0 & \cdots & 0 \\ \beta & -(\alpha+\beta) & \alpha & 0 & \cdots & 0 \\ 0 & \beta & -(\alpha+\beta) & \alpha & \cdots & 0 \\ \vdots & \vdots & \vdots & \vdots & & \vdots \\ 0 & 0 & 0 & 0 & \cdots & -(\alpha+\beta) \end{bmatrix},$$

i.e., $a_{ii} = -(\alpha+\beta)$, $a_{i+1,i} = \beta$, $a_{i,i+1} = \alpha$, and $a_{ij} = 0$ otherwise. By expansion by minors, it can be shown that

$$P_n(\lambda) = -(\alpha + \beta + \lambda) P_{n-1}(\lambda) - \alpha\beta P_{n-2}(\lambda), \tag{1}$$

with $P_0(\lambda) = 1$ and $P_1(\lambda) = -(\alpha + \beta + \lambda)$, where $P_n(\lambda)$ is the characteristic polynomial for the $n \times n$ tridiagonal matrix.

Assume that the solution to Eq. (1) is of the form

$$P_n = \gamma^n$$

and prove that

$$\gamma_\pm = -\frac{-(\alpha+\beta+\lambda) \pm \sqrt{(\alpha+\beta+\lambda)^2 - 4\alpha}}{2}$$

and that the general form of P_n is

$$P_n = c_1 \gamma_+^n + c_2 \gamma_-^n.$$

Evaluate c_1 and c_2 using the special cases $n = 0$ and $n = 1$.
Show that the eigenvalues $\mathbf{A}$ are

$$\lambda_j = -\alpha - \beta - \sqrt{4\alpha\beta} \cos \frac{\pi j}{n+1}, \qquad j = 1, \ldots, n,$$

and that the component x_{ij} of the eigenvector $\mathbf{x}_j$ is

$$x_{ij} = (-\beta)^{i-1} (\alpha\beta)^{(n-i)/2} \frac{\sin[\pi j(n-i+1)/(n+1)]}{\sin[\pi j/(n+1)]}, \qquad i = 1, \ldots, n.$$

Give the eigenvalues and eigenvectors of $\mathbf{A}^\dagger$.

30. Prove the following inequalities for *any* complex matrix $\mathbf{A}$:

(a)
$$\sum_{i=1}^n |\lambda_i|^2 \leq \sum_{i,j=1}^n |a_{ij}|^2.$$

(b)
$$\sum_{i=1}^n |\text{Re}(\lambda_i)|^2 \leq \sum_{i,j=1}^n \left| \frac{a_{ij} + a_{ji}^*}{2} \right|^2.$$

(c)
$$\sum_{i=1}^n |\text{Im}(\lambda_i)|^2 \leq \sum_{i,j=1}^n \left| \frac{a_{ij} - a_{ji}^*}{2} \right|^2.$$

31. Prove the Hölder inequality

$$\sum_{i=1}^n x_i y_i \leq \left(\sum_{i=1}^n x_i^p \right)^{1/p} \left(\sum_{i=1}^n y_i^q \right)^{1/q},$$

where x_i and y_i are real, $p > 1$, and $q = p/(p-1)$.

32. Prove the determinant inequality

$$|\lambda \mathbf{A} + (1-\lambda)\mathbf{B}| \geq |\mathbf{A}|^\lambda |\mathbf{B}|^{1-\lambda}$$

for $0 \leq \lambda \leq 1$, where **A** and **B** are $n \times n$ positive-definite matrices. *Hint:* Note that

$$\int_{-\infty}^{\infty} \cdots \int_{-\infty}^{\infty} \exp(-\lambda \mathbf{x}^\dagger \mathbf{A} \mathbf{x} - (1-\lambda)\mathbf{x}^\dagger \mathbf{B} \mathbf{x}) \, dx_1 \cdots dx_n$$
$$= \frac{\pi^{n/2}}{|\lambda \mathbf{A} + (1-\lambda)\mathbf{B}|^{1/2}}$$

and use the Hölder inequality on the left-hand side of the equation.

33. Use the results of Problem 31 to prove

$$\int_{\Omega_n} f(\vec{r}) g(\vec{r}) \, d^n r \leq \left(\int_{\Omega_n} f(\vec{r})^p \, d^n r \right)^{1/p} \left(\int_{\Omega_n} g(\vec{r})^q \, d^n r \right)^{1/q},$$

where f and g are real functions, $p > 1$, and $q = p/(p-1)$.

34. Show that if **A** is positive definite and **B** is self-adjoint, then

$$\int_{-\infty}^{\infty} \cdots \int_{-\infty}^{\infty} \exp(-\mathbf{x}^\dagger \mathbf{A} \mathbf{x} - i \mathbf{x}^\dagger \mathbf{B} \mathbf{x}) \, dx_1 \cdots dx_n = \frac{\pi^{n/2}}{|\mathbf{A} + i\mathbf{B}|^{n/2}}.$$

35. In the Fig. 6.P.35, a sequence of rotations is outlined by means of which a set of axes (x, y, z) is transferred to a set (x^*, y^*, z^*) by three two-dimensional rotations involving angles ψ, θ, ϕ, which are called the

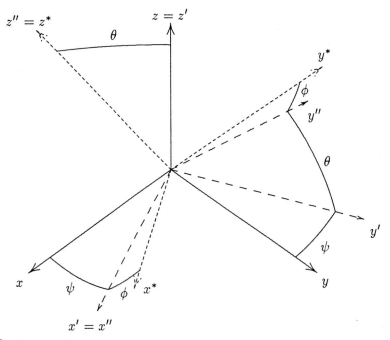

FIGURE 6.P.35

Eulerian angles. Show that

$$\begin{bmatrix} x^* \\ y^* \\ z^* \end{bmatrix} = \begin{bmatrix} \cos\phi & \sin\phi & 0 \\ -\sin\phi & \cos\phi & 0 \\ 0 & 0 & 1 \end{bmatrix} \begin{bmatrix} 1 & 0 & 0 \\ 0 & \cos\theta & \sin\theta \\ 0 & -\sin\theta & \cos\theta \end{bmatrix}$$

$$\times \begin{bmatrix} \cos\psi & \sin\psi & 0 \\ -\sin\psi & \cos\psi & 0 \\ 0 & 0 & 1 \end{bmatrix} \begin{bmatrix} x \\ y \\ z \end{bmatrix} = \mathbf{A} \begin{bmatrix} x \\ y \\ z \end{bmatrix},$$

where

$$\mathbf{A} = \begin{bmatrix} \cos\phi\cos\psi & \cos\phi\sin\psi & \sin\phi\sin\theta \\ -\sin\phi\cos\theta\sin\psi & +\sin\phi\cos\theta\cos\psi & \\ -\sin\theta\cos\psi & -\sin\phi\sin\psi & \cos\phi\sin\theta \\ -\cos\phi\cos\theta\sin\psi & +\cos\phi\cos\theta\cos\psi & \\ \sin\theta\sin\psi & -\sin\theta\cos\psi, & \cos\theta \end{bmatrix}.$$

Show (without forming $\mathbf{A}^T\mathbf{A}$ or $\mathbf{A}\mathbf{A}^T$) directly that $\mathbf{A}$ is orthogonal. The rotations illustrated in the Fig. 6.P.35 are:

(i) Positive rotation about the z axis by the angle ψ; new axes $x'y'z'$.
(ii) Positive rotation about the x' axis by the angle θ; new axes $x''y''z''$.
(iii) Positive rotation about the z'' axis by the angle ϕ; new axes $x^*y^*z^*$.

36. Prove that half of the eigenvalues of $\mathbf{A}$ are equal to those of $\mathbf{B}$ and the other half are equal to those of $\mathbf{C}$ if

$$\mathbf{A} = \begin{bmatrix} \mathbf{B} & \mathbf{O} \\ \mathbf{O} & \mathbf{C} \end{bmatrix}.$$

What is the relationship among the eigenvectors of $\mathbf{A}$, $\mathbf{B}$, and $\mathbf{C}$?

37. Consider the generalized eigenvalue problem

$$\mathbf{A}\mathbf{x} = \lambda\mathbf{B}\mathbf{x}.$$

Suppose

$$\mathbf{A} = \begin{bmatrix} 4 & -1 & 0 & 0 \\ -1 & 4 & -1 & 0 \\ 0 & -1 & 4 & -1 \\ 0 & 0 & -1 & 4 \end{bmatrix}$$

and

$$\mathbf{B} = \begin{bmatrix} 2 & -\frac{1}{2} & 0 & 0 \\ -\frac{1}{2} & 2 & -\frac{1}{2} & 0 \\ 0 & -\frac{1}{2} & 2 & -\frac{1}{2} \\ 0 & 0 & -\frac{1}{2} & 2 \end{bmatrix}.$$

Find the generalized eigenvalues λ_i and eigenvectors $\mathbf{x}_i$ for these matrices. Give the spectral decomposition of the matrix $\mathbf{B}^{-1}\mathbf{A}$.

38. A chemical reactor process may be described by the following equations:

$$\frac{dx}{dt} = -\frac{1}{\theta}x + \frac{\mu y}{K + y}$$

$$\frac{dy}{dt} = -\frac{1}{\theta}(y_0 - y) - \frac{\alpha \mu y}{K + y},$$

where θ is the holding time; y_0 is the constant-feed concentration of species y; and α, μ, and K are positive rate constants. Concentrations must lie in the ranges $x \geq 0$ and $0 \leq y \leq y_0$.

Find the two steady-state solutions to these equations. Determine the ranges for the quantities θ, y_0, α, μ, and K for which each of the steady-state solutions is stable to small perturbations. For a microbial population for which y_0 is nutrient and $\mu = 1/\text{h}$, $K = 0.2$ g/L, and $\alpha = 2$, find the holding time θ for which a feed of $y_0 = 0.3$ g/L will produce at steady state a concentration $x = 0.1$ g/L of microorganisms.

39. Evaluate the integral

$$I_{ij}^{2x,\, 2l} = \int_0^\infty \cdots \int_0^\infty \exp(-\mathbf{x}^T\mathbf{A}\mathbf{x}) x_i^{2r} x_j^{2l}\, dx_1 \cdots dx_n$$

in terms of appropriate derivatives of the determinant of $\mathbf{A}$ for r, l zero or positive integers. Assume $\mathbf{A}$ is an $n \times n$, positive-definite, real matrix.

40. If $\mathbf{A}$ and $\mathbf{B}$ are $n \times n$, real, positive-definite matrices, evaluate

$$I_n = \int_0^\infty \cdots \int_0^\infty \exp(-\mathbf{x}^T\mathbf{A}\mathbf{B}\mathbf{x})\, dx_1 \cdots dx_n.$$

Is $\mathbf{AB}$ positive definite? Prove your answer. What is the value of I_n for the special case

$$\mathbf{A} = \begin{bmatrix} 2 & 1 \\ 1 & 2 \end{bmatrix}, \quad \mathbf{B} = \begin{bmatrix} 1 & 1 \\ 2 & 3 \end{bmatrix}?$$

PROBLEMS

41. Consider two simultaneous reactions involving n components with stoichiometric equations

$$\sum_{i=1}^{n} \nu_{\alpha j} c_j = 0, \qquad \alpha = 1, 2,$$

where $\nu_{\alpha j}$ represents the stoichiometric coefficient of components j in the αth reaction. The rates of transformation of components are given by

$$\frac{dc_i}{dt} = \sum_{\alpha=1}^{2} \nu_{\alpha j} \left(k_{\alpha r} \prod_{j=1}^{n} c_j^{\beta_{\alpha j}} - k_{\alpha f} \prod_{j=1}^{n} c_j^{\gamma_{\alpha j}} \right), \qquad i = 1, \ldots, a,$$

where $k_{\alpha f}$ and $k_{\alpha r}$ are the forward and reverse reaction rate constants of reaction α and $\beta_{\alpha j}$ and $\gamma_{\alpha j}$ are constants. At equilibrium, $dc_1/dt = 0$ and the compositions $c_i = c_i^0$ can be determined from the above equations. Derive the conditions under which the equilibrium will be stable to small fluctuations.

42. (a) Consider the symmetric matrix

$$\mathbf{A} = \begin{bmatrix} 1 & 2 \\ 2 & 1 \end{bmatrix}.$$

Find the solution to the problem

$$\frac{d^2 \mathbf{x}}{dt^2} = \mathbf{A} \mathbf{x}, \qquad \mathbf{x}(0) = \begin{bmatrix} 1 \\ 2 \end{bmatrix}, \qquad \left. \frac{d\mathbf{x}}{dt} \right|_{t=0} = \begin{bmatrix} 0 \\ 0 \end{bmatrix}.$$

(b) Give the general form of the eigenvector solution to the equation

$$\frac{d^2 \mathbf{x}}{dt^2} = \mathbf{A} \mathbf{x}, \qquad \mathbf{x}(0) = \mathbf{x}_0, \qquad \left. \frac{d\mathbf{x}}{dt} \right|_{t=0} = \dot{\mathbf{x}}_0,$$

where $\mathbf{A}$ is an $n \times n$ self-adjoint matrix.

43. A chemical reactor process is described by the following equations:

$$V \frac{dc_A}{dt} = q c_{A_c} - q c_A - k_0 \exp\left(\frac{-\epsilon}{RT}\right) c_A$$

$$V c_p \rho \frac{dT}{dt} = q c_p \rho (T_f - T) - k_0 \exp\left(\frac{-\epsilon}{RT}\right) c_A V (\Delta H) - U(T - T_c)$$

Data:

$\epsilon/R = 10^4$ K $\qquad (-\Delta H/c_p \rho) = 200$ K L/gmol

$(U/V c_p \rho) = 1$ min^{-1} $\qquad q/V = 1$ min^{-1}

$k_0 = e^{25}$ min^{-1} $\qquad c_{A_0} = 1$ gmol/L

$T_f = T_c = 350$ K.

(a) Verify that the following are steady-state solutions of the problem:

$$T_s = 354 \text{ K} \qquad c_{A_s} = 0.964 \text{ gmol/L}$$
$$T_s = 400 \text{ K} \qquad c_{A_s} = 0.5 \text{ gmol/L}$$
$$T_s = 441 \text{ K} \qquad c_{A_s} = 0.0885 \text{ gmol/L}.$$

(b) Check for the stability of each of the three steady states.

(c) While the reactor is operating at the steady state $T_s = 441$ K and $c_{A_s} = 0.0885$ gmol/L, a small perturbation brings the system to a new state $(T_i = c_{A_i})$. Find T and c_A as a function of time after that small disturbance.

44. Two masses are suspended by springs as shown in Fig. 6.P.44. We can neglect gravity. If x_i is the displacement of mass i from its motionless or static position, the motion of the masses is governed by the equations

$$m_1 \frac{d^2 x_1}{dt^2} = -k_1 x_1 + k_2 (x_2 - x_1)$$
$$m_2 \frac{d^2 x_2}{dt^2} = -k_2 (x_2 - x_1).$$

(a) Let $m_1 = m_2 = m$. Show that the independent vibrational motions (independent modes) of this system can be deduced from the eigenproblem

$$\mathbf{Ax} = \lambda \mathbf{x},$$

where

$$\mathbf{A} = \frac{1}{m} \begin{bmatrix} k_1 + k_2 & -k_2 \\ -k_2 & k_2 \end{bmatrix}.$$

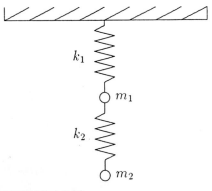

FIGURE 6.P.44

(b) Let $k_1 = \frac{3}{2}k_2$ and $k_1/m = 1$. Suppose that, initially, $x_1 = -0.5$ and $x_2 = 0.5$. Plot $x_1(t)$ and $x_2(t)$ versus time t. Also plot the positions $\zeta_1(t)$ and $\zeta_2(t)$ of the independent modes versus time t.

45. Consider the two-mass system shown in Fig. 6.P.45. This is a simplified schematic of a vibratory feeder for moving particulate material. The top mass ("feeder trough") is angled with respect to the floor and is designed to convey material through high-frequency oscillations induced by the lower mass ("exciter"). The set of springs fixing the feeder trough to the floor (isolation springs) are assumed to have a combined effective spring constant of K_1, and the springs attaching the exciter to the trough (reactor springs) have an effective spring constant of K_R. The feeder trough and exciter have masses of M_1 and M_2, respectively, and the heights of the centers of mass are labeled as a_1 and a_2 (as shown in the figure) for the case when there is zero tension on both sets of springs.

    (a) By defining the displacement of M_1 and M_2 from a_1 and a_2 by the vector
    $$\mathbf{x} = \begin{bmatrix} z_1 - a_1 \\ z_2 - a_2 \end{bmatrix},$$
    show that the equations of motion for the two masses can be written in the form
    $$\frac{d^2\mathbf{x}}{dt^2} = \mathbf{Ax} + \mathbf{b}. \tag{1}$$
    Express $\mathbf{A}$ and $\mathbf{b}$ in terms of the system masses, spring constants, and g (the acceleration due to gravity). *Note:* In this case, gravity cannot be neglected.

    (b) Transform the above coupled system in Eq. (1) to a decoupled system by introducing the transformation $\boldsymbol{\xi} = \mathbf{X}^{-1}\mathbf{x}$, where $\mathbf{X}^{-1}\mathbf{AX} = \mathbf{\Lambda}$.

    (c) Prove that the natural frequencies (eigenvalues of $-\mathbf{A}$) are real and positive for all real, positive values of M_1, M_2, K_1, and K_R.

    (d) Find the steady-state solution $\mathbf{x}^{(s)}$ to Eq. (1). What does this state represent physically?

    (e) Under normal operations, a motor-driven forcing function is applied to the exciter (lower mass) of the form $F(t) = \gamma \sin \omega t$ in the z direction. By defining a new displacement vector $\mathbf{u} = \mathbf{x} - \mathbf{x}^{(s)}$, solve for

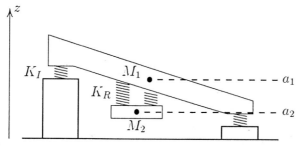

FIGURE 6.P.45

displacement of each mass, assumed to be initially at rest, as a function of time. What operation frequencies ω lead to instabilities in this system?

(f) In actual operation, oscillations are damped by air resistance and through the dynamics of the particulate material being transported. As an approximation, assume each vibrational mode is individually damped by a force factor of the form $-\kappa_i\, d\xi_i/dt$, $i = 1, 2$, in the decoupled system. Solve for the total stroke (defined as twice the amplitude) of the feeder trough in the large time limit (i.e., as $t \to \infty$). What effect does damping have on the natural frequencies and the trough stroke?

46. A damped pendulum in small oscillations obeys the equation

$$\frac{d^2\theta}{dt^2} + k_1 \frac{d\theta}{dt} + k_2 \theta = 0, \qquad \theta(t=0) = \alpha, \qquad \frac{d\theta}{dt}(t=0) = \beta,$$

where θ is the angle of displacement of the pendulum and k_1 and k_2 are positive constants. Setting $x_1 = \theta$ and $x_2 = dx_1/dt$, we can convert this to the system

$$\frac{d\mathbf{x}}{dt} = \mathbf{A}\mathbf{x}, \qquad \mathbf{A} = \begin{bmatrix} 0 & 1 \\ -k_2 & -k_1 \end{bmatrix},$$

where

$$\mathbf{x}(t=0) = \mathbf{x}_0 = \begin{bmatrix} \alpha \\ \beta \end{bmatrix}.$$

From the spectrum of $\mathbf{A}$, determine the relationship between k_1 and k_2 such that the motion is (a) periodic, (b) damped periodic, and (c) overdamped.

47. Consider the augmented matrix

$$[\mathbf{A}, \mathbf{b}] = \begin{bmatrix} 1 & 1 & -8 & -14 \\ 3 & -4 & -3 & 0 \\ 2 & -1 & -7 & -10 \end{bmatrix}.$$

(a) Use simple Gauss elimination to find the rank of $[\mathbf{A}, \mathbf{b}]$. What is the rank of $\mathbf{A}$?

(b) How many of the column vectors of $[\mathbf{A}, \mathbf{b}]$ are linearly independent? Why?

(c) If the problem

$$\mathbf{A}\mathbf{x} = \mathbf{b} \qquad (1)$$

has a solution, find the most general one. If it does not have a solution, why not?

(d) Evaluate the traces, $\operatorname{tr}_j \mathbf{A}$, $j = 1, 2, 3$, of the matrix $\mathbf{A}$.

(e) Find the eigenvalues of $\mathbf{A}$ and $\mathbf{A}^\dagger$.

(f) Is $\mathbf{A}$ perfect? Why? Can $\mathbf{A}$ be diagonalized by a similarity transformation? Why?

(g) Is there a solution to the homogeneous problem

$$\mathbf{A}^\dagger \mathbf{z} = \mathbf{0}?$$

What are the implications of the answer to this question to the solvability of Eq. (1)?

48. Consider the following predator–prey model for big fish with population x_p (predators) and little fish with population x_n (prey):

$$\frac{dx_n}{dt} = r\left(1 - \frac{x_n}{\kappa}\right)x_n - \alpha \frac{x_n^2}{\beta^2 + x_n^2} x_p,$$

$$\frac{dx_p}{dt} = k\left(1 - h\frac{x_p}{x_n}\right)x_p.$$

(a) Using the Newton–Raphson method, find the nontrivial steady-state populations $x_n^{(s)}$ and $x_p^{(s)}$ of little fish and big fish for the case where $\kappa = 1000, r = 100, \alpha = 50, \beta = 12, k = 10$, and $h = 2$.
(b) At steady state, would the population be stable to small perturbations, i.e., if at t_0 the population densities were perturbed to $x_n = x_n^{(s)} + \epsilon_n$ and $x_p = x_p^{(s)} + \epsilon_p$, ϵ_n and ϵ_p quite small, would the population return to steady state in time? Why?

49. The generalized eigenvalue problem, with two $n \times n$ matrices $\mathbf{A}$ and $\mathbf{B}$, is defined by the equation

$$\mathbf{A}\mathbf{x} = \lambda \mathbf{B} \mathbf{x}.$$

(a) Assume that the determinants of $\mathbf{A}$ and $\mathbf{B}$ are not 0 and prove that

$$\det(\mathbf{A}) = \det(\mathbf{B}) \prod_{i=1}^{n} \lambda_i.$$

(b) If $\mathbf{A}$ and $\mathbf{B}$ are positive definite, show that λ is positive.
(c) If

$$\mathbf{A} = \begin{bmatrix} 2 & 1 \\ 1 & 2 \end{bmatrix}, \quad \mathbf{B} = \begin{bmatrix} 2 & -1 \\ -1 & 2 \end{bmatrix},$$

find the eigenvalues and the eigenvectors.
(d) Solve the differential equation

$$\mathbf{B}\frac{d\mathbf{x}}{dt} = \mathbf{A}\mathbf{x}, \quad \mathbf{x}(t=0) = \begin{bmatrix} 1 \\ 0 \end{bmatrix},$$

where $\mathbf{A}$ and $\mathbf{B}$ are given in part (c).

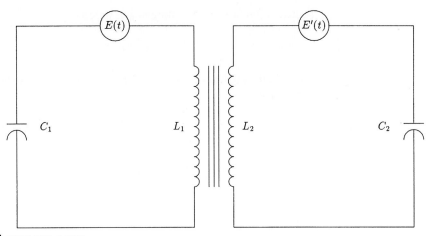

FIGURE 6.P.50

50. For the electric circuit shown in Fig. 6.P.50, the currents I_1 and I_2 obey the equations

$$L_1 \frac{d^2 I_1}{dt^2} + M \frac{d^2 I_2}{dt^2} + \frac{I_1}{C_1} = E_0 \sin \omega t$$

$$M \frac{d^2 I_1}{dt^2} + L_2 \frac{d^2 I_2}{dt^2} + \frac{I_2}{C_2} = E_0' \cos \nu t \tag{1}$$

where L_i and M are inductances, C_i are capacitances, E_0, E_0' are electric fields, and ω and ν are frequencies. Show that Eq. (1) can be expressed in the form

$$\frac{d\mathbf{x}}{dt} = \mathbf{A}\mathbf{x} + \mathbf{b},$$

where $\mathbf{x}$ and $\mathbf{b}$ are four-dimensional vectors and $\mathbf{A}$ is a 4×4 matrix.

Under what conditions will the currents I_1 and I_2 be purely oscillational in the case $E_0 = E_0' = 0$? Sketch the current versus time for the case $E_0 \neq 0$, $E_0' \neq 0$ under the same conditions.

51. A system whose transient behavior is described by

$$\frac{d\mathbf{x}}{dt} = \mathbf{A}\mathbf{x}$$

may be unstable. A forcing function $\mathbf{f}$ can be imposed on the system to yield

$$\frac{d\mathbf{x}}{dt} = -\mathbf{A}\mathbf{x} + \mathbf{f}.$$

In terms of eigenvalue analysis, describe how a "proportional" forcing function,

$$\mathbf{f} = \mathbf{K}_1 \mathbf{x},$$

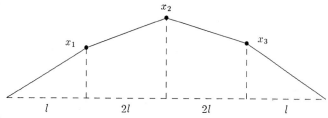

FIGURE 6.P.52

or a differential forcing function,

$$\mathbf{f} = \mathbf{K}_2 \frac{d\mathbf{x}}{dt},$$

can be used to stabilize the system. Can stability always be obtained if $\mathbf{K}_1$ and $\mathbf{K}_2$ are suitable constants times the unit matrix?

52. Consider the three beads on the string shown in Fig. 6.P.52. The string is under tension T, its total length is $6l$, and its mass is negligible. The mass of each bead is m and the ratio

$$\frac{ml}{T} = 1 \text{ s}^2.$$

For small displacements, the equations of motion for the beads are

$$m\frac{d^2 x_1}{dt^2} = -\frac{T}{l}x_1 - \frac{T}{2l}(x_2 - x_1)$$

$$m\frac{d^2 x_2}{dt^2} = -\frac{T}{2l}(x_2 - x_1) + \frac{T}{2l}(x_3 - x_2)$$

$$m\frac{d^2 x_3}{dt^2} = -\frac{T}{2l}(x_3 - x_2) - \frac{T}{l}x_3.$$

(a) Transform these equations into the set

$$\frac{d^2 y_i}{dt^2} = \omega_i^2 y_i, \quad i = 1, 2, 3,$$

by a linear transformation of the x_i. Give the relationship between x_i and y_i and the values of ω_i.

(b) If, initially, $x_1 = 0.02l$, $x_2 = 0.04l$, $x_3 = 0.02l$, and $dx_1/dt = dx_2/dt = dx_3/dt = 0$, find x_i as functions of time.

(c) Suppose each bead is subjected to an external force equal to $I \sin \omega t$ at $t = 0$. Solve this problem l with the same initial conditions. What happens if $\omega = \omega_i$?

53. Is the following matrix positive definite; i.e., is

$$Q = \mathbf{x}^\dagger \mathbf{A} \mathbf{x} > 0$$

for all real and complex vectors $\mathbf{x} \neq 0$?

$$\mathbf{A} = \begin{bmatrix} 1 & 2 & 1 \\ 1 & 3 & 2 \\ 2 & 1 & 4 \end{bmatrix}.$$

Is $Q = \mathbf{x}^T \mathbf{A} \mathbf{x} > 0$ for all real vectors $\mathbf{x} \neq \mathbf{0}$? Is $\mathbf{A}_s$ positive definite, where

$$\mathbf{A}_s = \frac{1}{2}[\mathbf{A} + \mathbf{A}^T]?$$

$\mathbf{A}_s$ is the symmetric part of $\mathbf{A}$.

FURTHER READING

Bellman, R. (1970). "Introduction to Matrix Analysis." McGraw-Hill, New York.
Noble, B. (1969). "Applied Linear Algebra." Prentice Hall International, Englewood Cliffs, NJ.
Noble B., and Daniel, J. W. (1977). "Applied Linear Algebra" Prentice Hall. Englewood Cliffs, NJ.
Parlett, B. N. (1980). "The Symmetric Eigenvalue Problem." Prentice Hall International, Englewood Cliffs, NJ.
Watkins, S. W. (1991). "Fundamentals of Matrix Computations." Wiley, New York.
Wilkinson, J. H. (1965). "The Algebraic Eigenvalue Problem." Clarendon, Oxford.

7
IMPERFECT OR DEFECTIVE MATRICES

7.1. SYNOPSIS

We have found that matrices with nondistinct eigenvalues may or may not be *perfect*. If the number of eigenvectors corresponding to the eigenvalue λ_i is less than the multiplicity p_{λ_i}, then we say that the matrix is *imperfect* or *defective*. Thus, by definition, an $n \times n$ matrix $\mathbf{A}$ is *imperfect* if it has fewer than n eigenvectors. In this chapter we examine the properties of imperfect matrices and, when possible, we will extend the theorems and properties, or present analogous relations, presented in Chapter 6 for perfect matrices.

We will begin by proving that for any square matrix $\mathbf{A}$ there exists a nonsingular matrix $\mathbf{Q}$ that, in a similarity transformation, transforms $\mathbf{A}$ to the Jordan canonical form, i.e., $\mathbf{Q}^{-1}\mathbf{A}\mathbf{Q} = \mathbf{\Lambda}_\mathbf{J}$, where the partitioned matrix $\mathbf{\Lambda}_\mathbf{J}$ has Jordan block matrices $\mathbf{J}_i$ along the main diagonal and 0's elsewhere. The Jordan block $\mathbf{J}$ is a square matrix whose elements are 0 except for those on the main diagonal, which are all equal, and those on the first diagonal above the main diagonal, which are all equal to 1.

We will see that, analogous to perfect matrices, if $f(t)$ can be expressed as a Taylor series in t, then $f(\mathbf{A}) = \mathbf{Q}f(\mathbf{\Lambda}_\mathbf{J})\mathbf{Q}^{-1}$, where $f(\mathbf{A})$ is a block diagonal matrix with the elements $f(\mathbf{J}_i)$ along the main diagonal and 0's elsewhere. The exponential of a $k \times k$ Jordan block is, therefore, of the form

$$\exp(t\mathbf{J}) = \exp(\alpha t)\mathbf{K}(t),$$

where α is the value of a main diagonal element of $\mathbf{J}$ and

$$\mathbf{K}(t) = \begin{bmatrix} 1 & t & \dfrac{t^2}{2!} & \cdots & \dfrac{t^{k-1}}{(k-1)!} \\ 0 & 1 & t & \cdots & \dfrac{t^{k-2}}{(k-2)!} \\ \vdots & \vdots & \vdots & & \vdots \\ 0 & 0 & 0 & \cdots & 1 \end{bmatrix}.$$

The general solution to $d\mathbf{x}/dt = \mathbf{A}\mathbf{x}$, $\mathbf{x}(0) = \mathbf{x}(t=0)$, is, therefore,

$$\mathbf{x} = \mathbf{Q} \begin{bmatrix} \exp(\lambda_1 t)\mathbf{K}_1 & 0 & \cdots & 0 \\ 0 & \exp(\lambda_r t)\mathbf{K}_2 & \cdots & 0 \\ 0 & 0 & \cdots & \exp(\lambda_s t)\mathbf{K}_s \end{bmatrix} \mathbf{Q}^{-1}\mathbf{x}(0),$$

where $\exp(\lambda_i t)\mathbf{K}_1 = \exp(t\mathbf{J}_i)$, $\mathbf{J}_i$ being the ith Jordan block of $\mathbf{A}$.

The column vectors of $\mathbf{Q}$ contain all of the eigenvectors of $\mathbf{A}$. If $\mathbf{A}$ is *imperfect*, then $\mathbf{Q}$ has more column vectors than there are eigenvectors of $\mathbf{A}$. The excess column vectors are called *generalized eigenvectors*. From the Jordan canonical transformation theorem, we can deduce an algorithm for calculating the generalized eigenvectors for any imperfect matrix. We will see that the Fredholm alternative theorem must be invoked in computing the generalized eigenvectors and examples of the calculation of $\mathbf{Q}$ will be given. We will also see an application of an initial value problem for which the companion matrix is imperfect.

Finally, we will show that any square matrix, *perfect* or *imperfect*, can be expressed in *Schmidt normal form* as

$$\mathbf{A} = \sum_{i=1}^{n} \kappa_i \mathbf{y}_i \mathbf{x}_i^{\dagger},$$

where $\{\mathbf{x}_1, \ldots, \mathbf{x}_n\}$ and $\{\mathbf{y}_1, \ldots, \mathbf{y}_n\}$ are orthonormal basis sets and κ_i are positive real numbers. The vectors $\{\mathbf{x}_i\}$ are eigenvectors of $\mathbf{A}^{\dagger}\mathbf{A}$ and $\mathbf{A}\mathbf{A}^{\dagger}$, respectively, and κ_i are the positive square roots of the eigenvalues of $\mathbf{A}^{\dagger}\mathbf{A}$ (or of $\mathbf{A}\mathbf{A}^{\dagger}$ since its eigenvalues are the same as those of $\mathbf{A}^{\dagger}\mathbf{A}$).

7.2. RANK OF THE CHARACTERISTIC MATRIX

The eigenvector or eigenvectors of the matrix $\mathbf{A}$, corresponding to the eigenvalue λ_i, are obtained by solving the characteristic equation

$$(\mathbf{A} - \lambda_i \mathbf{I})\mathbf{x}_i = \mathbf{0}. \tag{7.2.1}$$

According to the general solvability theorem, the number of solutions to Eq. (7.2.1) is $n - r_{\mathbf{A} - \lambda_i \mathbf{I}}$, where n is the number of columns of $\mathbf{A}$ (and thus the dimension of $\mathbf{x}_i$) and $r_{\mathbf{A} - \lambda_i \mathbf{I}}$ is the rank of the characteristic matrix $\mathbf{A} - \lambda_i \mathbf{I}$. In Section 6.4 we showed that if λ_i is a distinct eigenvalue, then Eq. (7.2.1) admits only one linearly independent eigenvector.

The starting point of the proof was the relationship

$$\frac{d^j P_n}{d\lambda^j}(\lambda) = (-1)^j j! \text{tr}_{n-j}(\mathbf{A} - \lambda \mathbf{I}), \qquad (7.2.2)$$

where $P_n(\lambda) = |\mathbf{A} - \lambda \mathbf{I}|$ is the characteristic polynomial. Suppose the multiplicity of the root λ_i is p_i. Then the characteristic polynomial can be factored as

$$P_n(\lambda) = (\lambda_i - \lambda)^{p_i} \prod_{\substack{j \neq i}}^{n-p_i} (\lambda_j - \lambda) \qquad (7.2.3)$$

from which it follows that

$$\left. \frac{d^k}{d\lambda^k} P_n(\lambda) \right|_{\lambda = \lambda_i} = \begin{cases} 0 & \text{if } k < p_i, \\ (-1)^{p_i} p_i! \prod_{\substack{j \neq i}}^{n-p_i} (\lambda_j - \lambda_i) & \text{if } k = p_i. \end{cases} \qquad (7.2.4)$$

From Eqs. (7.2.2) and (7.2.4), it follows that

$$\text{tr}_{n-p_i}(\mathbf{A} - \lambda_i \mathbf{I}) \neq 0, \qquad (7.2.5)$$

which proves that at least one $(n - p_i)$th-order principal minor is not 0. This, in turn, implies that

$$r_{\mathbf{A} - \lambda_i \mathbf{I}} \geq n - p_i. \qquad (7.2.6)$$

If the eigenvalue is distinct, i.e., $p_i = 1$, then Eq. (7.2.6), plus the fact that the rank of $\mathbf{A} - \lambda_i \mathbf{I}$ is less than n, requires that $r_{\mathbf{A} - \lambda_i \mathbf{I}} = n - 1$. If $p_i > 1$, then we only know that

$$n - r_{\mathbf{A} - \lambda_i \mathbf{I}} \leq p_i, \qquad (7.2.7)$$

which means that there might be fewer eigenvectors than p_i. Since $\sum_i p_i = n$, Eq. (7.2.7) implies that

$$\sum_i (n - r_{\mathbf{A} - \lambda_i \mathbf{I}}) \leq n, \qquad (7.2.8)$$

which means that a matrix might be imperfect if any eigenvalue is degenerate.

As pointed out in Chapter 5, having degenerate eigenvalues is a necessary but not a sufficient condition for a matrix to be imperfect. For example, the characteristic polynomial of $\mathbf{I}$ is $P_n(\lambda) = (1 - \lambda)^n$, and so $\lambda_1 = 1$ and $p_1 = n$. Nevertheless, $\mathbf{I}$ is a perfect matrix because any basis set $\{\mathbf{x}_i, \ldots, \mathbf{x}_n\}$ in E_n forms a linearly independent set of eigenvectors of $\mathbf{I}$.

On the other hand, consider the $k \times k$ matrix

$$\mathbf{J} = \begin{bmatrix} \alpha & 1 & 0 & \cdots & 0 & 0 \\ 0 & \alpha & 1 & \cdots & 0 & 0 \\ \vdots & \vdots & \vdots & & \vdots & \vdots \\ 0 & 0 & 0 & \cdots & \alpha & 1 \\ 0 & 0 & 0 & \cdots & 0 & \alpha \end{bmatrix}, \tag{7.2.9}$$

which has α's on the main diagonal, 1's on the diagonal just above the main diagonal, and 0's everywhere else. The characteristic polynomial for this matrix is

$$P_k(\lambda) = (\alpha - \lambda)^k, \tag{7.2.10}$$

and so $\lambda_1 = \alpha$ and $p_1 = k$. However, the rank of $\mathbf{J} - \lambda_1 \mathbf{I}$ is $k - 1$, which means that $\mathbf{J}$ has only one eigenvector, namely,

$$\mathbf{e}_1 = \begin{bmatrix} 1 \\ 0 \\ \vdots \\ 0 \end{bmatrix}. \tag{7.2.11}$$

A matrix of the form of $\mathbf{J}$ is known as a Jordan block matrix. In what follows, it will be shown that Jordan block matrices play a fundamental role in the theory of imperfect matrices.

7.3. JORDAN BLOCK DIAGONAL MATRICES

The $k \times k$ Jordan block matrix $\mathbf{J}$ was defined in Eq. (7.2.9). It is a square matrix whose elements are 0 everywhere except on the main diagonal, which are all equal to α, and those on the first diagonal above the main diagonal, which are all equal to 1. A Jordan block matrix has only one eigenvalue of multiplicity k, and only one eigenvector, namely, the k-dimensional unit vector $\mathbf{e}_1$.

We define a *Jordan block diagonal matrix* as a partitioned matrix of the form

$$\mathbf{\Lambda_J} = \begin{bmatrix} \mathbf{J}_1 & \mathbf{0} & \mathbf{0} & \cdots & \mathbf{0} \\ \mathbf{0} & \mathbf{J}_2 & \mathbf{0} & \cdots & \mathbf{0} \\ \vdots & \vdots & \vdots & & \vdots \\ \mathbf{0} & \mathbf{0} & \mathbf{0} & \cdots & \mathbf{J}_s \end{bmatrix}, \tag{7.3.1}$$

where the matrices $\mathbf{J}_i$ along the main diagonal of the partitioned matrix are Jordan blocks and all the other matrices are 0. An example of a Jordan block diagonal matrix is

$$\Lambda_J = \begin{bmatrix} \lambda_1 & 1 & 0 & 0 & 0 & 0 \\ 0 & \lambda_1 & 1 & 0 & 0 & 0 \\ 0 & 0 & \lambda_1 & 0 & 0 & 0 \\ 0 & 0 & 0 & \lambda_2 & 1 & 0 \\ 0 & 0 & 0 & 0 & \lambda_2 & 0 \\ 0 & 0 & 0 & 0 & 0 & \lambda_1 \end{bmatrix} = \begin{bmatrix} \mathbf{J}_1 & \mathbf{0}_{12} & \mathbf{0}_{13} \\ \mathbf{0}_{21} & \mathbf{J}_2 & \mathbf{0}_{23} \\ \mathbf{0}_{31} & \mathbf{0}_{32} & \mathbf{J}_3 \end{bmatrix}, \qquad (7.3.2)$$

where

$$\mathbf{J}_1 = \begin{bmatrix} \lambda_1 & 1 & 0 \\ 0 & \lambda_1 & 1 \\ 0 & 0 & \lambda_1 \end{bmatrix}, \quad \mathbf{0}_{12} = \begin{bmatrix} 0 & 0 \\ 0 & 0 \\ 0 & 0 \end{bmatrix}, \quad \mathbf{0}_{13} = \begin{bmatrix} 0 \\ 0 \\ 0 \end{bmatrix},$$

$$\mathbf{0}_{21} = \begin{bmatrix} 0 & 0 & 0 \\ 0 & 0 & 0 \end{bmatrix}, \quad \mathbf{J}_2 = \begin{bmatrix} \lambda_2 & 1 \\ 0 & \lambda_2 \end{bmatrix}, \quad \mathbf{0}_{23} = \begin{bmatrix} 0 \\ 0 \end{bmatrix}, \qquad (7.3.3)$$

$$\mathbf{0}_{31} = \begin{bmatrix} 0, 0, 0 \end{bmatrix}, \quad \mathbf{0}_{32} = \begin{bmatrix} 0, 0 \end{bmatrix} \quad \mathbf{J}_3 = \lambda_3.$$

In the above example, there are three Jordan blocks of dimension 3, 2, and 1, respectively. And since $|\Lambda_J - \lambda \mathbf{I}| = (\lambda_1 - \lambda)^3 (\lambda_2 - \lambda)^2 (\lambda_3 - \lambda)$, there is an eigenvalue λ_1 of multiplicity 3, λ_2 of multiplicity 2, and λ_3 of multiplicity 1. The characteristic matrix $\Lambda_J - \lambda_i \mathbf{I}$ has the rank 5 for all three eigenvalues. Thus, the 6×6 matrix has only three eigenvectors. These are

$$\mathbf{x}_1 = \begin{bmatrix} 1 \\ 0 \\ 0 \\ 0 \\ 0 \\ 0 \end{bmatrix}, \quad \mathbf{x}_2 = \begin{bmatrix} 0 \\ 0 \\ 0 \\ 1 \\ 0 \\ 0 \end{bmatrix}, \quad \mathbf{x}_3 = \begin{bmatrix} 0 \\ 0 \\ 0 \\ 0 \\ 0 \\ 1 \end{bmatrix}. \qquad (7.3.4)$$

This result points out an important aspect of a Jordan block diagonal matrix. Let us partition the eigenvectors of Λ_J and the unit matrix $\mathbf{I}$ to conform with the block diagonal form in Eq. (7.3.2); i.e., let us write

$$\mathbf{x} = \begin{bmatrix} \mathbf{x}^{(1)} \\ \mathbf{x}^{(2)} \\ \mathbf{x}^{(3)} \end{bmatrix}, \quad \mathbf{I} = \begin{bmatrix} \mathbf{I}^{(1)} & \mathbf{0} & \mathbf{0} \\ \mathbf{0} & \mathbf{I}^{(2)} & \mathbf{0} \\ \mathbf{0} & \mathbf{0} & \mathbf{I}^{(3)} \end{bmatrix}, \qquad (7.3.5)$$

where $\mathbf{x}^{(1)}$ is a three-dimensional vector, $\mathbf{x}^{(2)}$ a two-dimensional vector, and $\mathbf{x}^{(1)}$ a one-dimensional vector, and $\mathbf{I}^{(1)}$, $\mathbf{I}^{(2)}$, and $\mathbf{I}^{(3)}$ are 3×3, 2×2, and 1×1 unit matrices. Then the eigenproblem given by

$$(\Lambda_J - \lambda \mathbf{I})\mathbf{x} = \mathbf{0} \qquad (7.3.6)$$

now becomes

$$\begin{bmatrix} \mathbf{J}_1 - \lambda \mathbf{I}^{(1)} & \mathbf{0}_{12} & \mathbf{0}_{13} \\ \mathbf{0}_{21} & \mathbf{J}_2 - \lambda \mathbf{I}^{(2)} & \mathbf{0}_{23} \\ \mathbf{0}_{31} & \mathbf{0}_{31} & \mathbf{J}_3 - \lambda \mathbf{I}^{(3)} \end{bmatrix} \begin{bmatrix} \mathbf{x}^{(1)} \\ \mathbf{x}^{(2)} \\ \mathbf{x}^{(3)} \end{bmatrix} = \begin{bmatrix} \mathbf{0}^{(1)} \\ \mathbf{0}^{(2)} \\ \mathbf{0}^{(3)} \end{bmatrix} \qquad (7.3.7)$$

or

$$(\mathbf{J}_i - \lambda \mathbf{I}^{(i)}) \mathbf{x}^{(i)} = \mathbf{0}^{(i)}, \qquad i = 1, 2, 3. \qquad (7.3.8)$$

Three linearly independent solutions to these equations are

$$\lambda = \lambda_1, \quad \mathbf{x}_1^{(1)} = \mathbf{e}_1^{(1)} = \begin{bmatrix} 1 \\ 0 \\ 0 \end{bmatrix}, \quad \mathbf{x}_1^{(2)} = \mathbf{0}^{(2)} = \begin{bmatrix} 0 \\ 0 \end{bmatrix}, \quad \mathbf{x}_1^{(3)} = \mathbf{0}^{(3)} = 0,$$

$$\lambda = \lambda_2, \quad \mathbf{x}_2^{(1)} = \mathbf{0}^{(1)} = \begin{bmatrix} 0 \\ 0 \\ 0 \end{bmatrix}, \quad \mathbf{x}_2^{(2)} = \mathbf{e}_2^{(2)} = \begin{bmatrix} 1 \\ 0 \end{bmatrix}, \quad \mathbf{x}_2^{(3)} = \mathbf{0}^{(3)} = 0, \qquad (7.3.9)$$

$$\lambda = \lambda_3, \quad \mathbf{x}_3^{(1)} = \mathbf{0}^{(1)} = \begin{bmatrix} 0 \\ 0 \\ 0 \end{bmatrix}, \quad \mathbf{x}_3^{(2)} = \mathbf{0}^{(2)} = \begin{bmatrix} 0 \\ 0 \end{bmatrix}, \quad \mathbf{x}_3^{(3)} = \mathbf{e}_1^{(3)} = 1.$$

The three solutions

$$\mathbf{x}_i = \begin{bmatrix} \mathbf{x}_i^{(1)} \\ \mathbf{x}_i^{(2)} \\ \mathbf{x}_i^{(3)} \end{bmatrix}, \qquad i = 1, 2, 3, \qquad (7.3.10)$$

correspond to the three eigenvectors given by Eq. (7.3.4).

If, in the general case Eq. (7.3.1), we partition the eigenvectors as

$$\mathbf{x} = \begin{bmatrix} \mathbf{x}^{(1)} \\ \mathbf{x}^{(2)} \\ \vdots \\ \mathbf{x}^{(s)} \end{bmatrix}, \qquad (7.3.11)$$

we can reduce the eigenproblem

$$(\mathbf{\Lambda}_\mathbf{J} - \lambda \mathbf{I})\mathbf{x} = \mathbf{0} \qquad (7.3.12)$$

to

$$(\mathbf{J}_i - \lambda \mathbf{I}^{(i)})\mathbf{x}^{(i)} = \mathbf{0}^{(i)}, \qquad i = 1, \ldots, s, \qquad (7.3.13)$$

which yields

$$\mathbf{x}_i = \begin{bmatrix} \mathbf{0}^{(1)} \\ \mathbf{0}^{(2)} \\ \vdots \\ \mathbf{0}^{(i-1)} \\ \mathbf{e}_1^{(i)} \\ \mathbf{0}^{(i+1)} \\ \vdots \\ \mathbf{0}^{(s)} \end{bmatrix}, \quad \text{where } \mathbf{\Lambda_J x}_i = \lambda_i \mathbf{x}_i, \ i = 1, \ldots, s. \qquad (7.3.14)$$

What we have just seen is that the eigenvalues of a Jordan block diagonal matrix are the eigenvalues of each of the Jordan blocks and that if the $n \times n$ matrix $\mathbf{\Lambda_J}$ is composed of s Jordan block matrices, then $\mathbf{\Lambda_J}$ has exactly s linearly independent eigenvectors. These eigenvectors are n-dimensional unit vectors $\mathbf{e}_i$ of the form given by Eq. (7.3.14). If we designate the dimension of a Jordan block as k_i, then it follows that $\sum_{i=1}^{s} k_i = n$.

By carrying out the matrix multiplication, it is easy to show from Eq. (7.3.1) that positive powers of $\mathbf{\Lambda_J}$ are given by

$$\mathbf{\Lambda_J}^k = \begin{bmatrix} \mathbf{J}_1^k & 0 & 0 & \cdots & 0 \\ 0 & \mathbf{J}_2^k & 0 & \cdots & 0 \\ \vdots & \vdots & \vdots & & \vdots \\ 0 & 0 & 0 & \cdots & \mathbf{J}_s^k \end{bmatrix}, \qquad (7.3.15)$$

where k is a positive integer. If none of the eigenvalues is 0, then the inverse $\mathbf{\Lambda_J}^{-1}$ exists and is of the form

$$\mathbf{\Lambda_J}^{-1} = \begin{bmatrix} \mathbf{J}_1^{-1} & 0 & \cdots & 0 \\ 0 & \mathbf{J}_2^{-1} & \cdots & 0 \\ \vdots & \vdots & & \vdots \\ 0 & 0 & \cdots & \mathbf{J}_s^{-1} \end{bmatrix}, \qquad (7.3.16)$$

as can be verified by showing that the product $\mathbf{\Lambda_J}^{-1}\mathbf{\Lambda_J}$ equals the unit matrix. Similar to Eq. (7.3.15), the kth power of $\mathbf{\Lambda_J}^{-1}$ is

$$\mathbf{\Lambda_J}^{-k} = \begin{bmatrix} \mathbf{J}_1^{-k} & 0 & \cdots & 0 \\ 0 & \mathbf{J}_2^{-k} & \cdots & 0 \\ \vdots & \vdots & & \vdots \\ 0 & 0 & \cdots & \mathbf{J}_s^{-k} \end{bmatrix}. \qquad (7.3.17)$$

From Eq. (7.3.15) it follows that if $f(t) = \sum_{k=0}^{\infty} f_k t^k$, then

$$f(\mathbf{\Lambda_J}) = \sum_{k=0}^{\infty} f_k \begin{bmatrix} \mathbf{J}_1^k & \mathbf{0} & \cdots & \mathbf{0} \\ \mathbf{0} & \mathbf{J}_2^k & \cdots & \mathbf{0} \\ \vdots & \vdots & & \vdots \\ \mathbf{0} & \mathbf{0} & \cdots & \mathbf{J}_s^k \end{bmatrix} \tag{7.3.18a}$$

or

$$f(\mathbf{\Lambda_J}) = \begin{bmatrix} f(\mathbf{J}_1) & \mathbf{0} & \cdots & \mathbf{0} \\ \mathbf{0} & f(\mathbf{J}_2) & \cdots & \mathbf{0} \\ \vdots & \vdots & & \vdots \\ \mathbf{0} & \mathbf{0} & \cdots & f(\mathbf{J}_s) \end{bmatrix}. \tag{7.3.18b}$$

If $\mathbf{\Lambda_J}$ is not singular, then it follows from Eqs. (7.3.15) and (7.3.17) that if $f(t) = \sum_{k=-\infty}^{\infty} f_k t^k$, the function $f(\mathbf{\Lambda_J})$ is also of the form given by Eq. (7.3.18b).

We saw in the previous chapter that if $\mathbf{A}$ is perfect the solution to the equation $d\mathbf{x}/dt = \mathbf{A}\mathbf{x}$ is an exponential function of time. What if $\mathbf{A}$ is imperfect? The simplest such case is

$$\frac{d\mathbf{x}}{dt} = \mathbf{J}\mathbf{x}, \qquad \mathbf{x}(0) = \mathbf{x}(t=0), \tag{7.3.19}$$

where $\mathbf{J}$ is a $k \times k$ Jordan block. In terms of its components, this vector equation becomes

$$\begin{aligned} \frac{dx_1}{dt} &= \lambda x_1 + x_2 \\ \frac{dx_2}{dt} &= \lambda x_2 + x_3 \\ &\vdots \\ \frac{dx_{k-1}}{dt} &= \lambda x_{k-1} + x_k \\ \frac{dx_k}{dt} &= \lambda x_k. \end{aligned} \tag{7.3.20}$$

The solution to these equations can be obtained by solving first the kth equation, then the $(k-1)$th, etc. From this we obtain

$$x_k = \exp(\lambda t) x_k(0), \tag{7.3.21}$$

then

$$\frac{d}{dt}(\exp(-\lambda t) x_{k-1}) = \exp(-\lambda t) x_k = x_k(0)$$
$$x_{k-1} = \exp(\lambda t) x_{k-1}(0) + \exp(\lambda t) t x_k(0), \tag{7.3.22}$$

and so on until

$$x_{k-j} = \exp(\lambda t)x_{k-j}(0) + \exp(\lambda t)tx_{k-j+1}(0)$$
$$+ \exp(\lambda t)\frac{t^2}{2!}x_{k-j+2}(0) + \cdots + \exp(\lambda t)\frac{t^j}{j!}x_k(0). \quad (7.3.23)$$

In vector form these equations become

$$\mathbf{x} = \exp(\lambda t)\mathbf{K}\mathbf{x}(0), \quad (7.3.24)$$

where

$$\mathbf{K} = \begin{bmatrix} 1 & t & \frac{t^2}{2!} & \cdots & \frac{t^{k-3}}{(k-3)!} & \frac{t^{k-2}}{(k-2)!} & \frac{t^{k-1}}{(k-1)!} \\ 0 & 1 & t & \cdots & \frac{t^{k-4}}{(k-4)!} & \frac{t^{k-3}}{(k-3)!} & \frac{t^{k-2}}{(k-2)!} \\ \vdots & \vdots & \vdots & & \vdots & \vdots & \vdots \\ 0 & 0 & 0 & \cdots & 1 & t & t^2 \\ 0 & 0 & 0 & \cdots & 0 & 1 & t \\ 0 & 0 & 0 & \cdots & 0 & 0 & 1 \end{bmatrix}. \quad (7.3.25)$$

Since the formal solution to Eq. (7.3.19) is

$$\mathbf{x} = \exp(t\mathbf{J})\mathbf{x}(0) \quad (7.3.26)$$

for an arbitrary initial condition $\mathbf{x}(0)$, Eqs. (7.3.24) and (7.3.26) prove that

$$\exp(t\mathbf{J}) = \exp(\lambda t)\mathbf{K}(t). \quad (7.3.27)$$

For a Jordan block, then, the elements of the exponential function $\exp(t\mathbf{J})\mathbf{x}(0)$ are not linear combinations of exponential functions of time, but are rather linear combinations of $\exp(\lambda t)t^j$, $j = 0, \ldots, k-1$. In fact, asymptotically,

$$\|\exp(t\mathbf{J})\| \to \exp(\lambda t)t^{k-1}. \quad (7.3.28)$$

Although the asymptotic form of $\|\exp(t\mathbf{J})\|$ is not a simple exponential function of time, it still follows that whether

$$\mathbf{x} = \exp(t\mathbf{J})\mathbf{x}(0) \quad (7.3.29)$$

tends to 0 or ∞ will depend on whether the real part of the eigenvalue λ is positive or negative.

Since

$$\exp(t\mathbf{\Lambda_J}) = \begin{bmatrix} \exp(t\mathbf{J}_1) & 0 & \cdots & 0 \\ 0 & \exp(t\mathbf{J}_1) & \cdots & 0 \\ \vdots & \vdots & & \vdots \\ 0 & 0 & \cdots & \exp(t\mathbf{J}_s) \end{bmatrix}, \quad (7.3.30)$$

it follows that

$$\exp(t\mathbf{\Lambda_J}) = \begin{bmatrix} \exp(\lambda_1 t)\mathbf{K}_1 & 0 & \cdots & 0 \\ 0 & \exp(\lambda_2 t)\mathbf{K}_2 & \cdots & 0 \\ \vdots & \vdots & & \vdots \\ 0 & 0 & \cdots & \exp(\lambda_s t)\mathbf{K}_s \end{bmatrix}, \quad (7.3.31)$$

and so the solution to

$$\frac{d\mathbf{x}}{dt} = \mathbf{\Lambda_J}\mathbf{x}, \quad \mathbf{x}(0) = \mathbf{x}(t=0), \quad (7.3.32)$$

is

$$\mathbf{x} = \begin{bmatrix} \exp(\lambda_1 t)\mathbf{K}_1 & 0 & \cdots & 0 \\ 0 & \exp(\lambda_2 t)\mathbf{K}_2 & \cdots & 0 \\ \vdots & \vdots & & \vdots \\ 0 & 0 & \cdots & \exp(\lambda_s t)\mathbf{K}_s \end{bmatrix} \mathbf{x}(0). \quad (7.3.33)$$

If all of the Jordan blocks are 1×1 matrices, $\mathbf{\Lambda_J}$ is just a diagonal matrix and the diagonal elements of the above matrix become simple exponentials.

7.4. THE JORDAN CANONICAL FORM

Why was the Jordan block diagonal matrix discussed in such detail in Section 7.3? The answer is that this form, known as the Jordan canonical form, is the closest we can come to diagonalizing an *imperfect* matrix through a similarity transformation. The theorem stating this is:

THEOREM. *If $\mathbf{A}$ is an $n \times n$ square matrix, there exists a nonsingular matrix $\mathbf{Q}$ such that*

$$\mathbf{Q}^{-1}\mathbf{A}\mathbf{Q} = \mathbf{\Lambda_J} = \begin{bmatrix} \mathbf{J}_1 & 0 & \cdots & 0 \\ 0 & \mathbf{J}_2 & \cdots & 0 \\ \vdots & \vdots & & \vdots \\ 0 & 0 & \cdots & \mathbf{J}_s \end{bmatrix}, \quad (7.4.1)$$

where $\mathbf{J}_i$ are Jordan blocks. The same eigenvalue may occur in different blocks but the number of different blocks having the same eigenvalue is equal to the number of independent eigenvectors corresponding to that eigenvalue. The number of distinct eigenvectors of $\mathbf{A}$ equals the number of Jordan blocks in $\mathbf{\Lambda_J}$.

THE JORDAN CANONICAL FORM

The proof of this theorem is rather detailed and will not be reproduced here. Instead, the consequences of the theorem will be explored. In the special case that **A** is *perfect*, the theorem becomes the spectral resolution theorem and all Jordan blocks are 1×1 matrices (i.e., $\mathbf{J}_i = \lambda_i, i = 1, \ldots, n$).

From what we learned in the previous section, we know there are s linearly independent eigenvectors of $\mathbf{\Lambda}_\mathbf{J}$, one for each Jordan block. For a given eigenvalue λ_i, the rank of $\mathbf{\Lambda}_\mathbf{J} - \lambda_i \mathbf{I}$ will determine how many eigenvectors correspond to this eigenvalue. Since the rank of a matrix is not changed when it is multiplied by a nonsingular matrix, the rank of $\mathbf{Q}(\mathbf{\Lambda}_\mathbf{J} - \lambda_i \mathbf{I})\mathbf{Q}^{-1} = \mathbf{A} - \lambda_i \mathbf{I}$ is the same as the rank of $\mathbf{\Lambda}_\mathbf{J} - \lambda_i \mathbf{I}$. This establishes that if $\mathbf{Q}$ transforms $\mathbf{A}$, as shown in Eq. (7.4.1), then the number of distinct eigenvectors of $\mathbf{A}$ equals the number of Jordan blocks. In fact, if $\mathbf{y}_i$ is the eigenvector of $\mathbf{\Lambda}_\mathbf{J}$ arising from the ith Jordan block, i.e., if

$$\mathbf{y}_i = \begin{bmatrix} \mathbf{0}^{(1)} \\ \vdots \\ \mathbf{0}^{(i-1)} \\ \mathbf{e}_1^{(i)} \\ \mathbf{0}^{(i+1)} \\ \vdots \\ \mathbf{0}^{(s)} \end{bmatrix}, \tag{7.4.2}$$

then the eigenproblems,

$$(\mathbf{\Lambda}_\mathbf{J} - \lambda_i \mathbf{I})\mathbf{y}_i = \mathbf{0}, \tag{7.4.3}$$

can be transformed into

$$\mathbf{Q}(\mathbf{\Lambda}_\mathbf{J} - \lambda_i \mathbf{I})\mathbf{Q}^{-1}\mathbf{Q}\mathbf{y}_i = \mathbf{0} \quad \text{or} \quad (\mathbf{A} - \lambda_i \mathbf{I})\mathbf{x}_i = \mathbf{0}, \tag{7.4.4}$$

where

$$\mathbf{x}_i = \mathbf{Q}\mathbf{y}_i. \tag{7.4.5}$$

Thus, if $\mathbf{Q}$ is known, the eigenvectors of $\mathbf{A}$ can be obtained from the eigenvectors of $\mathbf{\Lambda}_\mathbf{J}$ by the simple relationship $\mathbf{x}_i = \mathbf{Q}\mathbf{y}_i$. Since the eigenvectors $\mathbf{y}_i$ are n-dimensional unit vectors $\mathbf{e}_i$, this relationship establishes that the eigenvectors $\mathbf{x}_i$ of $\mathbf{A}$ are some subset of s column vectors of $\mathbf{Q}$. The remaining $n - s$ column vectors of $\mathbf{Q}$ are called *generalized eigenvectors*.

In the next section, we will discuss how to calculate the generalized eigenvectors. First, however, let us consider some of the implications of the theorem. Note that

$$\mathbf{Q}^{-1}\mathbf{A}^k\mathbf{Q} = \mathbf{Q}^{-1}\mathbf{A}\mathbf{Q}\mathbf{Q}^{-1}\mathbf{A}\mathbf{Q} \cdots \mathbf{Q}^{-1}\mathbf{A}\mathbf{Q} = \mathbf{\Lambda}_\mathbf{J}^k, \tag{7.4.6}$$

where k is a positive integer. Thus, if $f(t)$ is a series function of t, $f(t) = \sum_{k=0}^{\infty} f_k t^k$, then

$$\mathbf{Q}^{-1} f(\mathbf{A}) \mathbf{Q} = f(\mathbf{\Lambda_J}) = \begin{bmatrix} f(\mathbf{J}_1) & 0 & \cdots & 0 \\ 0 & f(\mathbf{J}_2) & \cdots & 0 \\ \vdots & \vdots & & \vdots \\ 0 & 0 & \cdots & f(\mathbf{J}_s) \end{bmatrix} \quad (7.4.7)$$

or

$$f(\mathbf{A}) = \mathbf{Q} \begin{bmatrix} f(\mathbf{J}_1) & 0 & \cdots & 0 \\ 0 & f(\mathbf{J}_2) & \cdots & 0 \\ \vdots & \vdots & & \vdots \\ 0 & 0 & \cdots & f(\mathbf{J}_s) \end{bmatrix} \mathbf{Q}^{-1}. \quad (7.4.8)$$

If $\mathbf{A}$ is nonsingular, then Eq. (7.4.8) also holds if $f(t)$ is a Laurent series in t, i.e., $f(t) = \sum_{k=-\infty}^{\infty} f_k t^k$.

The general solution to

$$\frac{d\mathbf{x}}{dt} = -\mathbf{A}\mathbf{x}, \quad \mathbf{x}(0) = \mathbf{x}(t=0), \quad (7.4.9)$$

now becomes

$$\mathbf{x} = \mathbf{Q} \begin{bmatrix} \exp(\lambda_1 t)\mathbf{K}_1 & 0 & \cdots & 0 \\ 0 & \exp(\lambda_2 t)\mathbf{K}_2 & \cdots & 0 \\ \vdots & \vdots & & \vdots \\ 0 & 0 & \cdots & \exp(\lambda_s t)\mathbf{K}_s \end{bmatrix} \mathbf{Q}^{-1}\mathbf{x}(0), \quad (7.4.10)$$

where, according to Eq. (7.3.22),

$$\mathbf{K}_i = \begin{bmatrix} 1 & t & \frac{t^2}{2!} & \cdots & \frac{t^{k_i-2}}{(k_i-2)!} & \frac{t^{k_i-1}}{(k_i-1)!} \\ 0 & 1 & t & \cdots & \frac{t^{k_i-3}}{(k_i-3)!} & \frac{t^{k_i-2}}{(k_i-2)!} \\ \vdots & \vdots & \vdots & & \vdots & \vdots \\ 0 & 0 & 0 & \cdots & 1 & t \\ 0 & 0 & 0 & \cdots & 0 & 1 \end{bmatrix}. \quad (7.4.11)$$

Just as in the case of a *perfect* matrix, if all we want to know is whether $\|\mathbf{x}\| \to 0$ or ∞ as $t \to \infty$, then all we need to know is whether the real parts of the eigenvalues of $\mathbf{A}$ are negative or positive. The solution to Eq. (7.4.9) for an *imperfect* matrix is a linear combination of terms of the form $t^\nu \exp(\lambda_i t)$, where the integers ν depend on the sizes of the Jordan blocks.

■ **ILLUSTRATION 7.4.1** (Damped Oscillator). Consider the case of the damped oscillator depicted schematically in Fig. 7.4.1. This illustrates a problem in control theory involving a second-order system and is related to the way shock absorbers on an automobile work. We denote the position of the mass m by u. The restoring force acting on the mass when it is displaced from its equilibrium position (defined in the absence of gravity) is $-k(u-u_0)$, where k is the spring constant. The force of gravity is $-mg$, where g is the acceleration of gravity. In addition, a viscous force of $-\eta\, du/dt$ is exerted by a dash pot (a plunger being pulled through a viscous fluid). Here, the damping factor η is proportional to the viscosity of the fluid.

The equation of motion for the mass is

$$m\frac{d^2u}{dt^2} = -k(u-u_0) - \eta\frac{du}{dt} - mg. \qquad (7.4.12)$$

With the definitions

$$\theta^2 \equiv \frac{m}{k}, \qquad \zeta \equiv \frac{\eta}{2\sqrt{mk}}, \qquad \psi \equiv \frac{u_0 k}{mg} - 1, \qquad (7.4.13)$$

and

$$y \equiv \frac{uk}{mg}, \qquad \tau \equiv \frac{t}{\theta}, \qquad (7.4.14)$$

the equation of motion becomes in dimensionless form

$$\frac{d^2y}{d\tau^2} + 2\zeta\frac{dy}{d\tau} + y = \psi. \qquad (7.4.15)$$

We convert this to a first-order system of equations by defining

$$x_1 \equiv y$$
$$x_2 \equiv \frac{dx_1}{d\tau} = \frac{dy}{d\tau}, \qquad (7.4.16)$$

to obtain

$$\frac{dx_1}{d\tau} = x_2$$
$$\frac{dx_2}{d\tau} = -x_1 - 2\zeta x_2 + \psi \qquad (7.4.17)$$

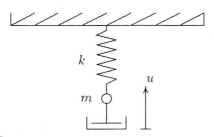

■ **FIGURE 7.4.1** Damped oscillator.

or

$$\frac{d\mathbf{x}}{d\tau} = \mathbf{Kx} + \mathbf{b}, \quad (7.4.18)$$

where

$$\mathbf{x} = \begin{bmatrix} x_1 \\ x_2 \end{bmatrix}, \quad \mathbf{b} = \begin{bmatrix} 0 \\ \psi \end{bmatrix}, \quad \mathbf{K} = \begin{bmatrix} 0 & 1 \\ -1 & -2\zeta \end{bmatrix}. \quad (7.4.19)$$

The eigenvalues of $\mathbf{K}$ are

$$\lambda_1 = -\zeta - \sqrt{\zeta^2 - 1} \quad \text{and} \quad \lambda_2 = -\zeta + \sqrt{\zeta^2 - 1}. \quad (7.4.20)$$

With the above solution, we note that if $\zeta > 1$ both eigenvalues are real and negative. Likewise, if $\zeta < 1$, both eigenvalues are complex, but have negative real parts. If $\zeta = 1$, $\lambda_1 = \lambda_2 = -1$.

The solution to the first-order system is

$$\mathbf{y} = \exp(\mathbf{K}\tau)\mathbf{y}_0, \quad (7.4.21)$$

where

$$\mathbf{y} = \mathbf{x} - \mathbf{x}_p \quad \text{and} \quad \mathbf{y}_0 = \mathbf{x}(0) - \mathbf{x}_p, \quad (7.4.22)$$

and

$$\mathbf{x}_p = -\mathbf{K}^{-1}\mathbf{b}. \quad (7.4.23)$$

(i) Discuss how a disturbance given by $\mathbf{y}(\tau = 0) = 1$ decays in time for the cases $\zeta > 1$, $\zeta < 1$, and $\zeta = 1$ in terms of the properties of $\mathbf{K}$, i.e., when $\mathbf{K}$ is perfect and when it is not.

(ii) Evaluate $\exp(\mathbf{K}\tau)$ for the three cases $\zeta < 1$, $\zeta > 1$, and $\zeta = 1$.

In what follows assume that the datum of the coordinate system is chosen so that $\psi = 0$ (and so $\mathbf{b} = \mathbf{0}$). Also, consider a disturbance such that $y_1 = 1$ and $y_2 = 0$ at $\tau = 0$.

(iii) For $\zeta < 1$ the system is an oscillatory, underdamped system. Give the solution for y_1 versus τ for this case.

(iv) For $\zeta > 1$ the system is a nonoscillatory, overdamped system. Give the solution for y_1 versus τ for this case.

(v) For $\zeta = 1$ the system is critically damped. Give the solution for y_1 versus τ for this case.

(vi) Plot y_1 versus τ for the values $\zeta = 0.2, 0.4, 0.6, 0.8, 1.0, 1.2, 1.4,$ ■ ■ ■ and 1.6.

The Jordan canonical form enables us to prove for an arbitrary square matrix the Hamilton–Cayley theorem:

HAMILTON–CAYLEY THEOREM. *An arbitrary square matrix satisfies its own characteristic equation.*

THE JORDAN CANONICAL FORM

To carry out the proof, we need to define the *nilpotent matrix*:

DEFINITION. *We say that the square matrix* **N** *is nilpotent of index p if* $\mathbf{N}^{p-1} \neq \mathbf{0}$ *and* $\mathbf{N}^p = \mathbf{0}$.

Consider the $k \times k$ matrix

$$\mathbf{H} = \begin{bmatrix} 0 & 1 & 0 & 0 & \cdots & 0 \\ 0 & 0 & 1 & 0 & \cdots & 0 \\ 0 & 0 & 0 & 1 & \cdots & 0 \\ \vdots & \vdots & \vdots & \vdots & & \vdots \\ 0 & 0 & 0 & 0 & \cdots & 0 \end{bmatrix} ; \tag{7.4.24}$$

i.e., all elements are 0 except on the diagonal just above the main diagonal, all of whose elements are unity. Consider what happens when a matrix is left-multiplied by **H**:

$$\mathbf{HA} = \begin{bmatrix} 0 & 1 & 0 & 0 & \cdots & 0 \\ 0 & 0 & 1 & 0 & \cdots & 0 \\ \vdots & \vdots & \vdots & \vdots & & \vdots \\ 0 & 0 & 0 & 0 & \cdots & 1 \\ 0 & 0 & 0 & 0 & \cdots & 0 \end{bmatrix} \begin{bmatrix} a_{11} & a_{12} & \cdots & a_{1k} \\ a_{21} & a_{22} & \cdots & a_{2k} \\ \vdots & \vdots & & \vdots \\ a_{k-1,1} & a_{k-1,2} & \cdots & a_{k-1,k} \\ a_{k1} & a_{k2} & \cdots & a_{kk} \end{bmatrix}$$

$$= \begin{bmatrix} a_{21} & a_{22} & \cdots & a_{2k} \\ a_{31} & a_{32} & \cdots & a_{3k} \\ \vdots & \vdots & & \vdots \\ a_{k1} & a_{k2} & \cdots & a_{kk} \\ 0 & 0 & \cdots & 0 \end{bmatrix}. \tag{7.4.25}$$

Multiplication of **H** by **A** eliminates the top row of **A**, raises each other row up one, and replaces the bottom row by 0's. Thus, the product $\mathbf{H}^2$ will have only 0's in the last two rows and the $(k-2)$th row will be $0, 0, \ldots, 0, 1$. Continuing the process $k-1$ times will place 0's in rows $2, \ldots, k$ and the first row will be $0, 0, \ldots, 0, 1$, i.e.,

$$\mathbf{H}^{k-1} = \begin{bmatrix} 0 & \cdots & 0 & 1 \\ 0 & \cdots & 0 & 0 \\ \vdots & & \vdots & \vdots \\ 0 & \cdots & 0 & 0 \end{bmatrix}. \tag{7.4.26}$$

Finally, multiplication of Eq. (7.4.26) by **H** yields

$$\mathbf{H}^k = \mathbf{0}, \tag{7.4.27}$$

which, since $\mathbf{H}^{k-1} \neq \mathbf{0}$, proves that the nilpotent index of a $k \times k$ matrix of the form of Eq. (7.4.24) is k.

Next, consider the characteristic polynomial of an arbitrary matrix $\mathbf{A}$. In factored form, it can be expressed as

$$P_n(\lambda) = \prod_{i=1}^{s}(\lambda_i - \lambda)^{k_i}, \qquad (7.4.28)$$

where k_i is the dimension of the Jordan block $\mathbf{J}_i$ appearing in the Jordan canonical form $\mathbf{\Lambda_J}$ given in Eq. (7.4.1). Let us consider the Jordan block $\mathbf{J}_j$ and examine the function $P_n(\mathbf{J}_j)$, which can be expressed as

$$P_n(\mathbf{J}_j) = \prod_{i=1}^{s}(\lambda_i \mathbf{I}^{(k_j)} - \mathbf{J}_j)^{k_i}, \qquad (7.4.29)$$

where $\mathbf{I}^{(k_j)}$ is a $k_j \times k_j$ unit matrix. The factor in the product corresponding to $i = j$ is

$$(\lambda_j \mathbf{I}^{(k_j)} - \mathbf{J}_j)^{k_j} = (-1)^{k_j} \mathbf{H}_j^{k_j}, \qquad (7.4.30)$$

where $\mathbf{H}_j$ is a $k_j \times k_j$ matrix of the form defined by Eq. (7.4.24). We have just proven that the nilpotency of such a matrix is k_j, and thus it follows that

$$P_n(\mathbf{J}_j) = \mathbf{0}, \qquad j = 1, \ldots, s. \qquad (7.4.31)$$

But, according to Eq. (7.4.8),

$$P_n(\mathbf{A}) = \mathbf{Q} \begin{bmatrix} P_n(\mathbf{J}_1) & 0 & \cdots & 0 \\ 0 & P_n(\mathbf{J}_2) & \cdots & 0 \\ \vdots & \vdots & & \vdots \\ 0 & 0 & \cdots & P_n(\mathbf{J}_s) \end{bmatrix} \mathbf{Q}^{-1}, \qquad (7.4.32)$$

which proves that

$$P_n(\mathbf{A}) = \mathbf{0} \qquad (7.4.33)$$

and establishes the Hamilton–Cayley theorem.

7.5. DETERMINATION OF GENERALIZED EIGENVECTORS

If $\mathbf{A}$ is a 2×2 *defective* matrix, the Jordan block diagonalization theorem gives

$$\mathbf{AQ} = \mathbf{Q} \begin{bmatrix} \lambda_1 & 1 \\ 0 & \lambda_1 \end{bmatrix} \qquad (7.5.1)$$

or

$$\mathbf{A}[\mathbf{q}_1, \mathbf{q}_2] = [\mathbf{q}_1, \mathbf{q}_2] \begin{bmatrix} \lambda_1 & 1 \\ 0 & \lambda_1 \end{bmatrix}, \qquad (7.5.2)$$

where $\mathbf{q}_1$ and $\mathbf{q}_2$ are the two-dimensional column vectors of $\mathbf{Q}$. Carrying out the matrix multiplication in Eq. (7.5.2), we obtain

$$[\mathbf{A}\mathbf{q}_1, \mathbf{A}\mathbf{q}_2] = [\lambda_1 \mathbf{q}_1, \lambda_1 \mathbf{q}_2 + \mathbf{q}_1]. \tag{7.5.3}$$

Equating the two column vectors, we find

$$\mathbf{A}\mathbf{q}_1 = \lambda_1 \mathbf{q}_1 \tag{7.5.4}$$

and

$$(\mathbf{A} - \lambda_1 \mathbf{I})\mathbf{q}_2 = \mathbf{q}_1. \tag{7.5.5}$$

We see that the column vector $\mathbf{q}_1$ is, in fact, the eigenvector of $\mathbf{A}$, whereas $\mathbf{q}_2$ satisfies an inhomogeneous equation. Recall that, according to the solvability theorem, Eq. (7.5.5) has a solution *if and only if* the rank of $\mathbf{A} - \lambda_i \mathbf{I}$ is the same as the rank of $[\mathbf{A} - \lambda_1 \mathbf{I}, \mathbf{q}_1]$. Or stated as the Fredholm alternative theorem, Eq. (7.5.5) has a solution *if and only if*

$$\mathbf{z}^\dagger \mathbf{q}_1 = 0, \qquad \text{where } \mathbf{z} \text{ satisfies } (\mathbf{A}^\dagger - \lambda_1^* \mathbf{I})\mathbf{z} = \mathbf{0}. \tag{7.5.6}$$

The matrix

$$\mathbf{A} = \begin{bmatrix} \dfrac{4}{3} & \dfrac{4}{3} \\ -\dfrac{1}{3} & \dfrac{8}{3} \end{bmatrix} \tag{7.5.7}$$

can help to illustrate the process. In this case, the characteristic polynomial is $P_2 = (-\lambda)^2 + 4(-\lambda) + 4$, whose root $\lambda_1 = 2$ is of multiplicity 2. The rank of $\mathbf{A} - 2\mathbf{I}$ is 1, and so there is only one eigenvector. We easily find it to be

$$\mathbf{x}_1 = \begin{bmatrix} 2 \\ 1 \end{bmatrix} = \mathbf{q}_1, \tag{7.5.8}$$

which we set equal to $\mathbf{q}_1$. The solution to $(\mathbf{A}^\dagger - 2\mathbf{I})\mathbf{z} = \mathbf{0}$ is

$$\mathbf{z} = \begin{bmatrix} -1 \\ 2 \end{bmatrix}. \tag{7.5.9}$$

By Eq. (7.5.6) we find that the required condition, $\mathbf{z}^\dagger \mathbf{q}_1 = 0$, is obeyed, and so the equation

$$\begin{bmatrix} \dfrac{4}{3} - 2 & \dfrac{4}{3} \\ -\dfrac{1}{3} & \dfrac{8}{3} - 2 \end{bmatrix} \begin{bmatrix} q_{12} \\ q_{22} \end{bmatrix} = \begin{bmatrix} 2 \\ 1 \end{bmatrix} \tag{7.5.10}$$

is ensured a solution. Either q_{12} or q_{22} can be chosen arbitrarily. With $q_{22} = 0$, the solution to Eq. (7.5.10) is

$$\mathbf{q}_2 = \begin{bmatrix} -3 \\ 0 \end{bmatrix} \qquad (7.5.11)$$

and

$$\mathbf{Q} = \begin{bmatrix} 2 & -3 \\ 1 & 0 \end{bmatrix}, \qquad \mathbf{Q}^{-1} = \begin{bmatrix} 0 & 1 \\ -\dfrac{1}{3} & \dfrac{2}{3} \end{bmatrix}. \qquad (7.5.12)$$

The reader can easily verify that

$$\mathbf{Q}^{-1}\mathbf{A}\mathbf{Q} = \begin{bmatrix} 2 & 1 \\ 0 & 2 \end{bmatrix}. \qquad (7.5.13)$$

Consider now the general case of an $n \times n$ matrix:

$$\mathbf{A}\mathbf{Q} = \mathbf{Q} \begin{bmatrix} \mathbf{J}_1 & 0 & \cdots & 0 \\ 0 & \mathbf{J}_2 & \cdots & 0 \\ \vdots & \vdots & & \vdots \\ 0 & 0 & \cdots & \mathbf{J}_s \end{bmatrix}, \qquad (7.5.14)$$

where the dimension of each Jordan block $\mathbf{J}_i$ will be denoted by k_i. It is convenient to repartition the matrix

$$\mathbf{Q} = [\mathbf{q}_1, \ldots, \mathbf{q}_n], \qquad (7.5.15)$$

as

$$\mathbf{Q} = \left[\mathbf{Q}^{(1)}, \ldots, \mathbf{Q}^{(s)}\right], \qquad (7.5.16)$$

where each matrix $\mathbf{Q}^{(i)}$ contains k_i of the n-dimensional column vectors of $\mathbf{Q}$, which will be relabeled as

$$\mathbf{Q}^{(i)} = \left[\mathbf{q}_1^{(i)}, \ldots, \mathbf{q}_{k_i}^{(i)}\right] \qquad (7.5.17)$$

for convenience. With this partitioning, Eq. (7.5.14) becomes

$$\mathbf{A}\left[\mathbf{Q}^{(1)}, \ldots, \mathbf{Q}^{(s)}\right] = \left[\mathbf{Q}^{(1)}, \ldots, \mathbf{Q}^{(s)}\right] \begin{bmatrix} \mathbf{J}_1 & 0 & \cdots & 0 \\ 0 & \mathbf{J}_2 & \cdots & 0 \\ \vdots & \vdots & & \vdots \\ 0 & 0 & \cdots & \mathbf{J}_s \end{bmatrix} \qquad (7.5.18)$$

or

$$\left[\mathbf{A}\mathbf{Q}^{(1)}, \ldots, \mathbf{A}\mathbf{Q}^{(s)}\right] = \left[\mathbf{Q}^{(1)}\mathbf{J}_1, \ldots, \mathbf{Q}^{(s)}\mathbf{J}_s\right]. \qquad (7.5.19)$$

DETERMINATION OF GENERALIZED EIGENVECTORS

Equating partitioned components of Eq. (7.5.19) yields

$$\mathbf{A}\mathbf{Q}^{(i)} = \mathbf{Q}^{(i)}\mathbf{J}_i, \qquad i = 1, \ldots, s. \tag{7.5.20}$$

The result is that the generalized eigenvectors of each Jordan block can be determined separately.

Using Eq. (7.5.17) in Eq. (7.5.20), we get the system of equations

$$\begin{aligned}
\mathbf{A}\mathbf{q}_1^{(i)} &= \lambda_i \mathbf{q}_1^{(i)} \\
(\mathbf{A} - \lambda_i \mathbf{I})\mathbf{q}_2^{(i)} &= \mathbf{q}_1^{(i)} \\
(\mathbf{A} - \lambda_i \mathbf{I})\mathbf{q}_3^{(i)} &= \mathbf{q}_2^{(i)} \\
&\vdots \\
(\mathbf{A} - \lambda_i \mathbf{I})\mathbf{q}_{k_i}^{(i)} &= \mathbf{q}_{k_i-1}^{(i)}, \qquad i = 1, \ldots, s.
\end{aligned} \tag{7.5.21}$$

Thus, as in the 2×2 case, the first column vector in $\mathbf{Q}^{(i)}$ is the eigenvector corresponding to λ_i and the other *generalized eigenvectors* have to be determined by solving the appropriate inhomogeneous equations.

If each Jordan block has a different eigenvalue, then the multiplicity of each eigenvalue determines the dimension of each Jordan block and the system represented by Eq. (7.5.21) can be solved for each block without further consideration. However, suppose we have the case

$$\mathbf{\Lambda}_\mathbf{J} = \begin{bmatrix} \mathbf{J}_1 & \mathbf{0} \\ \mathbf{0} & \mathbf{J}_2 \end{bmatrix}, \tag{7.5.22}$$

where the eigenvalues of $\mathbf{J}_1$ and $\mathbf{J}_2$ are the same, say λ_1. A priori, we know from the rank of $\mathbf{A} - \lambda_1\mathbf{I}$ that $\mathbf{A}$ has only two eigenvectors and, therefore, only two Jordan blocks. However, we do not know the dimension of the blocks. For an $n \times n$ matrix, $\mathbf{J}_1$ can have any dimension ranging from $k_1 = n - 1$ to 1 as long as $k_1 + k_2 = n$. The strategy for finding the appropriate Jordan blocks and the generalized eigenvectors has to take into account this uncertainty.

The following strategy is recommended:

Step 1. Find the eigenvalues λ_i and their multiplicities p_{λ_i}.

Step 2. For each distinct eigenvector λ_i ($p_{\lambda_i} = 1$), set $\mathbf{q}_i = \mathbf{x}_i$, the eigenvector corresponding to λ_i, and set $\mathbf{J}_i = \lambda_i$.

Step 3. For each case $p_{\lambda_i} > 1$, compute the rank $r_{\mathbf{A}-\lambda_i\mathbf{I}}$ of the characteristic matrix. The number of eigenvectors equals $n - r_{\mathbf{A}-\lambda_i\mathbf{I}}$. This is also the number of Jordan blocks corresponding to λ_i.

Step 4. For the case $n - r_{\mathbf{A}-\lambda_i\mathbf{I}} = p_{\lambda_i} > 1$, each Jordan block is a 1×1 matrix equal to λ_i. For each such case, compute the p_{λ_i} eigenvectors $\mathbf{x}_i$ and set them equal to the appropriate $\mathbf{q}_i$.

Step 5. For each case $p_{\lambda_i} > 1$ and $n - r_{\mathbf{A}-\lambda_i\mathbf{I}} = 1$, there is only one Jordan block and it is a $p_{\lambda_i} \times p_{\lambda_i}$ matrix. The generalized eigenvectors can be computed according to Eq. (7.5.21) with $k_i = p_{\lambda_i}$.

Step 6. When $p_{\lambda_i} > n - r_{A-\lambda_i I} > 1$, there will be $n - r_{A-\lambda_i I}$ Jordan blocks for the eigenvalue λ_i, but we will not usually know at the outset the dimension of the block (if $p_{\lambda_i} = 3$, $n - r_{A-\lambda_i I} = 2$, then one block will be 1×1 and the other 2×2, but for $p_{\lambda_i} > 3$, there is no unique choice).

(6a) The procedure for this case is to calculate the $n - r_{A-\lambda_i I}$ eigenvectors $\{x_i\}$.

(6b) Next, calculate the rank of the augmented matrix $[A - \lambda_i I, x]$ for each eigenvector x. If, for any eigenvector, the rank of the augmented matrix is not equal to $r_{A-\lambda_i I}$, then its Jordan block is simply λ_i.

(6c) If the ranks of the augmented matrix with some of the eigenvectors (say x_1 and x_2 for illustration) are equal to $r_A - \lambda_i I$, then the *generalized eigenvectors* $q_2^{(j)}$ and $q_2^{(k)}$ can be computed from

$$(A - \lambda_i I)q_2^{(j)} = x_1 \quad \text{and} \quad (A - \lambda_i I)q_2^{(k)} = x_2. \quad (7.5.23)$$

Next, try to find *generalized eigenvectors* by solving

$$(A - \lambda_i I)q_3^{(j)} = q_2^{(j)} \quad (7.5.24)$$

and

$$(A - \lambda_i I)q_3^{(k)} = q_2^{(k)}. \quad (7.5.25)$$

Keep going until no further solutions can be found.

(6d) Suppose Eqs. (7.5.24) and (7.5.25) do not admit solutions, but we know (from p_{λ_i}) that the Jordan blocks have to be bigger than two 2×2's. The failure to find a solution means that $z_i^\dagger q_2^{(j)}$ or $z_i^\dagger q_2^{(k)}$ are not all 0, where z_i, $i = 1, 2$, are the solutions to $(A^\dagger - \lambda_i^* I)z = 0$ corresponding to x_1 and x_2. We have to return to Eq. (7.5.23) and obtain a solution to

$$\begin{aligned}(A - \lambda_i I)q_2^{(j)} &= \alpha x_1 + \beta x_2 \\ (A - \lambda_i I)q_2^{(k)} &= \gamma x_1 + \delta x_2.\end{aligned} \quad (7.5.26)$$

$q_1^{(j)} = \alpha x_1 + \beta x_2$ and $q_1^{(k)} = \gamma x_1 + \delta x_2$ are still eigenvectors of A corresponding to λ_i. The new solutions are

$$\begin{aligned}\tilde{q}_2^{(j)} &= \alpha q_2^{(j)} + \beta q_2^{(k)} \\ \tilde{q}_2^{(k)} &= \gamma q_2^{(j)} + \delta q_2^{(k)},\end{aligned} \quad (7.5.27)$$

where $q_2^{(j)}$ and $q_2^{(k)}$ are the solutions found in Eq. (7.5.23). We now impose the conditions

$$z_i^\dagger \tilde{q}_2^{(j)} = 0 = \alpha z_i^\dagger q_2^{(j)} + \beta z_i^\dagger q_2^{(k)}, \quad i = 1, 2, \quad (7.5.28)$$

$$z_i^\dagger \tilde{q}_2^{(k)} = 0 = \gamma z_i^\dagger q_2^{(j)} + \delta z_i^\dagger q_2^{(k)}, \quad i = 1, 2. \quad (7.5.29)$$

DETERMINATION OF GENERALIZED EIGENVECTORS

Equation (7.5.28) or (7.5.29) will have a solution for α and β or γ and δ if a solution to Eq. (7.5.24) or (7.5.25) exists. We continue this process until all of the Jordan blocks are accounted for.

A couple of examples will perhaps help to illustrate the procedure for finding the transformation matrix $\mathbf{Q}$.

■ **EXAMPLE 7.5.1.** Define the matrix

$$\mathbf{A} = \frac{1}{5}\begin{bmatrix} 14 & 8 & 2 & 1 \\ 1 & 12 & 8 & 4 \\ -1 & -2 & 7 & 1 \\ -2 & -4 & -6 & 2 \end{bmatrix}, \quad (7.5.30)$$

and reduce it to its Jordan canonical form.

The eigenvalues of $\mathbf{A}$ are $\lambda_1 = 2$, $p_{\lambda_1} = 3$ and $\lambda_2 = 1$, $p_{\lambda_2} = 1$. The rank of $\mathbf{A} - \lambda_1 \mathbf{I}$ is 3 and the rank of $\mathbf{A} - \lambda_2 \mathbf{I}$ is 3. Thus, there is one eigenvector corresponding to each of the two eigenvalues. These are

$$\mathbf{x}_1 = \begin{bmatrix} -2 \\ 1 \\ 0 \\ 0 \end{bmatrix} \quad \text{and} \quad \mathbf{x}_2 = \begin{bmatrix} 0 \\ 0 \\ -\frac{1}{2} \\ 1 \end{bmatrix}. \quad (7.5.31)$$

Since λ_2 is distinct, its Jordan block is $\mathbf{J}_2 = \lambda_2 = 1$. The Jordan block of $\lambda_1 = 2$ is

$$\mathbf{J}_1 = \begin{bmatrix} 2 & 1 & 0 \\ 0 & 2 & 1 \\ 0 & 0 & 2 \end{bmatrix} \quad (7.5.32)$$

and the solution of $(\mathbf{A}^\dagger - \lambda_1 \mathbf{I})\mathbf{z}_1 = \mathbf{0}$ is

$$\mathbf{z}_1 = \begin{bmatrix} \frac{2}{3} \\ \frac{4}{3} \\ 2 \\ 1 \end{bmatrix}. \quad (7.5.33)$$

The generalized eigenvectors of $\lambda_1 = 2$ obey the equations

$$(\mathbf{A} - 2\mathbf{I})\mathbf{q}_2 = \mathbf{x}_1 \quad (7.5.34)$$

and

$$(\mathbf{A} - 2\mathbf{I})\mathbf{q}_3 = \mathbf{q}_2. \quad (7.5.35)$$

Since $\mathbf{z}_1^\dagger \mathbf{x}_1 = 0$, Eq. (7.5.34) is solvable and the solution is

$$\mathbf{q}_2 = \begin{bmatrix} -3 \\ 0 \\ 1 \\ 0 \end{bmatrix}. \tag{7.5.36}$$

Also, since $\mathbf{z}_1^\dagger \mathbf{q}_2 = 0$, Eq. (7.4.34) is solvable with the solution

$$\mathbf{q}_3 = \begin{bmatrix} -4 \\ 0 \\ 0 \\ 1 \end{bmatrix}. \tag{7.5.37}$$

The transformation matrix $\mathbf{Q} = [\mathbf{x}_1, \mathbf{q}_2, \mathbf{q}_3, \mathbf{x}_2]$ is given by

$$\mathbf{Q} = \begin{bmatrix} -2 & -3 & -4 & 0 \\ 1 & 0 & 0 & 0 \\ 0 & 1 & 0 & -\tfrac{1}{2} \\ 0 & 1 & 1 & 1 \end{bmatrix}, \tag{7.5.38}$$

and through the similarity transformation $\mathbf{Q}^{-1}\mathbf{A}\mathbf{Q}$, $\mathbf{A}$ is transformed into the Jordan canonical form. With

$$\mathbf{Q}^{-1} = \frac{1}{5}\begin{bmatrix} 2 & 4 & 6 & 8 \\ 0 & 1 & 0 & 0 \\ 1 & 2 & 8 & 4 \\ -2 & -4 & -6 & -3 \end{bmatrix}, \tag{7.5.39}$$

it is straightforward to show that

$$\mathbf{Q}^{-1}\mathbf{A}\mathbf{Q} = \begin{bmatrix} \mathbf{J}_1 & \mathbf{0} \\ \mathbf{0} & \mathbf{J}_2 \end{bmatrix}. \tag{7.5.40}$$

In this example, the determination of $\mathbf{Q}$ represents an easy application of the transformation theorem. The next example represents a more complex application.

It should be noted that, since the column vectors of $\mathbf{Q}$ are calculated from homogeneous equations and particular solutions to singular equations, $\mathbf{Q}$ is not unique. In fact, the matrix

$$\mathbf{P} = \begin{bmatrix} -2 & 1 & 0 & 0 \\ 1 & -2 & 1 & 0 \\ 0 & 1 & -2 & 1 \\ 0 & 0 & 1 & -2 \end{bmatrix} \tag{7.5.41}$$

DETERMINATION OF GENERALIZED EIGENVECTORS

also reduces the matrix $\mathbf{A}$ defined in Eq. (7.5.30) to the Jordan canonical form in a similarity transformation.

■ ■ ■

EXAMPLE 7.5.2. Find the Jordan canonical form of

$$\mathbf{A} = \frac{1}{7}\begin{bmatrix} -20 & 1 & 4 & -8 & 5 & -1 \\ -5 & -2 & 15 & -16 & 10 & -2 \\ -4 & 0 & -2 & -17 & 15 & -3 \\ -3 & 0 & 9 & -32 & 20 & -4 \\ -2 & 0 & 6 & -12 & -3 & 2 \\ -1 & 0 & 3 & -6 & 2 & -6 \end{bmatrix}. \qquad (7.5.42)$$

The eigenvalues of $\mathbf{A}$ are $\lambda_1 = -2$, $p_{\lambda_1} = 5$ and $\lambda_2 = -1$, $p_{\lambda_2} = 1$. The rank of $\mathbf{A} - \lambda_1\mathbf{I}$ is 4, and so there are two eigenvectors corresponding to $\lambda_1 = -2$. This means that there will be two Jordan blocks of dimension 4×4 and 1×1 or 3×3 and 2×2. There is one eigenvector for $\lambda_2 = -1$ and its Jordan block is λ_2.

The eigenvectors of $\mathbf{A}$ are

$$\mathbf{x}_1 = \begin{bmatrix} -6 \\ -3 \\ 0 \\ 3 \\ 2 \\ 1 \end{bmatrix}, \quad \mathbf{x}_2 = \begin{bmatrix} 3 \\ 2 \\ 1 \\ 0 \\ 0 \\ 0 \end{bmatrix}, \quad \mathbf{x}_3 = \frac{1}{6}\begin{bmatrix} 1 \\ 2 \\ 3 \\ 4 \\ 5 \\ 6 \end{bmatrix}, \qquad (7.5.43)$$

where $\mathbf{A}\mathbf{x}_1 = -2\mathbf{x}_1$, $\mathbf{A}\mathbf{x}_2 = -2\mathbf{x}_2$, and $\mathbf{A}\mathbf{x}_3 = -\mathbf{x}_3$. The solutions to $(\mathbf{A}^\dagger - \lambda_1^*\mathbf{I})\mathbf{z}_i = \mathbf{0}$, $i = 1, 2$, are

$$\mathbf{z}_1 = \begin{bmatrix} 0 \\ -1 \\ 2 \\ 0 \\ -2 \\ 1 \end{bmatrix} \quad \text{and} \quad \mathbf{z}_2 = \begin{bmatrix} 0 \\ 1 \\ -2 \\ 1 \\ 0 \\ 0 \end{bmatrix}. \qquad (7.5.44)$$

Since $\mathbf{z}_i^\dagger \mathbf{x}_j = 0$, $i = 1, 2$, the equations for the generalized eigenvectors

$$(\mathbf{A} - \lambda_1\mathbf{I})\tilde{\mathbf{q}}_2 = \mathbf{x}_1 \qquad (7.5.45)$$

and

$$(\mathbf{A} - \lambda_1\mathbf{I})\tilde{\mathbf{p}}_2 = \mathbf{x}_2 \qquad (7.5.46)$$

have solutions. These are

$$\tilde{\mathbf{q}}_2 = \begin{bmatrix} 17 \\ 4 \\ 0 \\ -4 \\ 0 \\ 0 \end{bmatrix} \quad \text{and} \quad \tilde{\mathbf{p}}_2 = \begin{bmatrix} -6 \\ -1 \\ 0 \\ 1 \\ 0 \\ 0 \end{bmatrix}. \tag{7.5.47}$$

Since the two generalized eigenvectors exist, the Jordan blocks must be of dimension 3×3 and 2×2. However, since $\mathbf{z}_1^\dagger \tilde{\mathbf{q}}_2 = -4$ and $\mathbf{z}_1^\dagger \tilde{\mathbf{p}}_2 = 1$, neither

$$(\mathbf{A} - \lambda_1 \mathbf{I})\tilde{\mathbf{q}}_3 = \tilde{\mathbf{q}}_2 \quad \text{nor} \quad (\mathbf{A} - \lambda_1 \mathbf{I})\tilde{\mathbf{p}}_3 = \tilde{\mathbf{p}}_2 \tag{7.5.48}$$

has a solution. Since any linear combination of $\mathbf{x}_1$ and $\mathbf{x}_2$ is also an eigenvector corresponding to λ_1, we replace the eigenvector $\mathbf{x}_1$ by the eigenvector

$$\mathbf{y}_1 = \mathbf{x}_1 + 4\mathbf{x}_2, \tag{7.5.49}$$

and compute $\mathbf{q}_2$ from

$$(\mathbf{A} - \lambda_1 \mathbf{I})\mathbf{q}_2 = \mathbf{y}_1 \tag{7.5.50}$$

to obtain

$$\mathbf{q}_2 = \begin{bmatrix} -7 \\ 0 \\ 0 \\ 0 \\ 0 \\ 0 \end{bmatrix} \tag{7.5.51}$$

for which $\mathbf{z}_1^\dagger \mathbf{q}_2 = \mathbf{z}_2^\dagger \mathbf{q}_2 = 0$. The linear combination of $\mathbf{x}_1$ and $\mathbf{x}_2$ chosen to form $\mathbf{y}_1$ thus satisfies Eq. (7.5.28). Now the equation

$$(\mathbf{A} - \lambda_1 \mathbf{I})\mathbf{q}_3 = \mathbf{q}_2 \tag{7.5.52}$$

has a solution, namely,

$$\mathbf{q}_3 = \begin{bmatrix} 0 \\ -7 \\ 0 \\ 0 \\ 0 \\ 0 \end{bmatrix}. \tag{7.5.53}$$

The eigenvectors and generalized eigenvectors form the matrix

$$\mathbf{Q} = [\mathbf{y}_1, \mathbf{q}_2, \mathbf{q}_3, \mathbf{x}_2, \mathbf{p}_2, \mathbf{x}_3] \tag{7.5.54}$$

or

$$\mathbf{Q} = \begin{bmatrix} 6 & -7 & 0 & 3 & -6 & 1 \\ 5 & 0 & -7 & 2 & -1 & 2 \\ 4 & 0 & 0 & 1 & 0 & 3 \\ 3 & 0 & 0 & 0 & 1 & 4 \\ 2 & 0 & 0 & 0 & 0 & 5 \\ 1 & 0 & 0 & 0 & 0 & 6 \end{bmatrix}. \tag{7.5.55}$$

With this matrix,

$$\mathbf{Q}^{-1}\mathbf{A}\mathbf{Q} = \begin{bmatrix} \mathbf{J}_1 & 0 & 0 \\ 0 & \mathbf{J}_2 & 0 \\ 0 & 0 & \mathbf{J}_3 \end{bmatrix}, \tag{7.5.56}$$

where

$$\mathbf{J}_1 = \begin{bmatrix} -2 & 1 & 0 \\ 0 & -2 & 1 \\ 0 & 0 & -2 \end{bmatrix}$$

$$\mathbf{J}_2 = \begin{bmatrix} -2 & 1 \\ 0 & -2 \end{bmatrix} \tag{7.5.57}$$

$$\mathbf{J}_3 = -1.$$

■ ■ ■

7.6. DYADIC FORM OF AN IMPERFECT MATRIX

For a *perfect* matrix, the spectral decomposition, $\mathbf{A} = \sum_{i=1}^{n} \lambda_i \mathbf{x}_i \mathbf{z}_i^\dagger$, amounts to the expression of $\mathbf{A}$ as a dyadic consisting of a linear combination of the dyads $\mathbf{x}_i \mathbf{z}_i^\dagger$, $i = 1, \ldots, n$, where $\{\mathbf{x}_i\}$ are the eigenvectors of $\mathbf{A}$ and $\{\mathbf{z}_i\}$ their reciprocal set. We are interested in searching for an equivalent decomposition of an *imperfect* matrix into a linear combination of n dyads.

Suppose $\mathbf{Q} = [\mathbf{q}_1, \ldots, \mathbf{q}_n]$ is a matrix that transforms $\mathbf{A}$ into the Jordan canonical form in a similarity transformation. Define

$$\mathbf{Z} = [\mathbf{z}_1, \ldots, \mathbf{z}_n] \equiv \left(\mathbf{Q}^{-1}\right)^\dagger. \tag{7.6.1}$$

The column vectors of $\mathbf{Z}$ are the reciprocal set to the column vectors of $\mathbf{Q}$, since $\mathbf{Z}^\dagger \mathbf{Q} = [\mathbf{z}_i^\dagger \mathbf{q}_j] = \mathbf{I} = [\delta_{ij}]$, or $\mathbf{z}_i^\dagger \mathbf{q}_j = \delta_{ij}$. If $\{\mathbf{q}_i\}$ is taken as a basis set in E_n, then the identity matrix can be expressed as

$$\mathbf{I} = \sum_{i=1}^{n} \mathbf{q}_i \mathbf{z}_i^\dagger. \tag{7.6.2}$$

From this, and the property $\mathbf{A} = \mathbf{AI}$, it follows that

$$\mathbf{A} = \sum_{i=1}^{n} \mathbf{y}_i \mathbf{z}_i^\dagger, \qquad (7.6.3)$$

where

$$\mathbf{y}_i = \mathbf{A}\mathbf{q}_i. \qquad (7.6.4)$$

Thus, any arbitrary matrix can be expressed as a dyadic consisting of a sum of n dyads.

If an *imperfect* matrix has s eigenvectors, denoting these by $\mathbf{x}_1, \mathbf{x}_2, \ldots, \mathbf{x}_s$ and noting that $\mathbf{y}_i = \mathbf{A}\mathbf{x}_i = \lambda_i \mathbf{x}_i$ for this set, we can rewrite Eq. (7.6.3) as

$$\mathbf{A} = \sum_{i=1}^{s} \lambda_i \mathbf{x}_i \mathbf{z}_i^\dagger + \sum_{i=s+1}^{n} \mathbf{y}_i \mathbf{z}_i^\dagger, \qquad (7.6.5)$$

where the $\mathbf{y}_i$ are linearly independent vectors related to the generalized eigenvectors through Eq. (7.6.4). The proof that the vectors $\mathbf{y}_i$ form a linearly independent set proceeds as follows. Assume that the $\mathbf{y}_i$ are linearly dependent; i.e., assume that there exists a set of numbers $\{\alpha_{s+1}, \ldots, \alpha_n\}$, not all 0, such that

$$\sum_{i=s+1}^{n} \alpha_i \mathbf{y}_i = \mathbf{0}. \qquad (7.6.6)$$

From Eq. (7.6.4), it follows that

$$\mathbf{A} \sum_{i=s+1}^{n} \alpha_i \mathbf{q}_i = \mathbf{0} \qquad (7.6.7)$$

or that $\sum_{i=s+1}^{n} \alpha_i \mathbf{q}_i$ is an eigenvector of $\mathbf{A}$. This is a contradiction, since the eigenvectors $\mathbf{x}_1, \ldots, \mathbf{x}_s$ are not included in the set of *generalized eigenvectors* generating the set $\{\mathbf{y}_{s+1}, \ldots, \mathbf{y}_n\}$ in Eq. (7.6.6).

Equation (7.6.5) indicates that a dyadic representation of $\mathbf{A}$ needs at most n dyads. If k of the eigenvalues of $\mathbf{A}$ are 0, then Eq. (7.6.5) will contain $n - k$ dyads.

■ **EXERCISE 7.5.1.** Suppose $\{\mathbf{x}_i\}$ is a basis set in E_n and $\{\mathbf{z}_i\}$ is its reciprocal set, and let

$$\mathbf{A} = \sum_{i=1}^{n-1} \mathbf{x}_i \mathbf{z}_{i+1}^\dagger.$$

Find the eigenvectors and eigenvalues of $\mathbf{A}$. Give the Jordan canonical form $\mathbf{\Lambda}_J$
■ ■ ■ of $\mathbf{A}$. What is the nilpotent index of $\mathbf{A}$?

7.7. SCHMIDT'S NORMAL FORM OF AN ARBITRARY SQUARE MATRIX

Since its proof is somewhat tedious, we will first state the theorem of interest in this section, examine its consequences, and then give its proof:

THEOREM. *For an arbitrary square matrix* $\mathbf{A}$, *the solutions* $\mathbf{x}_1, \mathbf{x}_2, \ldots, \mathbf{x}_n$ *and* $\mathbf{y}_1, \mathbf{y}_2, \ldots, \mathbf{y}_n$ *of the equations*

$$\mathbf{A}\mathbf{x}_i = \kappa_i \mathbf{y}_i \tag{7.7.1}$$

and

$$\mathbf{A}^\dagger \mathbf{y}_i = \kappa_i \mathbf{x}_i \tag{7.7.2}$$

form orthonormal basis sets in E_n *and the quantities* κ_i *are real numbers equal to or greater than* 0.

Note that this theorem is valid whether or not $\mathbf{A}$ is a perfect matrix.

By multiplying Eq. (7.7.1) by $\mathbf{A}^\dagger$ and Eq. (7.7.2) by $\mathbf{A}$ and then substituting Eqs. (7.7.2) and (7.7.1) into the resulting expressions, we find

$$\mathbf{A}^\dagger \mathbf{A} \mathbf{x}_i = \kappa_i^2 \mathbf{x}_i \tag{7.7.3}$$

and

$$\mathbf{A}\mathbf{A}^\dagger \mathbf{y}_i = \kappa_i^2 \mathbf{y}_i. \tag{7.7.4}$$

Thus, we see that the orthonormal sets $\{\mathbf{x}_1, \ldots, \mathbf{x}_n\}$ and $\{\mathbf{y}_1, \ldots, \mathbf{y}_n\}$ are, respectively, the eigenvectors of the self-adjoint matrices $\mathbf{A}^\dagger \mathbf{A}$ and $\mathbf{A}\mathbf{A}^\dagger$. In Chapter 5 we saw that the eigenvalues of $\mathbf{AB}$ and $\mathbf{BA}$ are the same, in agreement with Eqs. (7.7.3) and (7.7.4). It also follows from the equation

$$\langle \mathbf{A}\mathbf{x}_i, \mathbf{A}\mathbf{x}_i \rangle = \langle \mathbf{x}_i, \mathbf{A}^\dagger \mathbf{A} \mathbf{x}_i \rangle = \kappa_i^2 \|\mathbf{x}_i\|^2 \tag{7.7.5}$$

that κ_i^2 is positive. Thus, as an alternative to finding $\kappa_i, \mathbf{x}_i$, and $\mathbf{y}_i$ by solving Eqs. (7.7.1) and (7.7.2), we can solve the two self-adjoint eigenproblems posed by Eqs. (7.7.3) and (7.7.4) to obtain $\mathbf{x}_i$, $\mathbf{y}_i$, and $\kappa_i = \sqrt{\kappa_i^2}$.

An important implication of the above theorem is that the matrix $\mathbf{A}$ can be expressed in the biorthogonal diagonal form

$$\mathbf{A} = \sum_{i=1}^{n} \kappa_i \mathbf{y}_i \mathbf{x}_i^\dagger. \tag{7.7.6}$$

This is known as Schmidt's normal form of a square matrix. The proof of Eq. (7.7.6) is easy. Since $\{\mathbf{x}_1, \ldots, \mathbf{x}_n\}$ is an orthonormal basis set,

$$\mathbf{I} = \sum_{i=1}^{n} \mathbf{x}_i \mathbf{x}_i^\dagger. \tag{7.7.7}$$

But since $\mathbf{A} = \mathbf{AI} = \sum_{i=1}^{n} \mathbf{A}\mathbf{x}_i \mathbf{x}_i^\dagger$ and $\mathbf{A}\mathbf{x}_i = \kappa_i \mathbf{y}_i$, we immediately see that Eq. (7.7.6) is true.

As an application of Eq. (7.7.6), consider the inhomogeneous equation

$$\mathbf{A}\mathbf{z} = \mathbf{b} \tag{7.7.8}$$

for the case $|\mathbf{A}| \neq 0$, or $\kappa_i \neq 0$. Expanding $\mathbf{b}$ as $\mathbf{b} = \sum_i \langle \mathbf{y}_i, \mathbf{b} \rangle \mathbf{y}_i$, and noting that $\mathbf{Az} = \sum_{i=1}^n \kappa_i \langle \mathbf{x}_i, \mathbf{z} \rangle \mathbf{y}_i$, we obtain $\kappa_i \langle \mathbf{x}_i, \mathbf{z} \rangle = \langle \mathbf{y}_i, \mathbf{b} \rangle$. Thus, the solution to Eq. (7.7.8) is

$$\mathbf{z} = \sum_{i=1}^n \frac{\langle \mathbf{y}_i, \mathbf{b} \rangle}{\kappa_i} \mathbf{x}_i. \tag{7.7.9}$$

For the case $|\mathbf{A}| = 0$, or some $\kappa_i = 0$, Eq. (7.7.8) has a solution *if and only if* $\langle \mathbf{y}_i, \mathbf{b} \rangle = 0$ for the vectors $\mathbf{y}_i$ satisfying $\mathbf{A}^\dagger \mathbf{y}_i = 0$. The solution is then

$$\mathbf{z} = {\sum_i}' \frac{\langle \mathbf{y}_i, \mathbf{b} \rangle}{\kappa_i} \mathbf{x}_i + {\sum_i}'' \alpha_i \mathbf{x}_i, \tag{7.7.10}$$

where α_i are arbitrary constants. The primed sum $\sum'$ means summation over the vectors for which $\kappa_i \neq 0$ and the double primed sum $\sum''$ means summation over the vectors for which $\mathbf{A}\mathbf{x}_i = 0$, or $\kappa_i = 0$. This result is consistent with our earlier solvability theorem. What is new is that the solution is given in terms of the eigenvalues and eigenvectors of $\mathbf{A}\mathbf{A}^\dagger$ and $\mathbf{A}^\dagger \mathbf{A}$.

Let us now turn to the proof of the theorem. Consider the vectors satisfying

$$\mathbf{A}\mathbf{x}_i = \mathbf{0} \quad \text{and} \quad \mathbf{A}^\dagger \mathbf{y}_i = \mathbf{0}. \tag{7.7.11}$$

The vectors $\mathbf{x}_i$ and $\mathbf{y}_i$ are, by definition, the null vectors of $\mathbf{A}$ and $\mathbf{A}^\dagger$. Furthermore, we know that there will be $n - r_A$ vectors $\mathbf{x}_i$ and $\mathbf{y}_i$ each, where r_A is the rank of $\mathbf{A}$ (and of $\mathbf{A}^\dagger$). These eigenvectors can, of course, be orthonormalized by the Gram–Schmidt procedure. The next step is to hunt the maximum of $|\langle \mathbf{y}, \mathbf{A}\mathbf{x} \rangle|^2$ among all normalized vectors $\mathbf{x}, \mathbf{y}$ in E_n such that $\mathbf{x}$ is orthogonal to the null vectors of $\mathbf{A}$, and $\mathbf{y}$ is orthogonal to the null vectors of $\mathbf{A}^\dagger$. Since

$$|\langle \mathbf{y}, \mathbf{A}\mathbf{x} \rangle|^2 \leq \|\mathbf{A}\|^2 \|\mathbf{x}\| \|\mathbf{y}\| = \|\mathbf{A}\|^2, \tag{7.7.12}$$

we know that the quantity $|\langle \mathbf{y}, \mathbf{A}\mathbf{x} \rangle|^2$ has a maximum M (which is, in fact, equal to $\|\mathbf{A}\|^2$). We can, therefore, find $\mathbf{x}_1$ and $\mathbf{y}_1$ such that

$$|\langle \mathbf{y}_1, \mathbf{A}\mathbf{x}_1 \rangle|^2 = M. \tag{7.7.13}$$

This is because there exist sequences of normalized vectors $\mathbf{x}^{(n)}$ and $\mathbf{y}^{(n)}$ such that $\lim |\langle \mathbf{y}^{(n)}, \mathbf{A}\mathbf{x}^{(n)} \rangle|^2 = M$, and so $\mathbf{x}_1 = \lim \mathbf{x}^{(n)}$ and $\mathbf{y}_1 = \lim \mathbf{y}^{(n)}$. We then let

$$\langle \mathbf{y}_1, \mathbf{A}\mathbf{x}_1 \rangle = \kappa_R + i\kappa_I = \kappa, \tag{7.7.14}$$

where κ_R and κ_I are real numbers and $|\kappa|^2 = \kappa_R^2 + \kappa_I^2 = M$.

Consider next a normalized vector $\mathbf{x}^*$ that is orthogonal to $\mathbf{x}_1$ and to the null vectors of $\mathbf{A}$. We can show that

$$\langle \mathbf{y}_1, \mathbf{A}\mathbf{x}^* \rangle = 0 \tag{7.7.15}$$

by assuming that $\langle \mathbf{y}_1, \mathbf{A}\mathbf{x}^* \rangle = \mu_1 + i\mu_2 = \mu$, where μ_1 and μ_2 are real numbers. We then consider the vector $c_1 \mathbf{x}_1 + c_2 \mathbf{x}^*$, where c_1 and c_2 are chosen so that $\|c_1 \mathbf{x}_1 + c_2 \mathbf{x}^*\| = 1$. This requires that $|c_1|^2 + |c_2|^2 = 1$ and now we have

$$|\langle \mathbf{y}_1, \mathbf{A}(c_1 \mathbf{x}_1 + c_2 \mathbf{x}^*) \rangle|^2 = |c_1 \kappa + c_2 \mu|^2 = \mathbf{c}^\dagger \mathbf{H} \mathbf{c}, \tag{7.7.16}$$

where

$$\mathbf{c} = \begin{bmatrix} c_1 \\ c_2 \end{bmatrix} \quad \text{and} \quad \mathbf{H} = \begin{bmatrix} |\kappa|^2 & \mu^*\kappa \\ \mu\kappa^* & |\mu|^2 \end{bmatrix}. \quad (7.7.17)$$

Since $\mathbf{H}$ is a self-adjoint matrix, the maximum value of $\mathbf{c}^\dagger \mathbf{H}\mathbf{c}$ is attained when $\mathbf{c}$ is the normalized eigenvector $\mathbf{c}_m$ of $\mathbf{H}$, corresponding to the maximum eigenvalue λ_m (which equals $|\kappa|^2 + |\mu|^2$). Thus, choosing $\mathbf{c} = \mathbf{c}_m$, we obtain

$$|\langle \mathbf{y}_1, \mathbf{A}(c_1\mathbf{x}_1 + c_2\mathbf{x}^*)\rangle|^2 = \lambda_m \|\mathbf{c}_m\|^2 = |\kappa|^2 + |\mu|^2. \quad (7.7.18)$$

However, we already know that $|\langle \mathbf{y}_1, \mathbf{A}\mathbf{x}\rangle|^2 \le |\kappa|^2$ for any normalized vector $\mathbf{x}$. Therefore, it follows that $\mu = 0$, which proves Eq. (7.7.15) for all normalized vectors $\mathbf{x}^*$ that are orthogonal to $\mathbf{x}_1$ and the null vectors of $\mathbf{A}$.

We can rewrite Eq. (7.7.14) as

$$\langle \mathbf{A}^\dagger \mathbf{y}_1, \mathbf{x}_1 \rangle = \kappa, \quad (7.7.19)$$

using the property $\langle \mathbf{y}, \mathbf{A}\mathbf{x}\rangle = \langle \mathbf{A}^\dagger \mathbf{y}, \mathbf{x}\rangle$ of the adjoint matrix, and then prove in a similar way that $\langle \mathbf{A}^\dagger \mathbf{y}^*, \mathbf{x}_1\rangle = 0$ for all normalized vectors $\mathbf{y}^*$ that are orthogonal to $\mathbf{y}_1$ and the null vectors of $\mathbf{A}^\dagger$.

Next, we will prove that

$$\mathbf{A}\mathbf{x}_1 = \kappa \mathbf{y}_1. \quad (7.7.20)$$

First, assume that $\mathbf{A}\mathbf{x}_1 - \kappa \mathbf{y}_1 = \mathbf{x}$. Then, for any vector $\mathbf{y}_i$ in the null vector set of $\mathbf{A}^\dagger$ ($\mathbf{A}^\dagger \mathbf{y}_i = \mathbf{0}$), it follows that

$$\langle \mathbf{y}_i, \mathbf{x}\rangle = \langle \mathbf{y}_i, \mathbf{A}\mathbf{x}_1 - \kappa \mathbf{y}_1\rangle = \langle \mathbf{A}\mathbf{y}_i, \mathbf{x}_1\rangle - \kappa \langle \mathbf{y}_i, \mathbf{y}_1\rangle = 0 \quad (7.7.21)$$

and

$$\langle \mathbf{y}_1, \mathbf{x}\rangle = \langle \mathbf{y}_1, \mathbf{A}\mathbf{x}_1\rangle - \kappa \langle \mathbf{y}_1, \mathbf{y}_1\rangle = \kappa - \kappa = 0. \quad (7.7.22)$$

Also, since $\langle \mathbf{A}^\dagger \mathbf{y}^*, \mathbf{x}_1\rangle = 0$ for all normalized vectors orthogonal to $\mathbf{y}_1$ and the null vectors of $\mathbf{A}^\dagger$, we find

$$\langle \mathbf{y}^*, \mathbf{x}\rangle = \langle \mathbf{A}^\dagger \mathbf{y}^*, \mathbf{x}_1\rangle - \kappa \langle \mathbf{y}^*, \mathbf{y}_1\rangle = 0. \quad (7.7.23)$$

Thus, we find that $\mathbf{x}$ is orthogonal to a complete orthonormal basis set and so $\mathbf{x} = \mathbf{0}$, which proves Eq. (7.7.20). A similar proof establishes that

$$\mathbf{A}^\dagger \mathbf{y}_1 = \kappa^* \mathbf{x}_1. \quad (7.7.24)$$

Let us express the complex number κ in the form $\kappa = \exp(i\theta)\kappa_1$, where $\kappa_1 = |\kappa|$ and $\kappa_R = \kappa_1 \cos\theta$ and $\kappa_I = \kappa_1 \sin\theta$. If the vectors $\mathbf{x}_1$ and $\mathbf{y}_1$ are replaced by $\exp(-i\theta/2)\mathbf{x}_1$ and $\exp(i\theta/2)\mathbf{y}_1$, they remain unit vectors. Thus, without loss of generality, Eqs. (7.7.20) and (7.7.24) can be rewritten as

$$\mathbf{A}\mathbf{x}_1 = \kappa_1 \mathbf{y}_1 \quad \text{and} \quad \mathbf{A}^\dagger \mathbf{y}_1 = \kappa_1 \mathbf{x}_1, \quad (7.7.25)$$

where $\kappa_1 > 0$ and $\|\mathbf{x}_1\| = \|\mathbf{y}_1\| = 1$.

The process of obtaining $\mathbf{x}_1, \mathbf{y}_1$, and κ_1 can be repeated to find $\mathbf{x}_2, \mathbf{y}_2$, and κ_2 with the constraint that $\mathbf{x}_2$ is orthogonal to $\mathbf{x}_1$ and the null vectors of $\mathbf{A}$, and $\mathbf{y}_2$ is orthogonal to $\mathbf{y}_1$ and the null vectors of $\mathbf{A}^\dagger$. The result is

$$\mathbf{A}\mathbf{x}_2 = \kappa_2 \mathbf{y}_2 \quad \text{and} \quad \mathbf{A}^\dagger \mathbf{y}_2 = \kappa_2 \mathbf{x}_2, \quad 0 < \kappa_2 \leq \kappa_1. \qquad (7.7.26)$$

Continuing this process will generate two sets of vectors corresponding to nonzero values of κ_i. The vectors $\mathbf{x}_1, \mathbf{x}_2, \ldots, \mathbf{x}_{r_A}$ plus the null vectors of $\mathbf{A}$ form an orthonormal basis set in E_n. Likewise, the vectors $\mathbf{y}_1, \mathbf{y}_2, \ldots, \mathbf{y}_{r_A}$ plus the null vectors of $\mathbf{A}^\dagger$ form an orthonormal basis set in E_n. This completes the proof of our theorem.

There is a subtlety that might be perplexing without a little thought. Suppose $\mathbf{A}$ is a self-adjoint matrix whose eigenproblem is

$$\mathbf{A}\mathbf{z}_i = \lambda_i \mathbf{z}_i. \qquad (7.7.27)$$

The eigenvalues λ_i are real and can be either negative or positive. Clearly, $\mathbf{z}_i$ are the eigenvectors of $\mathbf{A}\mathbf{A}^\dagger$ and $\mathbf{A}^\dagger\mathbf{A}$ with eigenvalues λ_i^2. Thus, κ_i^2 in Eqs. (7.7.3) and (7.7.4) must equal λ_i^2. However, we cannot simply set $\mathbf{x}_i = \mathbf{y}_i = \mathbf{z}_i$ in Eqs. (7.7.1) and (7.7.2) because this would imply that $\lambda_i = \kappa_i$, and so all of the eigenvalues of a self-adjoint matrix would have to be positive. The subtlety is that $\mathbf{x}_i$ and $\mathbf{y}_i$ differ by the factor $\exp(i\theta)$, where $\lambda_i = \exp(i\theta_i)\kappa_i$. If λ_i is negative, then $\theta_i = \pi$. Thus, in Eqs. (7.7.1) and (7.7.2), we set $\mathbf{x}_i = \mathbf{z}_i$ and $\mathbf{y}_i = \exp(i\theta_i)\mathbf{z}_i$ for which they correctly reduce to $\mathbf{A}\mathbf{z} = \lambda_i \mathbf{z}_i$ and $\mathbf{A}^\dagger \mathbf{z}_i = \lambda_i^* \mathbf{z}_i$. Of course, $\lambda_i^* = \lambda_i$ for self-adjoint matrices.

7.8. THE INITIAL VALUE PROBLEM

The solutions to the initial value problems (IVPs) given by Eqs. (6.10.8) and (6.10.24) are valid whether or not $\mathbf{A}$ or $\mathbf{T}$ is perfect. However, in the general case, the simplest form that the equation

$$\mathbf{x} = \exp(\mathbf{A}t)\mathbf{x}_0 + \int_0^t \exp(\mathbf{A}(t-\tau))\mathbf{b}(\tau)\,d\tau \qquad (7.8.1)$$

can take is

$$\mathbf{x} = \mathbf{Q} \begin{bmatrix} \exp(t\lambda_1)\mathbf{K}_1 & 0 & \cdots & 0 \\ 0 & \exp(t\lambda_2)\mathbf{K}_2 & \cdots & 0 \\ \vdots & \vdots & & \vdots \\ 0 & 0 & \cdots & \exp(t\lambda_s)\mathbf{K}_s \end{bmatrix} \mathbf{Q}^{-1}\mathbf{x}_0$$

$$+ \mathbf{Q} \int_0^t d\tau \begin{bmatrix} \exp((t-\tau)\lambda_1)\mathbf{K}_1 & 0 & \cdots & 0 \\ 0 & \exp((t-\tau)\lambda_2)\mathbf{K}_2 & \cdots & 0 \\ \vdots & \vdots & & \vdots \\ 0 & 0 & \cdots & \exp((t-\tau)\lambda_s)\mathbf{K}_s \end{bmatrix}$$

$$\times \mathbf{Q}^{-1}\mathbf{b}(\tau). \qquad (7.8.2)$$

THE INITIAL VALUE PROBLEM

The following example illustrates the value of the Jordan canonical form in solving the IVP.

EXAMPLE 7.8.1. The quantity $u(t)$ obeys the second-order equation

$$\frac{d^2u}{dt^2} + 2\frac{du}{dt} + u = \cos\left(\frac{t}{2}\right), \tag{7.8.3}$$

with

$$u(t) = 2, \quad \frac{du}{dt} = -1 \text{ at } t = 0. \tag{7.8.4}$$

The problem transforms (see Section 6.10) to

$$\frac{d\mathbf{x}}{dt} = \mathbf{A}\mathbf{x} + \mathbf{f}, \quad \mathbf{x}(0) = \begin{bmatrix} 2 \\ -1 \end{bmatrix}, \tag{7.8.5}$$

where the companion matrix is

$$\mathbf{A} = \begin{bmatrix} 0 & 1 \\ -1 & -2 \end{bmatrix}. \tag{7.8.6}$$

Recall that

$$\mathbf{f} = \begin{bmatrix} 0 \\ \cos\left(\dfrac{t}{2}\right) \end{bmatrix} \tag{7.8.7}$$

and

$$\mathbf{x} = \begin{bmatrix} u \\ \dfrac{du}{dt} \end{bmatrix}. \tag{7.8.8}$$

Thus, $\mathbf{A}$ is imperfect; it has one eigenvalue, $\lambda_1 = -1$, and one eigenvector,

$$\mathbf{x}_1 = \begin{bmatrix} 1 \\ 1 \end{bmatrix}. \tag{7.8.9}$$

The solution $\mathbf{q}_2$ to

$$(\mathbf{A} - \lambda_1 \mathbf{I})\mathbf{q}_2 = \mathbf{x}_1 \tag{7.8.10}$$

is

$$\mathbf{q}_2 = \begin{bmatrix} 1 \\ 0 \end{bmatrix}, \tag{7.8.11}$$

and so
$$Q = \begin{bmatrix} 1 & 1 \\ -1 & 0 \end{bmatrix}, \quad Q^{-1} = \begin{bmatrix} 0 & -1 \\ 1 & 1 \end{bmatrix}, \tag{7.8.12}$$

and
$$A = QJQ^{-1} = Q \begin{bmatrix} -1 & 1 \\ 0 & -1 \end{bmatrix} Q^{-1}. \tag{7.8.13}$$

The solution to Eq. (7.8.5) is
$$\begin{aligned} x &= \exp(tA)x(0) + \int_0^t \exp((t-\tau)A)f(\tau)\,d\tau \\ &= Q\exp(tJ)Q^{-1}x(0) + Q\int_0^t \exp((t-\tau)J)Q^{-1}f(\tau)\,d\tau. \end{aligned} \tag{7.8.14}$$

When the matrix products and integrations are carried out in Eq. (7.8.14), the final result is

$$x = \begin{bmatrix} u \\ \dfrac{du}{dt} \end{bmatrix}$$

$$= e^{-t}\begin{bmatrix} 2+t \\ -1-t \end{bmatrix} + \frac{4}{25}e^{-t}\begin{bmatrix} -3-5t+e^t\left(3\cos\dfrac{t}{2} + 4\sin\dfrac{t}{2}\right) \\ 2+5t+e^t\left(2\cos\dfrac{t}{2} - \dfrac{3}{2}\sin\dfrac{t}{2}\right) \end{bmatrix}. \tag{7.8.15}$$

Thus,
$$u = (2+t)e^t + \frac{4}{25}\left[-(3+5t)e^{-t} + 3\cos\frac{t}{2} + 4\sin\frac{t}{2}\right]. \tag{7.8.16}$$

The first part of the solution comes from the initial conditions $x(0)$ and the second part comes from the inhomogeneous part, or the driving force, f. Note that, for $t \gg 1$,

$$u(t) \approx \frac{4}{25}\left(3\cos\frac{t}{2} + 4\sin\frac{t}{2}\right); \tag{7.8.17}$$

■ ■ ■ i.e., asymptotically, the solution to Eq. (7.8.3) is an oscillating function of time.

PROBLEMS

1. Find the Jordan block form and transformation matrix Q for the following imperfect matrices:

   (a)
   $$\begin{bmatrix} 1 & 3 \\ 4 & 2 \end{bmatrix}.$$

(b)
$$\begin{bmatrix} 5 & 1 & 2 \\ 0 & 3 & 0 \\ 2 & 1 & 5 \end{bmatrix}.$$

(c)
$$\begin{bmatrix} 1 & 1 & 2 \\ -1 & 3 & 4 \\ 0 & 0 & 2 \end{bmatrix}.$$

(d)
$$\begin{bmatrix} \pi & \pi/3 & -\pi \\ 0 & \pi & \pi/2 \\ 0 & 0 & \pi \end{bmatrix}.$$

(e)
$$\begin{bmatrix} -1 & -4 & 4 \\ -2 & -2 & 3 \\ -3 & -5 & -6 \end{bmatrix}.$$

2. Find $\exp(\mathbf{A}t)$ for the matrices $\mathbf{A}$ given in parts (a)–(e) of Problem 1.

3. Suppose $\mathbf{N}$ is an $n \times n$ nilpotent matrix of index p, i.e., $\mathbf{N}^{p-1} \neq \mathbf{0}$ and $\mathbf{N}^p = \mathbf{0}$. Prove that if $\mathbf{u}$ is a vector such that $\mathbf{N}^{p-1}\mathbf{u} \neq \mathbf{0}$, then the vectors $\mathbf{u}, \mathbf{Nu}, \ldots, \mathbf{N}^{p-1}\mathbf{u}$ form a linearly independent set.

4. (a) Suppose $\mathbf{x}_1, \mathbf{x}_2, \ldots, \mathbf{x}_n$ is an orthonormal basis set in E_n. Prove that the matrix
$$\mathbf{A} = \mathbf{x}_i \mathbf{x}_j^\dagger, \qquad i \neq j,$$
is an *imperfect* matrix.

(b) Suppose $\{\mathbf{x}_1, \ldots, \mathbf{x}_n\}$ is a basis set in E_n and $\{\mathbf{z}_1, \ldots, \mathbf{z}_n\}$ is the reciprocal set ($\mathbf{z}_i^\dagger \mathbf{x}_j = \delta_{ij}$). Prove that the matrix
$$\mathbf{A} = \mathbf{x}_i \mathbf{z}_j^\dagger, \qquad i \neq j,$$
is an *imperfect* matrix.

5. Consider the fourth-order equation
$$\frac{d^4u}{dt^4} + 5\frac{d^3u}{dt^3} + 9\frac{d^2u}{dt^2} + 7\frac{du}{dt} + 2u = \sin t,$$

with the initial conditions ($t = 0$)
$$\frac{d^3u}{dt^3} = 2, \qquad \frac{d^2u}{dt^2} = 0, \qquad \frac{du}{dt} = -3, \qquad u = 4.$$

Convert this problem into a first-order system and, using the Jordan canonical transformation, find and plot u and du/dt as a function of time.

6. Consider the matrix

$$A = \begin{bmatrix} 1 & 0 & 0 & 0 & 0 & 0 & 0 \\ 0 & 2 & 0 & 0 & 0 & 0 & 0 \\ 0 & 0 & 1 & 0 & 0 & 0 & 0 \\ 0 & 0 & 0 & 1 & 0 & 1 & 0 \\ 0 & 0 & 0 & 0 & 1 & 0 & 0 \\ 1 & 0 & 0 & 0 & 0 & 1 & 0 \\ 0 & 0 & 0 & 0 & 0 & 2 & 0 \end{bmatrix}.$$

(a) Determine the eigenvalues of A and their respective multiplicities.
(b) Find the rank of $A - \lambda_i I$. How many eigenvectors does each eigenvalue have?
(c) Find the matrix Q that Jordan block diagonalizes A in a similarity transformation.
(d) Consider the problem

$$\frac{d^2 x}{dt^2} = A^2 x, \qquad x(t=0) = x_0, \qquad \frac{dx}{dt}(t=0) = 0,$$

and its solution

$$x = \cos(At)\, x_0.$$

Give the solution in terms of the Jordan block diagonalization.

7. Transform the matrix

$$A = \begin{bmatrix} 5 & 4 & 3 \\ -1 & 0 & -3 \\ 1 & -2 & 1 \end{bmatrix}$$

into its Jordan block form using a similarity transformation. Give the solution to

$$\frac{dx}{dt} = Ax, \qquad x(t=0) = \begin{bmatrix} -1 \\ 2 \\ 1 \end{bmatrix},$$

and plot x_i, $i = 1, 2, 3$, versus t.

8. Find the eigenvalues of **A** and the matrix **Q** that transforms **A** into Jordan block diagonal form in a similarity transformation.

$$A = \begin{bmatrix} \frac{17}{6} & \frac{5}{3} & \frac{1}{2} & \frac{1}{3} & \frac{1}{6} \\ \frac{1}{3} & \frac{8}{3} & 2 & \frac{4}{3} & \frac{2}{3} \\ -1 & -2 & -1 & -3 & -2 \\ \frac{5}{6} & \frac{5}{3} & \frac{5}{2} & \frac{16}{3} & \frac{13}{6} \\ -\frac{1}{6} & -\frac{1}{3} & -\frac{1}{2} & -\frac{2}{3} & \frac{13}{6} \end{bmatrix}.$$

9. A galvanometer is an electromagnetic device for measuring electric current via the deflection of a needle through some angle θ. In a typical galvanometer, the deflection obeys the equation of motion

$$J\frac{d^2\theta}{dt^2} + D\frac{d\theta}{dt} + S\theta = \frac{EG}{R}, \tag{1}$$

where G is a calibration coefficient, D is a damping coefficient, S is the effective force constant of a spring opposing the deflection, J is the moment of inertia of the needle about its rotation axis, and R is the resistance of the galvanometer. The quantity E is the applied initial voltage (at $t = 0$) when $\theta = d\theta/dt = 0$.

Depending on the parameters J, D, and S, the galvanometer can undergo undamped

$$\theta = \theta_p(1 - \cos \omega_0 t),$$

underdamped

$$\theta = \theta_p[1 - K\exp(-at)\sin(\omega t + \phi)],$$

and critical damped motion

$$\theta = \theta_p[1 - (1 + \omega_0 t)\exp(-\omega_0 t)].$$

(a) Convert the second-order problem in Eq. (1) into the first-order system

$$\frac{d\mathbf{x}}{dt} = \mathbf{A}\mathbf{x}, \quad \mathbf{x} = \begin{bmatrix} \theta \\ \frac{d\theta}{dt} \end{bmatrix}. \tag{2}$$

(b) From the properties of **A**, determine the necessary conditions on D, J, and S needed to satisfy each type of motion described above.

(c) Solve Eq. (2) and give the formulas for θ_p, ω_0, ω, K, a, and θ in terms of the constants J, D, S, E, G, and R.

FURTHER READING

Bellman, R. (1970). "Introduction to Matrix Analysis." McGraw-Hill, New York.
Noble, B. (1969). "Applied Linear Algebra," Prentice Hall International, Englewood Cliffs, NJ.
Noble B., and Daniel, J. W. (1977). "Applied Linear Algebra." Prentice Hall International, Englewood Cliffs, NJ.
Watkins, S. W. (1991). "Fundamentals of Matrix Computations." Wiley, New York.

8
INFINITE-DIMENSIONAL LINEAR VECTOR SPACES

8.1. SYNOPSIS

The remaining chapters of this text will be devoted to the theory of infinite-dimensional linear vector spaces and function spaces. We will find, however, that a great deal of what we learned about finite-dimensional spaces will still apply—with a few added stipulations. Therefore, we will, whenever possible, draw upon the theorems and definitions presented in Chapters 1–7 with the aim of forming a solid analogy between matrix theory and linear operator theory.

We will begin by showing with examples that function spaces are, in fact, infinite dimensional, where functions of the form $f(t)$ become analogous to vectors in finite-dimensional spaces. We will quickly see, however, that the concept of infinite dimensionality allows for many types of function spaces. We will consequently define several of the more common ones.

In the finite-dimensional vector space, E_n, n linearly independent vectors $\mathbf{x}_i$, $i = 1, \ldots, n$, always form a basis set; i.e., any vector in E_n can be expressed as a linear combination of the $\mathbf{x}_i$. In an infinite-dimensional space, however, an infinite number of linearly independent vectors is not necessarily a basis set. Hence, infinite dimensionality requires us to develop the concept of the *completeness* of a vector space. This, in turn, leads to the need to enlarge the concept of an integral from the definition of Riemann to the definition of Lebesgue.

The abstract linear vector spaces having the closest analog to finite linear vector spaces are Hilbert spaces. A Hilbert space $\mathcal{H}$ is a linear vector space in which an inner or scalar product is defined, and in which every Cauchy sequence converges to a vector contained in $\mathcal{H}$. A Cauchy sequence is a sequence of vectors $\mathbf{v}_1, \mathbf{v}_2, \ldots$ such that the length of the difference $\mathbf{v}_n - \mathbf{v}_m$ converges in the infinite limit. Namely,

$$\|\mathbf{v}_n - \mathbf{v}_m\| < \epsilon$$

if n and m are greater than some integer $N(\epsilon)$ that depends on ϵ. In a Hilbert space, the vector $\mathbf{v} = \lim_{n \to \infty} \mathbf{v}_n$ always lies in the space if $\mathbf{v}_n, n = 1, 2, \ldots$, is a sequence converging in the Cauchy sense.

A typical function space that forms a Hilbert space is $\mathcal{L}_2(a, b)$, whose functions are square integrable, i.e.,

$$\int_a^b f^*(t) f(t) \, dt < \infty.$$

The need for square integrability arises directly from the definition of the inner product of $\mathbf{f}$ and $\mathbf{g}$:

$$\langle \mathbf{f}, \mathbf{g} \rangle = \int_a^b f^*(t) g(t) \, dt,$$

where $f^*(t)$ is the complex conjugate of $f(t)$. As we shall see, the integrals must be defined according to the measure theory of Lebesgue.

We will introduce the concept of linear operators and compare their properties with the finite-dimensional analogs of matrices. We will see, for instance, that, for a vector $\mathbf{x} \in E_n$, the matrix product $\mathbf{A}\mathbf{x}$ is also contained in E_n. However, this is not necessarily the case in Hilbert spaces. We will show that the *domain* of a linear operator in $\mathcal{H}$ is actually, in general, only a subset of $\mathcal{H}$.

We will define a special class of operators called dyadic operators represented in the form

$$\mathbf{K} = \sum_{i=1}^{k} \mathbf{a}_i \mathbf{b}_i^{\dagger}.$$

We will discuss the solutions of the linear equations, $\mathbf{L}\mathbf{u} = \mathbf{f}$, in which the operator $\mathbf{L}$ is either a k-term dyadic or the sum of the identity operator and a k-term dyadic. These problems can be solved using the solvability theorem for linear systems in a finite-dimensional vector space.

Finally, we will show that perfect operators in a Hilbert space can be represented by a dyadic operator just as in finite linear vector spaces—although the dyadic may contain an infinite number of dyads. This representation is a spectral decomposition totally analogous to perfect matrices in finite-dimensional vector spaces. In function spaces, the implication of this is that, whether the operator is an integral or a differential operator, the spectral representation is with certainty an integral operator.

8.2. INFINITE-DIMENSIONAL SPACES

The definition of a linear vector space is the same whether the space is finite or infinite dimensional. As described in Chapter 2, a vector space S is a collection of elements $\mathbf{x}, \mathbf{y}, \mathbf{z} \ldots$ having the operation of addition (denoted by $+$) and possessing

INFINITE-DIMENSIONAL SPACES

the following properties:

1. If $\mathbf{x}, \mathbf{y} \in S$, then

$$\mathbf{x} + \mathbf{y} \in S. \tag{8.2.1}$$

2.
$$\mathbf{x} + \mathbf{y} = \mathbf{y} + \mathbf{x}. \tag{8.2.2}$$

3. There exists a zero vector $\mathbf{0}$ such that

$$\mathbf{x} + \mathbf{0} = \mathbf{x}. \tag{8.2.3}$$

4. For every $\mathbf{x} \in S$, there exists $-\mathbf{x}$ such that

$$\mathbf{x} + (-\mathbf{x}) = \mathbf{0}. \tag{8.2.4}$$

5. If α and β are complex numbers, then

$$\begin{aligned} \alpha(\beta \mathbf{x}) &= (\alpha \beta) \mathbf{x} \\ (\alpha + \beta) \mathbf{x} &= \alpha \mathbf{x} + \beta \mathbf{x} \\ \alpha(\mathbf{x} + \mathbf{y}) &= \alpha \mathbf{x} + \alpha \mathbf{y}. \end{aligned} \tag{8.2.5}$$

Since a real vector space is contained within a complex vector space, we will deal with complex vector spaces throughout the rest of this text.

In both E_n and E_∞, a vector is denoted by a boldface, lowercase letter, e.g., $\mathbf{x}$, and its ith component is denoted by a lowercase letter with the subscript i, e.g., x_i. We have concentrated in previous chapters on the vector space E_n in which

$$\mathbf{x} = \begin{bmatrix} x_1 \\ x_2 \\ \vdots \\ x_n \end{bmatrix}, \tag{8.2.6}$$

and we defined the adjoint $\mathbf{x}^\dagger$ of $\mathbf{x}$ as

$$\mathbf{x}^\dagger \equiv [x_1^*, \ldots, x_n^*]. \tag{8.2.7}$$

Analogously to E_n, a vector in the space E_∞ is of the form

$$\mathbf{x} = \begin{bmatrix} x_1 \\ x_2 \\ \vdots \end{bmatrix}, \tag{8.2.8}$$

and its adjoint is given by

$$\mathbf{x}^\dagger = [x_1^*, x_2^*, \ldots]. \tag{8.2.9}$$

We say that $\mathbf{x}$ has a denumerably infinite number of components.

Function spaces can also form a linear vector space. For example, $C^0(0, 1)$, the space of all continuous functions defined on the interval $(0, 1)$, is a linear vector space. We will denote a vector in $C^0(0, 1)$ by a boldface, lowercase letter, e.g., **f**. This vector represents a continuous function $f(t)$, $0 < t < 1$, in our function space. Whereas **f** represents all values of the function (analogous to **x** in E_∞), the quantity $f(t)$ is the value of the function at t and is analogous to a component x_i of **x**. In fact, in a discretized representation of the function $f(t)$, we can introduce the approximation

$$\mathbf{f} \rightarrow \begin{bmatrix} f_1 \\ f_2 \\ \vdots \\ f_n \end{bmatrix}, \qquad (8.2.10)$$

where $f_i = f(t_i)$ and t_i, $i = 1, \ldots, n$, are the values of the continuous variable t at which we estimate $f(t)$.

In function spaces, we also introduce the concept of an adjoint vector. Thus, if **f** is a vector in the function space $C^0(0, 1)$ representing the function $f(t)$, $0 < t < 1$, then its adjoint vector $\mathbf{f}^\dagger$ represents all values of the function $f^*(s)$, $0 < s < 1$. Here $f^*(s)$ is just the complex conjugate of $f(s)$.

We need not restrict ourselves to one-dimensional arguments. For example, we can define another vector space $C^0(\Omega_D)$ consisting of the continuous functions $f(\vec{r})$ of the D-dimensional vector $\vec{r}$ in the volume Ω_D. If the vector **f** belongs to $C^0(\Omega_D)$, then the component of the adjoint vector $\mathbf{f}^\dagger$ is the function $f^*(\vec{s})$, where, again, $f^*(\vec{s})$ is the complex conjugate of $f(\vec{s})$ and $\vec{s}$ is a D-dimensional vector in Ω_D. For the purposes of this book, $f(\vec{r})$ need not be a geometrical vector. It can be any continuous variable with D components (such as concentrations of chemical species in solution) with Ω_D denoting a D-dimensional volume over which the values of the components range.

We shall define the dimension of a space as the number of linearly independent vectors the space contains. It is then easy to establish that function spaces are infinite dimensional. For example, in the space $C^0(0, 1)$, the functions $f_n(t) = t^n$, $n = 0, 1, 2, \ldots$, are linearly independent vectors and so $C^0(0, 1)$ is an infinite-dimensional vector space.

One can define numerous function spaces. A few examples are:

1. $C^n(a, b)$: the space of functions $f(t), g(t), \ldots$ whose first n derivatives are continuous in the interval (a, b).
2. $C^n(\Omega_D)$: the space of functions $f(\vec{r}), g(\vec{r}), \ldots$ whose first- through nth order derivatives are continuous functions of the D-dimensional Euclidean vector $\vec{r}$ in the volume Ω_D.
3. $R(a, b)$: the space of functions whose Riemann integral exists; i.e., if $\mathbf{f} \in R(a, b)$, then

$$\int_a^b f(t)\, dt \qquad (8.2.11)$$

exists, where the above integral is the Riemann integral.

RIEMANN AND LEBESGUE INTEGRATION

4. $\mathcal{L}_1(a,b)$: the space of functions whose Lebesgue integral exists; i.e., if $\mathbf{f} \in \mathcal{L}_1(a,b)$, then

$$\int_a^b f(t)\,dt \qquad (8.2.12)$$

exists, where the above integral is the Lebesgue integral. *The difference between the Riemann and the Lebesgue definitions of an integral will be explained in the next section.*

5. $\mathcal{L}_2(\Omega_D)$: the space of functions, of D variables, whose squared absolute values are Lebesgue integrable; i.e., if $\mathbf{f} \in \mathcal{L}_2(\Omega_D)$, then

$$\int_{\Omega_D} \cdots \int f^*(\vec{r}) f(\vec{r})\, d^D r \qquad (8.2.13)$$

exists, where the D-dimensional integral is a Lebesgue integral.

In the subsequent sections, we shall utilize the concept of a complete vector space. In essence, this property will ensure that a set of basis functions will exist for that space. Thus, if a space S is complete, then we can find a basis set $\mathbf{f}_1, \mathbf{f}_2, \ldots$ such that any vector $\mathbf{f}$ in S can be represented by the series

$$\mathbf{f} = \sum_n \alpha_n \mathbf{f}_n, \qquad (8.2.14)$$

where the scalars α_n are unique for a given vector $\mathbf{f}$.

For example, E_n forms a complete vector space, but of the function spaces defined above, only $\mathcal{L}_1$ and $\mathcal{L}_2$ are complete. This is why Lebesgue integrability is important in the general theory of linear vector spaces and why we shall take it up in the next section.

8.3. RIEMANN AND LEBESGUE INTEGRATION

The integral of a continuous function $f(t)$ in the interval $a < t < b$ is the area under the curve representing $f(t)$ as shown in Fig. 8.3.1. The way we define a Riemann integral is to discretize the interval (a,b) into the subintervals $(t_i, t_i + \Delta t_i)$, $i = 1, \ldots, n$, and then construct the upper and lower Darboux sums D_u^n and D_l^n. In the interval $(t_i, t_i + \Delta t_i)$, we define the upper bound on $f(t)$ as $f_i^u = \max_t f(t)$

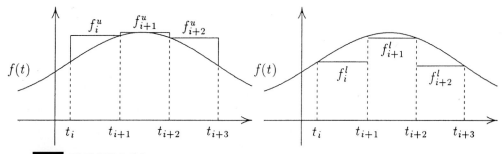

FIGURE 8.3.1

and the lower bound as $f_i^l = \min_t f(t)$. These are illustrated in Fig. 8.3.1. The upper Darboux sum is the sum of the areas of the rectangles of height f_i^u and width Δt_i, i.e.,

$$D_u^n = \sum_{i=1}^n f_i^u \, \Delta t_i. \tag{8.3.1}$$

Likewise, the lower Darboux sum is defined as the sum of the areas of the rectangles of height f_i^l and width Δt_i, i.e.,

$$D_l^n = \sum_{i=1}^n f_i^l \, \Delta t_i. \tag{8.3.2}$$

If, as the discretization is refined, i.e., $n \to \infty$ and $\Delta t_i \to 0$, the Darboux sums converge and are equal to each other, we say the Riemann integral exists and has the value

$$\int_a^b f(t) \, dt = \lim_{n \to \infty} D_u^n = \lim_{n \to \infty} D_l^n. \tag{8.3.3}$$

It is easy to show that continuous and piecewise-continuous functions are Riemann integrable, but what about the situation when $f(t)$ is continuous except at a set of points t_j, $j = 1, 2, \ldots$, where $f(t_j) \neq \lim_{t \to t_j} f(t)$? The area under an isolated point is 0, and so one does not expect the area under a curve representing $f(t)$ to be affected by these isolated points. However, the Riemann integral of a function can fail to exist even though the area under its curve exists. For example, consider the interval (a, b) and the function

$$\begin{aligned} f(t) &= 1, & t \text{ an irrational number} \\ &= 0, & t \text{ a rational number.} \end{aligned} \tag{8.3.4}$$

In any interval Δt_i, there will be both rational and irrational numbers, and so

$$D_u^n = 1 \quad \text{and} \quad D_l^n = 0. \tag{8.3.5}$$

Since D_u^n and D_l^n have different limits, the function is not Riemann integrable. The Lebesgue measure theory introduced early in the 20th century leads to a definition of the integral of a function that captures the concept of the integral as the area under the curve, avoiding the problem caused by a function having discontinuities such as in Eq. (8.3.4).

In Lebesgue's theory, one considers the class of functions $f(t)$, which can be approximated by a sequence of step functions $\psi_n(t)$ almost everywhere, i.e., everywhere but on a denumerable set of isolated points (denumerable means the number can be infinite, but countably infinite in the sense that the set can be mapped one to one onto the set of all positive integers—the irrational numbers are an example of an uncountable set). The step functions are defined in the function $\psi_n(t)$ as

$$\psi_n(t) \equiv \bar{f}_i^n, \qquad t_i < t < t_i + \Delta t_i, \tag{8.3.6}$$

where $\bar{f}_i^n$ is the value of $f(t)$ within the interval i *not* including the isolated points in question. The integral of the step functions is then

$$\int_a^b \psi_n(t)\,dt = \sum_{i=1}^n \bar{f}_i^n \Delta t_i. \tag{8.3.7}$$

The requirement for the existence of a Lebesgue integral of $f(t)$ is that

$$\left| \int_a^b \psi_n(t)\,dt \right| < A \neq \infty \tag{8.3.8}$$

and that

$$f(t) = \lim_{n \to \infty} \psi_n(t) \tag{8.3.9}$$

almost everywhere. Consequently, the the Lebesgue integral is given by

$$\int_a^b f(t)\,dt = \lim_{n \to \infty} \int_a^b \psi_n(t)\,dt. \tag{8.3.10}$$

Clearly, when a Riemann integral exists, the Lebesgue integral exists. For example, we could sufficiently choose as $\psi_n(t)$ either

$$\psi_n(t) = f_i^u, \qquad t_i < t < t_i + \Delta t_i, \tag{8.3.11}$$

or

$$\psi_n(t) = f_i^l, \qquad t_i < t < t_i + \Delta t_i. \tag{8.3.12}$$

On the other hand, while the Riemann integral of the function defined by Eq. (8.3.4) does not exist, the step function

$$\psi_n(t) = 1, \qquad a < t < b, \tag{8.3.13}$$

is equal to $f(t)$ almost everywhere, and so the Lebesgue integral of $f(t)$ does exist and gives the result

$$\int_a^b f(t)\,dt = \lim_{n \to \infty} \int_a^b 1\,dt = b - a. \tag{8.3.14}$$

In practical problems, functions are generally Riemann integrable and so the subtleties involved in Lebesgue integration need not concern us except in as much as they guarantee the completeness of certain function spaces. The need for completeness of a linear vector space will be addressed in the next section. Those who are interested in a deeper appreciation of the measure theory and the Lebesgue integral are encouraged to explore references recommended at the end of this chapter.

8.4. INNER PRODUCT SPACES

In the space E_n, the simplest inner or scalar product we can define is $\mathbf{x}^\dagger \mathbf{y} = \sum_{i=1}^n x_i^* y_i$. Analogously, in E_∞, we can define the inner product $\mathbf{x}^\dagger \mathbf{y} \equiv \sum_{i=1}^\infty x_i^* y_i$. More generally, we defined in Section 5.2 the inner product for an arbitrary abstract linear vector space $\mathcal{S}$ by the properties

$$\langle \mathbf{x}, \mathbf{y} \rangle = \langle \mathbf{y}, \mathbf{x} \rangle^* \tag{8.4.1}$$

$$\langle \alpha \mathbf{x} + \beta \mathbf{y}, \mathbf{z} \rangle = \alpha^* \langle \mathbf{x}, \mathbf{z} \rangle + \beta^* \langle \mathbf{y}, \mathbf{z} \rangle \tag{8.4.2}$$

and

$$\langle \mathbf{x}, \mathbf{x} \rangle > 0 \quad \text{for any } \mathbf{x} \neq \mathbf{0} \text{ in } \mathcal{S}. \tag{8.4.3}$$

The length (or norm) of $\mathbf{x}$ in $\mathcal{S}$ was defined by

$$\|\mathbf{x}\| \equiv \sqrt{\langle \mathbf{x}, \mathbf{x} \rangle}, \tag{8.4.4}$$

and we also proved in Section 5.2 that the properties in Eqs. (8.4.1)–(8.4.3) imply the triangle inequality

$$\|\mathbf{x} + \mathbf{y}\| \leq \|\mathbf{x}\| + \|\mathbf{y}\| \tag{8.4.5}$$

and the Schwarz inequality

$$|\langle \mathbf{x}, \mathbf{y} \rangle| \leq \|\mathbf{x}\| \, \|\mathbf{y}\|. \tag{8.4.6}$$

These proofs are valid for an arbitrary inner product space. Some examples of inner products in linear vector spaces include:

(i) The space E_n, with

$$\langle \mathbf{x}, \mathbf{y} \rangle = \mathbf{x}^\dagger \mathbf{y} = \sum_{i=1}^n x_i^* y_j. \tag{8.4.7}$$

(ii) The space E_n, with

$$\langle \mathbf{x}, \mathbf{y} \rangle = \mathbf{x}^\dagger \mathbf{A} \mathbf{y} = \sum_{i=1}^n a_{ij} x_i^* y_j, \tag{8.4.8}$$

where $\mathbf{A}$ is a positive-definite matrix.

(iii) The space E_∞, with

$$\langle \mathbf{x}, \mathbf{y} \rangle = \mathbf{x}^\dagger \mathbf{y} = \sum_{i=1}^\infty x_i^* y_j. \tag{8.4.9}$$

(iv) The space E_∞, with

$$\langle \mathbf{x}, \mathbf{y} \rangle \equiv \mathbf{x}^\dagger \mathbf{A} \mathbf{y} = \sum_{i=1}^\infty a_{ij} x_i^* y_j, \tag{8.4.10}$$

where $\mathbf{A}$ is a positive-definite matrix.

INNER PRODUCT SPACES

(v) The space $\mathcal{L}_2(0, 1)$ of functions square integrable in the interval $(0, 1)$, with

$$\langle \mathbf{f}, \mathbf{g} \rangle = \mathbf{f}^\dagger \mathbf{g} = \int_0^1 f^*(t) g(t) \, dt. \tag{8.4.11}$$

(vi) The space $\mathcal{L}_2(0, 1; k(t))$, with

$$\langle \mathbf{f}, \mathbf{g} \rangle \equiv \int_0^1 f^*(t) g(t) k(t) \, dt, \tag{8.4.12}$$

where $k(t) > 0$ almost everywhere.

(vii) The space $\mathcal{L}_2(\Omega_D)$ of functions square integrable in the D-dimensional volume Ω_D, with

$$\langle \mathbf{f}, \mathbf{g} \rangle = \mathbf{f}^\dagger \mathbf{g} \equiv \int \cdots \int_{\Omega_D} f^*(\vec{r}) g(\vec{r}) \, d^D r. \tag{8.4.13}$$

(viii) The space $\mathcal{L}_2(\Omega_D, k)$, with

$$\langle \mathbf{f}, \mathbf{g} \rangle \equiv \int \cdots \int_{\Omega_D} f^*(\vec{r}) g(\vec{r}) k(\vec{r}) \, d^D r, \tag{8.4.14}$$

where $k(\vec{r}) > 0$ almost everywhere in Ω_D.

In Eqs. (8.4.11) and (8.4.13), we have generalized the notation $\mathbf{x}^\dagger \mathbf{y}$ of the vector space E_n to function spaces. In E_n, the notation $\mathbf{x}^\dagger \mathbf{y}$ means the sum over all i of the elements $x_i^* y_i$. Likewise, in a function space $\mathcal{S}$, the notation $\mathbf{f}^\dagger \mathbf{g}$ means the integral of the elements $f^*(\vec{r}) g(\vec{r})$ over the volume (or interval in one dimension) in which the functions are defined.

Notice in the function spaces $\mathcal{L}_2(0, 1)$ and $\mathcal{L}_2(\Omega_D)$ we can define the inner product with or without a weighting function k. We can even define an inner product analogously to Eqs. (8.4.8) and (8.4.10) in which $\mathbf{A}$ is replaced by a positive-definite integral operator. In this and in subsequent chapters, we will use the notation $\langle \mathbf{x}, \mathbf{y} \rangle$ and $\mathbf{x}^\dagger \mathbf{y}$ interchangeably as inner products. In handling dyadic expressions, $\mathbf{x}^\dagger \mathbf{y}$ and $\mathbf{y} \mathbf{x}^\dagger$ turn out to be quite suggestive and will aid us significantly in generalizing our thinking from finite-dimensional vector spaces to function spaces.

Another inner product that further illustrates the generality of the concept is

(ix) The space $C^1(0, 1)$ of functions with continuous first derivatives in the closed interval $(0, 1)$, with

$$\langle \mathbf{f}, \mathbf{g} \rangle = \int_0^1 \frac{df^*(t)}{dt} \frac{dg(t)}{dt} \, dt + f^*(0) g(0). \tag{8.4.15}$$

Note that the temptingly simpler definition

$$\langle \mathbf{f}, \mathbf{g} \rangle = \int_0^1 \frac{df^*(t)}{dt} \frac{dg(t)}{dt} \, dt \tag{8.4.16}$$

does *not* define an inner product. This is because the vector $\mathbf{f} = \mathbf{1}$ (or the function $f(t) = 1$) is a nonzero vector belonging to $C^1(0, 1)$, and, if the inner product is defined by Eq. (8.4.16),

$$\langle \mathbf{f}, \mathbf{f} \rangle = \int_0^1 0 \times 0 \, dt = 0, \tag{8.4.17}$$

which violates the property required in Eq. (8.4.3).

8.5. HILBERT SPACES

We say a vector belongs to a *normed* linear vector space $\mathcal{S}$ if every vector $\mathbf{x}$ in $\mathcal{S}$ has a finite norm or length, i.e., if $\|\mathbf{x}\| < \infty$. We claimed in earlier chapters that norms can be defined independently of inner products. For example, in E_n, the $p = \infty$ norm of a vector is given by $\|\mathbf{x}\|_\infty = \max_{1 \leq i \leq n} |x_i|$. In the remainder of this text, however, we shall restrict ourselves to spaces whose norm is defined as an inner product, i.e., $\|\mathbf{x}\|^2 = \langle \mathbf{x}, \mathbf{x} \rangle$. This is because we want to exploit as much as possible what we have learned about the theory of matrix operators in E_n in understanding linear operators in abstract linear vector spaces.

The space E_∞, with the inner product $\langle \mathbf{x}, \mathbf{y} \rangle = \sum_i x_i^* y_i$, is not an example of a normed linear vector space. This is because vectors exist in E_∞ that are of infinite length; e.g., if

$$\mathbf{x} = \begin{bmatrix} 1 \\ \frac{1}{\sqrt{2}} \\ \frac{1}{\sqrt{3}} \\ \vdots \end{bmatrix}, \tag{8.5.1}$$

then

$$\|\mathbf{x}\|^2 = \sum_{i=1}^\infty \frac{1}{i} = \infty. \tag{8.5.2}$$

On the other hand, the space $\mathcal{S} = E_\infty$ with the condition $\|\mathbf{x}\|^2 = \langle \mathbf{x}, \mathbf{x} \rangle < \infty$ does form a normed linear vector space. Interestingly, if we chose the $p = \infty$ norm, $\|\mathbf{x}\|_\infty = 1$ for this example. However, the vector $\mathbf{x}$ with components $x_i = i$ has an infinite $p = \infty$ norm.

Unlike finite-dimensional spaces, in an infinite-dimensional space we must concern ourselves with the issues of convergence (of a sequence of vectors) and completeness of the space. A sequence of vectors $\mathbf{x}_1, \mathbf{x}_2, \ldots$ in $\mathcal{S}$ is said to converge to a vector $\mathbf{x}$ if, for any number $\epsilon > 0$, there exists an integer $N(\epsilon)$ such that

$$\|\mathbf{x} - \mathbf{x}_n\| < \epsilon \qquad \text{for } n > N(\epsilon). \tag{8.5.3}$$

The vector $\mathbf{x}$ is said to be the limit of the sequence, namely,

$$\mathbf{x} = \lim_{n \to \infty} \mathbf{x}_n. \tag{8.5.4}$$

From the triangle inequality, it follows that if Eq. (8.5.3) is true, then

$$\|\mathbf{x}_m - \mathbf{x}_n\| = \|\mathbf{x} - \mathbf{x}_n - (\mathbf{x} - \mathbf{x}_m)\| \leq \|\mathbf{x} - \mathbf{x}_n\| + \|\mathbf{x} - \mathbf{x}_m\|$$
$$< \epsilon \qquad \text{for } n, m > N\left(\frac{\epsilon}{2}\right). \tag{8.5.5}$$

Such a sequence of vectors is said to converge in the Cauchy sense; i.e., we say that if a sequence $\mathbf{x}_n$ in $\mathcal{S}$ converges to a vector $\mathbf{x}$ in $\mathcal{S}$, then the sequence converges in the Cauchy sense.

If the sequence $\mathbf{x}_1, \mathbf{x}_2, \ldots$ converges in the Cauchy sense, for vectors in the finite-dimensional vector space E_n, then

$$\mathbf{x} = \lim_{m \to \infty} \mathbf{x}_m \in E_n; \tag{8.5.6}$$

i.e., the sequence converges to a vector in E_n if it converges in the Cauchy sense. This is because of the well-known property that if, for the sequence of complex numbers $\alpha_1, \alpha_2, \ldots$, there exists an integer $N(\epsilon)$ such that

$$|\alpha_m - \alpha_p| < \epsilon \qquad \text{for } m, p > N(\epsilon), \tag{8.5.7}$$

then $\lim_{m \to \infty} \alpha_m$ exists and is equal to a complex number.

On the other hand, in an infinite-dimensional vector space S, Cauchy convergence of a sequence does not always imply that the limit of the sequence is a vector in the space S. For example, consider the space $C^0(0, 1)$ of continuous functions defined on the interval $(0, 1)$. Defining the inner product as

$$\langle \mathbf{f}, \mathbf{g} \rangle = \int_0^1 f^*(t) g(t)\, dt, \tag{8.5.8}$$

we can define the following sequence of continuous functions:

$$f_n(t) = \begin{cases} 0, & 0 \le t \le \dfrac{1}{2} - \dfrac{1}{2n}, \\ n\left(t - \dfrac{1}{2}\right) + \dfrac{1}{2}, & \dfrac{1}{2} - \dfrac{1}{2n} \le t \le \dfrac{1}{2} + \dfrac{1}{2n}, \\ 1, & \dfrac{1}{2} + \dfrac{1}{2n} \le t \le 1. \end{cases} \tag{8.5.9}$$

These functions are plotted in Fig. 8.5.1 for a few values of n.

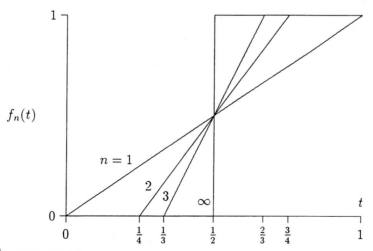

FIGURE 8.5.1

EXERCISE 8.5.1. For a sequence defined by Eq. (8.5.9), show that, for any number $\epsilon > 0$, there exists an integer $N(\epsilon)$ such that

$$\|\mathbf{f}_n - \mathbf{f}_m\| = \int_0^1 |f_n(t) - f_m(t)|^2 \, dt < \epsilon, \qquad n, m > N(\epsilon). \qquad (8.5.10)$$

Equation (8.5.10) establishes that $\mathbf{f}_n$ converges in the Cauchy sense. However, the limit of this sequence ($n \to \infty$) is the function

$$f(t) = \begin{cases} 0, & 0 \le t \le \dfrac{1}{2}, \\ 1, & \dfrac{1}{2} < t \le 1. \end{cases} \qquad (8.5.11)$$

This function is not continuous in the interval $C^0(0, 1)$, and so the limit vector $\mathbf{f} = \lim_{n \to \infty} \mathbf{f}_n$ is not in the vector space of continuous functions.

We say that a vector space $\mathcal{S}$ is *complete* if the limit vector of any sequence converging in the Cauchy sense is a vector in the space $\mathcal{S}$. Clearly, if a given space is not complete, it can be made complete by adding to it all of the limit vectors of Cauchy sequences. When this is done for the space of continuous functions with the inner product defined by Eq. (8.5.8), the resulting complete space is $\mathcal{L}_2(0, 1)$—the space of functions square integrable in the sense of Lebesgue. The need to complete a vector space forced analysts to abandon the Riemann integral in favor of the Lebesgue integral.

Without proving it, let us remark that the linear spaces (i)–(viii), with the inner products indicated by Eqs. (8.4.7)–(8.4.14), and with the requirement that $\|\mathbf{x}\|^2 = \langle \mathbf{x}, \mathbf{x} \rangle < \infty$, form complete vector spaces. If a sequence in any of these spaces converges in the Cauchy sense, then it converges to a vector in the space. Another way of saying this is that if the sequence $\mathbf{x}_n$ converges in the Cauchy sense in a complete space $\mathcal{S}$, then the limit $\mathbf{x}$ of the sequence exists and is in the space $\mathcal{S}$.

A complete linear vector space with an inner product norm is called a Hilbert space. In the remainder of this text, the theory of linear operators will be conducted in an appropriate Hilbert space. We shall denote a Hilbert space generically by $\mathcal{H}$, although more suggestive notation such as E_n, $\mathcal{L}_2(0, 1)$, and $\mathcal{L}_2(\Omega_D)$ will be used at times.

8.6. BASIS VECTORS

Just as in finite-dimensional vector spaces, all Hilbert spaces possess basis sets. We say that the linearly independent vectors $\mathbf{u}_1, \mathbf{u}_2, \ldots$ form a basis set in $\mathcal{H}$ if, for any vector $\mathbf{f}$ in $\mathcal{H}$, there exists a unique set of complex numbers α_1, α_2 such that

$$\mathbf{f} = \sum_n \alpha_n \mathbf{u}_n. \qquad (8.6.1)$$

In function spaces, the basis sets are denumerably infinite; i.e., the number of basis vectors is infinite, but the vectors are countable in the sense that they are in one-to-one correspondence with the positive integers.

As a simple example of a basis set, consider the set of all analytic functions $f(t)$ in the interval $(-1, 1)$. Since the functions are analytic, the series expansion

$$f(t) = \sum_{n=0}^{\infty} f_n t^n \tag{8.6.2}$$

always exists. This means that the functions $u_n = t^n$, $n = 0, 1, 2, \ldots$, form a basis set for analytic functions in the interval $(-1, 1)$. Although the functions $u_n = t^n$, $n = 0, 1, \ldots$, are linearly independent, they are not orthogonal since

$$\int_{-1}^{1} u_n^* u_m \, dt = \frac{2}{n+m+1}. \tag{8.6.3}$$

However, the Gram–Schmidt procedure can be used to orthogonalize a linearly independent set in any Hilbert space. Recall the procedure: if $\mathbf{u}_1, \mathbf{u}_2, \ldots$ is a linearly independent set, then the set $\mathbf{v}_1, \mathbf{v}_2, \ldots$, where

$$\begin{aligned}
\mathbf{v}_1 &= \mathbf{u}_1 \\
\mathbf{v}_2 &= \mathbf{u}_2 - \frac{\langle \mathbf{v}_1, \mathbf{u}_2 \rangle}{\langle \mathbf{v}_1, \mathbf{v}_1 \rangle} \mathbf{v}_1 \\
&\vdots \\
\mathbf{v}_j &= \mathbf{u}_j - \sum_{i=1}^{j-1} \frac{\langle \mathbf{v}_i, \mathbf{u}_j \rangle}{\langle \mathbf{v}_i, \mathbf{v}_i \rangle} \mathbf{v}_i,
\end{aligned} \tag{8.6.4}$$

is an orthogonal set, i.e.,

$$\langle \mathbf{v}_i, \mathbf{v}_i \rangle \neq 0 \quad \text{and} \quad \langle \mathbf{v}_i, \mathbf{v}_j \rangle = 0, \quad i \neq j. \tag{8.6.5}$$

EXAMPLE 8.6.1. Consider the vectors defined by the set of functions $u_n = t^n$, $n = 0, 1, 2, \ldots$, in the space $\mathcal{L}_2(-1, 1)$. Orthogonalize the set.

By the Gram–Schmidt procedure, the first three orthogonal vectors are defined by the functions v_0, v_1, and v_2:

$$\begin{aligned}
v_0(t) &= u_0(t) = 1 \\
v_1(t) &= u_1(t) - \frac{\langle v_0, u_1 \rangle}{\langle v_0, v_0 \rangle} v_0(t) \\
&= t - \frac{\int_{-1}^{1} t \, dt}{\int_{-1}^{1} dt} = t \\
v_2(t) &= u_2(t) - \frac{\langle v_0, u_2 \rangle}{\langle v_0, v_0 \rangle} v_0(t) - \frac{\langle v_1, u_2 \rangle}{\langle v_1, v_1 \rangle} v_1(t) \\
&= t^2 - \frac{\int_{-1}^{1} t^2 \, dt}{\int_{-1}^{1} dt} - \frac{\int_{-1}^{1} t^3 \, dt}{\int_{-1}^{1} t^2 \, dt} \\
&= t^2 - \frac{1}{3}.
\end{aligned} \tag{8.6.6}$$

Aside from multiplicative factors, these functions are the first three Legendre polynomials defined by

$$P_n(t) = \frac{1}{2^n n!} \frac{d^n}{dt^n} (t^2 - 1)^n. \tag{8.6.7}$$

■ ■ ■ The Legendre polynomials are known (e.g., from quantum mechanics) to be a basis set in the Hilbert space $\mathcal{L}_2(-1, 1)$.

From the theory of Fourier series, we know that the set of functions

$$v_n(t) = \sqrt{2} \sin n\pi t, \qquad n = 1, 2, \ldots, \tag{8.6.8}$$

form an orthonormal basis set in the Hilbert space $\mathcal{L}_2(0, 1)$. We shall use this set to illustrate what is meant when we say that vectors in a Hilbert space are equal. Consider the function $f(t)$. We define the vector $\tilde{\mathbf{f}}$ by the expansion

$$\tilde{f}(t) = \sum_{n=1}^{\infty} \tilde{\alpha}_n v_n(t) = \sum_{n=1}^{\infty} \tilde{\alpha}_n \sqrt{2} \sin n\pi t \tag{8.6.9}$$

and claim it is equal to the vector $\mathbf{f}$ if

$$\tilde{\alpha}_n = \int_0^1 v_n^*(t) f(t) \, dt = \sqrt{2} \int_0^1 \sin n\pi t f(t) \, dt \tag{8.6.10}$$

for all n. In other words, two vectors are equivalent if they have the same Fourier coefficients.

In $\mathcal{L}_2(0, 1)$, vector equality means

$$\|\mathbf{f} - \tilde{\mathbf{f}}\|^2 = \int_0^1 |f(t) - \tilde{f}(t)|^2 \, dt = 0. \tag{8.6.11}$$

If $f(t)$ is a continuous function, the theory of Fourier series teaches us that $\tilde{f}(t) = f(t)$ at every point in the interval $(0, 1)$. Thus, Eq. (8.6.11) follows since $f(t) - \tilde{f}(t) = 0$ for all values of t in $(0, 1)$. On the other hand, if $f(t)$ is piecewise continuous, $f(t) - \tilde{f}(t) = 0$ everywhere $f(t)$ is continuous, but, at a point of discontinuity t',

$$\tilde{f}(t') = \frac{1}{2}[f(t'_-) + f(t'_+)], \tag{8.6.12}$$

where $f(t'_-)$ is the value of $f(t)$ as t' is approached from below and $f(t'_+)$ is the value of $f(t)$ as t' is approached from above. For example, if

$$f(t) = \begin{cases} 0, & 0 \le t \le \frac{1}{2}, \\ 1, & \frac{1}{2} < t \le 1, \end{cases} \tag{8.6.13}$$

its Fourier expansion is

$$\tilde{f}(t) = \sum_{n=0}^{\infty} \tilde{\alpha}_n \sqrt{2} \sin n\pi t, \tag{8.6.14}$$

with

$$\tilde{\alpha}_n = \frac{\sqrt{2}}{n\pi}\left\{\frac{1}{2}[1+(-1)^n]^{n/2} - (-1)^n\right\}. \tag{8.6.15}$$

Then $\tilde{f}(t) = 0$, $0 \le t < \frac{1}{2}$; $\tilde{f}(t) = 1$, $\frac{1}{2} < t \le 1$, but

$$\tilde{f}\left(t = \frac{1}{2}\right) = \frac{1}{2}(0+1) = \frac{1}{2}. \tag{8.6.16}$$

Thus, $f(t) - \tilde{f}(t) \ne 0$ at $t = \frac{1}{2}$. However, even though $f(t)$ and $\tilde{f}(t)$ are not equal at every point in the interval t, since they differ only on a set of measure 0, it follows that $\|\mathbf{f} - \tilde{\mathbf{f}}\|^2 = \int_0^1 |f(t) - \tilde{f}(t)|^2 \, dt = 0$, and so the vector $\mathbf{f}$ is equal to the vector $\tilde{\mathbf{f}}$ in the Hilbert space $\mathcal{L}_2(0, 1)$.

Another point to make with the expansion in Eq. (8.6.14) is that the sequence

$$\tilde{f}_m(t) = \sum_{n=1}^{m} \tilde{\alpha}_n \sqrt{2} \sin n\pi t, \qquad m = 1, 2, \ldots, \tag{8.6.17}$$

with $\tilde{\alpha}_n$ given by Eq. (8.6.15), is a sequence of continuous functions, i.e., functions in $C^0(0, 1)$, which converges to a piecewise-continuous function in the limit $m \to \infty$. This points out the need to enlarge the space to include functions outside $C^0(0, 1)$ in order to complete it.

This next example is indicative of the need to use Lebesgue instead of Riemann integration in completing function spaces. Consider again the orthonormal basis set defined by Eq. (8.6.8). We define the piecewise-continuous function in $(0, 1)$:

$$f(t) = \frac{1}{i}, \qquad \frac{1}{i+1} < t \le \frac{1}{i}, \qquad i = 1, 2, \ldots. \tag{8.6.18}$$

The expansion of $f(t)$ in the basis set (8.6.8), i.e., the Fourier expansion, yields

$$\tilde{f}(t) = \sum_{n=0}^{\infty} \frac{2}{i}\left[\cos\left(\frac{n\pi}{i+1}\right) - \cos\left(\frac{n\pi}{i}\right)\right]\sin n\pi t, \qquad \frac{1}{i+1} < t \le \frac{1}{i}, \ i = 1, 2, \ldots, \tag{8.6.19}$$

and from the properties of Fourier series it follows that

$$f(t) - \tilde{f}(t) = 0, \qquad \frac{1}{i+1} < t < \frac{1}{i}, \ i = 1, 2, \ldots, \tag{8.6.20}$$

and

$$f(t) - \tilde{f}(t) = \frac{1}{2}\left(\frac{1}{i} - \frac{1}{i+1}\right) \quad \text{at } t = \frac{1}{i}, \ i = 2, 3, \ldots. \tag{8.6.21}$$

If we try to take the upper and lower Darboux sums for the function $f(t) - \tilde{f}(t)$ for arbitrary discretizations of the interval $(0, 1)$, we find that the Riemann integral does not exist because the upper and lower Darboux sums differ. However, since $f(t)$ and $\tilde{f}(t)$ only differ on a denumerable set of measure 0, the Lebesgue integral of $f(t) - \tilde{f}(t)$ is 0, and so $\|\mathbf{f} - \tilde{\mathbf{f}}\| = 0$ and $\mathbf{f} = \tilde{\mathbf{f}}$ in the Hilbert space $\mathcal{L}_2(0, 1)$.

We should note that, although n linearly independent vectors in E_n always form a basis set, an infinite number of linearly independent vectors do not necessarily form a basis set in an infinite-dimensional space. For example, if $v_1 = \sqrt{2}\sin\pi t$ is left out of the set defined by Eq. (8.6.8), then the remaining infinite number of vectors do not form a basis. This follows since v_1 cannot be expanded in terms of them since v_1 is linearly independent of v_n, $n > 1$.

8.7. LINEAR OPERATORS

Analogous to matrices in E_n, we define a linear operator $\mathbf{L}$ in a Hilbert space $\mathcal{H}$ as follows: (1) if $\mathbf{x}, \mathbf{y} \in \mathcal{H}$ and $\mathbf{Lx}, \mathbf{Ly} \in \mathcal{H}$ and (2) if

$$\mathbf{L}(\alpha\mathbf{x} + \beta\mathbf{y}) = \alpha\mathbf{Lx} + \beta\mathbf{Ly}, \tag{8.7.1}$$

where α and β are complex numbers, then we say $\mathbf{L}$ is a linear operator in $\mathcal{H}$.

An $n \times n$ matrix $\mathbf{A}$ is, of course, a linear operator in E_n and, with an appropriate inner product, an $\infty \times \infty$ matrix is a linear operator in the space E_∞. For example, suppose

$$\langle \mathbf{x}, \mathbf{y} \rangle = \mathbf{x}^\dagger \mathbf{y} = \sum_{i=1}^{\infty} x_i^* y_i \tag{8.7.2}$$

is the inner product in E_∞ and the matrix $\mathbf{A}$ is given by

$$\mathbf{A} = \begin{bmatrix} 1 & 0 & 0 & 0 & \cdots \\ 0 & 2^{1/2} & 0 & 0 & \cdots \\ 0 & 0 & 3^{1/2} & 0 & \cdots \\ \cdot & \cdot & \cdot & \cdot & \cdots \\ \cdot & \cdot & \cdot & \cdot & \cdots \\ \cdot & \cdot & \cdot & \cdot & \cdots \end{bmatrix}, \tag{8.7.3}$$

i.e., $a_{ij} = i^{1/2}\delta_{ij}$, $i = 1, 2, \ldots$. The vectors

$$\mathbf{x} = \begin{bmatrix} 1 \\ \frac{1}{2} \\ \frac{1}{3} \\ \vdots \end{bmatrix}, \quad \mathbf{y} = \begin{bmatrix} 1 \\ \frac{1}{2^{3/2}} \\ \frac{1}{3^{3/2}} \\ \vdots \end{bmatrix}, \quad \mathbf{z} = \begin{bmatrix} 1 \\ \frac{1}{2^2} \\ \frac{1}{3^2} \\ \vdots \end{bmatrix} \tag{8.7.4}$$

are in E_∞, and since $\|\mathbf{x}\|^2 = \sum_{i=1}^{\infty} 1/i^2 < \infty$, $\|\mathbf{y}\|^2 = \sum_{i=1}^{\infty} 1/i^3 < \infty$, and $\|\mathbf{z}\|^2 = \sum_{i=1}^{\infty} 1/i^4 < \infty$, $\mathbf{x}$, $\mathbf{y}$, and $\mathbf{z}$ belong to the Hilbert space formed by imposing the inner product $\mathbf{x}^\dagger \mathbf{y}$ on E_∞. The vectors

$$\mathbf{Ay} = \begin{bmatrix} 1 \\ \frac{1}{2} \\ \frac{1}{3} \\ \vdots \end{bmatrix} \quad \text{and} \quad \mathbf{Az} = \begin{bmatrix} \frac{1}{2^{3/2}} \\ \frac{1}{3^{3/2}} \\ \vdots \end{bmatrix} \qquad (8.7.5)$$

also belong to the Hilbert space because $\|\mathbf{Ay}\| < \infty$ and $\|\mathbf{Az}\| < \infty$. However, $\mathbf{Ax}$ does not belong to the Hilbert space since

$$\|\mathbf{Ax}\|^2 = \sum_{i=1}^{\infty} \frac{1}{i} = \infty. \qquad (8.7.6)$$

Indeed, $\mathbf{A}$ is a linear operator in the space $\mathcal{H}$ for all $\mathbf{x} \in E_n$ for which $\|\mathbf{x}\| < \infty$, but for some vectors $\mathbf{x} \in \mathcal{H}$, the vector $\mathbf{Ax}$ lies outside the Hilbert space. This property of linear operators did not arise in our analysis of finite-dimensional vector spaces. Because of this property, in infinite-dimensional Hilbert spaces we must define the *domain* $\mathcal{D}$ of a linear operator.

DEFINITION. *We say that a vector $\mathbf{x}$ in $\mathcal{H}$ belongs to the domain $\mathcal{D}$ of the linear operator $\mathbf{L}$ in $\mathcal{H}$ if $\mathbf{Lx}$ belongs to $\mathcal{H}$.*

For the operator $\mathbf{A}$ defined by Eq. (8.7.3), and the inner product $\mathbf{x}^\dagger \mathbf{y} = \sum_i x_i^* y_i$, it follows that $\mathbf{y}$ and $\mathbf{z}$ belong to the domain $\mathcal{D}$ of $\mathbf{A}$, but $\mathbf{x}$ does not.

As another example, consider the differential operator $\mathbf{L}$ operating on the vectors $\mathbf{u} \in \mathcal{L}_2(0, 1)$. By definition, the domain $\mathcal{D}$ of $\mathbf{L}$ consists of the vectors $\mathbf{v}$ in $\mathcal{L}_2(0, 1)$ for which $\mathbf{Lv}$ belongs to $\mathcal{L}_2(0, 1)$. Consider, in particular, the operator defined by the differential expression

$$Lv(t) = -\frac{d^2 v(t)}{dt^2}, \qquad (8.7.7)$$

with the boundary conditions

$$v(0) = \gamma_1, \qquad v(1) = \gamma_2. \qquad (8.7.8)$$

Note that a differential operator must be defined by its differential expression *and* its boundary conditions. The vectors forming the domain of $\mathbf{L}$ in $\mathcal{L}_2(0, 1)$ must be square-integrable functions whose second derivatives exist. Furthermore, the functions must satisfy the boundary conditions in Eq. (8.7.8).

The eigenvalue problem for a linear operator follows analogously from matrix theory as

$$\mathbf{Lv} = \lambda \mathbf{v}, \qquad (8.7.9)$$

where **v** belongs to the domain of **L**. If γ_1 and γ_2 equal 0 in the differential example above, then the eigenvalue equation becomes

$$\frac{d^2v}{dt} = -\lambda v, \tag{8.7.10}$$

whose general solution is

$$v = a\sin\sqrt{\lambda}\, t + b\cos\sqrt{\lambda}\, t. \tag{8.7.11}$$

Upon applying the boundary conditions $v(0) = v(1) = 0$, it follows that $b = 0$ and $\sin\sqrt{\lambda} = 0$, requiring that $\lambda = (n\pi)^2$ for $n = 1, 2, \ldots$. The solutions to the eigenproblem are then given by

$$v_n = a\sin n\pi t, \qquad n = 1, 2, \ldots. \tag{8.7.12}$$

We know from the theory of Fourier series that the functions $a\sin n\pi t$, $n = 1, 2, \ldots$, form a basis set for $\mathcal{L}_2(0, 1)$. Thus, the operator **L** defined by Eq. (8.7.7) with $v(0) = v(1) = 0$ is a perfect operator in the sense that its eigenvectors form a complete set (i.e., a basis set) in $\mathcal{L}_2(0, 1)$. In fact, the set is a complete, orthonormal basis set if a is set to $\sqrt{2}$. In Chapter 9 we will explore the spectral theory of integral and differential operators and we will find that there are perfect as well as imperfect (defective) operators.

Another example of a differential operator is the operator **L** in $\mathcal{L}_2(\Omega_D)$, where **Lv** is defined by the differential expression

$$\nabla^2 v = \frac{\partial^2 v}{\partial x^2} + \frac{\partial^2 v}{\partial y^2} + \frac{\partial^2 v}{\partial z^2}, \tag{8.7.13}$$

with the boundary condition

$$v(\vec{r}) = \gamma(\vec{r}) \qquad \text{for } \vec{r} \text{ on } \partial\Omega_D, \tag{8.7.14}$$

where $\gamma(\vec{r})$ is a given function for values of $\vec{r}$ lying on the boundary $\partial\Omega_D$ of the volume Ω_D.

An integral operator **K** in $\mathcal{L}_2(a, b)$ is defined by the expression

$$\mathbf{K}v = \int_a^b k(t, s) v(s)\, ds. \tag{8.7.15}$$

$k(t, s)$ is referred to as the kernel of the integral operator **K** and it is analogous to the component a_{ij} of the matrix operator. Thus, if $\mathbf{y} = \mathbf{A}\mathbf{x}$, then the ith component of **y** is $y_i = \sum_j a_{ij} x_j$, and if $\mathbf{f} = \mathbf{K}v$, then the "component" of **f** corresponding to the index t is $f(t) = \int_a^b k(t, s) v(s)\, ds$.

What about the domain of an integral operator **K**? If **K** is an integral operator in $\mathcal{L}_2(a, b)$, if a and b are finite, and if the kernel $k(t, s)$ is a continuous function of t and s, then $f(t) = \int_a^b k(t, s) v(s)\, ds$ is a continuous function for any function $v(s)$ in $\mathcal{L}_2(a, b)$. Since any continuous function belongs to $\mathcal{L}_2(a, b)$, every vector in $\mathcal{L}_2(a, b)$ belongs to the domain **K**.

On the other hand, if **K** is an operator in $\mathcal{L}_2(0, 1)$ and if $k(t, s) = 1/(s+t)^2$, $v(s) = 1$, and $\mathbf{f} = \mathbf{Kv}$, then

$$f(t) = \frac{1}{t} - \frac{1}{1+t}. \tag{8.7.16}$$

Since $f(t)$ is not square integrable, the function $v = 1$ does not define a vector in the domain of **K**. However, if $v(s) = s$ and $\mathbf{f} = \mathbf{Kv}$, then

$$f(t) = -1 + \frac{t}{1+t} + \ln\left(\frac{1+t}{t}\right), \tag{8.7.17}$$

which is square integrable in $\mathcal{L}_2(0, 1)$:

$$\int_0^1 f(t)^2 \, dt = \frac{1}{2} + (\ln 2)^2. \tag{8.7.18}$$

Thus, the function $v = s$ does define a vector in the domain of **K**. In fact, the functions $v_n(s) = s^n$, $n = 1, 2, \ldots$, all belong to the domain of **K**.

Perhaps the simplest kind of integral operator is the one with the kernel $k(t, s) = g(t)h^*(s)$. Symbolically, the operator corresponding to this kernel is a dyad, i.e.,

$$\mathbf{K} = \mathbf{gh}^\dagger. \tag{8.7.19}$$

Application of such an operator to **v** yields

$$\mathbf{f} = \mathbf{Kv} = \mathbf{gh}^\dagger \mathbf{v} = \mathbf{g}\langle \mathbf{h}, \mathbf{v} \rangle \tag{8.7.20}$$

or

$$f(t) = \langle \mathbf{h}, \mathbf{v} \rangle g(t), \tag{8.7.21}$$

where

$$\langle \mathbf{h}, \mathbf{v} \rangle = \int_a^b h^*(s)v(s) \, ds. \tag{8.7.22}$$

We can define the dyad operator for an arbitrary Hilbert space $\mathcal{H}$ by Eq. (8.7.19) and the expression

$$\mathbf{Kv} = \mathbf{g}\langle \mathbf{h}, \mathbf{v} \rangle, \tag{8.7.23}$$

where $\langle \mathbf{h}, \mathbf{v} \rangle$ denotes the inner product defining the Hilbert space $\mathcal{H}$. For example, suppose $\mathcal{H} = \mathcal{L}_2(\Omega_D, k)$, such that

$$\langle \mathbf{h}, \mathbf{v} \rangle = \int \cdots \int_{\Omega_D} h^*(\vec{r})v(\vec{r})k(\vec{r}) \, d^D r. \tag{8.7.24}$$

The kernel of $\mathbf{K} = \mathbf{gh}^\dagger$ is given by $g(\vec{r})h^*(\vec{s})$, but the action of **K** on **v** is to take the inner product of $\mathbf{h}^\dagger$ and **v** in $\mathcal{L}_2(\Omega_D, k)$, and so if $\mathbf{f} = \mathbf{Kv}$, it follows that

$$f(\vec{r}) = \langle \mathbf{h}, \mathbf{v} \rangle g(\vec{r}), \tag{8.7.25}$$

where $\langle \mathbf{h}, \mathbf{v} \rangle$ is given by Eq. (8.7.24).

A dyadic operator is, by definition, just a sum of dyads, i.e.,

$$\mathbf{K} = \sum_{i=1}^{n} \mathbf{g}_i \mathbf{h}_i^{\dagger}. \tag{8.7.26}$$

The dyadic form of an operator figured prominently in the spectral theory of matrix operators in finite-dimensional vector spaces. As we shall see in Chapter 9, this form will also become important in the spectral theory (i.e., the eigenanalysis) of operators in infinite-dimensional vector spaces.

If $\mathbf{f}$ is a vector in a Hilbert space $\mathcal{H}$, then $\alpha \mathbf{f}$, $|\alpha| < \infty$, also belongs to the space $\mathcal{H}$. Moreover, the adjoint $(\alpha \mathbf{f})^{\dagger}$ of $\alpha \mathbf{f}$ obeys the relationship

$$(\alpha \mathbf{f})^{\dagger} = \alpha^* \mathbf{f}^{\dagger}, \tag{8.7.27}$$

where α^* is the complex conjugate of α. Correspondingly, the adjoint $\mathbf{K}^{\dagger}$ of the k-term dyadic $\mathbf{K} = \sum_{i=1}^{k} \alpha_i \mathbf{a}_i \mathbf{b}_i^{\dagger}$ is

$$\mathbf{K}^{\dagger} = \sum_{i=1}^{k} \alpha_i^* \mathbf{b}_i \mathbf{a}_i^{\dagger}. \tag{8.7.28}$$

Dyadics provide the simplest examples of perfect and imperfect operators in Hilbert spaces. For example, consider the operator

$$\mathbf{K} = \mathbf{P}_0 \mathbf{P}_0^{\dagger} + \mathbf{P}_1 \mathbf{P}_1^{\dagger} \tag{8.7.29}$$

in $\mathcal{L}_2(-1, 1)$, where $P_0(t)$ and $P_1(t)$ are the zeroth- and first-degree Legendre polynomials $P_0(t) = 1$ and $P_1(t) = t$. The solutions to the eigenproblem

$$\mathbf{K}\mathbf{u} = \lambda \mathbf{u} \tag{8.7.30}$$

are $u_n = P_n$ for $n = 0, 1, \ldots$, with the eigenvalues $\lambda_0 = 2$, $\lambda_1 = \frac{2}{3}$, and $\lambda_n = 0$ for $n > 1$. This results from the orthonormality of the Legendre polynominals and the nature of the dyadic operator. Namely,

$$\mathbf{K}\mathbf{P}_n = \langle \mathbf{P}_0, \mathbf{P}_n \rangle \mathbf{P}_0 + \langle \mathbf{P}_1, \mathbf{P}_n \rangle \mathbf{P}_1, \tag{8.7.31}$$

and $\langle \mathbf{P}_m, \mathbf{P}_n \rangle = 0$ if $m \neq n$. Since the Legendre polynomials form a basis set in $\mathcal{L}_2(-1, 1)$, the dyadic defined by Eq. (8.7.27) is a perfect operator.

Consider, however, the operator

$$\mathbf{K} = \mathbf{P}_1 \mathbf{P}_0^{\dagger} + \mathbf{P}_1 \mathbf{P}_1^{\dagger} \tag{8.7.32}$$

and the corresponding eigenproblem

$$\mathbf{K}\mathbf{u} = \langle \mathbf{P}_0, \mathbf{u} \rangle \mathbf{P}_1 + \langle \mathbf{P}_1, \mathbf{u} \rangle \mathbf{P}_1 = \lambda \mathbf{u}. \tag{8.7.33}$$

For this case, $\mathbf{u}_1 = \mathbf{P}_1$, $\lambda_1 = \frac{2}{3}$ and $\mathbf{u}_n = \mathbf{P}_n$, $\lambda_n = 0$ for $n = 2, 3, \ldots$. However, $\mathbf{P}_0$ is not an eigenvector of $\mathbf{K}$, and so the eigenvectors of $\mathbf{K}$ do not form a basis set. The operator defined by Eq. (8.7.32) is, therefore, imperfect.

■ **EXERCISE 8.7.1.** Determine whether or not the operator

$$\mathbf{K} = \mathbf{P}_1\mathbf{P}_0^\dagger + \mathbf{P}_0\mathbf{P}_1^\dagger \tag{8.7.34}$$

■ ■ ■ is perfect, where $\mathbf{P}_0$ and $\mathbf{P}_1$ are the first two Legendre polynomials. Find all of the eigenvectors and eigenvalues of $\mathbf{K}$.

We end this section by deriving an explicit expression for the inverse of the operator

$$\mathbf{L} = \mathbf{I} + \mathbf{K}, \tag{8.7.35}$$

where $\|\mathbf{K}\| < \gamma$, $0 < \gamma < 1$. Heuristically, we expect the inverse, denoted by

$$(\mathbf{I} + \mathbf{K})^{-1} \quad \text{or} \quad \frac{1}{\mathbf{I} + \mathbf{K}}, \tag{8.7.36}$$

to be expandable in the series

$$\mathbf{S} = \mathbf{I} - \mathbf{K} + \mathbf{K}^2 - \mathbf{K}^3 + \cdots. \tag{8.7.37}$$

Thus, we must first prove that this series converges. Consider the finite sum

$$\mathbf{S}_n = \mathbf{I} + \sum_{i=1}^{n-1}(-\mathbf{K})^i. \tag{8.7.38}$$

Recalling the definition of the norm of an operator in the Hilbert space $\mathcal{H}$:

$$\|\mathbf{L}\|^2 = \max_{\mathbf{u} \in \mathcal{H}} \frac{\langle \mathbf{Lu}, \mathbf{Lu} \rangle}{\langle \mathbf{u}, \mathbf{u} \rangle}, \tag{8.7.39}$$

we observe that

$$\|\mathbf{Lu}\| \leq \|\mathbf{L}\| \, \|\mathbf{u}\|, \tag{8.7.40}$$

where the length $\|\mathbf{u}\|$ of a vector is defined by $\|\mathbf{u}\|^2 = \langle \mathbf{u}, \mathbf{u} \rangle$. Thus, for any vector $\mathbf{u} \in \mathcal{H}$, it follows that

$$\|(\mathbf{S}_n - \mathbf{S}_m)\mathbf{u}\| = \left\|\sum_{i=m}^{n-1}(-\mathbf{K})^i\mathbf{u}\right\| \leq \sum_{i=m}^{n-1}\|(-\mathbf{K})^i\mathbf{u}\|. \tag{8.7.41}$$

Observing that $\|\mathbf{K}^i\mathbf{u}\| < \gamma^i\|\mathbf{u}\|$, Eq. (8.7.41) becomes

$$\|(\mathbf{S}_n - \mathbf{S}_m)\mathbf{u}\| < \frac{\gamma^m - \gamma^n}{1 - \gamma}\|\mathbf{u}\|. \tag{8.7.42}$$

Since $\gamma < 1$, there exists an $m(\epsilon)$ such that $\gamma^m < \epsilon(1 - \gamma)$, and so

$$\|(\mathbf{S}_n - \mathbf{S}_m)\mathbf{u}\| < \epsilon\left(1 - \gamma^{n-m}\right) < \epsilon, \qquad n > m > m(\epsilon). \tag{8.7.43}$$

This implies that $\mathbf{S}_n\mathbf{u}$ is a Cauchy sequence and so $\lim_{n \to \infty} \mathbf{S}_n\mathbf{u}$ converges to a vector $\mathbf{v}$ in $\mathcal{H}$. Thus, we define the operator $\mathbf{S}$ as

$$\mathbf{v} = \mathbf{S}\mathbf{u} = \lim_{n \to \infty} \mathbf{S}_n\mathbf{u}. \tag{8.7.44}$$

The next question is whether $\mathbf{S} = (\mathbf{I}+\mathbf{K})^{-1}$. Note that

$$\|\mathbf{S}_n(\mathbf{I}+\mathbf{K})\mathbf{u} - \mathbf{S}(\mathbf{I}+\mathbf{K})\mathbf{u}\| = \left\|\left(\sum_{i=n}^{\infty}\mathbf{K}^i\right)(\mathbf{I}+\mathbf{K})\mathbf{u}\right\|$$
$$< \left(\sum_{i=n}^{\infty}\gamma^i\right)\|(\mathbf{I}+\mathbf{K})\mathbf{u}\| \qquad (8.7.45)$$
$$\leq \frac{\gamma^n}{1-\gamma}(1+\gamma)\|\mathbf{u}\|.$$

Since the right-hand side of Eq. (8.7.45) goes to 0 as $n \to \infty$, it follows that $\lim_{n\to\infty} \mathbf{S}_n(\mathbf{I}+\mathbf{K}) = \mathbf{S}(\mathbf{I}+\mathbf{K})$. Similar details establish that $\lim_{n\to\infty}(\mathbf{I}+\mathbf{K})\mathbf{S}_n = (\mathbf{I}+\mathbf{K})\mathbf{S}$.

In summary, we have proved that:

Theorem. *If $\mathbf{L} = \mathbf{I}+\mathbf{K}$ and $\|\mathbf{K}\| < \gamma$, $0 < \gamma < 1$, then the inverse of $(\mathbf{I}+\mathbf{K})$ exists and can be represented as*

$$(\mathbf{I}+\mathbf{K})^{-1} = \mathbf{I} + \sum_{i=1}^{\infty}(-\mathbf{K})^i. \qquad (8.7.46)$$

This also leads to the conclusion that the solution of

$$(\mathbf{I}+\mathbf{K})\mathbf{u} = \mathbf{f} \qquad (8.7.47)$$

always exists and is unique. It could be computed from the series

$$\mathbf{u} = \mathbf{f} + \sum_{i=1}^{\infty}(-1)^i \mathbf{K}^i \mathbf{f}, \qquad (8.7.48)$$

although this might not be the most economical way to obtain the solution. Alternatively, an iterative solution to the problem could be obtained from the series

$$\begin{aligned}\mathbf{u}_1 &= \mathbf{f} \\ \mathbf{u}_2 &= \mathbf{f} - \mathbf{K}\mathbf{u}_1 \\ &\vdots \\ \mathbf{u}_n &= \mathbf{f} - \mathbf{K}\mathbf{u}_{n-1},\end{aligned} \qquad (8.7.49)$$

since the theorem at Eq. (8.7.46) guarantees convergence of the series.

8.8. SOLUTIONS TO PROBLEMS INVOLVING k-TERM DYADICS

In the previous section we encountered the k-term dyadic operator of the form

$$\mathbf{K} = \sum_{i=1}^{k} \mathbf{a}_i \mathbf{b}_i^{\dagger}, \qquad (8.8.1)$$

where the sets $\{\mathbf{a}_1, \ldots, \mathbf{a}_k\}$ and $\{\mathbf{b}_1, \ldots, \mathbf{b}_k\}$ are linearly independent sets. (*Note*: Some texts refer to operators of the form of Eq. (8.8.1) as *degenerate* operators.) If the sets $\{\mathbf{a}_i\}$ and $\{\mathbf{b}_i\}$ are not linearly independent, then the dyadic operator in Eq. (8.8.1) can be transformed into another dyadic operator composed of smaller sets $\{\tilde{\mathbf{a}}_i\}$ and $\{\tilde{\mathbf{b}}_i\}$ that are linearly independent. For example, suppose $\mathbf{a}_k$ is linearly dependent on the vectors $\mathbf{a}_1, \ldots, \mathbf{a}_{k-1}$. Then there exists a set of numbers $\{\alpha_1, \ldots, \alpha_{k-1}\}$ such that

$$\mathbf{a}_k = \sum_{i=1}^{k-1} \alpha_i \mathbf{a}_i. \tag{8.8.2}$$

Inserting this expression into Eq. (8.8.1) gives

$$\mathbf{K} = \sum_{i=1}^{k-1} \mathbf{a}_i \mathbf{b}_i^\dagger + \sum_{i=1}^{k-1} \mathbf{a}_i \alpha_i \mathbf{b}_k^\dagger \tag{8.8.3}$$

$$= \sum_{i=1}^{k-1} \mathbf{a}_i \tilde{\mathbf{b}}_i^\dagger,$$

a $(k-1)$-term dyadic in which

$$\tilde{\mathbf{b}}_i = \mathbf{b}_i + \alpha_i^* \mathbf{b}_k, \qquad i = 1, \ldots, k-1. \tag{8.8.4}$$

If one vector in the set $\{\tilde{\mathbf{b}}_i\}$ is linearly dependent on the rest, then a similar process will enable reduction of Eq. (8.8.3) to

$$\mathbf{K} = \sum_{i=1}^{k-2} \tilde{\mathbf{a}}_i \tilde{\mathbf{b}}_i^\dagger. \tag{8.8.5}$$

The process can be continued until $\mathbf{K}$ is reduced to a sum of dyads $\mathbf{a}_i \mathbf{b}_i^\dagger$ whose vectors $\{\mathbf{a}_i\}$ and $\{\mathbf{b}_i\}$ form linearly independent sets. We say such a dyadic is irreducible. In what follows, we will always assume that the k-term dyadic is irreducible.

Consider next the inhomogeneous equation

$$\mathbf{K}\mathbf{u} = \mathbf{f}, \tag{8.8.6}$$

where $\mathbf{K}$ is a k-term irreducible dyadic. Since Eq. (8.8.6) can be expressed as

$$\sum_{i=1}^{k} \langle \mathbf{b}_i, \mathbf{u} \rangle \mathbf{a}_i = \mathbf{f}, \tag{8.8.7}$$

it follows that Eq. (8.8.6) will have a solution *only if* $\mathbf{f}$ is a linear combination of the set $\{\mathbf{a}_i\}$, i.e., if

$$\mathbf{f} = \sum_{i=1}^{k} \alpha_i \mathbf{a}_i. \tag{8.8.8}$$

If **f** is, indeed, of the form of Eq. (8.8.8), it follows from Eq. (8.8.7) that

$$\langle \mathbf{b}_i, \mathbf{u} \rangle = \alpha_i, \qquad i = 1, \ldots, k. \tag{8.8.9}$$

To find a particular solution $\mathbf{u}_p$, we assume that

$$\mathbf{u}_p = \sum_{j=1}^{k} \beta_j \mathbf{b}_j, \tag{8.8.10}$$

and substitute this into Eq. (8.8.9) to yield the matrix problem

$$\mathbf{A}\boldsymbol{\beta} = \boldsymbol{\alpha}, \tag{8.8.11}$$

where $\mathbf{A}$ is a self-adjoint, $k \times k$ matrix given by

$$\mathbf{A} = [\langle \mathbf{b}_i, \mathbf{b}_j \rangle], \tag{8.8.12}$$

and $\boldsymbol{\beta}$ and $\boldsymbol{\alpha}$ are k-dimensional vectors with components β_i and α_i, respectively.

From the linear independence of the set $\{\mathbf{b}_1, \ldots, \mathbf{b}_k\}$, it follows that the matrix $\mathbf{A}$ is nonsingular, i.e., $|\mathbf{A}| \neq 0$. To prove this, let us begin with the hypothesis that $\mathbf{A}$ is singular; i.e., its rank is less than k. From the theory of matrices, this implies that at least one column, say column k, of $\mathbf{A}$ is a linear combination of the other columns; i.e., there exists a set of numbers $c_1, \ldots, c_{k-1}$, not all 0, such that

$$\begin{bmatrix} \langle \mathbf{b}_1, \mathbf{b}_k \rangle \\ \vdots \\ \langle \mathbf{b}_k, \mathbf{b}_k \rangle \end{bmatrix} = \sum_{j=1}^{k-1} c_j \begin{bmatrix} \langle \mathbf{b}_1, \mathbf{b}_j \rangle \\ \vdots \\ \langle \mathbf{b}_k, \mathbf{b}_j \rangle \end{bmatrix} \tag{8.8.13}$$

or

$$\langle \mathbf{b}_i, \mathbf{b}_k \rangle = \sum_{j=1}^{k-1} c_j \langle \mathbf{b}_i, \mathbf{b}_j \rangle, \qquad i = 1, \ldots, k. \tag{8.8.14}$$

Rearrangement gives

$$\left\langle \mathbf{b}_i, \sum_{j=1}^{k} \tilde{c}_j \mathbf{b}_j \right\rangle = 0, \qquad i = 1, \ldots, k, \tag{8.8.15}$$

where $\tilde{c}_j = -1$ if $j = k$ and $\tilde{c}_j = c_j$ otherwise. Equation (8.8.15) implies that the vector $\mathbf{b}_{k+1} \equiv \sum_{j=1}^{k} \tilde{c}_j \mathbf{b}_j$ is orthogonal to the linearly independent set $\mathbf{b}_1, \ldots, \mathbf{b}_k$, and thus the set $\mathbf{b}_1, \ldots, \mathbf{b}_{k+1}$ is linearly independent. This is a contradiction since the equation

$$\sum_{i=1}^{k+1} \delta_i \mathbf{b}_i = \mathbf{0} \tag{8.8.16}$$

has the solution $\delta_{k+1} = 1$ and $\delta_i = -\tilde{c}_i$ for $i = 1, \ldots, k$. Thus, we have proven that if the set $\{\mathbf{b}_i\}$ is linearly independent, then $\mathbf{A}$ must be nonsingular.

SOLUTIONS TO PROBLEMS INVOLVING k-TERM DYADICS

We can restate this point in the following theorem:

THEOREM. *If $\mathbf{b}_1, \ldots, \mathbf{b}_k$ is a linearly independent set in an abstract linear vector space $\mathcal{H}$ (of arbitrary dimension as long as the dimension is greater than or equal to k), then the $k \times k$ matrix $\mathbf{A} = [\langle \mathbf{b}_i, \mathbf{b}_j \rangle]$ is nonsingular.*

Since $\mathbf{A}$ is nonsingular, the inverse of $\mathbf{A}$ exists, and so Eq. (8.8.11) has the unique solution $\boldsymbol{\beta} = \mathbf{A}^{-1}\boldsymbol{\alpha}$, yielding

$$\mathbf{u}_p = \sum_{i=1}^{k}\sum_{j=1}^{k} A_{ij}^{-1} \alpha_j \mathbf{b}_i. \tag{8.8.17}$$

We can assign the vectors $\mathbf{b}_1, \ldots, \mathbf{b}_k$ as the first k vectors of a basis set an infinite number of basis vectors. To complete the basis set, we can add to these the vectors $\mathbf{b}_{k+1}, \mathbf{b}_{k+2}, \ldots$, and with the aid of the Gram–Schmidt process, the added vectors can even be chosen to be orthogonal to the vectors $\mathbf{b}_1, \ldots, \mathbf{b}_k$. The added vectors then obey the homogeneous equation

$$\mathbf{K}\mathbf{b}_i = \mathbf{0}, \qquad i = k+1, \; k+2, \ldots. \tag{8.8.18}$$

Thus, the solvability theorem for the k-term dyadic problem reads as follows:

THEOREM. *The equation*

$$\mathbf{K}\mathbf{u} \equiv \sum_{i=1}^{k} \mathbf{a}_i \langle \mathbf{b}_i, \mathbf{u} \rangle = \mathbf{f} \tag{8.8.19}$$

has the solution

$$\mathbf{u} = \mathbf{u}_p + \sum_{i=k+1}^{\infty} c_i \mathbf{b}_i, \tag{8.8.20}$$

where $\mathbf{u}_p$ is a particular solution to Eq. (8.8.19) and the $\mathbf{b}_i$, $i = k+1, k+2, \ldots$, are solutions to the homogeneous equation $\mathbf{K}\mathbf{u} = 0$.

Equation (8.8.11) guarantees the existence of a particular solution, and the solutions to the homogeneous problem, $\{\mathbf{b}_{k+1}, \mathbf{b}_{k+2}, \ldots\}$, can always be found by constructing basis vectors orthogonal to the set $\{\mathbf{b}_1, \ldots, \mathbf{b}_k\}$. The theorem, of course, assumes that the k-term dyadic has been reduced to a sum of linearly independent dyads.

■ **EXAMPLE 8.8.1.** Consider the space $\mathcal{L}_2(-1, 1)$ and the two-term dyadic operator $\mathbf{K}$ with the kernel

$$k(t, s) = 2\exp(-t) + s \sin t \tag{8.8.21}$$

and the equation

$$\mathbf{K}\mathbf{u} = \mathbf{f}, \tag{8.8.22}$$

where

$$f(t) = 3\exp(-t) + 5\sin t. \tag{8.8.23}$$

Equation (8.8.22) is given explicitly as

$$2\exp(-t)\int_{-1}^{1} u(s)\,ds + \sin t \int_{-1}^{1} su(s)\,ds = 3\exp(-t) + 5\sin t. \qquad (8.8.24)$$

The operator $\mathbf{K}$ is represented by Eq. (8.8.1) with $b_1(t) = 2$, $b_2(t) = t$, $a_1(t) = e^{-t}$, and $a_2 = \sin t$. The vector $\mathbf{f}$ yields the coefficients $\alpha_1 = 3$ and $\alpha_2 = 5$, and the relevant matrix elements are given by $\langle \mathbf{b}_1, \mathbf{b}_1 \rangle = 4\int_{-1}^{1} dt = 8$, $\langle \mathbf{b}_1, \mathbf{b}_2 \rangle = \langle \mathbf{b}_2, \mathbf{b}_1 \rangle = 2\int_{-1}^{1} t\,dt = 0$, and $\langle \mathbf{b}_2, \mathbf{b}_2 \rangle = \int_{-1}^{1} t^2\,dt = \frac{2}{3}$. Correspondingly, the equation $\mathbf{A}\boldsymbol{\beta} = \boldsymbol{\alpha}$ Eq. (8.8.11) becomes

$$\begin{bmatrix} 8 & 0 \\ 0 & 2/3 \end{bmatrix} \begin{bmatrix} \beta_1 \\ \beta_2 \end{bmatrix} = \begin{bmatrix} 3 \\ 5 \end{bmatrix} \qquad (8.8.25)$$

whose solution is $\beta_1 = \frac{3}{8}$ and $\beta_2 = \frac{15}{2}$. Thus, the particular solution is

$$u_p(t) = \frac{3}{4} + \frac{15}{2}t. \qquad (8.8.26)$$

Since $b_1(t) = 2P_0(t)$ and $b_2(t) = P_1(t)$, it follows that

$$\mathbf{K}\mathbf{P}_l = \mathbf{0}, \qquad l = 2, 3, \ldots, \qquad (8.8.27)$$

where $\mathbf{P}_l$ is the Legendre polynomial:

$$P_l(t) = \frac{1}{2^l l!} \frac{d^l}{dt^l}(t^2 - 1)^l. \qquad (8.8.28)$$

The general solution to the problem is, therefore,

$$u(t) = \frac{3}{4} + \frac{15}{2}t + \sum_{l=2}^{\infty} c_l P_l(t), \qquad (8.8.29)$$

where the coefficients c_l are arbitrary and

$$\int_{-1}^{1} P_l(t) P_m(t)\,dt = 0, \qquad l \neq m. \qquad (8.8.30)$$

■ ■ ■ The set $\{\mathbf{P}_l\}$ is known to be an orthogonal basis set in $\mathcal{L}_2(-1, 1)$.

Another interesting problem involves the operator of the form $\mathbf{L} = \mathbf{I} + \mathbf{K}$, where $\mathbf{I}$ is the identity operator and $\mathbf{K}$ is a k-term dyadic. The corresponding inhomogeneous equation is

$$(\mathbf{I} + \mathbf{K})\mathbf{u} = \mathbf{f} \quad \text{or} \quad \mathbf{u} + \mathbf{K}\mathbf{u} = \mathbf{f}, \qquad (8.8.31)$$

or

$$\mathbf{u} + \sum_{i=1}^{k} \mathbf{a}_i \langle \mathbf{b}_i, \mathbf{u} \rangle = \mathbf{f}. \qquad (8.8.32)$$

SOLUTIONS TO PROBLEMS INVOLVING k-TERM DYADICS

Clearly, if the quantities $\langle \mathbf{b}_i, \mathbf{u} \rangle$ can be found, a solution of Eq. (8.8.32) is

$$\mathbf{u} = \mathbf{f} - \sum_{i=1}^{k} \langle \mathbf{b}_i, \mathbf{u} \rangle \mathbf{a}_i. \tag{8.8.33}$$

To generate a set of equations for the coefficients $\langle \mathbf{b}_i, \mathbf{u} \rangle$, we take the inner product of Eq.(8.8.32) with $\mathbf{b}_i$, $i = 1, \ldots, k$. The result is the matrix problem

$$\mathbf{A}\boldsymbol{\beta} = \boldsymbol{\alpha}, \tag{8.8.34}$$

where $\mathbf{A}$ is a $k \times k$ matrix with elements

$$a_{ii} = 1 + \langle \mathbf{b}_i, \mathbf{a}_i \rangle \quad \text{and} \quad a_{ij} = \langle \mathbf{b}_i, \mathbf{a}_j \rangle, i \neq j, \tag{8.8.35}$$

and $\boldsymbol{\beta}$ and $\boldsymbol{\alpha}$ are k-dimensional vectors with elements

$$\beta_i = \langle \mathbf{b}_i, \mathbf{u} \rangle \quad \text{and} \quad \alpha_i = \langle \mathbf{b}_i, \mathbf{f} \rangle. \tag{8.8.36}$$

According to the theory of the solvability of linear algebraic equations, Eq. (8.8.34) has a solution *if and only if* the rank of $\mathbf{A}$ is the same as the rank of the augmented matrix $[\mathbf{A}, \boldsymbol{\alpha}]$. When Eq. (8.8.34) is solvable, we know from Chapter 4 that the solutions are of the form

$$\boldsymbol{\beta} = \boldsymbol{\beta}^{\mathrm{p}} + \sum_{j=1}^{k-r} c_j \boldsymbol{\beta}_j^{\mathrm{h}}, \tag{8.8.37}$$

where r is the rank of $\mathbf{A}$, $\boldsymbol{\beta}^{\mathrm{p}}$ is a particular solution, and $\mathbf{A}\boldsymbol{\beta}_j^{\mathrm{h}} = \mathbf{0}$, $j = 1, \ldots, k-r$. The general solution to Eq. (8.8.32) then becomes

$$\mathbf{u} = \mathbf{f} - \sum_{i=1}^{k} \beta_i^{\mathrm{p}} \mathbf{a}_i - \sum_{j=1}^{k-r} c_j \sum_{i=1}^{k} \beta_{ji}^{\mathrm{h}} \mathbf{a}_i. \tag{8.8.38}$$

EXAMPLE 8.8.2. For the space $\mathcal{L}_2(0, 1)$, find the conditions of the solution of

$$u(t) + \lambda \int_0^1 s u(s) \, ds = f(t). \tag{8.8.39}$$

Here $l(t, s) = \delta(t - s) + \lambda s$, where $l(t, s)$ is the kernel of $\mathbf{L} = \mathbf{I} + \mathbf{K}$, and so $a_1(t) = \lambda$ and $b_1(s) = s$. We can multiply Eq. (8.8.39) by t and integrate to get

$$\int_0^1 t u(t) \, dt + \lambda \int_0^1 t \, dt \int_0^1 s u(s) \, ds = \int_0^1 t f(t) \, dt, \tag{8.8.40}$$

or

$$\left(1 + \frac{\lambda}{2}\right) \langle \mathbf{b}_1, \mathbf{u} \rangle = \int_0^1 t f(t) \, dt. \tag{8.8.41}$$

We see that Eq. (8.8.40) has a unique solution only if $\lambda \neq -2$. Namely,

$$u(t) = f(t) - \lambda \left(1 + \frac{\lambda}{2}\right)^{-1} \int_0^1 t f(t) \, dt \tag{8.8.42}$$

for arbitrary $f(t)$ in $\mathcal{L}_2(0, 1)$.

If $\lambda = -2$, Eq. (8.8.40) requires that $f(t)$ satisfy the equation

$$0 = \int_0^1 tf(t)\,dt. \tag{8.8.43}$$

The resulting homogeneous solution is $u(t) = \beta$, where β is an arbitrary complex number. The solution to Eq. (8.8.39) then becomes $u(t) = f(t) + \beta$. On the other hand, if $\lambda = -2$ and $f(t)$ does not satisfy Eq. (8.8.43), the problem does not have a solution. ∎ ∎ ∎

We will end this section with the following theorem:

THEOREM. *A k-term dyadic operator in $\mathcal{H}$,*

$$\mathbf{L} = \sum_{i=1}^k \mathbf{u}_i \mathbf{v}_i^\dagger, \tag{8.8.44}$$

can always be transformed into a p-term dyadic of the form

$$\mathbf{L} = \sum_{i,j=1}^p w_{ij} \boldsymbol{\phi}_i \boldsymbol{\phi}_j^\dagger, \tag{8.8.45}$$

where $p \leq 2k$ and the $\boldsymbol{\phi}_i$ are orthonormal vectors, i.e.,

$$\langle \boldsymbol{\phi}_i, \boldsymbol{\phi}_j \rangle = \delta_{ij}. \tag{8.8.46}$$

The proof of this is simple. By the definition of a k-term dyadic, the sets $\{\mathbf{u}_1, \ldots, \mathbf{u}_k\}$ and $\{\mathbf{v}_1, \ldots, \mathbf{v}_k\}$ are both linearly independent. The vectors $\mathbf{v}_1, \ldots, \mathbf{v}_k$, however, are not necessarily linearly independent of the vectors $\mathbf{u}_1, \ldots, \mathbf{u}_k$. If they are, the set $\{\mathbf{u}_1, \ldots, \mathbf{u}_k, \mathbf{v}_1, \ldots, \mathbf{v}_k\}$ forms a set of $2k$ linearly independent vectors and the Gram–Schmidt procedure can be used to transform this set into the orthonormal set $\{\boldsymbol{\phi}_i, \ldots, \boldsymbol{\phi}_{2k}\}$. However, if some of the vectors $\mathbf{v}_1, \ldots, \mathbf{v}_k$ are linearly dependent on the vectors $\mathbf{u}_1, \ldots, \mathbf{u}_k$, then the Gram–Schmidt procedure yields the orthonormal set $\{\boldsymbol{\phi}_1, \ldots, \boldsymbol{\phi}_p\}$ with $p < 2k$. In either case, $\mathbf{u}_i$ and $\mathbf{v}_i$ can be expressed as the linear combinations

$$\mathbf{u}_i = \sum_{j=1}^p \alpha_{ij} \boldsymbol{\phi}_j \quad \text{and} \quad \mathbf{v}_i = \sum_{j=1}^p \beta_{ij} \boldsymbol{\phi}_j, \tag{8.8.47}$$

and so $\mathbf{L}$ becomes

$$\mathbf{L} = \sum_{i=1}^k \sum_{j=1}^p \sum_{l=1}^p \alpha_{ij} \beta_{il}^* \boldsymbol{\phi}_j \boldsymbol{\phi}_l^\dagger$$
$$= \sum_{j,l=1}^p \omega_{jl} \boldsymbol{\phi}_j \boldsymbol{\phi}_l^\dagger, \tag{8.8.48}$$

where

$$\omega_{jl} = \sum_{i=1}^k \alpha_{ij} \beta_{il}^*. \tag{8.8.49}$$

Note that if $\mathbf{L}$ is self-adjoint, i.e., if $\mathbf{L} = \mathbf{L}^\dagger$, then

$$\sum_{i,j=1}^{p} \omega_{ij} \boldsymbol{\phi}_i \boldsymbol{\phi}_j^\dagger = \sum_{i,j=1}^{p} \omega_{ij}^* \boldsymbol{\phi}_j \boldsymbol{\phi}_i^\dagger = \sum_{i,j}^{p} \omega_{ji}^* \boldsymbol{\phi}_i \boldsymbol{\phi}_j^\dagger, \qquad (8.8.50)$$

and, therefore, $\omega_{ij} = \omega_{ji}^*$.

8.9. PERFECT OPERATORS

We call an operator in $\mathcal{H}$ *perfect* if its eigenvectors form a basis set. For example, the operator $\mathbf{L}$ in $\mathcal{L}_2(0, 1)$ defined by the differential expression

$$Lv(t) = -\frac{d^2 v}{dt^2}, \qquad (8.9.1)$$

with the boundary conditions

$$v(0) = v(1), \qquad (8.9.2)$$

is perfect since its eigenvectors

$$v_r = \sqrt{2} \sin n\pi t, \qquad n = 1, 2, \ldots, \qquad (8.9.3)$$

with the corresponding eigenvalues $\lambda_n = (n\pi)^2$ for $n = 1, 2, \ldots$, form a basis set in $\mathcal{L}_2(0, 1)$. In fact, not only does (8.9.3) form a basis set, it is an orthonormal basis set since we observe that

$$\langle \mathbf{v}_n, \mathbf{v}_m \rangle = \mathbf{v}_n^\dagger \mathbf{v}_m = 2 \int_0^1 \sin n\pi t \, \sin m\pi t \, dt = \delta_{nm}. \qquad (8.9.4)$$

Thus, any vector $\mathbf{f}$ in $\mathcal{L}_2(0, 1)$ can be expressed as

$$\mathbf{f} = \sum_{n=1}^{\infty} \alpha_n \mathbf{v}_n, \qquad (8.9.5)$$

where

$$\alpha_n = \langle \mathbf{v}_n, \mathbf{f} \rangle = \mathbf{v}_n^\dagger \mathbf{f}. \qquad (8.9.6)$$

By substituting Eq. (8.9.6) and rearranging, Eq. (8.9.5) becomes

$$\mathbf{f} = \sum_{n=1}^{\infty} \mathbf{v}_n \mathbf{v}_n^\dagger \mathbf{f} = \left(\sum_{n=1}^{\infty} \mathbf{v}_n \mathbf{v}_n^\dagger \right) \mathbf{f}. \qquad (8.9.7)$$

Since Eq. (8.9.7) is valid for any arbitrary vector $\mathbf{f}$ in $\mathcal{L}_2(0, 1)$, it follows that the expression

$$\mathbf{I} \equiv \sum_{n=1}^{\infty} \mathbf{v}_n \mathbf{v}_n^\dagger \qquad (8.9.8)$$

is the identity operator in $\mathcal{L}_2(0, 1)$. The individual dyads $\mathbf{v}_n\mathbf{v}_n^\dagger$ are integral operators with the kernels $v_n(t)v_n^*(s) = \sqrt{2}\sin n\pi t \sqrt{2}\sin n\pi s$, and so $\mathbf{I}$ is an integral operator with the kernel

$$i(t, s) = \sum_{n=1}^\infty v_n(t)v_n^*(s). \tag{8.9.9}$$

Therefore, Eq. (8.9.7) reads $\mathbf{f} = \mathbf{I}\mathbf{f}$, or

$$f(t) = \int_0^1 i(t, s)f(s)\,ds \tag{8.9.10}$$

for an arbitrary Lebesgue integrable function in $\mathcal{L}_2(0, 1)$. From this it follows that the kernel $i(t, s)$ must actually be the Dirac delta function $\delta(t - s)$, i.e.,

$$\delta(t - s) = \sum_{i=1}^\infty v_n(t)v_n^*(s). \tag{8.9.11}$$

Equation (8.9.8) represents the resolution of the identity and Eq. (8.9.11) provides a representation of the Dirac delta function in terms of the functions of an orthonormal basis set in $\mathcal{L}_2(0, 1)$. Equation (8.9.11) is sometimes referred to as the *completeness relation*. For the purpose of illustration, we considered the basis set given in Eq. (8.9.3). However, for any orthonormal basis set in $\mathcal{L}_2(0, 1)$, Eqs. (8.9.8) and (8.9.11) follow.

Similar arguments show that if $\mathbf{v}_n, n = 1, 2, \ldots$, is an orthonormal basis set in $\mathcal{L}_2(\Omega_D)$, then the resolution of the identity is given by

$$\mathbf{I} = \sum_{n=1}^\infty \mathbf{v}_n\mathbf{v}_n^\dagger, \tag{8.9.12}$$

where $\mathbf{I}$ is an integral operator with the kernel

$$\delta(\vec{r} - \vec{s}). \tag{8.9.13}$$

In this case, $\delta(\vec{r} - \vec{s})$ is the Dirac delta function in the D-dimensional Euclidean space.

Suppose the eigenvectors $\mathbf{v}_n, n = 1, 2, \ldots$, of the operator $\mathbf{L}$ in $\mathcal{H}$ form an orthonormal basis set. Then, since $\mathbf{f} = \sum_{n=1}^\infty \mathbf{v}_n\mathbf{v}_n^\dagger\mathbf{f}$ and $\mathbf{L}\mathbf{v}_n = \lambda_n\mathbf{v}_n$, it follows that

$$\mathbf{L}\mathbf{f} = \sum_{n=1}^\infty \lambda_n\mathbf{v}_n\mathbf{v}_n^\dagger\mathbf{f} \tag{8.9.14}$$

for arbitrary $\mathbf{f}$. From this it follows that the operator $\mathbf{L}$ can be represented by

$$\mathbf{L} = \sum_{n=1}^\infty \lambda_n\mathbf{v}_n\mathbf{v}_n^\dagger. \tag{8.9.15}$$

We have just derived the spectral resolution theorem for perfect operators in $\mathcal{H}$ that have orthonormal eigenvectors. We will explore this theorem in greater detail in Chapter 9.

PERFECT OPERATORS

The significance of Eq. (8.9.15) is tremendous. Consider the case of the second-order differential operator defined by Eqs. (8.9.1) and (8.9.2). According to Eq. (8.9.15), this differential operator can be represented in the Hilbert space $\mathcal{L}_2(0, 1)$ by an integral operator whose kernel is given by

$$l(t, s) = \sum_{n=1}^{\infty} \lambda_n v_n(t) v^*(s)$$
$$= \sum_{n=1}^{\infty} 2(n\pi)^2 \sin n\pi t \, \sin n\pi s. \tag{8.9.16}$$

It is straightforward to show that

$$\mathbf{L}^k = \sum_{n=1}^{\infty} \lambda_n^k \mathbf{v}_n \mathbf{v}_n^\dagger, \tag{8.9.17}$$

and, from that,

$$f(\mathbf{L}) = \sum_{n=1}^{\infty} f(\lambda_n) \mathbf{v}_n \mathbf{v}_n^\dagger, \tag{8.9.18}$$

where $f(t)$ is any function that can be represented by a series in t.

In the notation of quantum mechanics, a dyadic $\mathbf{K} = \sum_n \mathbf{a}_n \mathbf{b}_n^\dagger$ is denoted by

$$\mathbf{K} = \sum_n |\mathbf{a}_n\rangle \langle \mathbf{b}_n|. \tag{8.9.19}$$

The meaning of the dyadic is that the operation of $\mathbf{K}$ on $\mathbf{u}$ gives

$$\mathbf{Ku} = \sum_n |\mathbf{a}_n\rangle \langle \mathbf{b}_n, \mathbf{u}\rangle; \tag{8.9.20}$$

i.e., $\mathbf{Ku}$ is a linear combination of the vectors $\mathbf{a}_n$ whose coefficients are the inner products $\langle \mathbf{b}_n, \mathbf{u}\rangle$. Our notation means the same thing. We write

$$\mathbf{Ku} = \sum_n \mathbf{a}_n \mathbf{b}_n^\dagger \mathbf{u} \tag{8.9.21}$$

in which $\mathbf{b}_n^\dagger \mathbf{u}$ denotes the inner product in $\mathcal{H}$. In the space $\mathcal{L}_2(\Omega_D)$, the inner product is simply

$$\mathbf{b}_n^\dagger \mathbf{u} = \int \cdots \int_{\Omega_D} b_n^*(\vec{r}) u(\vec{r}) \, d^D r, \tag{8.9.22}$$

whereas, in the space $\mathcal{L}_2(\Omega_D, k)$, the inner product is

$$\mathbf{b}_n^\dagger \mathbf{u} = \int \cdots \int_{\Omega_D} b_n^*(\vec{r}) u(\vec{r}) k(\vec{r}) \, d^D r. \tag{8.9.23}$$

Thus, the operation of the dyad $\mathbf{a}_n \mathbf{b}_n^\dagger$ on $\mathbf{u}$ is defined in this case as taking the $k(\vec{r})$-weighted inner product between $\mathbf{b}_n^\dagger$ and $\mathbf{u}$. We prefer our dyad notation $\mathbf{ab}^\dagger$ to the so-called "bra" and "ket" notation of quantum mechanics, $|\mathbf{a}\rangle\langle\mathbf{b}|$, because the former notation is more suggestive of the analogy between operators in abstract linear vector spaces and matrix operators in E_n.

As was true for basis sets in E_n, basis sets in an arbitrary Hilbert space need not be orthogonal. In such cases, the notion of a reciprocal basis set becomes important. If $\mathbf{v}_n$, $n = 1, 2, \ldots$, is a basis set, the reciprocal set $\mathbf{u}_n$, $n = 1, 2, \ldots$, is defined by the conditions

$$\langle \mathbf{v}_n, \mathbf{u}_m \rangle = \delta_{nm}. \qquad (8.9.24)$$

How we go about finding the reciprocal basis set is an issue that we will address in later chapters. If $\mathbf{v}_n$ and $\mathbf{u}_n$, $n = 1, 2, \ldots$, denote reciprocally related sets in $\mathcal{H}$, then, since the coefficients α_n exist such that

$$\mathbf{f} = \sum_n \alpha_n \mathbf{v}_n \qquad (8.9.25)$$

for any $\mathbf{f}$, it follows from Eq. (8.9.24) that $\alpha_n = \langle \mathbf{u}_n, \mathbf{f} \rangle$. Equation (8.9.25) can, thus, be written as

$$\mathbf{f} = \sum_n \mathbf{v}_n \langle \mathbf{u}_n, \mathbf{f} \rangle = \left(\sum_n \mathbf{v}_n \mathbf{u}_n^\dagger \right) \mathbf{f}, \qquad (8.9.26)$$

and so the resolution of the identity becomes

$$\mathbf{I} = \sum_n \mathbf{v}_n \mathbf{u}_n^\dagger \qquad (8.9.27)$$

for a nonorthogonal basis set in $\mathcal{H}$.

If the set $\{\mathbf{v}_n\}$, $n = 1, 2, \ldots$, are eigenvectors of a perfect operator $\mathbf{L}$, it follows from Eq. (8.9.27) that

$$\mathbf{L} = \sum_n \lambda_n \mathbf{v}_n \mathbf{u}_n^\dagger \qquad (8.9.28)$$

and, similarly,

$$f(\mathbf{L}) = \sum_n f(\lambda_n) \mathbf{v}_n \mathbf{u}_n^\dagger, \qquad (8.9.29)$$

where $f(t)$ is a function that can be represented by a series in t.

An important property of the dyad $\mathbf{ab}^\dagger$, as we have defined it, is that its adjoint is given by

$$(\mathbf{ab}^\dagger)^\dagger = \mathbf{ba}^\dagger. \qquad (8.9.30)$$

Thus, if $\mathcal{H} = E_n$, the quantity $\mathbf{ab}^\dagger$ is a matrix whose components are $a_i b_j^*$ and whose adjoint matrix, $\mathbf{ba}^\dagger$, has components $b_i a_j^*$. If $\mathcal{H}$ is a function space, say $\mathcal{L}_2(\Omega_D)$, then $\mathbf{ab}^\dagger$ is an integral operator with the kernel $a(\vec{r})b^*(\vec{s})$. In this case, the adjoint $\mathbf{ba}^\dagger$ is also an integral operator with the kernel $b(\vec{r})a^*(\vec{s})$. This property has important implications for a *perfect* operator whose eigenvectors are not orthogonal. The spectral resolution of such an operator is given by Eq. (8.9.28). The adjoint of this operator is

$$\mathbf{L}^\dagger = \sum_n \lambda_n^* \mathbf{u}_n \mathbf{v}_n^\dagger \tag{8.9.31}$$

from which it follows that

$$\mathbf{L}^\dagger \mathbf{u}_n = \lambda_n^* \mathbf{u}_n, \qquad n = 1, 2, \ldots. \tag{8.9.32}$$

Hence, the eigenvectors of the adjoint $\mathbf{L}^\dagger$ of a perfect operator $\mathbf{L}$ form the reciprocal basis set for the eigenvectors of $\mathbf{L}$. Moreover, the eigenvalues of $\mathbf{L}^\dagger$ are the complex conjugates of the eigenvalues of $\mathbf{L}$.

We see that the analogy between linear operators in abstract linear vector spaces and matrix operators in E_n is strikingly close, especially for perfect operators. The challenge, however, is to determine what classes of operators are perfect. In the following chapters we will investigate certain classes of integral and differential operators and we will try to provide an answer to this question.

Next, let us consider the following theorem:

THEOREM. *A self-adjoint, k-term dyadic is a perfect operator, has a complete set of orthonormal eigenvectors, and has real eigenvalues.*

To prove this, suppose

$$\mathbf{L} = \sum_{l=1}^{k} \mathbf{u}_l \mathbf{v}_l^\dagger, \tag{8.9.33}$$

where the sets $\{\mathbf{u}_1, \ldots, \mathbf{u}_k\}$ and $\{\mathbf{v}_1, \ldots, \mathbf{v}_k\}$ are linearly independent sets in a Hilbert space $\mathcal{H}$. We define the adjoint of $\mathbf{L}$ by

$$\mathbf{L}^\dagger = \sum_{l=1}^{k} \mathbf{v}_l \mathbf{u}_l^\dagger. \tag{8.9.34}$$

The operations $\mathbf{L}\mathbf{u}$ and $\mathbf{L}^\dagger \mathbf{u}$ can then be composed as follows:

$$\mathbf{L}\mathbf{u} = \sum_{l=1}^{k} \mathbf{u}_l \mathbf{v}_l^\dagger \mathbf{u} = \sum_{l=1}^{k} \langle \mathbf{v}_l, \mathbf{u} \rangle \mathbf{u}_l \tag{8.9.35}$$

and

$$\mathbf{L}^\dagger \mathbf{u} = \sum_{l=1}^{k} \mathbf{v}_l \mathbf{u}_l^\dagger \mathbf{u} = \sum_{l=1}^{k} \langle \mathbf{u}_l, \mathbf{u} \rangle \mathbf{v}_l, \tag{8.9.36}$$

where $\mathbf{v}_l^\dagger \mathbf{u}$ and $\langle \mathbf{v}_l, \mathbf{u} \rangle$ are, again, our two different notations for the inner product defining the Hilbert space. Using the Gram–Schmidt procedure, a set of orthonormal vectors $\{\boldsymbol{\phi}_1, \ldots, \boldsymbol{\phi}_k\}$ can be constructed as a linear combination of $\{\mathbf{v}_1, \ldots, \mathbf{v}_k\}$ and, thus, we can find coefficients γ_{lj} such that

$$\mathbf{v}_l = \sum_{j=1}^{k} \gamma_{lj} \boldsymbol{\phi}_j, \qquad \text{where } \langle \boldsymbol{\phi}_i, \boldsymbol{\phi}_j \rangle = \delta_{ij}. \tag{8.9.37}$$

Recall that the orthonormal functions $\boldsymbol{\phi}_{k+1}, \boldsymbol{\phi}_{k+2}, \ldots$ can always be found to complete an orthonormal basis set $\boldsymbol{\phi}_i$, $i = 1, 2, \ldots$, in $\mathcal{H}$.

Insertion of the expression for $\mathbf{v}_l$ in Eq. (8.9.37) into Eq. (8.9.33) yields

$$\mathbf{L} = \sum_{j=1}^{k} \boldsymbol{\psi}_j \boldsymbol{\phi}_j^\dagger, \tag{8.9.38}$$

where

$$\boldsymbol{\psi}_j \equiv \sum_{l=1}^{k} \gamma_{lj}^* \mathbf{u}_l. \tag{8.9.39}$$

Since the set $\{\boldsymbol{\phi}_i\}$ forms a basis set, we can express the vectors $\boldsymbol{\psi}_j$ as

$$\boldsymbol{\psi}_j = \sum_{i=1}^{\infty} \alpha_{ij} \boldsymbol{\phi}_i, \qquad j = 1, \ldots, k. \tag{8.9.40}$$

The operator $\mathbf{L}$ and its adjoint can now be represented as

$$\mathbf{L} = \sum_{j=1}^{k} \sum_{i=1}^{\infty} \alpha_{ij} \boldsymbol{\phi}_i \boldsymbol{\phi}_j^\dagger \tag{8.9.41}$$

and

$$\mathbf{L}^\dagger = \sum_{j=1}^{k} \sum_{i=1}^{\infty} \alpha_{ij}^* \boldsymbol{\phi}_j \boldsymbol{\phi}_i^\dagger. \tag{8.9.42}$$

Operating on the vector $\boldsymbol{\phi}_n$ these expressions become

$$\mathbf{L}\boldsymbol{\phi}_n = \begin{cases} \sum_{i=1}^{\infty} \alpha_{in} \boldsymbol{\phi}_i, & n \leq k, \\ 0, & n > k. \end{cases} \tag{8.9.43}$$

and

$$\mathbf{L}^\dagger \boldsymbol{\phi}_n = \sum_{i=1}^{k} \alpha_{ni}^* \boldsymbol{\phi}_i. \tag{8.9.44}$$

Since $\mathbf{Lu} = \mathbf{L}^\dagger \mathbf{u}$, comparison of Eqs. (8.9.43) and (8.9.44) leads to the conclusions $\alpha_{in} = \alpha_{ni}^*$ for $n \leq k$ and $\alpha_{ij} = 0$ for $n \geq k$. So the $k \times k$ matrix $\mathbf{A} = [\alpha_{ij}]$ has the property

$$\mathbf{A} = \mathbf{A}^\dagger; \tag{8.9.45}$$

i.e., $\mathbf{A}$ is self-adjoint. We know from Chapter 6 that a self-adjoint matrix can be expressed in the form

$$\mathbf{A} = \mathbf{U}\boldsymbol{\Lambda}\mathbf{U}^\dagger, \tag{8.9.46}$$

where $\mathbf{\Lambda}$ is a diagonal matrix whose diagonal elements are the real eigenvalues of $\mathbf{A}$, and $\mathbf{U}$ is a unitary matrix, i.e.,

$$\mathbf{U}^\dagger \mathbf{U} = \mathbf{I} \quad \text{or} \quad \sum_{l=1}^{k} u_{li}^* u_{lj} = \delta_{ij}. \tag{8.9.47}$$

Thus, we can write the coefficients of $\mathbf{A}$ as

$$\alpha_{ij} = \sum_{l=1}^{k} \lambda_l u_{il} u_{jl}^* \tag{8.9.48}$$

and express the operator $\mathbf{L}$ as

$$\mathbf{L} = \sum_{i,j,l=1}^{k} \lambda_l u_{il} u_{jl}^* \boldsymbol{\phi}_i \boldsymbol{\phi}_j^\dagger. \tag{8.9.49}$$

With the definitions

$$\boldsymbol{\chi}_l = \sum_{i=1}^{k} u_{il} \boldsymbol{\phi}_i \quad \text{and} \quad \boldsymbol{\chi}_l^\dagger = \sum_{i=1}^{k} u_{il}^* \boldsymbol{\phi}_i^\dagger, \tag{8.9.50}$$

Eq. (8.9.49) becomes

$$\mathbf{L} = \sum_{l=1}^{k} \lambda_l \boldsymbol{\chi}_l \boldsymbol{\chi}_l^\dagger. \tag{8.9.51}$$

Finally, we note that

$$\begin{aligned}\langle \boldsymbol{\chi}_l, \boldsymbol{\chi}_{l'} \rangle &= \sum_{i=1}^{k} \sum_{j=1}^{k} u_{il}^* u_{jl'} \langle \boldsymbol{\phi}_i, \boldsymbol{\phi}_j \rangle \\ &= \sum_{j=1}^{k} u_{jl}^* u_{jl'} = \delta_{ll'}.\end{aligned} \tag{8.9.52}$$

Thus, the vectors $\boldsymbol{\chi}_i$, $i = 1, \ldots, k$, are the orthonormal eigenvectors of $\mathbf{L}$ and the eigenvalues of $\mathbf{A}$ are the eigenvalues of $\mathbf{L}$ corresponding to $\boldsymbol{\chi}_i$. Moreover, since $\langle \boldsymbol{\phi}_i, \boldsymbol{\phi}_j \rangle = 0$ for $i \leq k$ and $j > k$, it follows that

$$\mathbf{L}\boldsymbol{\phi}_j = \mathbf{0}, \qquad j = k+1, k+2, \ldots; \tag{8.9.53}$$

i.e., the vectors $\boldsymbol{\phi}_j$, for which $j > k$, are eigenvectors of $\mathbf{L}$ with zero eigenvalue. We see that the eigenvectors $\boldsymbol{\chi}_1, \ldots, \boldsymbol{\chi}_k, \boldsymbol{\phi}_{k+1}, \boldsymbol{\phi}_{k+2}, \ldots$ form an orthonormal basis set in $\mathcal{H}$ and a k-term dyadic in an infinite-dimensional vector space will always have an infinite number of eigenvectors having a zero eigenvalue.

A similar proof can be given for the following theorem:

THEOREM. *A normal k-term dyadic operator (defined by the property $\mathbf{L}\mathbf{L}^\dagger = \mathbf{L}^\dagger \mathbf{L}$) in $\mathcal{H}$ has a complete set of orthonormal eigenvectors in $\mathcal{H}$.*

L can, again, be shown to be expressible in the form

$$\mathbf{L} = \sum_{j=1}^{k}\sum_{i=1}^{\infty}\alpha_{ij}\boldsymbol{\phi}_i\boldsymbol{\phi}_j^{\dagger}, \qquad (8.9.54)$$

where the set $\{\boldsymbol{\phi}_i\}$ forms an orthonormal set. The condition $\mathbf{LL}^{\dagger} = \mathbf{L}^{\dagger}\mathbf{L}$ and the inner products $\langle\boldsymbol{\phi}_r, \mathbf{LL}^{\dagger}\boldsymbol{\phi}_s\rangle$ and $\langle\boldsymbol{\phi}_r, \mathbf{L}^{\dagger}\mathbf{L}\boldsymbol{\phi}_s\rangle$ lead to the conditions

$$\sum_{j=1}^{k}\alpha_{rj}\alpha_{sj}^{*} = 0, \qquad r \text{ or } s > k, \qquad (8.9.55)$$

and

$$\sum_{j=1}^{k}\alpha_{rj}\alpha_{sj}^{*} = \sum_{j=1}^{\infty}\alpha_{js}\alpha_{jr}^{*}. \qquad (8.9.56)$$

Equation (8.9.55) for $r = s > k$ implies that $\alpha_{rj} = 0$ for $r > 0$. Thus, the $k \times k$ matrix $\mathbf{A} = [\alpha_{ij}]$, according to Eq. (8.9.56), obeys the condition

$$\mathbf{AA}^{\dagger} = \mathbf{A}^{\dagger}\mathbf{A}; \qquad (8.9.57)$$

i.e., $\mathbf{A}$ is a normal matrix. We proved previously that normal matrices can be expressed as

$$\mathbf{A} = \mathbf{U\Lambda U}^{\dagger}, \qquad (8.9.58)$$

where $\mathbf{U}$ is a unitary matrix and the diagonal matrix $\mathbf{\Lambda}$ contains the eigenvalues λ_i of $\mathbf{A}$. This result again allows us to transform $\mathbf{L}$ into the form of Eq. (8.9.51) in which the $\boldsymbol{\chi}_l$'s are orthonormal eigenvectors of $\mathbf{L}$ with eigenvalues λ_i. The λ_i are not necessarily real as in the case of a self-adjoint matrix.

It follows from the spectral theorem that, for any function $f(t)$ that can be expressed as a series in t,

$$f(\mathbf{L}) = \sum_{l=1}^{k} f(\lambda_l)\boldsymbol{\chi}_l\boldsymbol{\chi}_l^{\dagger} \qquad (8.9.59)$$

for a normal k-term dyadic operator (of which a self-adjoint operator is a special case). A further generalization is that, for any $f(t)$ that exists at $t = \lambda_i$, $i = 1, 2, \ldots$, Eq. (8.9.59) defines $f(\mathbf{L})$ for a normal operator. Examples of this are $f(t) = t^{1/2}$ and $f(t) = \ln t$, for which

$$f(\mathbf{L}) = \sum_{l=1}^{k}\lambda_l^{1/2}\boldsymbol{\chi}_l\boldsymbol{\chi}_l^{\dagger} \quad \text{and} \quad f(\mathbf{L}) = \sum_{\ell=1}^{k}\ln\lambda_l\boldsymbol{\chi}_l\boldsymbol{\chi}_l^{\dagger}. \qquad (8.9.60)$$

Let us close this chapter by noting that a linear problem in any Hilbert space can be mapped into a linear problem in the space E_{∞}. Consider the equation

$$\mathbf{Lu} = \mathbf{f}, \qquad \mathbf{u}, \mathbf{f} \in \mathcal{H}. \qquad (8.9.61)$$

Suppose $\mathbf{v}_n$, $n = 1, 2, \ldots$, is an orthonormal basis set in $\mathcal{H}$. Then it follows that $\mathbf{u} = \sum_n \mathbf{v}_n \langle \mathbf{v}_n, \mathbf{u} \rangle$ and $\mathbf{f} = \sum_n \mathbf{v}_n \langle \mathbf{v}_n, \mathbf{f} \rangle$ from which Eq. (8.9.61) becomes

$$\sum_n \mathbf{L}\mathbf{v}_n \langle \mathbf{v}_n, \mathbf{u} \rangle = \sum_n \mathbf{v}_n \langle \mathbf{v}_n, \mathbf{f} \rangle. \tag{8.9.62}$$

Taking the inner product of Eq. (8.9.62) with $\mathbf{v}_m$, we obtain

$$\sum_n \langle \mathbf{v}_m, \mathbf{L}\mathbf{v}_n \rangle \langle \mathbf{v}_n, \mathbf{u} \rangle = \langle \mathbf{v}_m, \mathbf{f} \rangle, \quad m = 1, 2, \ldots. \tag{8.9.63}$$

We can define the matrix $\mathbf{A}$ in E_∞ with components

$$a_{mn} = \langle \mathbf{v}_m, \mathbf{L}\mathbf{v}_n \rangle, \tag{8.9.64}$$

and the vectors $\mathbf{x}$ and $\mathbf{b}$ with components

$$x_n = \langle \mathbf{v}_n, \mathbf{u} \rangle \quad \text{and} \quad b_n = \langle \mathbf{v}_n, \mathbf{f} \rangle. \tag{8.9.65}$$

Equation (8.9.63) then becomes the matrix equation

$$\mathbf{A}\mathbf{x} = \mathbf{b}, \quad \mathbf{x}, \mathbf{b} \in E_\infty. \tag{8.9.66}$$

Hence, the problem posed in Eq. (8.9.61) has been converted into an equivalent problem in E_∞. Of course, handling vectors in the infinite-dimensional space E_∞ is not necessarily easier than dealing directly with differential and integral equations. However, the equivalence of any $\mathcal{H}$ to E_∞ does provide encouragement to seek analogies between problems in $\mathcal{H}$ and known results in E_n.

In the usual solution of practical problems, one almost always converts differential and integral problems to algebraic problems of finite dimension. Finite-element analysis is a popular example. The unknowns are approximated by a finite number of basis functions, and then the methods and theory laid out in Chapters 1–7 for solving algebraic systems are employed to get answers. This being the case, why pursue the theory of abstract vector spaces further? The answer is that the solvability of differential and integral problems is an important issue underlying any approximate solution. Also, there exists a great body of theory from classical analysis that often provides analytical solutions to important, illustrative problems. These solutions often identify similarities and important differences between finite- and infinite-dimensional problems and can be used to test approximating computer codes.

PROBLEMS

1. Solve the dyadic equation

$$\int_{-1}^{1} \sum_{i=0}^{3} (st)^i u(s) \, ds = 1 + 3t^2 + t^3.$$

2. Discuss the solution of the equation

$$\int_0^\pi \cos 3(t-s) u(t)\, dt = \sin 3t, \cos 3t$$

in $\mathcal{L}_2(0, \pi)$.

3. Consider the dyadic operators

$$\mathbf{L}_1 = \sum_{i=1}^\infty \lambda_i \mathbf{v}_i \mathbf{v}_i^\dagger,$$

$$\mathbf{L}_2 = \sum_{i=1}^\infty \mu_i \mathbf{v}_i \mathbf{v}_i^\dagger,$$

and

$$\mathbf{L}_3 = \sum_{i=1}^\infty \nu_i \mathbf{v}_i \mathbf{v}_i^\dagger,$$

where $\{\mathbf{v}_i\}$ is an orthonormal basis set in a Hilbert space $\mathcal{H}$ and $\lambda_i = i$, $\mu_i = 1/\sqrt{i}$, and $\nu_i = 1/i$. The square of the norm of an operator in $\mathcal{H}$ is

$$\|\mathbf{L}\|^2 = \max_{\mathbf{u} \in \mathcal{H}} \frac{\langle \mathbf{Lu}, \mathbf{Lu} \rangle}{\langle \mathbf{u}, \mathbf{u} \rangle}.$$

(a) Prove that $\mathbf{L}_1$ and $\mathbf{L}_2$ are unbounded and that $\mathbf{L}_3$ is bounded, i.e., $\|\mathbf{L}_1\| = \|\mathbf{L}_2\| = \infty$ and $\|\mathbf{L}_3\| < \infty$.

(b) Which of the three operators are perfect?

4. Consider the operator $\mathbf{L}$, defined by

$$Lu = -\frac{d^2 u}{dx^2},$$

$$u(1) = u(0) \quad \text{and} \quad \frac{du(1)}{dx} = \frac{du(0)}{dx}$$

in $\mathcal{L}_2(0, 1)$.

(a) Find the eigenvectors and eigenvalues of $\mathbf{L}$.
(b) Normalize the eigenvectors.
(c) As will be proved in a later chapter, $\mathbf{L}$ is a perfect operator. Give its dyadic form; i.e., give its spectral representation. Note that this enables you to express a differential operator as an integral operator.
(d) Is $\mathbf{L}$ a bounded operator in the sense described in Problem 3?

5. Prove that if $\mathbf{f}$ belongs to $\mathcal{L}_2(0, 1)$, then it belongs to $\mathcal{L}_2(0, 1; e^{-t})$; i.e., prove that if

$$\int_0^1 |f(t)|^2\, dt < \infty,$$

then

$$\int_0^1 |f(t)|^2 \exp(-t)\, dt < \infty,$$

where the integrals are Lebesgue integrals. The proof is simple.

6. The Hermite functions

$$h_n(t) = \frac{(-1)^n}{\sqrt{2^n n! \sqrt{\pi}}} \exp\left(\frac{t^2}{2}\right) \frac{d^n}{dt^n} \exp(-t^2),$$

$n = 0, 1, 2, \ldots$, form a basis set in $\mathcal{L}_2(-\infty, \infty)$. Prove that they form an orthonormal basis set.

7. Suppose $\mathbf{f}_n \in \mathcal{L}_2(0, 1)$, where $f_n(t) = t^n$. Show by direct calculation that

$$|\langle \mathbf{f}_n, \mathbf{f}_m \rangle| \le \|\mathbf{f}_n\| \, \|\mathbf{f}_m\|.$$

8. Consider the vector space $\mathcal{L}_2(-\infty, \infty)$ and the sequence of linearly independent vectors

$$v_n(t) = t^n \exp\left(\frac{-t^2}{2}\right), \quad n = 0, 1, 2, \ldots.$$

Use the Gram–Schmidt procedure to orthonormalize this sequence for $n = 0, 1, 2$. Compare the orthonormalized vectors $u_n, n = 0, 1, 2$, with the Hermite functions defined in Problem 6.

9. Find the eigenvectors of the differential operator $\mathbf{L}$ in $\mathcal{L}_2(0, 1)$, where

$$Lv(t) = -\frac{d^2 v(t)}{dt^2},$$

$v(0) = 0$, and $dv/dt = v$ at $t = 1$.

10. Consider the equation

$$\mathbf{u} + \mathbf{K}\mathbf{u} = \mathbf{f},$$

where $\mathbf{u}$ and $\mathbf{f}$ are vectors in $\mathcal{L}_2(0, 1)$ and $\mathbf{K}$ is an integral operator with the kernel

$$k(t, s) = \frac{1}{2} \exp(-(t + s)).$$

(a) Prove that $\|\mathbf{K}\| < 1$, and so the equation has a unique solution.
(b) Find the solution when $f(t) = 1$. Note that $\mathbf{K}$ is a dyad operator.
(c) Find the solution when $f(t) = \sin t$.

FURTHER READING

Bear, H. S. (1995). *A Primer of Lebesgue Integration*, Academic Press, San Diego.
Bell, D. J. (1990) "Mathematics of Linear and Nonlinear Systems: for Engineers and Applied Scientists." Clarendon Press, Oxford.
Churchill, R. V. (1958). "Operational Mathematics," McGraw-Hill, New York.
Friedman, B. (1956). "Principles and Techniques of Applied Mathematics." Wiley, New York.
Kantorovich, L. V. and Akilov, G. P. (1964). "Functional Analysis in Normed Spaces." Pergamon, New York.
Kolmogorov, A. N. and Fomin, S. V. (1961). "Measure, Lebesque Integrals, and Hilbert Spaces." Academic, New York.
Kolmogorov, A. N. and Fomin, S. V. (1970). "Introductory Real Analysis." Prentice Hall International, Englewood Cliffs, NJ.
Riesz, F. and Nagy, B. S. (1965). "Functional Analysis." Ungar, New York.
Schmeidler, W. (1965). Linear Operators in Hilbert Spaces. Academic, New York.

9
LINEAR INTEGRAL OPERATORS IN A HILBERT SPACE

9.1. SYNOPSIS

In this chapter we will study linear integral operators in Hilbert spaces. The emphasis here will be primarily on bounded operators (i.e., operators whose norms $\|\mathbf{K}\|$ are less than ∞). The domain of an unbounded operator is a subset of vectors in the Hilbert space. In the class of bounded operators, *completely continuous operators* are of special interest since almost everything we have learned about operators in finite-dimensional vector spaces is true for these operators. A completely continuous operator is bounded and can be approximated as closely as we please by a finite, n-term dyadic. Finally, we will consider the class of Hilbert–Schmidt operators that satisfy the condition $\sum_n \|\mathbf{K}\boldsymbol{\psi}_n\|^2 < \infty$, where $\{\boldsymbol{\psi}_n\}$ is a complete orthonormal set in the Hilbert space. We will see that Hilbert–Schmidt operators are actually a subclass of completely continuous operators.

As we did in Chapter 4 for finite-dimensional vector spaces, we will prove the Fredholm solvability theorems as they apply to the above classes of operators—allowing us to explore the solvability of linear integral equations. In particular, we will show that the equation $\mathbf{Lu} = \mathbf{f}$ has a unique solution *if and only if* the only solution to the homogeneous equation $\mathbf{Lu} = \mathbf{0}$ is $\mathbf{u} = \mathbf{0}$. Furthermore, a solution to the inhomogeneous equation exists *if and only if* $\langle \mathbf{v}_i, \mathbf{f} \rangle = 0$, $i = 1, \ldots, m$, where $\{\mathbf{v}_i\}$ are all the solutions to the adjoint equation $\mathbf{L}^\dagger \mathbf{v} = \mathbf{0}$.

Section 4 of this chapter is devoted to Volterra equations of the first and second kind. This important class of equations is unusual in that solutions are generally unique and Volterra operators have no eigenvectors. We will show that one-dimensional Volterra equations can be solved by the method of Laplace transforms when the kernel $k(t, s)$ depends only on $t - s$. Volterra equations of the second kind can generally be solved by a method of iteration, which always converges for these equations. We will further show that one-dimensional initial value problems involving pth-order linear differential equations can be converted to the problem of solving corresponding Volterra equations.

The important and extremely useful *spectral theory of integral operators* will be presented and derived for the class of completely continuous operators. We will show that any completely continuous operator can be represented in Schmidt's normal form $\mathbf{K} = \sum_i \kappa_i \boldsymbol{\psi}_i \boldsymbol{\phi}_i^\dagger$, where $\kappa_i \neq 0$. The orthonormal sets $\{\boldsymbol{\phi}_1, \boldsymbol{\phi}_2, \ldots\}$ and $\{\boldsymbol{\psi}_1, \boldsymbol{\psi}_2, \ldots\}$ form orthonormal basis sets in $\mathcal{H}$ and are solutions of the self-adjoint equations $\mathbf{K}^\dagger \mathbf{K} \boldsymbol{\phi}_i = \kappa_i^2 \boldsymbol{\phi}_i$ and $\mathbf{K} \mathbf{K}^\dagger \boldsymbol{\psi}_i = \kappa_i^2 \boldsymbol{\psi}_i$. For the cases of *normal*, completely continuous operators (with $\mathbf{K}^\dagger \mathbf{K} = \mathbf{K} \mathbf{K}^\dagger$) and *self-adjoint* operators (with $\mathbf{K}^\dagger = \mathbf{K}$), the spectral decomposition becomes $\mathbf{K} = \sum_i \lambda_i \boldsymbol{\phi}_i \boldsymbol{\phi}_i^\dagger$, $\langle \boldsymbol{\phi}_i \boldsymbol{\phi}_j \rangle = \delta_{ij}$, where the eigenfunctions $\boldsymbol{\phi}_i$ form complete orthonormal basis sets in the appropriate Hilbert space. The eigenvalues $\{\lambda_i\}$, analogous to the case for matrices in the finite-dimensional theory, are always real numbers for self-adjoint operators, and, in general, are complex for normal operators. We will subsequently show that the spectral resolution for a function, $f(\mathbf{K}) = \sum_i f(\lambda_i) \boldsymbol{\phi}_i \boldsymbol{\phi}_i^\dagger$, follows quite naturally, provided $f(t)$ exists for $t = \lambda_i$, $i = 1, 2, \ldots$.

In our presentation of the spectral resolution theorem, we will call upon various definitions and theorems that are useful in their own right, including Bessel's inequality: $\sum_i |\langle \boldsymbol{\phi}_i, \mathbf{f} \rangle|^2 \leq \|\mathbf{f}\|^2$, where the set $\{\boldsymbol{\phi}_i\}$ is an arbitrary orthonormal set in $\mathcal{H}$. We will also show that completely continuous operators have a finite number of eigenvectors for any nonzero eigenvalue λ_i. We note that if $\boldsymbol{\phi}_i$ is an eigenvector of $\mathbf{K}$ of eigenvalue λ_i, then $\boldsymbol{\phi}_i$ is also an eigenvector of $f(\mathbf{K})$ of eigenvalue $f(\lambda_i)$ if the function $f(t)$ is defined for $t = \lambda_i$. We will also prove that if the operators $\mathbf{L}$ and $\mathbf{K}$ commute, i.e., if $\mathbf{LK} = \mathbf{KL}$, then $\mathbf{K}$ and $\mathbf{L}$ have at least one common eigenvector.

Virtually everything we have learned about matrix operators in finite-dimensional vector spaces also holds for completely continuous operators in a Hilbert space. The class of completely continuous integral operators contains operators with continuous or piecewise-continuous kernels in a closed, bounded domain, Hilbert–Schmidt operators in bounded and unbounded domains, and operators with weakly singular kernels (kernels diverging as $1/|\vec{r} - \vec{s}|^\alpha$, where $0 < \alpha < D/2$ and D is the number of independent components of $\vec{r}$ or $\vec{s}$).

9.2. SOLVABILITY THEOREMS

The linear integral equations we will study in this chapter will be either of the first kind

$$\mathbf{Ku} = \mathbf{f} \qquad (9.2.1)$$

or of the second kind

$$\mathbf{u} + \mathbf{Ku} = \mathbf{f}, \qquad (9.2.2)$$

where **K** is a linear integral operator. For the one-dimensional problem, these equations become

$$\int_a^b k(t,s)u(s)\,ds = f(t) \tag{9.2.3}$$

and

$$u(t) + \int_a^b k(t,s)u(s)\,ds = f(t), \tag{9.2.4}$$

where $k(t,s)$ is the kernel of the integral operator **K**, $k(t,s)$ and $f(t)$ are known functions, and $u(t)$ is to be determined.

For the D-dimensional problem, these equations take the form

$$\int_{\Omega_D} k(\vec{r},\vec{s})u(\vec{s})\,d^D s = f(\vec{r}) \tag{9.2.5}$$

and

$$u(\vec{r}) + \int_{\Omega_D} k(\vec{r},\vec{s})u(\vec{s})\,d^D s = f(\vec{r}), \tag{9.2.6}$$

where $\vec{r}$ and $\vec{s}$ represent the D independent variables, $d^D s$ represents a volume element in the D-dimensional space in which $\vec{r}$ and $\vec{s}$ are defined, and Ω_D is the volume of space over which these variables range. Whether $\vec{r}$ and $\vec{s}$ are D-dimensional Euclidean vectors or simply sets of D independent variables (such as concentrations of chemicals in solution) makes no difference. The kernel $k(\vec{r},\vec{s})$ and the quantity $f(\vec{r})$ are, again, known functions and $u(\vec{r})$ is to be determined.

In proving the solvability theorems presented in this section, we will assume that u, k, and f are Riemann integrable functions in the closed domains of the independent variables. Examples of such functions include continuous and piecewise-continuous functions.

We define the adjoint $\mathbf{K}^\dagger$ of the integral operator **K**, in a one-variable problem, by

$$\mathbf{K}^\dagger \mathbf{v} = \int_a^b k^*(s,t)v(s)\,ds; \tag{9.2.7}$$

i.e., if $k(t,s)$ is the kernel of **K**, then $k^*(s,t)$ is the kernel of $\mathbf{K}^\dagger$, where $k^*(s,t)$ is the complex conjugate of $k(s,t)$. The adjoint $(\mathbf{K}\mathbf{u})^\dagger$ of $\mathbf{K}\mathbf{u}$ is then given by

$$(\mathbf{K}\mathbf{u})^\dagger = \mathbf{u}^\dagger \mathbf{K}^\dagger = \int_a^b u^*(s)k^*(t,s)\,ds. \tag{9.2.8}$$

So in the Hilbert space $\mathcal{L}_2(a,b)$, with the inner product defined as

$$\langle \mathbf{v}, \mathbf{u} \rangle \equiv \int_a^b v^*(t)u(t)\,dt, \tag{9.2.9}$$

the inner product of $\mathbf{K}\mathbf{v}$ and $\mathbf{K}\mathbf{u}$ becomes

$$\langle \mathbf{K}\mathbf{v}, \mathbf{K}\mathbf{u} \rangle = \mathbf{v}^\dagger \mathbf{K}^\dagger \mathbf{K}\mathbf{u} = \int_a^b \int_a^b \int_a^b v^*(s_1)k^*(t,s_1)k(t,s_2)u(s_2)\,ds_1\,ds_2\,dt. \tag{9.2.10}$$

We define the norm of $\mathbf{K}$ by

$$\|\mathbf{K}\|^2 \equiv \max_{\mathbf{u} \neq 0} \frac{\langle \mathbf{Ku}, \mathbf{Ku}\rangle}{\langle \mathbf{u}, \mathbf{u}\rangle}$$

$$= \max_{\mathbf{u}\neq 0} \frac{\int_a^b \int_a^b \int_a^b k^*(t, s_1) k(t, s_2) u^*(s_1) u(s_2)\, ds_1\, ds_2\, dt}{\int_a^b |u(s)|^2\, ds}. \qquad (9.2.11)$$

Similarly, for the D-dimensional problem in the Hilbert space $\mathcal{L}_2(\Omega_D)$, the adjoint $\mathbf{K}^\dagger$ of $\mathbf{K}$ is defined as

$$\mathbf{K}^\dagger \mathbf{v} \equiv \int k^*(\vec{s}, \vec{r}) v(\vec{s})\, d^D s. \qquad (9.2.12)$$

Likewise, $(\mathbf{Kv})^\dagger$ is given by

$$\mathbf{v}^\dagger \mathbf{K}^\dagger = \int_{\Omega_D} v^*(\vec{s}) k^*(\vec{r}, \vec{s})\, d^D s, \qquad (9.2.13)$$

and the inner product $\langle \mathbf{Kv}, \mathbf{Ku}\rangle$ by

$$\langle \mathbf{Kv}, \mathbf{Ku}\rangle = \mathbf{v}^\dagger \mathbf{K}^\dagger \mathbf{Ku}$$
$$= \int_{\Omega_D}\int_{\Omega_D}\int_{\Omega_D} v^*(\vec{s}_1) k^*(\vec{r}, \vec{s}_1) k(\vec{r}, \vec{s}_2) u(\vec{s}_2)\, d^D s_1\, d^D s_2\, d^D r. \qquad (9.2.14)$$

With the above definitions in hand, we can now state the following theorem:

THEOREM. *If Ω_D is a finite volume (or interval in one dimension) and $k(\vec{r}, \vec{s})$ is a bounded function (which it certainly is if it is continuous on the closed domain Ω_D), then the norm of $\mathbf{K}$ is bounded; i.e., it is finite.*

To prove this, first note that

$$\left|\int_{\Omega_D} u(\vec{s})\, d^D \vec{s}\right| = |\langle \mathbf{u}, 1\rangle| \leq \langle|\mathbf{u}|, 1\rangle \leq \|\mathbf{u}\|\,\|1\| = \|\mathbf{u}\|\Omega_D^{1/2}. \qquad (9.2.15)$$

If $M = \max_{\vec{r},\vec{s}} k(\vec{r}, \vec{s})$, it follows that

$$|\langle \mathbf{Ku}, \mathbf{Ku}\rangle| \leq M^2 \Omega_D |\langle|\mathbf{u}|, 1\rangle\langle|\mathbf{u}|, 1\rangle| \leq M^2 \Omega_D^2 \|\mathbf{u}\|^2 \qquad (9.2.16)$$

or, using Eq. (9.2.11),

$$\|\mathbf{K}\|^2 = \max_{\mathbf{u}\neq 0} \frac{\langle \mathbf{Ku}, \mathbf{Ku}\rangle}{\|\mathbf{u}\|^2} \leq M^2 \Omega_D^2. \qquad (9.2.17)$$

Since $\|\mathbf{I} + \mathbf{K}\| \leq \|\mathbf{I}\| + \|\mathbf{K}\|$, it also follows from Eq. (9.2.17) that $\|\mathbf{I} + \mathbf{K}\| \leq (1+M)\Omega_D$, proving the theorem. All of the integral operators studied in this section will be of the type described above and will, therefore, have bounded norms.

Let us now turn to the solvability theory of equations of the first and second kind. We will extend the problem to include D independent variables since the one-dimensional case is included in this more general case. First, imagine that the volume Ω_D is filled with small volume elements $\Delta^D s$. We shall number these volume elements $1, 2, \ldots, n$ with the center of the ith element being located by

SOLVABILITY THEOREMS

the vector $\vec{r}_i$. Since the integral in Eq. (9.2.5) is Riemann integrable, the integral equations of the first and second kind can be approximated as

$$\sum_{j=1}^{n} k_{ij} \Delta^D s \, u_i = f_i, \qquad i = 1, \ldots, n, \qquad (9.2.18a)$$

and

$$\sum_{j=1}^{n} (\delta_{ij} + k_{ij} \Delta^D s) u_j = f_i, \qquad i = 1, \ldots, n, \qquad (9.2.18b)$$

where $k_{ij} = k(\vec{r}_i, \vec{r}_j)$, $u_i = u(\vec{r}_i)$, and $f_i = f(\vec{r}_i)$. The homogeneous adjoint equations, $\mathbf{K}^\dagger \mathbf{v} = \mathbf{0}$ and $(\mathbf{I}^\dagger \mathbf{K}^\dagger) \mathbf{v} = \mathbf{0}$, are similarly approximated as

$$\sum_{j=1}^{n} k_{ji}^* \Delta^D s \, v_j = 0, \qquad i = 1, \ldots, n, \qquad (9.2.19a)$$

and

$$\sum_{j=1}^{n} (\delta_{ij} + k_{ji}^* \Delta^D s) v_j = 0, \qquad i = 1, \ldots, n, \qquad (9.2.19b)$$

where $k_{ji}^* = k^*(\vec{r}_j, \vec{r}_i)$ and $v_i = v(\vec{r}_i)$. In the limit $\Delta^D s \to 0$, Eqs. (9.2.18) and (9.2.19) become the integral equations

$$\mathbf{K}\mathbf{u} = \mathbf{f} \qquad (9.2.20a)$$

$$(\mathbf{I} + \mathbf{K})\mathbf{u} = \mathbf{f} \qquad (9.2.20b)$$

and

$$\mathbf{K}^\dagger \mathbf{v} = \mathbf{0} \qquad (9.2.21a)$$

$$(\mathbf{I} + \mathbf{K}^\dagger)\mathbf{v} = \mathbf{0}. \qquad (9.2.21b)$$

From Section 4.6, we know that Eq. (9.2.18) has a solution *if and only if* the condition

$$\sum_{i=1}^{n} v_i^* f_i = 0 \qquad (9.2.22)$$

is obeyed for any solution to the homogeneous adjoint equation (9.2.19). Multiplying Eq. (9.2.22) by $\Delta^D s$, we can express the solvability condition as

$$\sum_{i=1}^{n} v_i^* f_i \, \Delta^D s = 0. \qquad (9.2.23)$$

Taking the limit $\Delta^D s \to 0$, Eq. (9.2.23) becomes

$$\int_{\Omega_D} v^*(\vec{r}) f(\vec{r}) \, d^D r = \langle \mathbf{v}, \mathbf{f} \rangle = 0. \qquad (9.2.24)$$

We also know from Section 4.6 that the number of solutions to the homogeneous equations

$$\sum_{j=1}^{n} k_{ij}\Delta^D s\, u_j = 0, \qquad i = 1,\ldots,n, \tag{9.2.25a}$$

and

$$\sum_{j=1}^{n}(\delta_{ij} + k_{ij}\Delta^D s)u_j = 0, \qquad i = 1,\ldots,n, \tag{9.2.25b}$$

are the same as the number of solutions to the homogeneous adjoint equations. And when there are no solutions to the homogeneous equation, there is a unique solution to Eq. (9.2.18) for arbitrary $f_1,\ldots,f_n$. In the limit $\Delta^D s \to 0$, Eq. (9.2.24) becomes

$$\mathbf{Ku} = \mathbf{0} \tag{9.2.26a}$$

and

$$(\mathbf{I} + \mathbf{K})\mathbf{u} = \mathbf{0}. \tag{9.2.26b}$$

Since the algebraic approximations to the integral equations converge to the continuum equations, it seems reasonable to expect the algebraic solutions to converge to solutions to the continuum equations. If u, k, and f are continuous functions in a closed finite domain of independent variables, this is indeed true. Detailed mathematical proofs can be found in W. V. Lovitt's *Linear Integral Equations*, Dover, 1950, and in R. Courant and D. Hilbert's *Methods of Mathematical Physics*, Interscience, 1953. Thus, for the continuum case, when $\mathbf{L} = \mathbf{K}$ or $\mathbf{I} + \mathbf{K}$ and u, k, and f are continuous and Ω_D is finite and closed (such that u, k, and f do not diverge on the boundary of Ω_D), we can summarize the Fredholm alternative theorems as follows:

FREDHOLM'S ALTERNATIVE THEOREM. (1) *The equation*

$$\mathbf{Lu} = \mathbf{f} \tag{9.2.27}$$

has a unique solution if and only if the only continuous solution to the homogeneous equation

$$\mathbf{Lu} = \mathbf{0} \tag{9.2.28}$$

is $\mathbf{u} = \mathbf{0}$. (2) *Alternatively, the homogeneous equation has at least one solution if the homogeneous adjoint equation*

$$\mathbf{L}^\dagger \mathbf{v} = \mathbf{0} \tag{9.2.29}$$

has at least one solution. The homogeneous equation (9.2.28) *and the adjoint homogeneous equation* (9.2.29) *have the same number of solutions.* (3) *When solutions to the homogeneous equation exist, the inhomogeneous equation* (9.2.27) *has a solution if and only if*

$$\langle \mathbf{v}_i, \mathbf{f} \rangle = 0, \qquad i = 1,\ldots,m, \tag{9.2.30}$$

where $\mathbf{v}_i$ is a solution to the homogeneous adjoint equation. The number m of solutions to the homogeneous equation can be infinite in infinite-dimensional vector spaces (as are the function spaces appropriate to integral equations).

Although we have outlined the proof of the theorem only for Riemann integrable kernels k and functions f and u in bounded (closed) domains Ω_D of independent variables, the solvability theorem often applies to singular problems—which involve unbounded kernels and/or cases in which one or more of the independent variables goes to ∞. Such cases will be illustrated in some of the examples to follow. In Section 9.3 we will examine the subclass of completely continuous operators for which the Fredholm solvability theorems apply.

EXAMPLE 9.2.1. Solve the equation $\mathbf{u} + \mathbf{K}\mathbf{u} = \mathbf{f}$, where $\mathbf{K}$ is a dyad operator, i.e., $\mathbf{K} = \mathbf{a}\mathbf{b}^\dagger$, and $\mathbf{u}, \mathbf{f} \in \mathcal{L}_2(0, 1)$. The equation is given by

$$u(x) - \frac{1}{2}\int_0^1 xtu(t)\,dt = \frac{5x}{6}. \qquad (9.2.31)$$

The methods for solving dyadic equations were described in Chapter 8. For the equation $\mathbf{u} + \mathbf{a}\mathbf{b}^\dagger\mathbf{u} = \mathbf{f}$, take the inner product with respect to $\mathbf{b}$ (to obtain $(1 + \mathbf{b}^\dagger\mathbf{a})\mathbf{b}^\dagger\mathbf{u} = \mathbf{b}^\dagger\mathbf{f}$) and solve for $\mathbf{b}^\dagger\mathbf{u}$. Here $a(x) = -\frac{1}{2}x$, $b(x) = x$, and $f(x) = 5x/6$. Multiplying Eq. (9.2.31) by x and integrating over dx, we obtain

$$\int_0^1 xu(x)\,dx - \frac{1}{2}\int_0^1 x^2\,dx \int_0^1 tu(t)\,dt = \frac{5}{6}\int_0^1 x^2\,dx \qquad (9.2.32)$$

or $\mathbf{b}^\dagger\mathbf{u} = \int_0^1 xu(x)\,dx = \frac{1}{3}$, and so $u(x) = x$ is the unique solution to Eq. (9.2.31)—unique because $u(x) = 0$ is the only solution to the homogeneous equation.

EXAMPLE 9.2.2. Solve the equation $\mathbf{u} + \mathbf{a}\mathbf{b}^\dagger\mathbf{u} = \mathbf{f}$, where $\mathbf{u}, \mathbf{f} \in \mathcal{L}_2(0, \infty)$. The equation is given by

$$u(x) + \int_0^\infty \exp(-(x+t))u(t)\,dt = x\exp(-x). \qquad (9.2.33)$$

Again, solve by taking the inner product of the equation with $\mathbf{b}$:

$$\int_0^\infty \exp(-x)u(x)\,dx + \int_0^\infty \exp(-2x)\,dx \int_0^\infty \exp(-t)u(t)\,dt$$
$$= \int_0^\infty x\exp(-2x)\,dx, \qquad (9.2.34)$$

and so $\mathbf{b}^\dagger\mathbf{u} = \int_0^\infty \exp(-x)u(x)\,dx = \frac{1}{6}$. Thus,

$$u(x) = \left(x - \frac{1}{6}\right)\exp(-x) \qquad (9.2.35)$$

is the unique solution to Eq. (9.2.33).

EXAMPLE 9.2.3. Solve the equation $\mathbf{K}\mathbf{u} = \mathbf{f}$, $\mathbf{u}, \mathbf{f} \in \mathcal{L}_2(-\infty, \infty)$, where

$$\int_{-\infty}^\infty \exp(-(x-t)^2)u(t)\,dt = \exp\left(\frac{-x^2}{2}\right). \qquad (9.2.36)$$

This equation is solved by the method of Fourier transforms. The Fourier transform $\tilde{u}(k)$ of a function $u(x)$ is, by definition,

$$\tilde{u}(k) = \frac{1}{\sqrt{2\pi}} \int_{-\infty}^{\infty} \exp(ikx) u(x)\, dx. \qquad (9.2.37)$$

The inverse Fourier transform of $\tilde{v}(k)$ is likewise

$$u(x) = \frac{1}{\sqrt{2\pi}} \int_{-\infty}^{\infty} \exp(-ikx) \tilde{u}(k)\, dk \qquad (9.2.38)$$

and nicely reciprocal in form to Eq. (9.2.37). According to the theory of Fourier transforms, the Fourier transform of $\int_{-\infty}^{\infty} k(x-y) u(y)\, dy$ is $\sqrt{2\pi}\, \tilde{k}(k) \tilde{u}(k)$. Thus, the Fourier transform of Eq. (9.2.36) is

$$\frac{\sqrt{2\pi}}{\sqrt{2}} \exp\left(\frac{-k^2}{4}\right) \tilde{u}(k) = \exp\left(\frac{-k^2}{2}\right), \qquad (9.2.39)$$

and so

$$\tilde{u}(k) = \sqrt{\frac{2}{\pi}} \frac{\exp(-k^2/4)}{\sqrt{2}}, \qquad (9.2.40)$$

which, since

$$\frac{1}{\sqrt{2\pi}} \int_{-\infty}^{\infty} \exp(ikx - x^2)\, dx = \frac{\exp(-k^2/4)}{\sqrt{2}}, \qquad (9.2.41)$$

inverts to

$$u(x) = \sqrt{\frac{2}{\pi}} \exp(-x^2). \qquad (9.2.42)$$

■ ■ ■ This is the only solution to Eq. (9.2.36).

EXAMPLE 9.2.4. Solve the equation $\mathbf{Ku} = \mathbf{f}$, where $\mathbf{u}, \mathbf{f} \in \mathcal{L}_2(R_3)$, and R_3 is an unbounded three-dimensional Euclidean vector space, i.e., $\vec{r} = x\hat{i} + y\hat{j} + z\hat{i}$, $-\infty < x, y, z < \infty$. In polar coordinates, $\vec{r} = r\sin\theta\cos\phi\hat{i} + r\sin\theta\sin\phi\hat{j} + r\cos\theta\hat{k}$, where $0 < r < \infty$, $0 < \theta < \pi$, and $0 < \phi < 2\pi$.

$$\int_{R_3} \exp(-(\vec{r}-\vec{s})^2) u(\vec{s})\, d^3s = \exp\left(\frac{-r^2}{2}\right). \qquad (9.2.43)$$

Again, the method of Fourier transforms is useful. In three dimensions, the Fourier transform is defined by

$$\tilde{u}(\vec{k}) = \frac{1}{(2\pi)^{3/2}} \int_{R_3} \exp(i\vec{k}\cdot\vec{r}) u(\vec{r})\, d^3r \qquad (9.2.44)$$

and has the inverse

$$u(\vec{r}) = \frac{1}{(2\pi)^{3/2}} \int_{R_3} \exp(-i\vec{k}\cdot\vec{r}) \tilde{u}(\vec{k})\, d^3k. \qquad (9.2.45)$$

SOLVABILITY THEOREMS

The three-dimensional Fourier transform of $\int_{R_3} k(\vec{r}-\vec{s})u(\vec{s})\,d^3s$ is $(2\pi)^{3/2}\tilde{k}(\vec{k})\tilde{u}(\vec{k})$, which yields for Eq. (9.2.43) the result

$$\tilde{u}(\vec{k}) = \frac{1}{(2\pi)^{3/2}} \frac{\tilde{f}(\vec{k})}{\tilde{k}(\vec{k})}. \tag{9.2.46}$$

The transforms of $\exp(-r^2)$ and $\exp(-r^2/2)$ are

$$\frac{1}{2^{3/2}} \exp\left(\frac{-k^2}{4}\right) \quad \text{and} \quad \exp\left(\frac{-k^2}{2}\right), \tag{9.2.47}$$

and so

$$\tilde{u}(\vec{k}) = \frac{1}{\pi^{3/2}} \exp\left(\frac{-k^2}{4}\right), \tag{9.2.48}$$

which inverts to

$$u(\vec{r}) = \left(\frac{2}{\pi}\right)^{3/2} \exp(-r^2). \tag{9.2.49}$$

■ ■ ■ This is the only solution to Eq. (9.2.43).

■ **EXAMPLE 9.2.5.** Solve the equation $\mathbf{Ku} = \mathbf{f}$ for $\mathbf{u}, \mathbf{f} \in \mathcal{L}_2(-1, 1)$, where $\mathbf{K}$ is the two-term dyadic

$$\mathbf{K} = \mathbf{P}_0\mathbf{P}_1^\dagger + \beta\mathbf{P}_1\mathbf{P}_1^\dagger, \tag{9.2.50}$$

β is a number, and $\mathbf{P}_l$ is a Legendre polynomial defined by

$$P_l(x) = \frac{1}{2^l l!} \frac{d^l}{dx^l}(x^2 - 1)^l. \tag{9.2.51}$$

Assume that $f(x) = 1 + 4x$ so that the integral equation is

$$\mathbf{P}_0\langle \mathbf{P}_1, \mathbf{u}\rangle + \beta\mathbf{P}_1\langle \mathbf{P}_1, \mathbf{u}\rangle = \mathbf{f} \tag{9.2.52}$$

or

$$\int_{-1}^{1} tu(t)\,dt + \beta x \int_{-1}^{1} tu(t)\,dt = 1 + 4x. \tag{9.2.53}$$

This has a solution only if

$$\int_{-1}^{1} tu(t)\,dt = 1 \quad \text{and} \quad \beta \int_{-1}^{1} tu(t)\,dt = 4. \tag{9.2.54}$$

Thus, only if $\beta = 4$ and $\int_{-1}^{1} tu(t)\,dt = 1$ is there a solution—for example, $u(t) = t$.

Since $\langle \mathbf{P}_l, \mathbf{P}_m\rangle = 0$ if $l \neq m$, the homogeneous equation $\mathbf{Ku} = \mathbf{0}$ has the solution

$$\mathbf{u}_h = \alpha_0 \mathbf{P}_0 + \sum_{i=2}^{\infty} \alpha_i \mathbf{P}_i, \tag{9.2.55}$$

where the α_i are arbitrary complex numbers. Thus, if $\beta = 4$, the general solution to Eq. (9.2.52) can be written as

$$\mathbf{u} = \sum_{i=0}^{\infty} \alpha_i \mathbf{P}_i, \qquad (9.2.56)$$

where $\alpha_1 = 1$ and the other α_i are arbitrary. If $\beta \neq 4$, there is no solution. Expressed as the Fredholm alternative theorem, the condition on $\mathbf{f}$ is

$$\langle \mathbf{v}, \mathbf{f} \rangle = 0, \qquad (9.2.57)$$

where $\mathbf{v}$ is any solution to $\mathbf{K}^\dagger \mathbf{v} = \mathbf{0}$ or any solution to

$$\mathbf{P}_1 \langle \mathbf{P}_0, \mathbf{v} \rangle + \beta \mathbf{P}_1 \langle \mathbf{P}_1, \mathbf{v} \rangle = \mathbf{0}. \qquad (9.2.58)$$

The solutions to this equation are $\mathbf{v}_i = \mathbf{P}_i$, $i = 2, 3, \ldots$, and $\mathbf{v}$ such that

$$\int_{-1}^{1} v(t) \, dt + \beta \int_{-1}^{1} t v(t) \, dt = 0. \qquad (9.2.59)$$

The solution to this equation that is linearly independent of $\mathbf{P}_i$, $i > 1$, is $v = b(t - \beta/2)$ or $\mathbf{v} = b(\mathbf{P}_1 - (\beta/2)\mathbf{P}_0)$, where b is arbitrary.

Since $\mathbf{f} = \mathbf{P}_0 + 4\mathbf{P}_1$, the Fredholm condition, Eq. (9.2.57), yields

$$b\left[-\frac{\beta}{2} \langle \mathbf{P}_0, \mathbf{P}_0 \rangle + 4\langle \mathbf{P}_1, \mathbf{P}_1 \rangle \right] = 0, \qquad (9.2.60)$$

or, as before, we find that $\beta = 4$ is the condition for solution to the inhomogeneous equation.

■ ■ ■

EXAMPLE 9.2.6. Consider an orthonormal basis set $\mathbf{v}_1, \mathbf{v}_2, \ldots$ in an arbitrary infinite-dimensional Hilbert space $\mathcal{H}$. The operator

$$\mathbf{K} = \sum_{i=1}^{\infty} \lambda_i \mathbf{v}_i \mathbf{v}_i^\dagger \qquad (9.2.61)$$

is a perfect operator for any set of numbers λ_i. Suppose $\lambda_i \neq 0$ and consider an arbitrary vector $\mathbf{f} \in \mathcal{H}$. The equation $\mathbf{K}\mathbf{u} = \mathbf{f}$, or

$$\sum_i \lambda_i \mathbf{v}_i \langle \mathbf{v}_i, \mathbf{u} \rangle = \mathbf{f}, \qquad (9.2.62)$$

has the unique solution

$$\mathbf{u} = \sum_i \lambda_i^{-1} \langle \mathbf{v}_i, \mathbf{f} \rangle \mathbf{v}_i. \qquad (9.2.63)$$

For this operator, a bound of the norm is simple to compute. Any vector $\mathbf{v}$ in $\mathcal{H}$ can be expressed as

$$\mathbf{v} = \sum_i \alpha_i \mathbf{v}_i, \qquad (9.2.64)$$

SOLVABILITY THEOREMS

and so

$$\mathbf{Kv} = \sum_i \alpha_i \lambda_i \mathbf{v}_i. \tag{9.2.65}$$

Then

$$\frac{\langle \mathbf{Kv}, \mathbf{Kv} \rangle}{\langle \mathbf{v}, \mathbf{v} \rangle} = \frac{\sum_i |\alpha_i|^2 |\lambda_i|^2}{\sum_i |\alpha_i|^2} \leq \sum_i |\lambda_i|^2. \tag{9.2.66}$$

Consider the three cases:

(a) $\lambda_i = 1/i$;
(b) $\lambda_i = 1/i^{1/2}$;
(c) $\lambda_i = i$.

In case (a), $\sum_i \lambda_i^2 = \sum_i 1/i^2 < \infty$, and so $\mathbf{K}_a$ is a bounded operator. In cases (b) and (c), $\sum_i |\lambda_i|^2 = \infty$, and so Eq. (9.2.66) does not establish a bound for $\mathbf{K}_b$ or $\mathbf{K}_c$. However, since

$$\frac{\langle \mathbf{Kv}_i, \mathbf{Kv}_i \rangle}{\langle \mathbf{v}_i, \mathbf{v}_i \rangle} = \lambda_i^2, \tag{9.2.67}$$

it follows that

$$\lim_{i \to \infty} \frac{\langle \mathbf{Kv}_i, \mathbf{Kv}_i \rangle}{\langle \mathbf{v}_i, \mathbf{v}_i \rangle} \tag{9.2.68}$$

is 0 for case (b) and is infinite for case (c). Thus, the operator

$$\mathbf{K}^a = \sum_{i=1}^\infty \frac{1}{i} \mathbf{v}_i \mathbf{v}_i^\dagger \tag{9.2.69}$$

is definitely bounded and

$$\mathbf{K}^c = \sum_{i=1}^\infty i \mathbf{v}_i \mathbf{v}_i^\dagger \tag{9.2.70}$$

is definitely unbounded, whereas

$$\mathbf{K}^b = \sum_{i=1}^\infty \frac{1}{i^{1/2}} \mathbf{v}_i \mathbf{v}_i^\dagger \tag{9.2.71}$$

is different from both. $\mathbf{K}^b$ has the property that there exists a sequence $\mathbf{K}_n$ of finite dyadics such that

$$\lim_{n \to \infty} \|(\mathbf{K}^b - \mathbf{K}_n)\mathbf{u}\| = 0 \qquad \text{for any } \mathbf{u} \in \mathcal{H}. \tag{9.2.72}$$

In particular, we can choose

$$\mathbf{K}_n = \sum_{i=1}^n \frac{1}{i^{1/2}} \mathbf{v}_i \mathbf{v}_i^\dagger, \tag{9.2.73}$$

so that

$$\|(\mathbf{K}^b - \mathbf{K}_n)\mathbf{u}\|^2 = \sum_{i=n+1}^{\infty} \frac{1}{i}|\langle \mathbf{v}_i, \mathbf{u}\rangle|^2$$

$$\leq \frac{1}{n+1} \sum_{i=n+1}^{\infty} |\langle \mathbf{v}_i, \mathbf{u}\rangle|^2 \qquad (9.2.74)$$

$$\leq \frac{1}{n+1} \sum_{i=1}^{\infty} |\langle \mathbf{v}_i, \mathbf{u}\rangle|^2 = \frac{1}{n+1}\|u\|^2.$$

■ ■ ■

Clearly, $\lim \|(\mathbf{K} - \mathbf{K}_n)\mathbf{u}\| = 0$ for the operator $\mathbf{K}^b$. Operators $\mathbf{K}$ obeying the condition that there exists a sequence of finite dyadic operators converging to $\mathbf{K}$ in the sense of Eq. (9.2.72) are called *completely continuous operators*. We shall see later that this special class of operators obeys most of the known theorems of finite linear vector spaces. For instance, the Fredholm alternative theorems of solvability are true for this class of operators, even though it is bigger than the class of integral operators having continuous kernels in a finite domain. $\mathbf{K}^a$ and $\mathbf{K}^b$ are both examples of completely continuous operators. However, the kernel of $\mathbf{K}^a$ is a continuous function in its domain of definition, whereas the kernel of $\mathbf{K}^b$ is not.

9.3. COMPLETELY CONTINUOUS AND HILBERT–SCHMIDT OPERATORS

A completely continuous operator $\mathbf{K}$ is one that can be uniformly approximated by a sequence of finite-term dyadic operators:

$$\mathbf{K}_n = \sum_{i=1}^{n} \mathbf{u}_i \mathbf{v}_i^\dagger. \qquad (9.3.1)$$

This means that, for any vector $\mathbf{u}$, there exists an integer $n(\epsilon)$ such that

$$\|(\mathbf{K} - \mathbf{K}_n)\mathbf{u}\| < \epsilon \|\mathbf{u}\| \qquad (9.3.2)$$

for all values of $n > n(\epsilon)$. Heuristically, if an operator is completely continuous, then it can always be well approximated by a finite n-term dyadic, and so all of the properties of matrix operators in finite-dimensional vector spaces can be expected to hold.

An example of a completely continuous operator is

$$\mathbf{K} = \sum_{i=1}^{\infty} \lambda_i \boldsymbol{\phi}_i \boldsymbol{\phi}_i^\dagger, \qquad (9.3.3)$$

where $\boldsymbol{\phi}_i$, $i = 1, 2, \ldots$, is a complete orthonormal basis set in a Hilbert space $\mathcal{H}$ and $\lambda_i \neq 0$, $\lambda_{i+1} < \lambda_i$, and $\lambda_i \to 0$ as $i \to \infty$. The n-term dyadic

$$\mathbf{K}_n = \sum_{i=1}^{n} \lambda_i \boldsymbol{\phi}_i \boldsymbol{\phi}_i^\dagger \qquad (9.3.4)$$

forms a sequence that approximates **K** uniformly. Since any vector **u** in $\mathcal{H}$ can be expressed as

$$\mathbf{u} = \sum_{i=1}^{\infty} \alpha_i \boldsymbol{\phi}_i, \tag{9.3.5}$$

it follows that

$$(\mathbf{K} - \mathbf{K}_n)\mathbf{u} = \sum_{i=n+1}^{\infty} \lambda_i \alpha_i \boldsymbol{\phi}_i, \tag{9.3.6}$$

and so

$$\|(\mathbf{K} - \mathbf{K}_n)\mathbf{u}\|^2 = \sum_{i=n+1}^{\infty} \lambda_i |\alpha_i|^2 < \lambda_{n+1} \sum_{i=n+1}^{\infty} |\alpha_i|^2 < \lambda_{n+1} \|\mathbf{u}\|^2. \tag{9.3.7}$$

Thus, if we choose $n(\epsilon)$ such that $\lambda_{n(\epsilon)+1}^2 = \epsilon$, then Eq. (9.3.2) holds for any $n > n(\epsilon)$. Actually, any perfect operator $\mathbf{K} = \sum_{i=1}^{\infty} \lambda_i \boldsymbol{\phi}_i \boldsymbol{\phi}_i^\dagger$ for which $\lambda_i \to 0$ as $i \to \infty$ is a completely continuous operator. As a counterexample, the operator

$$\mathbf{K} = \sum_{i=1}^{\infty} i \boldsymbol{\phi}_i \boldsymbol{\phi}_i^\dagger \tag{9.3.8}$$

is *not* completely continuous, even though it is perfect (perfect since its eigenvectors form a basis set in $\mathcal{H}$).

We define a Hilbert–Schmidt operator as follows:

DEFINITION. *An operator **K** is a Hilbert–Schmidt operator if **K** is bounded and if*

$$\sum_{i=1}^{\infty} \|\mathbf{K}\boldsymbol{\psi}_i\|^2 < \infty, \tag{9.3.9}$$

where $\boldsymbol{\psi}_i$, $i = 1, 2, \ldots$, is an orthonormal basis in $\mathcal{H}$ for which the following theorem holds:

THEOREM. *A Hilbert–Schmidt operator is completely continuous.*

To prove this, we note that $\mathbf{I} = \sum_{i=1}^{\infty} \boldsymbol{\psi}_i \boldsymbol{\psi}_i^\dagger$ and make use of the identity $\mathbf{u} = \mathbf{I}\mathbf{u}$, or

$$\mathbf{u} = \sum_{i=1}^{\infty} \boldsymbol{\psi}_i \boldsymbol{\psi}_i^\dagger \mathbf{u}, \tag{9.3.10}$$

giving

$$\mathbf{K}\mathbf{u} = \sum_{i=1}^{\infty} (\mathbf{K}\boldsymbol{\psi}_i) \boldsymbol{\psi}_i^\dagger \mathbf{u}. \tag{9.3.11}$$

The fact that **K** is bounded (i.e., $\|\mathbf{K}\| < M$, where $M < \infty$) ensures that the series in Eq. (9.3.11) converges to **Ku**. Since $\mathbf{K}\boldsymbol{\psi}_i$ is a vector in $\mathcal{H}$, the operator

$$\mathbf{K}_n = \sum_{i=1}^{n} (\mathbf{K}\boldsymbol{\psi}_i) \boldsymbol{\psi}_i^\dagger \tag{9.3.12}$$

is an n-term dyadic operator and

$$(\mathbf{K} - \mathbf{K}_n)\mathbf{u} = \sum_{i=n+1}^{\infty} \mathbf{K}\boldsymbol{\psi}_i \boldsymbol{\psi}_i^\dagger \mathbf{u}. \qquad (9.3.13)$$

However, by the triangle and Schwarz inequalities, it follows that

$$\left\| \sum_{i=n+1}^{\infty} \mathbf{K}\boldsymbol{\psi}_i (\boldsymbol{\psi}_i^\dagger \mathbf{u}) \right\| \leq \sum_{i=n+1}^{\infty} \|\mathbf{K}\boldsymbol{\psi}_i\| \, |\boldsymbol{\psi}_i^\dagger \mathbf{u}|$$

$$\leq \left(\sum_{i=n+1}^{\infty} \|\mathbf{K}\boldsymbol{\psi}_i\|^2 \sum_{i=n+1}^{\infty} |\boldsymbol{\psi}_i^\dagger \mathbf{u}|^2 \right)^{1/2} \qquad (9.3.14)$$

$$\leq \left(\sum_{i=n+1}^{\infty} \|\mathbf{K}\boldsymbol{\psi}_i\|^2 \right)^{1/2} \|\mathbf{u}\|,$$

and the convergence of the series in Eq. (9.3.9) implies that there exists an integer $n(\epsilon)$ such that

$$\sum_{i=n+1}^{\infty} \|\mathbf{K}\boldsymbol{\psi}_i\|^2 < \epsilon^2 \qquad (9.3.15)$$

for $n > n(\epsilon)$. Thus, from Eqs. (9.3.13) and (9.3.14), we conclude that

$$\|(\mathbf{K} - \mathbf{K}_n)\mathbf{u}\| < \epsilon \|\mathbf{u}\| \qquad (9.3.16)$$

for $n > n(\epsilon)$. This proves that $\mathbf{K}$ can be uniformly approximated by the sequence of n-term dyadics $\mathbf{K}_n$, and so $\mathbf{K}$ is completely continuous.

As an example, consider the operator

$$\mathbf{K} = \sum_{i=1}^{\infty} \frac{1}{i} \boldsymbol{\phi}_i \boldsymbol{\phi}_i^\dagger, \qquad (9.3.17)$$

where $\{\boldsymbol{\phi}_i\}$ is a complete orthonormal basis set. $\mathbf{K}$ is, by definition, a completely continuous operator. Furthermore, $\mathbf{K}$ is a Hilbert–Schmidt operator. This is easily seen by letting $\boldsymbol{\psi}_i = \boldsymbol{\phi}_i$. We then find that

$$\sum_{i=1}^{\infty} \|\mathbf{K}\boldsymbol{\phi}_i\|^2 = \sum_{i=1}^{\infty} \frac{1}{i^2} \langle \boldsymbol{\phi}_i, \boldsymbol{\phi}_i \rangle = \sum_{i=1}^{\infty} \frac{1}{i^2} < \infty. \qquad (9.3.18)$$

As we have shown, all Hilbert–Schmidt operators are completely continuous. However, Hilbert–Schmidt operators form a subclass of the class of completely continuous operators. As an example, consider the completely continuous operator defined by Eq. (9.3.3). It follows that

$$\sum_{i=1}^{\infty} \|\mathbf{K}\boldsymbol{\phi}_i\|^2 = \sum_{i=1}^{\infty} \frac{1}{i} = \infty, \qquad (9.3.19)$$

illustrating that $\mathbf{K}$ is an operator that is completely continuous but it is not a Hilbert–Schmidt operator. It is important to note that boundedness alone is not enough to make an operator either completely continuous or Hilbert–Schmidt. For example, the unit operator $\mathbf{I} = \sum_{i=1}^{\infty} \boldsymbol{\phi}_i \boldsymbol{\phi}_i^{\dagger}$ is bounded, since $\|\mathbf{I}\mathbf{u}\| = \|\mathbf{u}\|$, whereas the sum of its eigenvalues, $\lambda_i = 1$, is infinite and there is no sequence of finite-term dyadics that converges uniformly to $\mathbf{I}$. Thus, $\mathbf{I}$ is neither Hilbert–Schmidt nor completely continuous.

Let us now consider the following theorem:

THEOREM: *If the kernel $k(\vec{r}, \vec{s})$ of an integral operator $\mathbf{K}$ is Lebesgue square integrable in $\mathcal{H}$ with respect to the variables $\vec{r}$ and $\vec{s}$, then the operator $\mathbf{K}$ is completely continuous.*

This means that $k(\vec{r}, \vec{s})$ is square integrable in the two-dimensional domain spanned by $\vec{r}$ and $\vec{s}$, i.e., $\mathcal{H} = \mathcal{L}_2(\Omega_D \times \Omega_D)$. To prove the theorem, consider an orthonormal basis set $\boldsymbol{\psi}_i$. Recall that $\mathbf{I} = \sum_{i=1}^{\infty} \boldsymbol{\psi}_i \boldsymbol{\psi}_i^{\dagger}$, where $\mathbf{I}$ is the identity integral operator whose kernel $i(\vec{r}, \vec{s}) = \sum_{i=1}^{\infty} \psi_i(\vec{r})\psi_i^*(\vec{s}) = \delta(\vec{r} - \vec{s})$, the Dirac delta function. Therefore,

$$\sum_{i=1}^{\infty} \|\mathbf{K}\boldsymbol{\psi}_i\|^2 = \sum_{i=1}^{\infty} \int d^D r \int k^*(\vec{r}, \vec{s})\psi_i^*(\vec{s})\, d^D s \int k(\vec{r}, \vec{s}')\psi_i(\vec{s}')\, d^D s'$$

$$= \int d^D r\, d^D s\, d^D s'\, k^*(\vec{r}, \vec{s}) k(\vec{r}, \vec{s}') \sum_{i=1}^{\infty} \psi_i(\vec{s}')\psi_i^*(\vec{s}) \quad (9.3.20)$$

$$= \int d^D r\, d^D s\, d^D s'\, k^*(\vec{r}, \vec{s}) k(\vec{r}, \vec{s}')\delta(\vec{s}' - \vec{s})$$

$$= \int d^D r\, d^D s\, |k(\vec{r}, \vec{s}')|^2.$$

Since $\int d^D r\, d^D s\, |k(\vec{r}, \vec{s})|^2 < \infty$ by hypothesis, it follows from Eq. (9.3.20) that $\mathbf{K}$ is a Hilbert–Schmidt operator, and so $\mathbf{K}$ is also completely continuous. Equation (9.3.20) also proves that if $\mathbf{K}$ is a Hilbert–Schmidt operator, i.e., Eq. (9.3.2) holds, then $\int d^D r\, d^D s\, |k(\vec{r}, \vec{s})|^2 < \infty$. Thus:

THEOREM. *An integral operator $\mathbf{K}$ is a Hilbert–Schmidt operator if and only if $k(\vec{r}, \vec{s}) \in \mathcal{L}_2(\Omega_D \times \Omega_D)$.*

■ EXAMPLE 9.3.1. Let $k(t, s)$ be a continuous function of t and s in the finite interval $[a, b]$. Then

$$\int_a^b \int_a^b |k(t, s)|^2\, ds\, dt \leq M^2(b - a)^2, \quad (9.3.21)$$

■■■ where $M = \max_{a \leq s, t \leq b} k(t, s)$, and so $\mathbf{K}$ is a Hilbert–Schmidt operator in $\mathcal{L}_2(a, b)$.

■ EXAMPLE 9.3.2. Let $k(t, s) = \exp(-t^2 - s^2)$ and $\mathcal{H} = \mathcal{L}_2(-\infty, \infty)$. Then

$$\int_{-\infty}^{\infty} \int_{-\infty}^{\infty} |k(t, s)|^2 ds\, dt = \int_{-\infty}^{\infty} \int_{-\infty}^{\infty} \exp(-2t^2 - 2s^2)\, ds\, dt = \frac{\pi}{2}, \quad (9.3.22)$$

■■■ and so $\mathbf{K}$ is a Hilbert–Schmidt operator in $\mathcal{L}_2(-\infty, \infty)$.

■ EXAMPLE 9.3.3. Suppose $\mathbf{K}$ is an integral operator in $\mathcal{L}_2(a, b)$ with a kernel of the form $k(t, s) = \bar{k}(t, s)/|t - s|^{\alpha}$, where a and b are finite, $0 < \alpha < \frac{1}{2}$, and

$|k(t, s)|$ is a continuous function of t and s in the interval $[a, b]$. Prove that **K** is a Hilbert–Schmidt operator.

Since $k(t, s)$ is continuous for $t, s \in [a, b]$, it follows that $|\bar{k}(t, s)| < M < \infty$, and so

$$\int_a^b \int_a^b |k(t,s)|^2 dt\, ds = \int_a^b \int_a^b |\bar{k}(t,s)|^2 \frac{1}{|t-s|^{2\alpha}} dt\, ds$$

$$< M^2 \int_a^b \int_a^b \frac{1}{|t-s|^{2\alpha}} dt\, ds \qquad (9.3.23)$$

$$< \frac{M^2 2(b-a)^{2-2\alpha}}{(1-2\alpha)(2-2\alpha)} < \infty.$$

■ ■ ■ This proves that **K** is a Hilbert–Schmidt operator.

EXERCISE 9.3.1. Suppose **K** is an integral operator in $\mathcal{L}_2(\Omega_D)$, where Ω_D is a finite D-dimensional closed domain. Assume that the kernel of **K** is of the form

$$k(\vec{r}, \vec{s}) = \frac{\bar{k}(\vec{r}, \vec{s})}{|\vec{r} - \vec{s}|^\alpha}, \qquad \alpha < \frac{D}{2}, \qquad (9.3.24)$$

where $\bar{k}(\vec{r}, \vec{s})$ is a continuous function of $\vec{r}$ and $\vec{s}$ in Ω_D. Prove that **K** is a Hilbert–Schmidt operator. *Hint*: Use D-dimensional polar coordinates to establish that the integral $\int_{\Omega_D} \int_{\Omega_D} |\vec{r} - \vec{s}|^{-2\alpha} d^D r\, d^D s$ is bounded. $|\vec{r} - \vec{s}|$ is the distance between $\vec{r}$
■ ■ ■ and $\vec{s}$. In Cartesian coordinates, $|\vec{r} - \vec{s}| = [\sum_{i=1}^D (r_i - s_i)^2]^{1/2}$.

To prove the solvability of the equation $(\mathbf{I} + \mathbf{K})\mathbf{u} = \mathbf{f}$, where **K** is completely continuous, consider the equation

$$(\mathbf{V} + \mathbf{W})\mathbf{u} = \mathbf{f}, \qquad (9.3.25)$$

where **V** has an inverse and **W** is the k-term dyadic

$$\mathbf{W} = \sum_{i=1}^k \mathbf{a}_j \mathbf{b}_j^\dagger. \qquad (9.3.26)$$

Then Eq. (9.3.25) can be rearranged to give

$$(\mathbf{I} + \mathbf{T})\mathbf{u} = \mathbf{g}, \qquad (9.3.27)$$

where **I** is the identity operator, **T** is the k-term dyadic

$$\mathbf{T} = \sum_{j=1}^k \mathbf{c}_j \mathbf{b}_j^\dagger, \qquad (9.3.28)$$

with

$$\mathbf{c}_j = \mathbf{V}^{-1} \mathbf{a}_j \qquad (9.3.29)$$

and

$$\mathbf{g} = \mathbf{V}^{-1} \mathbf{f}. \qquad (9.3.30)$$

Taking the inner product of Eq. (9.3.27) with $\mathbf{b}_i$ yields the algebraic system

$$\sum_{j=1}^{k}(\delta_{ij}+\alpha_{ij})x_j = \beta_i, \qquad i=1,\ldots,k, \qquad (9.3.31)$$

which can be written as

$$\mathbf{A}\mathbf{x} = \boldsymbol{\beta}, \qquad (9.3.32)$$

where the components of $\mathbf{A}$, $\mathbf{x}$, and $\boldsymbol{\beta}$ are

$$\begin{aligned} a_{ij} &= \delta_{ij} + \alpha_{ij} \\ \alpha_{ij} &= \mathbf{b}_i^\dagger \mathbf{c}_j = \langle \mathbf{b}_i, \mathbf{c}_j \rangle \\ x_i &= \mathbf{b}_i^\dagger \mathbf{u} = \langle \mathbf{b}_i, \mathbf{u} \rangle \\ \beta_i &= \mathbf{b}_i^\dagger \mathbf{g} = \langle \mathbf{b}_i, \mathbf{u} \rangle. \end{aligned} \qquad (9.3.33)$$

The solvability theory of the algebraic problem was presented in great detail in Chapter 4 and will not be repeated here. We note only that if $\mathbf{L}$ is the sum of $\mathbf{I}$ and a k-term dyadic operator, the solution to $\mathbf{L}\mathbf{u} = \mathbf{g}$ reduces to solving the problem $\mathbf{A}\mathbf{x} = \boldsymbol{\beta}$, where $\mathbf{A}$ is a $k \times k$ matrix and $\mathbf{x}, \boldsymbol{\beta} \in E_k$. Furthermore, the problem has a solution *if and only if* $\mathbf{z}^\dagger \boldsymbol{\beta} = 0$, where $\mathbf{z}$ is any solution to $\mathbf{A}^\dagger \mathbf{z} = \mathbf{0}$. If there is a solution to the algebraic system, then the solution to Eq. (9.3.27) is

$$\begin{aligned} \mathbf{u} &= \mathbf{u}^p + \sum_{i=1}^{k-r} \gamma_i \mathbf{u}^{h_i} \\ &= \mathbf{g} - \sum_{j=1}^{k} x_j^p \mathbf{c}_j - \sum_{i=1}^{k-r} \gamma_i \sum_{j=1}^{k} x_j^{h_i} \mathbf{c}_j, \end{aligned} \qquad (9.3.34)$$

where x_j^p are components of a particular solution to Eq. (9.3.32), and $x_j^{h_i}$ are components of the ith solution to the homogeneous equation $\mathbf{A}\mathbf{x}^h = \mathbf{0}$. The coefficients γ_i are arbitrary, and r is the rank of the matrix $\mathbf{A}$.

As stated above, the purpose of discussing the solution to Eq. (9.3.25) is to recognize that the equation

$$(\mathbf{I} + \mathbf{K})\mathbf{u} = \mathbf{f} \qquad (9.3.35)$$

can be converted to the form of Eq. (9.3.32) if $\mathbf{K}$ is a completely continuous operator. If $\mathbf{K}$ is completely continuous, then there exists an n-term dyadic $\mathbf{K}_n$ such that if

$$\mathbf{R}_n = \mathbf{K} - \mathbf{K}_n, \qquad (9.3.36)$$

the remainder operator $\mathbf{R}_n$ has a norm less than unity, i.e.,

$$\|\mathbf{R}_n\| < 1. \qquad (9.3.37)$$

Accordingly, Eq. (9.3.35) can be written as

$$(\mathbf{I} + \mathbf{R}_n + \mathbf{K}_n)\mathbf{u} = \mathbf{f}, \qquad (9.3.38)$$

where

$$\mathbf{K}_n = \sum_{i=1}^{n} \mathbf{u}_i \mathbf{v}_i^\dagger. \qquad (9.3.39)$$

We showed in Section 8.7 that if $\|\mathbf{R}_n\| < 1$, then the inverse $(\mathbf{I} + \mathbf{R}_n)^{-1}$ of $\mathbf{I} + \mathbf{R}_n$ exists and can be expressed as

$$(\mathbf{I} + \mathbf{R}_n)^{-1} = \mathbf{I} + \sum_{j=1}^{\infty} (-\mathbf{R}_n)^j. \qquad (9.3.40)$$

Thus, by multiplying Eq. (9.3.38) by the inverse $(\mathbf{I} + \mathbf{R}_n)^{-1}$, we obtain

$$(\mathbf{I} + \mathbf{T})\mathbf{u} = \mathbf{g}, \qquad (9.3.41)$$

where

$$\begin{aligned} \mathbf{T} &= \sum_{i=1}^{n} \mathbf{w}_i \mathbf{v}_i^\dagger \\ \mathbf{w}_i &= (\mathbf{I} + \mathbf{R}_n)^{-1} \mathbf{u}_i \\ \mathbf{g} &= (\mathbf{I} + \mathbf{R}_n)^{-1} \mathbf{f}. \end{aligned} \qquad (9.3.42)$$

This proves that the solution of Eq. (9.3.35) can be reduced to solving a linear system in a finite-dimensional vector space if $\mathbf{K}$ is completely continuous.

EXAMPLE 9.3.4. Solve the equation of the form

$$\mathbf{u} + \mathbf{a}_1 \mathbf{b}_1^\dagger \mathbf{u} + \mathbf{a}_2 \mathbf{b}_2^\dagger \mathbf{u} = \mathbf{f}, \qquad (9.3.43)$$

given by

$$\begin{aligned} u(t) &+ \frac{1}{4}\sin(\pi t)\int_{-1}^{1}\cos(\pi s)u(s)\,ds \\ &+ \frac{1}{4}\sin(2\pi t)\int_{-1}^{1}(\cos 2\pi s)u(s)\,ds = t^2. \end{aligned} \qquad (9.3.44)$$

Define

$$x_1 = \int_{-1}^{1}\cos(\pi s)u(s)\,ds \quad \text{and} \quad x_2 = \int_{-1}^{1}\cos(2\pi s)u(s)\,ds. \qquad (9.3.45)$$

COMPLETELY CONTINUOUS AND HILBERT–SCHMIDT OPERATORS

Multiply Eq. (9.3.44) by $\cos(\pi t)\, dt$ and $\cos(2\pi t)\, dt$, respectively, and integrate to obtain

$$\left(1 + \frac{1}{4}\int_{-1}^{1} \cos(\pi t)\, dt\right) x_1 + \frac{1}{4}\left(\int_{-1}^{1} \cos(\pi t)\sin t\, dt\right) x_2$$

$$= \int_{-1}^{1} t\cos(\pi t)\, dt$$

$$\left(\frac{1}{4}\int_{-1}^{1} \cos(2\pi t)\cos t\, dt\right) x_1 + \left(1 + \frac{1}{4}\int_{-1}^{1} \cos(2\pi t)\sin(2\pi t)\, dt\right) x_2$$

$$= \int_{-1}^{1} t^2 \cos(2\pi t)\, dt. \tag{9.3.46}$$

The unique solution to this system is

$$x_1 = \frac{-4}{\pi^2} \quad \text{and} \quad x_2 = \frac{1}{\pi^2}, \tag{9.3.47}$$

and so the unique solution to Eq. (9.3.44) is

$$u(t) = t^2 + \frac{1}{\pi^2} \sin \pi t - \frac{1}{4\pi^2} \sin 2\pi t. \tag{9.3.48}$$

■ ■ ■
▬ **EXAMPLE 9.3.5.** Solve the equation

$$\mathbf{u} + \mathbf{K}\mathbf{u} = \mathbf{f}, \tag{9.3.49}$$

where $\mathbf{u}, \mathbf{f} \in \mathcal{L}_2(-\infty, \infty; \exp(-t^2))$ and

$$\mathbf{K} = \mathbf{a}\mathbf{a}^\dagger, \tag{9.3.50}$$

with

$$a(t) = t^2 \quad \text{and} \quad f(t) = t^4. \tag{9.3.51}$$

In this Hilbert space, $\mathbf{v}^\dagger \mathbf{u}$ denotes the inner product

$$\langle \mathbf{v}, \mathbf{u} \rangle = \int_{-\infty}^{\infty} v^*(t) u(t) \exp(-t^2)\, dt. \tag{9.3.52}$$

Thus, Eq. (9.3.49) corresponds to the expression

$$u(t) + t^2 \int_{-\infty}^{\infty} \exp(-s^2) s^2 u(s)\, ds = t^4. \tag{9.3.53}$$

Let

$$x_1 = \int_{-\infty}^{\infty} \exp(-s^2) s^2 u(s)\, ds. \tag{9.3.54}$$

Multiply Eq. (9.3.53) by $t^2 \exp(-t^2)\, dt$ and integrate (this is equivalent to taking the inner product between $\mathbf{a}$ and $\mathbf{u} + \mathbf{K}\mathbf{u} = \mathbf{f}$) to obtain

$$\left(1 + \int_{-\infty}^{\infty} \exp(-t^2) t^2\, dt\right) x_1 = \int_{-\infty}^{\infty} \exp(-t^2) t^4\, dt \tag{9.3.55}$$

or

$$x_1 = \frac{3\sqrt{\pi}/4}{1+\sqrt{\pi}/2}, \qquad (9.3.56)$$

since

$$\int_{-\infty}^{\infty} \exp(-t^2)t^2\,dt = \frac{\sqrt{\pi}}{2} \quad \text{and} \quad \int_{-\infty}^{\infty} \exp(-t^2)t^4\,dt = \frac{3}{4}\sqrt{\pi}. \qquad (9.3.57)$$

Thus, the unique solution to Eq. (9.3.53) is

$$u(t) = t^4 - \frac{3\sqrt{\pi}/4}{1+\sqrt{\pi}/2}t^2. \qquad (9.3.58)$$

■ ■ ■

Completely continuous operators are of special interest for two reasons. First, the Fredholm alternative theorem is obeyed for these operators. That is:

THEOREM. *Let* **K** *be a completely continuous operator. If* $\mathbf{L} = \mathbf{I} + \mathbf{K}$ *or* $\mathbf{L} = \mathbf{K}$, *then*

$$\mathbf{Lu} = \mathbf{f} \qquad (9.3.59)$$

has a solution if and only if

$$\langle \mathbf{v}, \mathbf{f}\rangle = 0 \qquad (9.3.60)$$

for any solution to the homogeneous adjoint equation

$$\mathbf{L}^\dagger \mathbf{v} = \mathbf{0}. \qquad (9.3.61)$$

Equation (9.3.61) and the homogeneous equation

$$\mathbf{Lu} = \mathbf{0} \qquad (9.3.62)$$

have the same number of solutions. If Eq. (9.3.61) has no solution, then a unique solution to Eq. (9.3.59) exists for any vector **f** *in the Hilbert space* $\mathcal{H}$ *in which* **K** *is defined. When Eqs. (9.3.61) and (9.3.62) have homogeneous solutions, the general solution to Eq. (9.3.59) is given by*

$$\mathbf{u} = \mathbf{u}^\mathrm{p} + \sum_{i=1}^{m} \gamma_i \mathbf{u}_i^\mathrm{h}, \qquad (9.3.63)$$

where $\mathbf{u}^\mathrm{p}$ *is a particular solution to the inhomogeneous equation (9.3.59) and* $\mathbf{u}_i^\mathrm{h}, i = 1, \ldots, m$, *are linearly independent solutions to the homogeneous equation (9.3.70).*

The proof of the above theorem is accomplished by proving it for the operators $\mathbf{L}_n = \mathbf{I} + \mathbf{K}_n$ or $\mathbf{L}_n = \mathbf{K}_n$, where $\mathbf{K}_n$, $n = 1, 2, \ldots$, is the n-term dyadic sequence that converges uniformly to $\mathbf{L} = \mathbf{I} + \mathbf{K}$ or $\mathbf{L} = \mathbf{K}$. The essence of the proof is to prove that the solutions $\mathbf{u}_n$ to $\mathbf{L}_n\mathbf{u}_n = \mathbf{f}$ converge to the solution $\mathbf{u}$ of $\mathbf{Lu} = \mathbf{f}$.

We will not present all of the details of the proof. However, the necessity of the condition expressed by Eq. (9.3.60) is easy to prove. We assume that a solution

$\mathbf{u}$ to $\mathbf{Lu} = \mathbf{f}$ exists and take the inner product of a solution $\mathbf{v}$ of $\mathbf{L}^\dagger \mathbf{v} = \mathbf{0}$ and $\mathbf{Lu}$, i.e., $\langle \mathbf{v}, \mathbf{Lu} \rangle$. By the definition of the adjoint operator, it follows that

$$\langle \mathbf{v}, \mathbf{Lu} \rangle = \langle \mathbf{L}^\dagger \mathbf{v}, \mathbf{u} \rangle \quad \text{for } \mathbf{v}, \mathbf{u} \in \mathcal{H}. \tag{9.3.64}$$

For example, consider the Hilbert space $\mathcal{L}_2(\Omega_D)$. Then Eq. (9.3.64) becomes

$$\begin{aligned}\langle \mathbf{v}, \mathbf{Lu} \rangle &= \int_{\Omega_D} d^3 r v^*(\vec{r}) \int_{\Omega_D} k(\vec{r}, \vec{s}) u(\vec{s}) \, d^D s \\ &= \int_{\Omega_D} \int_{\Omega_D} k(\vec{r}, \vec{s}) v^*(\vec{r}) u(\vec{s}) \, d^D r d^D s,\end{aligned} \tag{9.3.65}$$

whereas

$$\begin{aligned}\langle \mathbf{L}^\dagger \mathbf{v}, \mathbf{u} \rangle &= \int_{\Omega_D} d^3 r \left(\int_{\Omega_D} k^*(\vec{s}, \vec{r}) v(\vec{s}) d^D s \right)^* u(\vec{r}) \\ &= \int_{\Omega_D} \int_{\Omega_D} k(\vec{s}, \vec{r}) v^*(\vec{s}) v(\vec{r}) \, d^D r \, d^D s.\end{aligned} \tag{9.3.66}$$

The integrands of Eqs. (9.3.65) and (9.3.66) differ only in the interchange of dummy variables, and so the integrals are identical, proving that Eq. (9.3.64) is true. Thus, if we assume that $\mathbf{u}$ is a solution to $\mathbf{Lu} = \mathbf{f}$ and $\mathbf{v}$ is a solution to $\mathbf{L}^\dagger \mathbf{v} = \mathbf{0}$, the Fredholm condition, Eq. (9.3.60), follows from Eq. (9.3.64).

The second reason that completely continuous operators are of interest is that completely continuous operators that are self-adjoint or normal are perfect operators and so obey the spectral resolution theorem of integral operators—examined in Section 9.5. In the next section, however, we will examine Volterra equations, which form a special, and somewhat different, class of integral equations.

9.4. VOLTERRA EQUATIONS

A Volterra equation of the first kind is a linear integral equation of the special form $\mathbf{Ku} = \mathbf{f}$, or, in integral form,

$$\int_{\Omega_D} k(\vec{r}, \vec{s}) u(\vec{s}) \, d^D s = f(\vec{r}), \tag{9.4.1}$$

where

$$k(\vec{r}, \vec{s}) = 0 \quad \text{if any } s_i > r_i, \tag{9.4.2}$$

and s_i and r_i are the independent components of $\vec{s}$ and $\vec{r}$. In one dimension, the Volterra equation reads

$$\int_0^t k(t, s) u(s) ds = f(t). \tag{9.4.3}$$

The lower limit of s and t can be a nonzero constant a, $|a| < \infty$, but without loss of generality the variables can be redefined so that the lower limit becomes 0.

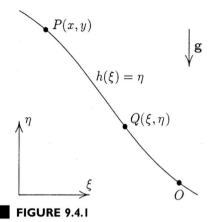

FIGURE 9.4.1

Volterra equations in more than one dimension are not so common, and so we will restrict ourselves to analyzing one-dimensional problems in the rest of this section.

An example of a Volterra equation of the first kind in which k is not continuous arises from Abel's problem. Consider a curve $h(\xi)$ in the $\xi - \eta$ plane as shown in Fig. 9.4.1. A particle beginning at rest at point P can slide freely under gravity (which is in the $-\eta$ direction) down the curve. From Newton's second law, it follows that the velocity of the particle at a point Q on the curve is given by

$$\frac{ds}{dt} = -\sqrt{2g(y-\eta)}, \tag{9.4.4}$$

where s represents the arc length along $h(\xi)$. Thus, the time it takes to go from point P at x, y to the origin O is given by

$$t = -\int_{s(y)}^{s(0)} \frac{ds}{\sqrt{2g(y-\eta)}}. \tag{9.4.5}$$

But since s is a function of ξ (or, equivalently, η through the relation $\eta = h(\xi)$ or $\xi = h'(\eta)$), we can write $ds = u(\eta)\,d\eta$, where $u(\eta) = \sqrt{1 + (dh'/d\eta)^2}$. Equation (9.4.5) then becomes

$$t = \int_0^y \frac{u(\eta)\,d\eta}{\sqrt{2g(y-\eta)}}. \tag{9.4.6}$$

Abel's problem was to find that curve $u(\eta)$ for which the time t to slide from the point x, y to the point $0, 0$ is a prescribed function $f(y)$ of y. Thus, Abel's problem is to solve the integral equation

$$f(y) = \int_0^y \frac{u(\eta)\,d\eta}{\sqrt{2g(y-\eta)}} \tag{9.4.7}$$

for $u(\eta)$. This is just a Volterra equation of the first kind with the kernel

$$k(y, \eta) = \frac{1}{\sqrt{2g(y-\eta)}}. \tag{9.4.8}$$

VOLTERRA EQUATIONS

A generalization of Eq. (9.4.7) is the Volterra equation

$$\int_0^y \frac{u(\eta)}{(y-\eta)^\alpha} d\eta = f(y), \qquad 0 < \alpha < 1. \tag{9.4.9}$$

This problem can be solved by the method of Laplace transforms.

The Laplace transform $\mathcal{L}g$ of a function $g(x)$ is defined as

$$\mathcal{L}g \equiv \tilde{g}(s) \equiv \int_0^\infty \exp(-sx) g(x)\, dx. \tag{9.4.10}$$

If $g(x)$ is continuously differentiable, then the Laplace transform of $dg(x)/dx = g'(x)$ is

$$\tilde{g}'(s) = \int_0^\infty \exp(-sx) \frac{dg}{dx}(x)\, dx = s\tilde{g}(s) - g(0), \tag{9.4.11}$$

as can be seen by integrating by parts. An important property of the Laplace transform is the convolution theorem, namely,

$$\mathcal{L}\left(\int_0^x k(x-\xi) u(\xi)\, d\xi\right) = \tilde{k}(s)\tilde{u}(s). \tag{9.4.12}$$

Thus, if a particular Volterra kernel $k(x,\xi)$ depends only on the difference $x-\xi$, i.e.,

$$\int_0^x k(x-\xi) u(\xi)\, d\xi = f(x), \tag{9.4.13}$$

then the Laplace transform of Eq. (9.4.13) yields the simple relation

$$\tilde{k}(s)\tilde{u}(s) = \tilde{f}(s), \tag{9.4.14}$$

and so the solution to Eq. (9.4.13) can be obtained by finding the inverse Laplace transform of $\tilde{u}(s) = \tilde{f}(s)/\tilde{k}(s)$.

Consider again Eq. (9.4.9). The Laplace transform of $1/x^\alpha$ is

$$\int_0^\infty \exp(-sx) \frac{1}{x^\alpha} dx = \frac{1}{s^{1-\alpha}} \int_0^\infty \exp(y) \frac{1}{y^\alpha} dy = \frac{\Gamma(1-\alpha)}{s^{1-\alpha}}, \tag{9.4.15}$$

where $\Gamma(1-\alpha)$ is a gamma function. From this result and the convolution theorem, we find by taking the Laplace transform of Eq. (9.4.9) that

$$\tilde{u}(s) = \frac{s^{1-\alpha}}{\Gamma(1-\alpha)} \tilde{f}(s). \tag{9.4.16}$$

Given the Laplace transform of $f(y)$, we can find $u(y)$ by inverting the Laplace transform in Eq. (9.4.16). For the special case that $f(y)$ is continuously differentiable, Eq. (9.4.11) implies that $s\tilde{f}(s) = \tilde{f}'(s) + f(0)$, allowing the rearrangement

$$\tilde{u}(s) = \frac{f(0)}{s^\alpha \Gamma(1-\alpha)} + \frac{1}{s^\alpha \Gamma(1-\alpha)} \tilde{f}'(s), \tag{9.4.17}$$

where $\tilde{f}'(s)$ denotes the Laplace transform of $f'(x) \equiv df(x)/dx$. Since the inverse Laplace transform of $1/s^\alpha$ is $x^{\alpha-1}/\Gamma(\alpha)$, it follows from Eq. (9.4.17) that the solution to Eq. (9.4.9) is

$$u(y) = \frac{\sin(\alpha\pi)}{\pi} \left[\frac{f(0)}{y^{1-\alpha}} + \int_0^y \frac{1}{(y-\eta)^{1-\alpha}} f'(\eta)\, d\eta \right], \qquad (9.4.18)$$

where we have made use of the relationship $1/[\Gamma(\alpha)\Gamma(1-\alpha)] = \sin(\alpha\pi)/\pi$ when $0 < \alpha < 1$.

Abel's problem corresponds to the special case $\alpha = \frac{1}{2}$ and $f(0) = 0$, and so the solution to Eq. (9.4.7) is

$$u(y) = \frac{\sqrt{2g}}{\pi} \int_0^y \frac{f'(\eta)}{\sqrt{y-\eta}}\, d\eta, \qquad (9.4.19)$$

when $f(y)$ is continuously differentiable. Even if $f(y)$ is not continuously differentiable, the solution can be found by Laplace inversion of Eq. (9.4.16). In any case, Eq. (9.4.16) implies that Abel's problem and its generalization has a unique solution since the only solution for $f(x) = 0$ is $\tilde{u}(s) = 0$.

A further generalization of Abel's problem is the Volterra equation

$$\int_0^y \frac{h(y,\eta)}{(y-\eta)^\alpha} u(\eta)\, d\eta = f(y), \qquad 0 < \alpha < 1, \qquad (9.4.20)$$

where $h(y, \eta)$ is a bounded function of y and η. Although more complicated, it can be shown that this equation also admits a unique solution.

Equations (9.4.7), (9.4.9), and (9.4.20) are all examples of singular equations. An integral equation is said to be singular if either its kernel becomes infinite for one or more points or any of the limits of integration are infinite. Being singular does not, of course, mean that the equation is unsolvable. In fact, all three variations of Abel's problem given above admit unique solutions. If the kernel $k(y, \eta)$ is itself a bounded function of y and η, then the solution to a Volterra equation of the first kind will be unique for continuous functions $f(y)$ for which $f(0) = 0$.

Volterra equations of the second kind are linear integral equations that can be expressed in the form

$$(\mathbf{I} + \mathbf{K})\mathbf{u} = \mathbf{f} \qquad (9.4.21)$$

or

$$u(t) + \int_0^t k(t,s)u(s)\, ds = f(t). \qquad (9.4.22)$$

If the kernel $k(t, s)$ and function $f(t)$ are differentiable, then the Volterra equation of the first kind, Eq. (9.4.3), can be differentiated with respect to t to yield

$$k(t,t)u(t) + \int_0^t k_t(t,s)u(s)\, ds = f'(t), \qquad (9.4.23)$$

where $k_t(t,s) \equiv dk/dt$ and $f'(t) \equiv df/dt$. If $k(t,t) \neq 0$, then Eq. (9.4.23) becomes a Volterra equation of the second kind. Similar to Eq. (9.4.13), if the

VOLTERRA EQUATIONS

kernel in Eq. (9.4.22) is of the form $k(t - s)$, then the equation can be solved by Laplace transforms. The Laplace transform of Eq. (9.4.22), in this case, is

$$\tilde{u}(z) = \frac{\tilde{f}(z)}{1 + \tilde{k}(z)}. \tag{9.4.24}$$

Turning now to the solvability of these equations, we begin by introducing the following theorem:

THEOREM. *If $f(t)$ is continuous and the kernel $k(t, s)$ is a continuous function of t and s in the interval $[a, b]$, then the Volterra equation of the second kind has one and only one continuous solution $u(t)$.*

To prove this theorem, consider the sequence of functions

$$\begin{aligned} u_1(t) &= f(t) \\ u_2(t) &= f(t) - \int_a^t k(t, s) u_1(s)\, ds \\ &\vdots \\ u_n(t) &= f(t) - \int_a^t k(t, s) u_{n-1}(s)\, ds. \end{aligned} \tag{9.4.25}$$

By successive substitution, the function $u_n(t)$ can be expressed in the form

$$\begin{aligned} u_n(t) = {}&f(t) - \int_a^t k(t, s_1) f(s_1)\, ds_1 \\ &+ \int_a^t k(t, s_1) \int_a^{s_1} k(s_1, s_2) f(s_2)\, ds_1\, ds_2 + \cdots \\ &+ (-1)^{n-1} \int_a^t k(t, s_1) \int_a^{s_1} k(s_1, s_2) \cdots \int_a^{s_{n-2}} k(s_{n-2}, s_{n-1}) \\ &\times f(s_{n-1})\, ds_1\, ds_2 \cdots ds_{n-1} \end{aligned} \tag{9.4.26}$$

or, in operator form,

$$\begin{aligned} \mathbf{u}_n &= \mathbf{f} - \mathbf{K}\mathbf{f} + \mathbf{K}^2\mathbf{f} + \cdots + (-1)^{n-1}\mathbf{K}^{n-1}\mathbf{f} \\ &= \mathbf{S}_n \mathbf{f}, \end{aligned} \tag{9.4.27}$$

where

$$\mathbf{S}_n = \mathbf{I} + \sum_{i=1}^{n-1} (-1)^i \mathbf{K}^i. \tag{9.4.28}$$

Since k and f are continuous, it follows that

$$\left| \int_a^t k(t, s_1) f(s_1)\, ds_1 \right| \leq AB \int_a^t ds_1 = AB(t - a), \tag{9.4.29}$$

where $A = \max_{a \leq t, s \leq b} |k(t,s)|$ and $B = \max_{a \leq t \leq b} |f(t)|$. Similarly,

$$\left| \int_a^t k(t, s_1) \int_a^{s_1} k(s_1, s_2) f(s_1) \, ds_1 \, ds_2 \right| \leq A^2 B \int_a^t ds_1 \int_a^{s_1} ds_2 = A^2 B \frac{(t-a)^2}{2!} \tag{9.4.30}$$

and, in general,

$$\left| \int_a^t k(t, s_1) \int_a^{s_1} k(s_1, s_2) \cdots \int_a^{s_n} k(s_{n-1}, s_n) f(s_n) \, ds_1 \cdots ds_n \right| \leq A^n B \frac{(t-a)^n}{n!}. \tag{9.4.31}$$

These inequalities imply that

$$|u_n(t)| \leq \sum_{i=0}^{n-1} B A^i \frac{(t-a)^i}{i!} < B \exp(A(t-a)), \tag{9.4.32}$$

which, in turn, implies that the series $u_1, u_2, \ldots$ converges absolutely and uniformly. Thus, the function $u(t) \equiv \lim_{n \to \infty} u_n(t)$ is a continuous function. Since

$$(\mathbf{I} + \mathbf{K})\mathbf{S}_n = \mathbf{I} + (-1)^{n-1} \mathbf{K}^n, \tag{9.4.33}$$

it follows that

$$(\mathbf{I} + \mathbf{K})\mathbf{u}_n = \mathbf{f} + (-1)^{n-1} \mathbf{K}^n \mathbf{f}. \tag{9.4.34}$$

However, according to the bound obtained at Eq. (9.4.30), $\lim_{n \to \infty} \mathbf{K}^n \mathbf{f} = \mathbf{0}$, and so taking the limit of Eq. (9.4.34) as $n \to \infty$, we find

$$(\mathbf{I} + \mathbf{K})\mathbf{u} = \mathbf{f}, \tag{9.4.35}$$

where $\mathbf{u} = \lim_{n \to \infty} \mathbf{u}_n$. This proves that the continuous function $u(t) = \lim_{n \to \infty} u_n(t)$ is a solution to the Volterra equation. To prove that the solution is unique, assume that $\mathbf{u}$ and $\mathbf{v}$ are two different solutions. The vector $\mathbf{w} = \mathbf{u} - \mathbf{v}$ then satisfies the homogeneous equation $(\mathbf{I} + \mathbf{K})\mathbf{w} = 0$, or

$$\mathbf{w} = -\mathbf{K}\mathbf{w}. \tag{9.4.36}$$

Successive substitution of $-\mathbf{K}\mathbf{w}$ for $\mathbf{w}$ on the right-hand side of this equation yields

$$\mathbf{w} = (-1)^n \mathbf{K}^n \mathbf{w} \tag{9.4.37}$$

for an arbitrary positive integer n. Since $w(t)$ is a continuous function, the argument leading to Eq. (9.4.31) leads to the conclusion

$$|w(t)| \leq A^n C \frac{(t-a)^n}{n!} \leq A^n C \frac{(b-a)^n}{n!}, \tag{9.4.38}$$

where $C = \max_{a \leq t \leq b} |w(t)|$. From this result, we find

$$|w(t)| \leq \lim_{n \to \infty} A^n C \frac{(b-a)^n}{n!} = 0, \tag{9.4.39}$$

requiring $\mathbf{w} = \mathbf{0}$, thus proving that the solution $\mathbf{u}$ to the Volterra equation is unique. Besides completing the proof, we have also established that no continuous function satisfies a homogeneous Volterra equation of the second kind.

We can extend our analysis of Volterra equations by repeating the above arguments for the general equation

$$\int_0^t k(t,s)u(s)\,ds + \alpha u(t) = f(t), \tag{9.4.40}$$

where α is a finite, arbitrary constant. The sequence analogous to Eq. (9.4.25) in this case is

$$\begin{aligned}
u_1(t) &= \frac{1}{\alpha} f(t) \\
u_2(t) &= \frac{1}{\alpha} f(t) - \frac{1}{\alpha^2} \int_a^t k(t,s) u_1(s)\,ds \\
&\vdots \\
u_n(t) &= \frac{1}{\alpha} f(t) - \frac{1}{\alpha^n} \int_a^t k(t,s) u_{n-1}(s)\,ds.
\end{aligned} \tag{9.4.41}$$

By using our previous definitions of $A = \max_{a \leq t, s \leq b} |k(t,s)|$ and $B = \max_{a \leq t \leq b} |f(t)|$, we can rederive Eq. (9.4.32) to get

$$|u_n(t)| \leq \frac{1}{|\alpha|} \sum_{i=0}^{n-1} BA^i \frac{(t-a)^i}{|\alpha|^i i!} < \frac{B}{|\alpha|} \exp\left(\frac{A(t-a)}{|\alpha|}\right). \tag{9.4.42}$$

We see that as long as α is non zero, the sequence $\{\mathbf{u}_n\}$ converges as $n \to \infty$. From here, the analysis follows exactly as above for the case $\alpha = 1$. Namely, we can prove that the continuous function $u(t) = \lim_{n \to \infty} u_n(t)$ is a unique solution to Eq. (9.4.40), and furthermore, there is no continuous function satisfying the homogeneous equation

$$\int_0^t k(t,s) u(s)\,ds + \alpha u(t) = 0 \tag{9.4.43}$$

for $\alpha \neq 0$.

Equation (9.4.43) is, in fact, simply the eigenequation (for negative α) for the Volterra operator $\mathbf{K}$. We have, therefore, shown that the Volterra operator does not have any non zero eigenvectors. There still remains the question of whether the equation

$$\int_0^t k(t,s) u(s)\,ds = 0 \tag{9.4.44}$$

has a solution. However, the only continuous function $u(s)$ that can satisfy Eq. (9.4.44) for any arbitrary value s is $u = 0$ since $k(t,s)$ is continuous. Combining this result with our above results allows us to form the following theorem:

THEOREM 9.4.2. *If $k(x,y)$ is continuous in the finite rectangle $a \leq x \leq b$, $a \leq y \leq b$, then the corresponding Volterra operators of the first and second kind have no eigenvectors.*

It does not, however, follow from our analysis that the general Volterra equation of the first kind,

$$\int_0^x k(x, y)u(y)\, ds = f(x), \tag{9.4.45}$$

always has a unique solution. For example, the equation

$$\int_0^x (x - y)u(y)\, ds = x \tag{9.4.46}$$

has no solution, whereas the equation

$$\int_0^x (x - y)u(y)\, ds = \frac{x^3}{6} \tag{9.4.47}$$

does have the unique solution $u(x) = x$.

One should note that a homogeneous Volterra equation of the second kind can have a *discontinuous* solution. For example, consider the equation

$$u(t) - \int_0^t s^{t-s} u(s)\, ds = f(t). \tag{9.4.48}$$

The kernel $k(t, s) = -s^{t-s}$ is continuous, and so, provided that $f(t)$ is continuous, the theorem proved above says that there is one and only one *continuous* solution to Eq. (9.4.39). This also means that $u(t) = 0$ is the only continuous function satisfying the homogeneous equation $\mathbf{u} + \mathbf{Ku} = \mathbf{0}$. However, the function

$$u(t) = ct^{t-1} \tag{9.4.49}$$

satisfies the homogeneous equation

$$u(t) - \int_0^t s^{t-s} u(s)\, ds = 0, \tag{9.4.50}$$

where c is an arbitrary constant. But ct^{t-1} is not a continuous function, since it diverges as $t \to 0$.

We will end this section with an example of the applicability of Volterra equations. It so happens that linear initial value problems can be converted into Volterra equations. Consider, for example, the equation

$$\frac{d^2 y}{dx^2} + a_1(x)\frac{dy}{dx} + a_2(x)y = f(x) \tag{9.4.51}$$

for $x \geq 0$. If we define

$$u(x) = \frac{d^2 y}{dx^2}, \tag{9.4.52}$$

then

$$\frac{dy}{dx} = \int_0^x u(x_1)\, dx_1 + c_1 \tag{9.4.53}$$

and

$$y = \int_0^x dx_2 \int_0^{x_2} u(x_1)\,dx_1 + c_1 x + c_2, \tag{9.4.54}$$

where c_1 and c_2 are constants. By interchanging the order of integration of x_1 and x_2 in Eq. (9.4.54) according to the rule

$$\int_0^x dx_2 \int_0^{x_2} dx_1\, A = \int_0^x dx_1 \int_{x_1}^x dx_2\, A, \tag{9.4.55}$$

Eq. (9.4.54) becomes

$$y(x) = \int_0^x (x - x_1) u(x_1)\,dx_1 + c_1 x + c_2. \tag{9.4.56}$$

Equation (9.4.51) can now be transformed into the Volterra equation

$$u(x) + \int_0^x \left[a_1(x) + a_2(x)(x-t) \right] u(t)\,dt = f(x) - \sum_{i=1}^2 c_i \alpha_i(x), \tag{9.4.57}$$

where

$$\alpha_1(x) = a_1(x) + a_2(x) x \qquad \text{and} \qquad \alpha_2(x) = a_2(x). \tag{9.4.58}$$

The constants c_1 and c_2 are fixed by the conditions $y(0) = \gamma_1$ and $y'(0) = \gamma_2$ in the initial value problem.

The pth-order initial value problem

$$\frac{d^p y}{dx^p} + a_1(x) \frac{d^{p-1} y}{dx^{p-1}} + \cdots + a_p(x) y = f(x) \tag{9.4.59}$$

can similarly be transformed into the Volterra equation

$$u(x) + \int_0^x \left[\sum_{i=1}^p a_i(x) \frac{(x-t)^{i-1}}{(i-1)!} \right] u(t)\,dt = f(x) - \sum_{i=1}^p c_i \alpha_i(x), \tag{9.4.60}$$

where the c_i are unknown constants, the α_i are given by

$$\alpha_i(x) = \sum_{j=0}^{p-i} a_{i+j}(x) \frac{x^j}{j!}, \tag{9.4.61}$$

and y and its derivatives are related to u through the equations

$$\frac{d^{p-i} y}{dx^{p-i}} = \int_0^x \frac{(x-t)^{i-1}}{(i-1)!} u(t)\,dt + \sum_{j=1}^i c_j \frac{x^{i-j}}{(i-j)!}. \tag{9.4.62}$$

The values of c_i have to be set by the initial conditions

$$\frac{d^j y(x=0)}{dx^j} = \gamma_{j+1}, \qquad j = 0, 1, \ldots, p-1. \tag{9.4.63}$$

When the coefficients a_i are all constant, the initial value problem is easier to solve by transformation to a problem in a finite vector space as given in Sections 6.10 and 7.7 rather than by transformation to a Volterra integral equation.

EXAMPLE 9.4.1. Find the Volterra equation corresponding to the initial value problem

$$\frac{d^2y}{dx^2} + \exp(-x)\frac{dy}{dx} + x^2 y = x, \tag{9.4.64}$$

where

$$\frac{dy}{dx} = 1 \quad \text{and} \quad y = 2 \quad \text{at } x = 0. \tag{9.4.65}$$

Set

$$\frac{d^2 y}{dx^2} = u. \tag{9.4.66}$$

Then

$$\frac{dy}{dx} = \int_0^x u(t)\, dt + 1 \tag{9.4.67}$$

and

$$y = \int_0^x (x-t) u(t)\, dt + x + 2, \tag{9.4.68}$$

and so

$$u(x) + \int_0^x \left[\exp(-x) + x^2(x-t) \right] u(t) = x - \exp(-x) - 2x^2 - x^3. \tag{9.4.69}$$

EXERCISE 9.4.1. Find the Volterra equation corresponding to the initial value problem

$$\frac{d^3 y}{dx^3} + x^2 \frac{d^2 y}{dx^2} + \sin x \frac{dy}{dx} + y = \exp(-x) \tag{9.4.70}$$

$$y = 1, \quad \frac{dy}{dx} = 3, \quad \frac{d^2 y}{dx^2} = 0 \quad \text{at } x = 0. \tag{9.4.71}$$

Write a computer program to solve the Volterra equation iteratively as given by Eq. (9.4.25).

EXAMPLE 9.4.2. Let us define p as the capillary pressure needed to force water out of a certain sample of porous rock until it occupies only the fraction s (saturation) of the pore space. How p depends on s is an important question in soil science and in the characterization of aquifers and oil reservoirs. To measure p versus s, one spins in a centrifuge a water-filled cylindrical sample of the porous rock and measures the volume of water removed (spun out) as a function of the spinning rate. The average saturation $\bar{s}$ of water left in the sample at the spin rate w is given by

$$\bar{s} = \frac{r_1 + r_2}{2 r_2 p} \int_0^p \frac{s(p')}{(1 - B p'/p)^{1/2}}\, dp', \tag{9.4.72}$$

where r_1 and r_2 are the distance of the two ends of the cylinder from the axis of rotation in the centrifuge and

$$p = \frac{1}{2} \Delta\rho \, w^2 (r_2^2 - r_1^2) \quad \text{and} \quad B = 1 - \frac{r_1^2}{r_2^2}, \qquad (9.4.73)$$

where $\Delta\rho$ is the difference between water and air densities.

Equation (9.4.72) is a Volterra equation of the first kind. One varies p to vary $\bar{s}$ and solves Eq. (9.4.72) for $s(p')$ versus p'. The problem is that experiments are accompanied by error and a Volterra equation of the first kind is ill conditioned (see Linz, 1982), which means that many solution techniques for solving integral equations are overly sensitive to error in the data ($\bar{s}$ and p). For example, if the data are available on a uniform mesh $h = p_{i+1} - p_i$, i being the ith measurement or the ith setting of rotation rate w, then the integral in Eq. (9.4.72) can be approximated numerically by

$$\bar{s}_i = \frac{h(r_1 + r_2)}{2 r_2 p_i} \left[\frac{s(p_0)}{2} + \sum_{j=1}^{i-1} \frac{s(p_j)}{\sqrt{1 - B p_j / p_i}} + \frac{s(p_i)}{2\sqrt{1-B}} \right] \qquad (9.4.74)$$

for $i = 1, \ldots, M$, where M is the total number of measurements made. This is a linear algebraic system that can be solved by forward substitution. However, since the left-hand side of Eq. (9.4.74) is multiplied by h, it follows that, upon grid refinement (smaller h), error in $\bar{s}_i$ is amplified as $1/h$ in the solution. The problem can be alleviated by using a least squares technique to smooth the data (Linz, 1982). Other techniques can also be used (Ayappa, 1989).

To illustrate the numerical problem in solving Eq. (9.4.74) when $\bar{s}_i$ contains error, consider the theoretical capillary pressure curve

$$s(p) = \begin{cases} 1, & 0 < p \leq 2, \\ \dfrac{1.5}{p} + 0.25, & p > 2, \end{cases} \qquad (9.4.75)$$

where p is in an appropriate set of units. Suppose $r_1/r_2 = 0.5$. Then $(r_1 + r_2)/2r_2 = 0.75$ and $B = 0.75$. From Eq. (9.4.72), it follows that

$$\bar{s}(p) = \frac{0.75}{p} \int_0^p \frac{dp'}{\sqrt{1 - 0.75 p'/p}}, \qquad 0 < p \leq 2, \qquad (9.4.76)$$

and

$$\bar{s}(p) = \frac{0.75}{p} \int_0^2 \frac{dp'}{\sqrt{1 - 0.75 p'/p}} + \frac{0.75}{p} \int_2^p \frac{1.5/p' + 0.25}{\sqrt{1 - 0.75 p'/p}} dp'$$

$$= \frac{7}{4} - \frac{3}{2}\sqrt{1 - \frac{1.5}{p}} \qquad (9.4.77)$$

$$+ \frac{2.25}{p} \left(\tanh^{-1} \sqrt{1 - \frac{1.5}{p}} - \tanh^{-1} \frac{1}{2} \right), \qquad p > 2,$$

and Eq. (9.4.74) becomes

$$\bar{s}_i = \frac{0.75}{i}\left[\frac{1}{2} + \sum_{j=1}^{i-1}\frac{s_j}{\sqrt{1-0.75j/i}} + s_i\right], \quad (9.4.78)$$

where $\bar{s}_i = \bar{s}(ih)$ and $s_i = s(ih)$. We can write Eq. (9.4.78) in matrix form as

$$\bar{\mathbf{s}} = \mathbf{As} + \mathbf{b}, \quad (9.4.79)$$

where $b_i = 0.75/2i$ and

$$a_{ij} = \begin{cases} \dfrac{0.75}{i\sqrt{1-0.75i/j}}, & j < i, \\ \dfrac{0.75}{2i\sqrt{1-0.75}}, & j = i, \\ 0, & j > i, \end{cases} \quad (9.4.80)$$

for $i = 1, \ldots, M$. Suppose now that we generate "experimental" values for $\bar{\mathbf{s}}$ by adding a fixed percentage of random fluctuation to the theoretical values from Eqs. (9.4.76) and (9.4.77). We can generate such data on a uniform mesh of values for p, resulting in the vector $\bar{\mathbf{s}}$, which can subsequently be used in Eq. (9.4.79) to solve for $\mathbf{s}$. A Mathematica program is provided in the Appendix for this example, where we show that even a fluctuation of 0.01% results in large deviations from the theoretical function in Eq. (9.4.75).

ILLUSTRATION 9.4.1 (Linear Viscoelastic Stress). The non-Newtonian stress/strain behavior of certain non-glassy polymers and plastics can be described by a viscoelastic model first conceived by Boltzmann. For an isotropic material (in which we assume the effects of deformation are independent of direction), the modulus G is defined as the ratio of linear stress, τ (force/area), to degree of strain, γ:

$$G = \frac{\tau}{\gamma}. \quad (9.4.81)$$

For perfectly elastic materials, this relation reflects the fact that an elastic response is essentially an instantaneous one. However, many materials, particularly polymer liquids and melts, exhibit time-dependent responses. The stress/strain behavior of these *viscoelastic* materials—as we call them—follows a generalization of Eq. (9.4.81) given by

$$G(\gamma, t) = \frac{\tau(t)}{\gamma(t)}, \quad (9.4.82)$$

where we recognize that the *relaxation modulus* G is a function of time. In general, the relaxation modulus decays with time under applied strain. Physically, we can attribute this to molecular rearrangement of weakly interacting macromolecules. Since these materials are typically highly disordered, the exact functional form of $G(\gamma, t)$ is very complicated. However, we can approximate the behavior by assuming an "instantaneously elastic" response of the form

$$d\tau = \gamma\, dG. \quad (9.4.83)$$

(i) By defining a *memory function* $M(t)$ such that

$$M(\gamma, t) = -\frac{dG(\gamma, t)}{dt}, \tag{9.4.84}$$

show that the isotropic stress at time t is given by

$$\tau(t) = -\int_{-\infty}^{t} M(\gamma, t - t')\gamma(t')\, ds. \tag{9.4.85}$$

(ii) One method of modeling the relaxation modulus involves imagining a linear collection of independent modes of relaxation. We write

$$G(t) = \sum_{k=1}^{n} G_k \exp\left(\frac{-t}{\alpha_k}\right), \tag{9.4.86}$$

where the sum is over "modes" (n in total), and the constants G_k and α_k represent the nominal modulus and relaxation time for mode k. Show that the operator $M(t - t')$ in Eq. (9.4.85) represented by Eq. (9.4.86) is a *completely continuous operator* in the Hilbert space $\mathcal{L}_2(-\infty, b)$, where b is some finite, yet large time.

(iii) Find an eigenfunction (and corresponding eigenvalue) for $M(t - t')$ derived from Eq. (9.4.86). What is the physical significance of this eigenpair? Justify why such an eigenfunction should exist in light of the theorem preceding Eq. (9.4.45). Is Eq. (9.4.85) a Volterra equation?

(iv) Consider a viscoelastic material with relaxation modulus given by the general expression Eq. (9.4.86). Derive an expression for the time-dependent stress resulting from an oscillatory strain of the form $\gamma = \gamma_0 \sin \omega t$. Notice that the stress oscillates with the same frequency as the strain but not generally with the same phase. Derive an expression for the phase shift in the stress for a single mode as a function of α_k and ω. What are the limits of the shift at low and high frequency?

9.5. SPECTRAL THEORY OF INTEGRAL OPERATORS

9.5.1. Bessel's Inequality

Throughout this section we will make important use of Bessel's inequality. Thus, it is appropriate to begin with its derivation. The inequality can be expressed in terms of the following theorem:

THEOREM. *If $\boldsymbol{\phi}_1, \boldsymbol{\phi}_2, \ldots$ is an orthonormal set (not necessarily a complete set or even an infinite set) in the Hilbert space $\mathcal{H}$, then, for any vector $\mathbf{f}$ in $\mathcal{H}$, the inequality*

$$\sum_i |\langle \boldsymbol{\phi}_i, \mathbf{f} \rangle|^2 \leq \|\mathbf{f}\|^2 \tag{9.5.1}$$

holds.

To prove the theorem, we define

$$\alpha_i = \langle \boldsymbol{\phi}_i, \mathbf{f} \rangle \tag{9.5.2}$$

and note that

$$\left\| \mathbf{f} - \sum_i \alpha_i \boldsymbol{\phi}_i \right\|^2 \geq 0. \qquad (9.5.3)$$

Specifically, the above expression will be equal to 0 only if the set is complete. Using the inner product property $\|\mathbf{x} + \alpha \mathbf{y}\|^2 = \|\mathbf{x}\|^2 + \alpha \langle \mathbf{x}, \mathbf{y} \rangle + \alpha^* \langle \mathbf{y}, \mathbf{x} \rangle + |\alpha|^2 \|\mathbf{y}\|^2$, we obtain from Eq. (9.5.3)

$$\|\mathbf{f}\|^2 - \sum_i \alpha_i^* \langle \boldsymbol{\phi}_i, \mathbf{f} \rangle - \sum_i \alpha_i \langle \mathbf{f}, \boldsymbol{\phi}_i \rangle + \sum_i |\alpha_i|^2 \geq 0, \qquad (9.5.4)$$

where the property $\|\boldsymbol{\phi}_i\|^2 = 1$ of an orthonormal function has been used. Since $\alpha_i = \langle \boldsymbol{\phi}_i, \mathbf{f} \rangle$ and $\alpha_i^* = \langle \mathbf{f}, \boldsymbol{\phi}_i \rangle$, Eq. (9.5.4) can be rearranged to yield Eq. (9.5.1)—Bessel's inequality.

9.5.2. Eigenvalue Degeneracy

Suppose the kernel $k(\vec{r}, \vec{s})$ of $\mathbf{K}$ is square integrable in the Hilbert space $\mathcal{H} = \mathcal{L}_2(\Omega_D \times \Omega_D)$ (i.e., $k(\vec{r}, \vec{s})$ is square integrable with respect to $\vec{r}$ and $\vec{s}$); then $\mathbf{K}$ is a Hilbert–Schmidt operator. Suppose also that $\boldsymbol{\psi}_i$, $i = 1, 2, \ldots, h$, are the eigenvectors of $\mathbf{K}$ corresponding to the same eigenvalue λ, i.e.,

$$\mathbf{K} \boldsymbol{\psi}_i = \lambda \boldsymbol{\psi}, \qquad i = 1, \ldots, h, \qquad (9.5.5)$$

where the number h is the degeneracy (or multiplicity in some texts) of λ. Without loss of generality, these eigenvectors can be assumed to be orthogonal (due to the Gram–Schmidt procedure) and normalized. Furthermore, they belong to the Hilbert space for $\mathcal{L}_2(\Omega_D)$. If $\boldsymbol{\psi}$ is an eigenvector in a function space, as is the case for integral operators, we say that $\psi(\vec{r})$ is an eigenfunction. The terms eigenfunction and eigenvector are often used interchangeably in function spaces. Strictly speaking, $\boldsymbol{\psi}$ denotes the eigenvector in $\mathcal{H}$, whereas $\psi(t)$ denotes a component of $\boldsymbol{\psi}$—analogous to $\mathbf{x}$ in E_n, and x_i a component of $\mathbf{x}$.

Consider now the following theorem:

THEOREM. *An operator $\mathbf{K}$, whose kernel is square integrable in the Hilbert space $\mathcal{H} = \mathcal{L}_2(\Omega_D \times \Omega_D)$, has only a finite number h of eigenvectors for any nonzero eigenvalue λ.*

To prove this, let us consider $k^*(\vec{r}, \vec{s})$ to be a function of $\vec{s}$ and apply Bessel's inequality to obtain

$$\sum_{i=1}^{h} |\langle k^*, \boldsymbol{\psi}_i \rangle|^2 \leq \|k^*\|^2 \qquad (9.5.6)$$

or, equivalently,

$$\sum_{i=1}^{h} \left| \int_{\Omega_D} k(\vec{r}, \vec{s}) \psi_i(\vec{s}) \, d^D s \right|^2 \leq \int_{\Omega_D} |k(\vec{r}, \vec{s})|^2 \, d^D s. \qquad (9.5.7)$$

However, $\int_{\Omega_D} k(\vec{r},\vec{s})\psi_i(\vec{s})\,d^Ds = \lambda\psi_i(\vec{s})$, and so Eq. (9.5.7) implies that

$$|\lambda|^2 \sum_{i=1}^{h} |\psi_i(\vec{r})|^2 \leq \int_{\Omega_D} k|(\vec{r},\vec{s})|^2\,d^Ds. \tag{9.5.8}$$

Integrating Eq. (9.5.8) with respect to d^Dr and using the property $\|\psi_i\|^2 = 1$, we obtain

$$|\lambda|^2 h \leq \int_{\Omega_D}\int_{\Omega_D} |k(\vec{r},\vec{s})|^2\,d^Ds\,d^Dr. \tag{9.5.9}$$

We know that the right-hand side of Eq. (9.5.9) is finite since $k(\vec{r},\vec{s})$ is a function in $\mathcal{L}_2(\Omega_D \times \Omega_D)$ by hypothesis. Thus, h must be finite as well, proving the theorem.

Even though $k(\vec{r},\vec{s})$ is a function in $\mathcal{L}_2(\Omega_D \times \Omega_D)$, the number of eigenvectors of $\mathbf{K}$ having *zero* eigenvalues can be infinite; i.e., a zero eigenvalue can have infinite degeneracy. For example, assume ψ_i, $i = 1, 2\ldots$, is an infinite orthonormal basis set in $\mathcal{H}$ and suppose $k(\vec{r},\vec{s}) = \psi_1(\vec{r})\psi_1^*(\vec{s})$. Then

$$\mathbf{K}\psi_i = \psi_1\langle\psi_1,\psi_i\rangle = \mathbf{0}, \qquad i = 2, 3, \ldots. \tag{9.5.10}$$

Thus, the vectors ψ_i, $i = 2, 3$, form an infinite number of eigenvectors of $\mathbf{K}$ having zero eigenvalue.

Actually, we can prove the even stronger theorem:

THEOREM. *If $\mathbf{K}$ is a completely continuous operator in $\mathcal{H}$ and λ is a nonzero eigenvalue of $\mathbf{K}$, then the number of eigenvectors corresponding to λ is finite.*

We proceed by assuming that there is an infinite number of eigenvectors $\phi_1, \phi_2, \ldots$ corresponding to λ. These can be assumed to make up an orthonormal set. We can add to this set the orthonormal set $\psi_1, \psi_2, \ldots$ required to form a complete set (or basis set) in $\mathcal{H}$. The ψ_i can be chosen to be orthogonal to the ϕ_i. Since $\phi_1, \phi_2, \ldots$ and $\psi_1, \psi_2, \ldots$ now form a complete orthonormal set, we can express the identity operator as

$$\mathbf{I} = \sum_{i=1}^{\infty} \phi_i \phi_i^\dagger + \sum_j \psi_i \psi_i^\dagger. \tag{9.5.11}$$

Note that the set $\{\psi_j\}$ may be finite or infinite. Since $\mathbf{KI} = \mathbf{K}$, it follows from Eq. (9.5.11) that

$$\mathbf{K} = \lambda \sum_{i=1}^{\infty} \phi_i \phi_i^\dagger + \sum_j (\mathbf{K}\psi_j)\psi_j^\dagger. \tag{9.5.12}$$

Because the quantity $\lambda \sum_{i=1}^{\infty} \phi_j \phi_j^\dagger$ cannot be approximated uniformly by a sequence of n-term dyadics, this is a contradiction to our hypothesis that $\mathbf{K}$ is a completely continuous operator. Therefore, there cannot be an infinite number of eigenvectors corresponding to a nonzero eigenvalue λ of a completely continuous operator.

9.5.3. Eigenvectors and Eigenvalues of the Function $f(\mathbf{K})$

Just as in the case of a matrix operator, if the function $f(t) = \sum_{k=1}^{\infty} a_k t^k$ exists, then the eigenvector $\boldsymbol{\phi}$ of the integral operator $\mathbf{K}$ is an eigenvector of $f(\mathbf{K})$. In fact, we can show that

$$f(\mathbf{K})\boldsymbol{\phi}_i = f(\lambda_i)\boldsymbol{\phi}_i. \tag{9.5.13}$$

If $\lambda_i \neq 0$, then, for a Laurent series $f(t) = \sum_{k=-\infty}^{\infty} a_k t^k$, it also follows that

$$f(\mathbf{K})\boldsymbol{\phi}_i = f(\lambda_i)\boldsymbol{\phi}_i. \tag{9.5.14}$$

Equations (9.5.13) and (9.5.14) follow from the properties

$$\begin{aligned} \mathbf{K}^2 \boldsymbol{\phi}_i &= \mathbf{K}(\lambda_i \boldsymbol{\phi}) = \lambda_i \mathbf{K} \boldsymbol{\phi}_i = \lambda_i^2 \boldsymbol{\phi}_i \\ \mathbf{K}^3 \boldsymbol{\phi}_i &= \lambda_i^3 \boldsymbol{\phi}, \ldots, \mathbf{K}^k \boldsymbol{\phi}_i = \lambda_i^k \boldsymbol{\phi}_i \end{aligned} \tag{9.5.15}$$

and, if $\lambda_i \neq 0$,

$$\mathbf{K}^{-1}\mathbf{K} = \mathbf{I}, \qquad \boldsymbol{\phi}_i = \mathbf{K}^{-1}\mathbf{K}\boldsymbol{\phi}_i = \mathbf{K}^{-1}\lambda_i \boldsymbol{\phi}_i = \lambda_i \mathbf{K}^{-1}\boldsymbol{\phi}_i$$

or

$$\mathbf{K}^{-1}\boldsymbol{\phi}_i = \lambda_i^{-1}\boldsymbol{\phi}_i, \ldots, \mathbf{K}^{-k}\boldsymbol{\phi}_i = \lambda_i^{-k}\boldsymbol{\phi}_i. \tag{9.5.16}$$

In order for Eqs. (9.5.13) and (9.5.14) to have meaning, it is, of course, required that $f(t)$ exists for $t = \lambda_i$. For example, if $f(t) = \sum_{k=1}^{\infty} t^{k-1}$, $f(\lambda_i) = 1/(1 - \lambda_i)$ for $|\lambda_i| < 1$ but $f(\lambda_i)$ does not exist if $|\lambda_i| > 1$.

We can extend Eqs. (9.5.13) and (9.5.14) to nonseries functions $f(t)$ of t if $f(t)$ exists for $t = \lambda_i$. For example, if $f(t) = t^{1/2}$, $t^{-1/2}$, or $\ln t$, respectively, then $\boldsymbol{\phi}_i$ is an eigenvector of $f(\mathbf{K})$ with eigenvalue $\lambda_i^{1/2}$, $\lambda_i^{-1/2}$, or $\ln \lambda_i$, respectively—again, so long as these functions exist for $t = \lambda_i$.

9.5.4. Some Special Properties of Spectral Operators

We saw in Section 9.5.3 that, from the definition of an integral operator given in Eq. (9.2.7) or (9.2.12), it follows that

$$\langle \mathbf{v}, \mathbf{K}\mathbf{u} \rangle = \langle \mathbf{K}^\dagger \mathbf{v}, \mathbf{u} \rangle \qquad \text{for every } \mathbf{v}, \mathbf{u} \in \mathcal{H}, \tag{9.5.17}$$

or that

$$\int d^D r \, v^*(\vec{r}) \int d^D s \, k(\vec{r}, \vec{s}) u(\vec{s}) = \int d^D s \left[\int d^D r \, k^*(\vec{s}, \vec{r}) v(\vec{r}) \right]^* u(\vec{s}). \tag{9.5.18}$$

A subtlety of Eq. (9.5.17) is that $\mathbf{K}\mathbf{u}$ and $\mathbf{K}^\dagger \mathbf{v}$ are defined for any function in the Hilbert space $\mathcal{H}$. This condition defines a large and interesting class of integral operators. However, for differential operators, the domains of $\mathbf{v}$ and $\mathbf{u}$ of $\mathbf{L}^\dagger$ and $\mathbf{L}$ are always subsets of the Hilbert space. In the general treatment of integral operators, the domains of $\mathbf{K}$ and $\mathbf{K}^\dagger$ can be different but are still subsets of the Hilbert space. This more general class of operators lies outside the scope of this text and the interested reader should consult the texts listed under Further Reading for a more general treatment.

SPECTRAL THEORY OF INTEGRAL OPERATORS

With the aid of Eq. (9.5.17), we can easily prove the following theorem:

THEOREM. *If $\mathbf{K}$ is self-adjoint, i.e., if $\mathbf{K} = \mathbf{K}^\dagger$, then*

(i) *the eigenvalues λ_i of $\mathbf{K}$ are real and*

(ii) *the eigenvectors $\boldsymbol{\phi}_i$ of $\mathbf{K}$ form an orthogonal set (which can be normalized).*

To prove this, we assume that $\mathbf{K}\boldsymbol{\phi}_i = \lambda_i \boldsymbol{\phi}_i$. We can, therefore, express Eq. (9.5.17) for these eigenvectors as

$$\langle \boldsymbol{\phi}_i, \mathbf{K}\boldsymbol{\phi}_i \rangle = \langle \mathbf{K}^\dagger \boldsymbol{\phi}_i, \boldsymbol{\phi}_i \rangle. \tag{9.5.19}$$

However $\langle \boldsymbol{\phi}_i, \mathbf{K}\boldsymbol{\phi}_i \rangle = \langle \boldsymbol{\phi}_i, \lambda_i \boldsymbol{\phi}_i \rangle = \lambda_i \langle \boldsymbol{\phi}_i \boldsymbol{\phi}_i \rangle$ and $\langle \mathbf{K}^\dagger \boldsymbol{\phi}_i, \boldsymbol{\phi}_i \rangle = \langle \mathbf{K}\boldsymbol{\phi}_i, \boldsymbol{\phi}_i \rangle = \langle \lambda_i \boldsymbol{\phi}_i, \boldsymbol{\phi}_i \rangle = \lambda_i^* \langle \boldsymbol{\phi}_i \boldsymbol{\phi} \rangle$. Thus, it follows from Eq. (9.5.19) that $\lambda_i \|\boldsymbol{\phi}_i\|^2 = \lambda_i^* \|\boldsymbol{\phi}_i\|^2$, meaning that λ_i must be equal to its complex conjugate λ_i^* and, therefore, must be real. To prove part (ii), we set $\mathbf{v} = \boldsymbol{\phi}_j$ and $\mathbf{u} = \boldsymbol{\phi}_i$ in Eq. (9.5.17) to obtain

$$(\lambda_i - \lambda_j)\langle \boldsymbol{\phi}_j, \boldsymbol{\phi}_i \rangle = 0. \tag{9.5.20}$$

It then follows that

$$\langle \boldsymbol{\phi}_j, \boldsymbol{\phi}_i \rangle = 0 \qquad \text{if } \lambda_i \neq \lambda_j. \tag{9.5.21}$$

Finally, we note that if the degeneracy of the eigenvalue λ_i is p, i.e., if there are p eigenvectors corresponding to λ_i, then these eigenvectors can be orthogonalized by the Gram–Schmidt procedure.

We can also prove Eq. (9.5.21) for a normal operator. However, we will need to establish another property first. Recall the following definition:

DEFINITION. *A normal operator is defined by the property*

$$\mathbf{K}\mathbf{K}^\dagger = \mathbf{K}^\dagger \mathbf{K}; \tag{9.5.22}$$

i.e., $\mathbf{K}$ commutes with its adjoint $\mathbf{K}^\dagger$.

First, we assume that $\mathbf{K}$ is a normal operator and then consider the inner products

$$\langle \mathbf{v}, \mathbf{K}^\dagger \mathbf{K} \mathbf{v} \rangle = \langle \mathbf{K}\mathbf{v}, \mathbf{K}\mathbf{v} \rangle \tag{9.5.23}$$

and

$$\langle \mathbf{v}, \mathbf{K}\mathbf{K}^\dagger \mathbf{v} \rangle = \langle \mathbf{K}^\dagger \mathbf{v}, \mathbf{K}^\dagger \mathbf{v} \rangle. \tag{9.5.24}$$

With the aid of Eq. (9.5.22), it follows that

$$\langle \mathbf{K}\mathbf{v}, \mathbf{K}\mathbf{v} \rangle = \langle \mathbf{K}^\dagger \mathbf{v}, \mathbf{K}^\dagger \mathbf{v} \rangle. \tag{9.5.25}$$

Now suppose $\boldsymbol{\phi}_i$ is an eigenvector of $\mathbf{K}$ with zero eigenvalue, i.e., $\mathbf{K}\boldsymbol{\phi}_i = \mathbf{0}$. It follows from Eq. (9.5.25) that $\mathbf{K}^\dagger \boldsymbol{\phi}_i = \mathbf{0}$; i.e., $\boldsymbol{\phi}_i$ is also an eigenvector of the adjoint operator $\mathbf{K}^\dagger$ corresponding to a zero eigenvalue. Suppose next that $\boldsymbol{\phi}_i$ is an eigenvector of $\mathbf{K}$ corresponding to the eigenvalue $\lambda_i \neq 0$, i.e., $\mathbf{K}\boldsymbol{\phi}_i = \lambda_i \boldsymbol{\phi}_i$ or $(\mathbf{K} - \lambda_i \mathbf{I})\boldsymbol{\phi}_i = \mathbf{0}$. We now define

$$\mathbf{L} \equiv \mathbf{K} - \lambda_i \mathbf{I}. \tag{9.5.26}$$

The adjoint of this operator is $\mathbf{L}^\dagger = \mathbf{K}^\dagger - \lambda_i^* \mathbf{I}$, and since $\mathbf{L}\mathbf{L}^\dagger = \mathbf{L}^\dagger \mathbf{L}$, this operator is also a normal operator for which $\mathbf{L}\boldsymbol{\phi}_i = \mathbf{0}$. Thus, according to Eq. (9.5.25) for normal operators, it follows that $\mathbf{L}^\dagger \boldsymbol{\phi}_i = \mathbf{0}$, or

$$\mathbf{K}^\dagger \boldsymbol{\phi}_i = \lambda_i^* \boldsymbol{\phi}_i. \tag{9.5.27}$$

We have, in fact, proven the theorem:

THEOREM. *If $\mathbf{K}$ is a normal operator, every eigenvector $\boldsymbol{\phi}_i$ of $\mathbf{K}$ with eigenvalue λ_i is an eigenvector of $\mathbf{K}^\dagger$ with eigenvalue λ_i^*.*

It is now easy to prove the following:

THEOREM. *If $\mathbf{K}$ is a normal operator and if $\lambda_i \neq \lambda_j$, then from*

$$\langle \boldsymbol{\phi}_i, \mathbf{K}\boldsymbol{\phi}_i \rangle = \langle \mathbf{K}^\dagger \boldsymbol{\phi}_i, \boldsymbol{\phi}_i \rangle \tag{9.5.28}$$

it follows that

$$(\lambda_j - \lambda_i)\langle \boldsymbol{\phi}_i, \boldsymbol{\phi}_j \rangle = 0, \tag{9.5.29}$$

and so $\langle \boldsymbol{\phi}_i, \boldsymbol{\phi}_j \rangle = 0$, proving that if $\lambda_i \neq \lambda_j$ the eigenvectors $\boldsymbol{\phi}_i$ and $\boldsymbol{\phi}_j$ are orthogonal.

If λ_i has a degeneracy p, i.e., if there are p eigenvectors corresponding to the eigenvalue λ_i, these eigenvectors can be orthogonalized by the Gram–Schmidt procedure and thus we can summarize the above in the following theorem:

THEOREM. *If $\mathbf{K}$ is a normal operator, its eigenvectors form an orthogonal set (which, of course, can be normalized). The eigenvectors $\boldsymbol{\phi}_i$ of $\mathbf{K}$ are also eigenvectors of the adjoint operator $\mathbf{K}^\dagger$ and the eigenvalues of $\mathbf{K}^\dagger$ are λ_i^*, the complex values of the eigenvalues λ_i of $\mathbf{K}$.*

An example of a normal operator is

$$\mathbf{L} = \exp(it\mathbf{K}), \tag{9.5.30}$$

where $i = \sqrt{-1}$, t is a real number, and $\mathbf{K}$ is a self-adjoint operator. $\mathbf{L}$ is also a unitary operator since $\mathbf{L}\mathbf{L}^\dagger = \mathbf{L}^\dagger \mathbf{L} = \mathbf{I}$.

Next, consider the linear operators $\mathbf{K}$ and $\mathbf{L}$, which commute; i.e., they have the property

$$\mathbf{KL} = \mathbf{LK}. \tag{9.5.31}$$

Such operators obey the theorem:

THEOREM. *If* **K** *and* **L** *commute and if one of the operators, say* **L**, *has an eigenvalue* λ *of finite degeneracy h, then* **K** *and* **L** *have a common eigenvector—i.e., there exists at least one vector* **x** *such that*

$$\mathbf{Lx} = \lambda \mathbf{x} \quad \text{and} \quad \mathbf{Kx} = \mu \mathbf{x}. \quad (9.5.32)$$

To prove this theorem, we introduce the concept of the *null space* of an operator **L**. The null space $\mathcal{N}_L$ is the collection of all vectors that are linear combinations of the zero-eigenvalue eigenvectors of **L**; i.e., any vector $\mathbf{x} \in \mathcal{N}_L$ can be expressed as a linear combination of the $\mathbf{x}_i$, where

$$\mathbf{Lx}_i = \mathbf{0}, \quad i = 1, 2, \ldots. \quad (9.5.33)$$

$\mathcal{N}_L$ can be of finite or infinite dimension. For example, suppose $\boldsymbol{\psi}_i$, $i = 1, 2, \ldots$, is a complete orthonormal basis set, then the null space of

$$\mathbf{L}_1 = \boldsymbol{\psi}_1 \boldsymbol{\psi}_1^\dagger \quad (9.5.34)$$

is the infinite set $\boldsymbol{\psi}_i$, $i = 2, 3, \ldots$, and all linear combinations thereof. The null space of the operator

$$\mathbf{L}_2 = \sum_{i=2}^\infty \lambda_i \boldsymbol{\psi}_i \boldsymbol{\psi}_i^\dagger, \quad \lambda_i \neq 0, \quad (9.5.35)$$

on the other hand, consists of all scalar multiples of $\boldsymbol{\psi}_1$ and the null space of

$$\mathbf{L}_3 = \sum_{i=1}^\infty \lambda_i \boldsymbol{\psi}_i \boldsymbol{\psi}_i^\dagger, \quad \lambda_i \neq 0, \quad (9.5.36)$$

is empty.

With this definition, we can now prove the following theorem:

THEOREM. *If* **L** *and* **K** *commute, then the null space of* **L** (*or* **K**) *is an invariant manifold of* **K** (*or* **L**).

What this means is that if $\mathbf{x} \in \mathcal{N}_L$ (or $\mathcal{N}_K$), then $\mathbf{Kx}$ (or $\mathbf{Lx}$) also belongs to $\mathcal{N}_L$ (or $\mathcal{N}_K$). Still another way to say this is that if **x** is a vector that **L** (or **K**) maps to the zero vector **0**, then the vector $\mathbf{Kx}$ (or $\mathbf{Lx}$) is also mapped to **0** by **L** (or **K**).

To prove the theorem, assume that $\mathbf{x}_i \in \mathcal{N}_L$. Then, since $\mathbf{Lx}_i = \mathbf{0}$ and $\mathbf{KL} = \mathbf{LK}$, it follows that

$$\mathbf{0} = \mathbf{KLx}_i = \mathbf{L}(\mathbf{Kx}_i). \quad (9.5.37)$$

Thus, $\mathbf{Kx}_i$ belongs to $\mathcal{N}_L$ for all vectors $\mathbf{x}_i$ in $\mathcal{N}_L$.

Now we can return to the theorem expressed at Eq. (9.5.32). The h eigenvectors $\mathbf{x}_1, \ldots, \mathbf{x}_h$, corresponding to the eigenvalue λ of **L**, form a finite-dimensional manifold for the null space $\mathcal{N}_{L-\lambda I}$ of $\mathbf{L} - \lambda \mathbf{I}$. According to the theorem just proved, the null space of $\mathcal{N}_{L-\lambda I}$ is invariant to $\mathbf{K}$ because $(\mathbf{L} - \lambda\mathbf{I})\mathbf{K} = \mathbf{K}(\mathbf{L} - \lambda\mathbf{I})$. Thus, if $\mathbf{L}\mathbf{x}_i = \lambda\mathbf{x}_i$, then

$$\mathbf{K}\mathbf{x}_i = \sum_{j=1}^{h} \alpha_{ij}\mathbf{x}_j, \qquad i = 1,\ldots,h. \tag{9.5.38}$$

Equation (9.5.38) can be summarized in the form

$$\mathbf{K}[\mathbf{x}_1,\ldots,\mathbf{x}_h] = [\mathbf{x}_1,\ldots,\mathbf{x}_h]\mathbf{A}, \tag{9.5.39}$$

where $[\mathbf{x}_1,\ldots,\mathbf{x}_h]$ is a row vector with elements $\mathbf{x}_i$, and $\mathbf{A}$ is the square matrix

$$\mathbf{A} = \begin{bmatrix} \alpha_{11} & \cdots & \alpha_{h1} \\ \alpha_{12} & \cdots & \alpha_{h2} \\ \vdots & & \vdots \\ \alpha_{1h} & \cdots & \alpha_{hh} \end{bmatrix}. \tag{9.5.40}$$

Suppose $\boldsymbol{\xi}$ is an eigenvector (a column vector) of $\mathbf{A}$ corresponding to the eigenvalue μ, i.e.,

$$\mathbf{A}\boldsymbol{\xi} = \mu\boldsymbol{\xi}. \tag{9.5.41}$$

Multiplying each side of Eq. (9.5.39) from the right by $\boldsymbol{\xi}$, we obtain

$$\mathbf{K}\sum_{i=1}^{h} \xi_i \mathbf{x}_i = [\mathbf{x}_1,\ldots,\mathbf{x}_h]\mathbf{A}\boldsymbol{\xi} = \mu \sum_{i=1}^{h} \xi_i \mathbf{x}_i \tag{9.5.42}$$

or

$$\mathbf{K}\mathbf{y} = \mu\mathbf{y}, \qquad \mathbf{y} = \sum_{i=1}^{h} \xi_i \mathbf{x}_i. \tag{9.5.43}$$

Since $(\mathbf{L} - \lambda\mathbf{I})\mathbf{x}_i = 0$, it follows that

$$\mathbf{L}\mathbf{y} = \lambda\mathbf{y}, \tag{9.5.44}$$

thus proving that $\mathbf{L}$ and $\mathbf{K}$ possess a common eigenvector. In fact, the number of eigenvectors $\mathbf{L}$ and $\mathbf{K}$ have in common depends on the rank of $\mathbf{A}$.

■ **EXAMPLE 9.5.1.** Suppose $\boldsymbol{\psi}_i$, $i = 1, 2, \ldots$, is a complete orthonormal set in $\mathcal{H}$ and define

$$\mathbf{L} = \boldsymbol{\psi}_1 \boldsymbol{\psi}_1^\dagger + \boldsymbol{\psi}_2 \boldsymbol{\psi}_2^\dagger + \sum_{i=3}^{\infty} \lambda_i \boldsymbol{\psi}_i \boldsymbol{\psi}_i^\dagger, \qquad \lambda_i \neq 1,\ i > 2, \tag{9.5.45}$$

and

$$\mathbf{K} = 2\boldsymbol{\psi}_1 \boldsymbol{\psi}_2^\dagger + 2\boldsymbol{\psi}_2 \boldsymbol{\psi}_1^\dagger. \tag{9.5.46}$$

Note that

$$\mathbf{LK} - \mathbf{KL} = (2\boldsymbol{\psi}_1\boldsymbol{\psi}_2^\dagger + 2\boldsymbol{\psi}_2\boldsymbol{\psi}_1^\dagger) - (2\boldsymbol{\psi}_1\boldsymbol{\psi}_2^\dagger + 2\boldsymbol{\psi}_2\boldsymbol{\psi}_1^\dagger) = \mathbf{0}, \qquad (9.5.47)$$

and so $\mathbf{K}$ and $\mathbf{L}$ commute. Since

$$\mathbf{L}\boldsymbol{\psi}_i = \boldsymbol{\psi}_i, \qquad i = 1, 2, \qquad (9.5.48)$$

it follows from the theorem at Eq. (9.5.32) that $\mathbf{K}$ must possess at least one eigenvector in the manifold spanned by $\boldsymbol{\psi}_1$ and $\boldsymbol{\psi}_2$. Indeed, this is true, since

$$\mathbf{K}(\boldsymbol{\psi}_1 + \boldsymbol{\psi}_2) = 2(\boldsymbol{\psi}_1 + \boldsymbol{\psi}_2) \qquad (9.5.49)$$

and

$$\mathbf{L}(\boldsymbol{\psi}_1 + \boldsymbol{\psi}_2) = (\boldsymbol{\psi}_1 + \boldsymbol{\psi}_2). \qquad (9.5.50)$$

$\mathbf{K}$, however, has only one eigenvector, $\boldsymbol{\psi}_1 + \boldsymbol{\psi}_2$, whereas $\mathbf{L}$ has two, which can be chosen as $\boldsymbol{\psi}_1$ and $\boldsymbol{\psi}_1 + \boldsymbol{\psi}_2$. ■ ■ ■

Finally, we note that if $\mathbf{K}\boldsymbol{\phi}_i = \lambda_i \boldsymbol{\phi}_i$ and $\mathbf{K}^\dagger \boldsymbol{\psi}_j = \nu_j \boldsymbol{\psi}_j$ and $\lambda_i \neq \nu_j^*$, then $\langle \boldsymbol{\psi}_j, \boldsymbol{\phi}_i \rangle = 0$. To prove this, note that, from the property of the adjoint operator $\langle \mathbf{v}, \mathbf{Ku} \rangle = \langle \mathbf{K}^\dagger \mathbf{v}, \mathbf{u} \rangle$, it follows that $(\lambda_i - \nu_j^*)\langle \boldsymbol{\psi}_j, \boldsymbol{\phi}_i \rangle = 0$. Thus, we have proved the following theorem:

THEOREM. *If $\boldsymbol{\phi}_i$ is an eigenvector of $\mathbf{K}$ of eigenvalue λ_i and $\boldsymbol{\psi}_j$ is an eigenvector of $\mathbf{K}^\dagger$ of eigenvalue ν_j, then $\boldsymbol{\phi}_i$ is orthogonal to $\boldsymbol{\psi}_j$ if $\lambda_i \neq \nu_j^*$.*

9.5.5. Completely Continuous Self-Adjoint Operators

We shall begin by stating the following main theorem for the operators of interest in this section:

THEOREM. *If $\mathbf{K}$ is a completely continuous integral operator in a Hilbert space and is self-adjoint, i.e. if $\mathbf{K} = \mathbf{K}^\dagger$, then the eigenvectors $\boldsymbol{\phi}_1, \boldsymbol{\phi}_2, \ldots$ of $\mathbf{K}$ form a complete orthonormal set (i.e., an orthonormal basis set) in $\mathcal{H}$ and the eigenvalues of $\mathbf{K}$ are real.*

Incidentally, this theorem is valid for any completely continuous *self-adjoint* linear operator in a Hilbert space. For example, $\mathbf{K}$ could be a matrix operator in E_∞. As noted before, a linear operator in a function space can be mapped into a matrix operator in E_∞.

The proof of this theorem will be addressed in what follows. We will first, however, review the implications of the theorem and define the spectral resolution of an operator and its functions. Of course, the above theorem means that self-adjoint, completely continuous operators in $\mathcal{H}$ are perfect; i.e., their eigenvectors form a basis set in $\mathcal{H}$. If $\mathbf{y}$ is an arbitrary vector in $\mathcal{H}$ and $\mathbf{K}$ is a normal, completely continuous operator, then there exists a set of numbers $\alpha_1, \alpha_2, \ldots$ such that $\mathbf{y} = \sum_i \alpha_i \boldsymbol{\phi}_i$, where

$$\boldsymbol{\phi}_i, \qquad i = 1, 2, \ldots, \qquad (9.5.51)$$

are the eigenvectors of **K**. Since

$$\langle \boldsymbol{\phi}_i, \boldsymbol{\phi}_j \rangle = \delta_{ij}, \tag{9.5.52}$$

it follows that $\alpha_i = \langle \boldsymbol{\phi}_i^\dagger, \mathbf{y} \rangle = \boldsymbol{\phi}_i^\dagger \mathbf{y}$, or

$$\mathbf{y} = \sum_i \boldsymbol{\phi}_i (\boldsymbol{\phi}_i^\dagger \mathbf{y}) = \left(\sum_i \boldsymbol{\phi}_i \boldsymbol{\phi}_i^\dagger \right) \mathbf{y} \tag{9.5.53}$$

for arbitrary $\mathbf{y} \in \mathcal{H}$. This implies that the identity operator can be resolved as

$$\mathbf{I} = \sum_i \boldsymbol{\phi}_i \boldsymbol{\phi}_i^\dagger, \tag{9.5.54}$$

which, in turn, implies the *spectral resolution theorem*:

SPECTRAL RESOLUTION THEOREM

$$\mathbf{K} = \sum_i \lambda_i \boldsymbol{\phi}_i \boldsymbol{\phi}_i^\dagger, \qquad \langle \boldsymbol{\phi}_i, \boldsymbol{\phi}_j \rangle = \delta_{ij}, \tag{9.5.55}$$

for any normal, completely continuous operator **K**.

Of course, the function $f(\mathbf{K})$ of **K** also obeys the spectral resolution theorem:

$$f(\mathbf{K}) = \sum_i f(\lambda_i) \boldsymbol{\phi}_i \boldsymbol{\phi}_i^\dagger \tag{9.5.56}$$

if $f(t)$ is defined at $t = \lambda_i$, $i = 1, 2, \dots$.

Let us now outline the proof of the theorem that a completely continuous, self-adjoint operator has a complete set of orthonormal eigenvectors. As a first step, we will prove that a normalized vector **y** is an eigenvector of a linear self-adjoint operator **K** in $\mathcal{H}$ if it is the vector satisfying the maximum condition

$$\max_{\mathbf{y}} \langle \mathbf{y}, \mathbf{Ky} \rangle, \qquad \|\mathbf{y}\|^2 = 1. \tag{9.5.57}$$

Equation (9.5.57) is a constrained maximum and by introducing the Lagrange multiplier λ, the unconstrained maximum of

$$f = \langle \mathbf{y}, \mathbf{Ky} \rangle - \lambda \langle \mathbf{y}, \mathbf{y} \rangle \tag{9.5.58}$$

gives the vector satisfying Eq. (9.5.57).

Expanding **y** in terms of an arbitrary orthonormal basis set $\boldsymbol{\psi}_1, \boldsymbol{\psi}_2, \dots$

$$\mathbf{y} = \sum_i \alpha_i \boldsymbol{\psi}_1, \tag{9.5.59}$$

and inserting this into Eq. (9.5.58), we obtain

$$f = \sum_{i,j} k_{ij} \alpha_i^* \alpha_j - \lambda \sum_i \alpha_i^* \alpha_i, \tag{9.5.60}$$

where $k_{ij} = \langle \boldsymbol{\psi}_i, \mathbf{K}\boldsymbol{\psi}_j \rangle$. The maximum can be found from the conditions $\partial f / \partial \alpha_i^R = \partial f / \partial \alpha_i^I = 0$, where α_i^R and α_i^I are the real and imaginary parts of α_i. These conditions yield the eigenproblem

$$\sum_j k_{ij} \alpha_j = \lambda \alpha_i, \qquad i = 1, 2, \ldots. \tag{9.5.61}$$

However, Eq. (9.5.61) is equivalent to the eigenproblem

$$\mathbf{K}\mathbf{y} = \lambda \mathbf{y}, \tag{9.5.62}$$

whose equivalence can be proved by inserting Eq. (9.5.59) into Eq. (9.5.62) and taking the inner product of the result with $\boldsymbol{\psi}_i, i = 1, 2, \ldots$. Thus, the solution to Eq. (9.5.57) is a solution to the eigenproblem in Eq. (9.5.62). Actually, Eq. (9.5.62) is the solution only to the extremal problem. If $\mathbf{K}$ has no positive eigenvalues, then we would consider the operator $-\mathbf{K}$ so as to find a positive maximum. This is permissible since $\mathbf{K}$ and $-\mathbf{K}$ have the same eigenvectors.

The next step in the proof is to recall that a completely continuous operator is bounded, i.e., $\|\mathbf{K}\|^2 = \max_{\mathbf{x} \neq 0} \langle \mathbf{K}\mathbf{x}, \mathbf{K}\mathbf{x} \rangle / \langle \mathbf{x}, \mathbf{x} \rangle = M^2 < \infty$. This means that, for any normalized vector $\mathbf{y}$ in $\mathcal{H}$,

$$\langle \mathbf{y}, \mathbf{K}\mathbf{y} \rangle \leq \|\mathbf{y}\| \|\mathbf{K}\mathbf{y}\| \leq M \|\mathbf{y}\|^2 = M. \tag{9.5.63}$$

If the largest magnitude eigenvalue of $\mathbf{K}$ is positive (as we assume, or otherwise we consider $-\mathbf{K}$), then there exists a sequence of normalized vectors $\mathbf{y}_1, \mathbf{y}_2, \ldots$ such that

$$\lim_{n \to \infty} \langle \mathbf{y}_n, \mathbf{K}\mathbf{y}_n \rangle = M. \tag{9.5.64}$$

Recall that a property of completely continuous, self-adjoint operators is that $\mathbf{y}_n \to \mathbf{y}$; i.e., there is a vector $\mathbf{y} \in \mathcal{H}$ such that

$$\langle \mathbf{y}, \mathbf{K}\mathbf{y} \rangle = M. \tag{9.5.65}$$

Thus, $\mathbf{y}$ is an eigenvector (say $\boldsymbol{\phi}_1$) of $\mathbf{K}$ and M is the maximum eigenvalue (say λ_1) of $\mathbf{K}$.

Next, we hunt maxima of $\langle \mathbf{y}, \mathbf{K}\mathbf{y} \rangle$ among the normalized vectors in $\mathcal{H}$ that are orthogonal to $\boldsymbol{\phi}_1$. If there is more than one eigenvector corresponding to λ_1, then this step will generate a second eigenvector $\boldsymbol{\phi}_2$. Continuing the process will generate all h eigenvectors corresponding to λ_1. Recall that we proved earlier that h has to be finite for a nonzero eigenvalue of a completely continuous operator. The next step in the process will generate the eigenvectors of the second largest positive eigenvalue of $\mathbf{K}$. Continuing in this manner will eventually generate all the nonzero positive eigenvalues and eigenvectors $\boldsymbol{\phi}_1, \boldsymbol{\phi}_2, \ldots$ of $\mathbf{K}$. Carrying out the same process for $-\mathbf{K}$ will generate all the nonzero negative eigenvalues and eigenvectors of $\mathbf{K}$. As we can see, the eigenvalues of a normal, completely continuous operator form a countable or denumerable (though possibly infinite) set. This property arises from the defining property that a completely continuous operator is the limit of a sequence of n-term dyadic operators.

Let us label all of the nonzero eigenvalues of $\mathbf{K}$ by the sequence $\lambda_1 \leq \lambda_2 \leq \lambda_3 \leq \cdots$ and the corresponding eigenvectors by $\boldsymbol{\phi}_1, \boldsymbol{\phi}_2, \boldsymbol{\phi}_3, \ldots$. We next define a new self-adjoint operator

$$\tilde{\mathbf{K}} = \mathbf{K} - \sum_i \lambda_i \boldsymbol{\phi}_i \boldsymbol{\phi}_i^\dagger, \qquad (9.5.66)$$

and suppose that $\tilde{\mathbf{K}}$ is nonzero. There then exist a nonzero eigenvalue μ and its eigenvector $\boldsymbol{\psi}$, i.e.,

$$\tilde{\mathbf{K}}\boldsymbol{\psi} = \mu \boldsymbol{\psi}. \qquad (9.5.67)$$

However, the eigenvectors $\boldsymbol{\phi}_1, \boldsymbol{\phi}_2, \ldots$ of $\mathbf{K}$ are also eigenvectors of $\tilde{\mathbf{K}}$ of zero eigenvalue. Thus, $\langle \boldsymbol{\phi}_i, \boldsymbol{\psi} \rangle = 0$, $i = 1, 2, \ldots$, which implies that

$$\tilde{\mathbf{K}}\boldsymbol{\psi} = \mathbf{K}\boldsymbol{\psi} = \mu \boldsymbol{\psi}, \qquad (9.5.68)$$

and, therefore, $\boldsymbol{\psi}$ must be an eigenvector of $\mathbf{K}$ of nonzero eigenvalue. This is, of course, a contradiction since the process we described above generated all of the eigenvectors of $\mathbf{K}$ having nonzero eigenvalues. Thus, it follows that $\tilde{\mathbf{K}} = \mathbf{0}$, or

$$\mathbf{K} = \sum_i \lambda_i \boldsymbol{\phi}_i \boldsymbol{\phi}_i^\dagger, \qquad (9.5.69)$$

proving the spectral resolution theorem for self-adjoint, completely continuous operators. If $\mathbf{K}$ has zero eigenvalues, then the set of eigenvectors in Eq. (9.5.69) does not form a basis set in $\mathcal{H}$. However, we can add to the eigenvectors $\boldsymbol{\phi}_1, \boldsymbol{\phi}_2, \ldots$, for which $\lambda_i \neq 0$, an orthonormal set $\boldsymbol{\psi}_1, \boldsymbol{\psi}_2, \ldots$ needed to form a basis set in $\mathcal{H}$. Since the $\boldsymbol{\psi}_i$ can be chosen to be orthogonal to the $\boldsymbol{\phi}_i$, it follows from Eq. (9.5.69) that $\mathbf{K}\boldsymbol{\psi}_i = 0$. Thus, the basis set $\boldsymbol{\phi}_1 \boldsymbol{\phi}_2, \ldots, \boldsymbol{\psi}_1 \boldsymbol{\psi}_2, \ldots$ in $\mathcal{H}$ are all eigenvectors of $\mathbf{K}$. Since $\langle \boldsymbol{\phi}_i, \mathbf{K}\boldsymbol{\phi}_i \rangle = \langle \mathbf{K}^\dagger \boldsymbol{\phi}_i, \boldsymbol{\phi}_i \rangle$ and $\mathbf{K} = \mathbf{K}^\dagger$ for a self-adjoint operator, it follows that $(\lambda_i - \lambda_i^*)\|\boldsymbol{\phi}_i\|^2 = 0$, and so the eigenvalues of $\mathbf{K}$ are real.

■ **EXAMPLE 9.5.2.** Suppose the vectors $\boldsymbol{\phi}_1 \boldsymbol{\phi}_2, \ldots$ form an orthonormal basis set in an infinite-dimensional Hilbert space $\mathcal{H}$.

(a) Then

$$\mathbf{K} = \sum_{i=1}^n \frac{1}{\sqrt{i}} \boldsymbol{\phi}_i \boldsymbol{\phi}_i^\dagger \qquad (9.5.70)$$

is a self-adjoint, completely continuous operator with eigenvalues and eigenvectors

$$\lambda_i = \frac{1}{\sqrt{i}}, \qquad \boldsymbol{\phi}_i, \qquad i = 1, \ldots, n, \qquad (9.5.71)$$

and

$$\lambda_i = 0, \qquad \boldsymbol{\phi}_i, \qquad i = n+1, n+2, \ldots. \qquad (9.5.72)$$

For this case, there is an infinite number of eigenvectors with $\lambda_i = 0$.

(b) On the other hand, if

$$\mathbf{K} = \sum_{i=n}^{\infty} \frac{1}{\sqrt{i}} \boldsymbol{\phi}_i \boldsymbol{\phi}_i^{\dagger}, \tag{9.5.73}$$

then $\mathbf{K}$ is a self-adjoint operator with eigenvalues and eigenvectors

$$\lambda_i = 0, \qquad \boldsymbol{\phi}_i, \qquad i = 1, \ldots, n-1, \tag{9.5.74}$$

and

$$\lambda_i = \frac{1}{\sqrt{i}}, \qquad \boldsymbol{\phi}_i, \qquad i = n, n+1, \ldots. \tag{9.5.75}$$

■ ■ ■ In this case, there is a finite number of eigenvectors with $\lambda_i = 0$.

EXAMPLE 9.5.3. Suppose $k(t,s) = \exp(-i\pi t + i\pi s)$, $-1 \le s, t \le 1$. Then $k(t,s) = k^*(s,t)$, and so $\mathbf{K}$ is self-adjoint. Since $\int_{-1}^{1} \int_{-1}^{1} |k(t,s)|^2 \, dt \, ds = 4$, it follows that $\mathbf{K}$ is completely continuous. The eigenvectors of $\mathbf{K}$ obey the equation

$$\exp(-i\pi t) \int_{-1}^{1} \exp(i\pi s) v(s) \, ds = \lambda v(t). \tag{9.5.76}$$

$v_0 = 1/\sqrt{2}$ is an eigenvector with $\lambda_0 = 0$ since $\int_{-1}^{1} \exp(i\pi s) \, ds = (\exp(i\pi) - \exp(-i\pi))/i\pi = 0$. $v_1 = \exp(-i\pi t)/\sqrt{2}$ is an eigenvector with $\lambda_2 = 2$. Since $\int_{-1}^{1} \exp(i\pi m s) \, ds = 0$, $m = \pm 1, \pm 2, \ldots$, it follows that all of the eigenvectors and eigenvalues of $\mathbf{K}$ are

$$\begin{aligned}
v_0 &= \frac{1}{\sqrt{2}}, & \lambda &= 0 \\
v_1 &= \frac{\exp(-i\pi t)}{\sqrt{2}}, & \lambda_1 &= 2 \\
v_{-1} &= \frac{\exp(i\pi t)}{\sqrt{2}}, & \lambda_{-1} &= 0 \\
v_n &= \frac{\exp(-in\pi t)}{\sqrt{2}}, & \lambda_n &= 0, \quad n = \pm 2, \pm 3, \pm 4, \ldots.
\end{aligned} \tag{9.5.77}$$

According to our theorem, the set $v_n = \exp(-in\pi t)/\sqrt{2}$, $n = 0, \pm 1, \pm 2, \ldots$, forms a complete orthonormal set in $\mathcal{L}_2(-1, 1)$, confirming that any function $f(t)$, $-1 \le t \le 1$, in $\mathcal{L}_2(-1, 1)$ can be expanded in the series

$$f(t) = \sum_n \alpha_n v_n(t) = \frac{\alpha_0}{\sqrt{2}} + \sum_{n=\pm 1, \pm 2, \ldots} \alpha_n \frac{\exp(-in\pi t)}{\sqrt{2}}. \tag{9.5.78}$$

■ ■ ■ This is well known from the theory of Fourier series.

EXAMPLE 9.5.4. Consider in $\mathcal{L}_2(-\infty, \infty)$ the operator

$$\mathbf{K}v = \int_{-\infty}^{\infty} \exp\left(\frac{-(x^2 + y^2)}{2}\right) v(y) \, dy. \tag{9.5.79}$$

(a) Prove that **K** is self-adjoint.
(b) Prove that **K** is completely continuous.
(c) Find the eigenvectors of **K**.

Since $k(x, y) = \exp(-(x^2 + y^2)/2) = k^*(y, x)$, the operator is self-adjoint. Also,

$$\int_{-\infty}^{\infty} \int_{-\infty}^{\infty} [k(x, y)]^2 \, dx \, dy = \int_{-\infty}^{\infty} \exp(-x^2) \, dx \int_{-\infty}^{\infty} \exp(-y^2) \, dy = \pi < \infty, \tag{9.5.80}$$

which is a sufficient condition that **K** is completely continuous.

The eigenproblem is

$$\exp\left(\frac{-x^2}{2}\right) \int_{-\infty}^{\infty} \exp\left(\frac{-y^2}{2}\right) v(y) \, dy = \lambda v(x). \tag{9.5.81}$$

One eigenvector is $v_0(x) = \alpha \exp(-x^2/2)$, for which Eq. (9.5.81) yields

$$\exp\left(\frac{-x^2}{2}\right) \alpha \int_{-\infty}^{\infty} \exp(-y^2) \, dy = \lambda_0 \alpha \exp\left(\frac{-x^2}{2}\right) \tag{9.5.82}$$

or $\lambda_0 = \sqrt{\pi}$. Normalizing v_1, we obtain

$$1 = \|\mathbf{v}_1\|^2 = \alpha^2 \int_{-\infty}^{\infty} \exp\left(\frac{-x^2}{2}\right) dx = \alpha^2 \sqrt{\pi} \tag{9.5.83}$$

or

$$v_0(x) = \frac{1}{\pi^{1/4}} \exp\left(\frac{-x^2}{2}\right), \qquad \lambda_0 = \pi^{1/2}. \tag{9.5.84}$$

This is the only eigenfunction of **K** with a nonzero eigenvalue. The eigenfunctions corresponding to $\lambda = 0$ obey the equation

$$\int_{-\infty}^{\infty} \exp\left(\frac{-y^2}{2}\right) v(y) \, dy = 0. \tag{9.5.85}$$

The first two of these are

$$v_1(x) = \frac{1}{\sqrt{2\sqrt{\pi}}} 2x \exp\left(\frac{-x^2}{2}\right) \tag{9.5.86}$$

and

$$v_2(x) = \frac{1}{2\sqrt{\sqrt{\pi}}} (4x^2 - 2) \exp\left(\frac{-x^2}{2}\right). \tag{9.5.87}$$

In general,

$$v_n(x) = \frac{1}{\sqrt{2^n n! \sqrt{\pi}}} H_n(x) \exp\left(\frac{-x^2}{2}\right), \tag{9.5.88}$$

where $H_n(x)$ is a Hermite polynomial generated by the formula

$$H_n(x) = (-1)^n \exp(x^2) \frac{d^n}{dx^n}\left(\exp(-x^2)\right). \qquad (9.5.89)$$

From the theorem for completely continuous self-adjoint operators, it follows that $v_0(x), v_1(x), v_2(x), \ldots$ form an orthonormal basis set in $\mathcal{L}_2(-\infty, \infty)$. The functions

$$\frac{H_n(x)}{\sqrt{2^n n! \sqrt{\pi}}}, \qquad n = 0, 1, 2, \ldots, \qquad (9.5.90)$$

■ ■ ■ therefore, form an orthonormal basis set in $\mathcal{L}_2(-\infty, \infty; \exp(-x^2))$.

9.5.6. Schmidt's Normal Form for Completely Continuous Operators

The theorem of interest in this section is:

THEOREM. *If $\mathbf{K}$ is a completely continuous operator in $\mathcal{H}$, then it can be expressed in the form*

$$\mathbf{K} = \sum \kappa_i \boldsymbol{\psi}_i \boldsymbol{\phi}_i^\dagger, \qquad (9.5.91)$$

where the κ_i are real numbers greater than 0 and the vectors $\boldsymbol{\phi}_i$ and $\boldsymbol{\psi}_i$, $i = 1, 2, \ldots$, are orthonormal sets in $\mathcal{H}$ satisfying the equations

$$\mathbf{K}\boldsymbol{\psi}_i = \kappa_i \boldsymbol{\phi}_i \qquad (9.5.92)$$

and

$$\mathbf{K}^\dagger \boldsymbol{\phi}_i = \kappa_i \boldsymbol{\psi}_i. \qquad (9.5.93)$$

From Eqs. (9.5.92) and (9.5.93), it follows that $\boldsymbol{\psi}_i$ and $\boldsymbol{\phi}_i$ are eigenvectors of the self-adjoint operators $\mathbf{K}^\dagger \mathbf{K}$ and $\mathbf{K}\mathbf{K}^\dagger$; namely, they satisfy the eigenequations

$$\mathbf{K}^\dagger \mathbf{K} \boldsymbol{\phi}_i = \kappa_i^2 \boldsymbol{\phi}_i \qquad (9.5.94)$$

and

$$\mathbf{K}\mathbf{K}^\dagger \boldsymbol{\psi}_i = \kappa_i^2 \boldsymbol{\psi}_i. \qquad (9.5.95)$$

A completely continuous operator can be approximated as closely as we please by an n-term dyadic. With this approximation, linear equations involving completely continuous operators can be transformed into matrix equations in a finite-dimensional vector space, in which case, the theorem has already been proven in Section 7.7. Thus, heuristically, we could anticipate the validity of the theorem for completely continuous operators without further work. Those with little patience for theorem proving can ignore the rest of this section as we will outline the details of the rigorous proof in what follows.

Completely continuous operators have the property that if $\mathbf{x}_n \to \mathbf{x}$ and $\mathbf{y}_n \to \mathbf{y}$, then $\langle \mathbf{y}_n, \mathbf{K}\mathbf{x}_n \rangle \to \langle \mathbf{y}, \mathbf{K}\mathbf{x} \rangle$ as $n \to \infty$. By definition, the null spaces $\mathcal{N}_K$ and $\mathcal{N}_{K^\dagger}$ of $\mathbf{K}$ and $\mathbf{K}^\dagger$ are subspaces of $\mathcal{H}$ spanned by the linearly independent solutions of

$$\mathbf{K}\mathbf{x} = \mathbf{0} \tag{9.5.96}$$

and

$$\mathbf{K}^\dagger \mathbf{y} = \mathbf{0}. \tag{9.5.97}$$

We note that the solutions of Eqs. (9.5.96) and (9.5.97) are the zero-eigenvalue eigenvectors of $\mathbf{K}$ and $\mathbf{K}^\dagger$, respectively, and these eigenvectors can be orthonormalized. We also note that every vector in $\mathcal{N}_K$ or $\mathcal{N}_{K^\dagger}$ can be expressed as a linear combination of the corresponding orthonormal set. However, these orthonormal sets do not form a complete set because $\mathbf{K}$ and $\mathbf{K}^\dagger$ have at least one nonzero eigenvalue unless $\mathbf{K} = \mathbf{K}^\dagger = \mathbf{0}$.

We consider now all vectors $\mathbf{x}$ and $\mathbf{y}$ in $\mathcal{H}$ such that $\|\mathbf{x}\| = \|\mathbf{y}\| = 1$. Since $\mathbf{K}$ is bounded, it follows that

$$|\langle \mathbf{y}, \mathbf{K}\mathbf{x} \rangle|^2 \leq B, \tag{9.5.98}$$

where B, the least upper bound of $|\langle \mathbf{y}, \mathbf{K}\mathbf{x} \rangle|^2$, is positive and finite. From Eq. (9.5.98) it follows that there exist weakly convergent sequences $\mathbf{x}^{(n)}$ and $\mathbf{y}^{(n)}$ such that

$$\lim_{n \to \infty} |\langle \mathbf{y}^{(n)}, \mathbf{K}\mathbf{x}^{(n)} \rangle|^2 = B. \tag{9.5.99}$$

We denote the limits of $\mathbf{x}^{(n)}$ and $\mathbf{y}^{(n)}$ by $\mathbf{x}_1$ and $\mathbf{y}_1$, i.e., $\lim \mathbf{x}^{(n)} = \mathbf{x}_1$ and $\lim \mathbf{y}^{(n)} = \mathbf{y}_1$ (recall that weakly converging sequences—Cauchy sequences—converge to a limit in a Hilbert space). Because $\mathbf{K}$ is completely continuous, it follows that

$$\lim_{n \to \infty} |\langle \mathbf{y}^{(n)}, \mathbf{K}\mathbf{x}^{(n)} \rangle|^2 = |\langle \mathbf{y}_1, \mathbf{K}\mathbf{x}_1 \rangle|^2 = B. \tag{9.5.100}$$

Without loss of generality, we can assume that $\mathbf{x}_1$ is orthogonal to all of the vectors in the null space $\mathcal{N}_K$, and $\mathbf{y}_1$ is orthogonal to all of the vectors in the null space $\mathcal{N}_{K^\dagger}$. We now set

$$\langle \mathbf{y}_1, \mathbf{K}\mathbf{x}_1 \rangle = \kappa_R + i\kappa_I = \kappa, \tag{9.5.101}$$

where κ_R and κ_I are the real and imaginary parts of κ. It immediately follows that $B = \kappa_R^2 + \kappa_I^2$. Consider next a normalized vector $\tilde{\mathbf{x}}$ that is orthogonal to $\mathbf{x}_1$ and the null vectors of $\mathbf{K}$. We assume that $\langle \mathbf{y}, \mathbf{K}\tilde{\mathbf{x}} \rangle \neq 0$ and set

$$\langle \mathbf{y}_1, \mathbf{K}\tilde{\mathbf{x}} \rangle = \mu_R + i\mu_I = \mu. \tag{9.5.102}$$

Choosing c_1 and c_2 such that $|c_1|^2 + |c_2|^2 = 1$, we consider the vector $c_1 \mathbf{x}_1 + c_2 \tilde{\mathbf{x}}$, noting that $\|c_1 \mathbf{x}_1 + c_2 \tilde{\mathbf{x}}\|^2 = |c_1|^2 \|\mathbf{x}_1\|^2 + |c_2|^2 \|\tilde{\mathbf{x}}\|^2 = 1$, and, therefore,

$$|\langle \mathbf{y}_1, \mathbf{K}(c_1 \mathbf{x}_1 + c_2 \tilde{\mathbf{x}}) \rangle|^2 = |c_1 \kappa + c_2 \mu|^2 = \mathbf{c}^\dagger \mathbf{H} \mathbf{c}, \tag{9.5.103}$$

where

$$\mathbf{c} = \begin{bmatrix} c_1 \\ c_2 \end{bmatrix} \quad \text{and} \quad \mathbf{H} = \begin{bmatrix} |\kappa|^2 & \mu^*\kappa \\ \mu\kappa^* & |\mu|^2 \end{bmatrix}. \tag{9.5.104}$$

The maximum value of $\mathbf{c}^\dagger \mathbf{H} \mathbf{c}$ is attained when $\mathbf{c}$ is the normalized eigenvector $\mathbf{c}_m$ of $\mathbf{H}$ corresponding to the largest eigenvalue $\lambda_m (= |\kappa|^2 + |\mu|^2)$. Setting $\mathbf{c} = \mathbf{c}_m$, we obtain

$$|\langle \mathbf{y}_1, \mathbf{K}(c_1 \mathbf{x}_1 + c_2 \tilde{\mathbf{x}}) \rangle|^2 = \lambda_m \|\mathbf{c}_m\|^2 = |\kappa|^2 + |\mu|^2. \tag{9.5.105}$$

However, we already know that

$$|\langle \mathbf{y}_1, \mathbf{A}\mathbf{x} \rangle|^2 \leq |\kappa|^2, \qquad \|\mathbf{x}\| = 1, \tag{9.5.106}$$

and so it follows that μ must be equal to 0, and, therefore,

$$\langle \mathbf{y}_1, \mathbf{K}\tilde{\mathbf{x}} \rangle = 0. \tag{9.5.107}$$

This proves that Eq. (9.5.107) is true for all vectors $\tilde{\mathbf{x}}$ in $\mathcal{H}$ that are orthogonal to $\mathbf{x}_1$ and to the vectors in the null space $\mathcal{N}_K$ of $\mathbf{K}$.

We can rewrite Eq. (9.5.101) as

$$\langle \mathbf{K}^\dagger \mathbf{y}_1, \mathbf{x}_1 \rangle = \kappa \tag{9.5.108}$$

using the general property, $\langle \mathbf{K}^\dagger \mathbf{v}, \mathbf{u} \rangle = \langle \mathbf{v}, \mathbf{K}\mathbf{u} \rangle$, of the adjoint operator, and prove by similar considerations that

$$\langle \mathbf{K}^\dagger \tilde{\mathbf{y}}, \mathbf{x}_1 \rangle = 0 \tag{9.5.109}$$

for all vectors $\tilde{\mathbf{y}}$ orthogonal to $\mathbf{y}_1$ and to the vectors in the null space $\mathcal{N}_{K^\dagger}$ of $\mathbf{K}^\dagger$.

Next, we assume that

$$\mathbf{K}\mathbf{x}_1 - \kappa \mathbf{y}_1 = \mathbf{x}. \tag{9.5.110}$$

It follows that, for any vector $\mathbf{y}$ in $\mathcal{N}_{K^\dagger}$ (all vectors such that $\mathbf{K}^\dagger \mathbf{y} = \mathbf{0}$),

$$\begin{aligned} \langle \mathbf{y}, \mathbf{x} \rangle &= \langle \mathbf{y}, \mathbf{K}\mathbf{x}_1 \rangle - \kappa \langle \mathbf{y}, \mathbf{y}_1 \rangle \\ &= \langle \mathbf{K}^\dagger \mathbf{y}, \mathbf{x}_1 \rangle = 0. \end{aligned} \tag{9.5.111}$$

Also,

$$\begin{aligned} \langle \mathbf{y}_1, \mathbf{x} \rangle &= \langle \mathbf{y}_1, \mathbf{K}\mathbf{x}_1 \rangle - \kappa \langle \mathbf{y}_1, \mathbf{y}_1 \rangle \\ &= \kappa - \kappa = 0, \end{aligned} \tag{9.5.112}$$

and from Eq. (9.5.109) it follows that

$$\begin{aligned} \langle \tilde{\mathbf{y}}, \mathbf{x} \rangle &= \langle \tilde{\mathbf{y}}, \mathbf{K}\mathbf{x}_1 \rangle - \kappa \langle \tilde{\mathbf{y}}, \mathbf{y}_1 \rangle \\ &= \langle \mathbf{K}^\dagger \tilde{\mathbf{y}}, \mathbf{x}_1 \rangle = 0, \end{aligned} \tag{9.5.113}$$

where $\tilde{\mathbf{y}}$ is any vector orthogonal to $\mathbf{y}_1$ and to the vectors in $\mathcal{N}_{K^\dagger}$.

From Eqs. (9.5.111)–(9.5.113), we can conclude that $\mathbf{x}$ is orthogonal to an orthonormal basis set in the Hilbert space $\mathcal{H}$. Therefore, $\mathbf{x} = \mathbf{0}$, and

$$\mathbf{K}\mathbf{x}_1 = \kappa \mathbf{y}_1. \tag{9.5.114}$$

Similar considerations can be used to prove that

$$\mathbf{K}^\dagger \mathbf{y}_1 = \kappa^* \mathbf{x}_1. \tag{9.5.115}$$

Finally, if we express κ in the form $\kappa_1 \exp(i\theta)$, where $\kappa_1 = |\kappa|$, and we replace the unit vectors $\mathbf{x}_1$ and $\mathbf{y}_1$ with the unit vectors $\exp(-i\theta/2)\mathbf{x}_1$ and $\exp(i\theta/2)\mathbf{y}_1$, Eqs. (9.5.114) and (9.5.115) take the form

$$\mathbf{K}\mathbf{x}_1 = \kappa_1 \mathbf{y}_1 \quad \text{and} \quad \mathbf{K}^\dagger \mathbf{y}_1 = \kappa_1 \mathbf{x}_1, \tag{9.5.116}$$

where $\|\mathbf{x}_1\| = \|\mathbf{y}_1\| = 1$ and $0 < \kappa_1 < \infty$. Note that, for the vectors $\exp(-i\theta/2)\mathbf{x}_1$ and $\exp(i\theta/2)\mathbf{y}_1$, we have again used the symbols $\mathbf{x}_1$ and $\mathbf{y}_1$.

The procedure used above can be used again to find the maximum of $|\langle \mathbf{y}, \mathbf{K}\mathbf{x}\rangle|^2$ among the normalized vectors $\mathbf{x}$ and $\mathbf{y}$ in $\mathcal{H}$ such that $\mathbf{x}$ is orthogonal to $\mathbf{x}_1$ and the vectors in $\mathcal{N}_K$, and $\mathbf{y}$ is orthogonal to $\mathbf{y}_1$ and the vectors in $\mathcal{N}_{K^\dagger}$. The result is that there exist vectors $\mathbf{x}_2$ and $\mathbf{y}_2$ such that $\langle \mathbf{y}_2, \mathbf{K}\mathbf{x}_2\rangle = \kappa_2$, where $\kappa_2 \leq \kappa_1$ and

$$\mathbf{K}\mathbf{x}_2 = \kappa_2 \mathbf{y}_2 \quad \text{and} \quad \mathbf{K}^\dagger \mathbf{y}_2 = \kappa_2 \mathbf{x}_2. \tag{9.5.117}$$

It is, of course, possible that $\kappa_1 = \kappa_2$, but there will only be a finite number of pairs $(\mathbf{x}_i, \mathbf{y}_i)$ for which this is true.

Continuation of the procedure will yield the orthonormal sets $\{\mathbf{x}_1, \mathbf{x}_2, \ldots\}$ and $\{\mathbf{y}_1, \mathbf{y}_2, \ldots\}$ such that

$$\mathbf{K}\mathbf{x}_i = \kappa_i \mathbf{y}_i \quad \text{and} \quad \mathbf{K}^\dagger \mathbf{y}_i = \kappa_i \mathbf{x}_i, \qquad \kappa_1 \geq \kappa_2 \geq \kappa_3 \geq \cdots. \tag{9.5.118}$$

The vectors $\mathbf{x}_1, \mathbf{x}_2, \ldots$ constitute a basis set for all of the vectors in $\mathcal{H}$ that are orthogonal to $\mathcal{N}_K$. Likewise, the vectors $\mathbf{y}_1, \mathbf{y}_2, \ldots$ constitute a basis set for all of the vectors in $\mathcal{H}$ that are orthogonal to $\mathcal{N}_{K^\dagger}$. The union of $\{\mathbf{x}_i\}$ and an orthonormal set $\{\mathbf{x}_i^0\}$, spanning $\mathcal{N}_K$, forms an orthonormal basis set for $\mathcal{H}$ as does the union of $\{\mathbf{y}_i\}$ and an orthonormal set $\{\mathbf{y}_i^0\}$, spanning $\mathcal{N}_{K^\dagger}$. Thus, if we define the set $\boldsymbol{\phi}_1, \boldsymbol{\phi}_2, \ldots$ to be the set $\mathbf{x}_1, \mathbf{x}_2, \ldots, \mathbf{x}_1^0, \mathbf{x}_2^0, \ldots$ and the set $\boldsymbol{\psi}_1, \boldsymbol{\psi}_2, \ldots$ to be the set $\mathbf{y}_1, \mathbf{y}_2, \ldots, \mathbf{y}_1^0, \mathbf{y}_2^0, \ldots$, we have proved Eqs. (9.5.92) and (9.5.93), and since

$$\mathbf{I} = \sum_i \boldsymbol{\phi}_i \boldsymbol{\phi}_i^\dagger, \tag{9.5.119}$$

it follows from $\mathbf{K} = \mathbf{K}\mathbf{I} = \sum_i \mathbf{K}\boldsymbol{\phi}_i \boldsymbol{\phi}_i^\dagger = \sum_i \kappa_i \boldsymbol{\psi}_i \boldsymbol{\phi}_i^\dagger$ that Eq. (9.5.91) is true. This completes the proof of the theorem.

As an illustration of the use of Schmidt's form for $\mathbf{K}$, consider the integral equation

$$\mathbf{K}\mathbf{u} = \mathbf{f}. \tag{9.5.120}$$

This problem has a solution only if

$$\mathbf{f} = {\sum_i}' \alpha_i \boldsymbol{\psi}_i, \qquad (9.5.121)$$

where the prime on $\sum'$ means that the sum is restricted to the functions $\boldsymbol{\phi}_i$ for which $\mathbf{K}\boldsymbol{\phi}_i \neq \mathbf{0}$. When this condition is obeyed, the solution to Eq. (9.5.118) is

$$\mathbf{u} = {\sum_i}' \frac{\langle \boldsymbol{\psi}_i, \mathbf{f} \rangle}{\kappa_i} \boldsymbol{\phi}_i + {\sum_i}'' c_i \boldsymbol{\phi}_i, \qquad (9.5.122)$$

where the c_i are arbitrary and the double prime on $\sum''$ indicates that the sum is over the basis functions in $\mathcal{N}_K$. The quantities $\boldsymbol{\phi}_i$, $\boldsymbol{\psi}_i$, and κ_i can be computed from the self-adjoint equations (9.5.94) and (9.5.93). Another application of the theorem will be given in the next section.

9.5.7. Completely Continuous Normal Operators

Again, let us begin by stating the main theorem of interest in this section:

THEOREM. *If* $\mathbf{K}$ *is a completely continuous integral operator in a Hilbert space and is normal, i.e.,* $\mathbf{K}\mathbf{K}^\dagger = \mathbf{K}^\dagger\mathbf{K}$, *then the eigenvectors* $\boldsymbol{\phi}_1, \boldsymbol{\phi}_2, \ldots$ *of* $\mathbf{K}$ *form a complete orthonormal basis set in* $\mathcal{H}$.

The theorem is actually valid for any completely continuous, normal, linear operator in a Hilbert space. For example, $\mathbf{K}$ could be a matrix operator in E_∞. As noted before, a linear operator in a function space can be mapped into a matrix operator in E_∞. As was the case for self-adjoint operators, this theorem immediately leads to the spectral resolution for the operator:

$$\mathbf{K} = \sum_i \lambda_i \boldsymbol{\phi}_i \boldsymbol{\phi}_i^\dagger, \qquad (9.5.123)$$

and for a function $f(\mathbf{K})$ of the operator:

$$f(\mathbf{K}) = \sum_i f(\lambda_i) \boldsymbol{\phi}_i \boldsymbol{\phi}_i^\dagger, \qquad (9.5.124)$$

as long as the function $f(t)$ is defined for $t = \lambda_i$, $i = 1, 2, \ldots$.

The proof of the theorem follows easily from Schmidt's normal form of $\mathbf{K}$. We first note that $\mathbf{K} = \sum_i \kappa_i \boldsymbol{\psi}_i \boldsymbol{\phi}_i^\dagger$, where $\boldsymbol{\phi}_i$ and $\boldsymbol{\psi}_i$ satisfy the equations $\mathbf{K}^\dagger \mathbf{K} \boldsymbol{\phi}_i = \kappa_i^2 \boldsymbol{\phi}_i$ and $\mathbf{K}\mathbf{K}^\dagger \boldsymbol{\psi}_i = \kappa_i^2 \boldsymbol{\psi}_i$. Since $\mathbf{K}^\dagger\mathbf{K} = \mathbf{K}\mathbf{K}^\dagger$ for normal operators, $\boldsymbol{\phi}_i$ and $\boldsymbol{\psi}_i$ can differ only by a multiplier of modulus 1, i.e., $\boldsymbol{\psi}_i = \exp(i\theta_i)\boldsymbol{\phi}_i$, where θ_i is real. Thus, Schmidt's normal form becomes

$$\mathbf{K} = \sum_i \kappa_i \exp(i\theta_i) \boldsymbol{\phi}_i \boldsymbol{\phi}_i^\dagger, \qquad (9.5.125)$$

where $\{\boldsymbol{\phi}_i\}$ is an orthonormal set and the eigenvalues of $\mathbf{K}$ are $\lambda_i = \kappa_i \exp(i\theta_i)$. We note that, in the case of a normal operator, λ_i can be either real or imaginary. Also, the vectors $\{\boldsymbol{\phi}_i\}$ in Eq. (9.5.92), plus an orthonormal set spanning the null space $\mathcal{N}_K$, form an orthonormal basis set in the Hilbert space $\mathcal{H}$—completing the proof.

The spectral resolution theorem for self-adjoint operators is, of course, a subcase of the theorem for normal operators, and so, in actuality, Section 9.5.5 could have been skipped.

EXAMPLE 9.5.5. Consider the equation

$$\frac{d\mathbf{u}}{dt} = i\mathbf{K}\mathbf{u}, \qquad \mathbf{u}(t=0) = \mathbf{u}_0, \qquad (9.5.126)$$

where $i = \sqrt{-1}$ and $\mathbf{K}$ is a completely continuous, normal operator in $\mathcal{H}$. Assume that the imaginary part of the eigenvalue λ_n goes as $1/n^{1/2}$ and examine the asymptotic behavior of $\mathbf{u}(t)$.

The formal solution to Eq. (9.5.126) is

$$\mathbf{u} = \exp(it\mathbf{K})\mathbf{u}_0 = \sum_n \exp(it\lambda_n)\langle \boldsymbol{\phi}_n, \mathbf{u}_0\rangle \boldsymbol{\phi}_n. \qquad (9.5.127)$$

The length of $\mathbf{u}$ is given by

$$\|\mathbf{u}\|^2 = \sum_n \exp(it(\lambda_n - \lambda_n^*))|\langle \boldsymbol{\phi}_n, \mathbf{u}_0\rangle|^2$$

$$= \sum_n \exp\left(\frac{-2t}{n^{1/2}}\right)|\langle \boldsymbol{\phi}_n, \mathbf{u}_0\rangle|^2 \qquad (9.5.128)$$

$$\leq \sum_n |\langle \boldsymbol{\phi}_n, \mathbf{u}_0\rangle|^2 = \|\mathbf{u}_0\|^2.$$

Thus, $\mathbf{u}$ remains a vector in the Hilbert space $\mathcal{H}$ as $t \to \infty$. In fact, its length is bounded by the initial length $\|\mathbf{u}_0\|$.

PROBLEMS

1. (a) What are the eigenvalues and eigenvectors of the integral $\mathbf{K}$ whose kernel is

$$k(t,s) = \sum_{n=1}^{\infty} \frac{\sin nt \sin ns}{n}$$

   in the Hilbert space $\mathcal{L}_2(0, \pi)$? Note that $\mathbf{K}$ is self-adjoint.

   (b) Give the solution to the problem

$$\frac{d\mathbf{u}}{dt} = -\mathbf{K}\mathbf{u},$$

   where

$$u(t=0) = 1.$$

2. (a) Convert the equation

$$\frac{d^2u}{dt^2} + \int_0^1 \sin 2(s-t)u(s)\,ds = \cos t,$$

   with $u(0) = du(0)/dt = 0$, into an integral equation.

   (b) Solve the equation.

PROBLEMS

3. The exact solution of the equation

$$u(t) + \int_0^t \exp((t-s))u(s)\,ds = \exp(t)$$

is

$$u(t) = 1.$$

Use the trapezoidal method to find an approximation $\tilde{u}(t)$ to the equation. Find the discretization interval Δt below which

$$\|\mathbf{u} - \tilde{\mathbf{u}}\|^2 < 10^{-4},$$

where $\|\mathbf{u}\|^2 = \int_0^{10} (u(t))^2\,dt$.

4. Use the method of Laplace transforms to solve the equation

$$u(t) - \int_0^t \sin(t-s)u(s)\,ds = 1.$$

5. Consider the Abel-type equation of the form

$$\int_0^t (t-s)^{-\alpha} h(t,s) u(s)\,ds = f(t), \qquad 0 < \alpha < 1,$$

where $h(t,s)$ is a continuous function. If the variable s is discretized such that $t_i \le s \le t_{i+1}$, and $h(t,s)u(s)$ is approximated in the interval (t_i, t_{i+1}) as $k(t, t_{i+1/2})u(t_{i+1/2})$, the integral equation becomes

$$\sum_{i=0}^{n-1} w_{ni} h(t_n, t_{i+1/2}) u_{i+1/2} = f(t_n), \qquad n = 1, 2, \ldots,$$

where

$$w_{ni} = \int_{t_i}^{t_{i+1}} (t_n - s)^{-\alpha}\,ds$$

and $u_{i+1/2} = u(t = \tfrac{1}{2}(t_i + t_{i+1}))$. This is known as the midpoint method. Assume that $h(t,s) = 1 + s$, $\alpha = \tfrac{1}{2}$, $f = \tfrac{16}{15} t^{5/2} + \tfrac{4}{3} t^{3/2}$, and $t_{i+1} - t_i = h$. Solve the equation for $h = \tfrac{1}{10}, \tfrac{1}{50}$, and $\tfrac{1}{100}$ for $0 < t < 2$. Plot the midpoint solutions versus the analytical solution, which is $u(t) = t$ (verify).

6. Prove that the operator

$$\mathbf{K}u = \int_{-1}^{1} \frac{1}{(t-s)^\alpha} u(s)\,ds, \qquad 0 < \alpha < 1,$$

is completely continuous in $\mathcal{L}_2(-1, 1)$.

7. Prove that the operator

$$\mathbf{K}u = \int_{-1}^{1} \frac{h(t,s)}{(t-s)^\alpha} u(s)\,ds, \qquad 0 \le \alpha < 1,$$

is completely continuous in $\mathcal{L}_2(-1, 1)$, where $h(t,s)$ is a continuous function with respect to t and s in the closed interval $[-1, 1]$.

8. Suppose $u(x)$ satisfies the equation

$$\frac{du}{dx} = \alpha u + \beta \int_0^x \exp(-(x-y))u(y)\,dy,$$

where $u(0) = \gamma$. Integrate the equation to obtain the Volterra equation

$$u(x) = \gamma + \int_0^x k(x-y)u(y)\,dy. \tag{$*$}$$

What is $k(x-y)$? Solve $(*)$ by the method of Laplace transforms for $\alpha = \beta = \gamma = 1$.

9. In a certain dielectric material, the electric displacement $D(t)$ is related to the electric field $E(t)$ by

$$D(t) = \epsilon E(t) + \int_0^t \phi(t-s)E(s)\,ds,$$

where $D(t=0) = \epsilon E_0$. ϵ is the dielectric constant and $\phi(t-s)$ is the "memory function" of the material. If the memory is exponential, i.e.,

$$\phi(t) = \alpha \exp\left(\frac{-t}{\tau_r}\right),$$

use the method of Laplace transforms to find $E(t)$ as a function of $D(t)$. Suppose $\epsilon = 10$, $\alpha = 5$, and $\tau_r = 1$ and assume that $E(t) = E_0 \sin \pi t$. Plot D/E_0 and E/E_0 versus t.

10. Consider the equation

$$\frac{d\mathbf{x}}{dt} = \lambda \mathbf{x} + \int_0^t \mathbf{A}(t-s)\mathbf{x}(s)\,ds, \tag{$*$}$$

where $\mathbf{x}(t)$ is an n-dimensional vector function of t and $\mathbf{A}(t)$ is an $n \times n$ matrix function of t. Suppose

$$\mathbf{x}(t=0) = \begin{bmatrix} x_1^0 \\ \vdots \\ x_n^0 \end{bmatrix}$$

and

$$a_{ij}(t) = \alpha_{ij}\exp(-\nu_{ij}t),$$

where α_{ij} and ν_{ij} are constants ($\nu_{ij} > 0$). Moreover, assume that $\mathbf{A}$ is real and $\alpha_{ij} = \alpha_{ji}$ and $\nu_{ij} = \nu_{ji}$.

Use the method of Laplace transforms to find the formal solution to $(*)$. For the special case $n = 2$, $\alpha_{11} = \alpha_{22} = -2$, $\alpha_{12} = \alpha_{21} = \nu_{ij} = 1$, $\alpha = 1$, and $x_1^0 = x_2^0 = 1$. Find $x_1(t)$ and $x_2(t)$. Plot the results for $0 < t < 5$.

11. Consider the following kernels of self-adjoint operators in $\mathcal{L}_2(-\pi, \pi)$:

(a) $k(t, s) = 4\cos(t-s)$.
(b) $k(t, s) = 1 + \cos(t-s)$.

(c) $k(t, s) = 1 + \cos(t - s) + \sin 2(t + s)$.
(d) $k(t, s) = \sin 3(t - s)$.

Determine all of the nonzero eigenvalues and their eigenvectors for each of these operators. Give the spectral decomposition of operator $\mathbf{K}$. Give the spectral decomposition of $\exp(\alpha \mathbf{K})$.

12. Find the Schmidt's normal form for the operator $\mathbf{K}$ in $\mathcal{L}_2(-\pi, \pi)$ if the kernel of $\mathbf{K}$ is given by

$$k(t, s) = \sin t \cos s.$$

Give the spectral decomposition of $\mathbf{K}\mathbf{K}^\dagger$ and $\mathbf{K}^\dagger \mathbf{K}$.

13. Solve

$$\frac{d^2 u}{dt^2} + \int_0^1 \sin \alpha (t - s) u(s)\, ds = f(t),$$

with the conditions $u(t) = du(t)/dt = 0$ at $t = 0$. Plot the solution $u(t)$ versus t for $\alpha = 1$ and $f(t) = \exp(-t)$.

14. Consider the operator $\mathbf{K}$ such that

$$\mathbf{K} u = \int_{-1}^1 k(t - s) u(s)\, ds,$$

where $k(t) = t^{-\nu}$, $0 \leq t \leq 1$, and $k(t) = 0$, $t < 0$. Prove that $\mathbf{K}$ is a completely continuous operator in $\mathcal{L}_2(-1, 1)$ when $0 \leq \nu < 1$. *Hint*: Show that $\mathbf{K}$ is a Hilbert–Schmidt operator using the orthonormal basis set $\phi_n = \frac{1}{2} \exp(in\pi t)$, $n = 0, \pm 1, \pm 2, \ldots$, in $\mathcal{L}_2(-1, 1)$.

15. Consider the operator $\mathbf{K}$ with the kernel

$$k(x, y) = \sum_{n=1}^\infty \frac{1}{n^2} \sin(n + 1)x \sin ny, \qquad 0 \leq x, y \leq \pi.$$

(a) Prove that $\mathbf{K}$ is a Hilbert–Schmidt operator and that $\mathbf{K}$ has no eigenvalue.
(b) Solve

$$(\mathbf{K} + \mathbf{I})u = f,$$

where

$$f(x) = x.$$

16. Consider the eigenproblem

$$\int_{-\infty}^\infty k(x, y) u(y)\, dy = \lambda u(x), \qquad -\infty < x < \infty,$$

where

$$k(x, y) = (1 - \nu^2)^{-1/2} \exp\left(\frac{x^2 + y^2}{2}\right) \exp\left(-\frac{x^2 + y^2 - 2\nu x y}{1 - \nu^2}\right),$$

where ν is a real number lying between 0 and 1.

(a) Show by substitution that

$$u_0(x) = \exp\left(\frac{-x^2}{2}\right)$$

is an eigenfunction corresponding to the eigenvalue $\lambda_0 = \sqrt{\pi}$.

(b) Let

$$u_n(x) = \exp\left(\frac{-x^2}{2}\right) H_n(x),$$

where $H_n(x)$ is a Hermite polynomial, given by

$$H_n(x) = (-1)^n \exp(x^2) \frac{d^n}{dx^n}\left[\exp(-x^2)\right].$$

Assume that $\mathbf{K}\mathbf{u}_n = \lambda_n \mathbf{u}_n$ and show that $\mathbf{K}\mathbf{u}_{n+1} = \lambda_{n+1}\mathbf{u}_{n+1}$, where $\lambda_{n+1} = \nu\lambda_n$. Thus, by induction, establish that $\mathbf{u}_n$, $n = 0, 1, 2, \ldots$, are eigenfunctions of $\mathbf{K}$ of eigenvalue $\lambda_n = \nu^n \sqrt{\pi}$.

(c) Prove that

$$\int_{-\infty}^{\infty} u_n^2\, dx = 2^n n! \sqrt{\pi}$$

and give the spectral decomposition of $\mathbf{K}$ (the vectors $\{\mathbf{u}_n\}$ form an orthogonal basis set in $\mathcal{L}_2(-\infty, \infty)$).

17. Consider the self-adjoint operator $\mathbf{K}$ in $\mathcal{L}_2(-\infty, \infty)$ whose kernel is

$$k(t, s) = \frac{1}{\sqrt{\pi}} \exp\left(\frac{t^2 + s^2}{2}\right) \int_{-\infty}^{t} \exp(-\zeta^2)\,d\zeta \int_{s}^{\infty} \exp(-\xi^2)\,d\xi, \qquad t \leq s.$$

(a) Show that the Hermite functions

$$h_n(t) = \exp\left(\frac{t^2}{2}\right) \frac{d^n \exp(-t^2)}{dt^n}$$

are eigenfunctions of $\mathbf{K}$ with eigenvalues $\lambda_n = 2n + 2$. These Hermite functions form a complete orthogonal basis set in $\mathcal{L}_2(-\infty, \infty)$.

(b) Give the solution to the equation $d\mathbf{u}/dt = -\mathbf{K}\mathbf{u}$, $u(t = 0) = 1$.

18. Show that the eigenfunctions of the self-adjoint operator $\mathbf{K}$ in $\mathcal{L}_2(0, \infty)$ with the kernel

$$k(t, s) = \exp\left(\frac{t + 2}{2}\right) \int_{s}^{\infty} \frac{\exp(-\tau)}{\tau}\, d\tau, \qquad 0 \leq t \leq s,$$

are the orthogonal Laguerre functions

$$\exp\left(\frac{-t}{2}\right) \frac{\partial^n}{\partial z^n}\left[\frac{\exp(-tz/(1 - z))}{1 - z}\right]_{z=0},$$

with eigenvalues $\lambda_n = n + 1$.

FURTHER READING

Akhiezer, N. I. and Glazman, I. M. (1963). "Theory of Linear Operators in Hilbert Space." Vol. II, Ungar, New York.

Akhiezer, N. I. and Glazman, I. M. (1966). "Theory of Linear Operators in Hilbert Space." Vol. I, Ungar, New York.

Courant, R. and Hilbert, D. (1953). "Methods of Mathematical Physics." Interscience Pub., Inc., New York.

Green, C. D. (1969). "Integral Equation Methods." Barnes & Noble, New York.

Hochstadt, H. (1973). "Integral Equations." Interscience, New York.

Korevaar, J. (1968). "Mathematical Methods." Vol. I, Academic, New York.

Linz, P. (1985). "Analytical and Volterra Methods for Volterra Equations." Soc. for Industr. & Appl. Math., Philadelphia.

Lovitt, W. V. (1950). Linear Integral Equations, Dover, New York.

Petrovskii, I. G. (1957). "Integral Equations." Graylock Press, Rochester, New York.

Porter, D. and Stirling, D. S. G. (1990). "Integral Equations, a Practical Treatment, from Spectral Theory to Applications." Cambridge Univ. Press, Cambridge.

Riesz, F. and Nagy, B. S. (1965). "Functional Analysis." Ungar, New York.

Schmeidler, W. (1965). "Linear Operators in Hilbert Space." Academic, New York.

Schwabik, S. and Turdy, M. (1979). "Differential and Integral Equations, Boundary Value Problems," Reidel, Dordrecht.

Smithies, F. (1958). Integral Equations. Cambridge Univ. Press, Cambridge.

10
LINEAR DIFFERENTIAL OPERATORS IN A HILBERT SPACE

10.1. SYNOPSIS

In this chapter we will extend our analysis of operators by considering differential operators in a Hilbert space. The primary emphasis will be on ordinary differential equations and their differential operators. However, much of our analysis can be applied to partial differential equations and operators, for a certain class of problems, and we will touch upon this in the last section of the chapter.

A differential operator differs from the integral operator discussed in Chapter 9 in that it requires specification of boundary conditions. Furthermore, a differential operator **L** has a domain $\mathcal{D}_L$ that is generally a subset of the Hilbert space $\mathcal{H}$ in which it is defined, and a differential operator is always unbounded.

Like integral operators, the adjoint operator of a differential operator exists and is itself a differential operator. We say a differential operator **L** is self-adjoint if its differential expression L is the same as the differential expression $L^\dagger$ of the adjoint operator $\mathbf{L}^\dagger$ *and* if the domain $\mathcal{D}_L$ of **L** is the same as the domain $\mathcal{D}_{L^\dagger}$ of $\mathbf{L}^\dagger$. With the required specification of boundary conditions in $\mathcal{D}_L$ and $\mathcal{D}_{L^\dagger}$, self-adjointness is not as straightforward to assess as it was for integral operators or matrices.

We will show that if L is a pth-order differential expression,

$$Lu = \sum_{i=0}^{p} a_i(x) \frac{d^i u}{dx^i}, \qquad (10.1.1)$$

and $a_i(x)$ are continuous and $a_p(x) \neq 0$, then the inhomogeneous equation $Lu = f$, with the initial conditions

$$\frac{d^i u}{dx^i} = \gamma_i, \quad x = a, \quad i = 0, \ldots, p-1, \tag{10.1.2}$$

has a unique solution for any piecewise continuous function $f(x)$. This is the initial value problem (IVP). The equation

$$Lu = 0, \tag{10.1.3}$$

where L is defined in Eq. (10.1.1), has p linearly independent solutions $\{u_1(x), \ldots, u_p(x)\}$, where $u_j(x)$ corresponds to the initial conditions

$$\frac{d^i u_j}{dx^i} = \gamma_{ij}, \quad x = a, \quad i = 0, \ldots, p-1, \tag{10.1.4}$$

such that the vectors $\boldsymbol{\gamma}_j^{\mathrm{T}} = (\gamma_{0j}, \ldots, \gamma_{p-1,j})$, $j = 1, \ldots, p$, form a linearly independent set. Thus, the functions $\{u_1, \ldots, u_p\}$ are solutions to the homogeneous IVP corresponding to a p linearly independent set of initial values of $u(x)$ and its first $p-1$ derivatives. The set $\{u_i\}$ is referred to as a fundamental system. Fundamental systems are useful for solving boundary value problems and eigenvalue problems.

The boundary value problem (BVP)

$$Lu = f, \quad a < x < b,$$

for a pth-order differential equation places constraints on u and its first $p-1$ derivatives at both ends of the interval $[a, b]$. These conditions can be summarized as

$$B_i u = \gamma_i, \quad i = 1, \ldots, m, \tag{10.1.5}$$

where $B_i u$ is some linear combination of u and its first $p-1$ derivatives at $x = a$ and b. The BVP may have a unique solution, a nonunique solution for certain functions f, or no solution at all. If, for the operator defined in Eq. (10.1.1), $a_i(x) \in C^i(a, b)$, i.e., a_i has a continuous ith derivative, then the following Fredholm alternative theorem holds: the equation

$$Lu = f, \quad B_i u = 0, \quad i = 1, \ldots, m, \tag{10.1.6}$$

has a solution *if and only if*

$$\langle \mathbf{v}, \mathbf{f} \rangle = 0, \tag{10.1.7}$$

where $\mathbf{v}$ is any solution to the homogeneous adjoint equation $L^\dagger v = 0$. If there is no solution to $L^\dagger v = 0$, then the solution to Eq. (10.1.6) is unique. If $L^\dagger v = 0$ has ν solutions, then the solution to Eq. (10.1.6) will be $u_p + \sum_{i=1}^{\nu} \alpha_i u_{h_i}$, where α_i are arbitrary, $Lu_p = f$, and $Lu_{h_i} = 0$, $i = 1, \ldots, \nu$. The functions u_{h_i} are solutions to the homogeneous problem $Lu = 0$, $B_i u = 0$, $i = 1, \ldots, m$. If $B_i u = \gamma_i$, $i = 1, \ldots, m$, then the necessary and sufficient condition for a solution to $Lu = f$ is that $\langle \mathbf{v}, \mathbf{f} \rangle$ equal a particular linear combination of γ_i for each solution v to the homogeneous adjoint equation $L^\dagger v = 0$. Note that this theorem is analogous to the Fredholm theorem in Chapter 4 for finite vector spaces.

We will prove that the inverse of a differential operator **L** is an integral operator **G**, where the kernel of **G** is called Green's function. If the homogeneous equation $Lu = 0$ has no solution for the given boundary conditions, then we will show that **G** exists. For the pth-order differential operator defined in Eq. (10.1.1), the operator **G** is a completely continuous integral operator. This property enables us to use what we learned in the previous chapter for completely continuous integral operators to show that differential operators are often closely analogous to matrix operators with respect to spectral (eigen) properties.

If $g(x, y)$ is Green's function corresponding to the inverse of **L**, then $g^*(y, x)$ is equal to the kernel $g^\dagger(x, y)$ corresponding to the inverse of $\mathbf{L}^\dagger$. We will show that the inverse **G** of **L** always exists for the IVP and it may or may not exist for the BVP. In many cases, **L** can be transformed to an operator that does have an inverse and that has the same spectral (eigen) properties as **L**.

We say that **L** is a perfect operator if its eigenvectors $\{\boldsymbol{\phi}_n\}$ form a complete set. For such an operator, the resolution of the identity is $\mathbf{I} = \sum_n \boldsymbol{\phi}_n \boldsymbol{\psi}_n^\dagger$ and the spectral decomposition of **L** is $\mathbf{L} = \sum_n \lambda_n \boldsymbol{\phi}_n \boldsymbol{\psi}_n^\dagger$, where the set $\{\boldsymbol{\psi}_n\}$ is the reciprocal of the set $\{\boldsymbol{\phi}_n\}$ (i.e., they obey the biorthonormality condition $\langle \boldsymbol{\phi}_m, \boldsymbol{\psi}_n \rangle = \delta_{nm}$) and are the eigenvectors of the adjoint $\mathbf{L}^\dagger$. If $f(t)$ is defined for $t = \lambda_n, n = 1, 2, \ldots$, then the spectral decomposition of $f(\mathbf{L})$ is $f(\mathbf{L}) = \sum_n f(\lambda_n) \boldsymbol{\phi}_n \boldsymbol{\psi}_n^\dagger$. We see that much of our analysis of perfect operators from Chapters 6 and 9 apply also to differential operators. We will show that if **L** is a pth-order differential operator, then the number of eigenvectors corresponding to a given eigenvalue must be less than or equal to p.

If the differential operator **L** is a regular self-adjoint operator, then its eigenvectors $\{\boldsymbol{\phi}_n\}$ form a complete orthonormal set in the Hilbert space of the operator. Furthermore, its eigenvalues are real. We require that the coefficients in the differential expression of a regular differential operator have no singularities, a_p is strictly positive, and the domain of definitions of the variable $\vec{x}$ of $u(\vec{x})$ (interval $[a, b]$ or volume Ω_n) is finite. We will see that some regular operators whose differential expression is normal also have a complete orthonormal set of eigenvalues.

Second-order differential operators, whose boundary conditions make them self-adjoint, are called self-adjoint *Sturm–Liouville operators*. With certain boundary conditions, self-adjoint Sturm–Liouville operators are bounded from below and so the eigenvalues obey the conditions $|\lambda_n| > M < \infty$ and $|\lambda_n| \to \infty$ as $n \to \infty$. We will see that regular self-adjoint *Sturm–Liouville operators* have a complete orthonormal set of eigenvectors. Furthermore, the eigenfunctions of these operators obey an oscillation theorem; i.e., if we label the eigenfunctions by increasing value of their corresponding eigenvalues, the nth eigenfunction will have exactly n zeros in the open interval (a, b).

Singular self-adjoint Sturm–Liouville operators (i.e., operators having singular coefficients in their differential expression, having zeros in $a_p(\vec{x})$, or being defined on an infinite interval $[a, b]$ or volume Ω_n) may have a complete set of eigenfunctions or may possess continuous spectra. In the case of singular self-adjoint operators, the spectral decomposition of **L** is $\mathbf{L} = \sum_m \lambda_m \boldsymbol{\phi}_m \boldsymbol{\phi}_m^\dagger + \int \lambda_k \mathbf{u}_k \mathbf{u}_k^\dagger \, dk$ in one dimension and $\mathbf{L} = \sum_\mathbf{m} \lambda_\mathbf{m} \boldsymbol{\phi}_\mathbf{m} \boldsymbol{\phi}_\mathbf{m} + \int \lambda_{\vec{k}} u_{\vec{k}} u_{\vec{k}}^\dagger \, d^n k$ in n dimensions. Here, the $\boldsymbol{\phi}$

are eigenvectors and the vectors **u** obey $\mathbf{Lu} = \lambda \mathbf{u}$. However, the **u** are not eigenvectors since they do not belong to the Hilbert space. Specific examples are given for the various types of singular self-adjoint operators.

The reason that these self-adjoint Sturm–Liouville operators are given so much attention is that they occur often in problems of heat and mass transfer, structural vibrational problems, fluid mechanics, and the quantum mechanics of particles. Several examples of applications to these problems are given in this chapter.

10.2. THE DIFFERENTIAL OPERATOR

Matrices in finite-dimensional vector spaces, and the classes of integral operators of the types considered in Chapter 9, form linear operators for any vector in the respective Hilbert spaces. For this reason, we have not yet made a distinction between the domain $\mathcal{D}_L$ of an operator **L** and the Hilbert space $\mathcal{H}$ in which the operator is defined. The domain $\mathcal{D}_L$ is composed of the subset of vectors **u** in $\mathcal{H}$ for which the object **Lu** is defined. For example, if the differential expression for the operator **L** is

$$Lu = -\frac{d^2 u(x)}{dx^2} + q(x)u(x) \qquad (10.2.1)$$

and $\mathcal{H} = \mathcal{L}_2(a, b)$, then the domain of **L** will include only those functions in $\mathcal{L}_2(a, b)$ that are twice differentiable. If we further restrict the domain of functions to those for which **Lu** belongs to the Hilbert space $\mathcal{L}_2(a, b)$, then $\mathcal{D}_L$ contains only functions such that $\|\mathbf{Lu}\| < \infty$. Thus, for the example given by Eq. (10.2.1), we required that

$$\langle \mathbf{Lu}, \mathbf{Lu} \rangle = \int_a^b \left(-\frac{d^2 u^*}{dx^2} + q^* u^* \right) \left(-\frac{d^2 u}{dx^2} + qu \right) dx < \infty. \qquad (10.2.2)$$

The definition of an operator is completed by specifying the boundary conditions that the functions $u(x)$ obey.

Consider the inhomogeneous equation

$$Lu(x) = f(x), \qquad a < x < b, \qquad (10.2.3)$$

where the interval $[a, b]$ is finite and L is a pth-order differential expression of the form

$$Lu(x) = a_p(x) \frac{d^p u(x)}{dx_p} + a_{p-1}(x) \frac{d^{p-1} u(x)}{dx^{p-1}} + \cdots$$
$$+ a_1(x) \frac{du(x)}{dx} + a_0(x)u(x). \qquad (10.2.4)$$

In this chapter we assume that the boundary conditions will be set by fixing the values of linear functionals of the form

$$B_1 u \equiv \sum_{j=1}^{p} \alpha_{1j} u^{(j-1)}(a) + \sum_{j=1}^{p} \alpha_{1, p+j} u^{(j-1)}(b)$$
$$\vdots$$
$$B_m u \equiv \sum_{j=1}^{p} \alpha_{mj} u^{(j-1)}(a) + \sum_{j=1}^{p} \alpha_{m, p+j} u^{(j-1)}(b),$$
(10.2.5)

where $u^{(i)}\xi, \xi = a, b$, denotes the ith derivative of u (i.e., $d^i u/d\xi^i$, evaluated at $\xi = a$ and b). We assume that these functionals are linearly independent, which requires that the rank of the matrix

$$\mathbf{B} = \begin{bmatrix} \alpha_{11} & \cdots & \alpha_{1, 2p} \\ \vdots & & \\ \alpha_{m1} & \cdots & \alpha_{m, 2p} \end{bmatrix}$$
(10.2.6)

is m, or, equivalently, that the row vectors

$$\boldsymbol{\rho}_i^T = [\alpha_{i1}, \ldots, \alpha_{i, 2p}], \qquad i = 1, \ldots, m,$$
(10.2.7)

are linearly independent. If the boundary conditions are set by $B_i u = \gamma_i$, $i = 1, \ldots, m$, where the γ_i are given numbers, then the domain of the operator defined by Eqs. (10.2.4) and (10.2.5) is expressed as

$$\mathcal{D}_L = \{\mathbf{u}, \mathbf{L}\mathbf{u} \in \mathcal{H}; \; B_i u = \gamma_i, \; i = 1, \ldots, m\}.$$
(10.2.8)

A pth-order differential equation with p linearly independent boundary conditions is called a "balanced problem." Only for a balanced problem is there any hope of obtaining a unique solution. On the other hand, an unbalanced problem can have a solution and a balanced one can fail to have one. For example, the equation

$$-\frac{d^2 u}{dx^2} = x,$$
(10.2.9)

with no conditions on u, i.e., $m = 0$, has the solution

$$u(x) = \alpha + \beta x - \frac{x^3}{6}$$

for arbitrary α and β. Thus, there are many solutions to this problem. However, the balanced problem

$$-\frac{d^2 u(x)}{dx^2} = x, \qquad u'(0) = u'(1) = 0$$
(10.2.10)

has no solution because the boundary condition $u'(0) = 0$ implies $\beta = 0$, whereas the condition $u'(1) = 0$ implies $\beta = \frac{1}{2}$, a contradiction. On the other hand, the balanced problem

$$-\frac{d^2 u(x)}{dx^2} = x, \qquad u(0) = u(1) = 0$$
(10.2.11)

has the unique solution

$$u(x) = \frac{1}{6}x(1-x^2). \tag{10.2.12}$$

It is important to realize that a differential operator **L** is defined by its differential expression L and its domain $\mathcal{D}_L$. For example, even though L is the same in the following cases, the operator **L** is different in each case:

(i) $\quad Lu = -\dfrac{d^2u}{dx^2}, \qquad u(0) = 0 \text{ and } u(1) = 0, \tag{10.2.13}$

(ii) $\quad Lu = -\dfrac{d^2u}{dx^2}, \qquad u(0) = u(1) \text{ and } u'(1) = 0, \tag{10.2.14}$

(iii) $\quad Lu = -\dfrac{d^2u}{dx^2}, \qquad u(0) = u(1) \text{ and } u'(0) = u'(1), \tag{10.2.15}$

(iv) $\quad Lu = -\dfrac{d^2u}{dx^2}, \qquad u(0) = 0 \text{ and } u'(1) = 0, \tag{10.2.16}$

where $u'(x) = du(x)/dx$.

In addition to the need to identify the domain $\mathcal{D}_L$ of a differential operator as a subset of a Hilbert space $\mathcal{H}$, the differential operator differs in another fundamental way from integral operators. Whereas some classes of integral operators are bounded, there are no bounded differential operators. A simple example illustrates this point. Suppose

$$Lu = -\frac{d^2u}{dx^2}, \qquad u(0) = u(1) = 0. \tag{10.2.17}$$

Then the sequence of functions

$$u_n = \frac{n^{3/2}}{2^{1/2}} x(x-1) \exp\left(\frac{-nx}{2}\right) \tag{10.2.18}$$

belongs to the domain $\mathcal{D}_L$ defining the operator **L** in $\mathcal{L}_2(0, 1)$. It is straightforward to show that

$$\|\mathbf{u}_n\|^2 \to 1 \qquad \text{as } n \to \infty \tag{10.2.19}$$

and

$$\|\mathbf{L}\mathbf{u}_n\|^2 \simeq \frac{n^4}{16} \to \infty \qquad \text{as } n \to \infty. \tag{10.2.20}$$

Thus, since

$$\max_{\mathbf{u} \in \mathcal{D}_L} \frac{\|\mathbf{L}\mathbf{u}\|}{\|\mathbf{u}\|} = \infty, \tag{10.2.21}$$

the norm of **L** is unbounded.

If **L** is a pth-order differential operator, it can be easily proven that **L** is unbounded. Consider the sequence of functions

$$u_n(x) = h_n(x) \exp\left(\frac{-n(x-a)}{2}\right) \qquad (10.2.22)$$

in $\mathcal{D}_L$, where $\mathcal{D}_L \subset \mathcal{L}_2(a, b)$ and the interval $[a, b]$ is finite. Assume, moreover, that the functions $h_n(x)$ are chosen such that the functions u_n obey the boundary conditions of **L** and that

$$\|\mathbf{u}_n\|^2 = 1. \qquad (10.2.23)$$

Such functions can always be constructed, as we did in Eq. (10.2.18). For this sequence of functions, it follows that

$$\frac{\|\mathbf{Lu}\|^2}{\|\mathbf{u}\|^2} \to \infty \qquad \text{(as } n^{2p}\text{) as } n \to \infty, \qquad (10.2.24)$$

proving that any pth-order differential operator is unbounded in a finite interval $[a, b]$.

A similar proof can be constructed if the interval is infinite, say $[-\infty, \infty]$. In the Hilbert space $\mathcal{L}_2(-\infty, \infty)$, the boundary conditions on the vectors in the domain of a differential operator are that $\|\mathbf{u}\| < \infty$ and $\|\mathbf{Lu}\| < \infty$ if $\mathbf{u} \in \mathcal{D}_L$. We can construct functions

$$u_n(x) = h_n(x)\exp(-nx^2), \qquad n = 1, 2, \ldots, \qquad (10.2.25)$$

where $\|\mathbf{u}_n\|^2 = 1$, $h_n(x)$ has the necessary differentiability, and $|h_n(x)| < \exp(-x^2)$ as $|x| \to \infty$. These functions belong to $\mathcal{D}_L$ and it is straightforward to show that

$$\|\mathbf{Lu}\|^2 \to \infty \qquad \text{(as } n^p\text{) as } n \to \infty. \qquad (10.2.26)$$

Thus, we see that all differential operators are unbounded.

To anticipate what we will learn later, let us consider a differential operator **L** whose eigenvectors $\boldsymbol{\phi}_n$, $n = 1, 2, \ldots$, form a complete orthonormal set. Then the resolution of the identity takes the form

$$\mathbf{I} = \sum_{n=1}^{\infty} \boldsymbol{\phi}_n \boldsymbol{\phi}_n^\dagger \qquad \text{or} \qquad i(x, y) = \delta(x - y) = \sum_n \phi_n(x)\phi_n^*(y) \qquad (10.2.27)$$

if $\boldsymbol{\phi}_n \in \mathcal{L}_2(a, b)$. $\mathbf{L}\boldsymbol{\phi}_n = \lambda_n \boldsymbol{\phi}_n$, and $\mathbf{LI} = \mathbf{L}$, and so it follows that

$$\mathbf{L} = \sum_n \lambda_n \boldsymbol{\phi}_n \boldsymbol{\phi}_n^\dagger. \qquad (10.2.28)$$

This has two implications. The first is that the differential operator L can be represented by an integral operator with the kernel

$$l(x, y) = \sum_n \lambda_n \phi_n(x) \phi_n^*(y). \qquad (10.2.29)$$

The second is that $|\lambda_n| \to \infty$ as $n \to \infty$ (where the labeling of eigenvalues is chosen such that $|\lambda_{n+1}| \geq |\lambda_n|$). This conclusion follows from the property that if $|\lambda_n| < \infty$ for all n, then $\mathbf{u} = \sum_n \alpha_n \boldsymbol{\phi}_n$ and

$$\frac{\|L\mathbf{u}\|^2}{\|\mathbf{u}\|^2} = \frac{\sum_n |\lambda_n|^2 |\alpha_n|^2}{\sum_n |\alpha_n|^2} \leq \max |\lambda_n|^2 < \infty, \qquad (10.2.30)$$

contradicting the fact that all differential operators are unbounded.

In the Hilbert space $\mathcal{L}_2(a, b; s)$, $s(x) > 0$, there is a technical detail that should be kept in mind as regards the resolution of the identity in terms of a basis set $\boldsymbol{\phi}_n$ in the space. Suppose $\boldsymbol{\phi}_n$ is an orthonormal set. Then $\mathbf{u} = \mathbf{I}\mathbf{u}$ is written as

$$\mathbf{u} = \sum_n \boldsymbol{\phi}_n \langle \boldsymbol{\phi}_n^\dagger, \mathbf{u} \rangle \qquad (10.2.31)$$

or

$$u(x) = \sum_n \int_a^b \phi_n(x) \phi_n^*(y) s(y) u(y) \, dy. \qquad (10.2.32)$$

Since Eq. (10.2.32) holds for arbitrary $u(x)$, it follows that

$$\delta(x - y) = s(x) \sum_n \phi_n(x) \phi_n^*(y). \qquad (10.2.33)$$

Thus, in the space $\mathcal{L}_2(a, b; s)$, the kernel of the identity operator is the product of the sum $\sum_n \phi_n(x) \phi_n^*(y)$ and the weighting factor $s(x)$. Similarly, if $\{\boldsymbol{\phi}_n\}$ is the set of eigenvectors of $\mathbf{L}$, the relationship

$$\mathbf{L}\mathbf{u} = \sum_n \lambda_n \boldsymbol{\phi}_n \langle \boldsymbol{\phi}_n, \mathbf{u} \rangle = \sum_n \int_a^b \lambda_n \phi_n(x) \phi_n^*(y) s(y) u(y) \, dy \qquad (10.2.34)$$

implies that the kernel of the integral operator representing $\mathbf{L}$ is

$$l(x, y) = s(x) \sum_n \lambda_n \phi_n(x) \phi_n^*(y). \qquad (10.2.35)$$

10.3. THE ADJOINT OF A DIFFERENTIAL OPERATOR

Consider the pth-order differential expression of the form

$$Lu(x) = a_p(x) \frac{d^p u(x)}{dx^p} + a_{p-1}(x) \frac{d^{p-1} u(x)}{dx^{p-1}} + \cdots + a_0(x) u(x), \qquad (10.3.1)$$

where $a_p(x) \neq 0$ and $a_i(x) \in C^{(i)}(a, b)$, $i = 1, \ldots, p$. $C^{(i)}(a, b)$ is the space of functions whose ith derivative is continuous in the finite interval $[a, b]$. The boundary conditions for the problem will be determined by setting the m linearly independent functionals

$$B_1 u = \alpha_{11} u(a) + \cdots + \alpha_{1p} u^{(p-1)}(a) + \alpha_{1,p+1} u(b) + \cdots + \alpha_{1,2p} u^{(p-1)}(b)$$

$$\vdots \qquad (10.3.2)$$

$$B_m u = \alpha_{m1} u(a) + \cdots + \alpha_{mp} u^{(p-1)}(a) + \alpha_{m,p+1} u(b) + \cdots + \alpha_{m,2p} u^{(p-1)}(b),$$

where $u^{(i)}(\xi) = d^i u(\xi)/d\xi^i$, $\xi = a$ or b. Accompanying the differential expression L is its formal adjoint differential expression $L^\dagger$. We can derive the form of the adjoint by considering the expression

$$\zeta(x) \equiv \int_a^x v^*(y) Lu(y)\, dy, \qquad (10.3.3)$$

or, in expanded form,

$$\begin{aligned}\zeta(x) = &\int_a^x v^*(y)\, a_0(y) u(y)\, dy + \int_a^x v^*(y)\, a_1(y) \frac{du(y)}{dy}\, dy \\ &+ \int_a^x v^*(y)\, a_2(y) \frac{d^2 u(y)}{dy^2}\, dy + \cdots.\end{aligned} \qquad (10.3.4)$$

Integrating the second term by parts once:

$$\begin{aligned}\int_a^x v^*(y)\, a_1(y) \frac{du(y)}{dy}\, dy = &\left. v^*(y) a_1(y) u(y) \right|_a^x \\ &- \int_a^x \frac{d}{dy}\bigl[v^*(y) a_1(y)\bigr] u(y)\, dy,\end{aligned} \qquad (10.3.5)$$

the third term by parts twice:

$$\begin{aligned}\int_a^x v^*(y)\, a_2(y) \frac{d^2 u(y)}{dy^2}\, dy = &\left. v^*(y) a_2(y) \frac{du(y)}{dy} \right|_a^x \\ &- \left.\frac{d}{dy}\bigl[v^*(y) a_2(y)\bigr] u(y)\right|_a^x \\ &+ \int_a^x \frac{d^2}{dy^2}\bigl[v^*(y) a_2(y)\bigr] u(y)\, dy,\end{aligned} \qquad (10.3.6)$$

etc., we obtain

$$\begin{aligned}\zeta(x) = &\int_a^x a_0(y) v^*(y)\, u(y)\, dy - \int_a^x \frac{d}{dy}\bigl[a_1(y) v^*(y)\bigr] u(y)\, dy \\ &+ \int_a^x \frac{d^2}{dy^2}\bigl[a_2(y) v^*(y)\bigr] u(y)\, dy - \cdots + \left. a_1(y) v^*(y) u(y) \right|_a^x \\ &+ \left. a_2(y) v^*(y) \frac{du(y)}{dy} \right|_a^x - \left.\frac{d}{dy}\bigl[a_2(y) v^*(y)\bigr] u(y)\right|_a^x + \cdots.\end{aligned} \qquad (10.3.7)$$

By comparing Eqs. (10.3.3) and (10.3.7), we find

$$\int_a^x \bigl[v^* Lu - (L^\dagger v)^* u\bigr]\, dx = \left. J(\mathbf{u}, \mathbf{v}) \right|_a^x, \qquad (10.3.8)$$

where

$$L^\dagger v = \sum_{j=0}^p (-1)^j \frac{d^j (a_j^* v)}{dx^j} \qquad (10.3.9)$$

and

$$J(\mathbf{u}, \mathbf{v}) = \sum_{j=1}^{p}\sum_{k=1}^{j}(-1)^{k+1}\frac{d^{k-1}(a_j v^*)}{dx^{k-1}}\frac{d^{j-k}u}{dx^{j-k}}. \quad (10.3.10)$$

Equation (10.3.8) is known as Green's formula. If we differentiate this equation, we obtain Lagrange's identity:

$$v^*(x)Lu(x) - \left(L^\dagger v^*(x)\right)u(x) = \frac{d}{dx}J(\mathbf{u}, \mathbf{v}). \quad (10.3.11)$$

If we define the vectors

$$\mathbf{y} = \begin{bmatrix} v(a) \\ \vdots \\ v^{(p-1)}(a) \\ v(b) \\ \vdots \\ v^{(p-1)}(b) \end{bmatrix} \quad \text{and} \quad \mathbf{x} = \begin{bmatrix} u(a) \\ \vdots \\ u^{(p-1)}(a) \\ u(b) \\ \vdots \\ u^{(p-1)}(b) \end{bmatrix}, \quad (10.3.12)$$

we find from Eqs. (10.3.8) and (10.3.10) that

$$\int_a^b \left[v^* Lu - (L^\dagger v^*)u\right]dx = \sum_{i,j=1}^{2p} y_i^* p_{ij} x_j = \mathbf{y}^\dagger \mathbf{P}\mathbf{x}, \quad (10.3.13)$$

where the elements p_{ij} of the $2p \times 2p$ matrix $\mathbf{P}$ are various derivatives of $a_i(x)$ evaluated at $x = a$ or b.

The m functionals of Eq. (10.3.2) will be used to define the boundary conditions for the operator $\mathbf{L}$. In general, we assume that $m \leq 2p$. Thus, to these functionals we will add the $2p - m$ linearly independent functionals

$$B_{m+1}u = \alpha_{m+1,1}u(a) + \cdots + \alpha_{m+1,2p}u^{(p-1)}(b)$$

$$\vdots \quad (10.3.14)$$

$$B_{2p}u = \alpha_{2p,1}u(a) + \cdots + \alpha_{2p,2p}u^{(p-1)}(b).$$

These supplemental boundary functionals can be defined by any linearly independent set of vectors

$$\boldsymbol{\rho}_i^T = [\alpha_{i1}, \ldots, \alpha_{i,2p}], \quad i = m+1, \ldots, 2p \quad (10.3.15)$$

that is also linearly independent of the set

$$\boldsymbol{\rho}_i^T = [\alpha_{i1}, \ldots, \alpha_{i,2p}], \quad i = 1, \ldots, m. \quad (10.3.16)$$

Combining Eqs. (10.3.2) and (10.3.14), we can write

$$\mathbf{b} = \mathbf{A}\mathbf{x},$$

THE ADJOINT OF A DIFFERENTIAL OPERATOR

where $b_i = B_i u$, $i = 1, \ldots, 2p$; $\mathbf{x}$ is defined in Eq. (10.3.12); and the elements of $\mathbf{A}$ are α_{ij}. Since the $2p$ row vectors of $\mathbf{A}$ are linearly independent, $\mathbf{A}$ is nonsingular and so its inverse exists. Thus, $\mathbf{x} = \mathbf{A}^{-1}\mathbf{b}$, which, when inserted into $\mathbf{y}^\dagger \mathbf{P}\mathbf{x}$, yields

$$\mathbf{y}^\dagger \mathbf{P} \mathbf{A}^{-1} \mathbf{b} = \sum_{i,j=1}^{2p} y_i^* c_{ij} b_j = \sum_{i=1}^{2p} \left(B_{2p+1-i}^\dagger v\right)^* B_i u, \tag{10.3.17}$$

with $c_{ij} = (\mathbf{PA}^{-1})_{ij}$ and $B_{2p+1-i}^\dagger v \equiv \sum_{j=1}^{2p} c_{ij}^* y_j$. What we have found is that Eq. (10.3.13) can be expressed as

$$\langle \mathbf{v}, \mathbf{L}\mathbf{u}\rangle - \langle \mathbf{L}^\dagger \mathbf{v}, \mathbf{u}\rangle = (B_{2p}^\dagger v)^* B_1 u + \cdots + (B_{p+1}^\dagger v)^* B_p u \\ + (B_p^\dagger v)^* B_{p+1} u + \cdots + (B_1^\dagger v)^* B_{2p} u. \tag{10.3.18}$$

In solving the equation $Lu = f$, we have to consider the following problems:

(i) The inhomogeneous problem

$$Lu = f, \quad B_i u = \gamma_i, \quad i = 1, \ldots, m \tag{10.3.19}$$

(ii) The homogeneous problem

$$Lu = 0, \quad B_i u = 0, \quad i = 1, \ldots, m \tag{10.3.20}$$

(iii) The homogeneous adjoint problem

$$L^\dagger v = 0, \quad B_i^\dagger v = 0, \quad i = 1, \ldots, 2p - m \tag{10.3.21}$$

The boundary functionals for the adjoint operator are determined by requiring that

$$\langle \mathbf{v}, \mathbf{L}\mathbf{u}\rangle = \langle \mathbf{L}^\dagger \mathbf{v}, \mathbf{u}\rangle \tag{10.3.22}$$

for any vector $\mathbf{u}$ in the domain

$$\mathcal{D}_L = \{\mathbf{u}, \mathbf{L}\mathbf{u} \in \mathcal{H}, \; B_i u = 0, \; i = 1, \ldots, m\}. \tag{10.3.23}$$

From Eq. (10.3.18), it follows that, for such vectors $\mathbf{u}$,

$$\langle \mathbf{v}, \mathbf{L}\mathbf{u}\rangle = \langle \mathbf{L}^\dagger \mathbf{v}, \mathbf{u}\rangle + \left(B_{2p-m}^\dagger v\right)^* B_{m+1} u + \left(B_{2p-m-1}^\dagger v\right)^* B_{m+2} u + \cdots \\ + \left(B_1^\dagger v\right)^* B_{2p} u. \tag{10.3.24}$$

Since the boundary functionals $B_i u$, $i = 1, \ldots, m$, for $\mathbf{L}$ are linearly independent of the functionals $B_i u$, $i = m+1, \ldots, 2p$, the values of $B_i u$, $i = m+1, \ldots, 2p$, are arbitrary and so it follows from Eq. (10.3.21) that the boundary conditions for the adjoint operator $\mathbf{L}^\dagger$ are

$$B_i^\dagger v = 0, \quad i = 1, \ldots, 2p - m. \tag{10.3.25}$$

The domain of $\mathbf{L}^\dagger$ for the problem defined by Eq. (10.3.21) is

$$\mathcal{D}_{L^\dagger} = \{\mathbf{v}, \mathbf{L}^\dagger \mathbf{v} \in \mathcal{H}, \; B_i^\dagger v = 0, i = 1, \ldots, 2p - m\}. \tag{10.3.26}$$

Since the functionals $B_i u$, $i = m+1, \ldots, 2p$, are not unique, neither are the functionals $B_i^\dagger v$, $i = 1, \ldots, 2p-m$. However, the domain $\mathcal{D}_{L^\dagger}$ defined by Eq. (10.3.26) is unique because any linearly independent set $\tilde{B}_i u$, $i = m+1, \ldots, 2p$, is a linear combination of the set $B_i u$, $i = m+1, \ldots, 2p$, which means, in turn, that the set $\tilde{B}_i^\dagger v$, $i = 1, \ldots, 2p - m$, corresponding to $\tilde{B}_i u$, $i = m+1, \ldots, 2p$, will be a linear combination of the set $B_i^\dagger v$, $i = 1, \ldots, 2p - m$.

If we started with the adjoint problem

$$L^\dagger v = h, \quad B_i^\dagger v = \delta_i, \qquad i = 1, \ldots, 2p - m, \tag{10.3.27}$$

we would find that its formal adjoint differential expression is L and the boundary functionals of $\mathbf{L}$ are $B_i u$, $i = 1, \ldots, m$. This means that $(\mathbf{L}^\dagger)^\dagger = \mathbf{L}$, which is equivalent to the properties $(\mathbf{A}^\dagger)^\dagger = \mathbf{A}$ and $(\mathbf{K}^\dagger)^\dagger = \mathbf{K}$ of matrix and integral operators.

Let us study some simple examples. Consider the differential expression

$$Lu = -\frac{d^2 u}{dx^2} \tag{10.3.28}$$

for the following five sets of boundary conditions:

(i) $u(0) = 0, \qquad u(1) = 0,$

(ii) $u(0) - u(1) = 0, \qquad u'(1) = 0,$

(iii) $u(0) - u(1) = 0, \qquad u'(0) - u'(1) = 0,$ (10.3.29)

(iv) $u(0) = 0, \qquad u'(1) = 0,$

(v) $u'(0) - u'(1) = 0.$

For this operator, $L^\dagger = L$ and the boundary conditions for $L^\dagger$ are determined from the equation

$$\begin{aligned} J(\mathbf{u}, \mathbf{v})\Big|_0^1 &= v^*(0)u'(0) - v^*(1)u'(1) \\ &\quad + [v'(1)]^* u(1) - [v'(0)]^* u(0) = 0. \end{aligned} \tag{10.3.30}$$

We will see that, even though the adjoint differential *expression* $L^\dagger$ is the same as L, the adjoint *operator* $\mathbf{L}^\dagger$ is not the same as the *operator* $\mathbf{L}$. For case (i), Eq. (10.3.30) implies that

$$v^*(0)u'(0) - v^*(1)u'(1) = 0 \qquad \text{for all } \mathbf{u} \in \mathcal{D}_L. \tag{10.3.31}$$

Note that the boundary conditions on $\mathbf{L}$ put no constraint on the derivatives $u'(x)$ at $x = 0$ and $x = 1$. Thus, $u'(0)$ and $u'(1)$ are arbitrary and so the boundary conditions on $\mathbf{L}^\dagger$ deduced from Eq. (10.2.26) are

$$v(0) = v(1) = 0. \tag{10.3.32}$$

THE ADJOINT OF A DIFFERENTIAL OPERATOR

This implies that $\mathcal{D}_L = \mathcal{D}_{L^\dagger}$, and since the differential expressions L and $L^\dagger$ are the same and the domains of the operators are the same, the operator $\mathbf{L}$ of case (i) is self-adjoint.

Consider next case (ii). With the corresponding boundary conditions, Eq. (10.3.30) becomes

$$v^*(0)u'(0) + [v'(1) - v'(0)]^* u(0) = 0 \quad \text{for all } \mathbf{u} \in \mathcal{D}_L. \quad (10.3.33)$$

Again, the quantities $u'(0)$ and $u(0)$ are arbitrary since the boundary conditions for case (ii) put no constraints on them. Consequently, Eq. (10.2.29) implies the boundary conditions

$$v(0) = 0 \quad \text{and} \quad v'(1) = v'(0) \quad (10.3.34)$$

for the adjoint operator $\mathbf{L}^\dagger$. Since $\mathcal{D}_{L^\dagger} \neq \mathcal{D}_L$, the operator $\mathbf{L}$ for case (ii) is not equal to its adjoint $\mathbf{L}^\dagger$ even though their differential expressions are the same.

The boundary conditions for case (iii) give the condition

$$[v(0) - v(1)]u'(0) + [v'(0) - v'(1)]^* u(0) = 0 \quad \text{for all } \mathbf{u} \in \mathcal{D}_L. \quad (10.3.35)$$

Since $u'(0)$ and $u(0)$ are unconstrained, it follows that the boundary conditions of $\mathbf{L}^\dagger$ are

$$v(0) = v(1) \quad \text{and} \quad v'(0) = v'(1). \quad (10.3.36)$$

Thus, $\mathcal{D}_L = \mathcal{D}_{L^\dagger}$ and so the operator $\mathbf{L}$ is self-adjoint in this case.

The boundary conditions of case (iv) lead to the condition that

$$v^*(0)u'(0) + [v'(1)]^* u(1) = 0 \quad \text{for all } \mathbf{u} \in \mathcal{D}_L. \quad (10.3.37)$$

Since $u'(0)$ and $u(1)$ are unconstrained, the boundary conditions of $\mathbf{L}^\dagger$ become

$$v(0) = v'(1) = 0, \quad (10.3.38)$$

and so $\mathbf{L}$ is self-adjoint.

Finally, the boundary conditions for case (v) yield

$$[v^*(0) - v^*(1)]u'(1) + [v'(1)]^* u(1) - [v'(0)]^* u(0) = 0. \quad (10.3.39)$$

Since $u'(1)$, $u(1)$, and $u(0)$ can have arbitrary values, the boundary conditions for $\mathbf{L}^\dagger$ are

$$v(0) - v(1) = 0, \quad v'(1) = 0, \quad v'(0) = 0. \quad (10.3.40)$$

Thus, the operators of cases (i), (iii), and (iv) are self-adjoint, whereas the operators of cases (ii) and (v) are not.

Of course, if $L \neq L^\dagger$, then the operator $\mathbf{L}$ will not be self-adjoint for any conditions. For example, if

$$Lu = a_2(x)\frac{d^2 u(x)}{dx^2} + a_1(x)\frac{du(x)}{dx} + a_0(x)u(x), \quad (10.3.41)$$

then

$$L^\dagger u = \frac{d^2(a_2^*(x)u(x))}{dx^2} - \frac{d(a_1^*(x)u(x))}{dx} + a_0^*(x)u(x). \tag{10.3.42}$$

However, if $L = L^\dagger$, we say that an operator is *formally self-adjoint*. For the second-order case, formal self-adjointness requires that

$$a_2(x) = a_2^*(x), \qquad a_1(x) = 2\frac{da_2^*(x)}{dx} - a_1^*(x) \tag{10.3.43a}$$

and

$$a_0(x) = \frac{d^2 a_2^*(x)}{dx^2} - \frac{da_1^*(x)}{dx} + a_0^*(x). \tag{10.3.43b}$$

If the a_i are real, then formal self-adjointness requires that $da_2/dx = a_1$, or

$$Lu = \frac{d}{dx}\left(a_2(x)\frac{du(x)}{dx}\right) + a_0(x)u(x). \tag{10.3.44}$$

10.4. SOLUTION TO THE GENERAL INHOMOGENEOUS PROBLEM

Consider the homogeneous equation

$$Lu = \sum_{i=0}^{p} a_i(x)\frac{d^i u}{dx^i} = 0, \qquad a < x < b, \tag{10.4.1}$$

and its solution for the initial conditions

$$u^{(i)}(x) = \zeta_i, \quad x = c, \qquad i = 0, \ldots, p-1, \tag{10.4.2}$$

where c is some point in the interval $[a, b]$ (here a and b may be finite or infinite). We assume that $a_p(x) \neq 0$ for x in $[a, b]$ and that the coefficients $a_i(x)$, $i = 0, \ldots, p$, are continuous in the interval $[a, b]$. The following theorem can be proved:

THEOREM. *The solution to the pth-order homogeneous equation* (10.4.1) *with initial conditions Eq.* (10.4.2) *exists and is unique.*

The proof of the above theorem was given in Section 9.4, where we showed that the initial value problem can be transformed into the problem of solving a Volterra equation of the second kind—which was shown to possess a unique solution. In this section, however, we prefer to examine an alternative proof of the theorem since it allows us to introduce "fundamental solutions" to pth-order differential equations.

SOLUTION TO THE GENERAL INHOMOGENEOUS PROBLEM

First, we convert the pth-order equation to a first-order system by defining a_i, $i = 1, \ldots, p$, by

$$z_1 \equiv u$$

$$z_2 \equiv \frac{du}{dx} = \frac{dz_1}{dx}$$

$$z_3 \equiv \frac{d^2 u}{dx^2} = \frac{dz_2}{dx} \tag{10.4.3a}$$

$$\vdots$$

$$z_p \equiv \frac{d^{p-1} u}{dx^{p-1}} = \frac{dz_{p-1}}{dx}$$

and noting that

$$\frac{d^p u}{dx^{p-1}} = -\frac{1}{a_p} \sum_{i=0}^{p-1} a_i z_{i+1} = \frac{dz_p}{dx}. \tag{10.4.3b}$$

Thus, the problem of solving Eq. (10.4.1) with initial conditions in Eq. (10.4.2) is transformed into the problem of solving system

$$\frac{d\mathbf{z}}{dx} = \mathbf{A}\mathbf{z}, \tag{10.4.4}$$

with the initial condition

$$\mathbf{z}(x) = \boldsymbol{\zeta} \quad \text{for } x = c, \tag{10.4.5}$$

where

$$\mathbf{z} = \begin{bmatrix} z_1 \\ z_2 \\ \vdots \\ z_p \end{bmatrix}, \quad \boldsymbol{\zeta} = \begin{bmatrix} \zeta_1 \\ \zeta_1 \\ \vdots \\ \zeta_p \end{bmatrix} = \begin{bmatrix} u(c) \\ u^{(1)}(c) \\ \vdots \\ u^{(p-1)}(c) \end{bmatrix} \tag{10.4.6}$$

and

$$\mathbf{A} = \frac{1}{a_p(x)} \begin{bmatrix} 0 & 1 & 0 & 0 & \cdots & 0 \\ 0 & 0 & 1 & 0 & \cdots & 0 \\ \vdots & \vdots & \vdots & \vdots & & \vdots \\ -a_0(x) & -a_1(x) & -a_2(x) & -a_3(x) & \cdots & -a_{p-1}(x) \end{bmatrix}. \tag{10.4.7}$$

The theorem now becomes:

THEOREM. *The solution to Eq. (10.4.4) with the initial condition in Eq. (10.4.5) exists and is unique.*

If all of the coefficients a_i are constant, we already know from matrix theory that the unique solution to Eqs. (10.4.6) and (10.4.7) is

$$\mathbf{z}(x) = \exp((x - c)\mathbf{A})\boldsymbol{\zeta}. \tag{10.4.8}$$

To prove the theorem for the general case, let us integrate Eq. (10.4.4) between c and x to obtain

$$\mathbf{z} = \int_c^x \mathbf{A}(y)\mathbf{z}(y)\,dy + \boldsymbol{\zeta} \equiv \mathbf{Kz} + \boldsymbol{\zeta}. \tag{10.4.9}$$

This is a Volterra equation of the second kind. Writing Eq. (10.4.9) in the form

$$(\mathbf{I} - \mathbf{K})\mathbf{z} = \boldsymbol{\zeta}, \tag{10.4.10}$$

we next consider the sequence of vectors

$$\begin{aligned}\mathbf{z}_1 &= \boldsymbol{\zeta} \\ \mathbf{z}_2 &= \boldsymbol{\zeta} + \mathbf{K}\boldsymbol{\zeta} \\ &\vdots \\ \mathbf{z}_n &= \boldsymbol{\zeta} + \mathbf{K}\boldsymbol{\zeta}_{n-1}.\end{aligned} \tag{10.4.11}$$

By successive substitution, we find

$$\mathbf{z}_n = \mathbf{S}_n \boldsymbol{\zeta}, \tag{10.4.12}$$

where

$$\mathbf{S}_n = \mathbf{I} + \sum_{i=1}^{n-1} \mathbf{K}^i. \tag{10.4.13}$$

From the identity

$$(\mathbf{I} - \mathbf{K})\mathbf{S}_n = \mathbf{I} - \mathbf{K}^n, \tag{10.4.14}$$

it follows that

$$(\mathbf{I} - \mathbf{K})\mathbf{z}_n = \boldsymbol{\zeta} - \mathbf{K}^n \boldsymbol{\zeta}. \tag{10.4.15}$$

Note that the quantity $\mathbf{R}_n \equiv -\mathbf{K}^n \boldsymbol{\zeta}$ is a p-dimensional vector given by

$$\mathbf{R}_n = -\int_c^x dy_1\, \mathbf{A}(y_1) \int_c^{y_1} dy_2\, \mathbf{A}(y_2) \cdots \int_c^{y_n} dy_n\, \mathbf{A}(y_n) \boldsymbol{\zeta}. \tag{10.4.16}$$

We can examine the vector as $n \to \infty$ in the norm $\|\mathbf{v}\|_\infty = \max_{1 \leq i \leq p} |v_i|$. From the properties of matrix norms ($\|\mathbf{AB}\| \leq \|\mathbf{A}\|\,\|\mathbf{B}\|$ and $\|\sum_i \mathbf{A}_i\| \leq \sum_i \|\mathbf{A}_i\|$), it follows that

$$\|\mathbf{R}_n\|_\infty \leq \int_c^x dy_1 \int_c^{y_1} dy_2 \cdots \int_c^{y_n} dy_n \|\mathbf{A}(y_1)\|_\infty \cdots \|\mathbf{A}(y_n)\|_\infty \|\boldsymbol{\zeta}\|_\infty. \tag{10.4.17}$$

SOLUTION TO THE GENERAL INHOMOGENEOUS PROBLEM

From the structure of $\mathbf{A}(x)$, we find that either $\|\mathbf{A}(x)\|_\infty < 1$ or

$$\|\mathbf{A}(x)\|_\infty < \max_{1 \le i \le p} \max_{a \le x \le b} \left| \frac{a_i(x)}{a_p(x)} \right|, \qquad (10.4.18)$$

depending on whether the right-hand side of the inequality in Eq. (10.4.18) is larger than 1. In either case, $0 < \|\mathbf{A}(x)\|_\infty < \Gamma < \infty$. Thus, we find

$$\|\mathbf{R}_n\|_\infty \le \frac{(x-c)^n}{n!} \Gamma^n \max_{1 \le i \le p} |\boldsymbol{\zeta}| \qquad \text{as } n \to \infty \qquad (10.4.19)$$

for $a < x < b$, where the interval $[a, b]$ is finite. This proves that $(\mathbf{I} - \mathbf{K})\mathbf{z}_n - \boldsymbol{\zeta} = \mathbf{0}$ as $n \to \infty$, or that $\mathbf{z} = \lim_{n \to \infty} \mathbf{z}_n$ is a solution to Eq. (10.4.10) and the quantity

$$(\mathbf{I} - \mathbf{K})^{-1} = \sum_{i=0}^\infty \mathbf{K}^i \qquad (10.4.20)$$

is the inverse of $\mathbf{I} - \mathbf{K}$. To prove that the solution $\mathbf{z}$ is unique, we assume that there is another solution $\mathbf{w}$. Then $(\mathbf{I} - \mathbf{K})(\mathbf{z} - \mathbf{w}) = \mathbf{0}$, or $\mathbf{z} - \mathbf{w} = \mathbf{K}(\mathbf{z} - \mathbf{w})$, from which it follows that $(\mathbf{z} - \mathbf{w}) = \mathbf{K}^n(\mathbf{z} - \mathbf{w})$. As shown in Eqs. (10.4.17)–(10.4.19), this result implies that $\mathbf{z} - \mathbf{w} = \mathbf{0}$, and thus the solution is unique since $\lim_{n \to \infty} S_n \mathbf{0} = \mathbf{0}$.

Since the initial conditions can be represented as a p-dimensional vector $\boldsymbol{\zeta}$, it follows that there are only p possible sets of linearly independent initial conditions. Suppose $\{\boldsymbol{\zeta}_1, \ldots, \boldsymbol{\zeta}_p\}$ is a set of p linearly independent p-dimensional vectors. Then any other set of initial conditions can be represented as a linear combination in these p vectors. We define a fundamental set of solutions (also known as a fundamental system) to Eq. (10.4.5) as a set of solutions $\{\mathbf{z}_1, \ldots, \mathbf{z}_p\}$ corresponding to p linearly independent sets of initial conditions $\{\boldsymbol{\zeta}_1, \ldots, \boldsymbol{\zeta}_p\}$.

THEOREM. *The fundamental solutions of a pth-order linear differential equation are linearly independent.*

To prove this, we assume that the contrary is true, namely, that there exists a set of numbers $\{\beta_1, \ldots, \beta_p\}$, not all 0, such that $\sum_i \beta_i \mathbf{z}_i = \mathbf{0}$. However, when $x = 0$, $\mathbf{z}_i = \boldsymbol{\zeta}_i$, and so $\sum_i \beta_i \boldsymbol{\zeta}_i = \mathbf{0}$, which is a contradiction since the $\boldsymbol{\zeta}_i$ were chosen to be linearly independent.

We can now express solutions to arbitrary initial conditions in terms of the fundamental system by using the following theorem:

THEOREM. *If $\boldsymbol{\zeta}$ represents an arbitrary set of initial conditions, then there exists a unique set of numbers $\{\alpha_1, \ldots, \alpha_p\}$ such that $\boldsymbol{\zeta} = \sum_{i=1}^p \alpha_i \boldsymbol{\zeta}_i$ and the solution to*

$$\frac{d\mathbf{z}}{dx} = \mathbf{A}(x)\mathbf{z}, \qquad \mathbf{z}(c) = \boldsymbol{\zeta} \qquad (10.4.21)$$

is a linear combination of the fundamental solutions, namely,

$$\mathbf{z} = \sum_{i=1}^p \alpha_i \mathbf{z}_i, \qquad (10.4.22)$$

where $\mathbf{z}_i(c) = \boldsymbol{\zeta}_i$.

We can restate these conclusions for the equivalent pth-order problem in the following theorems:

THEOREM. *The homogeneous equation*

$$Lu = \sum_{i=0}^{p} a_i(x) \frac{d^i u}{dx^i} = 0 \tag{10.4.23}$$

has a fundamental system, i.e., a set of p fundamental solutions $\{u_1(x), \ldots, u_p(x)\}$ corresponding to p sets of linearly independent initial conditions

$$u_j^{(i)}(c) = \zeta_{ji}, \quad i = 0, \ldots, p-1, \; j = 1, \ldots, p. \tag{10.4.24}$$

THEOREM. *The solution $u(x)$ for an arbitrary set of initial conditions $u^{(i)}(c) = \zeta_i, i = 0, \ldots, p-1$, is a linear combination of the fundamental solutions $\{u_1(x), \ldots, u_p(x)\}$.*

A fundamental system is not unique, but any fundamental system can be expressed as a linear combination of the fundamental solutions of any other fundamental system.

It might not be obvious that linear independence of the fundamental solutions $\{\mathbf{z}_1(x), \ldots, \mathbf{z}_p(x)\}$ implies linear independence of the functions $\{u_1(x), \ldots, u_p(x)\}$, which are the first components of $\{\mathbf{z}_1, \ldots, \mathbf{z}_p\}$. Assume that the solutions $u_1(x)$ and $u_2(x)$ are linearly dependent while $\boldsymbol{\zeta}_1$ and $\boldsymbol{\zeta}_2$ are linearly independent. But if u_1 and u_2 are linearly dependent, $u_1(x) = \beta u_2(x)$, and so $u_1^{(i)}(x) = c u_2^{(i)}(x), i = 0, \ldots, p-1$ (remember $u^{(i)} \equiv d^i u/dx^i$). This, in turn, implies that $\mathbf{z}_i(x) = \beta \mathbf{z}_2(x)$ or, setting $x = c$, that $\boldsymbol{\zeta}_1 = \beta \boldsymbol{\zeta}_2$, contradicting the fact that $\boldsymbol{\zeta}_1$ and $\boldsymbol{\zeta}_2$ are linearly independent. Thus, the functions $\{u_1(x), \ldots u_p(x)\}$ must be linearly independent functions if the vectors $\{\boldsymbol{\zeta}_1, \ldots, \boldsymbol{\zeta}_p\}$ are linearly independent vectors.

EXAMPLE 10.4.1. The homogeneous equation

$$\frac{d^2 u}{dx^2} = 0 \tag{10.4.25}$$

has $u_1 = 1$ and $u_2 = x$ as fundamental solutions for the initial conditions $\{u(0) = 1, u'(0) = 0$ and $u(1) = 0, u'(1) = 1\}$, respectively. Thus,

$$\boldsymbol{\zeta}_1 = \begin{bmatrix} 1 \\ 0 \end{bmatrix}, \quad \boldsymbol{\zeta}_2 = \begin{bmatrix} 0 \\ 1 \end{bmatrix}. \tag{10.4.26}$$

For the arbitrary initial conditions,

$$\boldsymbol{\zeta} = \begin{bmatrix} \alpha \\ \beta \end{bmatrix} = \alpha \boldsymbol{\zeta}_1 + \beta \boldsymbol{\zeta}_2, \tag{10.4.27}$$

the solution is

$$u = \alpha u_1 + \beta u_2 = \alpha + \beta x. \tag{10.4.28}$$

SOLUTION TO THE GENERAL INHOMOGENEOUS PROBLEM

EXAMPLE 10.4.2. The homogeneous equation

$$\frac{d^2u}{dx^2} + u = 0 \tag{10.4.29}$$

has $u_1 = \sin x$ and $u_2 = \cos x$ as fundamental solutions for the initial conditions $\{u(0) = 0, u'(0) = 1 \text{ and } u(0) = 1, u'(0) = 0\}$, respectively, and so the solution for the arbitrary initial conditions $u(0) = \alpha$ and $u'(0) = \beta$ is

$$u(x) = \beta \sin x + \alpha \cos x. \tag{10.4.30}$$

EXAMPLE 10.4.3. The homogeneous equation

$$\frac{d^3u}{dx^3} - \frac{d^2u}{dx^2} - \frac{du}{dx} + u = 0 \tag{10.4.31}$$

has $u_1(x) = e^{-x}$, $u_2(x) = e^x$, and $u_3(x) = xe^x$ as fundamental solutions for the initial conditions $u(0) = 1, u'(0) = -1, u''(0) = 1; u(0) = u'(0) = u''(0) = 1;$ and $u(0) = 0, u'(0) = 1, u''(0) = 2$, respectively. The solution for the general initial value problem $u(0) = \alpha, u'(0) = \beta, u''(0) = \gamma$ is

$$\begin{aligned} u(x) = &\tfrac{1}{4}(\alpha - 2\beta + \gamma)e^{-x} + \tfrac{1}{4}(3\alpha + 2\beta - \gamma)e^x \\ &+ \tfrac{1}{2}(\alpha + \gamma)xe^x. \end{aligned} \tag{10.4.32}$$

Let us next consider the inhomogeneous initial value problem:

$$\sum_{i=0}^{p} a_i(x) \frac{d^i u}{dx^i} = f(x), \tag{10.4.33}$$

$$u^{(i)}(c) = \zeta_i, \qquad i = 0, \ldots, p-1, \tag{10.4.34}$$

where the functions $f(x)$ and $a_i(x)$ are continuous and $a_p(x) \neq 0$ in the interval $[a, b]$. The corresponding first-order system is

$$\frac{d\mathbf{z}}{dx} = \mathbf{A}(x)\mathbf{z} + \mathbf{b}(x), \tag{10.4.35}$$

where

$$\mathbf{b}(x) = \begin{bmatrix} f(x) \\ 0 \\ \vdots \\ 0 \end{bmatrix}. \tag{10.4.36}$$

We again integrate between c and x to obtain

$$(\mathbf{I} - \mathbf{K})\mathbf{z} = \boldsymbol{\zeta} + \int_c^x \mathbf{b}(y)\, dy. \tag{10.4.37}$$

The solution to this equation is

$$\mathbf{z} = (\mathbf{I} - \mathbf{K})^{-1}\boldsymbol{\zeta} + (\mathbf{I} - \mathbf{K})^{-1}\int_c^x \mathbf{b}(y)\,dy. \qquad (10.4.38)$$

We proved above that the integral operator $(\mathbf{I} - \mathbf{K})^{-1}$ exists. The solution is unique and is just the sum of the solution of the homogeneous equation with the inhomogeneous initial conditions $\mathbf{z}(c) = \boldsymbol{\zeta}$ and of the solution of the inhomogeneous equation with the homogeneous initial conditions $\mathbf{z}(c) = \mathbf{0}$.

We have learned so far that the solution to the general initial value problem exists and is unique. In the case of the general boundary value problem, however, things are more complicated. A solution might exist but not be unique or a solution might not even exist. The rest of this section will be devoted to the general boundary value problem.

For the pth-order differential equation, the fully inhomogeneous boundary value problem we want to consider is

$$Lu(x) = f(x), \qquad a < x < b, \qquad (10.4.39)$$

with balanced boundary conditions

$$\begin{aligned} B_1 u &= \gamma_1 \\ &\vdots \\ B_p u &= \gamma_p, \end{aligned} \qquad (10.4.40)$$

where not all γ_i are 0. The differential operator is defined in Eq. (10.3.2) and the functionals $B_i u$, $i = 1, \ldots, p$, are defined in Eq. (10.3.1). The difference between the operators defined in Eq. (10.3.1) and in Eq. (10.4.1) is that the former requires that $a_i(x) \in C^{(i)}(a,b)$, $i = 0, \ldots, p$.

Suppose $\hat{u}(x)$ is a solution to the homogeneous equation

$$L\hat{u}(x) = 0, \qquad a < x < b, \qquad (10.4.41)$$

obeying the inhomogeneous boundary conditions

$$B_i \hat{u} = \gamma_i, \qquad i = 1, \ldots, p. \qquad (10.4.42)$$

Suppose also that $\tilde{u}(x)$ is a solution to the inhomogeneous equation

$$L\tilde{u}(x) = f(x), \qquad a < x < b, \qquad (10.4.43)$$

obeying the homogeneous boundary conditions

$$B_i \tilde{u} = 0, \qquad i = 1, \ldots, p. \qquad (10.4.44)$$

Clearly, the function $u(x) = \hat{u}(x) + \tilde{u}(x)$ obeys the equation $Lu = f$ and the boundary conditions $B_i u = \gamma_i$. Thus, the problem of solving the fully inhomogeneous problem defined by Eqs. (10.4.39) and (10.4.40) can be divided into the problem of solving the homogeneous equation for inhomogeneous boundary conditions defined by Eqs. (10.4.41) and (10.4.42) and the problem of solving the

SOLUTION TO THE GENERAL INHOMOGENEOUS PROBLEM

inhomogeneous equation for the homogeneous boundary conditions defined by Eqs. (10.4.43) and (10.4.44). This is a special case of the *superposition principle*, which states

SUPERPOSITION PRINCIPLE. *If u is the solution to the problem $Lu = f$, $B_i u = \gamma_i$, $i = 1, \ldots, p$, and U is the solution to the problem $LU = F$, $B_i U = \Gamma_i$, $i = 1, \ldots, p$, then $c_1 u + c_2 U$ is the solution to the problem $Lw = c_1 f + c_2 F$, $B_i w = c_1 \gamma_1 + c_2 \Gamma_2$, $i = 1, \ldots, p$.*

The principle follows from the linearity of L and $B_i u$.

The *superposition principle* leads to the theorem:

THEOREM. *If the homogeneous problem $Lu = 0$, $B_i u = 0$, $i = 1, \ldots, p$, has only the trivial solution $u = 0$, then the fully inhomogeneous problem has at most one solution.*

The proof of the theorem is easy. Assume that u_1 and u_2 are solutions to the fully inhomogeneous problem. Then $u_2 - u_1$ obeys the equation $L(u_2 - u_1) = 0$ and the boundary conditions $B_i(u_2 - u_1) = 0$, implying that $u_2 - u_1 = 0$ or $u_2 = u_1$.

Let us examine a few simple examples before continuing with the general theory.

■ **EXAMPLE 10.4.4.** Consider the inhomogeneous equation

$$-\frac{d^2 u}{dx^2} = f(x), \qquad u(0) = \gamma_1, \ u(1) = \gamma_2. \qquad (10.4.45)$$

The general solution to the homogeneous equation $-d^2 \hat{u}/dx^2 = 0$ is $\hat{u} = \alpha + \beta x$, and from the boundary conditions $\hat{u}(0) = \gamma_1$ and $\hat{u}(1) = \gamma_2$ it follows that

$$\hat{u} = \gamma_1 + (\gamma_2 - \gamma_1)x. \qquad (10.4.46)$$

We will see in Section 10.5 that the solution to the inhomogeneous equation $-d^2 \tilde{u}/dx^2 = f(x)$, $u(0) = u(1) = 0$ is

$$\tilde{u} = \int_0^1 \bigl[\eta(x-y)y(1-x) + \eta(y-x)x(1-y)\bigr] f(y)\, dy, \qquad (10.4.47)$$

and, thus, the solution to Eq. (10.4.45) is

$$u = \gamma_1 + (\gamma_2 - \gamma_1)x \\ + \int_0^1 \bigl[\eta(x-y)y(1-x) + \eta(y-x)x(1-y)\bigr] f(y)\, dy. \qquad (10.4.48)$$

Since the only solution to $-d^2 u/dx^2 = 0$ with $u(0) = u(1) = 0$ is $u = 0$, it follows that Eq. (10.4.48) is the unique solution to Eq. (10.4.45). ■■■

■ **EXAMPLE 10.4.5.** Consider the inhomogeneous equation

$$-\frac{d^2 u}{dx^2} = 2, \qquad u'(0) = 1, \ u'(1) = -1. \qquad (10.4.49)$$

The general solution is

$$u = \alpha + \beta x - x^2, \qquad (10.4.50)$$

and the boundary conditions yield $\beta = 1$, and so the solution is

$$u = \alpha + x - x^2, \qquad (10.4.51)$$

where α is an arbitrary constant. In this case, the solution exists but is not unique.

■ ■ ■ **EXAMPLE 10.4.6.** Consider the inhomogeneous equation

$$-\frac{d^2 u}{dx^2} = f(x), \qquad u'(0) = \gamma_1, \; u'(1) = -\gamma_2, \qquad (10.4.52)$$

where the function $f(x)$ is such that

$$\int_0^1 f(x)\,dx \neq \gamma_1 + \gamma_2. \qquad (10.4.53)$$

Integrating Eq. (10.4.52) once yields

$$-u'(1) + u'(0) = \int_0^1 f(x)\,dx, \qquad (10.4.54)$$

which contradicts Eq. (10.4.53) since $-u'(1) + u'(0) = \gamma_2 + \gamma_1$. Thus, no solution exists for any $f(x)$ obeying Eq. (10.4.53). ■ ■ ■

To understand the general solvability conditions for the fully inhomogeneous problem, we need to return to Eq. (10.3.18), which is

$$\begin{aligned}\int_a^b \left[v^* Lu - (L^\dagger v)^* u\right] dx \\ = \sum_{i=1}^p (B^\dagger_{2p+1-i} v)^* B_i u + (B^\dagger_i v)^* B_{2p+1-i} u.\end{aligned} \qquad (10.4.55)$$

The domain $\mathcal{D}_L$ for the inhomogeneous problem is

$$\mathcal{D}_L = \{\mathbf{u},\; \mathbf{Lu} \in \mathcal{H},\; B_i \mathbf{u} = \gamma_i,\; i = 1, \ldots, p\}. \qquad (10.4.56)$$

The domain $\mathcal{D}_{L^\dagger}$ of the operator $\mathbf{L}^\dagger$ of the homogeneous adjoint problem is

$$\mathcal{D}_{L^\dagger} = \{\mathbf{v},\; \mathbf{L}^\dagger \mathbf{v} \in \mathcal{H},\; B_i^\dagger \mathbf{v} = 0,\; i = 1, \ldots, p\}. \qquad (10.4.57)$$

Thus, if we restrict ourselves to vectors $\mathbf{v} \in \mathcal{D}_{L^\dagger}$ and $\mathbf{u} \in \mathcal{D}_L$, Eq. (10.4.56) becomes

$$\langle \mathbf{v}, \mathbf{Lu}\rangle - \langle \mathbf{L}^\dagger \mathbf{v}, \mathbf{u}\rangle = \sum_{i=1}^p \gamma_i (B^\dagger_{2p+1-i} v)^*. \qquad (10.4.58)$$

We have now developed the ideas needed to state the alternative theorem for boundary value problems. Consider the three problems:

$$Lu = f, \qquad a < x < b, \qquad B_i u = \gamma_i, \qquad i = 1, \ldots, p, \qquad (10.4.59)$$

$$Lu = 0, \qquad a < x < b, \qquad B_i u = 0, \qquad i = 1, \ldots, p, \qquad (10.4.60)$$

$$L^\dagger v = 0, \qquad a < x < b, \qquad B_i^\dagger v = 0, \qquad i = 1, \ldots, p. \qquad (10.4.61)$$

where Lu and $B_i u$ are defined by Eqs. (10.3.1) and (10.3.2). The following alternative theorem holds:

ALTERNATIVE THEOREM. (a) *If the homogeneous problem, Eq. (10.4.60), possesses only the trivial solution, so does the homogeneous adjoint problem, Eq. (10.4.61).*

(b) *If the homogeneous problem, Eq. (10.4.60), has k linearly independent solutions, then the homogeneous adjoint problem, Eq. (10.4.61), has k linearly independent solutions.*

(c) *If the homogeneous problem, Eq. (10.4.60), has k linearly independent solutions $\mathbf{u}_h^i$, $i = 1, \ldots, k$, then the fully inhomogeneous problem, Eq. (10.4.59), has a solution if and only if*

$$\langle \mathbf{v}_h^i, \mathbf{f} \rangle = \sum_{j=1}^{p} \gamma_j \left(B^\dagger_{2p+1-j} v_h^i \right)^*, \qquad i = 1, \ldots, k, \tag{10.4.62}$$

where the $\mathbf{v}_h^i$, $i = 1, \ldots, k$, are the linearly independent solutions to the homogeneous adjoint problem, Eq. (10.4.61). If the solvability conditions hold, the solution to the fully homogeneous problem will be of the form

$$\mathbf{u} = \mathbf{u}_p + \sum_{i=1}^{k} c_i \mathbf{u}_h^i, \tag{10.4.63}$$

where $\mathbf{u}_p$ is a particular solution to Eq. (10.4.59); $\mathbf{u}_h^i$, $i = 1, \ldots, k$, are the linearly independent solutions to the homogeneous problem, Eq. (10.4.60); and the constants c_i, $i = 1, \ldots, k$, are arbitrary.

The necessity of the conditions in Eq. (10.4.62) follows easily from Eq. (10.4.58): if $\mathbf{u}$ is a solution to $L\mathbf{u} = \mathbf{f}$ and if $\mathbf{v}$ is a solution to $L^\dagger \mathbf{v} = \mathbf{0}$, then Eq. (10.4.58) reduces to Eq. (10.4.62). The proof that the conditions at Eq. (10.4.62) are sufficient to ensure solvability of the inhomogeneous problem is significantly more difficult and will not be given here. A more advanced textbook should be consulted for the proof.

When the solvability conditions are met, the solution to the fully inhomogeneous problem, Eq. (10.4.59), can be constructed in the following way. Let $\{u_1, \ldots, u_p\}$ denote a fundamental system of $Lu = 0$. Let u_I denote the solution to the initial value problem

$$Lu_I = f, \quad u^{(i)}(x = a) = 0, \qquad i = 0, \ldots, p-1, \ a < x < b. \tag{10.4.64}$$

The set $\{u_1, \ldots, u_p\}$ exists according to the theorem proved above. u_I exists and can be computed from a Volterra equation as indicated by Eqs. (9.4.59)–(9.4.63). The function

$$u(x) = \sum_{j=1}^{p} \alpha_j u_j(x) + u_I(x) \tag{10.4.65}$$

is, therefore, a solution to $Lu = f$ for arbitrary constants $\{\alpha_1, \ldots, \alpha_p\}$. If we can choose these constants to satisfy $B_i u = \gamma_i$, $i = 1, \ldots, p$, i.e.,

$$\sum_{j=1}^{P}(B_i u_j)\alpha_j = \gamma_i - B_i u_1, \qquad i = 1, \ldots, p, \qquad (10.4.66)$$

then Eq. (10.4.65) is the solution to the fully inhomogeneous problem, Eq. (10.4.59).

Let us illustrate this method of solving $Lu = f$, $B_i u = \gamma_i$, $i = 1, \ldots, p$, with the simple examples given above for the equation

$$-\frac{d^2 u}{dx^2} = f(x), \qquad 0 < x < 1. \qquad (10.4.67)$$

For this case, direct integration of Eq. (10.4.64) yields

$$u_1(x) = -\int_0^x (x - t)f(t)\, dt. \qquad (10.4.68)$$

Consider again the following examples.

EXAMPLE 10.4.7. Consider again the inhomogeneous equation

$$-\frac{d^2 u}{dx^2} = f(x), \qquad u(0) = \gamma_1, \ u(1) = \gamma_2. \qquad (10.4.69)$$

A linearly independent set of fundamental solutions to $Lu = 0$ (i.e., a fundamental system of $Lu = 0$) is

$$u_1(x) = 1, \qquad u_2(x) = x. \qquad (10.4.70)$$

(An equally acceptable system would be $u_1 = 1 + x$ and $u_2 = 1 - x$.) The solution to Eq. (10.4.69) is then

$$u = \alpha_1 u_1 + \alpha_2 u_2 + u_1, \qquad (10.4.71)$$

where $u(0) = \gamma_1$ and $u(1) = \gamma_2$, or

$$\begin{aligned}\alpha_1 &= \gamma_1, \\ \alpha_1 + \alpha_2 - \int_0^1 (1-t)f(t)\, dt &= \gamma_2.\end{aligned} \qquad (10.4.72)$$

Solving for α_1 and α_2, we obtain

$$\begin{aligned}u(x) = \gamma_1 + (\gamma_2 - \gamma_1)x + x\int_0^1 (1-t)f(t)\, dt \\ - \int_0^x (x-t)f(t)\, dt.\end{aligned} \qquad (10.4.73)$$

This result (when appropriately rearranged) is in agreement with the solution given earlier by Eq. (10.4.48).

SOLUTION TO THE GENERAL INHOMOGENEOUS PROBLEM

EXAMPLE 10.4.8. Consider again the inhomogeneous equation

$$-\frac{d^2u}{dx^2} = 2, \qquad u'(0) = 1, \ u'(1) = -1. \tag{10.4.74}$$

In this case,

$$u_I = -x^2, \tag{10.4.75}$$

and again we choose $u_1 = 1$ and $u_2 = x$ as the fundamental system of $Lu = 0$. With

$$u = \alpha_1 u_1 + \alpha_2 u_2 + u_I, \tag{10.4.76}$$

the boundary conditions require

$$\begin{aligned} \alpha_2 &= 1 \\ \alpha_2 - 2 &= -1 \end{aligned} \tag{10.4.77}$$

or the solution to Eq. (10.4.74) is

$$u = \alpha_1 + x - x^2$$

for arbitrary α_1 in agreement with the solution given by Eq. (10.4.49).

EXAMPLE 10.4.9. Consider again the inhomogeneous equation

$$-\frac{d^2u}{dx^2} = f(x), \qquad u'(0) = \gamma_1, \ u'(1) = -\gamma_2, \tag{10.4.78}$$

where

$$\int_0^1 f(x)\,dx \neq \gamma_1 + \gamma_2. \tag{10.4.79}$$

Then

$$u(x) = \alpha_1 + \alpha_2 x - \int_0^x (x-t) f(t)\,dt \tag{10.4.80}$$

and

$$u'(x) = \alpha_2 - \int_0^x f(t)\,dt. \tag{10.4.81}$$

We seek α_1 and α_2 such that

$$\begin{aligned} \alpha_2 &= \gamma_1 \\ \alpha_2 - \int_0^1 f(t)\,dt &= -\gamma_2, \end{aligned} \tag{10.4.82}$$

which requires that

$$\int_0^1 f(t)\,dt = \gamma_1 + \gamma_2. \tag{10.4.83}$$

Since we consider now $f(x)$ such that the inequality at Eq. (10.4.79) holds, there is no solution to this example.

■ **EXERCISE 10.4.1.** Find the adjoint operators for the operators defined in Examples 10.4.7–10.4.9 and demonstrate that the alternative theorem is obeyed in each example.

Let us close this section by showing that the solution to the general second-order problem

$$Lu = a_2 \frac{d^2u}{dx^2} + a_1 \frac{du}{dx} + a_0 u = f, \qquad a < x < b, \qquad (10.4.84)$$

with $B_i u = \gamma_i$, $i = 1, \ldots, p$, and $a_2(x) \neq 0$, can be constructed from a fundamental system u_1 and u_2, of $Lu = 0$. We define the determinant

$$W(u_1, u_2, x) \equiv \begin{vmatrix} u_1(x) & u_2(x) \\ u_1'(x) & u_2'(x) \end{vmatrix} = u_1(x)u_2'(x) - u_2(x)u_1'(x) \qquad (10.4.85)$$

(called the Wronskian). If u_1 and u_2 is a fundamental set, then $Lu_1 = Lu_2 = 0$. From this it follows that

$$\begin{aligned}
0 &= u_1 L u_2 - u_2 L u_1 \\
&= a_2(u_1 u_2'' - u_2 u_1'') + a_1(u_1 u_2' - u_2 u_1') \\
&= \frac{d}{dx} W + \frac{a_1}{a_2} W.
\end{aligned} \qquad (10.4.86)$$

Thus,

$$W(x) = W(x_0) \exp\left(-\int_{x_0}^{x} \left[\frac{a_1(y)}{a_2(y)}\right] dy\right), \qquad (10.4.87)$$

where $W(x_0)$ is an arbitrary point in the interval $[a, b]$. From this it follows that $W(x)$ is either identically 0 or nowhere 0 in $[a, b]$. Suppose $W(x)$ is identically 0. Then

$$u_1(x)u_2'(x) - u_2(x)u_1'(x) = 0, \qquad (10.4.88)$$

or $d \ln u_1/dx = d \ln u_2/dx = 0$, or $u_1(x) = Au_2(x)$, where A is an arbitrary constant. This is a contradiction since u_1 and u_2 form a fundamental system and so are linearly independent. Thus, $W(x)$ is nowhere 0 in the interval $[a, b]$. We assert that $u_1(x)$, the solution to Eq. (10.4.64) for this second-order case, is

$$u_I = -\int_a^x \frac{u_1(x)u_2(y) - u_2(x)u_1(y)}{a_2(y)\left[u_1(y)u_2'(y) - u_2(y)u_1'(y)\right]} f(y) dy. \qquad (10.4.89)$$

Since $a_2(y) \neq 0$ and $W(u_1, u_2; y) \neq 0$, the function $u_I(x) \in C^{(2)}(a, b)$, and by direct differentiation we can show that $Lu_I = f$ and $u_I(a) = u'(a) = 0$. The general solution to the second-order case, when it exists, can thus be expressed as

$$\begin{aligned}
u(x) &= \alpha_1 u_1(x) + \alpha_2 u_2(x) \\
&\quad - \int_a^x \frac{u_1(x)u_2(y) - u_2(x)u_1(y)}{a_2(y)\left[u_1(y)u_2'(y) - u_2(y)u_1'(y)\right]} f(y) dy,
\end{aligned} \qquad (10.4.90)$$

with α_1 and α_2 determined by the conditions $B_1 u = \gamma_1$ and $B_2 u = \gamma_2$.

EXAMPLE 10.4.10. Consider the problem

$$\frac{d^2u}{dx^2} + \pi^2 u = \cos \pi x, \qquad 0 < x < \pi, \qquad (10.4.91)$$

with $u(0) = u(1) = 0$. The adjoint operator $\mathbf{L}^\dagger$ equals $\mathbf{L}$ since $L = L^\dagger$ and $\mathcal{D}_L = \mathcal{D}_{L^\dagger}$. The solution to the homogeneous adjoint equation $L^\dagger v = 0$ is $v = \sin \pi x$. Since

$$\langle \mathbf{v}, \mathbf{f} \rangle = \int_0^\pi \sin \pi x \, \cos \pi x \, dx = 0, \qquad (10.4.92)$$

it follows that Eq. (10.4.91) has a solution. A fundamental system for $Lu = 0$ is $u_1 = \sin \pi x$ and $u_2 = \cos \pi x$. u_1 satisfies the initial conditions $u_1(0) = 0$, $u_1'(0) = -\pi$ and u_2 satisfies the initial conditions $u_2(0) = 1$, $u_2'(0) = 0$. Insertion of u_1 and u_2 into Eq. (10.4.90) yields

$$u(x) = \alpha_1 \sin \pi x + \alpha_2 \cos \pi x + \frac{1}{\pi} \sin \pi x \int_0^x \cos^2 \pi y \, dy \\ - \frac{1}{\pi} \cos \pi x \int_0^x \sin \pi y \, \cos \pi y \, dy. \qquad (10.4.93)$$

The boundary conditions $u(0) = u(1) = 0$ yield $\alpha_2 = 0$. Thus,

$$u(x) = \alpha_1 \sin \pi x + \frac{1}{\pi} \sin \pi x \int_0^x \cos^2 \pi y \, dy \\ - \frac{1}{\pi} \cos \pi x \int_0^x \sin \pi y \, \cos \pi y \, dy, \qquad (10.4.94)$$

where α_1 is arbitrary. Thus, $u = u_p + \alpha_1 u_h$ as required by the alternative theorem.

10.5. GREEN'S FUNCTION: INVERSE OF A DIFFERENTIAL OPERATOR

Let us consider again the equation

$$Lu(x) = f(x), \qquad a < x < b, \qquad (10.5.1)$$

where the interval $[a, b]$ is finite and L is a pth-order differential expression defined by

$$Lu(x) = a_p(x) \frac{d^p u(x)}{dx^p} + a_{p-1} \frac{d^{p-1} u(x)}{dx^{p-1}} + \cdots \\ + a_1(x) \frac{du(x)}{dx} + a_0(x) u(x). \qquad (10.5.2)$$

We assume that $a_p(x) \neq 0$ and that $a_i(x) \in C^{(i)}(a, b)$, $i = 0, \ldots, p$, where, as stated before, $C^{(i)}(a, b)$ is the space of functions whose ith derivative is continuous in $[a, b]$.

As pointed out in previous sections, in order to associate an operator $\mathbf{L}$ with the differential expression L, we must specify its domain $\mathcal{D}_L$. For example, we can say that $\mathbf{u} \in \mathcal{D}_L$ if $\mathbf{u}$ and $\mathbf{Lu} \in \mathcal{L}_2(a, b)$. Usually, p independent boundary conditions are also specified for a pth-order differential equation. This need not be done, but if p boundary conditions are not stated, a pth-order differential equation will certainly not have a unique solution. Of course, if the homogeneous equation $Lu = 0$ has a nontrivial solution, then the inhomogeneous equation will not have a unique solution even with p independent boundary conditions.

In this section we consider p linear, homogeneous boundary conditions of the form

$$B_1 u \equiv \sum_{j=1}^{p} \alpha_{1j} u^{(j-1)}(a) + \sum_{j=1}^{p} \alpha_{1,p+j} u^{(j-1)}(b) = 0$$
$$\vdots \qquad (10.5.3)$$
$$B_p u \equiv \sum_{j=1}^{p} \alpha_{pj} u^{(j-1)}(a) + \sum_{j=1}^{p} \alpha_{p,p+j} u^{(j-1)}(b) = 0,$$

where $u^{(i)}(\xi)$, $\xi = a, b$, denotes the ith derivative of u ($d^i u(\xi)/d\xi^i$) evaluated at $\xi = a$ and b. We assume that the row vectors

$$\boldsymbol{\rho}_i^{\mathrm{T}} = [\alpha_{i1}, \ldots, \alpha_{i,2p}], \qquad i = 1, \ldots, p, \qquad (10.5.4)$$

are linearly independent. We pointed out earlier that this assumption implies that the rank of the matrix $[\boldsymbol{\rho}_1, \ldots, \boldsymbol{\rho}_p]$ is p and ensures that Eq. (10.5.3) defines p linearly independent boundary conditions; thus, the problem is a balanced one.

In the special cases $\alpha_{ij} = 0$, $j = 1, \ldots, p$, or $\alpha_{ij} = 0$, $j = p+1, \ldots, 2p$, Eqs. (10.5.1) and (10.5.3) define the initial value problem (IVP). We know from Section 10.4 that, for the functions such that $a_p(x) \neq 0$ and $a_i(x) \in C^0(a, b)$, $i = 1, \ldots, p$, the IVP has a unique solution. On the other hand, if some of the α_{ij}, $1 \leq j \leq p$, and some of the α_{ij}, $p+1 \leq j \leq 2p$, are nonzero, Eqs. (10.5.1) and (10.5.3) define a boundary value problem (BVP). The BVP may or may not have a unique solution as stated in the *alternative theorem* in Section 10.4. For example, consider the equation

$$-\frac{d^2 u}{dx^2} = 1, \qquad 0 < x < 1. \qquad (10.5.5)$$

The general solution is

$$u(x) = \alpha + \beta x - \frac{x^2}{2}. \qquad (10.5.6)$$

With the boundary conditions $u'(0) = u'(1) = 0$, we find

$$0 = \beta \qquad \text{and} \qquad 0 = \beta - 1, \qquad (10.5.7)$$

which is clearly impossible. Thus, this simple BVP has no solution. Since the homogeneous equation $Lu = 0$ has the solution $u = \alpha$ for this BVP, we already knew the problem would not have a unique solution. It turns out that this problem has no solution, which can be confirmed from the general solvability conditions for a BVP—given in Section 10.4.

We will now restrict our attention to problems in which the homogeneous equations defined by $Lu = 0$ and $B_i u = 0$, $i = 1, \ldots, p$, have only the trivial solution $u = 0$. For this class of problems, the inverse of the differential operator $\mathbf{L}$ exists. In summary, the operator problem in this section becomes

$$\mathbf{Lu} = \mathbf{f}, \quad \text{where } \mathcal{D}_L = \{\mathbf{u}, \mathbf{Lu} \in \mathcal{H}; B_i u = 0, \ i = 1, \ldots, p\}; \tag{10.5.8}$$

the differential expression for $\mathbf{L}$ is defined by Eq. (10.5.2); and the boundary conditions are defined by Eq. (10.5.3). One choice of the Hilbert space $\mathcal{H}$ is $\mathcal{L}_2(a, b)$. However, since $a_p(x)$ never equals 0 on the interval $[a, b]$, it is either strictly positive or strictly negative. An alternative Hilbert space is $\mathcal{L}_2(a, b; \pm 1/a_p(x))$, where the expression $+1/a_p(x)$ is the weighting function if a_p is positive and $-1/a_p(x)$ is the weighting function if $a_p(x)$ is negative. If $\mathbf{u}, \mathbf{v} \in \mathcal{L}_2(a, b; 1/a_p(x))$, then the inner product $\langle \mathbf{u}, \mathbf{v} \rangle$ is, by definition,

$$\langle \mathbf{u}, \mathbf{v} \rangle = \int_a^b \frac{u^*(x) v(x)}{a_p(x)} \, dx. \tag{10.5.9}$$

Let us now consider the simplest of linear, inhomogeneous differential equations

$$\frac{du(x)}{dx} = f(x), \quad u(0) = 0, \ 0 < x < 1. \tag{10.5.10}$$

In operator form, the equation becomes

$$\mathbf{Lu} = \mathbf{f}. \tag{10.5.11}$$

If $\mathbf{L}^{-1}$ is defined as the inverse of $\mathbf{L}$, then $\mathbf{L}^{-1}\mathbf{L} = \mathbf{L}\mathbf{L}^{-1} = \mathbf{I}$, where $\mathbf{I}$ is the identity operator. Multiplication of Eq. (10.5.11) by $\mathbf{L}^{-1}$ yields the solution

$$\mathbf{u} = \mathbf{L}^{-1}\mathbf{f}. \tag{10.5.12}$$

This works, of course, only if $\mathbf{L}^{-1}$ exists and is useful only if we can construct $\mathbf{L}^{-1}$.
Integration of Eq. (10.5.10) yields

$$u(x) = \int_0^1 \eta(x - y) f(y) \, dy, \tag{10.5.13}$$

where η is the step function defined by

$$\eta(x) = \begin{cases} 0, & x < 0, \\ 1, & x > 0. \end{cases} \tag{10.5.14}$$

In order to accommodate operator theory, the simpler solution $u(x) = \int_0^x f(y) \, dy$ has been rewritten in the form given in Eq. (10.5.13).
In operator form, Eq. (10.5.13) becomes

$$\mathbf{u} = \mathbf{Gf}, \tag{10.5.15}$$

where $\mathbf{G}$ is an integral operator with the kernel $g(x, y) = \eta(x, y)$. The kernel of the operator $\mathbf{G}$ is called Green's function. Since $\mathbf{Lu} = \mathbf{LGf} = \mathbf{f}$ for arbitrary $\mathbf{f}$, it follows that $\mathbf{LG} = \mathbf{I}$. Since $\mathbf{I}$ is the identity operator in a function space, its kernel is $\delta(x - y)$, where δ is the Dirac delta function. Thus, for the operator $\mathbf{L}$, it follows that Green's function must satisfy the equation

$$Lg(x, y) = \frac{dg(x, y)}{dx} = \delta(x - y). \tag{10.5.16}$$

It is well known from functional analysis that the derivative of a step function is the Dirac delta function, i.e., $d\eta(x - y)/dx = \delta(x - y)$. To see that this is true, consider the basic properties of a Dirac delta function: $\delta(x) = 0$ if $x \neq 0$ and $\delta(x) \to \infty$ as $x \to 0$ such that

$$\int_a^b \delta(x)\, dx = \begin{cases} 1, & \text{if } a < x < b, \\ 0, & \text{otherwise.} \end{cases} \tag{10.5.17}$$

From this property it follows that

$$\int_0^x \delta(\xi - y)\, d\xi = \begin{cases} 0, & x < y, \\ 1, & x > y, \end{cases} \tag{10.5.18}$$

or

$$\int_0^x \delta(\xi - y)\, d\xi = \eta(x - y). \tag{10.5.19}$$

Thus, since $g(x, y) = \eta(x - y)$, it follows that $dg(x, y)/dx = \delta(x - y)$, as required by the condition that $\mathbf{LG} = \mathbf{I}$.

Let us next examine the slightly more complicated differential equation

$$Lu = -\frac{d^2 u(x)}{dx^2} = f(x), \quad u(0) = u(1) = 0. \tag{10.5.20}$$

The operator $\mathbf{L}$ in $\mathcal{L}_2(0, 1)$ corresponding to this equation has the domain

$$\mathcal{D}_L = \{u(0) = u(1) = 0, \ \mathbf{u}, \mathbf{Lu} \in \mathcal{L}_2(0, 1)\}. \tag{10.5.21}$$

Formal integration of $Lu = f$ yields

$$u(x) = a + bx - \int_0^x d\xi \int_0^\xi f(y)\, dy, \tag{10.5.22}$$

where a and b are constants. By interchange of variables of integration, we obtain

$$\int_0^x d\xi \int_0^\xi f(y)\, dy = \int_0^x f(y)\, dy \int_y^x d\xi = \int_0^x (x - y) f(y)\, dy. \tag{10.5.23}$$

From this and the boundary conditions for $\mathbf{L}$, it follows that

$$a = 0 \quad \text{and} \quad b = -\int_0^1 (1 - y) f(y)\, dy, \tag{10.5.24}$$

and so

$$u(x) = \int_0^1 \left[-\eta(x - y)(x - y) + x(1 - y)\right] f(y)\, dy. \tag{10.5.25}$$

The kernel of Eq. (10.5.25) possesses a symmetry that, though important to us in relation to the adjoint Green's function, is not apparent as written. Equation (10.5.25) can be rewritten in the form

$$u(x) = \int_0^x \left[-(x-y) + x(1-y)\right] f(y)\,dy + \int_x^1 x(1-y) f(y)\,dy$$

$$= \int_0^x y(1-x) f(y)\,dy + \int_x^1 x(1-y) f(y)\,dy \qquad (10.5.26)$$

$$= \int_0^1 \left[\eta(x-y) y(1-x) + \eta(y-x) x(1-y)\right] f(y)\,dy.$$

Thus, we have found that the solution to Eq. (10.5.20) is $\mathbf{u} = \mathbf{G}\mathbf{f}$, where Green's function is

$$g(x, y) = \eta(x-y) y(1-x) + \eta(y-x) x(1-y). \qquad (10.5.27)$$

The condition $\mathbf{LG} = \mathbf{I}$ requires that Green's function obey $Lg(x, y) = \delta(x-y)$. Differentiation of Eq. (10.5.27) with respect to x yields

$$\frac{dg(x,y)}{dx} = \delta(x-y) y(1-x) - \eta(x-y) y - \delta(y-x) x(1-y)$$
$$\qquad\qquad + \eta(y-x)(1-y) \qquad (10.5.28)$$
$$= -\eta(x-y) y + \eta(y-x)(1-y).$$

We arrived at this result by noting that $\delta(x)$ is an even function, i.e., $\delta(x-y) = \delta(y-x)$, and that $\delta(x-y) y(1-x) = \delta(x-y) x(1-y)$ since $\delta(x-y)$ is nonzero only at $x = y$. Differentiating Eq. (10.5.28) with respect to x, we find

$$-\frac{d^2 g(x,y)}{dx^2} = \delta(x-y) y + \delta(y-x)(1-y) = \delta(x-y), \qquad (10.5.29)$$

as required for Green's function.

Another property of the kernel $g(x, y)$ is that

$$g(0, y) = g(1, y) = 0 \qquad \text{for } 0 < y < 1; \qquad (10.5.30)$$

i.e., Green's function satisfies the boundary condition for its corresponding operator $\mathbf{L}$.

The simple examples just analyzed point out some general properties of Green's ffunctions. In terms of Green's function, the solution to $Lu = f$ is $u(x) = \int_a^b g(x, y) f(y)\,dy$ for arbitrary functions $f(y)$. From the boundary conditions for $\mathbf{L}$, it follows that

$$\int_a^b (B_i g) f(y)\,dy = 0, \qquad i = 1, \ldots, p. \qquad (10.5.31)$$

From this one concludes (since we can choose $f(y) = B_i g$) that Green's function $g(x, y)$ satisfies, with respect to the variable x, the boundary conditions of $\mathbf{L}$, i.e.,

$$B_i g = 0, \qquad i = 1, \ldots, p. \qquad (10.5.32)$$

Also, since $\mathbf{LG} = \mathbf{I}$, it follows that for the pth-order operator $g(x, y)$ satisfies the differential equation

$$\sum_{i=0}^{p} a_i(x) \frac{d^i g(x, y)}{dx^i} = \delta(x - y). \tag{10.5.33}$$

If $g(x, y)$ obeys Eq. (10.5.33), then it has to have $p - 2$ continuous derivatives at $x = y$. Otherwise, if the derivative $d^{p-2}g(x, y)/dx^{p-2}$ were discontinuous at $x = y$, then the derivative $d^{p-1}g(x, y)/dx^{p-1}$ would be a Dirac delta function of $x - y$ and the derivative $d^p g(x, y)/dx^p$ would be the derivative of a Dirac delta function of $x - y$, instead of just the delta function as in Eq. (10.5.33). Given that the first $p - 2$ derivatives of $g(x, y)$ are continuous, division of Eq. (10.5.33) by $a_p(x)$ and integration over x from $y - \epsilon$ to $y + \epsilon$ (where ϵ is an infinitesimal positive number), we obtain

$$\left.\frac{d^{p-1} g(x, y)}{dx^{p-1}}\right|_{x=y+\epsilon} - \left.\frac{d^{p-1} g(x, y)}{dx^{p-1}}\right|_{x=y-\epsilon}$$
$$+ \sum_{i=0}^{p-1} \int_{y-\epsilon}^{y+\epsilon} \frac{a_i(x) i\, d^i g(x, y) dx}{a_p(x)\, dx^i} = \frac{1}{a_p(y)}. \tag{10.5.34}$$

In the limit as $\epsilon \to 0^+$, the integrals in Eq. (10.5.34) vanish because the quantities (a_i/a_p) are continuous, and so

$$\lim_{\epsilon \to 0^+} \int_{y-\epsilon}^{y+\epsilon} \frac{a_i(x)}{a_p(x)} \frac{d^i g(x, y) dx}{dx^i} = \frac{a_i(y)}{a_p(y)} \lim_{\epsilon \to 0^+} \int_{y-\epsilon}^{y+\epsilon} \frac{d^i g(x, y) dy}{dx^i} = 0 \tag{10.5.35}$$

for $i = 0$ to $p - 1$. Thus, Eq. (10.5.34), in the limit as $\epsilon \to 0^+$, yields the following jump condition on $g(x, y)$:

$$\left.\frac{d^{p-1} g(x, y)}{dx^{p-1}}\right|_{x=y^+} - \left.\frac{d^{p-1} g(x, y)}{dx^{p-1}}\right|_{x=y^-} = \frac{1}{a_p(y)}. \tag{10.5.36}$$

Let us now examine Green's functions for some special cases.

10.5.1. The Initial Value Problem

Suppose the coefficients α_{ij}, $j = p + 1, \ldots, 2p$, in Eq. (10.5.3) are 0. Then, since the rank of $[\boldsymbol{\rho}_1, \ldots, \boldsymbol{\rho}_p] = [\alpha_{ij}]$ is p, we find

$$u(a) = u'(a) = \cdots = u^{(p-1)}(a) = 0 \tag{10.5.37}$$

for the initial value problem with *homogeneous boundary conditions*. Consider the solution $u_y(x)$ to the following pth-order IVP problem (called the causal fundamental solution in some texts):

$$Lu_y(x) = 0;$$
$$u_y(y) = u_y^{(1)}(y) = \cdots = u_y^{(p-2)}(y) = 0, \quad u_y^{(p-1)}(y) = \frac{1}{a_p(y)}. \tag{10.5.38}$$

GREEN'S FUNCTION: INVERSE OF A DIFFERENTIAL OPERATOR

Note that the boundary conditions in Eq. (10.5.38) are defined at the point $x = y$. The boundary condition for $u_y^{(p-1)}(y)$ was purposely chosen for reasons that will soon become clear. It turns out the function $u(x)$ defined by

$$u(x) \equiv \int_a^x u_y(x) f(y)\, dy = \int_a^b \eta(x-y) u_y(x) f(y)\, dy \qquad (10.5.39)$$

is a solution to $Lu = f$ with the initial conditions of Eq. (10.5.37). To see this, we simply apply L to Eq. (10.5.39) and use the initial conditions to obtain

$$Lu = \int_a^x Lu_y(x) f(y)\, dy + a_p(x) \left[\frac{d^{p-1}}{dx^{p-1}} u_y(x)\right]_{x=y} f(x) = f(x). \qquad (10.5.40)$$

Thus, Green's function for the IVP in this section is

$$g(x, y) = \eta(x - y) u_y(x), \qquad (10.5.41)$$

where $u_y(x)$ is the solution to Eq. (10.5.38). From this we can state the following theorem:

THEOREM. *The inverse operator* **G** *of the operator* **L** *for the IVP problem is a Volterra integral operator of the first kind.*

∎ EXAMPLE 10.5.1. Find Green's function for the problem

$$\frac{d^2u}{dx^2} + u = x, \qquad u(0) = u'(0) = 0. \qquad (10.5.42)$$

The formal solution to

$$\frac{d^2 u_y(x)}{dx^2} + u_y(x) = 0 \qquad (10.5.43)$$

is

$$u_y(x) = \alpha \sin(x + \delta) + \beta \cos(x + \gamma). \qquad (10.5.44)$$

With the boundary conditions

$$u_y(y) = 0 \quad \text{and} \quad \left.\frac{du_y(x)}{dx}\right|_{x=y} = 1, \qquad (10.5.45)$$

we find

$$u_y(x) = \sin(x - y). \qquad (10.5.46)$$

Thus,

$$g(x, y) = \eta(x - y) \sin(x - y) \qquad (10.5.47)$$

and the solution to Eq. (10.5.42) is

$$u(x) = \int_0^x \sin(x - y) y\, dy \qquad (10.5.48)$$
$$= x - \sin x.$$

∎∎∎

EXERCISE 10.5.1. Show by integration by parts that $\mathbf{GLu} = \mathbf{u}$ for the operators $\mathbf{G}$ and $\mathbf{L}$ defined in Example 10.5.1. Since $\mathbf{LG} = \mathbf{I}$ by construction, this completes the proof that $\mathbf{GL} = \mathbf{LG} = \mathbf{I}$, and so the inverse $\mathbf{L}^{-1}$ of $\mathbf{L}$ is the integral operator $\mathbf{G}$.

In the remainder of this section, we will restrict our attention to the BVP for second-order equations such that $a_2(x) \neq 0$, $a_i(x)$ are continuous, and $a < x < b$, where a and b are finite. The problem can be divided into two cases.

10.5.2. The Boundary Value Problem for Unmixed Boundary Conditions

In this case,
$$B_1 u = \alpha_{11} u(a) + \alpha_{12} u'(a) = 0$$
$$B_2 u = \alpha_{23} u(b) + \alpha_{24} u'(b) = 0, \tag{10.5.49}$$

and Green's function can be constructed from the solution of two IVPs, i.e., from a fundamental system of $Lu = 0$.

Consider first the function $u_1(x)$ satisfying the boundary condition $B_1 u_1 = 0$ and the differential equation

$$Lu_1 = a_2(x)\frac{d^2 u_1}{dx^2} + a_1(x)\frac{du_1}{dx} + a_0(x)u_1 = 0, \qquad a < x < y. \tag{10.5.50}$$

We are free to choose $u_1(a) = -\alpha_{12}$ and $u_1'(a) = \alpha_{11}$ to satisfy $B_1 u = 0$. The solution to this IVP exists and is nontrivial if at least one of the two coefficients α_{11} and α_{12} are nonzero—in order to ensure that the rank of the matrix

$$\begin{bmatrix} \alpha_{11} & \alpha_{12} & 0 & 0 \\ 0 & 0 & \alpha_{23} & \alpha_{24} \end{bmatrix} \tag{10.5.51}$$

is 2. Similarly, the solution to the IVP

$$Lu_2 = a_2(x)\frac{d^2 u_2}{dx^2} + a_1(x)\frac{du_2}{dx} + a_0(x)u_2 = 0, \qquad y < x < b. \tag{10.5.52}$$

exists when $u_2(b) = -\alpha_{24}$ and $u_2'(b) = \alpha_{23}$, ensuring that $B_2 u_2 = 0$. Again, at least one of the quantities α_{23} and α_{24} must be nonzero.

If we define the kernel

$$g(x, y) \equiv \begin{cases} c_1 u_1(x), & a < x < y, \\ c_2 u_2(x), & y < x < b, \end{cases} \tag{10.5.53}$$

where c_1 and c_2 are constants, it follows that

$$B_1 g = B_2 g = 0 \tag{10.5.54}$$

and

$$Lg(x, y) = 0 \qquad \text{if } x \neq y. \tag{10.5.55}$$

According to our arguments in Eqs. (10.5.32), (10.5.33), and (10.5.36), $g(x, y)$ will be Green's function for the second-order problem if Eqs. (10.5.54) and (10.5.55) hold, if $g(x, y)$ is continuous at $x = y$, and if

$$\left.\frac{dg(x, y)}{dx}\right|_{x=y^+} - \left.\frac{dg(x, y)}{dx}\right|_{x=y^-} = \frac{1}{a_2(y)}. \tag{10.5.56}$$

To enforce the continuity and jump conditions, we require that c_1 and c_2 be chosen to satisfy the equations

$$-c_1 u_1(y) + c_2 u_2(y) = 0$$
$$-c_1 u_1'(y) + c_2 u_2'(y) = \frac{1}{a(y)}. \tag{10.5.57}$$

Solving these equations for c_1 and c_2, we find

$$c_1 = \frac{u_2(y)}{a_2(y) W(u_1, u_2, y)} \quad \text{and} \quad c_2 = \frac{u_1(y)}{a_2(y) W(u_1, u_2, y)}, \tag{10.5.58}$$

where $W(u_1, u_2, y)$ is the Wronskian determinant defined in Eq. (10.4.85)

$$W(u_1, u_2, y) = \begin{vmatrix} u_1(y) & u_2(y) \\ u_1'(y) & u_2'(y) \end{vmatrix}. \tag{10.5.59}$$

We proved in Section 10.4 that the Wronskian is nowhere 0. We have, thus, found that, for the second-order differential equation with unmixed boundary conditions, Green's function is given by

$$g(x, y) = \begin{cases} \dfrac{u_2(y) u_1(x)}{a_2(y) W(u_1, u_2, y)}, & a < x < y, \\ \dfrac{u_1(y) u_2(x)}{a_2(y) W(u_1, u_2, y)}, & y < x < b, \end{cases} \tag{10.5.60}$$

where $u_1(x)$ and $u_2(x)$ are solutions of the initial value problems posed at Eqs. (10.5.50) and (10.5.52).

■ **EXAMPLE 10.5.2.** Find Green's function for the problem

$$Lu = -(1+x)^2 \frac{d^2 u}{dx^2} + 4(1+x)\frac{du}{dx} + 2u = f(x), \quad 0 < x < 1, \tag{10.5.61}$$

$$B_1 u = u(0) = 0, \quad B_2 u = u(1) = 1. \tag{10.5.62}$$

A solution to $Lu_1 = 0$ with $u_1(0) = 0$ and $u_1'(0) = 1$ is

$$u_1 = \frac{x}{(1+x)^2}. \tag{10.5.63}$$

A solution to $Lu_2 = 0$ with $u_2(1) = 1$ and $u_2'(1) = -1$ is

$$u_2 = \frac{4}{(1+x)^2}. \tag{10.5.64}$$

These fundamental solutions satisfy $Lu_1 = 0$, $B_1u_1 = 0$ and $Lu_2 = 0$, $B_2u_2 = 0$. Thus, Green's function for this example is (from Eq. (10.5.60))

$$g(x, y) = \begin{cases} -\dfrac{x}{(1+x)^2}, & 0 < x < y, \\ -\dfrac{y}{(1+x)^2}, & y < x < 1. \end{cases} \quad (10.5.65)$$

EXAMPLE 10.5.3. Find Green's function for the problem

$$Lu = \frac{d^2u}{dx^2} + 3\frac{du}{dx} + 2u = f(x), \quad -1 < x < 1, \quad (10.5.66)$$

$$\begin{aligned} u(-1) - 2u'(-1) &= 0 \\ u(1) + u'(1) &= 0 \end{aligned} \quad (10.5.67)$$

and solve the inhomogeneous problem for the cases

(a) $\quad f(x) = x \quad (10.5.68)$

and

(b) $\quad f(x) = \exp(-x^2). \quad (10.5.69)$

The general solution to $Lu = 0$ for Eq. (10.5.66) is

$$u = \alpha \exp(-2x) + \beta \exp(-x). \quad (10.5.70)$$

To obtain u_1, set $u(-1) = 2$ and $u'(-1) = 1$. This leads to $\alpha = -3/\exp(2)$ and $\beta = 5/e$, and so

$$u_1(x) = -3\exp(-2(x+1)) + 5\exp(-(x+1)). \quad (10.5.71)$$

Similarly, to find u_2, set $u(1) = 1$ and $u'(1) = -1$ to obtain $\alpha = 0$, $\beta = e$, giving

$$u_2(x) = \exp(-(x-1)). \quad (10.5.72)$$

From u_1 and u_2, we find

$$W(u_1, u_2, y) = -3\exp(-(3y+1)) \quad (10.5.73)$$

and

$$g(x, y) = \begin{cases} -\dfrac{1}{3}\Big[-3\exp(-2(x+1)) \\ \quad +5\exp(-(x+1))\Big]\exp(2(y+1)), & -1 < x < y, \\ -\dfrac{1}{3}\exp(-(x-1)) \\ \quad \times\Big[-3\exp(y-1)+5\exp(2y)\Big], & y < x < 1. \end{cases} \quad (10.5.74)$$

The solution to the inhomogeneous equation is then

$$u(x) = \int_{-1}^{1} g(x, y) f(y) \, dy, \qquad (10.5.75)$$

or, upon substitution of Eq. (10.5.74),

$$\begin{aligned} u(x) = &-\frac{1}{3} \exp(-(x-1)) \int_{-1}^{x} \left[-3\exp(y-1) + 5\exp(2y) \right] f(y) \, dy \\ &- \frac{1}{3} \left[-3\exp(-2(x+1)) + 5\exp(-(x+1)) \right] \\ &\times \int_{x}^{1} \exp(2(y+1)) f(y) \, dy. \end{aligned} \qquad (10.5.76)$$

For case (a), $f(y) = y$, and direct substitution yields

$$\begin{aligned} u(x) = &-\frac{1}{12} \left[-3\exp(-2(1+x)) + 5\exp(-(1+x)) \right] \\ &\times \left[\exp(4) - \exp(2(1+x))(2x-1) \right] \\ &- \frac{1}{12} \exp(1-x) \left\{ -9\exp(-2) + \exp(x-1) \right. \\ &\left. \times \left[12 - 5\exp(1+x) + 2x(5\exp(1+x) - 6) \right] \right\}. \end{aligned} \qquad (10.5.77)$$

For case (b), $f(y) = \exp(-y^2)$, and Eq. (10.5.76) does not yield an analytical solution. We can, however, solve for a solution by utilizing the numerical techniques of Chapter 3 for integration. Both solutions are plotted in Fig. 10.5.1 for the interval $-1 \le x \le 1$. (A Mathematica program has been provided in the Appendix for this example.) ∎ ∎ ∎

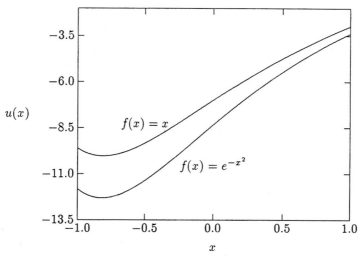

FIGURE 10.5.1

EXERCISE 10.5.2. Calculate and plot the solution $u(x)$ to Eqs. (10.5.66) and (10.5.67) when $f(x) = \sin \pi x$, $f(x) = 1/(2-x)$, and $f(x) = \sqrt{x+1}$.

10.5.3. The Boundary Value Problem for General Boundary Conditions

In this case, the boundary conditions are

$$B_1 u \equiv \alpha_{11} u(a) + \alpha_{12} u'(a) + \alpha_{13} u(b) + \alpha_{14} u'(b) = 0$$
$$B_2 u \equiv \alpha_{21} u(a) + \alpha_{22} u'(a) + \alpha_{23} u(b) + \alpha_{24} u'(b) = 0, \qquad (10.5.78)$$

with the rank of the matrix

$$\begin{bmatrix} \alpha_{11} & \alpha_{12} & \alpha_{13} & \alpha_{14} \\ \alpha_{21} & \alpha_{22} & \alpha_{23} & \alpha_{24} \end{bmatrix} \qquad (10.5.79)$$

equaling 2. The differential expression is again

$$Lu = a_2(x)\frac{d^2u}{dx^2} + a_1(x)\frac{du}{dx} + a_0(x)u, \qquad a < x < b, \qquad (10.5.80)$$

where the functions $a_i(x)$ are continuous, $a_2(x) \neq 0$, and $[a, b]$ is a finite interval. If we wish to define the adjoint differential expression $L^\dagger v$, we require that $a_i(x)$ belong to $C^{(i)}(a, b)$, i.e., that they have continuous ith-order derivatives.

Let u_1 be a nontrivial solution to $Lu_1 = 0$ with $B_1 u_1 = 0$ and let u_2 be a nontrivial solution to $Lu_2 = 0$ with $B_2 u_2 = 0$. Solutions to the second-order differential equation satisfying only one homogeneous boundary condition always exist and can always be computed from linear combinations of fundamental solutions. Recall also that the causal fundamental solution, $u_y(x)$, defined by Eq. (10.5.38) exists and satisfies the continuity and jump conditions for Green's function. Thus, if we set

$$g(x, y) = \eta(x - y)u_y(x) + \alpha u_1(x) + \beta u_2(x), \qquad (10.5.81)$$

then $Lg(x, y) = \delta(x - y)$ and $g(x, y)$ is the desired Green's function if α and β are chosen so that $g(x, y)$ satisfies the boundary conditions $B_1 g = B_2 g = 0$, or

$$\alpha_{13} u_y(b) + \alpha_{14} u'_y(b) + \alpha B_1 u_2 = 0$$
$$\alpha_{23} u_y(b) + \alpha_{24} u'_y(b) + \alpha B_2 u_1 = 0. \qquad (10.5.82)$$

The quantities $u_y(a)$ and $u'_y(a)$ are missing from these equations because $\eta(a - y) = 0$ for $a < y < b$. Note that neither $B_1 u_2$ nor $B_2 u_1$ can be 0 since, then, there would be a nontrivial solution to $Lu = 0$ with $B_1 u = B_2 u = 0$, a contradiction to the class of problems considered in this section. Thus, α and β are given by

$$\alpha = -[\alpha_{23} u_y(b) + \alpha_{24} u'_y(b)]/B_2 u_1$$
$$\beta = -[\alpha_{13} u_y(b) + \alpha_{14} u'_y(b)]/B_1 u_2. \qquad (10.5.83)$$

GREEN'S FUNCTION: INVERSE OF A DIFFERENTIAL OPERATOR

■ **EXAMPLE 10.5.4.** Use the method just described to find Green's function $g(x, y)$ for the differential expression

$$Lu = -\frac{d^2u}{dx^2}, \qquad (10.5.84)$$

with

$$\begin{aligned} B_1 u &\equiv u(0) = 0 \\ B_2 u &\equiv u(1) = 0. \end{aligned} \qquad (10.5.85)$$

We have already clearly determined $g(x, y)$ directly (Eq. (10.5.27)) and so we will only use the problem to illustrate the procedure for incorporating the general boundary conditions.

The solution to

$$-\frac{d^2}{dx^2} u_y(x) = 0, \qquad u_y(x)\bigg|_{x=y} = 0, \qquad \frac{du_y(x)}{dx}\bigg|_{x=y} = \frac{1}{a_2(y)} \qquad (10.5.86)$$

is

$$u_y(x) = -(x - y). \qquad (10.5.87)$$

Furthermore, the solution to $d^2u_1/dx^2 = 0$ with $u_1(0) = 0$ is $u_1(x) = c_1 x$ and to $d^2u_2/dx^2 = 0$ with $u_2(1) = 0$ is $u_2(x) = c_2(1-x)$, where c_1 and c_2 are arbitrary constants. From these results, it follows that $\alpha = -u_y(1)/c_1 = (1-y)/c_1$ and $\beta = 0$. Thus, Eq. (10.5.81) becomes

$$\begin{aligned} g(x, y) &= -\eta(x-y)(x-y) + x(1-y) \\ &= \eta(x-y)y(1-x) + \eta(y-x)x(1-y), \end{aligned} \qquad (10.5.88)$$

which is in agreement with Green's function (Eq. (10.5.27)) found by direct
■ ■ ■ integration.

Let us end this section with a few comments on Green's function for the adjoint $L^\dagger$ of the operator L. Remember that the boundary conditions for $L^\dagger$ are $B_i^\dagger u = 0$, $u = 1, \ldots, p$. The functionals $B_i^\dagger v$ can be derived from the functionals $B_i u$ and the condition $J(\mathbf{u}, \mathbf{v})|_a^b = 0$ for $\mathbf{u} \in \mathcal{D}_L$. If, however, we have Green's function $g(x, y)$ for L, we do not have to find the boundary conditions $B_i^\dagger v = 0$, $i = 1, \ldots, p$, to find Green's function $g^\dagger(x, y)$ for $L^\dagger$. This is because of the theorem:

THEOREM. *If $g(x, y)$ is Green's function for L, then*

$$g^\dagger(x, y) = g^*(y, x), \qquad (10.5.89)$$

where $g^\dagger(x, y)$ is Green's function for $L^\dagger$.

The proof of this theorem is straightforward. By construction of Green's function, we have

$$Lg(x, y) = \delta(x - y) \quad \text{and} \quad L^\dagger g^\dagger(x, y) = \delta(x - y). \qquad (10.5.90)$$

But from the property of the adjoint, $\langle \mathbf{v}, L\mathbf{u}\rangle = \langle \mathbf{L}^\dagger \mathbf{v}, \mathbf{u}\rangle$, it follows that

$$\int_a^b [g^\dagger(x,y)]^* L g(x,z)\,dx = \int_a^b [L^\dagger g^\dagger(x,y)]^* g(x,z)\,dx. \tag{10.5.91}$$

Together, Eqs. (10.5.90) and (10.5.91) imply

$$\int_a^b [g^\dagger(x,y)]^* \delta(x-z)\,dx = \int_a^b \delta(x-y) g(x,z)\,dx, \tag{10.5.92}$$

or

$$g^\dagger(z,y) = g^*(y,z), \tag{10.5.93}$$

in agreement with Eq. (10.5.89).

From the above theorem, it follows that the kernel of the adjoint $\mathbf{G}^\dagger$ of the integral operator $\mathbf{G}$ is, as expected, Green's function of the adjoint $\mathbf{L}^\dagger$ of the operator $\mathbf{L}$. Given Green's function $g(x,y)$ for $\mathbf{L}$, the solution to the equation

$$\mathbf{L}^\dagger \mathbf{v} = \mathbf{h} \tag{10.5.94}$$

is simply

$$\mathbf{v} = \mathbf{G}^\dagger \mathbf{h} \quad \text{or} \quad v(x) = \int_a^b g^*(y,x) h(y)\,dy \tag{10.5.95}$$

for the boundary conditions $B_i^\dagger v = 0$, $i = 1, \ldots, p$. What is nice is that we do not even have to determine the boundary conditions of $\mathbf{L}^\dagger$ to find its inverse $\mathbf{G}^\dagger$ or to solve Eq. (10.5.94).

■ ■ ■ **EXERCISE 10.5.3.** Use the result in Eq. (10.5.93) to find the boundary conditions for the adjoint $\mathbf{L}^\dagger$ of the operator $\mathbf{L}$ defined in Example 10.5.3.

10.6. SPECTRAL THEORY OF DIFFERENTIAL OPERATORS: SOME GENERAL PROPERTIES

The eigenproblem for differential operators is, of course, to find values of the scalar λ for which the equation

$$\mathbf{L}\boldsymbol{\phi} = \lambda\boldsymbol{\phi}, \quad \boldsymbol{\phi} \in \mathcal{D}_L, \tag{10.6.1}$$

has solutions—where L is a pth-order differential expression and

$$\mathcal{D}_L = \{\mathbf{u}, \mathbf{L}\mathbf{u} \in \mathcal{H}, \ B_i u = 0, \ i = 1, \ldots, p\}.$$

Certain things we know for matrix and integral operators follow immediately for differential operators. If $\mathbf{L}^\dagger$ is the adjoint operator with the domain $\mathcal{D}_L = \{\mathbf{v}, \mathbf{L}^\dagger \mathbf{v} \in \mathcal{H}, \ B_i^\dagger v = 0, \ i = 1, \ldots, p\}$ and obeying the relationship $\langle \mathbf{v}, \mathbf{L}\mathbf{u}\rangle = \langle \mathbf{L}^\dagger \mathbf{v}, \mathbf{u}\rangle$, it follows immediately that

THEOREM. *If $\boldsymbol{\phi}_i, \lambda_i$ are an eigenvector and an eigenvalue of $\mathbf{L}$ and $\boldsymbol{\psi}_j, v_j$ are an eigenvector and an eigenvalue of $\mathbf{L}$ and if $v_j^* \neq \lambda_i$, then $\boldsymbol{\phi}_i$ is orthogonal to $\boldsymbol{\psi}_j$.*

We can see this from

$$\langle \boldsymbol{\psi}_j, \mathbf{L}\boldsymbol{\phi}_i \rangle = \langle \mathbf{L}^\dagger \boldsymbol{\psi}_j, \boldsymbol{\phi}_i \rangle = \lambda_i \langle \boldsymbol{\psi}_j, \boldsymbol{\phi}_i \rangle = \nu_i^* \langle \boldsymbol{\psi}_j, \boldsymbol{\phi}_i \rangle. \tag{10.6.2}$$

Another property can be expressed by:

THEOREM. *If $\mathbf{L}$ is self-adjoint, then the eigenvalues λ_i are real, since* $\langle \boldsymbol{\phi}_i, \mathbf{L}\boldsymbol{\phi}_i \rangle = \langle \mathbf{L}\boldsymbol{\phi}_i, \boldsymbol{\phi}_i \rangle = \lambda_i \langle \boldsymbol{\phi}_i, \boldsymbol{\phi}_i \rangle = \lambda_i^* \langle \boldsymbol{\phi}_i, \boldsymbol{\phi}_i \rangle.$

Of course, if $\mathbf{L}$ is self-adjoint, it follows that $\langle \boldsymbol{\phi}_i, \boldsymbol{\phi}_j \rangle = 0$ if $\lambda_i \neq \lambda_j$.

If $\mathbf{L}$ is a perfect operator, i.e., if the eigenvectors $\boldsymbol{\phi}_i$, $i = 1, \ldots$, form a complete set, then the spectral resolution theorem holds:

$$\mathbf{L} = \sum_{i=1}^{\infty} \lambda_i \boldsymbol{\chi}_i^\dagger \boldsymbol{\phi}_i, \tag{10.6.3}$$

where $\langle \boldsymbol{\chi}_i^\dagger, \boldsymbol{\phi}_i \rangle = \delta_{ij}$. The set $\{\boldsymbol{\chi}_i\}$ is the reciprocal set to $\{\boldsymbol{\phi}_i\}$. Moreover, since the adjoint of Eq. (10.6.3) is

$$\mathbf{L}^\dagger = \sum_{i=1}^{\infty} \lambda_i^* \boldsymbol{\phi}_i^\dagger \boldsymbol{\chi}_i, \tag{10.6.4}$$

it follows that the vectors $\boldsymbol{\chi}_i$ are the eigenvectors of the adjoint $\mathbf{L}^\dagger$ of $\mathbf{L}$ and that the eigenvalues of $\mathbf{L}^\dagger$ are the complex conjugates of the eigenvalues of $\mathbf{L}$. Of course, when Eq. (10.6.3) holds, the function $f(\mathbf{L})$ obeys the decomposition theorem

$$f(\mathbf{L}) = \sum_{i=1}^{\infty} f(\lambda_i) \boldsymbol{\chi}_i^\dagger \boldsymbol{\phi}_i, \tag{10.6.5}$$

as long as the function $f(t)$ is defined at $t = \lambda_i$, $i = 1, 2, \ldots$.

A couple of simple examples of perfect differential operators are

$$\text{(i)} \quad Lu = i\frac{du}{dx}, \qquad 0 < x < 1, \ u(0) = u(1), \tag{10.6.6}$$

and

$$\text{(ii)} \quad Lu = -\frac{d^2 u}{dx^2}, \qquad 0 < x < 1, \ u(0) = u(1) = 0. \tag{10.6.7}$$

The reader can verify that these operators are self-adjoint. The eigenvectors and eigenvalues of case (i) are

$$\phi_n(x) = \exp(-i\lambda_n x), \qquad \lambda_n = 2\pi n, \ n = 0, \pm 1, \pm 2, \ldots, \tag{10.6.8}$$

and of case (ii) are

$$\phi_n(x) = \sqrt{2} \sin \sqrt{\lambda_n} x, \qquad \lambda_n = \pi n, \ n = 1, 2, \ldots. \tag{10.6.9}$$

We know from the theory of Fourier series that Eqs. (10.6.8) and (10.6.9) each define complete orthonormal sets in $\mathcal{L}_2(0, 1)$. We also see that for these perfect differential operators $|\lambda_n| \to \infty$ as $n \to \infty$.

We can also establish completeness of the eigenvectors of cases (i) and (ii) from the theory of integral equations. Consider first case (ii). The eigenproblem is

$$\mathbf{L}\boldsymbol{\phi} = \lambda\boldsymbol{\phi}, \qquad \boldsymbol{\phi} \in \mathcal{D}_L. \tag{10.6.10}$$

Since $\mathbf{L}\boldsymbol{\phi} = 0$ has only the trivial solution, the inverse $\mathbf{G}$ of $\mathbf{L}$ exists. Thus, since $\mathbf{GL} = \mathbf{I}$, multiplication of Eq. (10.6.10) by $\mathbf{G}$ yields (since $\lambda \neq 0$)

$$\mathbf{G}\boldsymbol{\phi} = \lambda^{-1}\boldsymbol{\phi} \tag{10.6.11}$$

or, recalling Green's function from Eq. (10.5.26), we obtain

$$\int_0^1 \left[\eta(x-y)y(1-x) + \eta(y-x)x(1-y)\right]\phi(y) = \lambda^{-1}\phi(x). \tag{10.6.12}$$

Since $\int_0^1 |g(x,y)|^2\, dx\, dy < \infty$ and $\mathbf{G} = \mathbf{G}^\dagger$, it follows that the integral operator $\mathbf{G}$ is a self-adjoint, Hilbert–Schmidt operator. In Chapter 9, we proved that such operators are perfect and have a complete orthonormal set of eigenvectors $\{\boldsymbol{\phi}_i\}$ in $\mathcal{H}$. Since these are also the eigenvectors of $\mathbf{L}$, it follows that $\mathbf{L}$ is perfect and that its eigenvalues λ_i are the reciprocal of the eigenvalues μ_i of $\mathbf{G}$. We also proved in Chapter 9 that the eigenvalues μ_i have a finite degeneracy (multiplicity) from which we conclude that so also do the eigenvalues λ_i.

In case (i), the homogeneous equation $Lu = 0$ has the solution $u = 1$, and so $\mathbf{L}$ has no inverse. However, we can define the operator $\tilde{\mathbf{L}}$ by

$$\tilde{\mathbf{L}}u \equiv \mathbf{L}u + u, \qquad u(0) = u(1) \tag{10.6.13}$$

for which solutions to

$$\tilde{\mathbf{L}}\boldsymbol{\phi} = \tilde{\lambda}\boldsymbol{\phi} \tag{10.6.14}$$

obey the equation

$$\mathbf{L}\boldsymbol{\phi} = (\tilde{\lambda} - 1)\boldsymbol{\phi}. \tag{10.6.15}$$

Thus, the eigenvectors of $\tilde{\mathbf{L}}$ are eigenvectors of $\mathbf{L}$ and the eigenvalues are related by $\lambda = \tilde{\lambda} - 1$. Green's function corresponding to $\tilde{\mathbf{L}}$ is

$$\tilde{g}(x,y) = \left[-\eta(x-y) + \frac{1}{1-\exp(-i)}\right] i\exp(i(x-y)). \tag{10.6.16}$$

Again, $\int_0^1 |\tilde{g}(x,y)|^2\, dx\, dy < \infty$ and $\tilde{\mathbf{L}}$ and $\tilde{\mathbf{G}}$ are self-adjoint so that Eq. (10.6.14) can be transformed to

$$\tilde{\mathbf{G}}\boldsymbol{\phi} = \tilde{\lambda}^{-1}\boldsymbol{\phi}, \tag{10.6.17}$$

where $\tilde{\mathbf{G}}$ is again a self-adjoint, Hilbert–Schmidt operator. Thus, the eigenvectors of $\tilde{\mathbf{G}}$ and, therefore, of $\tilde{\mathbf{L}}$ and $\mathbf{L}$, form a complete orthonormal set in $\mathcal{L}_2(0,1)$.

The device we just used in studying case (i) can be employed to obtain some general results for the spectral theory of differential operators. Suppose α is a number, real or complex, that is not an eigenvalue of a regular self-adjoint pth-order differential operator $\mathbf{L}$ (where $a_i(x)$ have continuous ith derivatives in the finite interval $[a, b]$ and $a_p(x) \neq 0$). Since $\mathbf{L}$ is self-adjoint, its eigenvalues are always real, and so any complex number α, $\text{Im}\,\alpha \neq 0$, will not be an eigenvalue of $\mathbf{L}$. Suppose the operator $\mathbf{L}$ is defined by

$$Lu = \sum_{i=0}^{p} a_i(x) \frac{d^i u}{dx^i}, \qquad B_j u = 0, \; j = 1, \ldots, p, \qquad (10.6.18)$$

with the requirement $\mathbf{u}, L\mathbf{u} \in \mathcal{H}$. We can define the new operator $\tilde{\mathbf{L}} \equiv \mathbf{L} + \alpha \mathbf{I}$ by

$$\tilde{L}u = \sum_{i=0}^{p} a_i(x) \frac{d^i u}{dx^i} + \alpha u, \qquad B_j u = 0, \; j = 1, \ldots, p, \qquad (10.6.19)$$

again with the requirement $\mathbf{u}, \tilde{L}\mathbf{u} \in \mathcal{H}$. $\mathbf{L}$ and $\tilde{\mathbf{L}}$ have the same eigenvectors because the equation

$$\tilde{\mathbf{L}}\boldsymbol{\phi}_i = \tilde{\lambda}_i \boldsymbol{\phi}_i \qquad (10.6.20)$$

can be rearranged to give

$$\mathbf{L}\boldsymbol{\phi}_i = (\tilde{\lambda}_i - \alpha)\boldsymbol{\phi}_i. \qquad (10.6.21)$$

Thus, the eigenvalues of $\mathbf{L}$ are $\lambda_i = \tilde{\lambda}_i - \alpha$.

Since α is not an eigenvalue of $\mathbf{L}$, the only solution to the homogeneous equation $\tilde{L}\mathbf{u} = \mathbf{0}$ is the trivial one $\mathbf{u} \equiv \mathbf{0}$, and so the inverse $\tilde{\mathbf{L}}^{-1}$ of $\mathbf{L}$ exists. This inverse is the integral operator $\tilde{\mathbf{G}}$ whose kernel $\tilde{g}(x, y)$ is Green's function for $\tilde{\mathbf{L}}$. The eigenvectors of $\tilde{\mathbf{L}}$ and $\tilde{\mathbf{G}}$ are the same because the equation

$$\tilde{\mathbf{G}}\boldsymbol{\phi}_i = \tilde{\gamma}_i \boldsymbol{\phi}_i \qquad (10.6.22)$$

can be rearranged to give

$$\tilde{\mathbf{L}}\boldsymbol{\phi}_i = \tilde{\gamma}_i^{-1} \boldsymbol{\phi}_i, \qquad (10.6.23)$$

where we used $\tilde{\mathbf{L}}^{-1} = \tilde{\mathbf{G}}$. Equations (10.6.21) and (10.6.23) demonstrate that $\mathbf{L}$ and $\tilde{\mathbf{G}}$ have the same eigenvectors. For the eigenvector $\boldsymbol{\phi}_i$, the eigenvalue λ_i of $\mathbf{L}$ is related to the eigenvalue $\tilde{\gamma}_i$ of $\tilde{\mathbf{G}}$ by the expression $\tilde{\gamma}_i = \tilde{\lambda}_i^{-1} = (\lambda_i + \alpha)^{-1}$. The eigenvalues $\tilde{\lambda}_i$ cannot be 0 since $\tilde{\mathbf{L}}$ is nonsingular. However, λ_i can be 0 and $\tilde{\gamma}_i$ may approach 0 as $i \to \infty$. Correspondingly, $|\lambda_i|$ may approach ∞ as $i \to \infty$, in keeping with the fact that a differential operator is unbounded (keep in mind that $\mathbf{L}$ may not even have an eigenvector—as, for example, in the case of initial value boundary conditions.) Note that if $\boldsymbol{\phi}_i$ is an eigenvector such that $\mathbf{L}\boldsymbol{\phi}_i = \mathbf{0}$, then $\boldsymbol{\phi}_i$ will be an eigenvector of $\tilde{\mathbf{G}}$ with eigenvalue $\tilde{\gamma}_i = 1/\alpha$. Thus, $\mathbf{L}$ and $\tilde{\mathbf{G}}$ have exactly the same eigenvectors.

The kernel $\tilde{g}(x, y)$ of $\tilde{\mathbf{G}}$ is a continuous function of x and y, and so, since $[a, b]$ is a finite interval, $\tilde{g}(x, y)$ is square integrable in the bounded rectangle $[a, b] \times [a, b]$. We proved in Section 9.5.2 that if the kernel of an integral operator $\tilde{\mathbf{G}}$ is square integrable, then $\tilde{\mathbf{G}}$ has only a finite number of eigenvectors for any nonzero eigenvalue. This enables us to state the theorem:

THEOREM. *If* **L** *is a pth-order differential operator in a finite interval* $[a, b]$, *if* $a_p(x) \neq 0$ *and* $a_i(x)$ *is continuous in* $[a, b]$, *and if there exists a number* α *not equal to an eigenvalue of* **L**, *then the number of eigenvectors corresponding to a given eigenvalue* λ_i *is finite.* λ_i *can be zero or nonzero.*

The property implied by this theorem distinguishes a differential operator from an integral operator. Consider, for example, the self-adjoint integral operator

$$\mathbf{K} = \boldsymbol{\phi}_1 \boldsymbol{\phi}_1^\dagger, \tag{10.6.24}$$

where $\boldsymbol{\phi}_1$ is the first of a complete orthonormal set $\{\boldsymbol{\phi}_1, \boldsymbol{\phi}_2, \ldots\}$. In this case, **K** has an infinite number of eigenvectors, $\boldsymbol{\phi}_2, \boldsymbol{\phi}_3, \ldots$, corresponding to the eigenvalue $\lambda = 0$. By the theorem just given, this cannot happen for a self-adjoint differential operator defined in a finite interval $[a, b]$.

There is another way to prove that the degeneracy of eigenvectors of a differential operator is finite. Consider the eigenproblem

$$Lu - \lambda u = 0, \tag{10.6.25}$$

where L is a pth-order differential operator with the boundary conditions

$$B_i u = 0, \quad i = 1, \ldots, p. \tag{10.6.26}$$

Suppose λ is an eigenvalue of L, 0 or otherwise. Equation (10.6.25) is a pth-order homogeneous equation,

$$\tilde{L}u = a_p(x)\frac{d^p u}{dx^p} + \cdots + a_1(x)\frac{du}{dx} + (a_0(x) - \lambda)u = 0. \tag{10.6.27}$$

If $\{u_1, \ldots, u_p\}$ is a fundamental system of $\tilde{L}u = 0$, a solution to Eq. (10.6.27)—which exists by the hypothesis that λ is an eigenvalue of L—can be expressed as

$$u = \sum_{i=1}^{p} \alpha_i u_i. \tag{10.6.28}$$

The boundary conditions, Eq. (10.6.26), determine the values of α_i. Insertion of Eq. (10.6.28) into Eq. (10.6.26) yields

$$\begin{bmatrix} B_1 u_1 & \cdots & B_1 u_p \\ \vdots & & \vdots \\ B_p u_1 & \cdots & B_p u_p \end{bmatrix} \begin{bmatrix} \alpha_1 \\ \vdots \\ \alpha_p \end{bmatrix} = \mathbf{0}. \tag{10.6.29}$$

The rank r of the matrix $[B_i u_j]$ determines how many linearly independent solutions $\{\alpha_1, \ldots, \alpha_p\}$ Eq. (10.6.29) admits. The rank r must lie between 0 and $p-1$, since we assumed at the outset that λ is an eigenvalue. The number of linearly independent solutions to Eq. (10.6.29), $p-r$, corresponds to the number of eigenfunctions L has for the eigenvalue λ. This yields the theorem:

THEOREM. *If* **L** *is a pth-order differential operator, then the number of eigenvectors corresponding to a given eigenvalue is less than or equal to p.*

Consider a pth-order differential operator $\mathbf{L}$ that has an adjoint, i.e., $a_p(x) \neq 0$ and $a_i(x) \in C^{(i)}(a, b)$. And suppose $\mathbf{L}$ is an operator for which there exists a real or complex number α that is not an eigenvalue. A self-adjoint operator is an example of such an operator. Since such an operator, if singular, can be converted to a nonsingular one by the addition of $\alpha \mathbf{I}$ to $\mathbf{L}$ (where α is not an eigenvalue of $\mathbf{L}$), and since $\tilde{\mathbf{L}}^\dagger \tilde{\mathbf{L}} = \tilde{\mathbf{L}} \tilde{\mathbf{L}}^\dagger$ if $\mathbf{L}^\dagger \mathbf{L} = \mathbf{L} \mathbf{L}^\dagger$, we can assume, for the purpose of proving the next theorem, that $\mathbf{L}$ is nonsingular. Consider the pth-order differential operator $\mathbf{L}$ and its adjoint with boundary conditions

$$B_i u = B_i^\dagger u = 0, \qquad i = 1, \ldots, p. \tag{10.6.30}$$

We assume that $\mathbf{L}\mathbf{u} = \mathbf{0}$ has only the trivial solution $\mathbf{u} = \mathbf{0}$ and that the differential expressions L and $L^\dagger$ obey the relationship

$$LL^\dagger w = L^\dagger L w. \tag{10.6.31}$$

We say $\mathbf{L}$ is a *normal differential operator* if it obeys Eqs. (10.6.30) and (10.6.31). We define the inverse of $\mathbf{L}$ as $\mathbf{G}$ and that of $\mathbf{L}^\dagger$ as $\mathbf{G}^\dagger$. From Eqs. (10.6.30) and (10.6.31), it follows that

$$\mathbf{L}\mathbf{L}^\dagger = \mathbf{L}^\dagger \mathbf{L} \quad \text{or} \quad (\mathbf{L}^\dagger)^{-1}\mathbf{L}^{-1} = \mathbf{L}^{-1}(\mathbf{L}^\dagger)^{-1}, \tag{10.6.32}$$

giving

$$\mathbf{G}^\dagger \mathbf{G} = \mathbf{G}\mathbf{G}^\dagger. \tag{10.6.33}$$

Since the kernel of $\mathbf{G}$ is square integrable in the bounded rectangle $[a, b] \times [a, b]$ and obeys Eq. (10.6.33), we conclude that $\mathbf{G}$ is a normal, completely continuous operator. According to the theorem proved in Section 9.5.7, such an operator possesses a complete set of orthonormal eigenvectors $\boldsymbol{\phi}_i, i = 1, 2, \ldots$, and since $\mathbf{L}\boldsymbol{\phi}_i = \nu_i^{-1}\boldsymbol{\phi}_i$ if $\mathbf{G}\boldsymbol{\phi}_i = \nu_i\boldsymbol{\phi}_i$, the following theorem is obeyed:

THEOREM. *If L is a normal differential expression (i.e., it obeys Eqs. (10.6.30) and (10.6.31)) and is nonsingular, or can be made nonsingular by the addition of $\alpha \mathbf{I}$, then the eigenvectors $\boldsymbol{\phi}_i$, $i = 1, 2, \ldots$, of the operator $\mathbf{L}$ form a complete orthonormal set.*

Since the eigenvalues of a self-adjoint operator are real, there always exists a complex number such that $\mathbf{L} + \alpha \mathbf{I}$ is not singular, and so the theorem always holds for self-adjoint operators.

The theorem yields the spectral decomposition of $\mathbf{L}$ and $f(\mathbf{L})$, namely,

$$\mathbf{L} = \sum_i \lambda_i \boldsymbol{\phi}_i \boldsymbol{\phi}_i^\dagger \quad \text{and} \quad f(\mathbf{L}) = \sum_i f(\lambda_i) \boldsymbol{\phi}_i \boldsymbol{\phi}_i^\dagger, \tag{10.6.34}$$

as long as $f(t)$ exists for $t = \lambda_i$, $i = 1, 2, \ldots$. The theorem generalizes to pth-order differential operators the theorem proved in Chapter 9 for normal, completely continuous operators. Self-adjoint operators are a subset of the class defined by Eqs. (10.6.30) and (10.6.31).

It is important to note that there are differential operators for which there does not exist a complex number that is not an eigenvalue. For example, consider

$$Lu = -\frac{d^2 u}{dx^2}, \qquad u(0) = -u(1), \quad u'(0) = u'(1). \tag{10.6.35}$$

The general solution to $Lu = \lambda u$ is

$$u = c_1 \sin \sqrt{\lambda} x + c_2 \cos \sqrt{\lambda} x, \tag{10.6.36}$$

where c_1 and c_2 are arbitrary constants ($\sin \sqrt{\lambda} x$ and $\cos \sqrt{\lambda} x$ form a fundamental system for the equation $Lu - \lambda u = 0$). Applying the boundary conditions to Eq. (10.6.36), we find

$$\begin{aligned} c_1 \sin \sqrt{\lambda} + c_2 (1 + \cos \sqrt{\lambda}) &= 0 \\ c_1 \sqrt{\lambda}(1 - \cos \sqrt{\lambda}) + c_2 \sqrt{\lambda} &= 0. \end{aligned} \tag{10.6.37}$$

Equation (10.6.37) has a nontrivial resolution for c_1 and c_2 if and only if

$$\begin{vmatrix} \sin \sqrt{\lambda} & 1 + \cos \sqrt{\lambda} \\ \sqrt{\lambda}(1 - \cos \sqrt{\lambda}) & \sqrt{\lambda} \sin \sqrt{\lambda} \end{vmatrix} = 0. \tag{10.6.38}$$

However, this determinant is

$$\sqrt{\lambda} \left[\sin^2 \sqrt{\lambda} + \cos^2 \sqrt{\lambda} - 1 \right] = 0 \qquad \text{for any } \lambda. \tag{10.6.39}$$

Thus, the eigenvectors of the operator **L** defined by Eq. (10.6.35) are given by Eq. (10.6.36), where the eigenvalues λ are any complex number. There is then no number α such that $\mathbf{L} + \alpha \mathbf{I}$ is nonsingular. Of course, in this case **L** is neither self-adjoint nor normal.

There are also non-self-adjoint operators **L** that have no eigenvectors or eigenvalues. For example,

$$Lu = -\frac{d^2 u}{dx^2}, \qquad u(0) = -2u(\pi), \quad u'(0) = 2u'(\pi). \tag{10.6.40}$$

The general solution to $Lu = \lambda u$ is again given by Eq. (10.6.34) and nonzero values of c_1 and c_2 can be found if

$$\begin{aligned} 0 &= \begin{vmatrix} 2 \sin \pi \sqrt{\lambda} & 1 + 2 \cos \pi \sqrt{\lambda} \\ \sqrt{\lambda}(1 - 2 \cos \pi \sqrt{\lambda}) & 2\sqrt{\lambda} \sin \pi \sqrt{\lambda} \end{vmatrix} \\ &= \sqrt{\lambda}(4 \sin^2 \pi \sqrt{\lambda} + 4 \sin^2 \pi \sqrt{\lambda} - 1) = 3\sqrt{\lambda}. \end{aligned} \tag{10.6.41}$$

From Eq. (10.6.41), we see that the only candidate for an eigenvalue is $\lambda = 0$. In this case, $u = c_2$. But the boundary condition $u(0) = -2u(\pi)$ yields $c_2 = -2c_2$ or $c_2 = 0$. Therefore, the operator defined by Eq. (10.6.40) has no eigenvalues.

10.7. SPECTRAL THEORY OF REGULAR STURM–LIOUVILLE OPERATORS

Consider the second-order differential expression

$$Lu = -\frac{1}{s(x)}\frac{d}{dx}\left(p(x)\frac{du}{dx}\right) + \frac{q(x)}{s(x)}u, \qquad a < x < b, \tag{10.7.1}$$

where $[a, b]$ is a finite interval; $s(x)$, $p(x)$, $p'(x)$, and $q(x)$ are real, continuous functions; and $s(x)$ and $p(x)$ are positive in the interval $[a, b]$. $s(x)$ and $p(x)$ are strictly positive; i.e., their minimum values for $x \in [a, b]$ are greater than a nonzero positive number. Such an operator is called a *Sturm–Liouville operator*. It will be convenient to define the operator corresponding to L in the Hilbert space $\mathcal{L}_2(a, b; s)$. The inner product in this Hilbert space is

$$\langle \mathbf{v}, \mathbf{u} \rangle \equiv \int_a^b s(x) v^*(x) u(x)\, dx. \tag{10.7.2}$$

The formal adjoint differential expression $L^\dagger$ corresponding to L, defined by the expression

$$\int_a^b s(x)\left[v^* Lu - (L^\dagger v)^* u\right] dx = p(x)\left[v^*(x) u'(x) - \left(v'(x)\right)^* u(x)\right]_a^b, \tag{10.7.3}$$

is the same as L. Thus, the Sturm–Liouville operator is formally self-adjoint. The operator $\mathbf{L}$ in $\mathcal{L}_2(a, b; s)$ is defined by the differential expression L and the boundary conditions

$$\begin{aligned} B_1 u &= \alpha_{11} u(a) + \alpha_{12} u'(a) + \alpha_{13} u(b) + \alpha_{14} u'(b) = 0 \\ B_2 u &= \alpha_{21} u(a) + \alpha_{22} u'(a) + \alpha_{23} u(b) + \alpha_{24} u'(b) = 0, \end{aligned} \tag{10.7.4}$$

where the coefficients α_{ij} are real numbers. The adjoint operator $\mathbf{L}^\dagger$ is determined by the differential expression L and the boundary conditions $B_1^\dagger v = B_2^\dagger v = 0$ derived from the condition

$$p(x)\left[v^*(x) u'(x) - \left(v'(x)\right)^* u(x)\right]_a^b = 0, \quad \text{for all } \mathbf{u} \in \mathcal{D}_L, \tag{10.7.5}$$

where

$$\mathcal{D}_L = \{\mathbf{u}, \mathbf{L}\mathbf{u} \in \mathcal{L}_2(a, b; s); \ B_1 u = B_2 u = 0\}. \tag{10.7.6}$$

The boundary functionals $B_i u$ of interest here are those for which $B_i^\dagger u = B_i u$, $i = 1, 2$, so that $\mathbf{L}^\dagger = \mathbf{L}$ (i.e., $\mathbf{L}$ is a self-adjoint operator). It is interesting to note that $\mathbf{L}$ cannot be self-adjoint if the functionals $B_i u$ define an initial value problem. To prove this, consider the case of initial conditions

$$\begin{aligned} B_1 u &= \alpha_{11} u(a) + \alpha_{12} u'(a) = 0 \\ B_2 u &= \alpha_{21} u(a) + \alpha_{22} u'(a) = 0. \end{aligned} \tag{10.7.7}$$

Since the matrix

$$\begin{bmatrix} \alpha_{11} & \alpha_{12} \\ \alpha_{21} & \alpha_{22} \end{bmatrix} \tag{10.7.8}$$

has rank 2, it follows that $u(a) = u'(a) = 0$ in this case. Equation (10.7.5) then becomes

$$p(b)\left[v^*(b)u'(b) - \left(v'(b)\right)^* u(b)\right] = 0 \tag{10.7.9}$$

for any $\mathbf{u} \in \mathcal{D}_L$. Since $u(b)$ and $u'(b)$ are arbitrary, it follows that the boundary conditions for $\mathbf{L}^\dagger$ are $v(b) = v'(b) = 0$—which are different from the boundary conditions of $\mathbf{L}$—and so $\mathbf{L}$ cannot be self-adjoint for the initial value problem.

The general condition for self-adjointness of the Sturm–Liouville operator is summarized in the following theorem:

THEOREM. *The Sturm–Liouville operator defined by Eqs.* (10.7.1) *and* (10.7.4) *is self-adjoint if and only if*

$$p(a)\begin{vmatrix} \alpha_{13} & \alpha_{14} \\ \alpha_{23} & \alpha_{24} \end{vmatrix} = p(b)\begin{vmatrix} \alpha_{11} & \alpha_{12} \\ \alpha_{21} & \alpha_{22} \end{vmatrix}. \tag{10.7.10}$$

Since $L = L^\dagger$, self-adjointness merely requires that $B_i^\dagger = B_i$, $i = 1, 2$, i.e., that v and u satisfy the same boundary conditions. This requirement can be summarized by the matrix equation

$$\begin{vmatrix} B_1 u & B_1 v^* \\ B_2 u & B_2 v^* \end{vmatrix} = \begin{vmatrix} 0 & 0 \\ 0 & 0 \end{vmatrix}, \tag{10.7.11}$$

which can be rearranged to give

$$\begin{bmatrix} \alpha_{11}u(a) + \alpha_{12}u'(a) & \alpha_{11}v^*(a) + \alpha_{12}\left(v'(a)\right)^* \\ \alpha_{21}u(a) + \alpha_{22}u'(a) & \alpha_{21}v^*(a) + \alpha_{22}\left(v'(a)\right)^* \end{bmatrix}$$
$$= -\begin{bmatrix} \alpha_{13}u(b) + \alpha_{14}u'(b) & \alpha_{13}v^*(b) + \alpha_{14}\left(v'(b)\right)^* \\ \alpha_{23}u(b) + \alpha_{24}u'(b) & \alpha_{23}v^*(b) + \alpha_{24}\left(v'(b)\right)^* \end{bmatrix}. \tag{10.7.12}$$

Taking the determinant of each side of Eq. (10.7.12) and using the elementary properties of determinants, we find that the expression

$$\begin{vmatrix} \alpha_{11} & \alpha_{12} \\ \alpha_{21} & \alpha_{22} \end{vmatrix} \begin{vmatrix} u(a) & v^*(a) \\ u'(a) & (v'(a))^* \end{vmatrix}$$
$$= \begin{vmatrix} \alpha_{13} & \alpha_{14} \\ \alpha_{23} & \alpha_{24} \end{vmatrix} \begin{vmatrix} u(b) & v^*(b) \\ u'(b) & (v'(b))^* \end{vmatrix} \tag{10.7.13}$$

is the condition that the boundary conditions of $\mathbf{L}$ and $\mathbf{L}^\dagger$ are the same. However, by the definition of $\mathbf{L}^\dagger$, in Eq. (10.7.3), u and v satisfy

$$p(a)\begin{vmatrix} u(a) & v^*(a) \\ u'(a) & (v'(a))^* \end{vmatrix} = p(b)\begin{vmatrix} u(b) & v^*(b) \\ u'(b) & (v'(b))^* \end{vmatrix}. \tag{10.7.14}$$

There are two possibilities to consider here. The first is that the determinants $\alpha_{11}\alpha_{22} - \alpha_{21}\alpha_{12}$ and $\alpha_{13}\alpha_{24} - \alpha_{23}\alpha_{14}$ are not 0. In this case, division of the left- and

SPECTRAL THEORY OF REGULAR STURM–LIOUVILLE OPERATORS

right-hand sides of Eq. (10.7.13) by the left- and right-hand sides, respectively, of Eq. (10.7.10) yields Eq. (10.7.14), proving that the condition at Eq. (10.7.10) yields self-adjoint boundary conditions for this case. The other possibility is that the determinants $\alpha_{11}\alpha_{22} - \alpha_{21}\alpha_{12}$ and $\alpha_{13}\alpha_{24} - \alpha_{23}\alpha_{14}$ are both 0. The only other alternative—that one of the determinants is 0 and the other is not—defines an initial value problem that we have already shown will not yield a self-adjoint operator. Since the rank of the matrix $[\alpha_{ij}]$ (for $i = 1, 2, j = 1, \ldots, 4$) is 2, at least one of the determinants

$$R_1 = \begin{vmatrix} \alpha_{11} & \alpha_{14} \\ \alpha_{21} & \alpha_{24} \end{vmatrix} \quad R_2 = \begin{vmatrix} \alpha_{12} & \alpha_{14} \\ \alpha_{22} & \alpha_{24} \end{vmatrix},$$

$$R_3 = \begin{vmatrix} \alpha_{11} & \alpha_{13} \\ \alpha_{21} & \alpha_{24} \end{vmatrix} \quad R_4 = \begin{vmatrix} \alpha_{12} & \alpha_{13} \\ \alpha_{22} & \alpha_{23} \end{vmatrix} \tag{10.7.15}$$

is nonzero since

$$\begin{vmatrix} \alpha_{11} & \alpha_{12} \\ \alpha_{21} & \alpha_{22} \end{vmatrix} = \begin{vmatrix} \alpha_{13} & \alpha_{14} \\ \alpha_{23} & \alpha_{24} \end{vmatrix} = 0 \tag{10.7.16}$$

for this case. From the linear combinations $\alpha_{24}B_1 u - \alpha_{14}B_2 u = 0$, $\alpha_{23}B_1 u - \alpha_{13}B_2 u = 0$, and the relation $\alpha_{13}\alpha_{24} - \alpha_{23}\alpha_{14} = 0$, we obtain

$$R_1 u(a) + R_2 u'(a) = 0$$
$$R_3 u(a) + R_4 u'(a) = 0 \tag{10.7.17}$$

The equations in Eq. (10.7.17) are also satisfied by v^*, i.e.,

$$R_1 v^*(a) + R_2 \bigl(v'(a)\bigr)^* = 0$$
$$R_3 v^*(a) + R_4 \bigl(v'(a)\bigr)^* = 0 \tag{10.7.18}$$

Consider $R_1, \ldots, R_4$ to be unknowns of the four equations in Eqs. (10.7.17) and (10.7.18). Since at least one of the R_i, $i = 1, \ldots, 4$, is nonzero, it follows that the determinant

$$\begin{vmatrix} u(a) & u'(a) & 0 & 0 \\ v^*(a) & \bigl(v'(a)\bigr)^* & 0 & 0 \\ 0 & 0 & u(a) & u'(a) \\ 0 & 0 & v^*(a) & \bigl(v'(a)\bigr)^* \end{vmatrix} \tag{10.7.19}$$

is 0, or, equivalently,

$$\begin{vmatrix} u(a) & u'(a) \\ v^*(a) & \bigl(v'(a)\bigr)^* \end{vmatrix}^2 = 0. \tag{10.7.20}$$

Analogously, the equations $\alpha_{22}B_1u - \alpha_{12}B_2u = 0$, $\alpha_{21}B_1u - \alpha_{11}B_2u = 0$, and the relation $\alpha_{11}\alpha_{22} - \alpha_{12}\alpha_{21} = 0$ lead to the condition

$$\begin{vmatrix} u(b) & u'(b) \\ v^*(b) & (v'(b))^* \end{vmatrix}^2 = 0. \qquad (10.7.21)$$

It follows from Eqs. (10.7.20) and (10.7.21) that Eq. (10.7.14) holds, and so $\mathcal{D}_L = \mathcal{D}_{L^\dagger}$ for this case. This completes the proof of the sufficiency part of the theorem. The proof of necessity requires the assumption that Eqs. (10.7.13) and (10.7.14) are true and then proving that Eq. (10.7.10) is implied. This part will be left to the reader.

It is important to note that there is a Hilbert space in which Eq. (10.7.10) guarantees self-adjointness for *any* real, regular second-order differential operator. Consider the general operator

$$Lu = -\left[a_2(x)\frac{d^2u}{dx^2} + a_1(x)\frac{dy}{dx} + a_0(x)u \right], \qquad a < x < b, \qquad (10.7.22)$$

where $a_i(x)$ are continuous and real, $a_2(x) > 0$, and $[a, b]$ is a finite interval. Next consider the differential operator defined by

$$L'u = -a_2(x)\exp\left(-\int_c^x \frac{a_1(\xi)}{a_2(\xi)}d\xi\right)\frac{d}{dx}\left\{\exp\left(\int_c^x \frac{a_1(\xi)}{a_2(\xi)}d\xi\right)\frac{du}{dx}\right\} + a_0(x)u, \qquad (10.7.23)$$

where c is an arbitrary point in $[a, b]$. Carrying out the differentiation on the right-hand side of Eq. (10.7.23), we find that $L' = L$. It follows then that if we set

$$p(x) = \exp\left(\int_c^x \frac{a_1(\zeta)}{a_2(\zeta)}d\zeta\right), \quad s(x) = \frac{a_2(x)}{p(x)} \quad \text{and} \quad q(x) = a_0(x)s(x), \qquad (10.7.24)$$

we can transform any real, regular second-order differential expression Lu into a self-adjoint form with the appropriate Hilbert space $\mathcal{L}_2(a, b; s(x))$. Thus, *if and only if* the boundary functions B_1u and B_2u obey Eq. (10.7.10), the operator defined by Eq. (10.7.22) is self-adjoint in the Hilbert space $\mathcal{L}_2(a, b; s)$, where $s(x)$ is given in Eq. (10.7.24). It follows from the transformation of Eq. (10.7.27) into Eq. (10.7.23) that Sturm–Liouville operators represent the entire class of real, regular second-order differential operators.

EXERCISE 10.7.1. Consider the operator **L** defined by

$$Lu = \frac{d^2u}{dx^2} - \frac{du}{dx}, \qquad 0 < x < 1, \ u(0) = u(1) = 0. \qquad (10.7.25)$$

(a) Find the Hilbert space in which Lu is formally self-adjoint.
(b) Is **L** self-adjoint for the boundary conditions given?
(c) Find the eigenvalues and eigenvectors of **L**.
(d) Solve the problem

$$\mathbf{L}u = \frac{\partial u}{\partial t}, \qquad (10.7.26)$$

where $u(x, 0) = 1$.

SPECTRAL THEORY OF REGULAR STURM–LIOUVILLE OPERATORS

(e) Solve the problem

$$Lu = f, \qquad (10.7.27)$$

■ ■ ■ where $f(x) = x$.

Some examples of self-adjoint boundary conditions for the operator

$$Lu = -\frac{d^2u}{dx^2}, \qquad 0 < x < \pi, \qquad (10.7.28)$$

are

(i) $u(0) = u(\pi) = 0$
(ii) $u(0) = u(\pi), \qquad u'(0) = u'(\pi)$
(iii) $u(0)\sin\theta + u'(0)\cos\theta = 0, \qquad \theta$ arbitrary
$u(\pi)\sin\phi + u'(\pi)\cos\phi = 0, \qquad \phi$ arbitrary. $\qquad (10.7.29)$

Consider now the eigenproblem

$$Lu = \lambda u \qquad (10.7.30)$$

for the operator in Eq. (10.7.28). A fundamental system for $Lu - \lambda u = 0$ is

$$u_1 = \sin\sqrt{\lambda}x, \qquad u_2 = \cos\sqrt{\lambda}x. \qquad (10.7.31)$$

The eigenproblem is solved by setting

$$u = \sum_{i=1}^{2} c_i u_i \qquad (10.7.32)$$

and applying the boundary conditions $B_i u = 0, i = 1, 2$. This leads to the equations

$$\begin{aligned} c_1 B_1 u_1 + c_2 B_1 u_2 &= 0 \\ c_1 B_2 u_1 + c_2 B_2 u_2 &= 0, \end{aligned} \qquad (10.7.33)$$

which have a solution *if and only if*

$$\begin{vmatrix} B_1 u_1 & B_1 u_2 \\ B_2 u_1 & B_2 u_2 \end{vmatrix} = 0. \qquad (10.7.34)$$

For the three examples in Eq. (10.7.29), we find the following equations for the eigenvalues of **L**:

(i) $\quad \begin{vmatrix} 0 & 1 \\ \sin \pi \sqrt{\lambda} & \cos \pi \sqrt{\lambda} \end{vmatrix} = 0 \qquad (10.7.35)$

(ii) $\quad \begin{vmatrix} -\sin \pi \sqrt{\lambda} & 1 - \cos \pi \sqrt{\lambda} \\ \sqrt{\lambda}(1 - \cos \pi \sqrt{\lambda}) & \sqrt{\lambda} \sin \pi \sqrt{\lambda} \end{vmatrix} = 0 \qquad (10.7.36)$

(iii) $\quad \begin{vmatrix} \sqrt{\lambda} \cos \theta & \sin \theta \\ \sin \phi \sin \pi \sqrt{\lambda} + \sqrt{\lambda} \cos \phi \cos \pi \sqrt{\lambda} & \sin \phi \cos \pi \sqrt{\lambda} - \sqrt{\lambda} \cos \phi \sin \pi \sqrt{\lambda} \end{vmatrix} = 0 \qquad (10.7.37)$

For case (i), we find that the eigenvalues must satisfy

$$\sin \pi \sqrt{\lambda} = 0. \qquad (10.7.38)$$

Candidate eigenvalues are then $\lambda_j = j^2$, $j = 0, 1, 2, \ldots$. Note, however, that the case $\lambda_0 = 0$ is not an eigenvalue since the eigenvector

$$u_j(x) = c_1 \sin \sqrt{\lambda_j} x + c_2 \cos \sqrt{\lambda_j} x \qquad (10.7.39)$$

yields $u_0 = c_2$ for $\lambda_0 = 0$. This does not satisfy the boundary condition $u(0) = u(\pi) = 0$. For $\lambda_j = j^2$, $j = 1, 2, \ldots$, $u_j(0) = u_j(\pi) = 0$ if $c_2 = 0$. Thus, the normalized eigenvectors of **L** for case (i) are

$$\phi_j = \sqrt{\frac{2}{\pi}} \sin jx, \qquad j = 1, 2, \ldots, \qquad (10.7.40)$$

with eigenvalues $\lambda_j = j^2$, $j = 1, 2, \ldots$. We know from the theorem given in the previous section that $\boldsymbol{\phi}_j$, $j = 1, 2, \ldots$, form a complete orthonormal set in $\mathcal{L}_2(0, \pi)$.

For case (ii), $\lambda = 0$ is an eigenvalue and corresponds to the eigenfunction $\phi_0 = 1/\sqrt{\pi}$. For $\lambda \neq 0$, Eq. (10.7.36) requires

$$-\sin^2 \pi \sqrt{\lambda} - 1 + 2\cos \pi \sqrt{\lambda} - \cos^2 \pi \sqrt{\lambda} = 0$$

or

$$\cos \pi \sqrt{\lambda} = 1. \qquad (10.7.41)$$

It follows from Eq. (10.7.41) that $\sqrt{\lambda} = j$, $j = 0, 1, 2, \ldots$. The rank of the determinant in Eq. (10.7.36) is 0, and so there are two linearly independent solutions corresponding to each eigenvalue $\lambda_j = 4j^2$, $j = 1, 2, \ldots$. For $\lambda_0 = 0$, there is only one eigenvector since $\sin \sqrt{\lambda_0} x = 0$. The orthonormal eigenfunctions for case (ii) are thus

$$\phi_0 = \frac{1}{\sqrt{\pi}}$$

$$\phi_j^{(1)} = \sqrt{\frac{2}{\pi}} \sin 2jx \qquad (10.7.42)$$

$$\phi_j^{(2)} = \sqrt{\frac{2}{\pi}} \cos 2jx, \qquad j = 1, 2, \ldots.$$

Again, these form a complete orthonormal set in $\mathcal{L}_2(0, \pi)$ and, as allowed by the theorem proven earlier, this second-order differential operator has two eigenvectors per eigenvalue (for the nonzero λ_j's).

If, for case (iii), we assume $\theta = 90°$ and $\phi = -45°$, we obtain the boundary conditions

$$u(0) = 0 \quad \text{and} \quad u(\pi) = u'(\pi). \qquad (10.7.43)$$

These conditions are appropriate for a heat transfer problem in a slab in which the slab is insulated at $x = 0$ and is in contact with a fluid at $x = \pi$. Again, $\lambda = 0$ is not an eigenvalue for this case and the eigenvalues obey the equation

$$\sin \pi \sqrt{\lambda} = \sqrt{\lambda} \cos \pi \sqrt{\lambda}. \qquad (10.7.44)$$

For this problem, the eigenvalues λ_j have to be found numerically. From the elementary properties of the equation $\tan x = x$, we can say that $\lambda_j \to \infty$ as j^2 as $j \to \infty$ and that, for large j, $\lambda_j \approx (2j + 1)^2/4$. The normalized eigenfunctions for this example are

$$\phi_j = \left(\frac{2\pi - \sin 2\pi \sqrt{\lambda_j}}{4} \right)^{1/2} \sin \sqrt{\lambda_j} x, \qquad j = 1, 2, \ldots. \qquad (10.7.45)$$

Although it is not apparent from Eq. (10.7.45) that these eigenfunctions are orthogonal in $\mathcal{L}_2(0, \pi)$, we know that they are from the basic theory of self-adjoint operators.

■ **EXAMPLE 10.7.1.** We can solve case (iii) above numerically for arbitrary values of θ and ϕ by using the Newton–Raphson method of Chapter 3 to solve for λ in Eq. (10.7.37). The eigenvalues of **L** defined in Eq. (10.7.28), in this case, must satisfy

$$\begin{aligned} 0 = &\sqrt{\lambda} \cos \theta \left[\sin \phi \cos(\pi \sqrt{\lambda}) - \sqrt{\lambda} \cos \phi \sin(\pi \sqrt{\lambda}) \right] \\ &- \sin \theta \left[\sqrt{\lambda} \cos \phi \cos(\pi \sqrt{\lambda}) + \sin \phi \sin(\pi \sqrt{\lambda}) \right]. \end{aligned} \qquad (10.7.46)$$

As an example, we will choose $\theta = 2$ and $\phi = 1.5$ (radians). Figure 10.7.1 contains a plot of the right-hand side of Eq. (10.7.46). We choose, as our initial guesses, estimates of the zeros read right off the graph. Using the set of initial guesses $\{1, 3, 7, 13, 21, 30, 42, 55, 71, 88\}$, the first 10 corresponding eigenvalues are

$$\{\lambda_n\} = \{0.741483, 3.04912, 7.07338, 12.9198, 20.6431, \\ 30.2711, 41.8202, 55.3014, 70.7238, 88.0946, \ldots\}. \qquad (10.7.47)$$

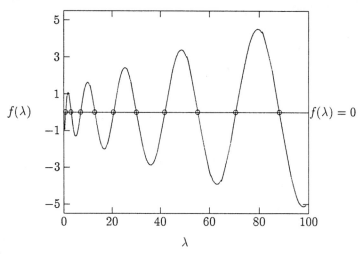

FIGURE 10.7.1

The corresponding eigenvectors can easily be determined from Eqs. (10.7.31) and (10.7.32) as

$$u_n(x) = c_{1,n} \sin \sqrt{\lambda_n} x + c_{2,n} \cos \sqrt{\lambda_n} x, \qquad (10.7.48)$$

where the coefficients are determined from

$$c_1 = -\frac{c_2}{\sqrt{\lambda}} \tan \theta, \qquad (10.7.49)$$

with the appropriate normalization.

Note that, although the solution $\lambda = 0$ satisfies Eq. (10.7.37), it is not an eigenvalue for this example. We can see this directly by noting that the coefficients c_1 and c_2 are solutions to the matrix equation

$$\begin{bmatrix} 0 & \sin \theta \\ 0 & \sin \phi \end{bmatrix} \begin{bmatrix} c_1 \\ c_2 \end{bmatrix} = \mathbf{0} \qquad (10.7.50)$$

when $\lambda = 0$. The only nontrivial solution possible is when $\sin \theta = \sin \phi = 0$, in which case $c_1 = 0$ (by default since $\sin(0) = 0$ in Eq. (10.7.48)) and $c_2 = 1/\sqrt{\pi}$ (after normalization).

EXAMPLE 10.7.2. In the classical physics of heat and mass transfer, one frequently encounters problems of the form

$$\frac{\partial u}{\partial t} = -DLu, \qquad B_1 u = B_2 u = 0, \ a < x < b, \qquad (10.7.51)$$

where $u(x, t)$ is a function of two variables (position x and time t). Typically, initial conditions of the type $u(x, 0) = f(x)$ are considered. In Eq. (10.7.51), D is a transport coefficient and is a positive scalar. In these problems, $Lu = -d^2u/dx^2$, and when the boundary conditions are such that **L** is self-adjoint, the solution to Eq. (10.7.51) is

$$\mathbf{u} = \exp(-t D \mathbf{L})\mathbf{f} = \sum_j \exp(-t D \lambda_j) \langle \boldsymbol{\phi}_j, \mathbf{f} \rangle \boldsymbol{\phi}_j, \qquad (10.7.52)$$

or

$$u(x) = \sum_j \exp(-t D \lambda_j) \phi_j(x) \int_a^b \phi_j^*(x) f(x)\, dx, \qquad (10.7.53)$$

where $\boldsymbol{\phi}_j$ and λ_j, $j = 1, 2, \ldots$, are the eigenvectors and eigenvalues of **L**. It is interesting to note that Eq. (10.7.53) yields a very general solution to Eq. (10.7.51) (owing to the spectral resolution theorem); of course, the particular eigenvectors and eigenvalues depend on the boundary conditions imposed.

EXERCISE 10.7.2. Choose units such that $D = 1$; let $f(x) = \exp(-x)$ and let $[a, b] = [0, 1]$. Plot the solution $u(x)$ versus x for $t = 0.1, 0.5, 1$, and 5 for the case analyzed in Example 10.7.2 (with $\theta = 2$ and $\phi = 1.5$). Plot the same curves in real units where $b = 5$ cm, $D = 0.245$ cm^2/s, and f and u are in kelvins. This corresponds to the cooling of thin iron rods insulated everywhere except the ends.

EXAMPLE 10.7.3. In quantum mechanics, one encounters the one-dimensional problem of finding the eigenvalues of the equation $Lu = \lambda u$, where the self-adjoint Sturm–Liouville operator L is the Schrödinger operator and the eigenvalue λ is the energy of the system. As an example, consider the wavefunction of a particle in a one-dimensional box obeying the equation

$$-\frac{\hbar^2}{2m} \frac{d^2 \psi}{dx^2} = E\psi, \qquad 0 < x < b, \qquad (10.7.54)$$

where $\hbar$ is Planck's constant (divided by 2π), m is the particle mass, b is the length of the box, and E is the energy of the particle. The fact that the particle is confined to the box is accounted for by the boundary conditions $\Psi(0) = \Psi(b) = 0$. According to quantum theory, the energy of a confined particle is quantized; i.e., it can only take on certain discrete real values. This property arises because the energy is the eigenvalue of a self-adjoint Sturm–Liouville operator. Introducing the notation

$$\lambda = \frac{2mEb^2}{\hbar^2} \qquad \text{and} \qquad \xi = \frac{x}{b}, \qquad (10.7.55)$$

Eq. (10.7.54) becomes

$$-\frac{d^2 \psi}{d\xi^2} = \lambda \psi, \qquad 0 < \xi < 1, \quad \psi(0) = \psi(1) = 0. \qquad (10.7.56)$$

A fundamental system for Eq. (10.7.56) is $u_1 = \sin\sqrt{\lambda}\xi$ and $u_2 = \cos\sqrt{\lambda}\xi$. Taking $\psi = c_1 u_1 + c_2 u_2$ and applying the boundary conditions $\psi(0) = \psi(1) = 0$, we find $c_2 = 0$ and the condition

$$\sin\sqrt{\lambda} = 0 \qquad (10.7.57)$$

for the eigenvalues λ. Thus, $\lambda_j = (j\pi)^2$, $j = 1, 2, \ldots$, and $\psi_j = c_1 \sin \pi j \xi$. Replacing ξ by x/b and determining c_1 from the normalization criterion, $\int_0^1 \psi_j^2 dx = 1$, the wavefunctions become

$$\psi_j = \sqrt{\frac{2}{b}} \sin \frac{\pi j x}{b}, \qquad j = 1, 2, \ldots, \qquad (10.7.58)$$

corresponding to the particle energies

$$E_j = \frac{\hbar^2}{2m} \frac{\pi^2 j^2}{b^2}, \qquad j = 1, 2, \ldots. \qquad (10.7.59)$$

EXAMPLE 10.7.4. The initial concentration of carbon in a certain long bar of steel is given by $c(x, t = 0) = c_0(x)$. The ends of the bar, say at $x = 0$ and $x = b$, are insulated, which means that $\partial c/\partial x = 0$ at $x = 0$ and $x = b$. The change in concentration in time is governed by the diffusion equation

$$\frac{\partial c}{\partial t} = D \frac{\partial^2 c}{\partial x^2}, \qquad (10.7.60)$$

where D is the diffusion coefficient. The problem of finding the time evolution concentration can be restated as the problem of solving

$$\frac{\partial \mathbf{c}}{\partial t} = -D\mathbf{L}\mathbf{c}, \qquad (10.7.61)$$

where $Lu = -d^2u/dx^2$, with boundary conditions $u'(0) = u'(b) = 0$, defines a self-adjoint operator. Thus, the solution to Eq. (10.7.61) is

$$\mathbf{c} = \exp(-t D\mathbf{L})\mathbf{c}_0 = \sum_j \exp(-t D\lambda_j) \langle \boldsymbol{\phi}_j, \mathbf{c}_0 \rangle \boldsymbol{\phi}_j, \qquad (10.7.62)$$

where λ_j and $\boldsymbol{\phi}_j$ obey the eigenequation $\mathbf{L}\boldsymbol{\phi}_j = \lambda_j \boldsymbol{\phi}_j$. Applying the boundary conditions to $u(x) = c_1 \sin \sqrt{\lambda} x + c_2 \cos \sqrt{\lambda} x$, we find that $c_1 = 0$ and the eigenvalues must satisfy

$$\sqrt{\lambda} \sin \sqrt{\lambda} b = 0, \qquad (10.7.63)$$

namely, $\lambda_0 = 0$ and $\sqrt{\lambda_j} b = j\pi$, $j = 1, 2, \ldots$. Thus, the eigenfunctions are

$$\begin{aligned} \phi_0 &= \frac{1}{\sqrt{b}}, \\ \phi_j &= \sqrt{\frac{2}{b}} \cos \frac{\pi j x}{b}, \qquad j = 1, 2, \ldots, \end{aligned} \qquad (10.7.64)$$

and the concentration $c(x, t)$ is given by

$$c(x, t) = \frac{1}{b} \int_0^b c_0(x) dx + \sum_{j=1}^\infty \exp\left(\frac{-t D \pi^2 j^2}{b^2}\right) \langle \boldsymbol{\phi}_j, \mathbf{c}_0 \rangle \sqrt{\frac{2}{b}} \cos \frac{\pi j x}{b}, \qquad (10.7.65)$$

with

$$\langle \phi_j, c_0 \rangle = \sqrt{\frac{2}{b}} \int_0^b c_0(x) \cos \frac{\pi j x}{b} \, dx. \tag{10.7.66}$$

Thus, we see that diffusion drives the concentration of carbon in an isolated bar toward a constant value $\bar{c} = b^{-1} \int_0^b c_0(x) \, dx$, which is the mean value of the initial concentration. The concentration approaches the mean value exponentially in time.

■ ■ ■ Consider now the operator $\mathbf{L}$ defined by

$$Lu = -\frac{d^2 u}{dx^2}, \quad 0 < x < 1, \tag{10.7.67}$$

with the following boundary conditions:

(a) $u(0) = u(1) = 0$ \hfill (10.7.68a)

(b) $u'(0) = u'(1) = 0$ \hfill (10.7.68b)

(c) $u(0) = u'(1) = 0$ \hfill (10.7.68c)

(d) $u(0) = u(1), \quad u'(0) = u'(1)$. \hfill (10.7.68d)

For each of the above conditions, the operator L is self-adjoint and the corresponding eigenvalues and eigenfunctions are

(a) $\lambda_n = (n\pi)^2, \quad \phi_n(x) = \sqrt{2} \sin n\pi x, \quad n = 1, 2, \ldots$ \hfill (10.7.69a)

(b) $\lambda_0 = 0, \quad \phi_0(x) = 1$

$\lambda_n = (n\pi)^2, \quad \phi_n(x) = \sqrt{2} \cos n\pi x, \quad n = 1, 2, \ldots$ \hfill (10.7.69b)

(c) $\lambda_n = \left(n + \frac{1}{2}\right)^2 \pi^2, \quad \phi_n(x) = \sqrt{2} \sin\left(n + \frac{1}{2}\right)\pi x, \quad n = 1, 2, \ldots$ \hfill (10.7.69c)

(d) $\lambda_n^{(1)} = (2n\pi)^2, \quad \phi_n^{(1)}(x) = \sqrt{2} \sin 2n\pi x, \quad n = 1, 2, \ldots$

$\lambda_0^{(2)} = 0, \quad \phi_0^{(2)}(x) = 1$

$\lambda_n^{(2)} = (2n\pi)^2, \quad \phi_n^{(2)}(x) = \sqrt{2} \cos 2n\pi x, \quad n = 1, 2, \ldots$ \hfill (10.7.69d)

■ **EXERCISE 10.7.3.** Show that the eigenvalues and eigenvectors in Eq. (10.7.69) are truly the eigenpairs of Eq. (10.7.67) for each of the boundary conditions in Eq. (10.7.68).

Since each of the examples above defines a self-adjoint operator, each set of eigenvectors forms a complete orthonormal set. Thus, the spectral resolution $\mathbf{I} = \sum_n \phi_n \phi_n^\dagger$ or $\delta(x - y) = \sum_n \phi_n(x) \phi_n^*(y)$ is valid for each of these cases. Moreover, $\mathbf{L} = \sum_n \lambda_n \phi_n \phi_n^\dagger$ and $f(\mathbf{L}) = \sum_n f(\lambda_n) \phi_n \phi_n^\dagger$, when $f(\lambda_n)$ exists. If we denote the kernel of $f(\mathbf{L})$ by $f(x, y)$, then the spectral resolution of $f(\mathbf{L})$ for cases (a)–(d) above becomes

(a) $\quad f(x, y) = \sum_{n=1}^{\infty} f(n^2\pi^2) 2 \sin \pi x \sin n\pi y \quad$ (10.7.70)

(b) $\quad f(x, y) = f(0) + \sum_{n=1}^{\infty} f(n^2\pi^2) 2 \cos \pi x \cos n\pi y \quad$ (10.7.71)

(c) $\quad f(x, y) = \sum_{n=1}^{\infty} f\left(\left(n+\frac{1}{2}\right)^2 \pi^2\right) 2 \sin\left(n+\frac{1}{2}\right)\pi x \sin\left(n+\frac{1}{2}\right)\pi y \quad$ (10.7.72)

(d) $\quad f(x, y) = f(0) + \sum_{n=1}^{\infty} f(4n^2\pi^2) 2 \big(\cos 2n\pi x \cos 2n\pi y$

$\qquad + \sin 2n\pi x \sin 2\pi n y\big) \quad$ (10.7.73)

By setting $f(x) = 1$, we recover the resolution of the identity operator $\mathbf{I}$ in $\mathcal{L}_2(0, 1)$ ($f(x, y) = \delta(x - y)$). When $F(t) = t$, $f(x, y)$ is simply the kernel $l(x, y)$ of the integral operator representing $\mathbf{L}$.

EXERCISE 10.7.4. Use the spectral resolution of $f(\mathbf{L})$ to solve the equation

$$\frac{\partial u}{\partial t} = -\mathbf{L}u, \qquad u(x, t = 0) = x, \ 0 < x < 1, \quad (10.7.74)$$

for cases (a)–(d) in Eq. (10.7.68). Plot the results as $u(x, t)$ versus x for various values of t.

ILLUSTRATION 10.7.1 (Case-hardening an Iron Bar). A flat bar of iron at 1100°C is case-hardened by putting it in a methane bath also at 1100° C. The bar, illustrated in Fig. 10.7.2, is 0.5 cm thick, 2 cm wide, and 20 cm long. Contact with the methane gas fixes the concentration of carbon at the surface of the bar at a mass fraction of $\theta = 0.012$. Carbon enters the bar by diffusion and converts the iron to steel, which is much harder than iron. At 1100°C, the diffusivity of carbon in iron is $D = 9.2 \times 10^{-5}$ cm²/s. For case-hardening, it is desired that the concentration of carbon be $\theta = 0.005$ at a depth of 0.05 cm. We want to determine how long the bar should be left in the methane bath before quenching it to room temperature in order to achieve the desired case-hardening.

The concentration in the bar obeys the diffusion equation

$$\frac{\partial \theta}{\partial t} = D \frac{\partial^2 \theta}{\partial x^2}, \qquad 0 < x < d, \quad (10.7.75)$$

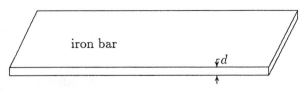

FIGURE 10.7.2 Case-hardened iron rod.

where t is time, x is distance from the flat surface of the bar, and $d = 0.5$ cm is the bar thickness. Edge effects will be neglected. The initial and boundary conditions are

$$\theta(x, t = 0) = 0, \quad \text{for all } x,$$
$$\theta(x = 0, t) = \theta_s, \quad t > 0, \quad (10.7.76)$$
$$\theta(x = d, t) = \theta_s, \quad t > 0,$$

where $\theta_s = 0.012$. Consider the operator

$$Lu = -\frac{d^2 u}{dx^2}, \quad u(0) = u(d) = 0. \quad (10.7.77)$$

This is a self-adjoint operator whose eigenvectors and eigenvalues are

$$u_n(x) = \sqrt{\frac{2n}{nd}} \sin\frac{n\pi}{d} x \quad \text{and} \quad \lambda_n = \left(\frac{n\pi}{d}\right)^2, \quad n = 1, 2, \ldots. \quad (10.7.78)$$

(i) Using the fact that u_n for $n = 1, \ldots$ forms a complete orthonormal basis set in $\mathcal{L}_2(0, d)$ space, solve the diffusion equation for the bar. Plot the concentration θ versus x at various times. Determine how long the bar should be left in the methane bath for the concentration of carbon to be 0.005 at $x = 0.05$ cm.

(ii) Suppose the concentration of carbon in the bar is initially

$$\theta(x, t = 0) = 0.004 - 0.016(x - 0.5 \text{ cm})^2/\text{cm}^2.$$

■ ■ ■ Find the concentration profiles (θ versus x) for various times for this case. Assume again that the surface concentration is fixed at $\theta_s = 0.012$ for $t > 0$.

The reader may have noticed that in all of the examples of self-adjoint Sturm–Liouville operators that we have studied, the eigenvalues λ_n are bounded from below by M, where $|M| < \infty$ and $\lambda_n \to \infty$ as $n \to \infty$. Boundedness from below is a general property for a large class of regular self-adjoint Sturm–Liouville operators. By bounded from below, we mean that, for any $\mathbf{u}$ in $\mathcal{D}_L$, the property

$$\langle \mathbf{u}, \mathbf{Lu} \rangle \geq M \langle \mathbf{u}, \mathbf{u} \rangle, \quad |M| < \infty, \quad (10.7.79)$$

holds.

Before stating the basic boundedness theorem and proving it, let us note that, for a self-adjoint Sturm–Liouville operator, one can, without loss of generality, limit the Hilbert space to real functions. This is because $\mathbf{L}$ is real and its eigenvalues are real. Suppose $\boldsymbol{\phi}_n$ is an eigenvector of $\mathbf{L}$ and suppose it is complex. Then $\boldsymbol{\phi}_n \equiv \boldsymbol{\theta}_n + i\boldsymbol{\psi}_n$, where $\boldsymbol{\theta}_n$ and $\boldsymbol{\psi}_n$ are real-valued functions and $i = \sqrt{-1}$. But $\mathbf{L}\boldsymbol{\phi}_n = \mathbf{L}\boldsymbol{\theta}_n + i\mathbf{L}\boldsymbol{\psi}_n = \lambda_n \boldsymbol{\theta}_n + i\lambda_n \boldsymbol{\psi}_n$, and since $\mathbf{L}\boldsymbol{\theta}_n$ and $\mathbf{L}\boldsymbol{\psi}_n$ are real it follows that $\mathbf{L}\boldsymbol{\theta}_n = \lambda_n \boldsymbol{\theta}_n$ and $\mathbf{L}\boldsymbol{\psi}_n = \lambda_n \boldsymbol{\psi}_n$. If the multiplicity of λ_n is 1, then $\boldsymbol{\theta}_n = \boldsymbol{\psi}_n$. If the multiplicity of λ_n is larger than 1, say p_n, then the p_n vectors $\boldsymbol{\phi}_n$ can always be chosen such that $\boldsymbol{\theta}_n = \boldsymbol{\psi}_n$. Thus, the eigenvectors of $\mathbf{L}$ are real except for the complex factor. Since an eigenvector is unchanged when multiplied by an arbitrary constant, it follows that the eigenvectors of $\mathbf{L}$ can be chosen to be real without loss of generality. Accordingly, we can prove the following theorem for real functions in the Hilbert space $\mathcal{L}_2(a, b; s)$.

THEOREM. *If $\mathbf{L}$ is a regular Sturm–Liouville operator in $\mathcal{L}_2(a,b;s)$, it is bounded from below if the boundary conditions are either*

$$u(a) = u(b) = 0 \tag{10.7.80}$$

or

$$u(a) = \alpha_{11} u(a) \quad \text{and} \quad u'(b) = -\alpha_{21} u(b), \tag{10.7.81}$$

where α_{11} and α_{21} are arbitrary real numbers.

Consider again the Sturm–Liouville operator defined in Eq. (10.7.1). The quantity $\langle \mathbf{u}, \mathbf{Lu} \rangle$, after integration by parts, yields

$$\langle \mathbf{u}, \mathbf{Lu} \rangle = p(a)u(a)u'(a) - p(b)u(b)u'(b) + \int_a^b p(x)[u'(x)]^2 \, dx + \int_a^b q(x)[u(x)]^2 \, dx. \tag{10.7.82}$$

For the boundary conditions $u(a) = u(b) = 0$ and $u'(a) = u'(b) = 0$ (the latter is a special case of the conditions in Eq. (10.7.81)), it follows immediately that $\mathbf{L}$ is bounded from below since

$$\int_a^b p(x)[u'(x)]^2 \, dx \geq 0 \tag{10.7.83}$$

and

$$\int_a^b q(x)[u(x)]^2 \, dx \geq M \int_a^b s(x)[u(x)]^2 \, dx = M \langle \mathbf{u}, \mathbf{u} \rangle, \tag{10.7.84}$$

where

$$M = \min_{a < x < b} \frac{q(x)}{s(x)}. \tag{10.7.85}$$

The quantity M is finite since $q(x)$ and $s(x)$ are continuous and $s(x)$ is strictly nonzero in $[a,b]$.

A little more work is needed to prove the theorem for the boundary conditions in Eq. (10.7.22). If we define

$$\nu = \min_{\{i,x\}} \alpha_{i1} p(x), \tag{10.7.86}$$

and use the boundary conditions $u'(a) = \alpha_{11} u(a)$ and $u'(b) = -\alpha_{21} u(b)$ in Eq. (10.7.86), we find

$$\langle \mathbf{u}, \mathbf{Lu} \rangle \geq \nu \left([u(a)]^2 + [u(b)]^2 \right) + \int_a^b p(u')^2 \, dx + \int_a^b q u^2 \, dx. \tag{10.7.87}$$

Moreover, if we define

$$p_m \equiv \max_{a < x < b} p(x) \quad \text{and} \quad \delta = \min(p_m, \nu), \tag{10.7.88}$$

SPECTRAL THEORY OF REGULAR STURM–LIOUVILLE OPERATORS

then it follows from Eq. (10.7.87) that

$$\langle \mathbf{u}, \mathbf{Lu} \rangle \geq \delta \left[u(a)^2 + u(b)^2 + \int_a^b (u')^2 dx \right] + \int_a^b q u^2 dx. \tag{10.7.89}$$

Next we need to find a bound for the quantity in the square brackets in Eq. (10.7.89). We do this by starting with the relation

$$\int_a^b \frac{d}{dx}(xu^2) dx = b[u(b)]^2 - a[u(a)]^2 \tag{10.7.90}$$

or

$$\int_a^b u^2 dx = -\int_a^b 2xuu' dx + b[u(b)]^2 - a[u(a)]^2. \tag{10.7.91}$$

Next consider the inequality

$$|\alpha \beta| \leq \frac{\gamma}{2} \alpha^2 + \frac{1}{2\gamma} \beta^2, \quad \gamma > 0 \ (\alpha, \beta \ \text{real}), \tag{10.7.92}$$

which is easily derived from the expression

$$\left(\sqrt{\gamma} |\alpha| - \frac{1}{\sqrt{\gamma}} \beta \right)^2 = |\gamma| \alpha^2 + \frac{1}{|\gamma|} \beta^2 - 2|\alpha \beta| \geq 0. \tag{10.7.93}$$

Letting $\alpha = u$ and $\beta = u'$, we can express the integrand on the right-hand side of Eq. (10.7.91) as

$$|-2xuu'| \leq 2\mu |uu'| \leq 2\mu \left(\frac{\gamma}{2} u^2 + \frac{1}{2\gamma} u'^2 \right), \tag{10.7.94}$$

where

$$\mu = \max_{a < x < b} |x|. \tag{10.7.95}$$

Next we set $\gamma = 1/(2\mu)$ and use the inequality in Eq. (10.7.94) to obtain

$$\int_a^b u^2 dx \leq 4\mu^2 \int_a^b (u')^2 dx + 2 \left(b[u(b)]^2 - a[u(a)]^2 \right). \tag{10.7.96}$$

But, since $b[u(b)]^2 - a[u(a)]^2 \leq \mu([u(b)]^2 + [u(a)]^2)$, Eq. (10.7.96) implies

$$\int_a^b u^2 dx \leq \epsilon \left[\int_a^b (u')^2 dx + u(a)^2 + u(b)^2 \right], \tag{10.7.97}$$

where

$$\epsilon = \max(4\mu^2, 2\mu). \tag{10.7.98}$$

Finally, combining Eqs. (10.7.89) and (10.7.97) yields

$$\langle \mathbf{u}, \mathbf{L}\mathbf{u} \rangle \geq \int_a^b s(x) \left[\frac{\delta}{\epsilon s(x)} + \frac{q(x)}{s(x)} \right] u^2 dx \tag{10.7.99}$$
$$\geq M \langle \mathbf{u}, \mathbf{u} \rangle,$$

where

$$M = \min_{a<x<b} \left[\frac{\delta}{\epsilon s(x)} + \frac{q(x)}{s(x)} \right] \tag{10.7.100}$$

and $\langle \mathbf{u}, \mathbf{u} \rangle = \int_a^b s(x) u(x)^2 dx$. Since $|M| < \infty$, this completes the proof.

The lower bound of $\mathbf{L}$, of course, does not have to be 0. For example, the operator

$$Lu = -\frac{d^2 u}{dx^2} - Mu, \qquad 0 < x < 1, \tag{10.7.101}$$

where M is an arbitrary real number and $u(0) = u(1) = 0$, has eigenvalues and eigenfunctions

$$\lambda_n = (n\pi)^2 + M \quad \text{and} \quad \phi_n = \sqrt{2} \sin n\pi x, \qquad n = 1, 2, \ldots . \tag{10.7.102}$$

We can let M be a large negative number and, thus, $\lambda_n \geq M$. However, $\lambda_n < 0$ until n is large enough such that $(n\pi)^2 > |M|$.

The various examples of regular, self-adjoint Sturm–Liouville operators bring out the special properties that are identified in the following theorem:

THEOREM. Let

$$Lu = -\frac{1}{s(x)} \frac{d}{dx} \left(p(x) \frac{du}{dx} \right) + \frac{q(x)}{s(x)} u, \tag{10.7.103}$$

where $p(x)$, $p'(x)$, $q(x)$, and $s(x)$ are real and continuous functions on the finite interval $[a, b]$; $p(x)$ and $s(x)$ are positive on $[a, b]$. Let

$$\begin{aligned} B_1 u &\equiv \alpha_{11} u(a) + \alpha_{12} u'(a) = 0 \\ B_2 u &\equiv \alpha_{23} u(b) + \alpha_{24} u'(b) = 0, \end{aligned} \tag{10.7.104}$$

where the coefficients α_{ij} are real and obey the conditions $\alpha_{11}^2 + \alpha_{12}^2 \neq 0$ and $\alpha_{23}^2 + \alpha_{24}^2 \neq 0$ (these boundary conditions are equivalent to those of case (iii) in Eq. (10.7.29)). Then $\mathbf{L}$ is self-adjoint and its eigenvalues can be arranged in the sequence

$$\lambda_0 < \lambda_1 < \lambda_2 \cdots < \lambda_n \cdots, \tag{10.7.105}$$

where

$$|\lambda_0| < \infty \quad \text{and} \quad \lim_{n \to \infty} \lambda_n = \infty. \tag{10.7.106}$$

Furthermore, the eigenfunction $\phi_n(x)$ corresponding to λ_n has exactly n zeros in the open interval (a, b).

Consider cases (a)–(d) given by Eq. (10.7.68). In case (a), $\lambda_n = (n+1)^2\pi^2$ and $\phi_n(x) = \sqrt{2}\sin(n+1)\pi x$, $n = 0, 1, 2\ldots$ (we have transformed the integer sequence labeling λ_n so that $n = 0$ gives the smallest eigenvalue). Notice that the eigenfunction $\phi_0 = \sqrt{2}\sin\pi x$ has no zeros in the interval $(0, 1)$. The zeros at $x = 0$ and $x = 1$ do not violate the theorem, because $x = 0$ and $x = 1$ do not belong to the open interval $(0, 1)$. The second eigenfunction $\phi_1 = \sqrt{2}\sin 2\pi x$ has a 0 at $x = \frac{1}{2}$; the third eigenfunction $\phi_2 = \sqrt{2}\sin 3\pi x$ has zeros at $x = \frac{1}{3}$ and $x = \frac{2}{3}$; etc. In this case, the eigenvalues order as $\lambda_0 < \lambda_1 < \lambda_2 < \cdots$. Furthermore, cases (b) and (c) can easily be seen to obey the theorem with the eigenvalue ordering $\lambda_0 < \lambda_1 < \lambda_2 < \cdots$.

The boundary conditions of case (d) are not covered by the functionals defined in Eq. (10.7.104). Nevertheless, the eigenvalues and eigenfunctions for this case behave similarly to those of cases (a)–(c). In case (d), the first eigenvalue is $\lambda_0 = 0$ and the corresponding eigenfunction, $\phi_0 = 1$, has no zeros in the open interval $(0, 1)$. The next eigenfunction is $\phi_1 = \sqrt{2}\sin 2\pi x$ (with eigenvalue $\lambda_1 = (2\pi)^2$) and has a 0 at $x = \frac{1}{2}$. The next eigenfunction is $\phi_2 = \sqrt{2}\cos 2\pi x$ (with eigenvalue $\lambda_2 = (2\pi)^2$) and has zeros at $x = \frac{1}{4}$ and $x = \frac{3}{4}$. This is an example in which the eigenvalue ordering is $\lambda_0 < \lambda_1 = \lambda_2 < \lambda_3 = \lambda_4 < \cdots$; i.e., there are two eigenfunctions corresponding to every nonzero eigenvalue. According to a theorem we proved earlier, a second-order differential operator can have at most two eigenvectors per eigenvalue. Thus, the behavior of cases (a)–(c) and case (d) exhausts the possibilities for a regular self-adjoint Sturm–Liouville operator.

Parts of the theorem given above have already been proved. That the eigenvalues form a denumerable set follows from an earlier proof that the eigenvectors of a regular self-adjoint differential operator form a complete orthonormal set—as such the set $\{\phi_n\}$ is denumerable and so also is the set $\{\lambda_n\}$. That $|\lambda_0| < \infty$ and $\lambda_n \to \infty$ as $n \to \infty$ follow from the facts that the Sturm–Liouville operator with the boundary conditions given by Eq. (10.7.104) is bounded from below and that a differential operator is unbounded as $n \to \infty$. The only thing we have not proved is that the eigenfunctions can be ordered such that ϕ_n has exactly n zeros in the open interval (a, b). This is the so-called *oscillation theorem* for Sturm–Liouville equations. The proof is a bit tedious, and so we refer the reader to texts on differential equations instead of reproducing it here. (Coddington and Levinson, *Theory of Ordinary Differential Equations*, is an excellent reference for this theorem.)

The theorem pertaining to case (d) is as follows:

THEOREM. *Let*

$$Lu = -\frac{1}{s(x)}\frac{d}{dx}\left(p(x)\frac{du}{dx}\right) + \frac{q(x)}{s(x)}u, \qquad (10.7.107)$$

where $p(x)$, $p'(x)$, $q(x)$, and $s(x)$ are real and continuous on the finite interval $(a, b]$ and $p(a) = p(b)$ (this last condition is needed to render $\mathbf{L}$ periodic in $\mathcal{L}_2(a, b; s)$). Suppose the boundary conditions are either periodic,

$$u(a) = u(b) \quad \text{and} \quad u'(a) = u'(b), \qquad (10.7.108)$$

or anti-periodic,

$$u(a) = -u(b) \quad \text{and} \quad u'(a) = -u'(b). \qquad (10.7.109)$$

With either set of boundary conditions, **L** is a self-adjoint operator. If $\lambda_n, n \geq 0$, denote the eigenvalues for periodic boundary conditions and $\tilde{\lambda}_n$, $n \geq 1$, denote those for anti-periodic boundary conditions, then the complete set of eigenvalues can be ordered into the following sequence:

$$\lambda_0 < \tilde{\lambda}_1 \leq \tilde{\lambda}_2 < \lambda_1 \leq \lambda_2 < \tilde{\lambda}_3 \leq \tilde{\lambda}_4 < \cdots, \tag{10.7.110}$$

where

$$|\lambda_0| < \infty \quad \text{and} \quad \lim_{n \to \infty} \tilde{\lambda}_n = \lim_{n \to \infty} \lambda_n = \infty. \tag{10.7.111}$$

For λ_0, there is a unique eigenfunction ϕ_0. When $\lambda_{2n+1} = \lambda_{2n+2}$ or $\tilde{\lambda}_{2n+1} = \tilde{\lambda}_{2n+2}$, there are two eigenfunctions, ϕ_{2n+1}, ϕ_{2n+2} or $\tilde{\phi}_{2n+1}, \tilde{\phi}_{2n+2}$, corresponding to the same eigenvalue. The function ϕ_0 has no zeros in $[a, b]$. ϕ_{2n+1} and $\phi_{2n+2}, n \geq 0$, have exactly $2n+2$ zeros in $(a, b]$ and $\tilde{\phi}_{2n+1}$ and $\tilde{\phi}_{2n+2}, n \geq 0$, have exactly $2n+1$ zeros in $[a, b)$.

Note that the semi-open interval $[a, b)$ includes $x = a$ but not $x = b$, whereas the semi-open interval $(a, b]$ contains $x = b$ but not $x = a$. We refer the reader to the literature on differential equations for the proof of this theorem. (Again, Coddington and Levinson, *Theory of Ordinary Differential Equations*, is an especially good text.)

Let us examine case (d) as an example of an operator obeying this theorem. For the periodic boundary conditions, we find from Eq. (10.7.69) that

$$\begin{aligned}
\lambda_0 &= 0, & \phi_0 &= 1 \\
\lambda_1 &= (2\pi)^2, & \phi_1 &= \sqrt{2} \sin 2\pi x \\
\lambda_2 &= (2\pi)^2, & \phi_2 &= \sqrt{2} \cos 2\pi x \\
\lambda_3 &= (4\pi)^2, & \phi_3 &= \sqrt{2} \sin 4\pi x \\
\lambda_4 &= (4\pi)^2, & \phi_4 &= \sqrt{2} \cos 4\pi x \\
&\vdots
\end{aligned} \tag{10.7.112}$$

while the eigenproblem for $u(0) = -u(\pi)$ and $u'(0) = -u'(\pi)$ with $Lu = -d^2u/dx^2$ yields

$$\begin{aligned}
\tilde{\lambda}_1 &= \pi^2, & \tilde{\phi}_1 &= \sqrt{2} \sin \pi x \\
\tilde{\lambda}_2 &= \pi^2, & \tilde{\phi}_2 &= \sqrt{2} \cos \pi x \\
\tilde{\lambda}_3 &= (3\pi)^2, & \tilde{\phi}_3 &= \sqrt{2} \sin 3\pi x \\
\tilde{\lambda}_4 &= (3\pi)^2, & \tilde{\phi}_4 &= \sqrt{2} \cos 3\pi x \\
&\vdots
\end{aligned} \tag{10.7.113}$$

The eigenvalues of the above two problems obey the sequence

$$\lambda_0 < \tilde{\lambda}_1 = \tilde{\lambda}_2 < \lambda_1 = \lambda_2 < \tilde{\lambda}_3 = \tilde{\lambda}_4 < \lambda_3 = \lambda_4 < \cdots, \tag{10.7.114}$$

and the functions ϕ_{2n+1} and $\phi_{2n+2}, n \geq 0$, have $2n + 2$ zeros in $[0, 1)$, and the functions $\tilde{\phi}_{2n+1}$ and $\tilde{\phi}_{2n+2}, n \geq 0$, have $2n + 1$ zeros in $(0, 1]$—as required by the theorem. According to the theorem, the eigenvalues of the periodic and the antiperiodic operators obey the separate sequences

$$-\infty < \lambda_0 < \lambda_1 \leq \lambda_2 < \lambda_3 \leq \lambda_4 < \cdots \qquad (10.7.115)$$

and

$$-\infty < \tilde{\lambda}_1 \leq \tilde{\lambda}_2 < \tilde{\lambda}_3 \leq \tilde{\lambda}_4 \leq \cdots, \qquad (10.7.116)$$

respectively. In the special case (d), the inequalities "$\leq$" are replaced by equalities.

10.8. SPECTRAL THEORY OF SINGULAR STURM–LIOUVILLE OPERATORS

The differential expression of a singular Sturm–Liouville operator is of the same form as that of a regular Sturm–Liouville operator. An operator is classified as singular if either $|a|$ or $|b|$ of the interval $[a, b]$ is infinite, if $s(x)$ or $p(x)$ has a zero in $[a, b]$, or if $|q(x)|$ is infinite at some point in $[a, b]$. The condition for self-adjointness is the same as for a regular operator, namely, that $L = L^\dagger$ and $\mathcal{D}_L = \mathcal{D}_{L^\dagger}$. The domain $\mathcal{D}_{L^\dagger}$ is defined as those vectors $\mathbf{v}$ in $\mathcal{H}$ such that $\mathbf{L}^\dagger \mathbf{v} \in \mathcal{H}$ and

$$\langle \mathbf{v}, \mathbf{L}\mathbf{u} \rangle = \langle \mathbf{L}^\dagger \mathbf{v}, \mathbf{u} \rangle \qquad \text{for all } \mathbf{u} \in \mathcal{D}_L. \qquad (10.8.1)$$

Two examples of singular, self-adjoint Sturm–Liouville equations are

(i) The Legendre equation:

$$L_1 u = -\frac{d}{dx}\left[(1 - x^2)\frac{du}{dx}\right] = \lambda u, \qquad -1 < x < 1. \qquad (10.8.2)$$

(ii) The Hermite equation:

$$L_2 u = -\frac{d^2 u}{dx^2} + x^2 u = \lambda u, \qquad -\infty < x < \infty. \qquad (10.8.3)$$

The Legendre equation arises in the quantum mechanical analysis of radially distributed potentials and angular momentum. The Hermite equation is encountered in the quantum mechanical analysis of the energy of a harmonic oscillator. In our analysis, u denotes the wavefunction and λ the energy expressed in appropriate units. The domain of the operator in Eq. (10.8.2) is

$$\mathcal{D}_{L_1} = \{\mathbf{u}, L_1\mathbf{u} \in \mathcal{L}_2(-1, 1), \; u(-1) = u(1), \; u'(-1) = u'(1)\}, \qquad (10.8.4)$$

and the domain of the operator in Eq. (10.8.3) is

$$\mathcal{D}_{L_2} = \{\mathbf{u}, L_2\mathbf{u} \in \mathcal{L}_2(-\infty, \infty)\}. \qquad (10.8.5)$$

It is easily shown that $L_1 = L_1^\dagger$ and $\mathcal{D}_{L_1} = \mathcal{D}_{L_1^\dagger}$ and that $L_2 = L_2^\dagger$ and $\mathcal{D}_{L_2} = \mathcal{D}_{L_2^\dagger}$. Thus, the operators L_1 and L_2 are self-adjoint. However, L_1 is singular because

$p(x) = 1 - x^2$ is 0 at $x = -1$ and $x = 1$. Likewise, L_2 is singular because it is defined in the infinite interval $(-\infty, \infty)$.

The eigenfunctions of Eq. (10.8.2) are the Legendre polynomials

$$P_l(x) = \frac{1}{2^l l!} \frac{d^l}{dx^l}(x^2 - 1)^l, \qquad l = 0, 1, 2, \ldots, \tag{10.8.6}$$

and the corresponding eigenvalues are $\lambda_l = l(l+1)$. The eigenfunctions of Eq. (10.8.3) are the Hermite functions

$$u_\nu(x) = H_\nu(x) \exp\left(\frac{-x^2}{2}\right), \qquad \nu = 0, 1, 2, \ldots, \tag{10.8.7}$$

where H_ν is the Hermite polynomial given by

$$H_\nu(x) = (-1)^\nu \exp(x^2) \frac{d^\nu}{dx^\nu}[\exp(-x^2)]. \tag{10.8.8}$$

The eigenvalue corresponding to u_ν is $\lambda_\nu = 2(\nu + \frac{1}{2})$.

In the theory of special functions, it can be proven that the Legendre polynomials form a complete orthogonal set in $\mathcal{L}_2(-1, 1)$ and the Hermite functions form a complete orthogonal set in $\mathcal{L}_2(-\infty, \infty)$ (P_l and u_ν are not normalized as given). By inspection, one can show that $P_l(x)$ has exactly l zeros in the interval $(-1, 1)$ and u_ν has exactly ν zeros in the interval $(-\infty, \infty)$. Moreover, the eigenvalues corresponding to P_l and u_ν are ordered in the sequence

$$-\infty < \lambda_0 < \lambda_1 < \lambda_2 < \ldots . \tag{10.8.9}$$

Thus, even though the Legendre and Hermite operators are singular self-adjoint Sturm–Liouville operators, they have a complete set of orthogonal eigenvectors and their eigenfunctions and eigenvalues obey the theorems proved for regular, self-adjoint Sturm–Liouville operators. Still another such case is given in Example 10.7.3. It is from examples like these that one says energy is *quantized* in the quantum mechanical formalism; i.e., it takes on only discrete values.

■ **EXAMPLE 10.8.1.** The vibrational modes λ of a circularly symmetric annular membrane with fixed boundaries satisfy the equation

$$Lu = -\frac{1}{r}\frac{d}{dr}\left(r\frac{du}{dr}\right) = \lambda u, \qquad a < x < b, \quad u(a) = u(b) = 0, \tag{10.8.10}$$

where $u(r)$ is the (time independent) amplitude of the vibration and a and b are finite with $b > a$. Physically, the eigenvalues in Eq. (10.8.10) are $\lambda = \omega^2/v^2$, where ω is the frequency of oscillation and v is the transverse wave velocity for the membrane (note that Eq. (10.8.10) derives from the one-dimensional wave equation in circular coordinates where $u(r, t) = u(r) \exp(i\omega t)$). The above equation is singular if $a = 0$, because $s(r) = p(r) = r$.

A fundamental system for $Lu - \lambda u = 0$ is $u_1 = J_0(\sqrt{\lambda}r)$ and $u_2 = N_0(\sqrt{\lambda}r)$, where J_0 is the Bessel function J_ν for $\nu = 0$ and N_0 is the Neumann function (Bessel function of the second kind) N_ν for $\nu = 0$. If we introduce the notation

$$\xi = \frac{r}{a}, \qquad \kappa = \frac{b}{a}, \qquad \mu = \lambda a^2, \tag{10.8.11}$$

Eq. (10.8.10) becomes

$$Lu = -\frac{1}{\xi}\frac{d}{d\xi}\left(\xi\frac{du}{d\xi}\right) = \mu u, \qquad 1 < \xi < \kappa,\ u(1) = u(\kappa) = 0. \qquad (10.8.12)$$

We can, therefore, express the eigenfunctions as the linear combination

$$u(\xi) = c_1 J_0(\sqrt{\mu}\xi) + c_2 N_0(\sqrt{\mu}\xi) \qquad (10.8.13)$$

for which the boundary conditions in Eq. (10.8.12) (for nonzero a) give the following conditions for the eigenvalues μ:

$$\begin{aligned} 0 &= c_1 J_0(\sqrt{\mu}) + c_2 N_0(\sqrt{\mu}) \\ 0 &= c_1 J_0(\sqrt{\mu}\kappa) + c_2 N_0(\sqrt{\mu}\kappa). \end{aligned} \qquad (10.8.14)$$

Equation (10.8.14) yields a nontrivial solution for the constants c_1 and c_2 if and only if

$$\begin{vmatrix} J_0(\sqrt{\mu}) & N_0(\sqrt{\mu}) \\ J_0(\sqrt{\mu}\kappa) & N_0(\sqrt{\mu}\kappa) \end{vmatrix} = 0. \qquad (10.8.15)$$

Notice that the redefined eigenvalues μ depend only on κ, the ratio of b/a. The explicit dependence on a comes in the definition of μ.

We can solve for the eigenvalues of a given κ by expanding Eq. (10.8.15):

$$J_0(\sqrt{\mu})N_0(\sqrt{\mu}\kappa) = N_0(\sqrt{\mu})J_0(\sqrt{\mu}\kappa). \qquad (10.8.16)$$

Consider, for example, the case where $\kappa = 1.5$. In Fig. 10.8.1, we have plotted both the left- and right-hand sides of Eq. (10.8.16). Here, as in Example 10.7.1, we will solve for the values of μ using the Newton–Raphson method, taking our

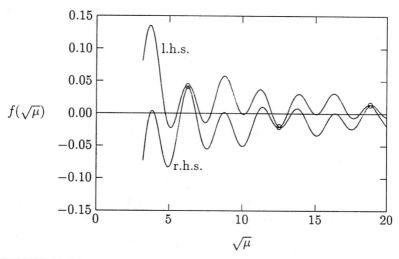

FIGURE 10.8.1

initial guesses directly from the intersecting curves in the figure. Using the initial guesses $\{6, 12.5, 19\}$ for $\sqrt{\mu}$, the first three resulting eigenvalues are

$$\{\sqrt{\mu_n}\} = \{6.27022, 12.5598, 18.8451, \ldots\}. \qquad (10.8.17)$$

The eigenfunctions are thus as in Eq. (10.8.13) with the coefficients c_1 and c_2 from Eq. (10.9.76) given by

$$c_2 = -c_1 \frac{J_0(\sqrt{\mu})}{N_0(\sqrt{\mu})}. \qquad (10.8.18)$$

For the special case $a = 0$, as mentioned above, Eq. (10.8.12) is singular. Physically, this case represents a circular membrane (drum). If we introduce the following new notation

$$\xi = \frac{r}{b} \quad \text{and} \quad \mu = \lambda b^2, \qquad (10.8.19)$$

Eq. (10.8.10) for the singular problem becomes

$$Lu = -\frac{1}{\xi}\frac{d}{d\xi}\left(\xi \frac{du}{d\xi}\right) = \mu u, \quad 0 < \xi < 1, \ u(0) = u(1) = 0, \qquad (10.8.20)$$

with the eigenfunctions $u(\xi)$ given in Eq. (10.8.13). Note, however, that the boundary condition $u(0) = 0$ can only be satisfied if $c_1 = c_2 = 0$ since $J_0(0) = 1$ and $N_0(0) \to -\infty$. Thus, with the singularity at $\xi = r = 0$, the operator L defined in Eq. (10.8.20) has no eigenvectors.

We can, however, alter the boundary conditions in Eq. (10.8.20) to make the problem nontrivial if we require that

$$u'(0) = 0 \quad \text{and} \quad u(1) = 0. \qquad (10.8.21)$$

This case represents physically the condition that the membrane at the center of the circle ($\xi = 0$) is free to move but is bounded at the edge ($\xi = 1$). Since $J_0'(0) = 0$ and $N_0'(0) \to \infty$, we can set $c_1 = 1$ and $c_2 = 0$ in Eq. (10.8.13). The second boundary condition in Eq. (10.8.21) then leads to the equation

$$J_0(\sqrt{\mu}) = 0. \qquad (10.8.22)$$

Thus, we see that the $\sqrt{\mu}$ can only have values of the zeros of J_0, namely,

$$\{\sqrt{\mu}\} = \{2.40483, 5.52008, 8.65373, 11.7915, 14.9309, \ldots\}. \qquad (10.8.23)$$

EXAMPLE 10.8.2. Consider the problem of diffusion of carbon into an iron cannonball initially free of carbon. This process is often conducted in order to case-harden (i.e., produce a steel casing on) cannonballs. Suppose the cannonball is a sphere of radius $r = b$. Initially, the cannonball contains no carbon, but at $t = 0$ it is contacted with a carbon source that keeps the surface concentration at the constant value C_s. We want to find the concentration distribution of carbon as a function of time.

Assuming that the concentration of carbon in the cannonball depends only on the radial distance r from the center of the ball, the concentration obeys the diffusion equation

$$\frac{\partial C}{\partial t} = D \frac{1}{r^2} \frac{\partial}{\partial r} \left(r^2 \frac{\partial C}{\partial r} \right), \tag{10.8.24}$$

where D is the diffusivity of carbon in iron. The initial and boundary conditions for this problem are

$$C(r, 0) = 0 \tag{10.8.25}$$

and

$$C(b, t) = C_s \quad \text{and} \quad \frac{\partial C}{\partial r}(0, t) = 0. \tag{10.8.26}$$

It is convenient to define $\tilde{C}(r, t) \equiv C(r, t) - C_s$, such that $\tilde{C}$ still satisfies Eq. (10.8.24) and the initial and boundary conditions

$$\tilde{C}(r, t=0) = -C_s, \quad \tilde{C}(r=b, t) = 0, \quad \frac{\partial \tilde{C}}{\partial r}(r=b, t) = 0. \tag{10.8.27}$$

The Sturm–Liouville operator

$$Lu = -\frac{1}{r^2} \frac{d}{dr} \left(r^2 \frac{du}{dr} \right) \tag{10.8.28}$$

is self-adjoint for the boundary conditions in Eq. (10.8.27). Thus, we wish to solve the eigenproblem

$$-\frac{1}{r^2} \frac{d}{dr} \left(r^2 \frac{du}{dr} \right) = \lambda u, \quad u \in \mathcal{L}_2(0, b; r^2), \tag{10.8.29}$$

with $u(b) = 0$ and $\partial u(0)/\partial r = 0$. Using the transformation

$$u(r) = \frac{w(r)}{r},$$

the eigenproblem becomes

$$-\frac{d^2 w}{dr^2} = \lambda w, \quad w \in \mathcal{L}_2(0, b), \tag{10.8.30}$$

with

$$w(b) = 0 \quad \text{and} \quad w(0) = 0. \tag{10.8.31}$$

The boundary condition for $w(0)$ comes from the physical requirement that $C < \infty$ at the center of the cannonball and is equivalent to the condition $\partial C/\partial r = 0$ at $r = 0$.

Equations (10.8.30) and (10.8.31) define a regular, self-adjoint Sturm–Liouville eigenproblem in $\mathcal{L}_2(0, b)$. Thus, the eigenfunctions ϕ_n, $n = 1, 2, \ldots$, for this problem form a complete orthonormal set in $\mathcal{L}_2(0, b)$. Likewise, since $\psi_n(r) = \phi_n(r)/r$ is an eigenfunction of the operator in Eq. (10.8.29), it follows that $\phi_n(r)/r$, $n = 1, 2, \ldots$, is a complete orthonormal set in $\mathcal{L}_2(0, b; r^2)$.

The eigenfunctions and eigenvalues of Eq. (10.8.30) are

$$\phi_n(r) = \sqrt{\frac{2}{b}} \sin \frac{n\pi r}{b}, \qquad \lambda_n = \left(\frac{n\pi}{b}\right)^2, \qquad n = 1, 2, \ldots. \qquad (10.8.32)$$

Thus, if Eq. (10.8.24) is expressed as

$$\frac{\partial \tilde{\mathbf{C}}}{\partial t} = -D\mathbf{L}\tilde{\mathbf{C}}, \qquad (10.8.33)$$

its solution is

$$\tilde{\mathbf{C}} = \exp(-tD\mathbf{L})\tilde{\mathbf{C}}_0, \qquad (10.8.34)$$

or

$$\begin{aligned}
C(r,t) &= C_s + \sum_{n=1}^{\infty} \exp\left(-tD\left(\frac{n\pi}{b}\right)^2\right) \frac{\sqrt{2/b}}{r} \sin \frac{n\pi r}{b} \\
&\quad \times \int_0^b \frac{\sqrt{2/b}}{r'} \sin \frac{n\pi r'}{b} (-C_s) r'^2 \, dr' \\
&= C_s + C_s \sum_{n=1}^{\infty} (-1)^n \exp\left(-\left(\frac{n\pi}{b}\right)^2 Dt\right) \frac{2b}{n\pi r} \sin \frac{n\pi r}{b}.
\end{aligned} \qquad (10.8.35)$$

In Fig. 10.8.2, we show C/C_s versus t (in units of b^2/D) for $r/b = 0.9$, 0.5, and 0. A Mathematica program is included in the Appendix for this example.

EXERCISE 10.8.1. Repeat Example 10.8.2, but for an initial concentration of $C(r, t = 0) = 0.2 C_s (r-b)^2/b^2$.

ILLUSTRATION 10.8.1 (Stokes' Oscillating Plate). Consider the problem of an incompressible Newtonian fluid between two parallel plates separated by a distance l as shown in Fig. 10.8.3. Keeping the top plate fixed, the bottom plate

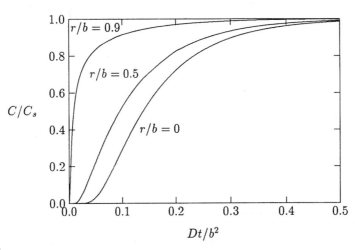

FIGURE 10.8.2 Time dependence of the concentration profile in Example 10.8.2 (Eq. (10.8.35)).

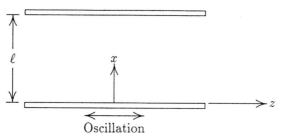

FIGURE 10.8.3

oscillates in the z direction (parallel to the top plate) with velocity $v = v_0 \sin \omega t$. We desire to find the velocity profile of the fluid as a function of its perpendicular distance away from the moving plate (x direction).

For a Newtonian fluid, the velocity in the x direction obeys the Navier–Stokes equation, which for this problem reduces to

$$\frac{\partial v}{\partial t} = \nu \frac{\partial^2 v}{\partial x^2}, \tag{10.8.36}$$

where the kinematic viscosity is given by $\nu = \mu/\rho$. The initial condition and boundary conditions are

$$\begin{aligned} v(t=0, x) &= 0 \\ v(t>0, x=0) &= v_0 \sin \omega t \\ v(t>0, x=l) &= 0. \end{aligned} \tag{10.8.37}$$

It is convenient to begin by defining the dimensionless quantities

$$\begin{aligned} \xi &\equiv \frac{x}{l} \\ u &\equiv \frac{v}{v_0} \\ \tau &\equiv \omega t, \end{aligned} \tag{10.8.38}$$

for which Eqs. (10.8.36) and (10.8.37) become

$$\frac{\partial u}{\partial \tau} = \alpha \frac{\partial^2 u}{\partial \xi^2}, \tag{10.8.39}$$

and

$$\begin{aligned} u(\tau=0, \xi) &= 0 \\ u(\tau>0, \xi=0) &= \sin \tau \\ u(\tau>0, \xi=1) &= 0, \end{aligned} \tag{10.8.40}$$

where we define $\alpha \equiv \nu/(\omega l^2)$.

The next step is to split the solution into a steady-state part and a transient part

$$u(\tau, \xi) = u_S(\tau, \xi) + u_T(\tau, \xi). \tag{10.8.41}$$

We now characterize the steady-state solution as a "pure oscillation" for which we assume a solution of the complex form

$$u_S = f(\xi) \exp(i\tau). \tag{10.8.42}$$

Here we specify the true velocity as the imaginary part of Eq. (10.8.42), i.e., $u_S = \text{Im}\{f(x) \exp(i\tau)\}$. This particular method is very common in physics and often, as in this case, offers the only viable method of obtaining a closed-form solution. We can now solve for the steady-state solution by noting that at large time the transient solution, by definition, is 0. Substituting Eq. (10.8.42) into Eq. (10.8.39) gives

$$\frac{d^2 f}{d\xi^2} = \frac{i}{\alpha} f(\xi). \tag{10.8.43}$$

(i) Show that the function

$$f(\xi) = \frac{\exp\bigl(-(1/\sqrt{2\alpha})(1+i)(1-\xi)\bigr) - \exp\bigl((1/\sqrt{2\alpha})(1+i)(1-\xi)\bigr)}{\exp\bigl(-(1/\sqrt{2\alpha})(1+i)\bigr) - \exp\bigl((1/\sqrt{2\alpha})(1+i)\bigr)} \tag{10.8.44}$$

is a solution to Eq. (10.8.43) and that the corresponding steady-state solution satisfies the two boundary conditions in Eq. (10.8.40). Derive an expression for the true steady state velocity profile $u_S(\tau, \xi)$. Plot this solution for several values of $0 \leq \tau \leq 2\pi$ and $\alpha = 1$.

To solve for the transient solution, we first note that it, too, is a solution to Eq. (10.8.39):

$$\frac{\partial u_T}{\partial \tau} = \alpha \frac{\partial^2 u_T}{\partial \xi^2}. \tag{10.8.45}$$

However, we need the transient solution to approach 0 as $\tau \to \infty$. We accomplish this by requiring the initial condition

$$u_T(\tau = 0, \xi) = -u_S(\tau = 0, \xi). \tag{10.8.46}$$

Equation (10.8.46) ensures that the initial condition in Eq. (10.8.37) is satisfied. Noting that at $\xi = 0$ the steady-state solution gives the correct velocity for all times, we can impose the boundary conditions

$$\begin{aligned} u_T(\tau > 0, \xi = 0) &= 0 \\ u_T(\tau > 0, \xi = 1) &= 0. \end{aligned} \tag{10.8.47}$$

(ii) Defining the operator

$$Lu = -\alpha \frac{d^2 u}{d\xi^2}, \tag{10.8.48}$$

we note that

$$Lu = -\alpha \frac{d^2 u}{d\xi^2} = \lambda u, \qquad u(0) = 0, \ u(1) = 0, \qquad (10.8.49)$$

defines a regular, self-adjoint Sturm–Liouville eigenproblem in $\mathcal{L}_2(0, 1)$. Show that the functions

$$u_n(\xi) = \sqrt{2} \sin n\pi \xi, \qquad \lambda_n = n^2 \pi^2 \alpha, \qquad (10.8.50)$$

are eigenfunctions of L with the corresponding eigenvalues given above for $n = 1, 2, \ldots$. Prove that there are no eigenvalues that are negative or 0.

(iii) In operator form, we can write Eq. (10.8.48) as

$$\frac{\partial \mathbf{U}}{\partial \tau} = -\mathbf{L}\mathbf{U}, \qquad (10.8.51)$$

which yields the formal solution

$$\mathbf{U}(\tau) = \exp(-\tau \mathbf{L}) \mathbf{U}(0). \qquad (10.8.52)$$

Using the spectral resolution theorem, show that

$$u_T(\tau, \xi) = \sum_{n=1}^{\infty} a_n u_n(\xi) \exp(-n^2 \pi^2 \alpha \tau). \qquad (10.8.53)$$

Use the initial condition in Eq. (10.8.46) to solve for the parameters a_n. Plot the transient solution for the case $\alpha = 1$ for several cycles of τ. After approximately how many cycles can we neglect the transient behavior?

With the four cases given by Eqs. (10.8.2), (10.8.3), (10.8.10), and (10.8.29), it seems tempting to conclude that singular self-adjoint Sturm–Liouville operators also always possess a complete set of orthogonal eigenvectors. This, alas, turns out not to be the case. Consider the self-adjoint problem

$$Lu = -\frac{d^2 u}{dx^2} = \lambda u, \qquad u \in \mathcal{L}_2(-\infty, \infty). \qquad (10.8.54)$$

This is the quantum mechanical problem for the energy λ and wavefunction u of a particle in free space. A fundamental system for $Lu - \lambda u = 0$ is

$$u_1 = \exp(i\sqrt{\lambda} x) \qquad \text{and} \qquad u_2 = \exp(-i\sqrt{\lambda} x) \qquad (10.8.55)$$

(an equally good set would be $u_1 = \sin \sqrt{\lambda} x$ and $u_2 = \cos \sqrt{\lambda} x$). The problem with Eq. (10.8.55) is that there is no linear combination,

$$u = c_1 \exp(i\sqrt{\lambda} x) + c_2 \exp(-i\sqrt{\lambda} x) \qquad (10.8.56)$$

for which u would be square integrable in $\mathcal{L}_2(-\infty, \infty)$. Thus, the operator $\mathbf{L}$ defined by Eq. (10.8.54) has no eigenvector. On the other hand, Eq. (10.8.56) satisfies Eq. (10.8.54) for $\sqrt{\lambda} = k$, where k is any real number. We say that $\lambda_k = k^2$ (k any real number) is the *continuous spectrum* of $\mathbf{L}$. Although the operator $\mathbf{L}$ does not have eigenvectors, it turns out it does have a spectral resolution.

Let us consider a function $h(x)$, which has a Fourier transform; i.e., the function $\tilde{h}(k)$ defined by

$$\tilde{h}(k) = \frac{1}{\sqrt{2\pi}} \int_{-\infty}^{\infty} \exp(ikx) h(x)\, dx \tag{10.8.57}$$

exists. From the theory of Fourier transforms, it is known that

$$h(k) = \frac{1}{\sqrt{2\pi}} \int_{-\infty}^{\infty} \tilde{h}(k) \exp(-ikx)\, dk. \tag{10.8.58}$$

Insertion of Eq. (10.8.57) into Eq. (10.8.58) yields

$$h(x) = \frac{1}{2\pi} \int_{-\infty}^{\infty} \left[\int_{-\infty}^{\infty} \exp(-ik(x-y))\, dk \right] h(y)\, dy, \tag{10.8.59}$$

and since Eq. (10.8.59) holds for any function having a Fourier transform, it follows that the quantity in square brackets must equal the Dirac delta function, i.e.,

$$\delta(x-y) = \frac{1}{2\pi} \int_{-\infty}^{\infty} \exp(-ik(x-y))\, dk. \tag{10.8.60}$$

Suppose $\mathbf{u} \in \mathcal{D}_L$. Then, since $\mathbf{u} = \mathbf{Iu}$, it follows that

$$\begin{aligned} u(x) &= \int_{-\infty}^{\infty} \delta(x-y) u(y)\, dy \\ &= \frac{1}{2\pi} \int_{-\infty}^{\infty} \int_{-\infty}^{\infty} \exp(-ik(x-y)) u(y)\, dk\, dy \end{aligned} \tag{10.8.61}$$

and

$$Lu = \int_{-\infty}^{\infty} \left[\frac{1}{2\pi} \int_{-\infty}^{\infty} k^2 \exp(-ik(x-y))\, dk \right] u(y)\, dy. \tag{10.8.62}$$

The quantity in square brackets on the right-hand side of Eq. (10.8.62) is a kernel of an integral operator representing the operator $\mathbf{L}$. If we denote this kernel by $l(x, y)$, we have

$$l(x, y) = \frac{1}{2\pi} \int_{-\infty}^{\infty} k^2 \exp(-ik(x-y))\, dk, \tag{10.8.63}$$

which is analogous to the spectral resolution

$$\mathbf{L} = \sum_n \lambda_n \boldsymbol{\phi}_n \boldsymbol{\phi}_n^\dagger \quad \text{or} \quad l(x, y) = \sum_n \lambda_n \phi_n(x) \phi_n^*(y) \tag{10.8.64}$$

of a regular self-adjoint Sturm–Liouville operator. (Note that the summation is replaced by an integral in Eq. (10.8.64).) The generalization of the spectral decomposition for the operator function $f(\mathbf{L})$ is

$$f(x, y) = \frac{1}{2\pi} \int_{-\infty}^{\infty} f(\lambda_k) \exp(-ik(x-y))\, dk \tag{10.8.65}$$

(where $\lambda_k = k^2$) for any function $f(t)$ that is defined for $f(\lambda_k)$. $f(x, y)$ denotes the kernel of the integral operator representing $f(\mathbf{L})$.

SPECTRAL THEORY OF SINGULAR STURM–LIOUVILLE OPERATORS

EXAMPLE 10.8.3. Consider the problem

$$Lu = -\frac{\partial^2 u}{\partial x^2} = i\frac{\partial u}{\partial t}, \qquad u \in \mathcal{L}_2(-\infty, \infty), \qquad (10.8.66)$$

and $u(x, 0) = u_0(x) \in \mathcal{L}_2(-\infty, \infty)$. The problem can be solved formally as

$$\mathbf{u} = \exp(-it\mathbf{L})\mathbf{u}_0, \qquad (10.8.67)$$

or with the aid of the spectral decomposition theorem

$$\begin{aligned} u(x) &= \frac{1}{2\pi} \int_{-\infty}^{\infty} \int_{-\infty}^{\infty} \exp\bigl[-i[k(x-y) + tk^2]\bigr] u_0(y) \, dk \, dy \\ &= \frac{1}{2\pi} \int_{-\infty}^{\infty} \exp(itk^2) \exp(-ikx) \tilde{u}_0(k) \, dk. \end{aligned} \qquad (10.8.68)$$

Thus, the solution to Eq. (10.8.66) is an integral transform. Interestingly, the solution corresponds to the process of Fourier transforming the initial concentration, followed by "time propagation" (through multiplication by the factor $\exp(itk^2)$) and concluded by a Fourier transformation back to "real space." If the boundary conditions for $\mathbf{L}$ were such that $\mathbf{L}$ is self-adjoint in a *finite* interval $[a, b]$, the solution to Eq. (10.8.66) would be

$$u(x) = \sum_n \exp(i\lambda_n t) \phi_n(x) \langle \phi_n, \mathbf{u}_0 \rangle. \qquad (10.8.69)$$

Equations (10.8.68) and (10.8.69) are quite similar. The eigenfunction $\phi_n(x)$ is analogous to $\exp(-ikx)$, the inner product $\langle \phi_n, \mathbf{u}_0 \rangle$ is analogous to $\tilde{u}_0(k) = (1/2\pi) \int_{-\infty}^{\infty} \exp(ikx) u_0(x) \, dx$, and the factors $\exp(itk^2)$ and $\exp(it\lambda_n)$ are equivalent. The major difference is that the summation in Eq. (10.8.69) is replaced by an integral in Eq. (10.8.68). In fact, the next exercise shows just how similar the spectral representations are in Eqs. (10.8.68) and (10.8.69).

EXERCISE 10.8.2. Consider the problem

$$Lu = -\frac{d^2x}{dx^2} = \lambda u, \qquad -a < x < a, \qquad (10.8.70)$$

with the periodic boundary conditions $u(-a) = u(a)$ and $u'(-a) = u'(a)$. Find the spectral representation of $\mathbf{L}$ for finite a. Show that in the limit that $a \to \infty$, the spectral decomposition of $\mathbf{L}$ becomes that given in Eq. (10.8.63).

For some other examples of the decomposition of the identity in terms of the continuous spectra of singular self-adjoint Sturm–Liouville equations, consider the following:

$$Lu = -\frac{d^2 u}{dx^2}, \qquad 0 < x < \infty, \ \mathbf{u} \in \mathcal{L}_2(0, \infty), \qquad (10.8.71)$$

for the cases

$$\begin{aligned} &\text{(a)} \quad u(0) = 0 \\ &\text{(b)} \quad u'(0) = 0 \qquad (10.8.72) \\ &\text{(c)} \quad u'(0) = \alpha u(0), \qquad \alpha \geq 0. \end{aligned}$$

The corresponding decompositions of the kernels of **I** are

(a) $\quad i(x,y) = \delta(x-y) = \dfrac{2}{\pi}\displaystyle\int_0^\infty \sin kx\,\sin ky\,dk$

(b) $\quad i(x,y) = \delta(x-y) = \dfrac{2}{\pi}\displaystyle\int_0^\infty \cos kx\,\cos ky\,dk$ \hfill (10.8.73)

(c) $\quad i(x,y) = \delta(x-y) = \dfrac{2}{\pi}\displaystyle\int_0^\infty \left(\cos kx + \dfrac{\alpha}{k}\sin kx\right)$

$$\times \left(\cos ky + \dfrac{\alpha}{k}\sin ky\right)\dfrac{k^2}{k^2+\alpha^2}\,dk.$$

■■■ **EXERCISE 10.8.3.** Derive from the results in Eq. (10.8.73) the spectral resolution of **L** for the above three cases.

From what we have examined thus far, one might hope that singular self-adjoint Sturm–Liouville operators have either only discrete spectra (when the eigenfunctions form a complete orthogonal set) or continuous spectra. Unfortunately, this hope is also soon dashed.

Consider case (c) in Eq. (10.8.72), but now assume that $\alpha < 0$. The function $\sqrt{2\alpha}\exp(\alpha x)$ is then an eigenfunction (with eigenvalue $\lambda_1 = \alpha^2$) of the operator defined by $Lu = -d^2u/dx^2$ and $u'(0) = \alpha u(0)$. Note that when $\alpha \geq 0$ the function $\exp(\alpha x)$ is not square integrable on $(0,\infty)$ and thus it is not an eigenfunction. The continuous spectra are the same as for case (c). Thus, the spectral decomposition of **I** when $\alpha < 0$ is the same as in Eq. (10.8.73) with an addition term:

$$i(x,y) = \delta(x-y) = 2\alpha\exp(\alpha(x+y))$$
$$+ \dfrac{2}{\pi}\int_0^\infty \left(\cos kx + \dfrac{\alpha}{k}\sin kx\right)\left(\cos ky + \dfrac{\alpha}{k}\sin ky\right)\dfrac{k^2}{k^2+\alpha^2}\,dk \quad (10.8.74)$$

Thus, the operator has both discrete and continuous spectra.

As another example, consider the following problem from quantum mechanics:

$$-\dfrac{\hbar^2}{2m}\dfrac{d^2u}{dx^2} + V(x)u = Eu, \qquad \mathbf{u}\in\mathcal{L}_2(-\infty,\infty), \qquad (10.8.75)$$

where the potential energy function,

$$\begin{aligned} V(x) &= V_0, & x &< -a \\ &= 0, & -a &< x < a \\ &= V_0, & x &> a, \end{aligned} \qquad (10.8.76)$$

is shown in Fig. 10.8.4.

a and V_0 are positive and the eigenvalue E is the energy of a particle. This is a singular self-adjoint problem corresponding to a particle in a space containing a potential well (often called a rectangular well because of its symmetry about $x=0$). The eigenvalues E are bounded from below by 0. Since $|u(x)|^2$ represents the probability density that the particle can be located between x and $x+dx$, the mathematical requirement that $\int_{-\infty}^\infty |u(x)|^2\,dx < \infty$, or $\mathbf{u}\in\mathcal{L}_2(-\infty,\infty)$, means

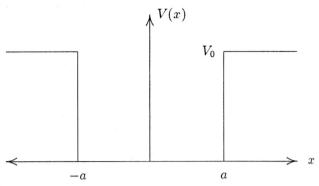

FIGURE 10.8.4

in quantum mechanics that the particle is in bound state. Formal solutions to Eq. (10.8.74) that are not in $\mathcal{L}_2(-\infty, \infty)$ are called continuum states.

Outside the potential well (i.e., when $|x| > a$), we need to solve the equation

$$\frac{d^2 u_1}{dx^2} - \kappa_1^2 u_1 = 0, \qquad |x| > a \text{ and } \kappa_1^2 = \frac{2m}{\hbar^2}(V_0 - E). \tag{10.8.77}$$

Inside the well, we need to solve the equation

$$\frac{d^2 u_2}{dx^2} + \kappa_2^2 u_1 = 0, \qquad |x| < a \text{ and } \kappa_2^2 = \frac{2m}{\hbar^2} E. \tag{10.8.78}$$

From these two equations, the solution to Eq. (10.8.76) is obtained by requiring $u_1(x) = u_2(x)$ and $u_1'(x) = u_2'(x)$ at $x = -a$ and a. Since the eigenfunctions must be square integrable in $(-\infty, \infty)$, i.e., $\mathbf{u} \in \mathcal{L}_2(-\infty, \infty)$, we can expect to find a solution to Eq. (10.8.76) only if $\kappa_1 > 0$, or $E < V_0$. In this case, a fundamental system for Eq. (10.8.77) is

$$\exp(\kappa_1 x) \qquad \text{and} \qquad \exp(-\kappa_1 x). \tag{10.8.79}$$

Likewise, a fundamental system for Eq. (10.8.78) is

$$\sin \kappa_2 x \qquad \text{and} \qquad \cos \kappa_2 x. \tag{10.8.80}$$

For $x > a$, only the fundamental solution $\exp(-\kappa_1 x)$ can be chosen since $\exp(\kappa_1 x) \to \infty$ as $x \to \infty$. Correspondingly, for $x < -a$, only the fundamental solution $\exp(\kappa_1 x)$ can be chosen since $\exp(-\kappa_1 x) \to \infty$ as $x \to -\infty$. Thus, one solution to Eq. (10.8.78) is

$$u^{(1)}(x) = \begin{cases} A \exp(\kappa_1 x), & -\infty < x < -a, \\ B \sin \kappa_2 x, & -a < x < a, \\ C \exp(-\kappa_1 x), & a < x < \infty. \end{cases} \tag{10.8.81}$$

Another solution is

$$u^{(II)}(x) = \begin{cases} D\exp(\kappa_1 x), & -\infty < x < -a, \\ F\cos\kappa_2 x, & -a < x < a, \\ G\exp(\kappa_1 x), & a < x < \infty. \end{cases} \quad (10.8.82)$$

These two solutions are equivalent to the more generally stated solution

$$u(x) = \begin{cases} c_1 \exp(\kappa_1 x), & -\infty < x < -a, \\ c_2 \sin\kappa_2 x + c_3 \cos\kappa_2 x, & -a < x < a, \\ c_4 \exp(-\kappa_1 x), & a < x < \infty. \end{cases} \quad (10.8.83)$$

Algebraically, the solution of Eq. (10.8.83) is a bit messier than the solutions at Eqs. (10.8.81) and (10.8.82). In quantum mechanics, $u^{(I)}$ is said to be the solution of odd parity ($u^{(I)}(x) = -u^{(I)}(-x)$) and $u^{(II)}$ is said to be of even parity ($u^{(II)}(x) = u^{(II)}(-x)$).

From the continuity conditions at $x = -a$ and a, it follows that $A = -C$ and $D = G$. Furthermore, from the continuity conditions of $u(x)$ and $u'(x)$ at $x = a$, it follows that, for solutions of odd parity,

$$C\exp(-\kappa_1 a) = B\sin\kappa_2 a \quad \text{and} \quad -\kappa_1 C\exp(-\kappa_1 a) = \kappa_2 B\cos\kappa_2 a, \quad (10.8.84)$$

and so the energy eigenvalues are determined by

$$\kappa_1 = -\kappa_2 \cot\kappa_2 a \quad (10.8.85)$$

and the coefficients C and B obey the equation

$$\frac{C}{B} = \exp(\kappa_1 a)\sin\kappa_2 a. \quad (10.8.86)$$

Similarly, we find for solutions of even parity that the eigenvalues are determined by

$$\kappa_1 = \kappa_2 \tan\kappa_2 a \quad (10.8.87)$$

and the coefficients obey

$$\frac{F}{G} = \exp(\kappa_1 a)\cos\kappa_2 a. \quad (10.8.88)$$

If we use the relation $\kappa_1^2 + \kappa_2^2 = (2m/\hbar^2)V_0$ to eliminate κ_1 from the squares of Eqs. (10.8.85) and (10.8.87), and define the dimensionless quantities

$$\alpha = \kappa_2 a \quad \text{and} \quad \beta^2 = \frac{2ma^2}{\hbar^2}V_0, \quad (10.8.89)$$

we can transform Eq. (10.8.85) into

$$\alpha^2 \csc^2\alpha = \beta^2 \quad (10.8.90)$$

and Eq. (10.8.87) into

$$\alpha^2 \sec^2 \alpha = \beta^2. \tag{10.8.91}$$

The positive roots of $\alpha_n^{(I)}$ of Eq. (10.8.90) determine the eigenvalues of the eigenfunctions $\phi_n^{(I)}$ of odd parity and the roots $\alpha_n^{(II)}$ of Eq. (10.8.90) determine the eigenvalues of the eigenfunctions $\phi_n^{(II)}$ of even parity. Acceptable roots of Eq. (10.8.78) have to meet the requirement that $\tan \kappa_2 a$ is positive, and those of Eq. (10.8.91) have to meet the requirement that $\cot \kappa_2 a$ is negative.

From the above definitions, the energy E_n is related to a root α_n by

$$E_n = \frac{\hbar^2}{2ma^2} \alpha_n^2. \tag{10.8.92}$$

If $E_n > V_0$, there is no solution to the problem because κ_1 would be imaginary and neither $\exp(\kappa_1 x)$ nor $\exp(-\kappa_1 x)$ would be in the Hilbert space $\mathcal{L}_2(-\infty, \infty)$ because it would not be square integrable. Thus, there are eigenfunctions only for solutions to Eqs. (10.8.90) and (10.8.91) for $\alpha \leq \beta$. Figure 10.8.5 shows how the quantum mechanical energies E vary with the depth V_0 of the potential well.

Notice that, for values β below $\pi/2$, there is only one eigenfunction or bound state for a particle in a potential well. With increasing β, or well depth, the number of eigenfunctions increases. However, for any finite value of V_0 there will be a finite number of eigenfunctions. Therefore, the eigenfunctions in a finite potential well cannot form a basis set in $\mathcal{L}_2(-\infty, \infty)$.

Consider next the situation when the particle energy E is greater than V_0. We then define $\kappa_1^2 = (2m/\hbar^2)(E - V_0)$ and $\kappa_2^2 = (2m/\hbar^2)E$ and construct the eigenproblems

$$\frac{d^2 u_1}{dx^2} + \kappa_1^2 u_1 = 0, \qquad |x| > a, \tag{10.8.93}$$

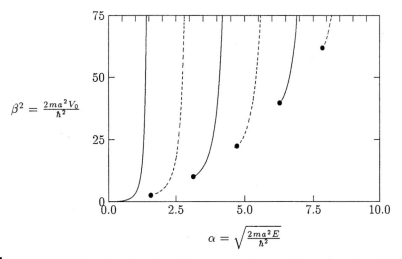

FIGURE 10.8.5 Energy levels for the bound states of a particle in a rectangular well. The full lines and broken lines refer to states of even and odd parity, respectively.

and

$$\frac{d^2u_2}{dx^2} + \kappa_2^2 u_1 = 0, \qquad |x| < a, \tag{10.8.94}$$

applying the continuity conditions $u_1(x) = u_2(x)$ and $u_1'(x) = u_2'(x)$ at $x = -a$ and a. In terms of the fundamental systems $\exp(\pm i\kappa_1 x)$ and $\exp(\pm i\kappa_2 x)$, we can express the solution as

$$u(x) = \begin{cases} c_1 e^{i\kappa_1 x} + c_2 e^{-i\kappa_1 x}, & -\infty < x < -a, \\ c_3 e^{i\kappa_2 x} + c_4 e^{-i\kappa_2 x}, & -a < x < a, \\ c_5 e^{i\kappa_1 x} + c_6 e^{-i\kappa_1 x}, & a < x < \infty. \end{cases} \tag{10.8.95}$$

For arbitrary c_3, c_4, κ_1, and κ_2, the continuity conditions at $x = -a$ can be used to determine c_1 and c_2 as linear combinations of c_3 and c_4. Furthermore, the continuity conditions at $x = a$ can be used to determine c_5 and c_6 as a linear combination of c_3 and c_4. Thus, Eq. (10.8.92) has a solution for any value of E greater than V_0. However, the solution is not square integrable, and so the values of E lying in the range $V_0 < E < \infty$ represent the continuum spectrum for the eigenproblem defined by the particle in a finite potential well.

The electronic states of the hydrogen atom represent another example. In that case, the potential has the form $V(r) = -e^2/r$, and there exist an infinite number of orthonormal eigenfunctions for $M < E < 0$, $|M| < \infty$, but they do not form a complete set. The energy range $0 < E < \infty$ represents the continuum spectrum for the electronic eigenstates (called the scattering states) of the hydrogen atom.

We see from the special cases examined in this section that a singular self-adjoint Sturm–Liouville operator may have a complete orthonormal set of eigenvectors, may have no eigenvectors (and thus only continuous spectra), or may have some orthonormal eigenvectors (even an infinite number) that are not a complete set and thus possess both continuous and dicrete spectra. So what is the most general theorem that can be stated for self-adjoint differential operators? At the end of this section, we will answer this question with a theorem that we will give without proof. Let us first reexamine the spectral decompositions given above for the operator $Lu = -d^2u/dx^2$ in $(-\infty, \infty)$ and $(0, \infty)$.

In the case of a self-adjoint Sturm–Liouville operator in $\mathcal{L}_2(0, \infty; s)$ or $\mathcal{L}_2(-\infty, \infty; s)$ having a complete set of eigenfunctions ϕ_n, the spectral resolution of $\mathbf{I}$ and $f(\mathbf{L})$ is

$$i(x, y) = \sum_{n=1}^{\infty} \phi_n(x)\phi_n^*(y) \tag{10.8.96}$$

$$f(x, y) = \sum_{n=1}^{\infty} f(\lambda_n)\phi_n(x)\phi_n^*(y), \tag{10.8.97}$$

where $i(x, y)$ is the kernel representing $\mathbf{I}$ and $f(x, y)$ is the kernel representing $f(\mathbf{L})$ when $f(t)$ exists for $t = \lambda_n$, $n = 1, 2, \ldots$. For the operators defined by Eq. (10.8.54) and Eqs. (10.8.71) and (10.8.72), the spectral decomposition of $\mathbf{I}$ and $f(\mathbf{L})$ is of the form

$$i(x, y) = \int_{k_0}^{\infty} u_k(x)u_k^*(y)\, dk \tag{10.8.98}$$

and

$$f(x, y) = \int_{k_0}^{\infty} f(\lambda_k) u_k(x) u_k^*(y) \, dk, \tag{10.8.99}$$

where $k_0 = -\infty$ for the operator defined by Eq. (10.8.54) and $k_0 = 0$ in the other three cases. The quantity λ_k corresponds to $Lu_k = \lambda_k u_k$, although the functions u_k are not eigenfunctions since they do not belong to $\mathcal{D}_L$ (which in these cases means they are not square integrable). For the self-adjoint operator having a complete set of eigenfunctions, the relation $\mathbf{u} = \mathbf{I}\mathbf{u} = \sum_n \mathbf{u}_n \mathbf{u}_n^\dagger \mathbf{u}$ corresponds to a Fourier expansion of $\mathbf{u}$ in terms of $\{\mathbf{u}_n\}$. For the self-adjoint operator having only continuous spectra, the relation $\mathbf{u} = \mathbf{I}\mathbf{u} = \int \mathbf{u}_k \mathbf{u}_k^\dagger \mathbf{u} \, dk$ is an integral transform.

In the examples given by Eq. (10.8.74) and by the particle in the potential well, the operators have both discrete spectra (eigenfunctions) and continuous spectra. These correspond to the most general case, which can be stated in the following theorem:

THEOREM. *If* $\mathbf{L}$ *is a linear self-adjoint operator in some domain* $\mathcal{D}_L$ *in a Hilbert space, then the identity operator* $\mathbf{I}$ *can be represented by the spectral decomposition*

$$i(x, y) = \sum_{n=1}^{n_u} \phi_n(x) \phi_n^*(y) + \int_{k_l}^{k_u} u_k(x) u_k^*(y) \, dk, \tag{10.8.100}$$

where the ϕ_n *are eigenfunctions of* $\mathbf{L}$ *and the* u_k *are functions satisfying* $Lu_k = \lambda_k u_k$ *(λ_k lying in the continuous spectra of* $\mathbf{L}$*). Depending on the operator,* k_u, $|k_n|$, *and* $|k_l|$ *can be finite or infinite.*

The proof of this theorem lies outside the scope of this text and advanced works on differential equations or functional analysis should be consulted for those interested.

Equation (10.8.100), through the relation $f(\mathbf{L}) = f(\mathbf{L})\mathbf{I}$, yields

$$f(x, y) = \sum_{n=1}^{n_u} f(\lambda_n) \phi_n(x) \phi_n^*(y) + \int_{k_l}^{k_u} f(\lambda_k) u_k(x) u_k^*(y) \, dk \tag{10.8.101}$$

if $f(t)$ exist for $t = \lambda_n$ and λ_k values. Here $f(x, y)$ is the spectral decomposition of the integral operator representing $f(\mathbf{L})$. It is important for the reader to remember that in the weighted Hilbert space $\mathcal{L}_2(a, b; s)$ the inner product is $\mathbf{u}^\dagger \mathbf{v} = \langle \mathbf{u}, \mathbf{v} \rangle = \int_a^b u^*(x) v(x) s(x) \, dx$, and so

$$\mathbf{I}\mathbf{u} = \int_a^b i(x, y) s(y) u(y) \, dy \tag{10.8.102}$$

and

$$f(\mathbf{L})\mathbf{u} = \int_a^b f(x, y) s(y) u(y) \, dy. \tag{10.8.103}$$

10.9. PARTIAL DIFFERENTIAL EQUATIONS

Although this chapter is primarily dedicated to ordinary differential equations, we will touch briefly in this section on partial differential equations. One reason for mentioning partial differential equations here is that frequently, by separation of variables, they can be solved by solution of ordinary differential equations. For example, diffusion or heat transfer in a solid rectangular parallelepiped obeys the equation

$$L u = -\frac{1}{D}\frac{\partial u}{\partial t}, \tag{10.9.1}$$

where

$$Lu = -\nabla^2 u = -\left(\frac{\partial^2 u}{\partial x^2} + \frac{\partial^2 u}{\partial y^2} + \frac{\partial^2 u}{\partial z^2}\right), \tag{10.9.2}$$

D is the molecular or thermal diffusivity (for physical reasons $D > 0$), and t is time. Suppose the boundary conditions on the solid are

$$u(0, y, z) = u(a, y, z) = 0$$
$$u(x, 0, z) = \frac{\partial u}{\partial y}(x, b, z) = 0 \tag{10.9.3}$$
$$\frac{\partial u}{\partial z}(x, y, 0) = \frac{\partial u}{\partial z}(x, y, c) = 0.$$

In a typical mass or heat transfer situation, the initial concentration or temperature is given, i.e.,

$$u(x, y, z, t = 0) = u_0(x, y, z). \tag{10.9.4}$$

Since the boundary conditions are separable (i.e., the constraints are on u in the x, y, and z directions separately), the method of separation of variables can be used. By this method, we choose the form of the solution as $u = X(x)Y(y)Z(z)$. We then define the operator **L** from the differential expression in Eq. (10.9.2) and the boundary conditions at Eq. (10.9.3), and seek the solution to the eigenproblem

$$\mathbf{Lu} = \lambda \mathbf{u}. \tag{10.9.5}$$

Inserting $u = XYZ$ into Eq. (10.9.5) and rearranging leads to

$$-\frac{X''(x)}{X(x)} - \frac{Y''(y)}{Y(y)} - \frac{Z''(z)}{Z(z)} = \lambda, \tag{10.9.6}$$

where the double primes on X, Y, and Z indicate second derivatives. Since the three quantities on the left-hand side of Eq. (10.9.6) depend only on x, y, and z, respectively, it follows that the three quantities are constant, i.e.,

$$-X'' = \lambda^{(x)} X, \qquad -Y'' = \lambda^{(y)} Y, \qquad -Z'' = \lambda^{(z)} Z, \tag{10.9.7}$$

where $\lambda = \lambda^{(x)} + \lambda^{(y)} + \lambda^{(z)}$. The boundary conditions for the three eigenproblems in Eq. (10.9.7) are

$$X(0) = X(a) = 0$$
$$Y(0) = Y'(b) = 0 \qquad (10.9.8)$$
$$Z'(0) = Z'(c) = 0,$$

and thus all three equations in Eq. (10.9.7) are regular, self-adjoint Sturm–Liouville equations.

The eigenfunctions and eigenvalues of these equations are

$$X_n = \sqrt{\frac{2}{a}} \sin \frac{n\pi x}{a}, \qquad \lambda_n^{(x)} = \left(\frac{n\pi}{a}\right)^2, \qquad n = 1, 2, \ldots$$

$$Y_n = \sqrt{\frac{2}{b}} \sin\left(n + \frac{1}{2}\right)\frac{\pi x}{b}, \qquad \lambda_n^{(y)} = \left(n + \frac{1}{2}\right)^2 \left(\frac{\pi}{b}\right)^2, \qquad n = 1, 2, \ldots$$

$$Z_0 = \frac{1}{\sqrt{c}}, \qquad \lambda_0^{(z)} = 0$$

$$Z_n = \sqrt{\frac{2}{c}} \cos \frac{n\pi x}{c}, \qquad \lambda_n^{(z)} = \left(\frac{n\pi}{c}\right)^2, \qquad n = 1, 2, \ldots.$$
$$(10.9.9)$$

Since $\{X_m\}$, $\{Y_n\}$, and $\{Z_p\}$ are complete orthonormal sets in $\mathcal{L}_2(0, a)$, $\mathcal{L}_2(0, b)$, and $\mathcal{L}_2(0, c)$, respectively, the set

$$\boldsymbol{\phi}_{mnp} = \mathbf{X}_m \mathbf{Y}_n \mathbf{Z}_p, \qquad m, n = 1, 2, \ldots, \quad p = 0, 1, 2 \ldots, \qquad (10.9.10)$$

forms a complete orthonormal set in $\mathcal{L}_2(\Omega_3)$, where Ω_3 is the volume of the rectangular parallelepiped defined by $0 < x < a$, $0 < y < b$, $0 < z < c$. Since $\boldsymbol{\phi}_{nmp}$ obeys the boundary conditions given by Eq. (10.9.3), we see that it is an eigenvector of Eq. (10.9.5) with the eigenvalue

$$\lambda_{mnp} = \left(\frac{m\pi}{a}\right)^2 + \left(n + \frac{1}{2}\right)^2 \left(\frac{\pi}{b}\right)^2 + \left(\frac{p\pi}{c}\right)^2. \qquad (10.9.11)$$

Thus, we have constructed the eigenvectors and eigenvalues of a regular, self-adjoint, three-dimensional Sturm–Liouville operator from the eigenvectors and eigenvalues of three regular, self-adjoint, one-dimensional Sturm–Liouville operators.

The formal solution to Eq. (10.9.1) is

$$\mathbf{u} = \exp(-t D\mathbf{L})\mathbf{u}_0, \qquad (10.9.12)$$

and so, from the spectral decomposition

$$\exp(-t D\mathbf{L}) \sum_{m, n, p} \exp(-t D\lambda_{mnp}) \boldsymbol{\phi}_{mnp} \boldsymbol{\phi}_{mnp}^{\dagger}, \qquad (10.9.13)$$

we find

$$u(x, y, z) = \sum_{m,n,p=1}^{\infty} \exp(-tD\lambda_{mnp})X_m(x)Y_n(y)Z_n(z) \qquad (10.9.14)$$
$$\times \int_0^a \int_0^b \int_0^c X_m(x_1)Y_n(y_1)Z_p(z_1)u_0(x_1,y_1,z_1)\,dx_1\,dy_1\,dz_1.$$

The operator **L** defined by Eqs. (10.9.2) and (10.9.3) is an example of a regular, selfadjoint, multidimensional Sturm–Liouville operator. For the general n-dimensional case, consider the finite volume Ω_n and surface $\partial\Omega_n$. $\bar{\Omega}_n$ will denote the volume Ω_n plus its boundary, i.e., $\bar{\Omega}_n = \Omega_n + \partial\Omega_n$. The differential expression defining the regular Sturm–Liouville operator is

$$Lu = \frac{1}{s(\vec{x})}\left\{-\sum_{i,j=1}^n \frac{\partial}{\partial x_i}\left[p_{ij}(\vec{x})\frac{\partial u}{\partial x_j}\right] + q(\vec{x})u\right\}, \qquad (10.9.15)$$

where $\vec{x}$ is an n-dimensional Euclidean vector with Cartesian components $\{x_1,\ldots,x_n\}$ and

(i) p_{ij}, s, and q are real-valued functions of $\vec{x}$ and $p_{ij} = p_{ji}$;
(ii) $p_{ij} \in C^1(\Omega_n)$, $s(\vec{x}), q(\vec{x}) \in C^0(\Omega_n)$;
(iii) $s(\vec{x}) > 0$ for $\vec{x} \in \bar{\Omega}_n$;
(iv) $\sum_{i,j=1}^n p_{ij}(\vec{x})\xi_i \geq c_0\sum_{i=1}^n \xi_i^2$ for all $\vec{x}$ in $\bar{\Omega}_n$ arbitrary real numbers $\xi_1,\ldots,\xi_n$; c_0 is a constant greater than 0. This condition is that the matrix $[p_{ij}(\vec{x})]$ is positive definite for all $\vec{x} \in \bar{\Omega}_n$.

This operator is formally self-adjoint in the Hilbert space $\mathcal{L}_2(\Omega_n;s)$, since it can be shown by integration by parts that $L^\dagger = L$.

The boundary conditions for the operator **L** are denoted by

$$Bu = 0, \qquad \vec{x} \in \partial\Omega_n. \qquad (10.9.16)$$

Thus, the domain $\mathcal{D}_L$ of the Sturm–Liouville operator defined by Eqs. (10.9.15)–(10.9.16) is

$$\mathcal{D}_L = \{\mathbf{u}, L\mathbf{u} \in \mathcal{L}_2(\Omega_n;s),\ \mathbf{u} \in C^2(\bar{\Omega}_n);\ Bu = 0,\ \vec{x} \in \partial\Omega_n\}. \qquad (10.9.17)$$

To derive the boundary conditions for the adjoint $L^\dagger$, we need to use the Gauss theorem

$$\int_{\partial\Omega_n} u(\vec{x})\hat{v}(\vec{x})\,dS_n = \int_{\Omega_n} \nabla u(\vec{x})\,d\Omega_n, \qquad (10.9.18)$$

where ∇ is the n-dimensional gradient operator and $\hat{v}(\vec{x})$ is the outward-pointing normal at the point $\vec{x}$ locating the element dS_n of the surface $\partial\Omega_n$. In Cartesian coordinates, the directional vectors are orthogonal and we can construct the Gauss theorem for individual coordinates as follows

$$\int\cdots\int_{\Omega_n} \frac{\partial u(\vec{x})}{\partial x_i}\,dx_1\cdots dx_n = \int\cdots\int_{\partial\Omega_n} u(\vec{x})v_i(\vec{x})\,dS_n, \qquad (10.9.19)$$

where v_i is the component of $\hat{v}$ in the direction of x_i. Using Eq. (10.9.20) and $\langle \mathbf{v}, \mathbf{u} \rangle \equiv \int_{\Omega_n} v^*(\vec{x}) u(\vec{x}) s(\vec{x}) \, d\Omega_n$, we find

$$\langle \mathbf{v}, \mathbf{L}\mathbf{u} \rangle - \langle \mathbf{L}^\dagger \mathbf{v}, \mathbf{u} \rangle = \int_{\partial \Omega_n} \sum_{i,j=1}^n p_{ij} v_i \left(v_{x_j}^* u - u_{x_j} v^* \right) dS_n, \tag{10.9.20}$$

where u_{x_j} and v_{x_j} denote the partial derivatives of u and v with respect to x_j. Thus, the domain $\mathcal{D}_{L^\dagger}$ is defined by the condition

$$\int_{\partial \Omega_n} \sum_{i,j=1}^n p_{ij} v_i \left(v_{x_j}^* u - u_{x_j} v^* \right) dS_n = 0 \quad \text{for all } \mathbf{u} \in \mathcal{D}_L. \tag{10.9.21}$$

If the domain $\mathcal{D}_{L^\dagger}$ of vectors $\mathbf{v}$ defined by Eq. (10.9.21) is equal to $\mathcal{D}_L$, the Sturm–Liouville operator will be self-adjoint since $L = L^\dagger$. Some examples for which the multidimensional Sturm–Liouville operator is self-adjoint are:

(i) $u(\vec{x}) = 0$ for $\vec{x} \in \partial \Omega_n$.
(ii) $u_{x_j}(\vec{x}) = 0$, $j = 1, \ldots, n$, for $\vec{x} \in \partial \Omega_n$.
(iii) $Ru = 0$ for $\vec{x} \in \partial \Omega_n$, where $Ru = \sum_{i,j=1}^n p_{ij}(\vec{x}) u_{x_j} v_j$.
(iv) $Ru + \sigma u = 0$ for $\vec{x} \in \partial \Omega_n$, σ real, and $\sigma(\vec{x}) \in C^0(\partial \Omega_n)$.

Since condition (i) places no constraint on u_{x_j}, it follows from Eq. (10.9.21) that $v = 0$ for $\vec{x} \in \partial \Omega_n$ for this case. Similarly, condition (ii) places no constraint on u and so Eq. (10.9.21) implies that $v_{x_j} = 0$, $j = 1, \ldots, n$, for $\vec{x} \in \partial \Omega_n$ for this case. Equation (10.9.21) can be expressed in the form

$$\int_{\partial \Omega_n} (uRv^* - v^* Ru) \, dS_n = 0 \tag{10.9.22}$$

from which it follows that $Ru = 0$ for $\vec{x} \in \partial \Omega_n$ for condition (iii) and $Ru + \sigma v = 0$ for $\vec{x} \in \partial \Omega_n$ for condition (iv). Thus, the regular, real Sturm–Liouville operators defined by Eqs. (10.9.15) and (10.9.16), with the boundary conditions (i)–(iv), are self-adjoint.

One can also prove the theorem:

THEOREM. *With the boundary conditions* (i)–(iv), *the regular, real Sturm–Liouville operator in* $\mathcal{L}_2(\Omega_n; s)$ *is bounded from below, i.e.,*

$$\langle \mathbf{u}, \mathbf{L}\mathbf{u} \rangle \geq M \langle \mathbf{v}, \mathbf{v} \rangle, \qquad |M| < \infty. \tag{10.9.23}$$

The proof of the theorem is more difficult than was the case in one dimension and will not be given here.

We will also give here without proof the following fundamental theorem for regular, real Sturm–Liouville operators:

THEOREM. *If the boundary conditions* $Bu = 0$ *for* $\vec{x} \in \partial \Omega_n$ *are such that the n-dimensional regular Sturm–Liouville operator* $\mathbf{L}$—*defined by Eqs.* (10.9.15) *and* (10.9.16)—*is a subset of* $\mathcal{L}_2(\Omega_n; s)$, *then the eigenvectors* $\{\boldsymbol{\phi}_\mathbf{m}\}$ *of* $\mathbf{L}$ *form a complete orthonormal set in* $\mathcal{L}_2(\Omega_n; s)$ *and the eigenvalues* $\lambda_\mathbf{m}$ *are real.*

Thus, the same spectral resolution theorem

$$\mathbf{I} = \sum_{\mathbf{m}} \boldsymbol{\phi}_{\mathbf{m}} \boldsymbol{\phi}_{\mathbf{m}}^{\dagger} \tag{10.9.24}$$

and

$$\mathbf{L} = \sum_{\mathbf{m}} \lambda_{\mathbf{m}} \boldsymbol{\phi}_{\mathbf{m}} \boldsymbol{\phi}_{\mathbf{m}}^{\dagger}, \qquad f(\mathbf{L}) = \sum_{\mathbf{m}} f(\lambda_{\mathbf{m}}) \boldsymbol{\phi}_{\mathbf{m}} \boldsymbol{\phi}_{\mathbf{m}}^{\dagger}, \tag{10.9.25}$$

when $f(\lambda_{\mathbf{m}})$ exists, holds for regular self-adjoint Sturm–Liouville operators in any dimension. As was evident in the example given by Eqs. (10.9.10) and (10.9.11), an eigenvector in n dimensions will generally be characterized by n indices. Thus, the boldface $\mathbf{m}$ in the above theorem indicates that n indices are required to specify the $\mathbf{m}$th eigenvector.

One route to proving the theorem is to prove that the Green's function operator, $\mathbf{G}$, for a regular Sturm–Liouville operator is a completely continuous integral operator. The kernel of the operator $\mathbf{G}$ is determined from

$$Lg(\vec{x}, \vec{y}) = \delta(\vec{x}, \vec{y}), \qquad Bg(\vec{x}, \vec{y}) = 0 \qquad \text{for } \vec{x} \in \partial\Omega_n. \tag{10.9.26}$$

Once it is well established that the Green's function for an n-dimensional regular Sturm–Liouville operator is completely continuous, then the proof of the theorem would be as easy as was the proof for the one-dimensional case. The interested reader should pursue more advanced texts for the complete proof.

The situation for singular self-adjoint operators in n dimensions is the same as that in one dimension. If $s(\vec{x}) = 0$ or $p(\vec{x}) = 0$ at some point, if $|q(\vec{x})|$ is infinite at some point, or if the domain Ω_n is infinite in any dimension, then a self-adjoint equation is singular. Singular self-adjoint operators can have exclusively discrete spectra (their eigenfunctions form a complete orthonormal set), exclusively continuous spectra, or a mixture of discrete and continuous.

Again, the simple example

$$Lu = -\sum_{i=1}^{n} \frac{\partial^2 u}{\partial x_i^2}, \qquad \mathbf{u} \in \mathcal{L}_2(-\infty < x_i < \infty, \ i = 1, \ldots, n) \tag{10.9.27}$$

provides us with a self-adjoint operator with no eigenfunctions. For this case, the functions

$$u_{\vec{k}} = \frac{1}{(2\pi)^{n/2}} \exp(i\vec{k} \cdot \vec{x}), \qquad \vec{k} \cdot \vec{x} = \sum_{i=1}^{n} k_i x_i, \tag{10.9.28}$$

obey the equation $Lu_{\vec{k}} = \lambda_{\vec{k}} u_{\vec{k}}$, where $\lambda_{\vec{k}} = k^2 = \sum_{i=1}^{n} k_i^2$ and k_i can be any real number. However, $u_{\vec{k}}$ is not square integrable and so does not belong to $\mathcal{L}_2(R_n)$, where R_n denotes the entire n-dimensional Euclidean space.

From the theory of Fourier transforms in R_n, it is known that if

$$\tilde{h}(\vec{k}) = \frac{1}{(2\pi)^{n/2}} \int_{R_n} \exp(i\vec{k} \cdot \vec{x}) h(\vec{x}) \, d^n x \tag{10.9.29}$$

(here $d\Omega_n$ and $d^n x$ denote the same thing), then

$$h(\vec{x}) = \frac{1}{(2\pi)^{n/2}} \int_{R_n} \exp(-i\vec{k} \cdot \vec{x})\tilde{h}(\vec{k}) \, d^n k \qquad (10.9.30)$$

from which it follows that

$$\delta(\vec{x} - \vec{y}) = \frac{1}{(2\pi)^n} \int_{R_n} \exp(i(\vec{x} - \vec{y}) \cdot \vec{k}) \, d^n k. \qquad (10.9.31)$$

But this means that

$$\mathbf{I} = \int_{R_n} \mathbf{u}_{\vec{k}} \mathbf{u}_{\vec{k}}^{\dagger} \, d^n k \qquad (10.9.32)$$

and

$$\mathbf{L} = \int_{R_n} \lambda_{\vec{k}} \mathbf{u}_{\vec{k}} \mathbf{u}_{\vec{k}}^{\dagger} \, d^n k. \qquad (10.9.33)$$

since $\mathbf{LI} = \mathbf{L}$ and $\mathbf{Lu}_{\vec{k}} = \lambda_{\vec{k}} \mathbf{u}_{\vec{k}}$. This is the n-dimensional version of the one-dimensional case of the spectral decomposition of a self-adjoint operator having only continuous spectra. The Schrödinger operator for the electronic states of the hydrogen atom is an example of a singular (because $\mathcal{H} = \mathcal{L}_2(R_n)$ and $|q(\vec{x})| = \infty$ at $|\vec{x}| = 0$) Sturm–Liouville operator having an infinite number of eigenvectors plus continuous spectra.

The general theorem for self-adjoint partial differential equations is the same as that for ordinary self-adjoint differential equations:

THEOREM. *If $\mathbf{L}$ is a linear self-adjoint operator in some domain $\mathcal{D}_L$ in a Hilbert space of functions $u(\vec{x})$ of the n-dimensional Euclidean vector $\vec{x}$, then the identity operator $\mathbf{I}$ can be represented by the spectral decomposition*

$$\mathbf{I} = \sum_m \boldsymbol{\phi}_m \boldsymbol{\phi}_m^{\dagger} + \int_{\Omega_n} \mathbf{u}_{\vec{k}} \mathbf{u}_{\vec{k}}^{\dagger} \, d^n k, \qquad (10.9.34)$$

where $L\phi_m(\vec{x}) = \lambda_m \phi_m^{\dagger}(\vec{x})$ and $Lu_{\vec{k}}(\vec{x}) = \lambda_{\vec{k}} u(\vec{x})$. The $\boldsymbol{\phi}_m$ are eigenvectors; i.e., they are square integrable in the Hilbert space. The $\mathbf{u}_{\vec{k}}$ are not square integrable and are therefore not eigenvectors and represent the continuous spectra. The λ_m and $\lambda_{\vec{k}}$ are real numbers, representing discrete and continuous spectra, respectively.

Again, the reader is referred to more advanced texts for the proof of this theorem.

■ ILLUSTRATION 10.9.1 (Heat Transfer in a Laminar-Flow Pipe). As an illustration of the separation of variables technique, consider a fluid flowing in a cylindrical pipe as shown in Figure 10.9.1. Assuming that the flow is laminar, the velocity in the z direction can be shown to be parabolic, satisfying the formula

$$v_z(r) = 2v_0 \left[1 - \left(\frac{r}{R}\right)^2\right], \qquad (10.9.35)$$

where v_0 is the average (or bulk) velocity. Assume that the fluid enters a length of pipe at $z = 0$ with a uniform temperature T_0. We would like to heat the fluid

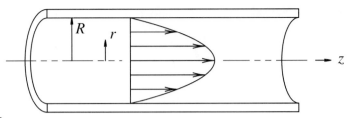

FIGURE 10.9.1

by holding the inside surface of the pipe (at $r = R$) fixed at some temperature T_s. Our goal is to find the temperature of the fluid as a function of r and z further downstream.

A differential energy balance on the appropriate control volume gives the well-known equation of change for the temperature profile

$$\rho \hat{C}_p \left(\frac{\partial T}{\partial t} + \vec{v} \cdot \vec{\nabla} T \right) = k \vec{\nabla}^2 T, \tag{10.9.36}$$

where ρ is the fluid density, $\hat{C}_p$ is the fluid heat capacity, and k is the fluid thermal conductivity. We wish to solve for the steady-state temperature profile in which the time derivative is 0. Further, we note that the velocity of the fluid only has a component in the z direction. In cylindrical coordinates, the above equation for steady-state conditions is thus given by

$$v_z(r) \frac{\partial T}{\partial z} = \alpha \left[\frac{1}{r} \frac{\partial}{\partial r} \left(r \frac{\partial T}{\partial r} \right) + \frac{\partial^2 T}{\partial z^2} \right], \tag{10.9.37}$$

where we define $\alpha \equiv k/(\rho \hat{C}_p)$. If the velocity is large enough or the pipe radius small enough, we can neglect the axial component of conduction; i.e., we can assume that $\partial^2 T/\partial z^2 \approx 0$. In this case, using the velocity profile in Eq. (10.9.35), the above equation becomes

$$2v_0 \left[1 - \left(\frac{r}{R} \right)^2 \right] \frac{\partial T}{\partial z} = \alpha \frac{1}{r} \frac{\partial}{\partial r} \left(r \frac{\partial T}{\partial r} \right) \tag{10.9.38}$$

and the boundary conditions are

$$\begin{aligned} T(z=0, r) &= T_0 \\ T(z, r=R) &= T_s \\ \left. \frac{\partial T}{\partial r} \right|_{r=0} &= 0. \end{aligned} \tag{10.9.39}$$

Before proceeding, it is convenient to define the following dimensionless variables:

$$\zeta \equiv \frac{z\alpha}{2v_0 R^2}, \quad \xi \equiv \frac{r}{R}, \quad \theta \equiv \frac{T_s - T}{T_s - T_0} \tag{10.9.40}$$

> # PARTIAL DIFFERENTIAL EQUATIONS

for which Eq. (10.9.38) becomes

$$(1-\xi^2)\frac{\partial \theta}{\partial \zeta} = \frac{1}{\xi}\frac{\partial}{\partial \xi}\left(\xi \frac{\partial \theta}{\partial \xi}\right). \tag{10.9.41}$$

The boundary conditions become

$$\begin{aligned}\theta(\zeta=0,\xi) &= 1 \\ \theta(\zeta,\xi=1) &= 0 \\ \left.\frac{\partial \theta}{\partial \xi}\right|_{\xi=0} &= 0.\end{aligned} \tag{10.9.42}$$

(i) We can solve this equation by using the separation of variables technique and assuming a solution of the form

$$\theta(\zeta,\xi) = Z(\zeta)\Xi(\xi). \tag{10.9.43}$$

Show that, with this solution form, Eq. (10.9.41) can be separated into the two coupled equations

$$\frac{dZ}{d\zeta} + \lambda^2 Z = 0 \tag{10.9.44}$$

$$\xi \frac{d^2\Xi}{d\xi^2} + \frac{d\Xi}{d\xi} + \lambda^2 \xi(1-\xi^2)\Xi = 0. \tag{10.9.45}$$

The solution to Eq. (10.9.44) is simply

$$Z(\zeta) = \exp(-\lambda^2 \zeta). \tag{10.9.46}$$

This exponential behavior in ζ seems reasonable in light of the boundary conditions for the problem. Thus, we are sure we chose the correct sign for λ^2. Next, we turn to the solution for Eq. (10.9.45).

(ii) First, show that the differential equation in Eq. (10.9.45) represents a Sturm–Liouville eigenproblem. What are the corresponding values of $s(\xi)$, $p(\xi)$, and $q(\xi)$? What are the boundary conditions?

(iii) Show that the polynomial functions

$$u_\lambda(\xi) = \sum_{n=0}^{\infty} c_n \xi^{2n}, \tag{10.9.47}$$

where

$$\begin{aligned} c_0 &= 1 \\ c_1 &= -\frac{\lambda^2}{4} \\ c_n &= \left(\frac{\lambda}{2n}\right)^2 (c_{n-2} - c_{n-1}), \end{aligned} \tag{10.9.48}$$

are solutions to Eq. (10.9.45). Write an expression representing the orthogonality conditions for the $u_\lambda(\xi)$.

(iv) Applying the boundary conditions for $\Xi(\xi)$, obtain an expression for the eigenvalues λ^2. Use the Newton–Raphson method to solve for the first 10 lowest eigenvalues.

(v) The spectral resolution theorem allows us to compose a solution from a superposition of eigenfunctions as

$$\theta(\zeta, \xi) = \sum_{n=1}^{\infty} a_n u_{\lambda_n}(\xi) \exp(-\lambda_n^2 \zeta). \qquad (10.9.49)$$

Derive an expression for the parameters a_n using the boundary condition when ∎∎∎ $\zeta = 0$.

PROBLEMS

1. Find the adjoint $L^\dagger$ of the operator L, where

$$Lu = \frac{d^3 u}{dx^3} - 2\frac{d^2 u}{dx^2} - \frac{du}{dx} + 2u, \qquad 0 < x < 1,$$

and $u(0) = 0$ and $u'(1) = 0$. Give $L^\dagger$ and $\mathcal{D}_{L^\dagger}$.

2. Find the adjoints $L^\dagger$ of the operators L, where

$$Lu = \frac{d^4 u}{dx^4}, \qquad 0 < x < 1,$$

and the boundary conditions are

(a) $u(0) = u(1) = u''(0) = u''(1) = 0$
(b) $u'(0) = u'(0) = u'''(0) = u'''(1) = 0$
(c) $u(0) = u(1) = u''(0) = u''(1) = 0$ and $u'(0) = u'(1)$
(d) $u(0) = u'''(0) = u(1) = u'''(1) = 0$.

For which set (or sets) of boundary conditions is the operator self-adjoint?

3. Consider the operator

$$Lu = \exp(r^2) \nabla \cdot (\exp(-r^2) \nabla u),$$

where ∇ is the gradient in three dimensions; i.e., in Cartesian coordinates, ∇ has the form

$$\nabla = \hat{i}\frac{\partial}{\partial x} + \hat{j}\frac{\partial}{\partial y} + \hat{k}\frac{\partial}{\partial z}.$$

Also $\vec{r} = x\hat{i} + y\hat{j} + z\hat{j}$ and $r^2 = x^2 + y^2 + z^2$ in these coordinates. Suppose the Euclidean domain for the problem is R_3, namely, all of the three-dimensional Euclidean space. In what Hilbert space will the operator L be self-adjoint? Is L a Sturm–Liouville operator? Is it regular or singular? Why?

4. Consider the differential expression

$$Lu = \frac{d^3u}{dx^3} - 2\frac{d^2u}{dx^2} - \frac{du}{dx} + 2u.$$

(a) Solve the initial value problems

$$Lu = 1, \quad u(0) = 1, \quad u'(0) = 0, \quad u''(0) = 2$$

and

$$L^\dagger v = x, \quad v(0) = 0, \quad v'(0) = 1, \quad v''(0) = 1,$$

where $L^\dagger$ is the differential expression of the formal adjoint of L.

(b) Plot $u(x)$ and $v(x)$ for $-2 < x < 2$.

5. Give fundamental systems for $Lu = 0$ and $L^\dagger v = 0$, where L and $L^\dagger$ are defined in Problem 4.

6. Give a fundamental system for $Lu = 0$, where

$$Lu = \frac{d^4u}{dx^4} - qu,$$

and q is a positive real constant.

7. Consider the differential operator

$$Lu = \frac{d^2u}{dx^2} + a(x)\frac{du}{dx} + b(x)u, \quad a < x < b,$$

where $a(x)$ and $b(x)$ are complex functions. Prove that if u_1 and u_2 are solutions to $Lu = 0$, they form a fundamental system (i.e., they are linearly independent) *if and only if* $W(x) \neq 0$, $x \in [a, b]$, where $W(x)$ is the Wronskian:

$$W(x) = \begin{vmatrix} u_1(x) & u_2(x) \\ u_1'(x) & u_2'(x) \end{vmatrix}.$$

8. Prove that if $u_1(x) \neq 0$ is a solution to $Lu = 0$, where Lu is given in Problem 7, then a fundamental system for $Lu = 0$ in $[a, b]$ is given by $u_1(x)$ and

$$u_2(x) = u_1(x) \int^x \frac{\exp\left[-\int^y a(z)\,dz\right]}{u_1^2(y)}\,dy.$$

For the special case $a(x) = 0$, $b(x) = -1$, use this method to find a fundamental system.

9. Give the functions $f(x)$ for which the equation

$$\frac{d^4u}{dx^4} - \pi^4 u = f(x), \quad 0 < x < 1,$$

with $u(0) = u(a) = u''(0) = u''(1) = 0$ has a solution.

10. Repeat Problem 9, but with the boundary conditions

$$u'(0) = u'(1) = u'''(0) = u'''(1).$$

11. Repeat Problem 9, but with the boundary conditions

$$u(0) = u'''(0) = u(1) = u'''(1) = 0.$$

12. Find the adjoint operator and the solvability condition for the problem

$$\frac{d^4u}{dx^4} = f(x), \qquad 0 < x < 1,$$

$$u(0) = u(1) = u''(0) = u''(1) = 0, \qquad u'(0) - u'(1) = 2.$$

13. Find the adjoint operator and the solvability conditions for the problem

$$-\frac{d^2u}{dx^2} - u = f(x), \qquad -\pi < x < \pi,$$

$$u(\pi) - u(-\pi) = \gamma_1, \qquad u'(\pi) - u'(-\pi) = \gamma_2.$$

14. Find Green's function for the operator

$$Lu = -\frac{d^2u}{dx^2} - 9u, \qquad 0 < x < 1.$$

15. For the operator given in Problem 14, solve the problem

$$Lu = f(x), \qquad 0 < x < 1,$$

with

$$u(0) = \gamma_1, \qquad u'(1) = \gamma_2.$$

Plot $u(x)$ versus x for $f(x) = x^3$, $\gamma_1 = 1$, and $\gamma_2 = 2$.

16. Find Green's function for the operator

$$Lu = -\frac{d^2u}{dx^2} - k^2u, \qquad 0 < x < 1,$$

with

$$u(0) = u(1) \qquad \text{and} \qquad u'(0) = u'(1).$$

17. Prove that Green's function $g(\vec{r}, \vec{r}')$ for the operator

$$Lu = \nabla^2 u + q^2 u, \qquad \mathbf{u}, \mathbf{Lu} \in \mathcal{L}_2(R_3)$$

is

$$G(\vec{r}, \vec{r}') = \frac{\exp(-q|\vec{r} - \vec{r}'|)}{|\vec{r} - \vec{r}'|},$$

where q is a positive constant and ∇^2 is the Laplacian in a three-dimensional Euclidean vector space. R_3 means that the domain is the entire space. *Hint:* Show that if

$$u(\vec{r}) = \int_{R_3} g(\vec{r} - \vec{r}') f(\vec{r}') d^3r',$$

then

$$Lu = f$$

for an arbitrary $\mathbf{f} \in \mathcal{L}_2(R_3)$.

18. Consider the eigenproblem

$$Lu = -\frac{d}{dx}\left(x^2 \frac{du}{dx}\right) = \lambda u, \qquad 1 < x < e, \ u(1) = u(e) = 0.$$

Show that the eigenfunctions of this equation are

$$\phi_n(x) = (2x)^{-1/2} \sin(n\pi \ln x), \qquad n = 1, 2, \ldots,$$

and that they form a complete set in $\mathcal{L}_2(1, e)$. What are the corresponding eigenvalues? Prove that $\langle \phi_n, \phi_m \rangle = 0$, $n \neq m$. Give the spectral decomposition of the Green's function operator $\mathbf{G}$ for $\mathbf{L}$. Solve the problem

$$Lu = x^2$$

and plot $u(x)$ versus x for $1 < x < e$.

19. Find the eigenvalues and normalized eigenfunctions of

$$Lu = -\frac{d^2 u}{dx^2}, \qquad -1 < x < 1,$$

where $u'(-1) = \cot \alpha\, u(-1)$ and $u'(1) = \cot \beta\, u(1)$, and $0 \le \alpha < \pi$, $0 \le \beta < \pi$. The situation when $\alpha = \beta = 0$ corresponds to $u(-1) = u(1) = 0$. Can an eigenvalue be less than 0? Prove your answer.

20. Find the eigenfunctions and eigenvalues of the operators in Problems 2 and 6 for the boundary conditions (a) and (b) in Problem 2. Do the eigenfunctions form a complete orthonormal set in these cases? Why?

21. Find the spectral decomposition of $\mathbf{L}$, where

$$Lu = \frac{d^4 u}{dx^4}, \qquad 0 < x < \infty,$$

for the boundary conditions

(a) $\qquad u(0) = u''(0) = 0.$
(b) $\qquad u'(0) = u'''(0) = 0.$

22. Suppose

$$Lu = -\frac{d^2 u}{dx^2}, \qquad 0 < x < \infty,$$

where $u(0) = 0$ and $\mathbf{u} \in \mathcal{L}_2(0, \infty)$. Use the spectral resolution of $\exp(-t\mathbf{L})$ to solve

$$\frac{\partial^2 u}{\partial x^2} = \frac{\partial u}{\partial t},$$

where $u(0, t) = 0$ and

$$u(x, 0) = \exp(-x^2).$$

23. Consider the operator

$$Lu = -\frac{d^2 u}{dx^2}, \qquad 0 < x < \infty,$$

where $\cos \alpha u(0) - \sin \alpha u'(0) = 0$, α is real, and $0 \leq \alpha < \pi$. This problem can have discrete and continuous spectra, depending on α. The Hilbert space for this operator is $\mathcal{L}_2(0, \infty)$.

(a) Give the spectral resolution of **I**, **L**, and $\mathbf{G} = \mathbf{L}^{-1}$ for the case $0 \leq \alpha < (\pi/2)$.

(b) Give the spectral resolution of **I**, **L**, and $\mathbf{G} = \mathbf{L}^{-1}$ for the case $(\pi/2) < \alpha < \pi$.

24. The temperature of a slab of material held between two plates and internally heated obeys the equation

$$\alpha \frac{\partial^2 T}{\partial x^2} = \frac{\partial T}{\partial t} + f, \qquad 0 < x < b,$$

where b is the length of the slab, α is the thermal diffusivity (thermal conductivity divided by the product of the density and the heat capacity), and $f(x)$ is the internally generated power divided by the product of the density and the heat capacity. Suppose one end of the slab is insulated, so that

$$\frac{\partial T}{\partial x}(0, t) = 0,$$

and the other end is at a fixed temperature, i.e.,

$$T(x = b, t) = T_b.$$

Suppose also that the initial temperature distribution is given by

$$T(x, t = 0) = T_0(x).$$

(a) Use the spectral decomposition of $\exp(-\alpha t \mathbf{L})$ to formally solve for $T(x, t)$, where **L** is the operator defined by

$$\mathbf{L}u = -\frac{d^2 u}{dx^2}, \qquad u(0) = 0, \ u'(b) = 0$$

and $\mathbf{u}, \mathbf{L}\mathbf{u} \in \mathcal{L}_2(0, b)$.

(b) Define a convenient set of units such that $\alpha = 1$ and $b = 1$. Suppose that

$$T_0(x) = 100 \qquad \text{and} \qquad f = 10x.$$

Plot T versus x for $t = 0, 1, 2, 5$, and 10.

25. Consider a long cylinder of radius $r = a$, which has an initial temperature distribution of

$$T(r, t = 0) = T_0(r).$$

Suppose the cylinder is immersed in a uniform heat bath that fixes the surface temperature at T_b, i.e.,

$$T(r = a, t) = T_b.$$

From symmetry, it follows that

$$\frac{\partial T}{\partial r} = 0 \quad \text{at } r = 0.$$

The differential equation for this problem is

$$\alpha \frac{1}{r} \frac{\partial}{\partial r}\left(r \frac{\partial T}{\partial r}\right) = \frac{\partial T}{\partial t}.$$

Using the fact that a solution to the eigenproblem

$$\frac{1}{r} \frac{d}{dr}\left(r \frac{du}{dr}\right) = \lambda u$$

is

$$J_0(\sqrt{\lambda} r),$$

find the temperature T of the cylinder as a function of r and t. Assuming that $r = 2$ in. $T_b = 600°C$ and $T_0 = 25°C$, and that the value of α is 3.3 ft²/hr (the value of the thermal diffusivity of aluminum), calculate $T(r, t)$. Plot T versus r for $t = 10$ sec, 10 min, 20 min, and 1 hr. Also, plot $T(r = 0, t)$ versus t.

26. Suppose a can of beer at room temperature (75°F) is put in a refrigerator at 40°F. The can is 6 in. long and $2\frac{1}{2}$ in. in diameter. Calculate how long the beer must be left in the refrigerator for its center to reach 60°F. Assume the following:

   (i) The temperature depends on the radial distance r from the center of the can and axial distance x from the bottom of the can, i.e., $T = T(r, x, t)$.

   (ii) The temperature of the surface of the can is 40°F, i.e.,

   $$T(r = 1.25 \text{ in.}, x, t)$$
   $$= T(r, x = 0, t)$$
   $$= T(r, x = 6 \text{ in.}, t) = 40°F.$$

   (iii) Because of symmetry, $\partial T/\partial r = 0$ at $r = 0$.

   (iv) The effect of the metal can on heat transfer is negligible and the thermal diffusivity of beer is the same as that of water, namely, $\alpha = 5.3 \times 10^{-3}$ ft²/hr.

The differential equation for this problem is

$$\alpha \left[\frac{1}{r} \frac{\partial}{\partial r}\left(r \frac{\partial T}{2r}\right) + \frac{\partial^2 T}{\partial x^2}\right] = \frac{\partial T}{\partial t}.$$

Hint: Use the method of separation of variables to solve the problem.

27. Suppose a particle is confined to a cube of length 5×10^{-8} cm on a side. The energy E of such a particle satisfies the Schrödinger equation

$$-\frac{\hbar^2}{2m}\left(\frac{\partial^2}{\partial x^2} + \frac{\partial^2}{\partial y^2} + \frac{\partial^2}{\partial z^2}\right)\psi = E\psi.$$

Confinement in quantum mechanics means $\psi = 0$ on all the faces of the cube (and outside the cube). Find the lowest energy of the confined particle for (a) a proton and (b) an electron. The value of Planck's constant $\hbar$ is 1.054×10^{-27} ergs. For an electron $m_e = 9.11 \times 10^{-28}$ g and for a proton $m_p = 1836 m_e$.

28. In organic molecules having conjugated double bonds, some electrons move freely from one end of the molecule to the other (or from one end of the conjugation sequence to the other). Thus, the electrons behave approximately like particles in a one-dimensional box. When white visible light shines on such molecules, photons with the frequency $\nu = (\epsilon_{2e} - \epsilon_{1e})/h$ will be absorbed, where ϵ_{ie} is the ith electronic energy state. Thus, molecules with conjugated bonds can be used as dyes. Using the quantum theory of a particle in a one-dimensional box, estimate the length δ that an organic molecule with conjugated bonds must have to absorb photons with wavelengths $\lambda = 6500$ or 4000 Å. These correspond to red and violet, respectively, in the spectrum of white light.

29. Consider the cannonball problem in Example 10.8.2. Suppose the initial concentration in the ball is 0 and that for $t > 0$ the concentration on the surface varies as $C_s = h(t)$. Find the concentration distribution $C(r, t)$ for $t \geq 0$. Assume that

$$h(t) = C_0\bigl(1 - \exp(-\beta t)\bigr).$$

and compute $3M/4\pi a^2 C_0$ versus Dt/b^2 for $\beta D/b^2 = 0.1, 0.5, 1$, and 2, where

$$M = 4\pi \int_0^b C(r,t) r^2 \, dr.$$

M is the total amount of carbon in the cannonball at time t.

30. Consider a string fixed at its ends ($x = 0$ and $x = L$) but free to move in the z direction everywhere else ($0 < x < L$). For small vibrations, the time-dependent displacement u of the string is given by the partial differential equation:

$$\frac{\partial^2 u}{\partial t^2} - v^2 \frac{\partial^2 u}{\partial x^2} = 0, \qquad 0 < x < L,$$

with the boundary conditions

$$u(0,t) = u(t,L) = 0, \qquad u(0,x) = u_0(x), \qquad u_t(0,x) = 0,$$

where u_t denotes the partial derivative of u with respect to t; v is the transverse wave velocity of the string and is assumed to be constant.

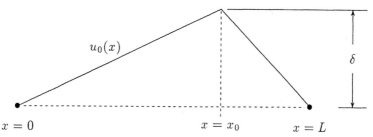

FIGURE 10.P.30

(a) Use the separation of variables technique to show that the vibrational motion of the string is given by the equation

$$u(x,t) = \sum_{n=1}^{\infty} \alpha_n \sin\frac{n\pi x}{L} \cos\frac{n\pi vt}{L}.$$

(b) Solve for the constants α_n when the string is initially deformed as $u_0(x)$ as shown in Fig. 10.P.30.

31. Solve the eigenproblem

$$-\frac{d^2u}{dx^2} = \lambda u, \qquad 0 < x < \pi,$$

$$u(0) = 0, \qquad u'(\pi) = \lambda u(\pi).$$

Do the eigenfunctions of this problem form a complete set? This is not easy to answer since the eigenvalue also appears in the boundary conditions.

Hint: Study the operator

$$\mathbf{LU} = \begin{pmatrix} -u''(x) \\ u'(\pi) \end{pmatrix}$$

for vector functions.

$$\mathbf{U} = \begin{pmatrix} u(x) \\ u_1 \end{pmatrix}, \qquad u(x) \in C^2(0,\pi) \text{ and } u_1 \text{ constant}.$$

Define the inner product

$$\langle \mathbf{U}, \mathbf{V} \rangle = \int_0^\pi u(x)v(x)\,dx + u_1 v_1$$

for this vector space. The domain of **L** is completed with the requirement that

$$u(0) = 0 \quad \text{and} \quad u_1 = u(\pi).$$

Is **L** self-adjoint? What conclusions about the eigenproblem posed above can be drawn from the spectral theory of **LU**?

FURTHER READING

Bates, D. R., editor, (1961). "Quantum theory." Academic, New York.
Carlslaw, H. S. and Jaeger, J. C. (1959). "Conduction of Heat in Solid." Oxford Univ. Press, Oxford.
Churchill, R. V. (1958). "Operational Mathematics." McGraw-Hill, New York.
Codington, E. H. and Levinson, N. (1955). "Theory of Ordinary Differential Equations." McGraw Hill, New York.
Conti, R. (1977). "Linear Differential Equations and Control." Academic, London.
Crank, J. (1975). "The Mathematics of Diffusion." Clarendon, Oxford.
Dym, H. and McKean, H. P. (1972). "Fourier Series and Integrals." Academic, New York.
Friedman, A. (1969). "Partial Differential Equations." Holt, Reinhart and Winston, New York.
Friedman, B. (1956). "Principles and Techniques of Applied Mathematics." Wiley, New York.
Garabedian, P. R. (1964). "Partial Differential Equations." Wiley, New York.
Hellwig, E. (1964). "Partial Differential Equations," Blaisdell, New York.
Hellwig, E. (1964). "Differential Operations of Mathematical Physics." Addison-Wesley, Reading, MA.
Kovach, L. D. (1984). "Boundary Value Problems." Addison-Wesley, Reading, MA.
Krein, M. G. (1983). "Topics in Differential and Integral Equations and Operator Theory," Birkhäuser, Basel.
Krieth, F. (1973). "Principles of Heat Transfer." Harper & Row, New York.
Lanczos, C. (1961). "Linear Differential Operators." Van Nostrand, London.
Merzbacher, E. (1970). "Quantum Mechanics." Wiley, New York.
Millen, K. S. (1963). "Linear Differential Equations in the Real Domain." W. W. Norton and Co., Inc., New York.
Morse, P. M. and Feshbach, H. (1953). "Methods of Theoretical Physics," McGraw-Hill Book Co., New York.
Naylor, A. W. and Sell, G. R. (1982). "Linear Operator Theory in Engineering and Science," Springer-Verlag, New York.
Ramkrishna, D. and Amundson, N. R. (1985). "Linear Operator Methods in Chemical Engineering with Applications to Transport and Chemical Reaction Systems," Prentice Hall, Englewood Cliffs, NJ.
Rapp, D. (1971). "Quantum Mechanics," Holt, Rinehart and Winston, Inc., New York.
Schwabik, S. and Turdy, M. (1979). "Differential and Integral Equations: Boundary Value Problems," Reidel, Dordrecht.
Stakgold, I. (1979). "Green's Functions and Boundary Value Problems," Wiley, New York.
Titchmarsh, E. C. (1946). "Eigenfunction Expansions Associated with Second Order Differential Equations," Oxford Univ. Press, Oxford.
Zwillinger, D. (1989). "Handbook of Differential Equations," Academic Press, San Diego.

APPENDIX

A.1. SECTION 3.2: GAUSS ELIMINATION AND THE SOLUTION TO THE LINEAR SYSTEM Ax = b

The following codes illustrate how to implement the Gauss elimination algorithms of Sections 3.2 and 3.3 to solve the linear system **Ax** = **b**. We include, in the following, implementations of simple Gauss elimination and Gauss elimination with partial and complete pivoting. The main structure of the routines follows that outlined in Section 3.2.

Auxiliary Functions

The following will be needed in the Gauss elimination routines that follow. The function RowSwap[**a**,i,j] is used to swap rows *i* and *j* in the matrix **a**. Likewise, the function ColumnSwap[**a**,i,j] swaps columns *i* and *j* of **a**. Note that the function RowSwap[] is called in ColumnSwap[].

```
RowSwap[a_,i_Integer,j_Integer] :=
    If[ i==j, a,
        If[ j>i,
            Delete[ ReplacePart[ Insert[a,
                Part[a,i],j], Part[a,j], i],
                j+1],
```

```
                RowSwap[a,j,i]
            ]
        ]
ColumnSwap[a_,i_Integer,j_Integer] := Transpose
    [ RowSwap[Transpose[a],i,j] ]
```

Simple Gauss Elimination

The simple Gauss elimination routine is defined below in the function SimpleGaussElim[**a**,**b**], where **a** is a square nonsingular matrix and **b** is a vector of equivalent dimension. The output contains the transformed augmented matrix $([a,b])_{tr}$.

```
SimpleGaussElim[a_,b_] :=
    Module[{aa,len,factor,i,j},
        len = Length[a];
        aa = Transpose[Append[Transpose[a],b]];
        For[ i=1, i≤len-1, i++,
            If[ aa[[i,i]]==0,
                For[ j=i+1, i≤len, j++,
                    If[ aa[[j,i]] != 0,
                        aa = RowSwap[aa,i,j];
                        Break[ ];
                    ];
                    Print["Error: singular matrix"];
                    Return[-1];
                ];
            ];
            For[ j=i+1, j≤len, j++,
                factor = aa[[j,i]]/aa[[i,i]];
                aa = ReplacePart[ aa, aa[[j]]-factor
                                        aa[[i]], j];
            ];
        ];
        aa
    ]
```

Solution of Linear Systems

The function SimpleGaussElim[**a**,**b**] defined above is used below in the function GaussSolve[**a**,**b**], which implements back substitution to generate the solution vector to the linear system **ax** = **b**.

```
GaussSolve[a_,b_] :=
    Module[ {aa,len,xsol,i,j},
        len = Length[a];
        aa = SimpleGaussElim[a,b];
        If[ aa == -1, Print["Error: no solution
```

```
                    found"];
                    Return[ ];
                ];
            xsol = Table[0,{len}];
            xsol[[len]] = aa[[len,len+1]]/aa[[len,len]];
            For[ i=len-1, i≥1, i--,
                    xsol[[i]] = aa[[i, len+1]];
                    For[ j=i+1, j≤len, j++,
                            xsol[[i]] -= aa[[i,j]] xsol[[j]];
                    ];
                    xsol[[i]] /= aa[[i,i]];
            ];
            xsol
        ]
```

Gauss Elimination with Pivoting

The Gauss elimination routine below (defined in the function GaussElim[**a**,**b**, pivot]) is an extended version of SimpleGaussElim[**a**,**b**] and includes both partial and complete pivoting. The value of the pivot can be set to (pivot=1) for partial (pivot=2) for complete pivoting. The default (pivot=0) means no pivoting, in which case the function is identical to SimpleGaussElim[**a**,**b**] defined above.

```
        GaussElim[a_,b_,pivot_:0] :=
            Module[{aa,len,factor,i,j,k,max,imax,jmax,piv},
                piv = pivot;
                If[ (piv ≠ 1 && piv ≠ 2), piv=0];
                len = Length[a];
                aa = Transpose[Append[Transpose[a],b]];
                For[ i=1, i≤len-1, i++,
                    If[ piv == 0,
                        If[ aa[[i,i]] == 0,
                            For[ j=i+1, i≤len, j++,
                                If[ aa[[j,i]] ≠ 0,
                                    aa = RowSwap[aa,i,j];
                                    Break[ ];
                                ];
                                Print["Error: singular
                                        matrix"];
                                Return[-1];
                            ];
                        ];
                    ];
                    If[ piv == 1,
                        max = Abs[aa[[i,i]]];
                        imax = i;
                        For[ j=i+1, j≤len, j++,
                            If[ Abs[aa[[j,i]]] ≥ max,
                                max = aa[[j,i]];
```

```
                    imax = j;
                ];
            ];
            If[ max == 0,
                Print["Error: singular matrix"];
                Return[-1];
            ];
            aa = RowSwap[aa,i,imax];
        ];
        If[ piv == 2,
            max = Abs[aa[[i,i]]];
            imax = i;
            jmax = i;
            For[ j=i+1, j≤len, j++,
                For[ k=i, k≤len, k++,
                    If[ Abs[aa[[j,k]]] ≥ max,
                        max = aa[[j,k]];
                        imax = j;
                        jmax = k;
                    ];
                ];
            ];
            If[ max == 0,
                Print["Error: singular matrix"];
                Return[-1];
            ];
            aa = RowSwap[aa,i,imax];
            aa = ColumnSwap[aa,i,jmax];
        ];
        For[ j=i+1, j≤len, j++,
            factor = aa[[j,i]]/aa[[i,i]];
            aa = ReplacePart[ aa, aa[[j]]-factor
                aa[[i]], j];
        ];
    ];
    aa
]
```

A.2. EXAMPLE 3.6.1: MASS SEPARATION WITH A STAGED ABSORBER

The following code is an example of how to include Gauss elimination in the definition of complicated functions in Mathematica. Below, we construct the function $x_{out}[n]$, which returns the outlet concentration of solute in an n-stage ideal absorber based on liquid and vapor mass flow rates (L and G), inlet liquid and vapor solute concentrations (xin and yin), and the vapor–liquid partition coefficient (K) defined by $y = Kx$.

Initialization of flow rates and partition coefficient: L and G are in units of lb$_m$/hr and K[=](weight fraction solute in vapor)/(weight fraction solute in liquid).

```
L = 3500;
G = 4000;
K = 0.876;
```

Initialization of inlet solute concentration (in units of weight fraction):

```
xin = 0.01;
yin = 0.0;
```

Program module for the function $x_{out}[n]$. The variables A and b are the corresponding matrix and vector used in Example 3.6.1 (Eqs. (3.6.37) and (3.6.38)). Here we use LinearSolve, the Mathematica built-in Gauss elimination routine for solving the system $\mathbf{A}\mathbf{x} = \mathbf{b}$. Alternately, the Gauss elimination routines above could be substituted.

```
x_out[n_Integer] := Module[{A, b, x, i, j},
            A = Table[
                    If[ i == j, -(L + GK),
                        If[ j == i+1, GK,
                            If[ j == i-1, L, 0]
                        ]
                    ],
                    {i, 1, n}, {j, 1, n}
                ];
            b = Table[
                    If[ i == 1, L xin,
                        If[ i == n, G yin, 0]
                    ],
                    {i, 1, n}
                ];
            x = LinearSolve[A, -b];
            x[[n]]
        ]
```

A.3. SECTION 3.7: ITERATIVE METHODS FOR SOLVING THE LINEAR SYSTEM Ax = b

The following codes illustrate how to implement the iterative methods of Section 3.7 to solve the linear system $\mathbf{A}\mathbf{x} = \mathbf{b}$. We include the Jacobi, Gauss–Seidel, and SOR methods in the form of user-defined functions in Mathematica. It should be noted that the control variables (TOLERANCE and Nmax) can be adjusted to the user's taste. TOLERANCE is the convergence criterion for the residual defined by $|x^{(k+1)} - x^{(k)}|$. Nmax is the maximum allowed number of iterations.

Jacobi Method

```
LinearJacobi[a_List, b_List, x0_List] :=
    Module[{len, xnew, xold, Residual, TOLERANCE,
            Nmax},
        TOLERANCE = 10^-15;
        Nmax = 100;
        len = Length[b];
        xnew = Table[0, {len}];
        xold = x0;
        Residual = 1.0;
        n = 0;
        While[Residual > TOLERANCE,
            n++;
            If[n ≥ Nmax,
                Print["Warning: Exceeded maximum
                    iterations."];
                Break[];
            ];
            For[i = 1, i ≤ len, i++,
                xnew[[i]] = - Sum[ a[[i, j]] xold[[j]]
                            // N, {j, 1, i-1}]
                          - Sum[ a[[i, j]] xold[[j]]
                            // N, {j, i+1, len}];
                xnew[[i]] += b[[i]] // N;
                xnew[[i]] /= a[[i, i]] // N;
            ];
            Residual = √((xnew − xold)·(xnew − xold)) // N;
            xold = xnew;
        ];
        xnew
    ]
```

Gauss–Seidel Method

```
LinearGaussSeidel[a_List, b_List, x0_List] :=
    Module[{len, xnew, xold, Residual, TOLERANCE,
            Nmax},
        TOLERANCE = 10^(-15);
        Nmax = 100;
        len = Length[b];
        xnew = Table[0, {len}];
        xold = x0;
        Residual = 1.0;
        n = 0;
        While[Residual > TOLERANCE,
            n++;
            If[n ≥ Nmax,
```

ITERATIVE METHODS FOR SOLVING THE LINEAR SYSTEM Ax = b

```
            Print["Warning: Exceeded maximum
                    iterations."];
            Break[];
         ];
         For[i = 1, i ≤ len, i++,
            xnew[[i]] = - Sum[ a[[i, j]] xnew[[j]]
                            // N, {j, 1, i-1}]
                        - Sum[ a[[i, j]] xold[[j]]
                            // N, {j, i+1, len}];
            xnew[[i]] += b[[i]] // N;
            xnew[[i]] /= a[[i, i]] // N;
         ];
         Residual = √((xnew − xold)·(xnew − xold)) // N;
         xold = xnew;
      ];
      xnew
   ]
```

Successive Overrelaxation Method

```
      LinearSOR[a_List, b_List, x0_List, ω_] :=
         Module[{len, xnew, xold, Residual, TOLERANCE,
               Nmax},
            TOLERANCE = 10⁻¹⁵;
            Nmax = 100;
            len = Length[b];
            xnew = Table[0, {len}];
            xold = x0;
            Residual = 1.0;
            n = 0;
            While[Residual > TOLERANCE,
               n++;
               If[n ≥ Nmax,
                  Print["Warning: Exceeded maximum
                          iterations."];
                  Break[];
               ];
               For[i = 1, i ≤ len, i++,
                  xnew[[i]] = - Sum[ a[[i, j]] xnew[[j]]
                                  // N, {j, 1, i-1}]
                              - Sum[ a[[i, j]] xold[[j]]
                                  // N, {j, i+1, len}];
                  xnew[[i]] += b[[i]] // N;
                  xnew[[i]] /= a[[i, i]] // N;
                  xnew[[i]] *= ω // N;
                  xnew[[i]] += (1 - ω) xold[[i]] // N;
               ];
               Residual = √((xnew − xold)·(xnew − xold)) // N;
```

```
            xold = xnew;
        ];
        xnew
    ]
```

A.4. EXERCISE 3.7.2: ITERATIVE SOLUTION TO Ax = b — CONJUGATE GRADIENT METHOD

The following code illustrates how to implement the conjugate gradient method of Exercise 3.7.2 to solve the linear system **Ax = b**. The routine is in the form of the user-defined Mathematica function, LinearConGard[ ]. The control variables (TOLERANCE and Nmax) can be adjusted to the user's taste. TOLERANCE is the convergence criterion for the residual defined by $|x^{(k+1)} - x^{(k)}|$. Nmax is the maximum allowed number of iterations.

Conjugate Gradient Method

The following routine, LinearConGrad[a,b,x0], solves for **ax = b** using the initial guess vector **x0**.

```
LinearConGrad[a_List, b_List, x0_List] :=
    Module[{xnew, xold, Rnew, Rold, TOLERANCE, Nmax,
            n, r1, r0, p0, α, β},
        TOLERANCE = 10^-15;
        Nmax = 500;
        r0 = b - a . x0;
        p0 = r0;
        xold = x0;
        Rold = √Conjugate[r0] . r0;
        n = 0;
        While[Rold > TOLERANCE,
            n++;
            If[n ≥ Nmax,
                Print["Warning: Exceeded maximum
                    iterations."];
                Break[];
            ];
```
$$\alpha = \frac{\text{Rold}^2}{(\text{Conjugate}[p0] \, . \, a \, . \, p0)};$$
```
            xnew = xold + α p0;
            r1 = r0 - αa · p0;
            Rnew = √Conjugate[r1].r1 // N;
```
$$\beta = \left(\frac{\text{Rnew}}{\text{Rold}}\right)^2;$$
```
            p0 = r1 + β p0;
            xold = xnew;
            Rold = Rnew;
```

```
            r0 = r1;
      ];
      xnew
]
```

A.5. EXAMPLE 3.8.1: CONVERGENCE OF THE PICARD AND NEWTON–RAPHSON METHODS

The following codes are examples of how to implement the Picard and Newton–Raphson methods. In both cases, the residual functions are defined externally and the solutions at each iteration are stored in the lists PSolution and NRSolution.

Picard Method

Define the Picard residual vector **g**:

$$g1[x1\_, x2\_] := x1 - x1\ \text{Tan}[x1] + \sqrt{x2^2 - x1^2}$$
$$g2[x1\_, x2\_] := x2 + \frac{x1 + \pi/4}{\text{Tan}[x1 + \pi/4]} + \sqrt{x2^2 - (x1 + \pi/4)^2}$$

Define the guess variables and provide initial guesses:

```
x1 = 1.0;
x2 = 1.0;
```

Initialize the solution lists. PSolution1 and PSolution2 contain the accumulated guesses for x1 and x2, respectively, at each step. PSolutionNorm contains the accumulated solution norms at each step.

```
PSolution1 = {x1};
PSolution2 = {x2};
PSolutionNorm = {};
```

Initialize the program parameters. TOLERANCE is the user-defined convergence criterion. Nmax is the user-defined maximum number of iterations.

```
TOLERANCE = 10^-5;
Nmax = 100;
```

Program module for the Picard method. The output values of the calculation are printed after the last iteration.

```
Module[{i, x1new, x2new, norm},
    norm = 1.0;
    i = 0;
    While[norm > TOLERANCE,
        x1new = g1[x1, x2];
        x2new = g2[x1, x2];
        norm = √((x1new − x1)² + (x2new − x2)²);
```

```
        AppendTo[PSolution1, x1new];
        AppendTo[PSolution2, x2new];
        AppendTo[PSolutionNorm, norm];
        x1 = x1new;
        x2 = x2norm;
        i++;
        If[ i ≥ Nmax,
            Print["WARNING: Exceeded maximum
                iterations."];
            Break[];
        ];
    ];
    Print[x1 " = x1"];
    Print[x1 " = x1"];
]
```

Newton–Raphson Method

Initialize the solution lists. NRSolution1 and NRSolution2 contain the accumulated guesses for x1 and x2, respectively, at each step. NRSolutionNorm1 and NRSolutionNorm2 contain the accumulated values of the norms, defined as the modulus of the difference in the update vectors and the modulus of the residuals, respectively.

```
NRSolution1 = {x1};
NRSolution2 = {x2};
NRSolutionNorm1 = {};
NRSolutionNorm2 = {};
```

Program module for the Newton–Raphson method. The output values of the calculation are printed after the last iteration.

```
Module[{i, jac, residual, xdel, norm1, norm2,
       TOLERANCE, Nmax},
    TOLERANCE = 10^-15;
    Nmax = 100;
    norm1 = 1.0;
    norm2 = 1.0;
    i = 0;
    While[norm1 > TOLERANCE && norm2 > TOLERANCE,
        jac = {{j11[x1, x2], j12[x1, x2]},
               {j21[x1, x2], j22[x1, x2]}};
        residual = {-f1[x1, x2], -f2[x1, x2]};
        xdel = LinearSolve[jac, residual];
        norm1 = √(xdel[[1]]² + xdel[[2]]²);
        norm2 = √(residual[[1]]² + residual[[2]]²);
        x1 += xdel[[1]];
        x2 += xdel[[2]];
        AppendTo[NRSolution1, x1];
```

```
                AppendTo[NRSolution2, x2];
                AppendTo[NRSolutionNorm1, norm1];
                AppendTo[NRSolutionNorm2, norm2];
                i++;
                If[i ≥ Nmax,
                    Print["WARNING: Exceeded maximum
                            iterations."];
                    Break[];
                ];
        ];
        Print[x1 " = x1"];
        Print[x2 " = x2"];
]
```

A.6. EXAMPLE 3.8.2: STEADY-STATE SOLUTIONS FOR A CONTINUOUSLY STIRRED TANK REACTOR

The following code is an example of how to implement the Newton–Raphson method to solve for the steady-state solutions (T and c_A) for an ideal, continuously stirred tank reactor (CSTR). Following Example 3.8.2, we first input the problem parameters, which include the inlet stream properties, reactor dimensions, kinetic constants, the heat transfer coefficient, and any other physical constants. We then define the reaction rate of component A as a user-defined function "reaction." Next, we define the residual functions and the Jacobian element functions and construct the function CSTRSolve as the Newton–Raphson routine for solving the problem.

Physical Constants

Inlet stream conditions:

```
        T0 = 273;           (* Inlet Temperature[deg. F] *)
        q = 1250;           (* Inlet Volumetric Flow Rate
                               [lbm/hr] *)
        cA0 = 0.15;         (* Inlet Concentration[moles/ft^3] *)
```

Reactor dimensions:

```
        V = 4500;           (* Volume of Reactor[ft^3] *)
        A = 25;             (* Effective Surface Area of
                               Reactor[ft^2] *)
```

Kinetic constants:

```
        k0 = 0.590;         (* Preexponential Rate Constant
                               [1/hr] *)
        E0 = 2500;          (* Activation Energy [Btu/mole] *)
        R = 8.314;          (* Gas Constant [Btu/lbmole---°R] *)
```

Thermodynamic and heat transfer constants:

```
H = - 33.5;        (* Heat of Reaction [Btu/lbmole] *)
U = 258;           (* Overall Heat Transfer
                      Coefficient *)
Tb = 298;          (* Ambient Temperature[deg. F] *)
ρ = 15.5;          (* Liquid Density of Solute
                      [lbm/ft³] *)
Cp = 4.54;         (* Heat Capacity of Solute
                      [Btu/lbmole---(deg. F)] *)
```

Auxiliary Functions

Define the function reaction $[T, c_A]$ as the production rate of component A via the reaction:

```
reaction[T_, c_] := - k0 Exp[-E0 / (RT)] c
```

Define the residual vector components:

```
f1[T_, c_] := q(cA0 - c) + reaction[T, c]V
f2[T_, c_] := q ρ Cp (T0 - T) + reaction[T, c]V H
              - A U(T - Tb)
```

Define the Jacobian matrix elements:

```
j11[x1_, x2_] := D[f1[x, y], x] /. {x → x1, y → x2}
j12[x1_, x2_] := D[f1[x, y], y] /. {x → x1, y → x2}
j21[x1_, x2_] := D[f2[x, y], x] /. {x → x1, y → x2}
j22[x1_, x2_] := D[f2[x, y], y] /. {x → x1, y → x2}
```

Newton–Raphson Method

Program module for the Newton–Raphson method. Ti and ci are the initial guesses for the outlet temperature and concentration.

```
CSTRSolve[Ti_, ci_] :=
    Module[{i, T, c, jac, residual, xdel, norm1,
            norm2, TOLERANCE, Nmax},
        TOLERANCE = 10⁻¹⁵;
        Nmax = 100;
        norm1 = 1.0;
        norm2 = 1.0;
        T = Ti;
        c = ci;
        i = 0;
        While[norm1 > TOLERANCE && norm2 > TOLERANCE,
            jac = {{j11[T, c], j12[T, c]},
                   {j21[T, c], j22[T, c]}};
```

```
                residual = {-f1[T, c], -f2[T, c]};
                xdel = LinearSolve[jac, residual] // N;
                norm1 = √xdel·xdel  // N;
                norm2 = √residual·residual  // N;
                T += xdel[[1]] // N;
                c += xdel[[2]] // N;
                i++;
                If[i ≥ Nmax,
                        Print["WARNING: Exceeded maximum
                                iterations.''"];
                        Break[];
                ];
        ];
        {T, c}
]
```

A.7. EXAMPLE 3.8.3: THE DENSITY PROFILE IN A LIQUID–VAPOR INTERFACE (ITERATIVE SOLUTION OF AN INTEGRAL EQUATION)

The following codes illustrate how to implement the iterative solution of an integral equation. We have included two iterative methods for solving Eq. (3.8.50). The first routine is a modification to the successive overrelaxation method in which the vector **g** is updated at each step using the previous value of **n**, the density profile. The second routine uses the Newton–Raphson method to handle the nonlinear function **g(n)**. It should be noted that in each routine the control variables (TOLERANCE and Nmax) can be adjusted to the user's taste. TOLERANCE is the convergence criterion for the residual defined by $|x^{(k+1)} - x^{(k)}|$. Nmax is the maximum allowed number of iterations.

Physical Constants

van der Waals coefficients:

```
a = 10.5;
b = 0.06;
```

Gas constant:

```
Rg = 0.082058;
T = 300;
```

Equilibrium conditions:

```
P = 2.096;
μ = -130.682;
nl = 13.8478;
ng = 0.0879781;
```

Auxiliary Functions

The following functions (and constants) are required in the routines below and are defined in Example 3.8.3.

$$K[x\_] := \frac{2\,a}{b^{1/3}} \mathrm{Exp}\left[-\frac{\mathrm{Abs}[x]}{b^{1/3}}\right]$$

$$\alpha = \mathrm{NIntegrate}[K[x], \{x, -\infty, \infty\}];$$

$$g[x\_, L\_] := -\mu - \mathrm{ng}\ \mathrm{NIntegrate}[K[y - x], \{y, L, \infty\}] - \mathrm{nl}\ \mathrm{NIntegrate}[K[y - x], \{y, -\infty, -L\}]$$

Equilibrium chemical potential:

$$\mathrm{u0}[n\_] := -\mathrm{Rg}\ T\ \mathrm{Log}\left[\frac{1}{n\,b} - 1\right] + \frac{n\,b\,\mathrm{Rg}\,T}{1 - n\,b} - 2\,n\,a$$

Successive Overrelaxation Method

The function DensitySOR[L,δ] uses the successive overrelaxation technique to solve Eq. (3.8.50) iteratively, where L is defined in Example 3.8.3 and δ is the user-input discretization length for the array **n**. The value of **g(n)** at each step is updated the previous step's value of **n**. As an initial value for **n**, a linear density profile is imposed across the interface.

```
DensitySOR[L_, δ_] :=
    Module[{},
        ω = 1.15;
        TOLERANCE = 10⁻¹⁵;
        Nmax = 100;
        Nstep = Ceiling[L / δ];
        delta = L / Nstep;
        Kmat =
    Table[K[delta (j - i)]delta, {i, -Nstep, Nstep},
            {j, -Nstep, Nstep}];
        For[ i = 1, i ≤ 2 Nstep + 1, i++,
            Kmat[[i, i]] -= α;
        ];
        bvec = Table[ g[i delta, L]
            - 1/2 (K[delta(-Nstep - 1 - i)]nl
                + K[delta (Nstep + 1 - i)] ng),
            {i, -Nstep, Nstep}];
        xold = Table[ng + (nl - ng)/2 (1 - delta i/L),
            {i, -Nstep, Nstep}];
        xnew = xold;
        Residual = 1.0;
```

THE DENSITY PROFILE IN A LIQUID–VAPOR INTERFACE

```
            n = 0;
            While[Residual > TOLERANCE,
                n++;
                If[n ≥ Nmax,
                    Print["Warning: Exceeded maximum
                        iterations."];
                    Break[];
                ];
                For[i = 1, i ≤ 2 Nstep + 1, i++,
                    xnew[[i]] = - Sum[Kmat[[i, j]] xnew[[j]]
                        // N, {j, 1, i - 1}]
                        -
        Sum[Kmat[[i, j]] xold[[j]]
                    // N, {j, i + 1, 2 Nstep + 1}];
                    xnew[[i]] += (bvec[[i]] + u0[xold[[i]]])
                        // N;
                    xnew[[i]] /= Kmat[[i, i]] // N;
                    xnew[[i]] *= ω // N;
                    xnew[[i]] += (1 - ω) xold[[i]] // N;
                ];
                Residual = √((xnew − xold) . (xnew − xold)) // N;
                xold = xnew;
            ];
            xnew
    ]
```

Newton–Raphson Method

The function Density[L,δ] uses the Newton–Raphson technique to solve Eq. (3.8.50) iteratively. NR and δ are as defined above for the SOR implementation. As above, for an initial value for **n**, a linear density profile is imposed across the interface.

```
    DensityNR[L_, δ_] :=
        Module[{n, i, jac, residual, x, xdel, norm1,
                norm2, TOLERANCE, Nmax, Kmat, delta, bvec,
                Nstep},
            TOLERANCE = 10⁻¹⁵;
            Nmax = 100;
            norm1 = 1.0;
            norm2 = 1.0;
            n = 0;
            Nstep = Ceiling[L / δ];
            delta = L / Nstep;
            Kmat =
        Table[K[delta (j - i)] delta, {i, -Nstep, Nstep},
                {j, -Nstep, Nstep}];
```

```
        For[ i = 1, i ≤ 2 Nstep + 1, i++,
            Kmat[[i, i]] -= α;
        ];
        bvec = Table[g[i delta, L]
                    - ½(K[delta (-Nstep - 1 - i)]nl
                    + K[delta (Nstep + 1 - i)] ng),
                            {i, -Nstep, Nstep}];
        x = Table[ng + (nl - ng)/2 (1 - delta i/L),
                 {i, -Nstep, Nstep}];
        While[norm1 > TOLERANCE && norm2 > TOLERANCE,
            If[n ≥ Nmax,
                Print["Warning: Exceeded maximum
                        iterations."];
                Break[];
            ];
            jac = Table[Kmat[[i, j]] - D[u0[z], z]
                    /. {z -> x[[i]]},
                    {i, 1, 2 Nstep + 1},
                    {j, 1, 2 Nstep + 1}];
            residual = Table[ Kmat · x - bvec];
            For[ i = 1, i ≤ 2 Nstep + 1, i++,
                residual[[i]] -= u0[x[[i]]];
            ];
            xdel = LinearSolve[jac, residual];
            norm1 = √(xdel · xdel);
            norm2 = √(residual · residual);
            x += xdel;
            n++;
        ];
        x
]
```

A.8. EXAMPLE 3.8.4: PHASE DIAGRAM OF A POLYMER SOLUTION

The following code is an example of how to implement the Newton–Raphson method with first-order continuation. The example involves calculation of the coexistence curves of two phases in a polymer solution using the Flory–Huggins theory. The residual functions for this case are defined externally.

Newton–Raphson Method

Define the residual vector **f**:

$$f1[x1\_, y1\_, \chi1\_, \nu\_] := \text{Log}[x1] - \text{Log}[y1] + \left(1 - \frac{1}{\nu}\right)(1 - x1)$$

$$f2[x1\_, y1\_, \chi1\_, \nu\_] := \mathrm{Log}[1 - x1] - \mathrm{Log}[1 - y1] \\ - \left(1 - \frac{1}{\nu}\right)(1 - y1) \\ + (1 - \nu) x1 - (1 - \nu) y1 \\ + \frac{(1-x1)^2}{\chi 1} - \frac{(1-y1)^2}{\chi 1} \\ + \frac{\nu\, x1^2}{\chi 1} - \frac{\nu\, y1^2}{\chi 1};$$

Define the Jacobian matrix elements:

$$j11[x1\_, y1\_, \chi1\_, \nu\_] := \partial_{xx} f1[xx, yy, \chi\chi, \nu\nu] \,/.\, \\ \{xx \to x1,\ yy \to y1, \\ \chi\chi \to \chi1,\ \nu\nu \to \nu\};$$

$$j12[x1\_, y1\_, \chi1\_, \nu\_] := \partial_{yy} f1[xx, yy, \chi\chi, \nu\nu] \,/.\, \\ \{xx \to x1,\ yy \to y1, \\ \chi\chi \to \chi1,\ \nu\nu \to \nu\};$$

$$j21[x1\_, y1\_, \chi1\_, \nu\_] := \partial_{xx} f2[xx, yy, \chi\chi, \nu\nu] \,/.\, \\ \{xx \to x1,\ yy \to y1, \\ \chi\chi \to \chi1,\ \nu\nu \to \nu\};$$

$$j22[x1\_, y1\_, \chi1\_, \nu\_] := \partial_{yy} f2[xx, yy, \chi\chi, \nu\nu] \,/.\, \\ \{xx \to x1,\ yy \to y1, \\ \chi\chi \to \chi1,\ \nu\nu \to \nu\};$$

$$d1[x1\_, y1\_, \chi1\_, \nu\_] := \partial_{\nu\nu} f1[xx, yy, \chi\chi, \nu\nu] \,/.\, \\ \{xx \to x1,\ yy \to y1, \\ \chi\chi \to \chi1,\ \nu\nu \to \nu\};$$

$$d2[x1\_, y1\_, \chi1\_, \nu\_] := \partial_{\nu\nu} f2[xx, yy, \chi\chi, \nu\nu] \,/.\, \\ \{xx \to x1,\ yy \to y1, \\ \chi\chi \to \chi1,\ \nu\nu \to \nu\};$$

Parameter initialization: νmax is the maximum value of ν to be calculated. νstep is the number of steps for incrementing ν in the calculation. χstep is the number of steps for incrementing χ, the dimensionless temperature.

```
νmax = 5;
νstep = 40;
χstep = 20;
```

Calculate the step size for incrementing ν.

$$\mathrm{delta}\nu = \frac{\nu\mathrm{max} - 1}{\nu\mathrm{step}};$$

Initialize the program parameters. TOLERANCE is the user-defined convergence criterion. Nmax is the user-defined maximum number of iterations.

$$\mathrm{TOLERANCE} = \frac{1}{10^7};$$
```
Nmax = 100;
```

Define the guess variables: x[n] is the solvent concentration for the solvent-rich phase for the nth value of χ (χ = n nustep). Likewise, y[n] is the solvent concentration for the polymer-rich phase for the nth value of v. χ[n] is the dimensionless temperature for the nth value of v.

```
x[0] = 0.0;
y[0] = 0.0;
χ[0] = 0.0;
```

Initialize all of the values of x[n], y[n], and χ[n] to their values at the critical point ($v = 1$). This will be the starting point for the continuation steps in χ.

$$\text{For}\Big[n=1, \; n \leq \chi\text{step} - 1, \; n++,$$
$$x[n] = \frac{n\;0.5}{\chi\text{step}};$$
$$y[n] = 1.0 - x[n];$$
$$\chi[n] = \frac{1.0 - 2\;x[n]}{\text{Log}\left[\frac{1.0-x[n]}{x[n]}\right]};$$
$$\Big];$$

Initialize the above value for the starting point when $\chi = 0.5$.

```
x[χstep] = 0.5;
y[χstep] = 0.5;
χ[χstep] = 0.5;
```

Program module. The calculation begins at $v = 1$ and, using first-order continuation, χ is decremented by χstep and x and y are calculated via the Newton–Raphson method. Again, χ is decremented using continuation for a total of χstep times. First-order continuation is then used to increment v and then x and y are recalculated at all values of χ using the Newton–Raphson method and first-order continuation in sequence. Each set of curves generated at a specific value of v (i.e., x[χ,v] and y[χ,v]) is a corresponding phase diagram (coexistence curves) for that value of v. Note that in order to calculate a phase diagram for some target $v \gg 1$, this continuation scheme must be used.

```
Module[{i, j, k, v, χc, jac, residual, xdel, delta,
        norm},
    v = 1.0;
    For[k = 1, k ≤ vstep, k++,
        For[i = 1, i ≤ χstep - 1, i++,
            residual = {-d1[x[i], y[i], χ[i], v]
                        deltav, -d2[x[i], y[i],
                        χ[i], v] deltav};
            jac = {{j11[x[i], y[i], χ[i], v],
                        j12[x[i], y[i], χ[i], v]},
                    {j21[x[i], y[i], χ[i], v],
                        j22[x[i], y[i], χ[i], v]}};
            xdel = LinearSolve[jac, residual];
```

```
            x[i] += xdel[[1]];
            y[i] += xdel[[2]];
        ];
        v = deltav;
        χc = 2.0 / (1.0 + 1/√v)^2;
        delta = χc / χstep;
        For[i = 1, i ≤ χstep - 1, i++,
            χ[i] = i delta;
            norm = 1.0;
            j = 0;
            While[norm > TOLERANCE,
                    jac = {{j11[x[i], y[i], χ[i], v],
                            j12[x[i], y[i], χ[i], v]},
                           {j21[x[i], y[i], χ[i], v],
                            j22[x[i], y[i], χ[i], v]}};
                    residual = {-f1[x[i], y[i], χ[i],
                                    v], -f2[x[i], y[i],
                                    χ[i], nv]};
                    xdel = LinearSolve[jac, residual];
                    norm = √(xdel[[1]]^2 + xdel[[2]]^2);
                    x[i] += xdel[[1]];
                    y[i] += xdel[[2]];
                    j++;
                    If[j ≥ Nmax,
                            Print["WARNING: Exceeded
                                    maximum iterations."];
                            Break[];
                    ];
            ];
        ];
        x[χstep] = 1.0 - 1.0 / (1 + √v);
        y[χstep] = x[χstep];
        χ[χstep] = χc;
    ];
]
```

A.9. SECTION 4.3: GAUSS–JORDAN ELIMINATION AND THE SOLUTION TO THE LINEAR SYSTEM Ax = b

The following code illustrates how to implement the Gauss–Jordan elimination algorithm of Section 4.3 to solve the linear system $\mathbf{Ax} = \mathbf{b}$. The main structure of the routine follows that outlined in Sections 3.2, 3.3, and 4.3.

Auxiliary Functions

The following functions will be needed in the Gauss–Jordan elimination routine below. The function RowSwap[a,i,j] is used to swap rows *i* and *j* in the matrix **a**. Likewise, the function ColumnSwap[a,i,j] swaps columns *i* and *j* of **a**. Note that the function RowSwap[ ] is called in ColumnSwap[ ].

```
RowSwap[a_, i_Integer, j_Integer] :=
    If[ i==j, a,
        If[ j>i,
            Delete[ ReplacePart
                [ Insert[a, Part[a,i], j],
                Part[a,j], i], j+1],
            RowSwap[a, j, i]
        ]
    ]
ColumnSwap[a_,i_Integer,j_Integer]
    := Transpose[RowSwap[Transpose[a], i, j] ]
```

Gauss–Jordan Elimination

The simple Gauss–Jordan elimination routine (without pivoting) is defined below in the function SimpleGaussJordan[**a**,**b**], where **a** is a square nonsingular matrix and **b** is a vector of equivalent dimension. The output contains the transformed augmented matrix $([a,b])_{tr}$. Note that the function ColumnSwap[ ] is not included in the routine below. We have included it in this section in case the reader wishes to include pivoting in a separate routine.

```
SimpleGaussJordan[a_,b_] :=
  Module[{len,aa,factor,i,j},
        len = Length[a];
        aa = Transpose[Append[Transpose[a],b]];
        For[ i=1, i≤len, i++,
            If[ aa[[i,i]] == 0,
                For[ j=i+1, i≤len, j++,
                    If[ aa[[j,i]] != 0,
                        aa = RowSwap[aa,i,j];
                        Break[ ];
                    ];
                    Print ["Error: singular
                        matrix"];
                    Return[-1];
                ];
            ];
            For[ j=i+1, j≤len, j++,
                factor = aa[[j,i]]/aa[i,i]];
                aa = ReplacePart[ aa, aa[[j]]
                    -factor aa[[i]], j];
            ];
```

```
            For[ j=i-1, j≥1, j--,
                factor = aa[[j,i]] /
                    aa[[i,i]];
                aa = ReplacePart[ aa, aa[[j]]
                    -factor aa[[i]],j];
            ];
        ];
        aa
    ]
```

A.10. SECTION 5.4: CHARACTERISTIC POLYNOMIALS AND THE TRACES OF A SQUARE MATRIX

The following codes are Mathematica programs for generating the traces and characteristic polynomial of a square matrix. The first routine is contained in the function Tr[j,**a**], which evaluates the *j*th trace of **a** where $1 < j < N$. The second routine is contained in the function CharPoly[**a**,λ], which expresses the characteristic polynomial in terms of the variable λ (i.e., $P_n(-\lambda)$).

Trace Routine

The following routine calculates the *j*th-order trace of **a** by summing up the determinants of the corresponding *j*th-order minors (as outlined in Section 1.7). The following function Tr[j,**a**] successively strikes out rows and columns using a recursive routine to nest For loops. The variable "case" contains the List of rows (and columns) to be stricken and is passed internally in the recursive cell. The routine is somewhat sophisticated and inexperienced programmers may have difficulty understanding the algorithm. It is, however, a good exercise to explore the routine and the interested reader may find Wolfram's textbook *The Mathematica Book* useful in doing so.

```
    Tr[n_Integer, a_List, case_: {}] :=
        Module[{ans, vec, dropit, start, i},
            vec = case;
            ans = 0;
            If[n > Length[a] || n < 1,
                Print["Error: Illegal integer value."];
                Return[];
            ];
            If[Length[vec] == Length[a] - n,
                dropit = Partition[vec, 1];
                ans = Det[Delete[Transpose[Delete[a,
                    dropit]], dropit]];,
                If[Length[vec] == 0, start = 0;, start
                    = vec[[Length[vec]]];];
                For[ i = start + 1, i ≤ n + 1
                    + Length[vec], i++,
```

```
            AppendTo[vec, i];
            ans += Tr[n, a, vec];
            vec = Drop[vec, -1];
         ];
      ];
      Return[ans];
   ]
```

Trace Routine

The following routine CharPoly[**a**,x] returns the characteristic polynomial of the matrix **a** with respect to the variable x. The output is in the form $P_n(-x)$ and derives from Eqs. (5.5.3) and (5.5.5). Note that the function Tr[ ] is called in CharPoly[ ].

```
CharPoly[a_, x_] :=
   Module[{},
      len = Length[a];
      ans = (-x) ^ len;
      For[i = 1, i ≤ len, i++,
         ans += Tr[i, a] (-x) ^ (len - i);
      ];
      ans
   ]
```

Example

The following is an example of the uses of Tr[ ] and CharPoly[ ]:

```
mat = {{1, 4, 3, 1},
       {2, 1, 0, 1},
       {1, 2, -1, 1},
       {0, 1, 1, 1}};

Tr[1, mat]

2

Tr[2, mat]

-13

Tr[4, mat]

14

CharPoly[mat, λ]

14 - 7 λ - 13 λ² - 2 λ³ + λ⁴
```

```
Solve[CharPoly[mat, λ] == 0, λ] // N
{{λ → 0.781784}, {λ → 4.8533},
  {λ → - 1.81754 - 0.621562 I},
  {λ → - 1.81754 + 0.621562 I}}
```

A.11. SECTION 5.6: ITERATIVE METHOD FOR CALCULATING THE EIGENVALUES OF TRIDIAGONAL MATRICES

The following code is the implementation of the iterative method described in Section 5.6 to calculate the eigenvalues of a tridiagonal matrix. The method involves the calculation of the polynomial expression $P_n(\lambda_0)$ for some initial guess eigenvalue λ_0, which is handled below in a separate function PolyTri. The Newton–Raphson method is then used to iteratively solve the equation $P_n(\lambda_0) = 0$, where n is the dimension of the matrix.

Polynomial Functions

The function PolyTri[a,λ,n] calculates the nth polynomial $P_n(\lambda)$ using the recursion formula Eq. (5.6.4). Notice that we must include three function definitions for the three separate cases where $n = 0$, $n = 1$, and $n \geq 1$.

```
PolyTri[a_List, λ_, n_Integer] :=
    (a[[n, n]] - λ) PolyTri[a, λ, n - 1] -
    a[[n - 1, n]] a[[n, n - 1]]
        PolyTri[a, λ, n - 2] /; n > 1

PolyTri[a_List, λ_, 1] := a[[1, 1]] - λ
PolyTri[a_List, λ_, 0] := 1.0
```

The function PolyTriD[**a**,λ,n] is the derivative with respect to λ of the corresponding function PolyTri defined above.

```
PolyTriD[a_List, λ_, n_Integer] :=
    (a[[n, n]] - λ) PolyTriD[a, λ, n - 1]
        - PolyTri[a, λ, n - 1] -
    a[[n - 1, n]] a[[n, n - 1]]
        PolyTriD[a, λ, n - 2] /; n > 1

PolyTriD[a_List, λ_, 1] := - 1.0
PolyTriD[a_List, λ_, 0] := 0.0
```

Newton–Raphson Method

We now define the function EigenvalueTri as an implementation of the one-dimensional Newton–Raphson scheme to solve for an eigenvalue of **a** given an initial guess λ_0. We use the above functions PolyTri and PolyTriD to calculate the residual and Jacobian below.

```
EigenvalueTri[a_List, λ0_] :=
    Module[{norm1, norm2, TOLERANCE, Nmax, jac,
                    residual, len, λnew, λdel, i},
        TOLERANCE = 10^-5;
        Nmax = 100;
        len = Length[a];
        norm1 = 1.0;
        norm2 = 1.0;
        λnew = λ0;
        residual = - PolyTri[a, λnew, len];
        i = 1;
        While[norm1 > TOLERANCE && norm2 > TOLERANCE
            jac = PolyTriD[a, λnew, len];
            λdel = residual / jac;
            λnew += λdel;
            residual = - PolyTri[a, λnew, len];
            norm1 = √λdel²;
            norm2 = √residual²;
            i++;
            If[i ≥ Nmax,
                Print["WARNING: Exceeded maximum
                        iterations."];
                Break[];
            ];
        ];
        λnew
    ]
```

A.12. EXAMPLE 5.6.1: POWER METHOD FOR ITERATIVE CALCULATION OF EIGENVALUES

The following code is an implementation of the power method in Section 5.6 to calculate the highest valued eigenvalue and eigenvector of a square matrix. The form of the program is a user-defined function called PowerMethod[**a**,x_0], where **a** is the input matrix and x_0 is the initial guess of the eigenvector. The function has been designed to handle complex eigenvalues and eigenvectors.

```
PowerMethod[a_List, x0_List] :=
    Module[{TOLERANCE, Nmax, u, λ, λold, norm, i},
        TOLERANCE = 10^-10;
        Nmax = 100;
        norm = 1.0;
        u = Chop[ x0 / √(Conjugate[x0] . x0) // N];
        λ = Chop[Conjugate[u] . a . u // N];
        i = 0;
        While[norm > TOLERANCE,
```

```
            i++;
            u = a . u // N;
            u = Chop[ u / Sqrt[Conjugate[u] . u] // N];
            λold = λ;
            λ = Chop[Conjugate[u] . a . u // N];
            norm = Chop[Sqrt[Conjugate[λ - λold](λ - λold)]
                       // N];
            If[i ≥ Nmax,
                Print["WARNING: Exceeded maximum
                        iterations."];
                Break[];
            ];
        ];
        {λ, u}
    ]
```

A.13. EXAMPLE 6.2.1: IMPLEMENTATION OF THE SPECTRAL RESOLUTION THEOREM—MATRIX FUNCTIONS

The following code is a Mathematica program for generating functions of matrices using the spectral resolution theorem of Chapter 6. Two built-in functions (Eigenvalues[] and Eigenvectors[]) are used to generate the required eigenanalysis; however, these routines can be easily replaced with the appropriate used-defined routines. The code is contained in the function MatrixFunction [f,a], where f is the Mathematica function head (e.g., Exp, Cos, Tanh), which can be a Mathematica built-in function or any user-defined function, and **a** is the matrix (or matrix expression). The function will only work for perfect matrices of a size feasible for handling by the functions Eigenvalues[] and Eigenvectors[]. Note that the function f(t) must be defined at the eigenvalues of **a**.

Routine

In the following program, we have made use of the built-in routine Map and the Mathematica implementation of "pure functions." The reader not familiar with such concepts is encouraged to consult Wolfram's text, *The Mathematica Book*, for a full description.

```
MatrixFunction[f_, a_] :=
    Module[{ans, eval, evec, revec, feval, i, j},
        ans = Table[0, {i, Length[a]}, {j, Length[a]}];
        eval = Eigenvalues[a];
        evec = Eigenvectors[a];
        evec = Map[(#/Sqrt[# . Conjugate[#]])&, evec];
        revec = Transpose[Inverse[Conjugate[evec]]];
```

```
        feval = Map[f, eval];
        For[i = 1, i ≤ Length[a], i++,
            ans += feval[[i]]
              Transpose[{evec[[i]]}] · {revec[[i]]}];
        ];
        ans
    ]
```

Examples

The following are examples of how to implement the above function: MatrixFunction[f,a]

```
a={{3, 1},
   {1, 2}};
```

$f(\mathbf{a})$:

```
MatrixFunction[f, a] // MatrixForm
```

$$\begin{pmatrix} -\dfrac{(1-\sqrt{5})\,f\left[\frac{1}{2}(5-\sqrt{5})\right]}{2\sqrt{5}} + \dfrac{(1+\sqrt{5})\,f\left[\frac{1}{2}(5+\sqrt{5})\right]}{2\sqrt{5}} & \dfrac{(1-\sqrt{5})(1+\sqrt{5})\,f\left[\frac{1}{2}(5-\sqrt{5})\right]}{4\sqrt{5}} - \dfrac{(1-\sqrt{5})(1+\sqrt{5})\,f\left[\frac{1}{2}(5+\sqrt{5})\right]}{4\sqrt{5}} \\ -\dfrac{f\left[\frac{1}{2}(5-\sqrt{5})\right]}{\sqrt{5}} + \dfrac{f\left[\frac{1}{2}(5+\sqrt{5})\right]}{\sqrt{5}} & \dfrac{(1+\sqrt{5})\,f\left[\frac{1}{2}(5-\sqrt{5})\right]}{2\sqrt{5}} - \dfrac{(1-\sqrt{5})\,f\left[\frac{1}{2}(5+\sqrt{5})\right]}{2\sqrt{5}} \end{pmatrix}$$

$f(\mathbf{a}) = e^{a}$:

```
MatrixFunction[Exp, a] // N // MatrixForm
```

$$\begin{pmatrix} 28.0655 & 14.8839 \\ 14.8839 & 13.1815 \end{pmatrix}$$

$f(\mathbf{a}) = \mathbf{J}_0(\mathbf{a})$:

```
MatrixFunction[(BesselJ[0, #])&, a ] // N //
    MatrixForm
```

$$\begin{pmatrix} -0.125315 & -0.433807 \\ -0.433807 & 0.308492 \end{pmatrix}$$

$f(\mathbf{a}) = \int_0^a \exp(-x^2)\sin(x)\,dx$:

```
MatrixFunction[(NIntegrate[Exp[-x^2] Sin[x],
   {x, 0, #}])&, a] // N // MatrixForm
```

$$\begin{pmatrix} 0.412416 & 0.0194503 \\ 0.0194503 & 0.392965 \end{pmatrix}$$

A.14. EXAMPLE 9.4.2: NUMERICAL SOLUTION OF A VOLTERRA EQUATION (SATURATION IN POROUS MEDIA)

The following codes are Mathematica programs for the numerical solution of Eq. (9.4.33), a Volterra equation of the first kind. One of the goals of Example 9.4.2 was to demonstrate the numerical difficulty involved with convergence of Volterra problems. Thus, the following codes have been designed to allow variable mesh sizes and variable inherent "data error." The user is urged to play with both these parameters and explore the stability of the resulting solutions.

Exact Solution

We begin by defining the functions that represent the "exact" solution to some physical system. The function sx[p] represents the exact physical saturation in a given sample porous media. The goal of our experiment is to back out this function by measuring the average saturation and a function of p, the capillary pressure. We will therefore eventually compare our numerical solution to sx[p]:

```
sx[p_] := 1.0 /; p <= 2.0
sx[p_] := 1.5/p + 0.25 /; p > 2.0
```

Next, we define the theoretical average saturation, ssx[p], by formally integrating Eq. (9.4.29) giving Eqs. (9.4.30) and (9.4.31):

```
ssx[p_] := 1.0 /; p <= 2.0
ssx[p_] := 1.75 - 1.5 Sqrt[1.0 - 1.5/p] -
           3.00.75
           ------ (ArcTanh[Sqrt[0.25]]
              p
           - ArchTanh[Sqrt[1 - 1.5/p]])/;
    p > 2.0
```

Numerical Solution

The routine NSat[pmax, error, n] is the numeric solution implemented by Eq. (9.4.32). The variable pmax is the user-specified maximum capillary pressure for evaluation (note that pmax must be greater than 0). The variable "error" is the requested inherent error to be imposed on the data. The data (representing values of the average saturation on a regular grid of capillary pressures) is generated by evaluating the theoretical value (via ssx[p]) at each pressure and then adding or subtracting a random fluctuation whose magnitude does not exceed the fraction "error" of the theoretical value. The variable n is simply the number of discretizations of capillary pressure on the interval ($0 < p \leq pmax$). The output of NSat is a list of points {p,s} representing the computed saturation (s) versus pressure (p).

```
NSat[pmax_, error_, n_: 200]:=
    Module[{d, ss, a, b, ans},
    If[
pmax ≤ 0.0, Print["Error: pmax must be greater
                    than zero"]; Return[];];
    d = pmax/n;
    ss = Table[ssx[dj](1 + error Random[Real,
                {-1, 1}]), {j, 1, n}];
    a = Table[ If[j ≤ i, 0.75/(i√(1.0 − 0.75 j/i)), 0.0]
            If[i == j, 0.5, 1.0],
            {i, 1, n}, {j, 1, n}];
    b = Table[0.375/i, {i, 1, n}];
    ans = LinearSolve[a, ss - b];
    ans = Transpose[{Table[ i d, {i, 1, n}], ans}];
    Return[ans];
]
```

Plotting Function

The function PlotSolution[data] uses the Mathematica function Show to generate a combined plot of the numerical solution (in data) and the theoretical solution (via ss[p]). The function List is used to plot the theoretical curve and ListPlot is used for the numerical data points. The option DisplayFunction is used to suppress the graphical output until both plots have been combined by Show.

```
PlotSolution[data_List]:=
    Show[ {Plot[sx[p],
    {p, 0.0, data[[Length[data], 1]]},
        DisplayFunction -> Identity],
        ListPlot[data, DisplayFunction -> Identity]},
        DisplayFunction -> $DisplayFunction]
```

Examples

The following examples show the sensitivity of the Volterra equation to random noise in the data:

NUMERICAL SOLUTION OF A VOLTERRA EQUATION (SATURATION IN POROUS MEDIA)

First, the solution with zero noise:

```
data = NSat[4.0, 0, 200];
PlotSolution[data]
```

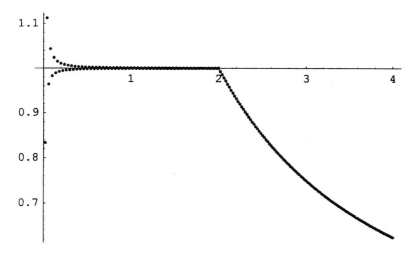

- Graphics -

The numerical solution with 0.001% noise:

```
data = NSat[4.0, 0.00001, 200];
PlotSolution[data]
```

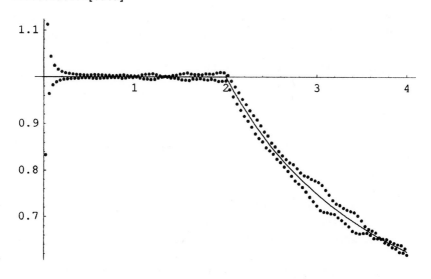

- Graphics -

The numerical solution with 0.01% noise:

```
data = NSat[4.0, 0.0001, 200];
PlotSolution[data]
```

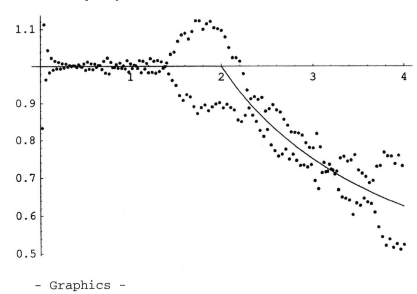

- Graphics -

A.15. EXAMPLE 10.5.3: NUMERICAL GREEN'S FUNCTION SOLUTION TO A SECOND-ORDER INHOMOGENEOUS EQUATION

The following codes illustrate how to implement the techniques of numerical integration (from Chapter 3) to define a Green's function solution to an inhomogeneous differential equation. The function u[x] is defined as the integral operation in Eq. (10.5.76) and ($0 \leq x \leq 1$).

Auxiliary Functions

We first define the function f[x], which is the forcing function in Eq. (10.5.66):

```
f[x_] := Exp[-x^2]
```

We then define a routine for numerical integration. The function NInt uses the rhombohedral method of integration where "integrand" contains the expression of the integrand; the list "limits" contain (in order) the variable of integration in "integrand," the lower integration limit, and the higher integration limit; and "delta" is the requested step size of the integration (dx). Note that the actual step size will be less than or equal to "delta" in actuality.

```
NInt[integrand_, limits_List, delta_] :=
    Module[{y, ymin, ymax, n, dy, ans, i},
        y = limits[[1]];
        ymin = limits[[2]];
        ymax = limits[[3]];
        n = Ceiling[(ymax - ymin)/delta - 1];
        dy = (ymax - ymin)/(n + 1);
        ans = 0.5 dy (integrand /.
            y -> ymin + integrand /.
            y -> ymax) // N;
        ans += Sum[ dy integrand /.
            y -> (ymin + i dy), {i, n}] // N
    ]
```

Solution

Finally, we define the module for the function u[x], incorporating the routine NInt above.

```
u[x_] := Module[ {z, int, ans, dx},
        dx = 0.001;
        z = -1/3 Exp[-(x - 1)];
        int = NInt[(-3 Exp[y - 1] + 5 Exp[2 y]) f[y],
            {y, -1, x}, dx];
        ans = z int;
        z = -1/3 (-3 Exp[-2 (x + 1)]
            + 5 Exp[-(x + 1)]);
        int = NInt[ Exp[2 (y + 1)] f[y],
            {y, x, 1}, dx];
        ans += z int // N
    ]
```

Note that we could have used the built-in Mathematica function NIntegrate instead of NInt, in which case the function u[x] would be redefined as:

```
u2[x_] := Module[ {z, int, ans},
        z = -1/3 Exp[-(x - 1)];
        int = NIntegrate[(-3 Exp[y - 1]
            + 5 Exp[2 y]) f[y], {y, -1, x} ];
        ans = z int;
        z = -1/3 (-3 Exp[-2 (x + 1)]
            + 5 Exp[-(x + 1)]);
        int = NIntegrate[ Exp[2 (y + 1)] f[y],
            {y, x, 1}];
        ans += z int // N
    ]
```

A.16. EXAMPLE 10.8.2: SERIES SOLUTION TO THE SPHERICAL DIFFUSION EQUATION (CARBON IN A CANNONBALL)

The following code illustrates how to define a series solution (with an imposed cutoff) to the diffusion equation in spherical coordinates. In Example 10.8.2, the boundary conditions are such that the problem is spherically symmetric (i.e., the spatial solution only depends on the radius r and not on the angular orientation). We proceed by defining the function f[r,t] as the solution C/C_s in Eq. (10.8.22), where r is the dimensionless radius (r/b) and t the dimensionless time (tD/b^2).

```
f[r_, t_] :=
    Module[{TOLERANCE, Nmax, norm, ans, del, n},
        TOLERANCE = 10^-15;
        Nmax = 100;
        norm = 1.0;
        ans = 1.0;
        n = 0;
        While[norm > TOLERANCE,
            n++;
            If[ r == 0.0,
                del = (-1)^n 2 Exp[ -(Pi n)^2 t];,
                del = (-1)^n 2 Exp[ -(Pi n)^2 t] Sin[Pi n r]/(Pi n r);
            ];
            ans += del;
            norm = √del^2;
            If[n ≥ Nmax,
                Print["WARNING: Exceeded maximum
                    iterations."];
                Break[];
            ];
        ];
        ans
    ]
```

INDEX

Abel's problem, 376
Adiabatic temperature profile, 106
Adjoint matrix, 164; *see also* Matrix
Adjoint of a differential operator, 413, 420–426
Adjugate, *see* Matrix
Alien cofactor expansion, 14
Anti-Hermitian matrix, 265
Arrhenius formula, 89
Atomic matrix, 147
Augmented matrix, 53–54, 55, 58, 123, 130

Backward substitution, 49, 53–55
Band matrix, 66–78
Basis set, 170–179; *see also* specific type
   biorthogonal, 199
   orthonormal, 199
   reciprocal, 171, 175–177
Basis vectors, 326–330
Bessel's inequality, 356, 387
Bilinear function, 206
Biorthogonal basis sets, 199
Biorthogonal set, perfect matrix, 205
Boundary conditions, 416–418
   homogeneous, 444–446
   self-adjoint, 463
   unmixed, 446–450
Boundary functionals, adjoint operator, 423
Boundary value problem, 414, 440
   general boundary conditions, 450–452
   for unmixed boundary conditions, 446–450
Bounded operators, 355

Cartesian unit vectors, 43
Cauchy convergence, 324–326
Cauchy sequence, 316
Centered difference formulas, 73
Characteristic polynomial equation, 165
Chemical reaction, 146–147, 210
Chemical stability, conditions of, 242
Cofactor
   definition, 9
   expansions, 9–14
Cofactor expansion theorem, 10–14
Commuting matrix, 239–240
Companion matrix, 255–258
Completely continuous normal operators, 405–406

Completely continuous operators, 355, 366–375, 387
   Schmidt's normal form of, 401–405
   self-adjoint, 395
Completeness, vector space, 315
Completeness relation, 344
Complete vector space, 326
Conformable partitioned matrix, 35–36
Conjugate gradient method, 84–85
   iterative methods for solving linear system $\mathbf{Ax} = \mathbf{b}$, 518–519
Continuation strategy, 92
Continuously stirred tank reactor (CSTR), 89
   multireaction at constant temperature, 104–106
   steady-state solutions for, 521–523
Continuous solutions, 382
Continuous spectrum, 485
Convergence rates, 80–81
Cooling, graphic electrode, 102–104
Countercurrent separation process, 75–76
Cramer's rule, 1, 14–16, 18–22, 29, 47

Damped oscillator, 291
Damped pendulum, 274
Darboux sum, 320, 329
Defective matrix, 279–314
Degenerate operators, 337
Determinants, 1–23
  addition, 1
  column addition or subtraction, 1
  definition, 3–5
  differentiation, 9
  elementary properties of, 6–9
  Vandermonde, 22
Diagonalization, similarity transformation, 213–215
Diagonal matrix, 30, 186, 241
Differentiability, in linear vector space, 38
Differential equation, partial, 493–501
Differential expression, adjoint, 420–426
Differential operator, 331–332, 345, 416–420
  adjoint of, 413, 420–426
  inverse of, Green's function, 439–452
  normal, 457
  spectral theory of, 452–458
Diffusion, in 3 dimensions, 494
Diffusion equation, spherical, series solution to, 542
Dilute partition coefficient, 78
Dirac delta function, 344, 442–444, 486
Discontinuous solutions, 382
Dyadic imperfect matrix, 303–304
Dyadic operator, 316, 333–334, 352
  finite, 366

Eigenfunction, 388
  function f(K), 390
Eigenproblem, 163–204
  generalized, 269, 275
Eigenvalues
  analysis, 179–184
  calculation of, 189–196
  degeneracy, 388–389
  distinct, 219–220
  iterative calculation, power method for, 534–535
  multiplicity, 193
  special properties of, 184–188
Eigenvector, 164
  adjoint matrix, 164
  function f(K), 390
  generalized, 280, 289, 294–303

Electric circuit, 143
Elimination process, 54–55
Entropy maximum principle, 242
Equations of state
  Peng–Robinson equation, 111
  Redlich–Kwong equation, 111
  Soave equation, 111
Euclidean space, 28, 40, 43
Euclidean vector space, 151, 167
  unbounded three-dimensional, 362
Eulerian angles, 268
Euler method, 116
  implicit or backward, 117

Finite-difference approximation, 71–76, 108
Finite-difference method, 71–76
Finite-element analysis, 351
First-order continuation, 92, 98
Floating-point arithmetic, 55–56
Flory-Huggins theory, polymeric solutions, 93–99
Forward substitution, 50
Fourier series, 328–330
Fourier transform, 362, 486
Fredholm's alternative theorem, 124, 155–159, 295, 360, 364
Fredholm's solvability theorems, 355
Frobenius norm, 45
Function spaces, 315, 318
Fundamental solutions, 426–439
Fundamental system, 503

Galvanometer, 313
Gas mixture, viral coefficients of, 148–150
Gauss elimination, 5, 47–70, 108–110, 511–514
  with pivoting, 55–58, 513–514
  simple, 48–55, 512
  solution of linear systems, 512–513
Gaussian distribution, 266
Gauss-Jordan elimination, solution to the linear system $Ax = b$, 529–531
Gauss-Jordan transformation, matrix, 129–132
Gauss-Seidel iteration, 47, 81
Gauss-Seidel method, 79–80
  iterative methods for solving linear system $Ax = b$, 516
Gauss theorem, 496
General boundary conditions, boundary value problem for, 450–452

Generalized eigenproblems, 269, 275
Generalized eigenvectors, 280, 289, 294–303
Gerschgorin's theorem, 165, 195
Gibbs free energy, 243
Gram-Schmidt procedure, 172–173, 197, 327, 347–350, 353
  orthogonalization, 229, 306
Green's formula, 422
Green's function, 415, 502–504
  inverse of a differential operator, 439–452
  numerical solution to second-order inhomogeneous equation, 540–541

Hadamard's inequality, 265
Hamilton-Cayley theorem, 199–201, 292–293
Harmonic oscillator, 206
Heat exchanger
  performance, 119–120
  profile, 215–216
Hermite functions, 353, 477–478
Hessian matrix, 232
Hilbert matrix, 117, 262
Hilbert-Schmidt operator, 355, 366–375, 388, 454
Hilbert space, 315–316, 324–336, 345
  linear differential operators in, 413–510
  linear integral operators in, 355–410
Holder inequality, 268
Homogeneous solutions, 141–143

Identity matrix, 174
  resolution, 174
Imperfect matrix, 169, 179, 279–314
  dyadic form, 303–304
Inertia tensor, 264
Infinite-dimensional linear vector spaces, 315–353
Inhomogeneous equation, second-order, numerical Green's function solution, 540–541
Inhomogeneous problem, general, 426–439
Initial value problem, 254–258, 308–313, 414, 440, 444–450
Inner product, 165
Inner product space, 163, 165, 322–323

Integrability, in linear vector space, 38
Integral operator, 369
  linear, in a Hilbert space, 355–410
  spectral theory of, 387–406
  Volterra, 375–387
Integral transform, 487
Inverse, see also Matrix
  computation of, 58–60
Inverse matrix, 28–33; see also Matrix
Isomerization reactions, 218
Iterative methods, 78–85
  solving the linear system $Ax = b$, 515–518
Iterative solutions, convergence
  Gauss–Seidel method, 81
  Jacobi method, 81
  Newton–Raphson method, 81
  SOR method, 81

Jacobian elements, 96
Jacobian matrix, 86
Jacobi iteration, 47, 78–79, 81
Jacobi method, iterative methods for solving linear system $Ax = b$, 516
Jordan block diagonal matrix, 282–288, 294
Jordan block matrix, 279, 282–288, 294–303
Jordan canonical form, 288–294, 299, 303
  solving initial value problem, 309–310

Kronecker delta function, 14
  self-adjoint, 347
  solutions to problems, 336–343

Lagrange multiplier, 396
Lagrange's identity, 422
Laplace transform, 377
Laurent series, 208
Least-squares data analysis, 2, 110, 149
Lebesgue integral, 316, 319–321, 329, 344, 352–353
Lebesgue square integral, 369
Legendre polynomial, 327, 334, 363, 477–478
Linear differential operators, Hilbert space, 413–510

Linear equations, 85–121
  solution by Cramer's rule, 14–16
Linear independence, 38–39, 44, 150–155
Linearly independent vectors, 124
Linear operator, 163, 330–336
  in normed linear vector space, 165–170
Linear operator theory, 315
Linear vector space, 26, 38–43, 163
  normed, linear operators in, 165–170
Linear-vector space,
  finite-dimensional, 170
Linear viscoelastic stress, 386
Liquid-vapor interface
  density profile, 90–92
  density profile in, 523–526
Liquid-vapor phase diagram, 99–102
Logarithm of perfect matrices, 209
Lower triangular matrix, 49, 61
LU-decomposition, 47, 61–65

Mass separation, 76
  with a staged absorber, 514–515
Matrix, 2-3, 25–46; see also specific type
  addition, 3, 25–28
  adjoint, 25, 33–35, 164
  adjugate, 25, 29, 43
  atomic, 147
  augmented, 123, 130
  commuting, 239–240
  companion, 255–258
  defective, 279–314
  diagonal, 30, 186, 241
  diagonally dominant, 65, 186
  with distinct eigenvalues, 219–220
  Gauss-Jordan transformation of, 129–132
  Hessian, 232
  identity, 174
  imperfect, 169, 179, 279–314
    dyadic form, 303–304
  implementation of spectral resolution theorem, 535–537
  inverse, 28–33, 43, 58–60
  linear operator norm, 38
  lower triangular, 49, 61
  multiplication, 26–28
  negative-definite, 206, 228, 237
  negative matrix, 228, 237
  nilpotent, 289, 293
  nonsingular, 288
  norm, 41–43

  normal, 206, 219–220, 245–249, 265
  orthogonal, 206, 219, 220–224
  partitioning, 35–37
  perfect, 164, 169, 191, 205–278
    logarithm of, 209
  positive-definite, 40, 62, 170, 206, 219, 228, 249
  positive-semidefinite, 264–265
  pseudo-inverse, 26
  rank of, linear dependence of a vector set and, 150–155
  scalar multiplication, 3
  self-adjoint, 190, 206, 219, 227–245, 249, 265, 348–350
  skew-symmetric, 161, 219
  square, 225, 288, 531–533
    arbitrary, Schmidt's normal form, 304–308
  symmetric, 271
  transpose, 26, 33–35
  transpositions, 3
  tridiagonal, 67, 72, 165, 189
    calculating eigenvalues of, 533–534
  unitary, 205, 206, 219, 220–224, 225–227, 349–350
  upper triangular, 49, 59, 61
Matrix products, determination of, 124–129
Minors, 16–18
Multicomponent diffusion system, 243–244, 251
Multicomponent reaction system, 217–219
Multiplicity, 252–253
  eigenvalues, 193

Negative-definite matrix, 206, 228, 237
Newton–Raphson iteration, 47, 81
Newton–Raphson method, 85–87, 197, 465
  continuously stirred tank reactor (CSTR), 521–523, 525–526
  convergence with Picard method, 519–521
  eigenvalues of tridiagonal matrix, 533–534
  polymer solution, 526–529
Newton–Raphson scheme, 210
Nilpotent matrix, 289, 293
Nonlinear systems, spectral resolution theorem in, 209–210
Norm, 39, 41–43
Normal matrix, 206, 219–220, 245–249, 265
Normal operator, 390–393

Normed linear vector space, 26, 39, 44, 45, 165–170, 324
  basis sets in, 170–179
Null space, 393

Ohm's law, 143
Onsager matrix, 251
Operator norm, 38
Orthogonalization, Gram–Schmidt, 172–173
Orthogonal matrix, 206, 219, 220–224
Orthonormal, 36–37
Orthonormal basis set, 171, 199, 206, 223, 305
  normal matrices, 246

Peng–Robinson equation, 111
Perfect matrix, 164, 169, 191, 205–278
Perfect operators, 179, 343–351, 347
Perrin's theorem, 202
Perturbation theory, 259–261
Phase diagram, polymer solution, 93–99, 526–529
Phase separator, process control of, 211–213
Picard iteration, 42–43, 44, 47
Picard method, 85–87
  convergence with Newton–Raphson method, 519–521
Pivoting, 55–58
  complete, 57
  partial, 56–58
Plug flow reactor, temperature profile in, 106–108
Polymerization, reversible anionic, 118
Polymer solutions, phase diagram of, 526–529
Polynomial functions, eigenvalues of tridiagonal matrix, 533
Positive-definite matrix, 40, 62, 170, 206, 219, 228, 249
Positive-semidefinite matrix, 264–265
Power method, 165, 190–194
  iterative calculation of eigenvalues, 534–535
Predator-prey, 275
Principal minors, 281
Process control, 34
Projection operators, 211
Pseudo-inverse, *see* Matrix

Quadratic convergence, Newton–Raphson method, 88
Quantum mechanics, 260–261, 488–492

Rank of matrices, 16–18
Reaction diffusion, thin film, 243–244
Reactor stability analysis, 34
Reciprocal basis set, 175, 207
Reciprocal vectors, 175–177
  perfect matrix, 205
Redlich–Kwong equation, 111
Residual, 97
Resolution, identity matrix, 174
Riemann integral, 318–321, 329
Runge–Kutta method, 115

Scalar multiplication, 27
Schmidt normal form, 280
Schmidt's normal form of arbitrary square matrix, 304–308
Schwarz inequality, 163, 166, 322
Self-adjoint integral operators, 356
  completely continuous, 395–401
Self-adjoint matrix, 190, 206, 219, 227–245, 249, 265, 348–350
Self-adjoint operators, 457
Semidiagonalization theorem, 225–227, 246
Separation of variables, 493–495, 499
Series solution, spherical diffusion equation, 542
Similarity transformation, 186
  diagonalization by, 213–219
  perfect matrix, 205
Simple perfect matrix, 208–209
Skew-symmetric matrix, 219
Soave equation, 111
Solutions
  homogeneous, 140–143
  particular, 140–143
Solvability theorem, 123
$\mathbf{Ax} = \mathbf{b}$, 133–150
  linear integral equations, 356–366
  for linear systems, 316
Spectral decomposition, 457–458, 486–487, 505
  completely continuous normal operators, 405–406
  self-adjoint operators, 405–406
Spectral operators, special properties, 390

Spectral resolution, 485
Spectral resolution theorem, 164, 177, 191, 205, 344
  implications of, 206–213
  integral operators, 375
  matrix functions, 535–537
  normal, completely continuous operator, 396
  normal matrix, 206
  partial differential equations, 493–501
  perfect matrix, 205
Spectral theorem
  differential operators, 452–458
  integral operators, 356, 387–406
  regular Sturm–Liouville operators, 459–477
  singular Sturm–Liouville operators, 477–493
Square matrix, 2, 31–32, 225, 288
  arbitrary, Schmidt's normal form, 304–308
  characteristic polynomials, 531–533
Staged absorber, 76
Stirling's approximation, 5
Stress tensor, 264
Sturm–Liouville operators, 415
  partial differential equations, 493–501
  regular, spectral theory of, 459–477
  singular, spectral theory of, 477–493
Successive overrelaxation method, 47, 79–81
  continuously stirred tank reactor, 524–525
  iterative methods for solving linear system $\mathbf{Ax} = \mathbf{b}$, 516
Sylvester's formula, 206, 254
Sylvester's theorem, 123–129
Symmetric matrix, 271

Taylor series, 208
Thermal diffusivity, 494
Tolerance, 96–99, 191–192
Trace routine, characteristic polynomials and traces of square matrix, 531–533
Traces, matrix, 17, 182–187
Traction vector, 201
Transpose, *see* Matrix
Trapezoidal rule, 108

INDEX

Triangle inequality, 163, 166, 322, 324
  vector, 40
Tridiagonal matrix, 67, 72, 165, 189
  calculating eigenvalues of, 533–534

Unbounded operators, 355
Unitary matrix, 205–206, 206, 219, 220–224, 225–227, 349–350
Unitary transformation, 205–206, 220
Upper triangular matrix, 49, 59, 61

van der Waals model, 91, 99
Vectors, 25–46
  column *versus* row, 2–3
  definition, 2
  length, 38
  linearly independent, 124
  orthonormal set, 36–37
  reciprocal, perfect matrix, 205
  scalar product, 27, 43
  space, 26
Vector spaces
  finite-dimensional, 372
  infinite-dimensional linear, 315–353

Vibratory feeder, 274–275
Volterra equation, 375–387
  first and second kind, 355–356
  numerical solution, 537–540

Wronskian, 438, 447, 503

Young–Laplace equation, 114

Zeroth-order continuation, 92